Molecular biophysics

Molecular biophysics

Structures in motion

Michel Daune

Honorary Professor, University of Strasbourg

Former Director of the Centre for Molecular Biophysics, Orléans

Translated from the French by

W. J. Duffin

With a foreword by

David Blow, FRS

Imperial College, London

OXFORD
UNIVERSITY PRESS

OXFORD
UNIVERSITY PRESS

Great Clarendon Street, Oxford OX2 6DP

Oxford University Press is a department of the University of Oxford.
It furthers the University's objective of excellence in research, scholarship,
and education by publishing worldwide in

Oxford New York

Athens Auckland Bangkok Bogotá Buenos Aires Calcutta
Cape Town Chennai Dar es Salaam Delhi Florence Hong Kong Istanbul
Karachi Kuala Lumpur Madrid Melbourne Mexico City Mumbai
Nairobi Paris São Paulo Singapore Taipei Tokyo Toronto Warsaw

Oxford is a registered trade mark of Oxford University Press
in the UK and in certain other countries

Published in the United States
by Oxford University Press Inc., New York

Reprinted 2003, 2004

Published with the help of the Ministère de la Culture

A catalogue record for this book is available from the British Library

Library of Congress Cataloging in Publication Data

ISBN 0 19 857783 4 (Hbk)
ISBN 0 19 857782 6 (Pbk)

3 5 7 9 10 8 6 4

Typeset by Newgen Imaging Systems (P) Ltd., Chennai, India

Printed in Great Britain
on acid-free paper by
CPI Bath

Foreword

by David Blow, FRS

Michel Daune's book fills a gap in the annals of Biophysics, linking several strands of science in a different way.

Under Desmond Bernal, Max Perutz, and Dorothy Hodgkin, biomolecular structural analysis was born, and is now strongly established world-wide. Also in Britain, study of neural transmission by Alan Hodgkin and Andrew Huxley led to the growth of neurobiology, a hugely growing study whose biophysical aspects have been developed in great detail. Meanwhile Francis Crick and Jim Watson's discovery pushed 'molecular biology' rapidly forward, so that cell biology was reorganized as a branch of molecular science. Between these three aspects, there is a wide realm of biophysical study associated with polymers, water, and their thermodynamic properties, which was built up most strongly in the United States by scientists like Paul Flory, Walter Kauzmann, and Charles Tanford. These workers were based in Chemistry departments. The difficult task of extending their insights into biological problems still lags behind, and is sometimes avoided by teachers. It also tends to be ignored in textbooks. As understanding of biological processes extends towards modelling them by strict physical rules, the weakness of this area of biophysical analysis becomes more embarrassingly apparent.

Daune has addressed the challenging problem of bringing all these strands together, and of developing each to the contemporary level of achievement. His book has a recognizably French flavour. The treatment is rigorously built on basic physical principles and a mathematical analysis is made of every problem. Each concept is developed into the dreadful complications of real biological systems. The role of experimental data is to demonstrate that such development is soundly based, or to show where further analysis is required.

Several familiar texts start from phenomena and experimental data, leading thence into examination of underlying biophysical principles. The contrast with Daune's bottom-up approach is not only fascinating but revealing.

Although the flavour may be French, the expression is clear and English. Bill Duffin's translation makes straightforward reading, and it needs careful comparison to see how skilfully the nuances of expression have been preserved. I thank Oxford University Press for the opportunity of editing Bill Duffin's excellent translation.

Preface

This book is addressed mainly to first degree and master's degree students in biochemistry and biophysics. It should also be accessible to master's degree students in physics, chemistry and physical chemistry as long as they have been able, in addition to their normal courses, to acquire some knowledge of biochemistry and molecular biology: it is assumed throughout the book that certain concepts relating to these two disciplines are already familiar.

The level reached by biochemistry students at the beginning of their studies is often quite varied, and for this reason the mathematical treatments in the body of the text are limited to what is strictly necessary. The detail of such treatments is confined to 'boxes', which will enable students with a substantial mathematical background to study some of the treatments and calculations in greater depth. A Mathematical Appendix is also included to provide further help. However, neither the boxes nor the appendix are essential for a general understanding of the text. The level of physics and chemistry required throughout the book is approximately that attained at the end of the second year of a university course.

The book reflects the experience gained by the author during more than 20 years' teaching of biophysics, mainly at the University of Strasbourg but also at the universities of Marseille–Luminy, Quebec (Laval), Orléans and Paris VI. Discussions with colleagues, suggestions from them, enquiries and criticisms from students, and the almost daily interaction between teaching and laboratory practice have all meant that the text has been 'reworked' many times. The various courses have in turn benefited from this, both in content and in the way the essential features of the subject are brought out.

Like biophysics itself, the text is interdisciplinary in the sense that the laws and methods of physics and chemistry are used to study the constituents of the cell on a molecular scale. Both in the content of the book and in the approach adopted, the author has tried to explain the bases of the biophysical treatment of the living world in the five main Parts, dealing in turn with the conformation, dynamics and hydration of biopolymers, with the properties of polyelectrolytes, and with association between biopolymers.

All detailed and specific descriptions of physical methods that are at present widely used in biophysics have deliberately been excluded. In particular, no treatment is provided of the three major techniques: X-ray crystallography, nuclear magnetic resonance and electron microscopy. All of these are best described and discussed by specialists, and the reader wishing to gain a thorough knowledge of them is referred to specialized monographs dealing with them. However, some physical methods are described, often very simply, in Part II of the book on the dynamics of biopolymers: the treatments are limited to the basic features essential for an understanding and interpretation of the results from experiments designed to study intermolecular and intramolecular motion.

At the end of each Part, references to further reading are provided so that students can add to their knowledge and deepen it. The list may also form a starting point for any new PhD student wishing to begin research work in a particular field.

Apart from the hundreds of students who, over the years, have gradually built up this book through their questions and reactions, I must not forget to acknowledge the contribution of the numerous discussions with all my colleagues, both teachers and researchers (whether they are biophysicists or are, on the contrary, resistant to the problems of biophysics because they are too much the physicist … or too much the biologist). I hope, in return for all this, that the book will help in the training of new students, will incite further exchanges and criticisms, and perhaps will even convert a few!

Word processors now make the task of producing a book considerably lighter, but the present work would not have been possible without the patience, efficiency and thoroughness of Liliane Diebold and Jeannine Florian, who fought their way from one correction to another and enabled the final text to see the light of day.

To all those who have, over many years and in various capacities, helped me to bring this work to fruition, I express my heartfelt thanks.

Preface to the English edition

The original edition of this book was published in France at the beginning of 1993 and the references listed in it included none beyond 1992. Over the six years since then, the field of molecular biophysics has undergone considerable development. The organization and morphogenesis of proteins, the development of molecular dynamics, the construction of maps of electric potential around proteins, and the formulation of new schemes for assembly are only a few examples of areas in which substantial progress has been made, both theoretically and experimentally.

Since almost every Part of the book has been affected by these advances, a thorough updating has had to be undertaken. Part IV dealing with polyelectrolytes has even had to be recast in order to reflect changes in the relative significance of several chapters and to alter the order in which they are presented.

In addition, some of the comments made on the book and my own critical re-reading of the text have led me to prune some of the more cumbersome calculations in it, either omitting them entirely or transferring them to 'boxes'. The contents of these are not essential for an overall understanding: they either show proofs of mathematical formulae or supplement the material in the main body of the book.

Finally, my re-reading of the original edition revealed a large number of errors of all kinds, distributed throughout the text, the formulae and even the figures! I have been helped considerably in the task of correcting these by Bill Duffin: I have particularly appreciated his competence, his care, his questioning, his demand for quality ... and also his typically British humour and kindness.

After a year's friendly and fruitful collaboration, I am sure that this English version is more complete, more balanced and often clearer than the original, and I hope that all the errors have been eliminated.

Michel Daune
Strasbourg, March 1998

Translator's note
I wish to express my gratitude and thanks to Professor David Blow for reading the whole manuscript, for help with the English terminology in the field, and for suggestions that have enhanced the clarity of the translation. I have also greatly appreciated the author's willing cooperation (and patience) in replying to my many queries.

Bill Duffin

Contents

General introduction

As we attempt to gain an ever deeper understanding of the modern biological world, the traditional differences between physics, chemistry and biology are tending to disappear. On the molecular or cellular scale, too, it is becoming increasingly pointless to set biochemist against geneticist, biophysicist against immunologist. All the specialists contribute their own technologies, their knowledge of processes and above all their scientific rigour.

When talking of biophysics, therefore, exactly what significance does the 'physics' part of the term have? There are two aspects to the answer.

Firstly, it characterizes the development of new methods of investigation. Nobody today questions the fundamental contributions made by crystallography to an understanding of the function of macromolecules: structures of entities as large as viruses, nucleosomes and ribosomes can now be determined. Two- or three-dimensional nuclear magnetic resonance (NMR) is a complementary approach independent of the production of crystals and giving information about intramolecular dynamics. Electron microscopy, coupled with new methods of preparing specimens (amorphous ice) and with image processing, enables complex structures to be visualized. Molecular spectroscopy allied to methods using fast-reaction kinetics means that we can tackle the dynamics of systems. This formidable arsenal of techniques will undoubtedly be supplemented by new methods before long.

Secondly, physics, more than any other discipline, provides thought processes indispensable in dealing with any 'natural' phenomenon. One reason for this is that its laws are universal on space and time scales that are generally larger than those of the living world. Another reason is its ability to reduce complex systems to models that can be treated theoretically and thus to make constant comparisons between theory and experiment. Testable models in a mathematical framework are now being introduced into an increasingly quantitative biology.

From its very beginning, molecular biology has adopted an approach in which biological phenomena, reduced to the molecular scale, appear as a set of algorithms forming a consistent logical whole into which any living system must fit. From this conceptual viewpoint, it is similar to thermodynamics, which, through a small number of laws, provides a logical framework for the existence of all physical phenomena. In both cases, however, the mechanism remains unknown. To say that all heat engines must obey the second law of thermodynamics gives no indication either of its mechanism or of the practical consequences. To say that a protein is synthesized by using an almost universal genetic code provides no information about the exact development of the transcription and translation processes. However, it is certainly true that molecular biology has developed its methodology by using tools provided directly by the living world to verify this or that model. Cloning is essential to an understanding of the modes of expression and regulation of genes and is one of the main biotechnological tools, yet it is achieved by using restriction enzymes of which more than 99 per cent have a completely unknown mechanism. Only the result counts, i.e. the cutting of the DNA at a precise point.

As a result of all this, biophysics appears as a new interface between two apparently different views of the world around us, those of physics and biology. Like any interface, there must be two interacting surfaces, one turned towards biological problems and their description in terms of molecular biology, and the other turned towards the concepts, models and techniques of physics. There must also of course be a constant flow between the two surfaces. Emerging from this are two ways of looking at the biophysics laboratory:

1. It displays only one of the surfaces (generally the physical or physico-chemical surface) but this interacts closely with the biological surface of another group. It is what could be called 'associative interdisciplinarity'.

2. It displays both of the interacting surfaces, and the flow between them is internal to the group of researchers forming it. This could be called 'constitutive interdisciplinarity'.

These two conceptions are equally valid, but the corresponding realities present different problems. In the first case, the biological problem more often remains under the control of the biological group, and the physics group must avoid being simply a technical extension, a subcontractor or a service provider. Dialogue on both sides of the interface can only be carried out through the respective apprenticeships in the language and experimental difficulties encountered by each group. In the second case, the dual experimental approach implies a relatively heavy infrastructure and the complete training of each researcher. The exchanges, on the other hand, form the very lifeblood of the group. In both cases, however, the prefix 'bio' implies that the biological problem is still the single reference point for the experimental approaches.

Formulating a question correctly often leads directly to the answer: thus, like other methodological approaches, biophysics is capable of explaining the living world provided that biophysicists know how to discern the type of problem suitable for their specific approach, know how to understand the approach and the language of all those who tackle it in another way, can if necessary use their own methodologies, and finally understand the significance of the biological problem they have decided to study.

Molecular biophysics asserts that it is possible to provide a quantitative explanation for all the biological phenomena considered on the molecular scale using the laws of physics established for the inanimate world. Such an approach is only possible at present, however, outside living matter. For it would be a mistake to attempt to explain the functioning of the cell in its entirety straight away. It would, however, be equally mistaken to study 'well-behaved' physico-chemical systems, lending themselves admirably to calculation and experiment but unfortunately having nothing in common with the mechanisms operating in a living being. In other words, biophysics attempts to study systems that have a limited number of parameters and are perfectly reproducible under given experimental conditions.

All such systems are simplified models of what happens in the cell, considered as possessing in itself all the attributes of a living being.

The biophysicist therefore works with a representation of the living world that is as faithful as possible. Biophysics not only has an explanatory value but a predictive value as well: its results, obtained for a given system, can be generalized to any biological system capable of being represented by an identical model. To carry out its plan, molecular biophysics uses a large number of standard physical methods, adapting them to the specific problems being posed.

The present treatise on molecular biophysics deliberately avoids any detailed approach to physical methods in order to deal mainly with the physico-chemical properties and concepts characteristic of the biological world on the molecular scale. It emphasizes the dynamic aspect of biological processes where molecular structures are continually evolving with time and where the significance of random phenomena increases with the complexity of the system. In the four first Parts of the book, the author shows that the physical and physico-chemical laws apply to the players in the biological game (water, ions, molecules and macromolecules).

Part I considers the problems of the structure and conformation of biopolymers, particularly the polymorphism of nucleic acids and the search for protein-building rules. Part II tackles the majority of dynamic problems, from the simple harmonic oscillator model to Brownian motion and current methods of molecular dynamics. Part III describes the behaviour of the water molecule and its interaction with molecules and polymers in a complex process defined by the general term 'hydration'. Part IV considers the problems of the potential and electric field created by distributions of free charge or carried by polyelectrolytes, which form the great majority of the macromolecules in the cell.

These four Parts cover the major aspects of the knowledge required for dealing with more specialized problems, such as the specificity of recognition between molecules or the morphogenesis of subcellular complexes (ribosomes, chromosomes, etc.), which are dealt with in Part V. The whole book also forms a starting point for an approach to properties of the membrane or of the entire cell. It could usefully be supplemented by a series of monographs on the various techniques that are now in general use.

The spirit in which this book has been written echoes that underlying the words of Erwin Schrödinger:

> Isolated knowledge obtained by a group of specialists in a narrow field has no value in itself. It only has value in the theoretical system which unites it with the rest of knowledge and even then only insofar as it really contributes to this synthesis by answering the question: 'who are we?'

Part I

Conformation of biopolymers

Introduction

A polymer in its simplest form (as occurs in synthetic polymers) consists of a covalent assembly of a sometimes very large number N of identical units. Each of the units is called a *monomer* or *monomeric unit*: *polyethylene* for example is a succession of CH_2 groups, but since it is produced by the polymerization of ethylene ($CH_2{=}CH_2$), the monomeric unit here is the $-CH_2-CH_2-$ group. Each end of the long CH_2 chain is attached to a methyl (CH_3) group, so that the formula for the polymer is:

$$CH_3-(CH_2-CH_2)_n-CH_3$$

It is clear from this that, in any synthetic or natural polymer, the repetitive element (the repeating unit or monomeric motif) is different from the two terminal elements.

One of the hydrogen atoms in the monomer may be replaced by a group R: for example, R is C_6H_5 in *styrene*, Cl in *vinyl chloride*, CN in *acetonitrile* or CH_3 in *propylene*. In such cases, the succession of monomeric units in the corresponding polymer may not be regular, since the R group may be located randomly on either side of the C–C backbone. Each monomeric unit can therefore adopt either of two conformations, leading to a large number of possible configurations for the whole molecule. A polymer with 1000 monomers in its chain, for example, can have 2^{1000} or about 10^{300} different configurations. In such materials, the probability of finding two identical polymer molecules is very small. However, a suitable choice of conditions for polymerization can produce what are called *stereoregular* polymers, in which a particular sequence is regularly repeated. The molecules can then pack together to form a regular array and can thus crystallize.

It is also possible to carry out polymerization from a mixture of two different monomers and produce a regular or random alternation of the two monomeric units, producing what is known as a *copolymer*. We shall see later that proteins and nucleic acids are copolymers. Proteins are covalent assemblies of 20 different *amino acids*. Nucleic acids are built from *nucleotides*, which themselves contain two different sugars (*ribose* and *deoxyribose*) and five different major bases (*adenine*, *guanine*, *thymine*, *cytosine* and *uracil*), thus giving rise to four distinct monomeric units both in *polydeoxyribonucleotides* and in *polyribonucleotides*. Although the majority of proteins and nucleic acids are extracted from living tissue, we have known for some years how to synthesize copolymers of amino acids or deoxyribonucleotides chemically with a perfectly predetermined chain sequence.

Moreover, it is now relatively easy to determine the sequence of monomeric units by chemical methods, provided that all the molecules in a solution of a given protein or nucleic acid are identical. The correspondence between the sequences in proteins and nucleic acids follows the rules of the *genetic code*. Such a perpetuation of the sequence (also called the *primary structure*) means that these molecules carry precise information. For proteins, this information is in principle sufficient to determine the three-dimensional structure of molecules when interactions with the solvent are taken into account.

The goal in *conformational analysis* is thus to find the rules or algorithms making it possible to predict the conformation adopted by the atoms in space from their linear succession in the polymer. We shall proceed step by step, first attempting to define the geometrical properties that characterize any monomeric chain. We shall then survey all the interaction energies between atoms or groups of atoms. It will then be possible to begin organizing certain parts of a polymer into certain types of *secondary structure*. Finally, we shall look for more complex spatial organizations leading to the final three-dimensional structure (or *tertiary structure*) of the polymer.

1

Geometry of a polymer chain

To represent a polymer in space, we shall first consider only the main backbone of the macromolecule: for example, the chain of carbon atoms in polyethylene, the peptide chain in proteins or the phosphodiester chain in nucleic acids. This amounts to saying that all such polymers have in common a small number of parameters and intrinsic properties characteristic of their geometry. For this purpose, the sequence of monomers is represented by a set of vectors $\boldsymbol{l}_1, \boldsymbol{l}_2, \ldots$ joining the atoms or groups of atoms represented by points $A_0, A_1, A_2, \ldots$ (Fig. 1.1). The ith bond links the $(i-1)$th atom and the ith atom and is characterized by the vector $\boldsymbol{l}_i$ with magnitude a_i. The set of n vectors $\boldsymbol{l}_i$ characterizes the molecule consisting of $(n+1)$ atoms at a given instant. In the absence of any data about the relations existing between these various vectors, the macromolecule can only be described statistically by general geometrical parameters.

1.1 Distance between the ends of a chain

The vector $\boldsymbol{r}$ connecting the two ends of the chain is given by

$$\boldsymbol{r} = \sum_i \boldsymbol{l}_i \tag{1.1}$$

For a purely statistical description, we require a scalar mean value, such as the square of the distance between the ends:

$$r^2 = \boldsymbol{r} \cdot \boldsymbol{r} = \sum_{i,j} \boldsymbol{l}_i \cdot \boldsymbol{l}_j$$

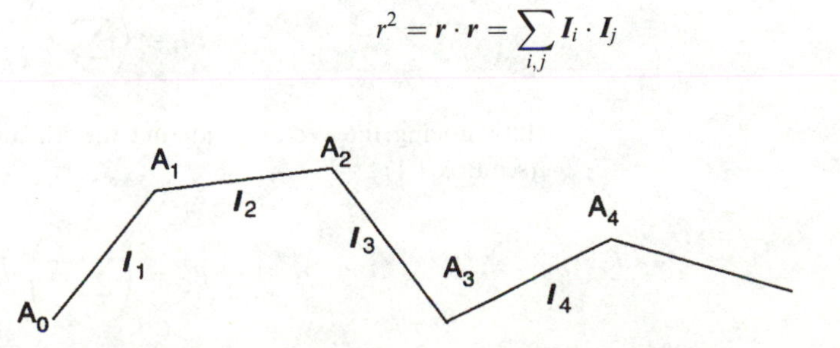

Figure 1.1 Representation of a polymer by a succession of atomic groups $A_0, A_1, A_2, \ldots$

Alongside the squared terms $\boldsymbol{l}_i \cdot \boldsymbol{l}_i = a_i^2$ in the scalar product, the cross-products $\boldsymbol{l}_i \cdot \boldsymbol{l}_j$ appear twice, so that

$$r^2 = \sum a_i^2 + 2\sum_{i<j} \boldsymbol{l}_i \cdot \boldsymbol{l}_j \tag{1.2}$$

by including only one (i, j) pair, that for which $i < j$.

For an assembly of molecules at a given instant, or for the whole set of configurations taken up by the same molecule as time passes, a distribution function for $\boldsymbol{r}$ can be defined whose mean square from eqn 1.2 is

$$\langle r^2 \rangle = \sum \langle a_i^2 \rangle + 2\sum_{i<j} \langle \boldsymbol{l}_i \cdot \boldsymbol{l}_j \rangle \tag{1.3}$$

The symbol $\langle \ldots \rangle$ indicates an average taken over a given molecule for a sufficiently long time, or an average at a given instant taken over a large number of molecules. If a^2 is the mean square of the distribution of bond lengths, we have $\sum \langle a_i^2 \rangle = na^2$ and so

$$\langle r^2 \rangle = na^2 + 2\sum_{i<j} \langle \boldsymbol{l}_i \cdot \boldsymbol{l}_j \rangle \tag{1.4}$$

If the orientations of the various bonds are completely independent of each other, the mean $\langle \boldsymbol{l}_i \cdot \boldsymbol{l}_j \rangle$ is that of a cosine, i.e. zero. In this case, we have

$$\langle r^2 \rangle = na^2 \tag{1.5}$$

The mean length between the ends, i.e. $\langle r^2 \rangle^{1/2}$, thus varies as the square root of the number of bonds.

This ideal chain model (called a *Gaussian chain*) assumes that the bonds are completely free to take any direction whatsoever in space, as if each atom were a universal joint allowing any possible orientation. This is the same as saying that each atom is separated from the previous one by a bond of mean length a but with a random direction. It is hardly surprising therefore to find for $\langle r^2 \rangle$ an expression exactly like that obtained for three-dimensional Brownian motion (see Chapter 7 in Part II).

1.2 Radius of gyration

The radius of gyration R_{G} is defined as the square root of the mean square of the distances ρ between the atoms and the centre of gravity of the chain. In other words

$$R_{\mathrm{G}}^2 = \left\langle \sum_{i=0}^{n} \rho_i^2 \right\rangle \Big/ (n+1) \tag{1.6}$$

Introducing the vector $\boldsymbol{r}_{ij}$ joining the ith and jth atoms, it can be shown that (see Box 1.1)

$$R_{\mathrm{G}}^2 = \left(\sum_{i<j} r_{ij}^2 \right) \Big/ (n+1)^2 \tag{1.7}$$

Box 1.1 Calculation of the radius of gyration

Consider two atoms *i* and *j* at respective distances ρ_i and ρ_j from the centre of gravity G. We have the standard formulae

$$r_{ij}^2 = \rho_i^2 + \rho_j^2 - 2\rho_i\rho_j \cos\phi_{ij}$$

and

$$\sum_i \sum_j r_{ij}^2 = \sum_j \sum_i \rho_i^2 + \sum_i \sum_j \rho_j^2 - 2\sum_i \sum_j \rho_i\rho_j \cos\phi_{ij}$$

The last term on the right is zero by definition of the centre of gravity and, in addition, from eqn 1.6:

$$\sum_i \rho_i^2 = \sum_j \rho_j^2 = (n+1)R_G^2$$

Carrying out a second summation simply amounts to multiplying by the number of atoms involved. However, $r_{ij} = 0$ for $i = j$. We must therefore multiply each sum not by $(n+1)$ but by $(n+1)/2$. We then obtain

$$\sum_i \sum_j r_{ij}^2 = (n+1)^2 R_G^2$$

which indeed gives eqn 1.7.

This definition of the radius of gyration is important since R_G is a characteristic length of the molecule, which remains significant whatever the shape of the polymer. Moreover, several experimental methods lead directly to a determination of R_G without any assumptions being made about the molecular shape.

In the case of an *ideal* or *Gaussian chain*, it can also be shown that the mean distance between the ends and the radius of gyration satisfy the simple relationship

$$R_G^2 = \langle r^2 \rangle / 6 \qquad (n \to \infty) \tag{1.8}$$

1.3 Constraints on the valence bond

The simplified geometrical approach used so far to describe the conformation of a macromolecule avoids the need to calculate the second term on the right-hand side of eqn 1.4. However, we know that the situation is not as simple as this and that the valence bonds are restricted in the directions they can assume (*steric constraints*). We also know that any atom in the macromolecule cannot occupy the space already occupied by another but is restricted to being a certain distance from such an atom determined by the short-range repulsive forces.

The scalar product $\sum_{i<j} \langle \boldsymbol{l}_i \cdot \boldsymbol{l}_j \rangle$ is a term denoting the correlation between bonds. Its calculation requires, amongst other things, a definition of the respective orientations of any two bonds *i* and *j* by taking into account all those between the *i*th and *j*th.

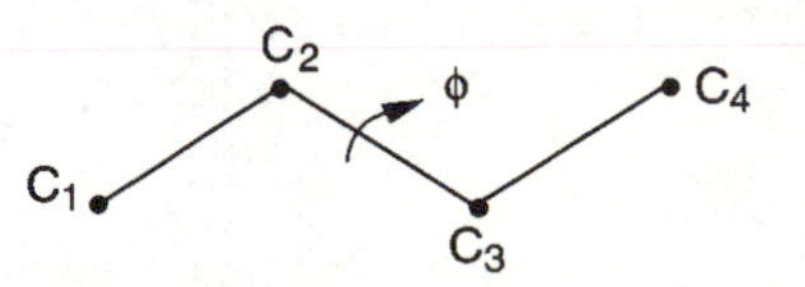

Figure 1.2 The orientation of the C_1C_2 bond with respect to the C_3C_4 bond defined by the rotation angle ϕ around

As a simple example of this, consider four successive carbon atoms C_1, C_2, C_3, C_4 in a chain (Fig. 1.2). The relative orientation of the C_3C_4 and C_1C_2 bonds is described by a rotation through an angle ϕ around C_2C_3. Because the valence angles $C_1C_2C_3$ and $C_2C_3C_4$ must be held constant, we keep the three atoms C_1, C_2 and C_3

arbitrarily in a plane and the C_3C_4 bond will then describe a cone, the *valence cone* (Fig. 1.3) around the axis defined by C_2C_3. The rotation is described by the angle ϕ made by OC_4 with a direction OM taken as origin. The angle ϕ is also the dihedral angle formed by the planes $C_1C_2C_3$ and $C_2C_3C_4$. This is the standard way of proceeding in organic chemistry in which the *Newman projection* makes it possible to define the conformations taken up by a molecule (e.g. n-butane, when the four carbon atoms belong to a saturated aliphatic hydrocarbon).

In order to define the origin M on the circle described by C_4, we can choose one of the two points at which the plane $C_1C_2C_3$ intersects the circle. If C_4 and C_1 are on the same side of C_2C_3, the position is described as *cis*. If, on the other hand, C_1 and C_4 are on opposite sides of C_2C_3 in the plane $C_1C_2C_3C_4$, the position is described as *trans* (Fig. 1.4).

It is convenient to take the *cis* position ($\phi = 0$) as the origin M and to count ϕ as positive if C_4 moves clockwise around the circle when looking in the direction from C_2 to C_3 (Fig. 1.5). Of course, if C_1 moves round its circle, the observation is made from C_4, i.e. in the direction from C_3 to C_2, and for a given relative rotation of C_1 and C_4 the value and sign of ϕ remain the same.

In both cases (*cis* and *trans*), the four atoms C_1, C_2, C_3, C_4 are in the same plane. As well as these two *eclipsed* positions, two *staggered* positions are possible, as in organic chemistry. With the same sign conventions, these will be called *gauche-plus* (*gauche+* or g^+) or *gauche-minus* (*gauche−* or g^-) depending on whether the rotation from the position *cis* ($\phi = 0$) is made towards the right or towards the left.

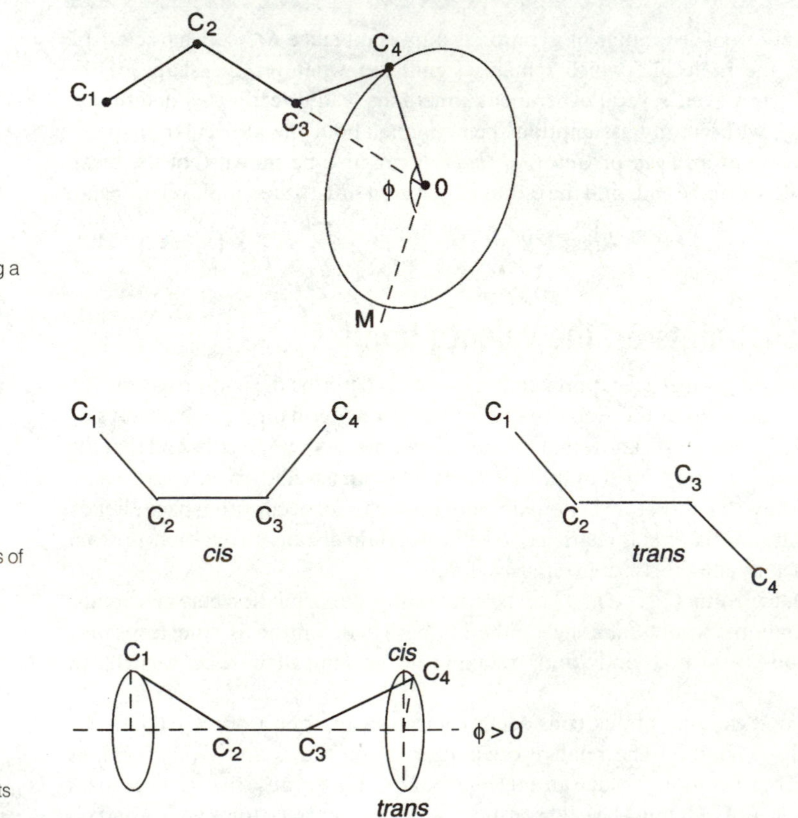

Figure 1.3 The C_3C_4 bond describing a valence cone of semi-vertical angle ϕ around C_2C_3.

Figure 1.4 The *cis* and *trans* positions of atoms C_1 and C_4.

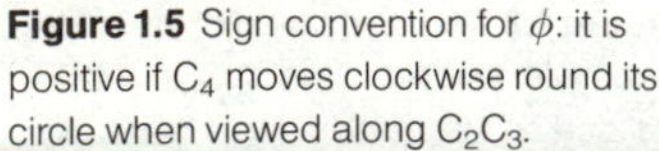

Figure 1.5 Sign convention for ϕ: it is positive if C_4 moves clockwise round its circle when viewed along C_2C_3.

If the angles made by successive bonds are known, the scalar product $\sum_{i<j} \boldsymbol{l}_i \cdot \boldsymbol{l}_j$ can be determined.

To illustrate the method, consider the case in which the valence angles between the different atoms are constant but it is assumed for simplicity that rotation occurs freely on the valence cone (there is no privileged value for the angle ϕ). Since each vector $\boldsymbol{l}$ has a length a, it is clear that $\boldsymbol{l}_i \cdot \boldsymbol{l}_{i+1} = a^2 \cos\phi$. When the $\boldsymbol{l}_{i+2}$ bond is made to rotate on its valence cone around $\boldsymbol{l}_{i+1}$, we have as before that on average $\boldsymbol{l}_{i+1} \cdot \boldsymbol{l}_{i+2} = a^2 \cos\phi$ and hence $\boldsymbol{l}_i \cdot \boldsymbol{l}_{i+2} = a^2 \cos^2\phi$. For the two bonds i and j we therefore have (with $j > i$)

$$\boldsymbol{l}_i \cdot \boldsymbol{l}_j = a^2 \cos^{j-i}\phi \tag{1.9}$$

When substituting eqn 1.9 into eqn 1.4, we have to take into account the number of segments separated by the same number $j - i = k$. If the chain has n segments there are $n - k$ pairs of this type. Carrying out the substitution and putting $x = \cos\phi$ gives

$$\langle r^2 \rangle = na^2 + 2a^2 \sum_k x^k (n-k) \tag{1.10}$$

The calculation of $\langle r^2 \rangle$ (Box 1.2) leads to the final expression:

$$\langle r^2 \rangle = na^2[(1+x)/(1-x) - 2x(1-x^n)/n(1-x)^2] \tag{1.11}$$

If n becomes large, the second term on the right becomes negligible compared with the first. In that case, if we assume that $x = 0.5$ ($\phi = 60°$), the term $(1+x)/(1-x) = 3$ and the mean square of the distance between the ends of the chain is now three times greater than that of a Gaussian chain of the same length.

In order to get a closer approximation to a real macromolecular chain, we must include the fact that there are limits to the rotation around the valence cone.

Box 1.2 Chain with limited rotation

In order to calculate $\sum_{k=1}^{n-1} x^k(n-k)$, we must find the two sums $\sum_{k=1}^{m} x^k$ and $\sum_{k=1}^{m} kx^k$ (with $m = n-1$). Putting $S_m = \sum_{k=1}^{m} x^k$, we have $S_{m+1} = \sum_{k=1}^{m+1} x^k = S_m + x^{k+1}$, but we can also write

$$S_{m+1} = x\sum(1 + x + \ldots + x^m) = x(1 + S_m)$$

Equating the two values of S_{m+1}, we obtain

$$S_m = x(1-x^m)/(1-x) \tag{1}$$

Differentiating S_m with respect to x, we obtain $\sum_{k=1}^{m} kx^{k-1}$ and hence

$$\sum_{k=1}^{m} kx^k = x\,\mathrm{d}S_m/\mathrm{d}x \tag{2}$$

From eqn 1 we have

$$\mathrm{d}S_m/\mathrm{d}x = [(1-x)(1-x^m-mx^m) + x(1-x^m)]/(1-x)^2$$

and hence

$$\sum_{k=1}^{m} kx^k = [x-(m+1)x^{m+1}]/(1-x) + x^2(1-x^m)/(1-x)^2$$

In terms of $n = 1 + m$, we obtain the two terms of eqn 1.10:

$$n\sum_{k=1}^{n-1} x^k = nx(1-x^{n-1})/(1-x)$$

$$\sum_{k=1}^{n-1} kx^k = (x-nx^n)/(1-x) - 2n^{-1}(x-nx^n)/(1-x) + x^2(1-x^{n-1})/(1-x)^2$$

Substituting these two terms into r^2/na^2 and regrouping the terms in $1/(1-x)^2$ we finally obtain

$$r^2/na^2 = (1+x)/(1-x) - 2x(1-x^n)/n(1-x)^2$$

The probability of a conformation characterized by angles $\phi_1, \phi_2, \phi_3, \ldots$ will depend on the potential energy $E_P(\phi_1, \phi_2, \phi_3, \ldots)$ of the molecule for this configuration, and more specifically on the Boltzmann factor $\exp[-E_P(\phi_1, \phi_2, \phi_3, \ldots)/k_BT]$, where k_B is Boltzmann's constant (a notation used throughout the book to distinguish it from other uses of the symbol k).

1.4 Torsional potential

Here again, a simplified model will enable us to gain a better understanding of the way to tackle the problem. Take the case of a chain of four carbon atoms. The rotation of C_4 with respect to C_1 is described by an angle ϕ around C_2C_3 (or again by the dihedral angle between the two planes $C_1C_2C_3$ and $C_2C_3C_4$). Experiment indicates that all values of ϕ are not equally likely. In other words, the potential function $E_P(\phi)$ has one or more minima separated by potential barriers. The Boltzmann factor $\exp[-E_P(\phi)/k_BT)]$ is a maximum in the potential wells. As a result, the molecule will spend most of its time in one of the configurations corresponding to a minimum in E_P. There are thus several *rotation isomers* or *rotamers*. The probability of finding the molecule outside one of these states is virtually negligible.

The situation is obviously more complex if the four carbon atoms form part of a long chain. Because of interactions between atomic groups, the minimum values of E_P are different but the general result remains. The function E_P representing the potential energy of rotation or the *torsional potential* can therefore be described by combinations of trigonometric functions.

The expression generally used is of the form

$$E_P(\phi) = (V/2)(1 \pm \cos n\phi) \quad (1.12)$$

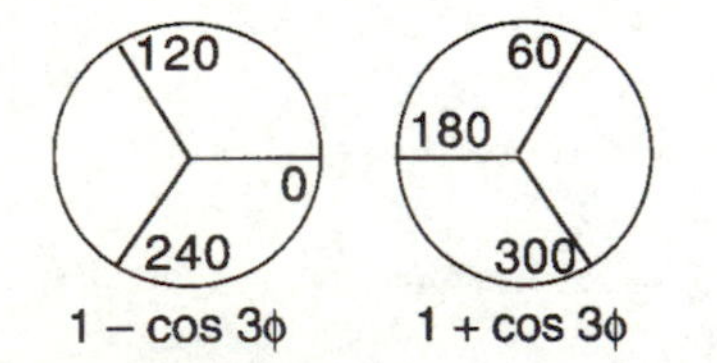

Figure 1.6 The positions of the minima in the functions $1 - \cos 3\phi$ and $1 + \cos 3\phi$.

where n is the multiplicity of potential wells and the + or − sign corresponds to a favoured eclipsed or staggered position respectively. V is the height of the potential barrier (e.g. in kcal mol^{-1}).

The function $1 + \cos 3\phi$ has three minima at $\phi = 60°$, 180° and 270°. For $1 - \cos 3\phi$, the minima are at 0°, 120° and 240°. The representation in Fig. 1.6, similar to the Newman projection, indicates the privileged positions, i.e. the locations of the various rotamers.

In order to analyse what happens with a large number of bonds and atoms, we must examine in more detail the various interaction energies between the atoms of a given macromolecule.

2

Intermolecular forces

2.1 Introduction

The simplest way to represent intermolecular forces is to start by considering two atoms A and B capable of forming a molecule AB, such as Cl and H to form HCl. When A and B are sufficiently far from each other (theoretically at infinity), the energy of the two-atom system is the sum of the energies E_A and E_B of the atoms taken separately. When A and B are separated by a distance r, a potential energy of interaction $E_P(r)$ must be added to the total energy. $E_P(r)$ is the work done in bringing the atoms from infinity to their final positions and is therefore given by

$$E_P(r) = \int_{\infty}^{r} F(r)\,\mathrm{d}r = -\int_{r}^{\infty} F(r)\,\mathrm{d}r \tag{2.1}$$

where $F(r)$ is the force exerted between A and B. By convention, a repulsive force is counted positive and an attractive force negative.

Differentiating both sides of eqn 2.1 leads to the usual relationship defining force in terms of potential energy:

$$F(r) = -\,\mathrm{d}E_P(r)/\,\mathrm{d}r \tag{2.2}$$

or, more generally,

$$\boldsymbol{F} = -\,\mathrm{grad}\,E_P \tag{2.3}$$

The components of the vector gradient (grad) are the partial derivatives of E_P with respect to each of the coordinates being used in the general problem. Equation 2.2 is appropriate when the two atoms or two molecules (forming two assemblies of atoms) have spherical symmetry. As soon as the relative orientations of the molecules themselves are involved, only eqn 2.3 is valid.

2.2 Electrical origin of the interaction energy

The interaction energy between molecules is due to electric forces. If the charge distribution in a molecule is known from wavefunction solutions of Schrödinger's equation, it is possible in principle to calculate the energy by superposition of

Coulomb potentials using the *Hellmann–Feynman theorem*:

$$V_{1,2} = \iint r_{ij}^{-1} \rho_1(\boldsymbol{r}_i)\rho_2(\boldsymbol{r}_j)\, \mathrm{d}\boldsymbol{r}_i\, \mathrm{d}\boldsymbol{r}_j \tag{2.4}$$

where $\boldsymbol{r}_i$ and $\boldsymbol{r}_j$ are the position vectors of the elementary charges $\rho(\boldsymbol{r}_i)\,\mathrm{d}\boldsymbol{r}_i$ distributed over molecules 1 and 2.

This type of calculation can be carried out by computer provided we have a precise knowledge of both the atomic structure of each molecule and their relative geometrical configuration. Several factors are, however, ignored in such a calculation:

(1) the deformation of the orbitals of one molecule in the electric field created by a neighbouring one;

(2) Brownian motion, which changes the respective positions of the molecules and leads to the replacement of instantaneous magnitudes by mean values;

(3) charge fluctuations on each molecule.

Because of these complications, classical formulae for interaction energies between charges and dipole moments (permanent, induced and spontaneous) are still used, and have the advantage of providing an analytical expression for the energy.

Our task is therefore to calculate the potential Ψ produced by a general charge distribution. It can be shown (see Box 2.1) that this is given by the sum of the potentials produced by a charge, a dipole, a quadrupole, etc. To begin with, we shall restrict ourselves to the charge–charge, charge–dipole and dipole–dipole interactions.

Charges will represent free ions or ionized groups in a biopolymer. Dipoles correspond to asymmetrical charge distributions such as those in H_2O, a peptide, a purine or pyrimidine base or an amino acid. In all these cases, the molecule is said to be *polar* and has a permanent electric dipole moment $\boldsymbol{p}$.

The potential energy of a charge q is given by the product of the charge and the potential Ψ at the position it occupies:

$$E_\mathrm{P} = q\Psi \tag{2.5}$$

The potential energy of a dipole is the negative scalar product of the dipole moment and the electric field $\boldsymbol{E}$ at the dipole:

$$E_\mathrm{P} = -\boldsymbol{p} \cdot \boldsymbol{E} \tag{2.6}$$

(see eqn 15.15).

2.3 Interactions between charges and permanent dipoles

2.3.1 Ion–ion

The interaction energy E_P between two ions with charges q_1 and q_2 separated by a distance r is given by *Coulomb's law*:

$$E_\mathrm{P} = q_1 q_2 / 4\pi\varepsilon\varepsilon_0 r \tag{2.7}$$

E_P is greater than 0 (repulsive energy) if q_1 and q_2 have the same sign, and less than 0 (attractive energy) if they have opposite signs; ε is the *dielectric constant* of the medium.

Box 2.1: **Equivalent distribution of any charge distribution**

Consider a set of charges $q_1, q_2, \ldots, q_i, \ldots$ (positive or negative) located at points $M_1, M_2, \ldots, M_i, \ldots$ at distances $\rho_1, \rho_2, \ldots, \rho_i, \ldots$ from an origin O at the centre of gravity of the molecule.

At a point P outside the charge system at a distance r from O and at distances $r_1, r_2, \ldots, r_i, \ldots$ from the charges, the potential Ψ_P is given by Coulomb's law as

$$\Psi_P = (1/4\pi\varepsilon_0)\sum_i q_i/r_i = (1/4\pi\varepsilon_0)\sum_i q_i/(r^2 + \rho_i^2 - 2\rho_i r \cos\theta_i)^{1/2}$$

where θ_i is the angle between OP and OM_i.

If we assume that $\rho_i < r$ so that $(-2\rho_i \cos\theta_i/r + \rho_i^2/r^2) < 1$, then Ψ_P can be obtained from the expansion of $(1 + \rho_i^2/r^2 - 2\rho_i \cos\theta_i/r)^{-1/2}$, i.e.

$$1 + \rho_i \cos\theta_i/r + \rho_i^2(3\cos^2\theta_i - 1)/2r^2 + \rho_i^3(5\cos^3\theta_i - 3\cos\theta_i)/2r^3 + \cdots$$

We therefore have

$$\Psi_P = (1/4\pi\varepsilon_0)\left[\sum q_i/r + \sum q_i\rho_i \cos\theta_i/r^2 + \sum q_i\rho_i^2(3\cos^2\theta_i - 1)/2r^3 + \cdots\right]$$

The *first term* is the potential that would be created by all the charges being located at the centre of gravity of the molecule. This will be zero if the total charge is zero.

In the *second term*, the quantity $\rho_i \cos\theta_i = a_i$ is simply the projection of the segments OM_i on to OP. The products $q_i a_i$ (a charge multiplied by a length) are dipole moments. If the sum is not zero, the *second term* is the potential due to a dipole. If $\sum q_i = 0$ (a molecule with the same number of positive and negative charges), we proceed differently. First we define a centre of gravity G_+ of the positive charges by

$$OG_+ = \sum q_i^+ \rho_i \Big/ \sum q_i^+$$

and similarly for a centre of gravity of the negative charges

$$OG_- = \sum q_i^- \rho_i \Big/ \sum q_i^-$$

If the points G_+ and G_- do not coincide but are separated by a distance l, and if we let Q be the sum of the charges of the same sign, then the quantity $p = Ql$ defines the equivalent dipole moment of the distribution, still with the assumption that $r > \rho_i$. If the direction from G_- to G_+ is given to the length l, then the dipole moment is a vector $\boldsymbol{p}$ making an angle θ with OP. The corresponding potential is

$$\Psi_{P,\text{dipole}} = (p\cos\theta)/4\pi\varepsilon_0 r^2$$

The *third term* is the potential due to a *quadrupole*, i.e. that of a charge distribution whose total charge and equivalent dipole moment are both zero.

Note: the potential Ψ_P can also be written in the following form for each value of i:

$$\Psi_P = (q/4\pi\varepsilon_0 r)\sum_{i=0}^{\infty} P_i(\cos\theta)\cdot(\rho/r)^i$$

where $P_0 = 1$, $P_1 = \cos\theta$, $P_2 = 3\cos^2\theta - 1$, $P_3 = 5\cos^3\theta - 3\cos\theta$, etc. These polynomials in $\cos\theta$ are *Legendre polynomials* (see Mathematical Appendix).

In many cases, it is difficult to assign a value to ε, particularly near charges. In the rest of this chapter, therefore, all expressions will be written with $\varepsilon = 1$ (vacuum), knowing that for other media we only have to replace ε_0 by $\varepsilon\varepsilon_0$ (see Part IV).

2.3.2 Ion–dipole

If a molecule with a *permanent* dipole moment $\boldsymbol{p}$ is located at a point where the electric field due to an ion is $\boldsymbol{E}$, then it has an energy

$$E_{\mathrm{P}} = -\boldsymbol{p} \cdot \boldsymbol{E} = -pE\cos\theta$$

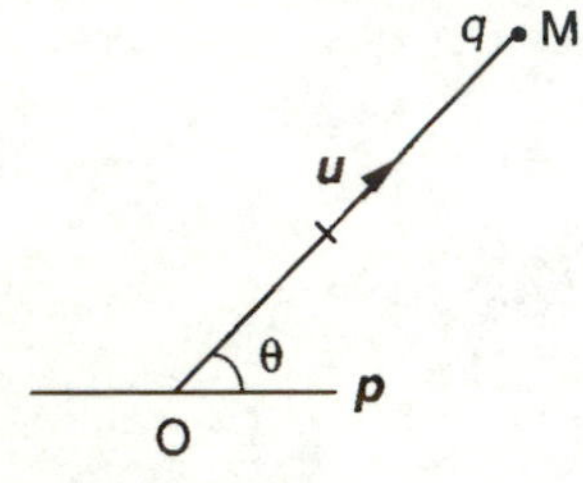

Figure 2.1 Geometry for the interaction between a dipole at O and an ion at M.

where θ is the angle between the direction of the dipole moment and the line joining the centre of the dipole O to the ion of charge q at the point M (see Fig. 2.1).

Since the field $\boldsymbol{E}$ produced by a charge q at a distance r is

$$E = \boldsymbol{u}q/4\pi\varepsilon_0 r^2$$

where $\boldsymbol{u}$ is a unit vector along OM, then

$$E_{\mathrm{P}} = -pq\cos\theta/4\pi\varepsilon_0 r^2 \tag{2.8}$$

If the molecule is undergoing Brownian motion, then all possible orientations of $\boldsymbol{p}$ are possible. The statistical weight of dipoles making an angle θ is given by the *Boltzmann factor*:

$$\exp(-E_{\mathrm{P}}/k_{\mathrm{B}}T) = \exp(pE\cos\theta/k_{\mathrm{B}}T)$$

The probability that the vector $\boldsymbol{p}$ lies between two cones with semi-vertical angles θ and $\theta + \mathrm{d}\theta$ is $\sin\theta\,\mathrm{d}\theta/2$. The projection $p\cos\theta$ of the dipole moment on to the direction of the field $\boldsymbol{E}$ thus has the mean value

$$\langle p \rangle = \frac{\int_0^{\pi} \exp(pE\cos\theta/k_{\mathrm{B}}T)p\cos\theta\sin\theta\,\mathrm{d}\theta/2}{\int_0^{\pi} \exp(pE\cos\theta/k_{\mathrm{B}}T)\sin\theta\,\mathrm{d}\theta/2}$$

Putting $pE/k_{\mathrm{B}}T = x$, this becomes a standard calculation (see Box 15.2) and gives

$$\langle p \rangle = p\mathcal{L}(x)$$

where $\mathcal{L}(x) = \coth x - 1/x$.

If $\boldsymbol{p} \cdot \boldsymbol{E} \ll k_{\mathrm{B}}T$ and $x \ll 1$, $\mathcal{L}(x)$ reduces to the first term of its power expansion, $x/3$, and hence

$$\langle p \rangle \approx p^2 E/3k_{\mathrm{B}}T$$

The mean energy is therefore

$$E_{\mathrm{P}} = -\langle \boldsymbol{p} \cdot \boldsymbol{E} \rangle = -p^2E^2/3k_{\mathrm{B}}T \tag{2.9}$$

Substituting the expression for E in eqn 2.9 gives, finally:

$$E_{\mathrm{P}} = -p^2q^2/(4\pi\varepsilon_0)^2 3k_{\mathrm{B}}Tr^4 \tag{2.10}$$

The energy varies as $1/r^4$ (and no longer as $1/r^2$ as it did in the absence of any motion).

If, on the other hand, $pE > k_B T$, the dipole takes up a direction near the position corresponding to minimum energy, i.e. $\theta = 0$, as in:

$$\overset{+\qquad\quad -}{\underset{p}{\longrightarrow}} \quad \bullet q^+$$

This is a situation encountered in the hydration of ions. The permanent dipole moment of H_2O takes up a mean position in the radial direction of the field of the ion.

2.3.3 Dipole–dipole

It can be shown (see Box 2.2) that the interaction energy between two dipoles $\boldsymbol{p}_1$ and $\boldsymbol{p}_2$ separated by a distance r is

$$E_P = p_1 p_2 K / 4\pi\varepsilon_0 r^3 \qquad (2.11)$$

where $K = \sin\theta_1 \sin\theta_2 \cos(\phi_1 - \phi_2) - 2\cos\theta_1 \cos\theta_2$. The angles θ_1, θ_2, ϕ_1, ϕ_2 are defined in Fig. 2.2.

If both dipoles are undergoing Brownian motion affecting their rotation (the angle θ) but not the distance r, then a mean value for K must be calculated.

Assuming that $\langle E_P \rangle \ll k_B T$, it is found that the mean value $\langle E_P \rangle$ is given by

$$\langle E_P \rangle = -2p_1^2 p_2^2 / 3k_B T (4\pi\varepsilon_0)^2 r^6 \qquad (2.12)$$

The interaction energy thus varies as $1/r^6$ (and no longer as $1/r^3$ as it does in the absence of motion).

2.4 Induced dipoles

So far, we have been considering interactions between charges, dipoles, etc., that exist in the molecules from the start without any external electrical influence. However, as soon as a collection of atoms is placed in an electric field $\boldsymbol{E}$, their charges (mainly electrons) are displaced and the molecule is polarized, producing an induced dipole (see section 15.3). The moment $\boldsymbol{p}$ of the dipole is related to the field $\boldsymbol{E}$ by

$$\boldsymbol{p} = \alpha\varepsilon_0 \boldsymbol{E} \qquad (2.13)$$

where α is the *polarizability* of the molecule, assumed to a first approximation to be a scalar quantity characteristic of the molecule and proportional to the molecular volume (α has the dimensions of a volume).

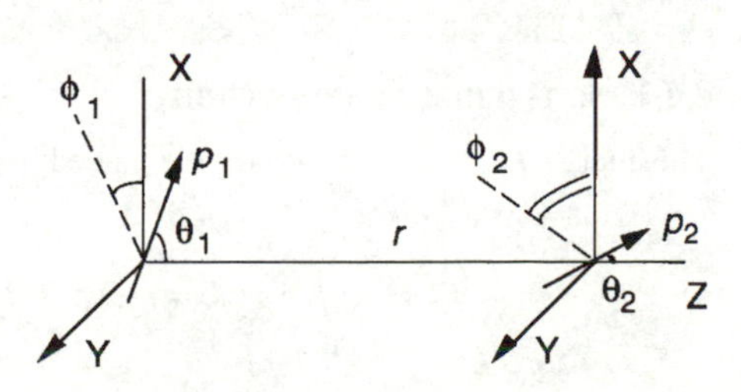

Figure 2.2 Geometrical parameters for dipole–dipole interaction.

Box 2.2 Interaction between two dipoles

The potential V at a point P situated at a distance r from a dipole of moment $\boldsymbol{p}$ at O is given by the standard expression

$$V = (p\cos\theta)/4\pi\varepsilon\varepsilon_0 r^2$$

where θ is the angle between $\boldsymbol{p}$ and the direction OP. The components of the electric field are therefore

$$E_r = -\partial V/\partial r = (2p\cos\theta)/4\pi\varepsilon\varepsilon_0 r^3$$
$$E_\theta = -r^{-1}\partial V/\partial\theta = (p\sin\theta)/4\pi\varepsilon\varepsilon_0 r^3$$

and hence, using the unit vectors $\boldsymbol{u}_r$ and $\boldsymbol{u}_\theta$ along and perpendicular to OP respectively, we have:

$$\boldsymbol{E}(r) = (2p\cos\theta\,\boldsymbol{u}_r + p\sin\theta\,\boldsymbol{u}_\theta)/4\pi\varepsilon\varepsilon_0 r^3$$

Similarly, the dipole moment $\boldsymbol{p}$ can be written as

$$\boldsymbol{p} = p(\boldsymbol{u}_r\cos\theta - \boldsymbol{u}_\theta\sin\theta)$$

so that

$$p\boldsymbol{u}_\theta\sin\theta = -\boldsymbol{p} + p\boldsymbol{u}_r\cos\theta = -\boldsymbol{p} + (\boldsymbol{p}\cdot\boldsymbol{u}_r)\boldsymbol{u}_r$$

Substituting this in the expression for $\boldsymbol{E}(r)$:

$$\boldsymbol{E}(r) = [3(\boldsymbol{p}\cdot\boldsymbol{u}_r)\boldsymbol{u}_r - \boldsymbol{p}]/4\pi\varepsilon\varepsilon_0 r^3$$

Now consider two dipoles $\boldsymbol{p}_1$ and $\boldsymbol{p}_2$ separated by a distance r (Fig. 2.2). The directions of the dipoles are defined by angles θ_1 and θ_2 made with the axis OZ and angles ϕ_1 and ϕ_2 that their projections on to the plane XOY make with OX. The interaction energy is

$$W_{1,2} = -\boldsymbol{p}_1\cdot\boldsymbol{E}_2$$
$$W_{1,2} = -\boldsymbol{p}_1.[3(\boldsymbol{p}_2\cdot\boldsymbol{u}_r)\boldsymbol{u}_r - \boldsymbol{p}_2]/4\pi\varepsilon\varepsilon_0 r^3$$

and putting $\boldsymbol{r} = r\boldsymbol{u}_r$

$$4\pi\varepsilon\varepsilon_0 W_{1,2} = \boldsymbol{p}_1\cdot\boldsymbol{p}_2/r^3 - 3(\boldsymbol{p}_1\cdot\boldsymbol{r})(\boldsymbol{p}_2\cdot\boldsymbol{r})/r^5$$

an expression symmetrical in 1 and 2. This expression can also be written as follows, introducing unit vectors $\boldsymbol{u}_1$ and $\boldsymbol{u}_2$ along the directions of the dipoles $\boldsymbol{p}_1$ and $\boldsymbol{p}_2$ respectively:

$$W_{1,2} = Kp_1p_2/4\pi\varepsilon\varepsilon_0 r^3$$

where $K = \boldsymbol{u}_1\cdot\boldsymbol{u}_2 - 3(\boldsymbol{u}_1\cdot\boldsymbol{u}_r)(\boldsymbol{u}_2\cdot\boldsymbol{u}_r)$ with $\boldsymbol{u}_1$ having components $\sin\theta_1\cos\phi_1$, $\sin\theta_1\sin\phi_1$, $\cos\theta_1$ and $\boldsymbol{u}_2$ having components $\sin\theta_2\cos\phi_2$, $\sin\theta_2\sin\phi_2$, $\cos\theta_2$. Using $\cos(\phi_1-\phi_2) = \cos\phi_1\cos\phi_2 + \sin\phi_1\sin\phi_2$, we obtain

$$K = \sin\theta_1\sin\theta_2\cos(\phi_1-\phi_2) - 2\cos\theta_1\cos\theta_2$$

2.4.1 Ion–molecule interaction

The energy E_P of the induced dipole placed in the inducing field $\boldsymbol{E}$ is

$$E_P = -\int_0^E \boldsymbol{p}\cdot d\boldsymbol{E} = -\varepsilon_0\alpha\int_0^E \boldsymbol{E}\cdot d\boldsymbol{E} = -\varepsilon_0\alpha E^2/2 \quad (2.14)$$

When the field E is due to a charge q, then

$$E = q/4\pi\varepsilon_0 r^2$$

giving

$$E_P = -\varepsilon_0 \alpha q^2/(4\pi\varepsilon_0)^2 2r^4 \quad (2.15)$$

This energy is always due to an attractive force and varies as $1/r^4$.

2.4.2 Dipole–molecule interaction

The field $\boldsymbol{E}$ produced at a point P (Fig. 2.3) by a dipole of moment $\boldsymbol{p}_O$ at a point O has two components (see section 15.2.1)

$$E_r = (2p_O \cos\theta)/4\pi\varepsilon_0 r^3 \qquad E_\theta = (p_O \sin\theta)/4\pi\varepsilon_0 r^3 \quad (2.16)$$

where r is the distance OP and θ is the angle between the dipole and OP.

Still assuming that the molecule is isotropic with a scalar polarizability α, the energy will depend on the field E given by

$$E^2 = (4\cos^2\theta + \sin^2\theta)p_O^2/(4\pi\varepsilon_0)^2 r^6 \quad (2.17)$$

and, from eqn 2.14, the energy will be

$$E_P = -\varepsilon_0 \alpha(3\cos^2\theta + 1)p_O^2/2(4\pi\varepsilon_0)^2 r^6 \quad (2.18)$$

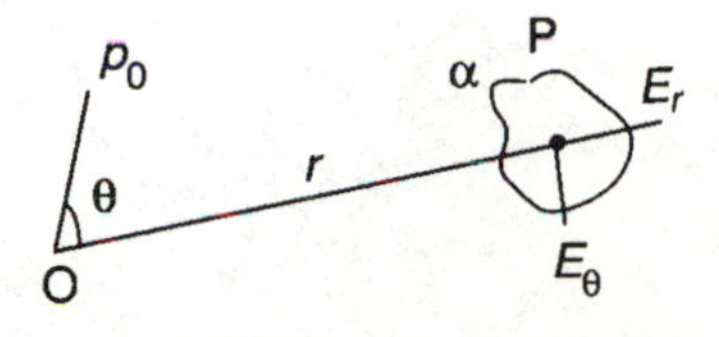

Figure 2.3 Geometrical parameters for dipole–molecule interaction.

If the molecule and/or the dipole are undergoing Brownian motion, the spatial mean of $\cos^2\theta$ (i.e. 1/3) is the appropriate value, and hence

$$E_P = -\varepsilon_0 \alpha p_O^2/(4\pi\varepsilon_0)^2 r^6 \quad (2.19)$$

If the molecule is anisotropic, the contributions from each component of the field and the value of the polarizability in the directions of each component must be used. However, although this calculation gives a different value for the numerical coefficient of E_P, it does not alter its dependence on r^{-6} and $p_O{}^2$.

2.4.3 Interaction between two induced dipoles

This type of interaction is present for all molecules, since it arises from fluctuations in electron density, which, at any given moment, create an instantaneous dipole whose external field induces a dipole in a neighbouring molecule. From the above calculation, the interaction energy is proportional to the inducing field and, since this varies as $1/r^3$, the energy will vary as $1/r^6$. An exact calculation of E_P can only be carried out by applying quantum mechanics. It should, however, be pointed out that the frequency of the fluctuations is very high and that, as a result, the polarizability is no longer a static quantity but is related to the motion of the electrons under the action of an electromagnetic field in the optical region of the spectrum.

Because of this, the interaction forces are also called *dispersion forces* since they can be related to interactions between radiation and matter, the province of the theory of dispersion of light in material media. By combining the classical theory of Drude with a quantum treatment of the harmonic oscillator, it is possible to calculate E_P. The result is:

$$E_P = -3h\nu_1\nu_2\alpha_1\alpha_2/2(\nu_1 + \nu_2)r^6 \quad (2.20)$$

where it is assumed that each molecule can be represented by a harmonic oscillator with a natural frequency ν_1 and ν_2 and static polarizabilities α_1 and α_2 (h is Planck's constant). If the two molecules are identical then

$$E_P = -3h\nu\alpha^2/4\pi^2 r^6 \tag{2.21}$$

As the distance r between the oscillators increases, the electromagnetic field produced by one of the instantaneous dipoles takes a time $t = r/c$ to propagate, where c is the speed of light. As a result, the interaction with the second induced dipole only occurs after a time $2r/c$. If the first dipole has had time to change its direction during this period of time, the interaction energy is reduced, which emphasizes the 'short-range' nature of this type of interaction.

A semi-empirical formula obtained from quantum mechanics is often used instead of eqn 2.20. If I_1 and I_2 are the ionization energies of the molecules, then it is found that

$$E_P = -3I_1 I_2 \alpha_1 \alpha_2 / 2(I_1 + I_2) r^6 \tag{2.22}$$

2.5 Forms of potential energy

2.5.1 General table

If we put $C = (4\pi\varepsilon_0)^{-1} = 9 \times 10^9$ SI units, the various forms taken by the potential energy can be summarized as in Table 2.1. It can be checked that every expression has the dimensions of energy, ML^2T^{-2}, remembering that ε_0 itself has dimensions of $M^{-1}L^3T^2Q^2$. It can also be seen that a $1/r$ law typical of a Coulomb potential is replaced by a $1/r^n$ law with $4 < n < 6$. In other words, long-range Coulomb interactions have only short-range influence when the system has no net charge, particularly under the averaging effect of Brownian motion.

It can be seen that, apart from the purely Coulomb energy, the value of E_P varies as r^{-4} or r^{-6} when Brownian motion occurs. The last three expressions in r^{-6} are included in what are known as *van der Waals interactions* resulting from *van der Waals forces*.

These formulae are only valid for small molecules with spherical symmetry. For elongated molecules (cylinders or discs) or for large spherical molecules, the energy variation with distance obeys an r^{-n} law with $n < 6$.

2.5.2 Spherically symmetric potential functions

Even if we restrict ourselves to van der Waals interactions for which the attractive potential energy varies as r^{-6}, many forms have been proposed for the functions $E_P(r)$ giving the total interaction energy between two uncharged molecules. In every case, the attractive potential is combined with a repulsive potential due to the very

Table 2.1 Expressions for interaction energies ($C = 1/4\pi\varepsilon_0$)

Type of interaction	E_P without Brownian motion	Mean value of E_P with Brownian motion
charge–charge (ion–ion)	Cq_1q_2/r	Cq_1q_2/r
charge–permanent dipole	$Cpq\cos\theta/r^2$	$C^2p^2q^2/3k_BTr^4$
permanent dipole–permanent dipole	Cp_1p_2k/r^3	$-2C^2p_1^2p_2^2/3k_BTr^6$
charge–molecule	$-C^2\varepsilon_0\alpha q^2/2r^4$	$-C^2\varepsilon_0\alpha q^2/2r^4$
permanent dipole–molecule	$-C^2\varepsilon_0\alpha(3\cos^2\theta+1)p_0^2/2r^6$	$-C^2\varepsilon_0\alpha p_0^2/r^6$
induced dipole–induced dipole	$-3h\nu_1\nu_2\alpha_1\alpha_2/2(\nu_1+\nu_2)r^6$	$-3h\nu_1\nu_2\alpha_1\alpha_2/2(\nu_1+\nu_2)r^6$

Box 2.3 **Lennard–Jones potential**

The potential function can be written in the following form, putting $x = r/r_m$ where r_m is the equilibrium distance at which the energy is a minimum:

$$U(x) = U_0(x^{-12} - 2x^{-6})$$

This passes through a minimum when $U'(x) = 0$, i.e. when $x = 1$.

If we examine the variation of $U(x)$ when $x < 1$, we obtain the figures in Table 2.2, showing a very rapid increase in the potential, which passes through zero for $x = 2^{-1/6} = 0.891$. Near $x = 1$, $U(x)$ can be expanded by putting $x = 1 + \varepsilon$. This gives $U(x) = -U_0(1 - 36\varepsilon^2)$.

The potential near $x = 1$ and $U = -U_0$ can therefore, to a first approximation, be described by the quadratic variation of the harmonic potential (see Chapter 6).

Table 2.2 Variation of U with x

x	U/U_0
0.891	0
0.8634	1
0.8458	2
0.8326	3
0.7937	8
0.77	13.6

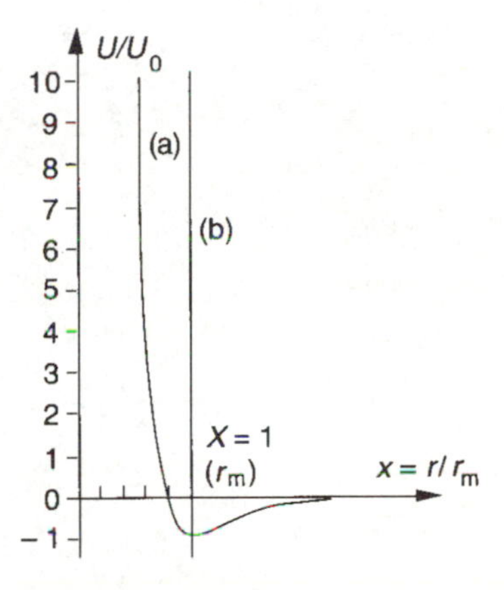

Figure 2.4 Potentials with spherical symmetry in reduced coordinates: (a) Lennard-Jones potential; (b) hard-sphere potential.

large repulsive forces between electron clouds coming into contact. The one most frequently used is the *Lennard–Jones potential* (Box 2.3):

$$E_P(r) = E_0(r_0/r)^{12} - 2E_0(r_0/r)^6 \qquad (2.23)$$

where the term in r^{-12} is a short-range repulsive potential. This function passes through a minimum ($dE_P/dr = 0$ and $E_P = -E_0$) for $r = r_0$. It is zero for $r = r_0{}^{-1/6} = 0.89r_0$ (Fig. 2.4)

In many cases, however, an even simpler form is used for $E_P(r)$ by assuming that the molecules can be modelled by hard spheres of radius r_0:

- if $r < r_0$ $\quad E_P(r) \to \infty$
- if $r > r_0$ $\quad E_P(r) = 0$

This is the potential implicitly used when so-called 'compact' molecular models are built, each atom being represented by a sphere or part of a sphere with a radius equal to the 'van der Waals radius'. The method for determining conformations (see Chapter 5) also depends on this simplified model of molecular contacts.

2.6 The hydrogen bond

Alongside the above interactions between molecules, which lead to an analytical form for E_P by classical methods, there is another type of intermolecular bond for which E_P can no longer be considered as a spherically symmetric function: the *hydrogen bond*. The interaction in this case is directional and it can only be studied completely by using quantum mechanics.

2.6.1 **Definition and properties**

The term *hydrogen bond* denotes a particular type of interaction occurring between a *proton donor* group D, which is strongly polar, such as FH, OH, NH, SH, etc., and a *proton acceptor* atom A, which is strongly electronegative, such as F, O, N, etc. The electronegativity in the latter case arises through the presence of electron pairs that, for most of the time, are not involved in a covalent bond, as in:

The bond is of the form D–H···A. The distance between D and A depends on the two atoms involved. For F–H···F it is 2.25 Å, for O–H···O it is 2.75 Å and for N–H···O it is 2.9–3.0 Å.

At first sight, this 'attraction' of the proton looks similar to a simple electrostatic type of force, which might provide a simple description of the bond. In such a simple form, each atom would be replaced by a positive charge and the electrons by negative charges for the D–H bond, and the electrons by two negative charges for the lone pair of A. Knowing the geometry of the system, the bond energy could be calculated by applying Coulomb's law. The values found in this way are close to experimental values and the decrease in energy in passing from F to N, i.e. with decreasing electronegativity, is also well verified. However, there is no correlation between this energy and the dipole moment of A or D. Moreover, the distance DA is always smaller than the sum of the van der Waals radii, which indicates the existence of forces other than electrostatic ones.

The quantum-mechanical treatment is difficult, but it brings out the fact that, in addition to an attractive energy of the Coulomb type, there are three other forms of energy:

1. The distance between the acceptor atom and H is less than the sum of the van der Waals radii (e.g. the H–O bond length in ice is 1.75 Å instead of 2.6 Å). It follows that there must be a short-range repulsive energy.
2. An attractive energy of the *charge-transfer* type (see Box 2.4) is involved but the hydrogen bond is not a simple charge-transfer complex since in this case it would require the quantity of charge transferred to increase with the strength of the complex, which is not observed.
3. The polarizabilities of A and D lead to a dispersion (London) energy due to the attractive coupling of the instantaneous dipoles produced by charge fluctuations.

The energies due to 1 and 2 are of the same order of magnitude as the electrostatic energy but, since they are opposite in sign, they will cancel each other, and this explains why the electrostatic theory explains the hydrogen bond so well in some cases.

At the moment, there is no analytical form for the potential energy corresponding to the hydrogen bond.

2.6.2 **Geometry and bifurcated bond**

The hydrogen bond plays an essential role in explaining the physical properties of water (see Part III) and in the binding of water molecules to biopolymers.

Box 2.4 **Electron donor–acceptor (D–A) complexes**

When two groups or two molecules A and D interact, the perturbation of the molecular orbitals of D by A may be such that the electron density becomes significant a long way from D, i.e. very near to A. The situation is then as if A had captured a fraction of D's electronic charge; hence the name *charge-transfer complex* is given to this type of interaction. An energy due to electrostatic attraction has to be added to the van der Waals type interactions between A and D, but it does not form the major part of the interaction energy.

Among donors are groups with lone pairs (amines, ethers, alcohols) or aromatic cyclic hydrocarbons (π donors). Among acceptors are ions with a free orbital such as Ag^+ and the π acceptors formed from cyclic aromatics with electrophilic groups such as NO_2, CN, Cl, Br, I and quinolic systems.

The classification into donors and acceptors is somewhat relative, with some molecules behaving as D or A depending on the partner. The complex AD is represented in the ground state by a wavefunction:

$$\Psi_F(AD) = a\Psi_0(A, D) + b\Psi_1(A^-D^+)$$

where Ψ_0 is the wavefunction describing the complex AD without a covalent bond and Ψ_1 describes the system when an electron has been transferred completely from D to A.

In the excited state, the contribution of the wavefunction Ψ_1 is greater. A transition can be observed between the ground state and this new excited state, which has a frequency and intensity characteristic of the complex. The various energy levels can in principle be calculated, but we shall be content with giving a simple diagram of a donor–acceptor system (Fig. 2.5).

A transition from the ground state to the excited state may be considered as causing an electron to pass from the highest occupied orbital of the donor to the lowest unoccupied orbital of the acceptor. When the distribution of energy levels in the two molecules is known, it is then possible to predict the existence of donor–acceptor pairs and thus the formation of bonds.

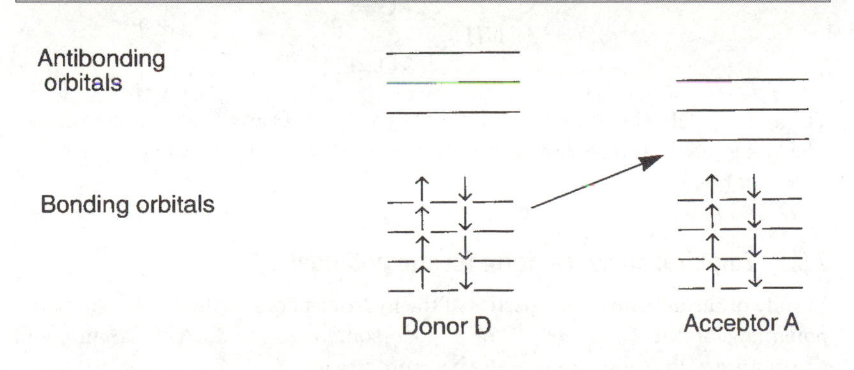

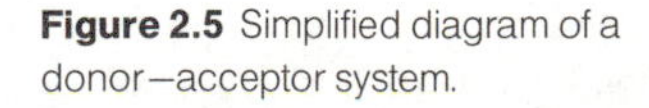
Figure 2.5 Simplified diagram of a donor–acceptor system.

In proteins, we shall see that hydrogen bonds 'lock' secondary structures such as α-helices, β-sheets and β-turns. The majority of hydrogen bonds are between NH and CO groups in the peptide chain. The oxygen of CO can, as in water, accept two hydrogens. A spatial description of this bond (Fig. 2.6) requires three

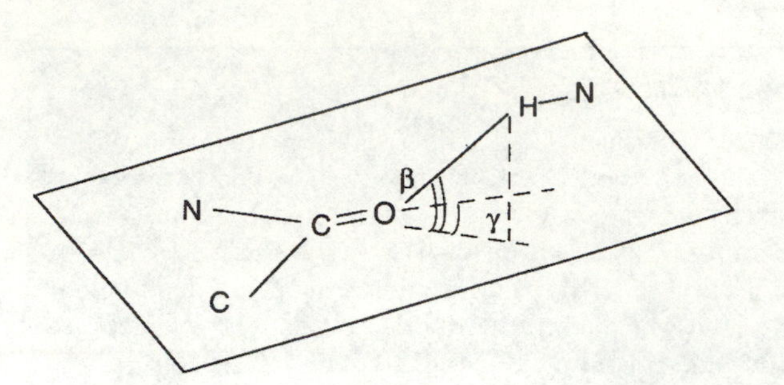

Figure 2.6 Spatial parameters in the peptide bond.

parameters:

(1) the N–H–O angle, which is a measure of the linearity of the hydrogen bond;

(2) the O–H distance, which is an indication of the bond energy;

(3) the C=O–H angle, or more precisely the angles β and γ, which are a measure of how far the three atoms deviate within the plane and out of it from the theoretically linear alignment of the C, O and H atoms.

For small molecules, the angle N–H–O is 165° on average and the O–H distance is 1.87 Å. The latter distance may reach 1.95 Å in proteins, with a higher mean deviation. The C=O–H angle has a mean value of 120° with values of β lying between 0 and 40°. A statistical survey of a large number of proteins with structures known to high resolution has made it possible to examine more than 1500 hydrogen bonds defined by H···O < 2.4 Å and C=O···H > 90°. There is a tendency for the hydrogen bond to be oriented in the direction of the sp_2 orbital in the hydrogen atom.

Similarly, in nucleic acids, we shall see hydrogen bonds NH···N or O···H_2N being established between the complementary bases (A and T/U, G and C) and contributing to a stabilization of double helices (type B, A or Z) and of some types of triple helix.

In the bond between water molecules and proteins or nucleic acids, a single proton may be shared between two acceptor atoms (*bifurcated* bond) as in:

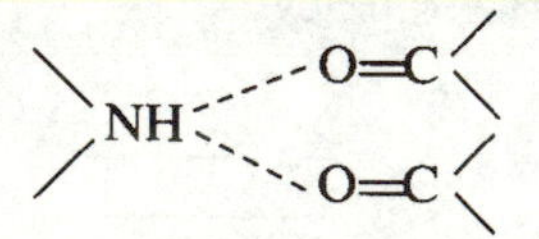

The angle O–H–O is close to 90° but the two N–H–O angles are often unequal. The largest angle also corresponds to the shortest O–H distance and hence to the strongest bond.

2.6.3 Empirical analytical form for the potential

In spite of the directional properties of the hydrogen bond, spherically symmetric potentials of the form $Ar^{-12} - Br^{-10}$ are sometimes used. A form more in accordance with reality has recently been proposed:

$$E_P = \sum_{ij} \cos\theta(-A' r_{ij}^{-6} + B' r_{ij}^{-12}) + (1 - \cos\theta)(-A r_{ij}^{-6} + B r_{ij}^{-12}) \qquad (2.24)$$

where the angle θ is a measure of the non-linearity of the hydrogen bond (for example, the angle N–H–O) and where the constants A, B, A', B' depend on the pairs of atoms involved.

3

Calculation of conformations

In determining the conformation of a macromolecular chain, calculating terms of the form $\boldsymbol{l}_i \cdot \boldsymbol{l}_j$ leads only to a value for the mean distance $\langle r \rangle$ or the radius of gyration. This is an inadequate description of the polymer, particularly where proteins are concerned.

If we wish to be able to give accurate spatial coordinates for each atom, we must use a set of interaction potentials between the atoms that takes into account:

- the valence bonds between bound atoms
- the forces of attraction and repulsion between non-bonded atoms.

3.1 Bound atoms

3.1.1 Valence bonds

Around the equilibrium value r_0, we assume a harmonic potential of the following type:

$$E_{\mathrm{P},1} = K_1(r - r_0)^2 \tag{3.1}$$

The values for K_1 in Table 3.1 are given by the AMBER program widely used in conformation calculations (Weiner *et al.*, 1984).

Similarly, around the equilibrium valence angle θ_0 we assume a harmonic potential of the type:

$$E_{\mathrm{P},2} = K_2(\theta - \theta_0)^2 \tag{3.2}$$

and using AMBER once again, the typical values in Table 3.2 are found for K_2.

In practice, the values must be adjusted for each type of monomer by using those obtained from small organic molecules whose structure is completely known, such as ethyl methyl ketone $CH_3{-}CO{-}C_2H_5$ or

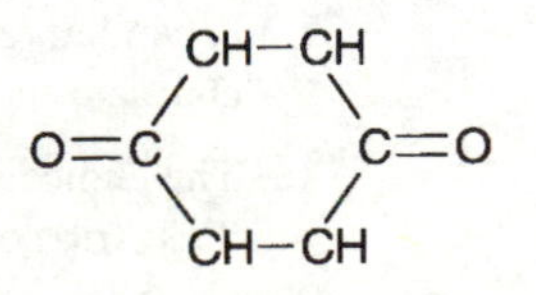

Table 3.1 Values of K_1 given by the AMBER program

Bond	r_0 (Å)	K_1 (kcal mol^{-1} Å^{-2})
C–C	1.507	317
C=C	1.336	570
C–N	1.449	337
C=N	1.273	570

Table 3.2 Values of K_2 given by the AMBER program

Angle	θ_0 (deg)	K_2 (kcal mol^{-1} rad^{-2})
C–C–C	112.4	63
C–N–C	121.9	50
C–C–N	111.2	80

3.1.2 Torsional potential

As we saw previously (eqn 1.12), we use an expression of the following type for the potential energy due to rotation around the valence cone:

$$E_{\mathrm{P},3} = (V/2)[1 + \cos(n\phi - \gamma)] \tag{3.3}$$

For example, X–C–C–X has $n = 3$, $\gamma = 0$ and $V/2 \approx 2$ kcal mol^{-1}, giving potential barriers of about 4 kcal mol^{-1}. The interactions involved in the rotation around the valence bond are mainly those between atoms bonded for example to two carbons, and those of electron orbitals surrounding the two carbons.

3.2 Non-bonded atoms

We saw previously that all the interactions between atoms or groups of atoms fell into three distinct groups: Coulomb interactions, interaction of the London–van der Waals type and hydrogen bonds.

The form taken by the potential energy has been determined for each group (see Table 2.1), so that overall the total interaction energy has the following general form summed over all atoms:

$$\begin{aligned} E_{\mathrm{P}} = {} & \sum_{\text{bonds}} K_1(r - r_0)^2 + \sum_{\text{angles}} K_2(\theta - \theta_0)^2 + \sum_{\text{dihedral angles}} (V/2)[1 - \cos(n\phi + \gamma)] \\ & + \sum_{i<j} q_i q_j/\varepsilon r_{ij} + \sum_{i<j} (A_{ij}/r^{12} - B_{ij}/r^6) + \sum_{\text{hydrogen bonds}} (C/r^{12} - D/r^{10}) \end{aligned} \tag{3.4}$$

and this is the basis of the AMBER program.

Several difficulties remain:

1. In Coulomb-type interactions, which are all the greater in that they obey the long-range $1/r$ law, it is not easy to choose values for q_1, q_2 and ε. A screening effect due to both the dielectric constant of water and the presence of free ions sometimes leads to a considerable reduction in the energy.
2. A similar effect occurs in the presence of ion pairs that reduce the net charge q.
3. The choice of a value for ε is made difficult because of the large variation in dielectric constant near a charge (see section 16.2.2 and Chapter 19).

3.3 Topological constraints

Covalent bonds between atoms a long way from each other along a chain can be involved in the building of three-dimensional structures. Examples include the disulphide bridges created in proteins by the oxidation of two SH groups whose position in the chain is indicated by the symbol >—:

$$\succ\!\!-SH \quad SH-\!\!\prec + \tfrac{1}{2}O_2 \rightarrow \succ\!\!-S{-}S-\!\!\prec + H_2O$$

A bond like this is not included in the previous calculations, but it plays a key role in maintaining a well-defined structure. Another example is provided by the phosphodiester bonds established between the two ends of a DNA chain to create closed circular structures whose topological properties we shall be studying later.

3.4 Helical structures

Despite the advances made both in the search for satisfactory analytical forms for the interaction potentials and in computer power, the exact calculation of a conformation is still a difficult problem. A first step therefore is to survey the experimentally observed solutions at the second organizational level of the macromolecular chain, also called its *secondary structure*.

A frequently occurring solution is the organization of the chain into a helix, described mathematically by a point that rotates at a given distance around an axis while moving parallel to that axis. In a macromolecular chain, the helix will be characterized by a form of organization in which we pass from one monomer to the next by rotation through an angle α around an axis and a translation p along the axis. The pitch P of the helix is the translation corresponding to one complete rotation ($\alpha = 2\pi$).

More precisely, the coordinates of an atom in the jth monomer can be expressed by the following equations in a coordinate system where the z axis lies along the axis of the helix:

$$\begin{aligned} z_j &= jp + z_0 \\ x_j &= r\cos(2\pi jP/p + \phi_0) \\ y_j &= r\sin(2\pi jP/p + \phi_0) \end{aligned} \tag{3.5}$$

where r is the distance of the atom from the axis, and z_0 and ϕ_0 are parameters depending on the selected atom in the monomer.

The ratio P/p is not necessarily an integer. The sense of the helix is determined by the sense of rotation needed to advance along the positive z direction. A helix is right-handed if, in order to move from one monomer to the next in the positive z direction, it is necessary to rotate it through a positive angle α (with the usual trigonometric convention) in the xy plane.

The sense of a helix is intrinsic (it does not depend on the end from which it is viewed!), a feature convincingly demonstrated by comparing right-handed and left-handed screw threads.

In proteins, the helicoidal shapes contribute to the rigidity of the molecule and play a role in recognition mechanisms with both nucleic acids (helix–turn–helix motif) and with membranes (transmembrane α-helices).

In nucleic acids, the double-stranded helix in DNA and in certain domains of RNA form the main structural features; regions with a triple-stranded helix may also be formed. Helicoidal structures also occur at various levels of organization in the nucleoprotein complex known as chromatin, and they also appear in a number of subcellular organizations (viruses, flagellae, tubules, collagen fibres, etc.). Their formation and stability are important problems in biology to which we shall return later. For the time being, we shall consider the helix–coil transition from a very general point of view.

3.5 The helix–coil conformational transition

The change from a helical structure to a much less organized configuration, which generally occurs in a narrow temperature range, has features in common with the melting of a pure substance, i.e. the abrupt transition between a solid and liquid phase at a well-defined temperature. The phase change in a pure substance is a *cooperative phenomenon*. In order to explain this term, consider a change in conformation that causes a molecule in state A to change reversibly into a state B: $A \rightleftharpoons B$.

In a dilute solution, the change in free energy $\Delta G = -RT \ln K$ (where K is the equilibrium constant) depends on the temperature and the internal coordinates of the states A and B.

Now consider the *condensed phase* and assume for example that the same scenario describes the passage of the molecule from the *solid* state A to the *liquid* state B. The value of ΔG will not depend very much on the internal coordinates, which often hardly change at all, but mainly on the *change in neighbours*. There is a regular order in state A but not in state B. The majority, if not the whole, of ΔG arises from changes in the energy of interactions with neighbours. The AA and BB pairs, which correspond to a lower energy than the AB pairs, will be favoured. The energy required to pass from an assemblage of A molecules to one of B molecules decreases as the reaction progresses. All these effects are typical of a cooperative phenomenon. In the case of the helix–coil transition, it is an *intra*molecular process.

3.5.1 Review of thermodynamics

As in any thermodynamic process, a study of the helix–coil transition involves first of all constructing a partition function from which all the other thermodynamic functions can be calculated. We recall that in a so-called *canonical* ensemble, N particles are enclosed in a volume V at a temperature T (there is no exchange of heat or matter with the external medium). If $n_1, n_2, n_3, \ldots$ are the numbers of particles in energy states U_1, U_2, $U_3, \ldots$, the partition function of this ensemble is the quantity Z defined by

$$Z = \sum_i \exp(-U_i/k_B T) \tag{3.6}$$

This indicates the distribution of the molecules according to the Boltzmann factor. It is, translating the German term literally, a *sum over states*. The probability p_i of finding n_i particles in the state U_i is

$$n_i/N = p_i = (1/Z) \exp(-U_i/k_B T) \tag{3.7}$$

and the mean value of a quantity X is $\sum p_i X_i$. If there are g_i states with the same energy U_i, there is said to be *degeneracy* and

$$Z = \sum_i g_i \exp(-U_i/k_{\mathrm{B}}T)$$

This definition can be extended to a macromolecule considered as a sum of microstates each with a Boltzmann factor assigned to them. We can assume in many cases that each monomer can only exist in two states denoted by '0' and '1'. For example, '0' would correspond to an amino acid in a disorganized peptide chain and '1' to the same amino acid forming part of an α-helix; or '0' could correspond to an open base pair in a denatured DNA molecule and '1' to a base pair in a B-type double-stranded helix. The change from '0' to '1' is accompanied by a change ΔG in the free energy and for the '0'$\rightleftharpoons$'1' equilibrium an equilibrium constant s is defined by

$$s = \exp(-\Delta G/RT) \tag{3.8}$$

3.5.2 'Zip' model

In the simplest model, the succession of '0's and '1's can only allow a single region of adjacent '1's. For example, it would be possible to have 00011111000 or 111111000, but not 111100011111 or 00011100111100011. The creation of a new '1' at the end of a sequence of '1's will be defined by the equilibrium constant s (*propagation* step):

$$s = 0011111\underline{1}00/0011111\underline{0}00$$

Creating a '1' in a succession of '0's (*nucleation* step) is much more difficult and will be defined by an equilibrium constant σs ($\sigma \ll 1$):

$$\sigma s = 000\underline{1}0000/000\underline{0}0000$$

The partition function is constructed by considering all the possible states of the chain each with an appropriate statistical weight. For example, 00011111000 contributes σs^5 to the partition function (the *sum* of the free energies corresponds to the *product* of the corresponding equilibrium constants). If the macromolecule has n elements, then

$$Z = 1 + \sum_{k=1}^{n} A_k \sigma s^k \tag{3.9}$$

where the 1 represents the total contribution of all the '0's, and A_k is the number of ways of arranging a set of k '1's after one of the other n elements. It is found (see Box 3.1) that the *helical fraction* θ is given by

$$\theta = s\,\partial Z/nZ\,\partial s \tag{3.10}$$

or

$$\theta = \partial \ln Z/n\,\partial \ln s \tag{3.11}$$

Box 3.1 Calculation of a partition function

In order to calculate Z we replace A_k by its expression

$$A_k = n - k + 1$$

and hence

$$Z = 1 + \sigma(n+1)\sum_{k=1}^{n} s^k - \sigma\sum_{k=1}^{n} ks^k$$

The second term on the right has the value $\sigma s(n+1)(s^n - 1)/(s-1)$. To calculate the third term, we note that $\sum ks^k = s\,\mathrm{d}(\sum s^k)/\mathrm{d}s$, so that we only have to differentiate the second term, which gives

$$\sum_{k=1}^{n} ks^k = s[ns^{n+1} - (n+1)s^n + 1]/(s-1)^2$$

and hence

$$Z = 1 + \sigma s[s^{n+1} - (n+1)s + n]/(s-1)^2$$

(The highest order of s in Z is n since there can be one sequence of n states '1' among all the possible states.) If we wish to calculate the fraction θ of monomers occurring in the '1' state (i.e. the helical fraction), we first calculate the probability $p(k)$ that there are k monomers in state '1'. By definition

$$p(k) = \sigma s^k(n-k+1)/Z$$

The mean value $\langle k \rangle$ is defined by $\sum kp(k)$ and so

$$\theta = \langle k \rangle/n = \sum kp(k)/n = \sum(n-k+1)k\sigma s^k/nZ$$

But $\sum k[(n-k+1)\sigma s^k] = s\,\mathrm{d}Z/\mathrm{d}s$, so that finally

$$\theta = (s/nZ)\,\mathrm{d}Z/\mathrm{d}s = (1/n)\,\mathrm{d}\log Z/\mathrm{d}\log s$$

This model fits short chains very well and leads either to the favouring of long helical sequences or to completely denatured molecules. This is even more so as σ becomes smaller.

with

$$Z = 1 + \sigma ss^{n+1}s - (n+1)s + n/(s-1)^2 \tag{3.12}$$

3.5.3 Matrix model (Zimm and Bragg)

The molecule will be described as a sequence of organized and disorganized regions, i.e. a set of '0's and '1's. We have to define (a) the statistical weight corresponding to any two successive states '0' or '1', and (b) a method of calculating the partition function Z describing the macromolecule.

Statistical weight

The statistical weight g is simply the Boltzmann factor $\exp(-\Delta G/RT)$, where ΔG is the change in free energy when a monomer in a given state is added to the preceding one. Thus we define

- $g = 1$ for any '0' that appears (after a '0' or '1')
- $g = s$ for any '1' that appears after another '1'
- $g = \sigma s$ for any '1' that appears after a '0'.

The change in free energy ΔG corresponding to s characterizes the 'reaction' of adding an organized monomer to a chain terminated by at least one organized monomer, compared with the addition of a '0', which by convention has a statistical weight of 1. The parameter σ introduces the cooperative idea, for if $\sigma = 1$ the probability of finding a '1' after a '0' is the same as that of finding a '1' after a '1': in this case, the fraction θ of organized monomers is

$$\theta = s/(1+s) \tag{3.13}$$

If, on the other hand, we give σ a very small value, it will be easier to supplement the sequences of '1's or '0's than to add a '1' after a sequence of '0's, i.e. to initiate an organized region. The quantity σ can be considered as a *nucleation* parameter (see note below), the phenomenon being more cooperative as σ becomes smaller. In the limit, when $\sigma = 0$, we have an 'all or nothing' system in which the molecule can only exist in one of two states: either completely disorganized or completely organized.

Note on nucleation centres

How can the change from a homogeneous metastable state to a homogeneous stable state be produced? Examples of this might be a vapour under a pressure and temperature at which it should be a liquid, or a superheated liquid above its boiling point that should be a vapour. The process is linked to the appearance of fluctuations that, at a certain point, create a small volume of the other phase (droplets of liquid in the vapour, for example). If the conditions of T and P are such that the gaseous phase is stable, such droplets are unstable and disappear. If the state is metastable, the droplets will be able to grow in volume. They become *nucleation centres* if their size is such that the energy gain corresponding to their existence is greater than the energy loss due to the creation of a surface of separation between the two phases. From these nucleation centres, the cooperative transition between one phase and another occurs rapidly and can, in some circumstances, take on a *catastrophic* character (in the mathematical sense of the term). This is what happens in the case of black ice, the explosive nature of delayed boiling or the phenomenon of 'tin plague', and is what was involved in the famous case of horses plunging into the supermelted Lake Ladoga.

The parameter σ can also be regarded as the inverse of a *surface tension* between the organized and non-organized regions. The limiting value $\sigma = 0$ would correspond to an infinite surface tension. It is only in this case that a true phase transition could occur (identical to that observed in the melting of a crystal) for a one-dimensional system. As soon as $\sigma \neq 0$, there is a mixture of two phases (sets of '0's and sets of '1's) and the transition is no longer a phase transition in the strictly thermodynamic sense.

Calculation of Z

The statistical weight of a given ith segment is represented by a column vector $\boldsymbol{a}_i$. If we restrict ourselves to *nearest-neighbour interactions*, the vector $\boldsymbol{a}_i$ has only two components, each representing the statistical weight of one of the states of the ith segment, *taking into account the state of the i−1 preceding monomers*. The move from one monomer to the next, i.e. the calculation of $\boldsymbol{a}_i$, is carried out by using a transition matrix $\boldsymbol{M}_i$ of the form

$$\boldsymbol{M}_i = \begin{pmatrix} 1 & 1 \\ \sigma s & s \end{pmatrix}$$

with $\boldsymbol{a}_{i+1} = \boldsymbol{M}_i \boldsymbol{a}_i$. The meaning of the elements of $\boldsymbol{M}$ follows from the definition of the 'g's. In the first row, the first term corresponds to a '0' following a '0' ($g = 1$) and the second term to a '0' following a '1' ($g = 1$). In the second row, the first term corresponds to a '1' following a '0' ($g = \sigma s$) and the second term to a '1' following a '1' ($g = s$).

If *more distant interactions* than those with nearest neighbours have to be included, then groups of μ segments must be considered and the column vector for the ith segment will then have 2μ components. Finally, for any μ, the last monomer in the chain is represented by a vector $\boldsymbol{a}_{i+1}$ whose components when summed give the partition function Z of the molecule. The whole problem reduces to a calculation of Z and the deduction from it of the characteristic parameters of the transition, which can then be determined experimentally.

From the calculation in Box 3.2, we have $Z = \lambda_1^n$ with

$$\lambda_1 = \{1 + s + [(1-s)^2 + 4\sigma s]^{1/2}\}/2 \tag{3.14}$$

and this gives a value for θ identical with that found with the first model, i.e.

$$\theta = (s/\lambda_1)\mathrm{d}\lambda_1/\mathrm{d}s \tag{3.15}$$

An examination of special cases gives a better understanding of the nature of the phenomenon:

Case 1: $\sigma = 1$

Here, $\lambda_1 = 1+s$, $\mathrm{d}\lambda_1/\mathrm{d}s = 1$ and $\theta = s/(1+s)$, so that we indeed obtain the formula for a non-cooperative transition.

Case 2: $\sigma = 0$

Here

$$\lambda_1 = [1 + s \pm (1-s)]/2 \quad \text{and} \quad \mathrm{d}\lambda_1/\mathrm{d}s = [1 \pm (s-1)/(1-s)]/2 \tag{3.16}$$

We must distinguish two domains depending on whether s is greater or less than 1.

- If $s < 1$, then the plus sign is appropriate and $\lambda_1 = 1$, so that

$$\mathrm{d}\lambda_1/\mathrm{d}s = 0 \quad \text{and} \quad \theta = 0 \tag{3.17a}$$

- If $s > 1$, then the minus sign is appropriate and $\lambda_1 = s$, so that

$$\mathrm{d}\lambda_1/\mathrm{d}s = 1 \quad \text{and} \quad \theta = 1 \tag{3.17b}$$

Box 3.2 **Calculation of Z and θ**

Taking the simplest example with only nearest-neighbour interactions, the column vector representing the first monomer in the chain is clearly

$$\begin{pmatrix} 1 \\ 1 \end{pmatrix}$$

The calculation of the successive products of vectors and matrices is considerably simplified if the matrix describing the passage is diagonalized. To diagonalize a matrix, we seek the eigenvalues, i.e. the values of the parameter λ satisfying the equation $Ax = \lambda x$ where A is the operator associated with the matrix. This can be written as $(A-\lambda I)x = 0$ where I is the unit matrix

$$\begin{pmatrix} 1 & 0 \\ 0 & 1 \end{pmatrix}$$

If this system of linear equations is to be satisfied, the determinant of the system must be zero. The parameter λ is thus the solution of the equation obtained by subtracting λ from each term of the main diagonal of A and then putting the determinant of the matrix equal to zero. We have

$$\begin{pmatrix} 1 & 1 \\ \sigma s & s \end{pmatrix} \rightarrow \begin{pmatrix} 1-\lambda & 1 \\ \sigma s & s-\lambda \end{pmatrix}$$

and a quadratic equation

$$(1-\lambda)(s-\lambda) = \sigma s$$

whose roots λ_1 and λ_2 are

$$\lambda_1 = \tfrac{1}{2}\{1 + s + [(1-s)^2 + 4\sigma s]^{1/2}\}$$

$$\lambda_2 = \tfrac{1}{2}\{1 + s - [(1-s)^2 + 4\sigma s]^{1/2}\}$$

The new transition matrix A_D is

$$A_D = \begin{pmatrix} \lambda_1 & 0 \\ 0 & \lambda_2 \end{pmatrix}$$

and the successive products are

$$A_D \begin{pmatrix} 1 \\ 1 \end{pmatrix} = \begin{pmatrix} \lambda_1 \\ \lambda_2 \end{pmatrix}$$

and

$$A_D \begin{pmatrix} \lambda_1 \\ \lambda_2 \end{pmatrix} = \begin{pmatrix} \lambda_1^2 \\ \lambda_2^2 \end{pmatrix}$$

The vector a_{n+1} will therefore have two components λ_1^n and λ_2^n and hence

$$Z = \lambda_1^n + \lambda_2^n$$

For large N, λ_2^n becomes negligible compared with λ_1^n and so

$$Z = \lambda_1^n$$

To calculate the fraction θ of organized regions, we use the following clue: if an average of k organized monomers ($0 < k < n$) occur in the macromolecule, their statistical weight in Z will be expressed by the factor s^k. From this, we deduce that

$$k = d(\log Z)/d(\log s)$$

and hence

$$\theta = k/n = d\log Z/n\, d\log s$$

an expression identical with that found above in the first model. Using the expression for Z, we obtain:

$$\theta = (s/\lambda_1)\, d\lambda_1/ds$$

and the derivative $d\lambda_1/ds$ is given by

$$d\lambda_1/ds = \tfrac{1}{2}\{1 + (s-1) + 2\sigma/[(1-s)^2 + 4\sigma s]^{1/2}\}$$

Combining all these expressions enables θ to be determined as a function of s.

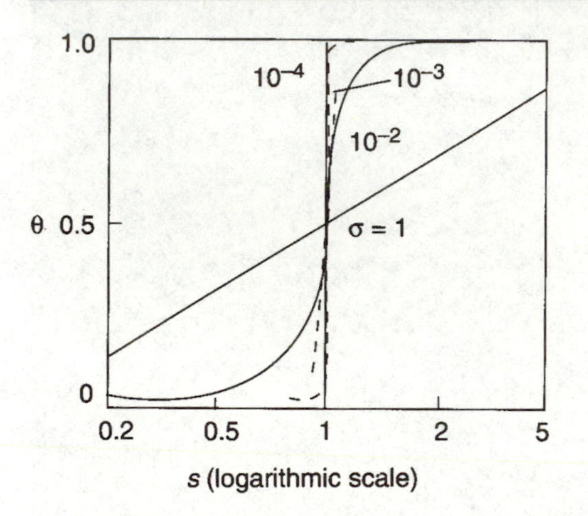

Figure 3.1 Transition curves for various values of σ.

Once again, we have the 'all or nothing' phenomenon already examined, which corresponds to a phase transition in the thermodynamic sense. The curves for $\theta(s)$ in Fig. 3.1 illustrate the general form of the transition. The slope at the origin ($s = 0$, $\theta = 0$) is σ. Only the curves for $\sigma < 1$ have a point of inflection for $s = 1$ and $\theta = 1/2$.

A transition width can be defined by

$$\Delta s = (d\theta/ds)^{-1}_{s=1} \tag{3.18}$$

(in Fig. 3.2, $\Delta s = AA' + BB'$). Differentiating θ with respect to s gives (Fig. 3.2):

$$d\theta/ds = (1/\lambda_1)\, d\lambda_1/ds - (s/\lambda_1^2)(d\lambda_1/ds)^2 + (s/\lambda_1)\, d^2\lambda_1/ds^2 \tag{3.19}$$

We only have to calculate $d^2\lambda_1/ds^2$ and replace s by 1 in this, to give

$$d\theta/ds = (1 + \sigma^{1/2} + \sigma)/4(\sigma + \sigma^{1/2}) \approx 1/(4\sigma^{1/2}) \tag{3.20}$$

(since $\sigma \approx 10^{-4}$) and hence

$$\Delta s = 4\sigma^{1/2} \tag{3.21}$$

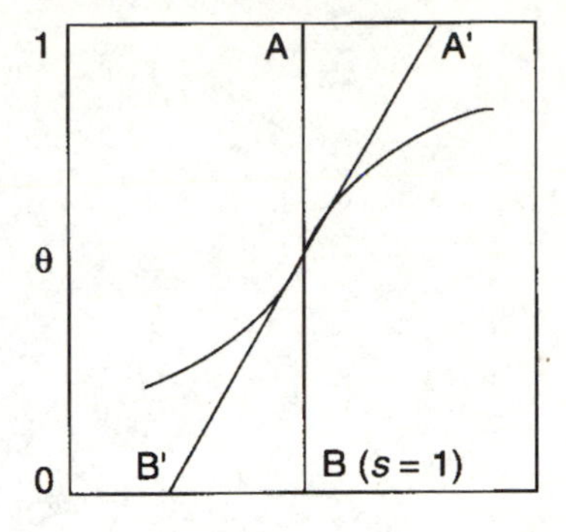

Figure 3.2 Curve for definition of transition width.

The width of the transition is therefore a direct measure of the cooperativity, i.e. the nucleation parameter. When the transition is caused by a rise in temperature, we have from the definition of s ($s = \exp(-\Delta G/RT)$) similar to an equilibrium constant:

$$\mathrm{d}(\ln s)/\mathrm{d}T = \Delta H/RT^2 \tag{3.22}$$

The transition width ΔT is defined in the same way as before by $\Delta T = (\mathrm{d}\theta/\mathrm{d}T)^{-1}$, and since $\mathrm{d}\theta/\mathrm{d}T = (\mathrm{d}\theta/\mathrm{d}s)(\mathrm{d}s/\mathrm{d}T)$, we obtain

$$\Delta T = 4\sigma^{1/2} RT_{\mathrm{m}}^2/\Delta H \tag{3.23}$$

since for $s = 1$, $T = T_{\mathrm{m}}$ (the transition temperature). ΔH is the change in enthalpy in passing from the helix to the coil. It should be pointed out that in this expression ΔH is an apparent enthalpy, which takes account of the cooperative nature of the process. Its value may be completely different from the change in enthalpy obtained directly by calorimetric measurements.

4

Conformation of nucleic acids

4.1 Introduction

The importance of the role played by nucleic acids in biological phenomena is now unquestioned. The deoxyribonucleic acid (DNA) in every cell behaves as a 'databank', and the information stored in it enables each cellular constituent to be synthesized, assembled and regulated. Ribonucleic acids (RNAs) are involved in the building of ribosomes (ribosomal RNA or rRNA), in protein synthesis (transfer RNA or tRNA), in carrying the 'genetic message' (messenger RNA or mRNA) and in regulatory mechanisms ('small' RNA). Viruses are primitive entities containing either a DNA or an RNA, while viroids are merely RNA fragments.

The structural model of DNA proposed by Watson and Crick (1953a,b) was a decisive event, triggering a dramatic development in molecular biology. Since then, and particularly over the last 20 years, the significance of nucleic acids in the living world has become even more evident, with important discoveries that have given molecular biology a second wind. The possibility of determining the sequence of several hundred nucleotides either by purely chemical methods (Maxam and Gilbert, 1977) or by chemical and enzymatic methods (Sanger *et al.*, 1977) has opened the way to a knowledge of the genome. The sequencing of the plasmid pBR322 (Sutcliffe, 1979) was the first great success using these methods. The DNA sequences of *E. coli*, of some other bacteria and of several simple eukaryotes are now completely determined, while the sequencing of human DNA has become a world-wide project.

Faster and more automated methods may accelerate the progress of the Human Genome Project. From DNA sequences determined by such methods (and stored with thousands of examples in a databank) it is very often possible to deduce sequences of proteins corresponding to an 'open reading frame' by applying the rules of the genetic code. The discovery of restriction enzymes (Arber and Linn, 1969) provided the essential tool for what is now called *genetic engineering*: the set of methods by which DNA fragments can be cut and assembled at will.

Alongside the limitless possibilities offered in molecular biology, in biotechnology and soon in genetic therapy, we should not forget the role of genetic engineering in the physico-chemical study of nucleic acids. The discovery of an enzymatic role for certain transfer RNAs, while extending the enzyme concept to non-protein structures (ribozymes), has thrown new light on the origin of life and has given a new impetus to methods for modelling the tertiary structures of RNA.

Finally, the many X-ray diffraction studies of synthesized crystalline DNA fragments have called into question the small-scale regularity of classical helical structures. This has necessitated the definition of new structural parameters, and has laid the foundations for establishing relations between sequence and structure, i.e. those involved in the specific recognition mechanisms between nucleic acids and other molecules. Quite recently, too, the development of specific chemical probes means that we can undertake detailed structural studies of *in vivo* DNA in a structure as complex as chromatin or the RNA within ribosomes.

Over the last 45 years, therefore, our knowledge of DNA and nucleic acids in general has been considerably enriched. Because of this, our examination of them requires an extremely detailed approach to their structure and conformation.

4.2 **Primary structure**

All macromolecules of nucleic acid are formed by the regular chaining of the same motif:

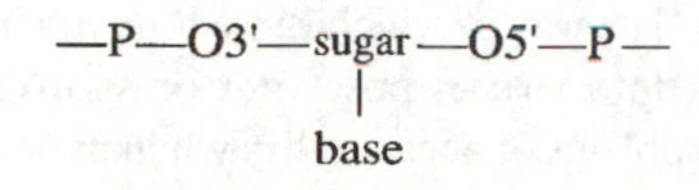

where the sugar is either a ribose or a deoxyribose (Fig. 4.1).

The corresponding monomeric unit is a nucleotide. However, the ester bond between the phosphate and the sugar may be made either with the C3′ atom or with the C5′ atom of the sugar, and this leads to the two structures shown in Fig. 4.2.

The primary structure or sequence of a nucleic acid is conventionally written as a set of bases going from left to right such that each phosphodiester bond is linked to the 3′ of the sugar on the left and to the 5′ of the sugar on the right. In a sequence such as GACTAGC..., for example, the guanine at the left-hand end therefore

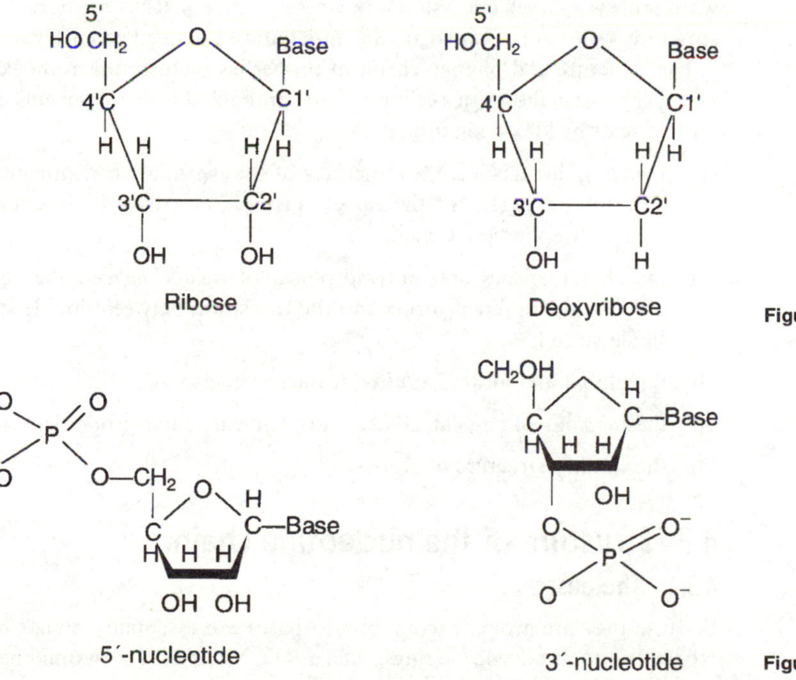

Figure 4.1 The sugars in RNA and DNA.

Figure 4.2 The nucleotides in RNA.

corresponds to the terminating 5′ atom and the last base on the right to the terminating 3′ atom. It should also be pointed out that, although the phosphate group of the nucleotide monomer can carry two negative charges (with respective p*K* values of about 1 and 6), there is only a single negative charge per motif (and hence per base) in the polymer. Since this corresponds to the lower p*K* (about 1), nucleic acids and polynucleotides carry a large number of negative charges. We shall see the importance of this in Part IV dealing with polyelectrolytes.

The base may be any one of those listed in Table 4.1. Only four bases occur in DNA: adenine (A), guanine (G), thymine (T) and cytosine (C), each linked to a deoxyribose. In messenger and ribosomal RNA, thymine is replaced by uracil (U) and the bases are linked to a ribose. This is also the case in transfer RNA, but here there is a greater variety of special bases.

The macromolecules of DNA are structures of immense length in which the motif is repeated between 10^7 times (in bacteria) and 10^{10} times (in eukaryotes). The human genome forms a thread about 2 nm in diameter that is altogether 1.5 m long! On a visible scale this would be equivalent to a steel wire with a diameter of 2 mm and a length of 1500 km. The way in which these long macromolecules are packed into cells, into nuclei or into viruses poses not only problems of flexibility and composition, but also problems of accessibility when material undergoing cellular replication is duplicated or when the genetic message is being read during transcription.

The various nucleic acids must also interact specifically with proteins. The recognition process involves the structure, the conformation, the charge, the hydration and the dynamics of all or part of the macromolecule. Eukaryotic DNA is packaged into a compact structure, *chromatin*, itself organized into *chromosomes* whose typical architecture is very visible at metaphase. Prokaryotic rRNA in ribosomes (which are veritable protein production lines) interacts with more than 50 different proteins whose organization we are beginning to determine in detail but whose role is less well understood. In viruses, DNA or RNA are buried in protein structures (*capsids*) of varied architecture but well-defined geometry.

The structure and physico-chemical properties of these macromolecules thus play a key role in the life of a cell. We shall approach the physico-chemical study of nucleic acids by looking in turn at:

(1) *structural* aspects, i.e. the structures of the two main constituent units, the base and the sugar, and the angles of rotation involved in the conformation of the phosphodiester chain;

(2) the characteristics and polymorphism of *double helices*, the equilibrium between the different forms and the transition between double strand and single strand;

(3) the global and local *flexibility* of nucleic acids;

(4) the *topology* of circular DNAs and their particular properties;

(5) the tertiary structure of RNAs.

4.3 Structure of the nucleotide chain

4.3.1 The bases

Because they are aromatic compounds, bases are essentially planar molecules. However, atoms outside the rings, such as O2, N4 and O4 in pyrimidines, and O6

Table 4.1 Nucleic acid bases

Base	Symbol	Formula
adenine	A	NH_2; N_1, 2, 3, 4, 5, 6, N_7, C_8, N_9
N^6-Δ^2-isopentenyladenosine	6iA	$(CH_3)_2$ —C=CH—CH_2—N(H)—6
N^6-Δ^2-isopentenyl-2-methylthioadenosine	2ms6iA	same group as above at 6 SCH3 at 2
guanine	G	O; HN_1, H_2N, 2, 3, 4, 5, 6, N_7, C_8, N_9
7-methylguanine	7meG	CH_3—N_7
N-2-dimethylguanine	2dmG	$(CH_3)_2$ —N—C 2
xanthine	X	O; HN_1, 2, N_3, 4, 5, 6, N_7, C_8, N_9
thymine	T	O; HN_3, 2, 1 N, O, 4, 5, 6, CH_3
uracil*	U	O; HN_3, 2, 1 N, O, 4, 5, 6
4-thiouracil	4sU	S=4

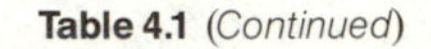

Table 4.1 (*Continued*)

Base	Symbol	Formula
dihydrouracil	D	
cytosine	C	

*Uracil may be bound to the sugar not at N1 but at C5 to give pseudouridine (ψU).

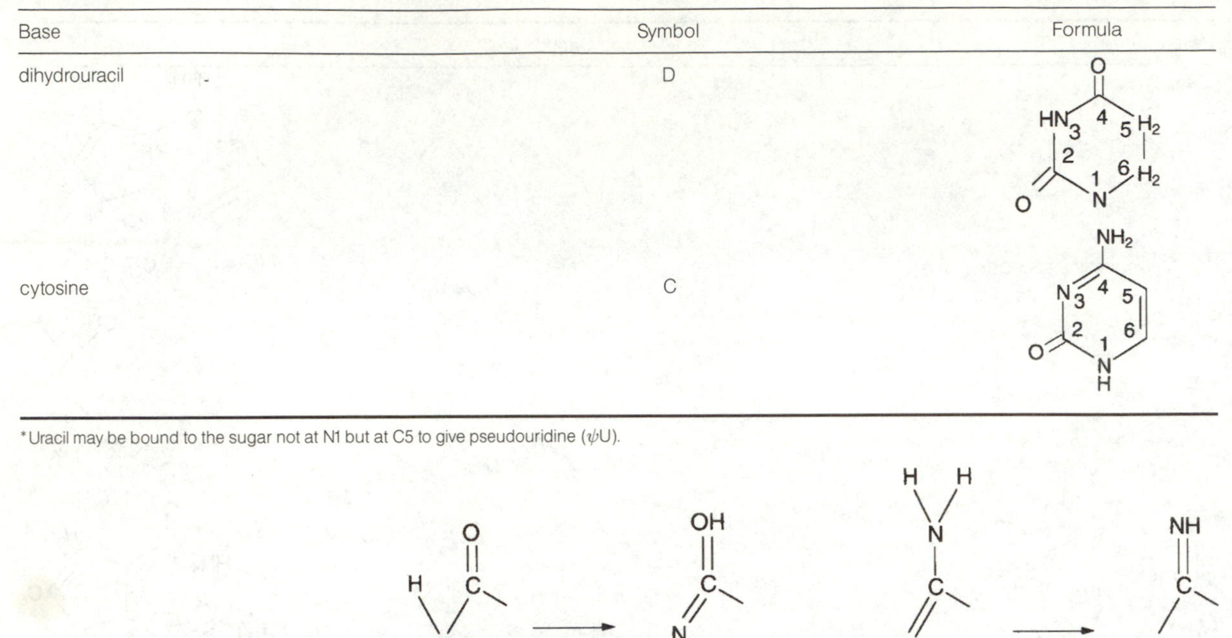

Figure 4.3 Tautomeric equilibrium between lactam and lactim and between amine and imine.

or N6 and N2 in purines, may lie at a distance of between 0.05 and 0.2 Å from the plane of the aromatic rings.

There is tautomeric equilibrium between the lactam and lactim forms (keto–enol equilibrium) and between the amine and imine forms (Fig. 4.3). The lactam form is stabilized with respect to the lactim form because of the resonance energy of the C=O group. The amine form is stabilized by the delocalization of free electron pairs of the amine group in the aromatic ring. The tautomeric keto and amine forms therefore form the large majority, if not all, of those found in solution. There is, however, a small probability that the lactim and imine forms will exist, which might explain certain mechanisms of mutagenesis.

4.3.2 The sugars

The sugar in nucleic acids is a furanose, i.e. one with a five-atom ring. The sugar in RNA is a β-D-ribose, and that in DNA is a 2′-deoxy-β-D-ribose. The asymmetry in the carbon atoms leads to D and L isomers (C4′ configuration) and to α and β isomers (or preferably *anomers*) (C1′ configuration). In a D-ribose, the substituent group at C4′ (the exocyclic C5′–O5′ group) points in the opposite direction to that of the oxygens O2′ and O3′. In the β anomer of ribose, the substituent group at C1′ (the base) points in the opposite direction to that of the oxygens O2′ and O3′ (Fig. 4.4).

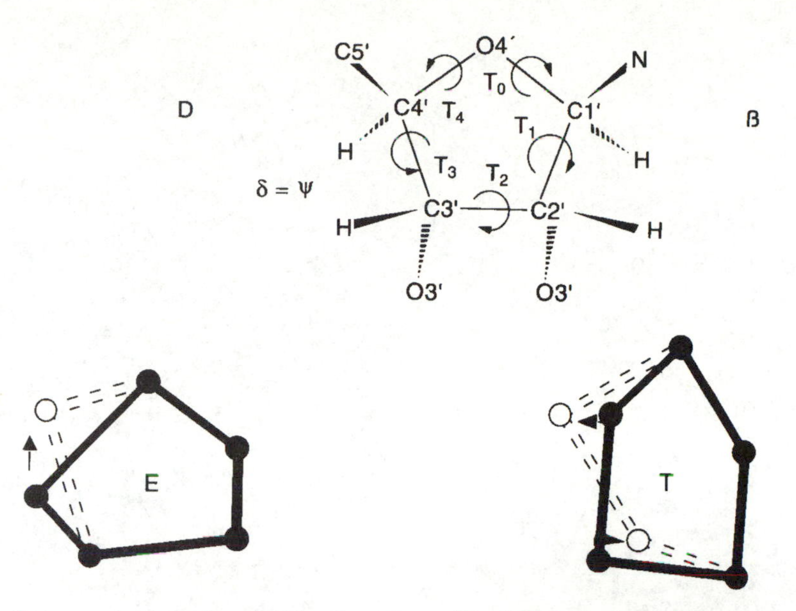

Figure 4.4 Configuration of β-D-ribose.

Figure 4.5 The E and T conformations of a furanose ring.

The furanose ring is puckered rather than planar (Fig. 4.5). It may have either one non-coplanar atom (the envelope or E form) or two atoms displaced from the plane of the other three (twist or T form).

Because there are five atoms in a furanose ring, there are 10 E conformations (five up and five down) and 10 T conformations (one atom up and the next down, and vice versa for each pair). In nucleic acids, 'up' means towards the exocylic group at C5′ and is called the *endo* form. 'Down' means in the other direction and is called the *exo* form. There are thus in principle 20 possible conformations for a furanose ring. To describe the puckering completely, we must specify not only the type or types of atom displaced, but their distance (of about 0.5 Å) from the plane defined by the others. The 20 different conformations can be placed on a circle whose radius is proportional to the amplitude of the puckering, each conformation being separated from the next by 360°/20 or 18° (Fig. 4.6).

Two parameters, the phase angle P and the amplitude of the pseudorotation τ_M, enable the five endocyclic (inside the ring) torsion angles τ_i $(i = 1, \ldots, 5)$ to be calculated. Each puckering can then be characterized using the relation

$$\tau_i = \tau_M \cos[P + 144(i - 1)] \qquad i = 1, \ldots, 5$$

where τ_1 is the torsion angle C1′–C2′–C3′–C4′, τ_2 is the torsion angle C2′–C3′–C4′–O4′, etc. Pseudorotation is a special property of sugars with five-membered rings. Through the pseudorotation, a furanose nucleus can adopt each of the 20 possible conformations without ever becoming planar. This is because the equation for the endocyclic torsion angles shows that they can vary continuously with the phase P. In nucleic acids, the phase is chosen so that $P = 0$ corresponds to the C3′-*endo*–C2′-*exo* puckering. $P = 180°$ then corresponds to C2′-*endo*–C3′-*exo* and $P = 90°$ to O4′-*endo*. The preferred conformations of sugars are C3′-*endo* (^3E or $P = 18°$) and C2′-*endo* (^2E or $P = 162°$). Sugars in solution are in a dynamic equilibrium between these two main conformations. That is why another way to approach this is to project the furanose ring on to the plane defined by the three atoms C1′, C4′ and O4′. The other two carbon atoms C2′ and C3′ are then displaced from the plane. In the *endo* conformation of the sugar, one of the two carbon atoms

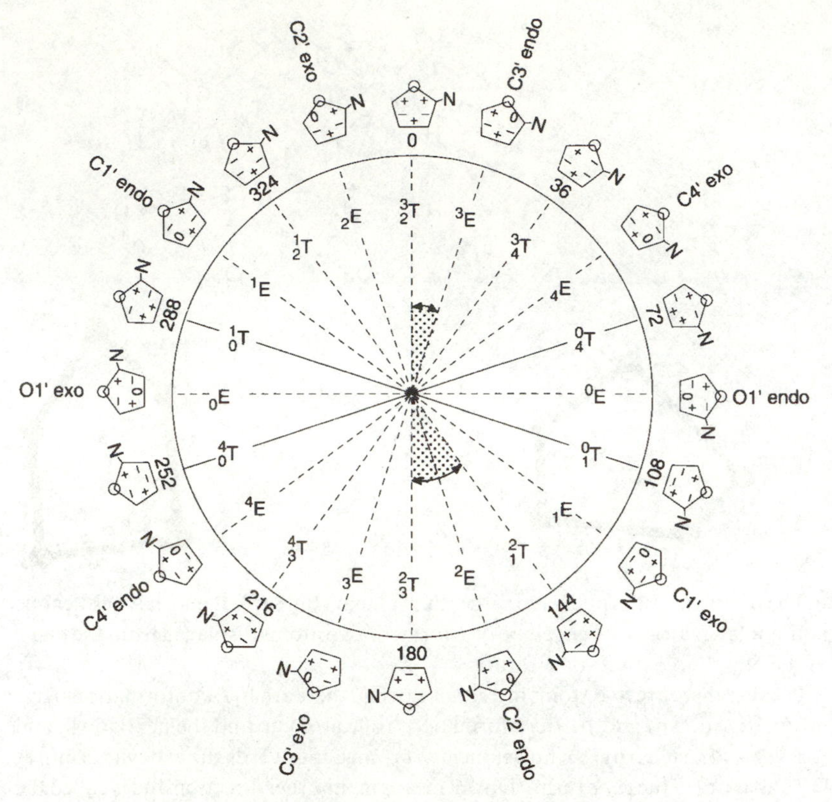

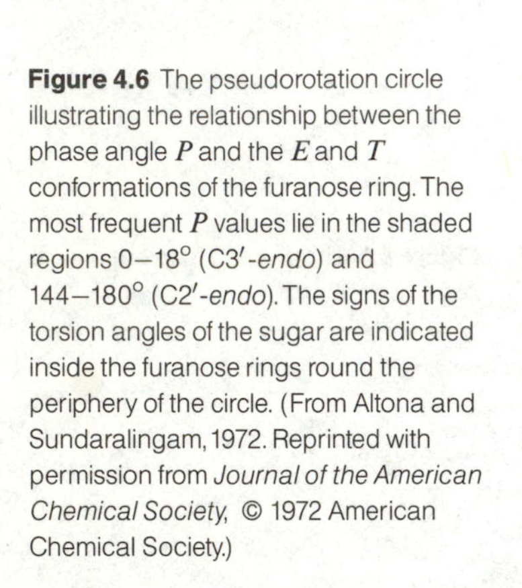

Figure 4.6 The pseudorotation circle illustrating the relationship between the phase angle P and the E and T conformations of the furanose ring. The most frequent P values lie in the shaded regions 0–18° (C3′-*endo*) and 144–180° (C2′-*endo*). The signs of the torsion angles of the sugar are indicated inside the furanose rings round the periphery of the circle. (From Altona and Sundaralingam, 1972. Reprinted with permission from *Journal of the American Chemical Society*, © 1972 American Chemical Society.)

is on the same side of the plane as the C5′ atom of the chain. Fig. 4.7 shows the two possibilities.

4.3.3 Rotation angles of the phosphodiester chain

As Fig. 4.8 shows, the set of rotations characterizing the nucleotide chain involves six independent angles, each type of C–C, C–O and P–O bond occurring twice because of the $3' \rightarrow 5'$ bond. Starting from one phosphate group and moving towards the other in the $5' \rightarrow 3'$ direction, we have in turn the angles α (rotation around P–O5′), β (around O5′–C5′), γ (around C5′–C4′), δ (around C4′–C3′), ε (around C3′–O3′) and ζ (around O3′–P). Added to these is a seventh angle χ measuring the rotation of the base with respect to the sugar around the N–C1′ bond. In the older notation proposed by Sundaralingam, the angles $\psi(\gamma)$ and $\psi'(\delta)$ related to the C–C bonds as in peptide chains and the angles $\phi(\beta)$ and $\phi'(\varepsilon)$ related to the C–O bonds by analogy with the C–N bond. The angles $\omega(\alpha)$ and $\omega'(\zeta)$ completed the description.

The seven angular parameters needed to describe the conformation of a nucleotide give some idea of the high degree of polymorphism in the phosphodiester chains of nucleic acids. In reality, steric problems and potential barriers associated with rotation limit the accessible ranges of all seven angles to a considerable extent. They also introduce correlations between the angles and we shall see later that these constraints mean, for example, that the possible types of double helix are limited in number. For the time being, however, it is worth simply examining the different angles of rotation and their effect on the conformation of the whole chain.

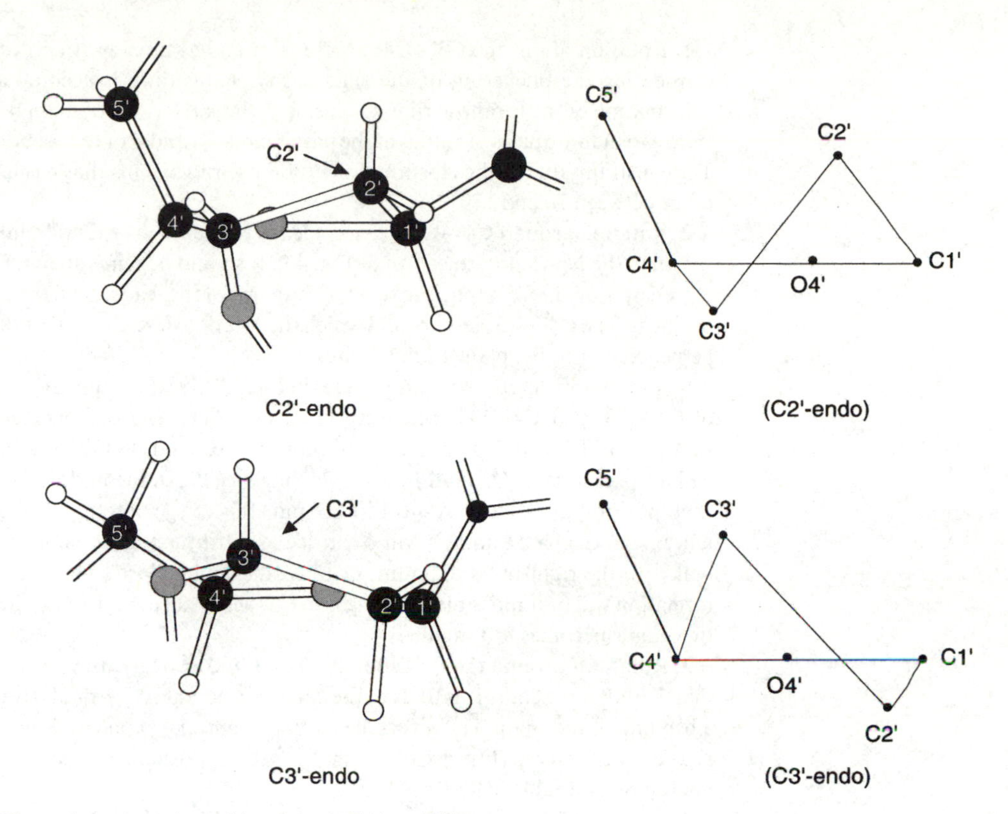

Figure 4.7 The *endo* conformations of the sugars in DNA and RNA.

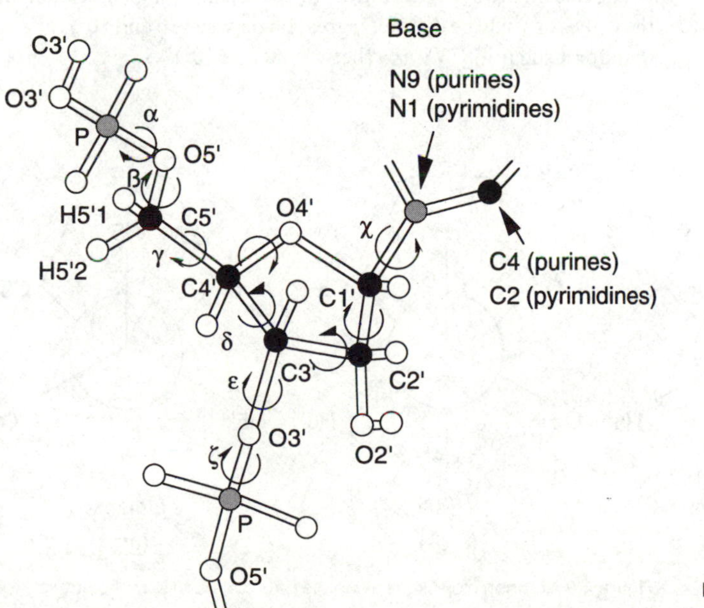

Figure 4.8 Possible rotations in the nucleotide chain.

1. Rotation about the C3′–C4′ bond (ψ' or δ) has already been dealt with in considering the puckering of the sugar. Any change in δ is accompanied by a deformation of the furanose ring. Conversely, the same value of δ can be obtained from two or more pairs of values of the parameters τ_M and P of the pseudorotation. The constraint due to the closure of the furanose ring means that δ is limited to a range between 70 and 170°.

2. Rotation around C4′–C5′ (ψ or γ) leads to the three types of conformation shown in the Newman projection of Fig. 4.9. It should be remembered that in this representation the C5′ atom is in front of the page for the reader, and the C4′ atom is behind it. The C4′–C5′ axis around which the rotation described by γ takes place is perpendicular to the plane of the figure.

The two conformations in parentheses in Fig. 4.9 give the respective orientations of C4′–C3′ and C4′–O4′ with respect to C5′–O5′. The commonest situation shown in the left-hand diagram, for example, corresponds to C4′–C3′ in *gauche+* with respect to C5′–O5′ and to C4′–O4′ in *gauche–*. In the middle diagram, the *trans* position of O5′ with respect to C3′ puts this oxygen atom at maximum distance from the base and this will occur if the substituent groups on the base are bulky. In the right-hand diagram, O5′ becomes quite close to C3′ and this conformation will be impossible for a sugar in C3′-*endo*, since the distance between the two atoms becomes too small.

3. Rotations around the P–O bonds (ω or α and ω' or ζ) are very important in that they alter the conformations of the chain considerably, particularly in creating a bending of the chain. However, the *gauche–*, *gauche–* conformation is the one usually encountered (Fig. 4.10). We shall see the importance of changes in α and ζ in the case of double helices.

4. χ is the angle of rotation around the C–N bond (C1′–N9 for purines and C4′–N1 for pyrimidines). It defines the orientation of the base with respect to the sugar. The range of χ around 180° (*trans*) is called *syn* and the range $-90° < \chi < +90°$ is called *anti* (Fig. 4.11).

The two angles δ (puckering of the sugar) and χ are correlated: in the *anti* region, the values of χ will be distinctly greater (between 30 and 80°) for a sugar in C2′-*endo* than for a sugar in C3′-*endo* (between 0 and 10°).

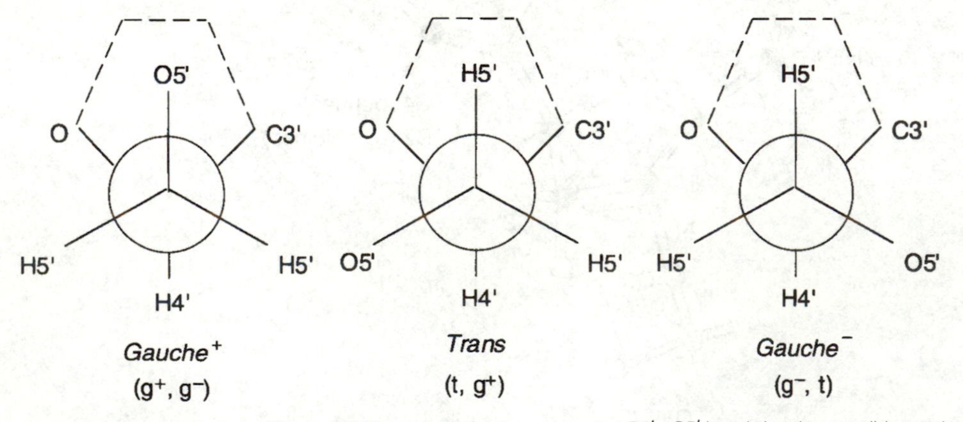

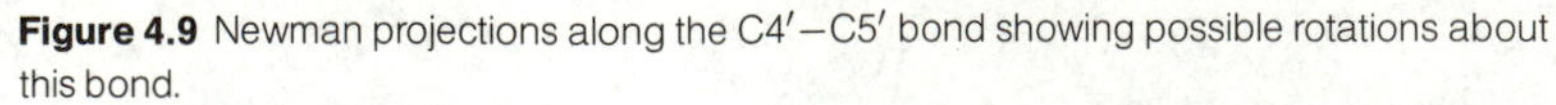

Figure 4.9 Newman projections along the C4′–C5′ bond showing possible rotations about this bond.

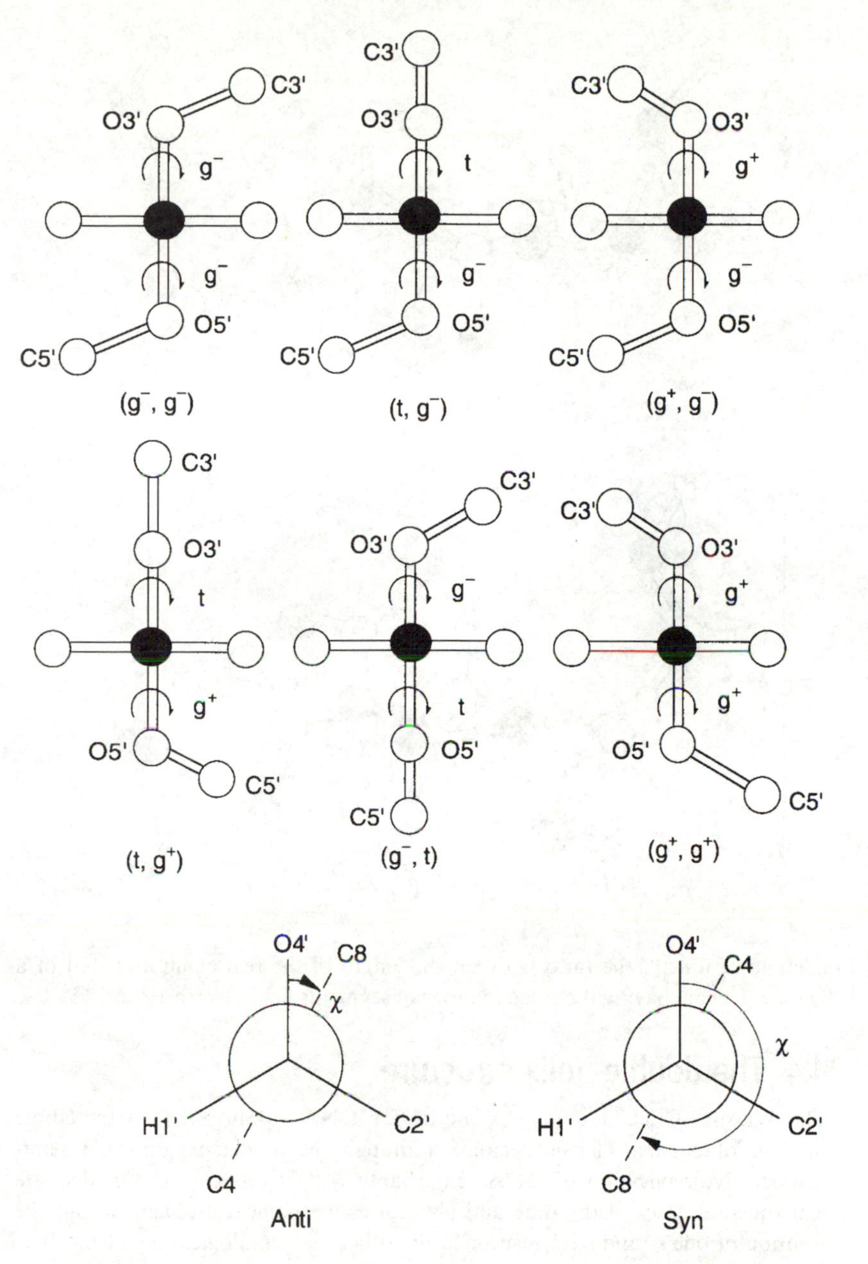

Figure 4.10 Conformations of the phosphodiester group (angles α and ζ). The first of these (g^-, g^-) is found in both A and B type helices. The others are important in single-stranded polynucleotides or in transient or permanent local perturbations. The (g^+, g^+) conformation cannot occur all along a chain for steric reasons. The (t, g^-) and (g^+, t) conformations produce kinks in the helix.

Figure 4.11 Definition of *anti* and *syn* conformations.

The most important data from those quoted above are summarized in Fig. 4.12, which shows the two most common nucleotide conformations, i.e. C3′-*endo* sugar, $\gamma = g^+$, $\chi = anti$ (Fig. 4.12a), and C2′-*endo* sugar, $\gamma = g^+/g^-$, $\chi = anti$ (Fig. 4.12b). Note particularly in these two cases that the distances between the phosphates and the orientations of the bases are different.

In general, the narrow ranges over which the rotation angles can vary and the correlation between them mean that the phosphodiester chain has great local rigidity. It is found experimentally that, for simple chains such as poly(rA) or poly(rU) (chaining of a single motif where the sugar is a ribose and the base an

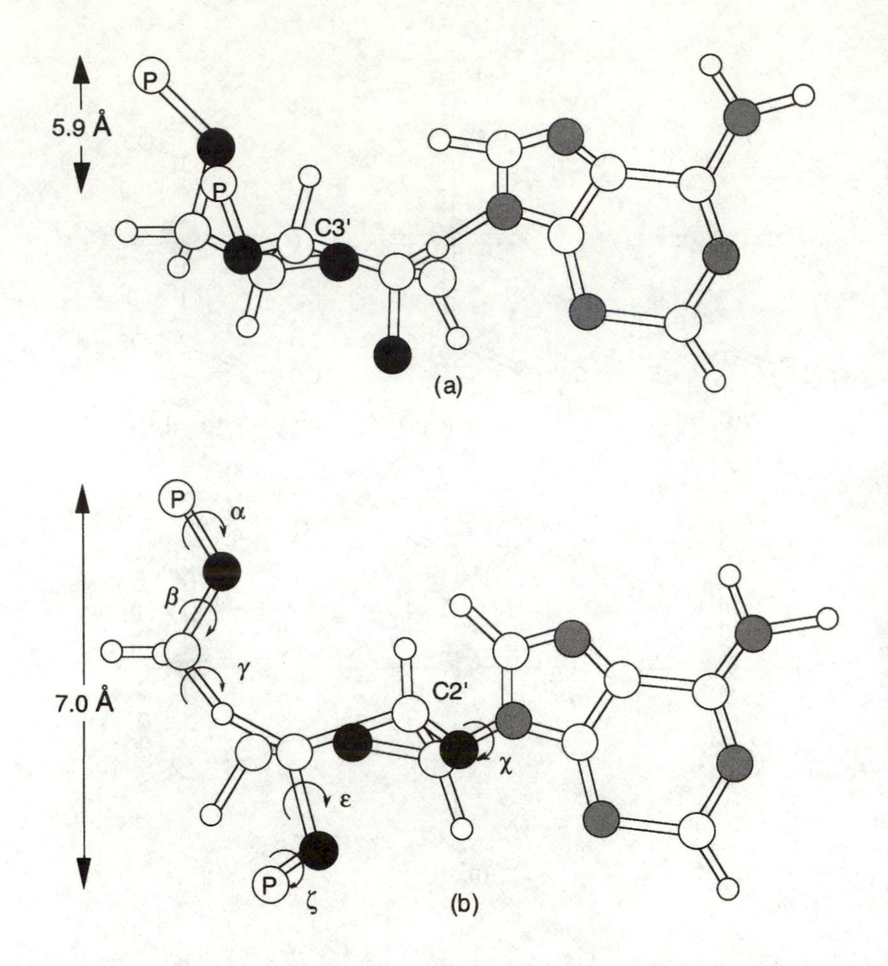

Figure 4.12 The most common nucleotide conformations: (a) C3′-*endo* and (b) C2′-*endo*.

adenine or uracil), the ratio between the length of the real chain and that of a Gaussian chain having the same number of segments lies between 10 and 15.

4.4 The double-helix structure

The Watson–Crick double-helix model for DNA has inspired an incredible number of papers. This is because, although the model explains the semi-conservative replication of DNA remarkably well, it relies on the inadequate experimental bases of the time and also raises mechanical problems about the rotation of one strand with respect to the other and topological problems. The structure has been determined in more detail and successively refined by recent crystallographic studies. As we shall see later, the double-helix parameters are only mean values and mask a fine structural modulation that is a function of the sequence. The discovery of topoisomerases and helicases has made it possible to solve the apparently mechanical problem of winding and unwinding.

4.4.1 The canonical B and A forms

The B form of the DNA double helix now appears as the typical conformation of *in vivo* DNA, whether naked or associated with proteins as in the bacterial chromosome or the chromatin of eukaryotes. What are its characteristics?

It is a right-handed double helix formed from two antiparallel helical phosphodiester chains: $3' \rightarrow 5'$ chaining occurs in one helix and $5' \rightarrow 3'$ chaining in the other (Fig. 4.13). The pairing of complementary bases (A and T, G and C) (Fig. 4.14) leads to an identical geometry for the AT and GC pairs. The four possible pairs (AT, TA, GC, CG) can occur in any arrangement whatsoever along the double helix without any appreciable local distortion.

The angle of *twist* around the double-helix axis defines the rotation around this axis when passing from one pair of bases to the next. This is similar to the angle between two successive stairs of a spiral (strictly, helical) staircase. The symmetry elements and the local reference axes are illustrated in Figs 4.15 and 4.16.

The angle of *tilt* defines the rotation of one base pair with respect to the next around the dyad axis. The angle of *roll* characterizes the rotational movement of one base pair with respect to the other around the C8–C6 axis in a way similar to the laths in a venetian blind.

Using the general criteria for the existence of a double helix, it is possible to define two main types of structure, called B and A, with the characteristics listed in Table 4.2. Both structures have a precise crystallographic definition in accordance with the various parameters and the seven characteristic angles given in the table. In fact, we must consider each of these structures as capable of generating a *family* by small variations in the geometrical parameters. The C structure discovered

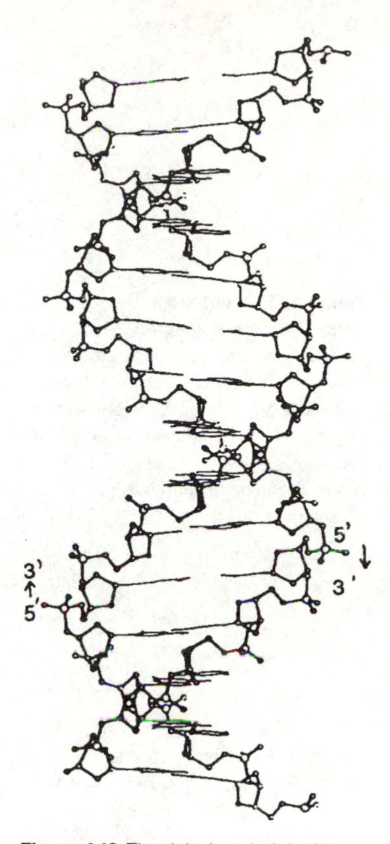

Figure 4.13 The right-handed double helix of B-DNA.

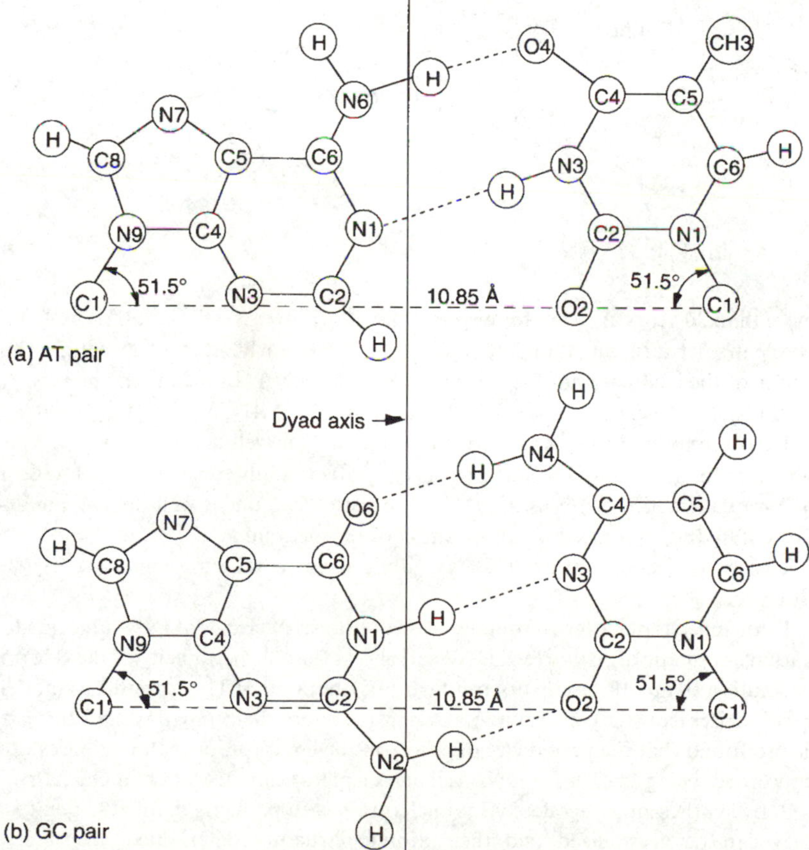

Figure 4.14 Watson–Crick pairing. A dyad axis (a 2-fold pseudosymmetry axis) in the plane of the bases brings the N9–C1 bond on the purine side into correspondence with the N1–C1 bond on the pyrimidine side.

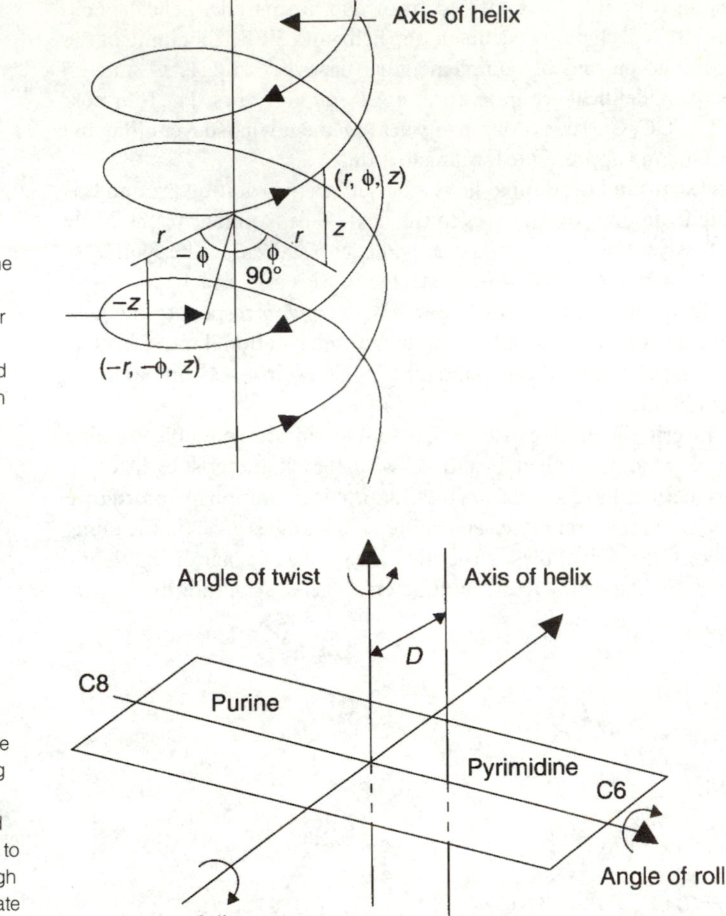

Figure 4.15 Since the two strands of the double helix are antiparallel, 2-fold symmetry occurs in the phosphodiester chain itself. At any point of the double helix, a phosphate group on one strand with cylindrical coordinates (r, ϕ, z) can be associated with a corresponding group on the other strand with coordinates $(r, -\phi, z)$.

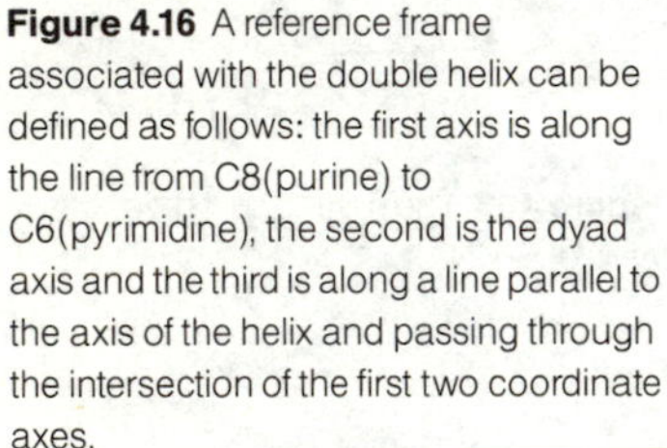

Figure 4.16 A reference frame associated with the double helix can be defined as follows: the first axis is along the line from C8(purine) to C6(pyrimidine), the second is the dyad axis and the third is along a line parallel to the axis of the helix and passing through the intersection of the first two coordinate axes.

more than 20 years ago can, for example, be considered as a variant of B with 9.3 base pairs per turn and a torsion angle of 38.6°. Similarly, in the A structure, the angles of the chain are not the same in DNA and RNA. In the latter, the type A double helix is the general rule for all double-stranded RNAs.

This description of DNAs in terms of three or four well-defined structures has prevailed for some 20 years. Such 'archetypes' arise mainly from the interpretation of X-ray diffraction patterns of DNA fibres, stretched under well-defined conditions of hydration and saline composition of the medium.

Two experimental approaches have obliged us to take another look at the situation:

1. An analysis of patterns from the electrophoresis of circular DNA subjected to partial enzymatic digestion (e.g. DNAseI) shows that the periodicity of the B form in solution is not 10 base pairs per turn but about 10.4. The 'canonical' model clearly arises from the constraints of packing the double helix into the DNA fibre. It is also found that the periodicity depends on the composition as far as bases are concerned, being smaller for DNA rich in AT pairs than DNA rich in GC pairs.

2. By synthesizing oligodeoxyribonucleotides with perfectly defined sequences, they can be crystallized and their atomic structure determined accurately,

Table 4.2 Some properties of B-DNA, A-DNA and RNA

Property	B-DNA	A-DNA
helix	right-handed	right-handed
sugar	C2′-*endo*	C3′-*endo*
base pairs per turn	10	11
pitch	34 Å	28 Å
axial distance between base pairs	3.4 Å	2.6 Å
torsion angle between base pairs	36°	32.7°
tilt angle	nearly 0°	about 20°
width of major groove*	11.4 Å	2.4 Å
width of minor groove*	6.0 Å	11.0 Å
distance of P atom from helical axis	9.3 Å	9.5 Å

	Rotation angles (deg)		
	B-DNA	A-DNA	RNA
$\alpha(\omega)$	−41	−90	−80
$\beta(\phi)$	136	211	175
$\gamma(\psi)$	38	47	49
$\delta(\psi')$	139	83	83
$\varepsilon(\phi')$	−133	−185	−147
$\zeta(\omega')$	−57	−45	−78
χ	78	27	13

*These widths correspond to the distance between two strands of the double helix measured in a direction perpendicular to the chain, reduced by 5.8 Å to allow for the van der Waals radii of the phosphates.

something that is impossible from a fibre diagram. The results from these experiments profoundly modify the regular primitive image of the double helix. When looked at more closely, the beauty of the harmonious Watson–Crick double helix, similar to that of the Leonardo da Vinci staircase in the Château de Chambord, is replaced by the imperfection of a spiral staircase with irregular steps that are all askew. Moreover, crystallization of the hexamer CGCGCG produces a left-handed double helix, called the Z form because of the zigzag shape of the phosphodiester chain.

4.4.2 The Z form

This form had in fact been found in solution many years previously on an alternating poly-D(CG)D(CG) placed in a strong saline concentration (Pohl and Jovin, 1972). A change in the sign of the circular dichroism that was observed in this case could be interpreted as a change in the chirality of the double helix. An X-ray analysis of the structure (Wang *et al.*, 1979) shows that the structural unit is formed from *two* base pairs. The *anti* form for C, which is linked to a C2′-*endo* sugar, is preserved. G on the other hand is *syn* and is linked to a C3′-*endo* deoxyribose. The major groove has disappeared and has been replaced by a convex surface. The minor groove is much deeper, and the double helix is more elongated and narrower than in the B form. Lastly, the bases in the Z form are much closer to the edge of the double helix (Fig. 4.17).

Table 4.3 lists the parameters of the Z form for comparison with those in Table 4.2. The figures quoted are typical values and variations are possible within the limits of the general orientations (g^+, g^-, t) defined in the text.

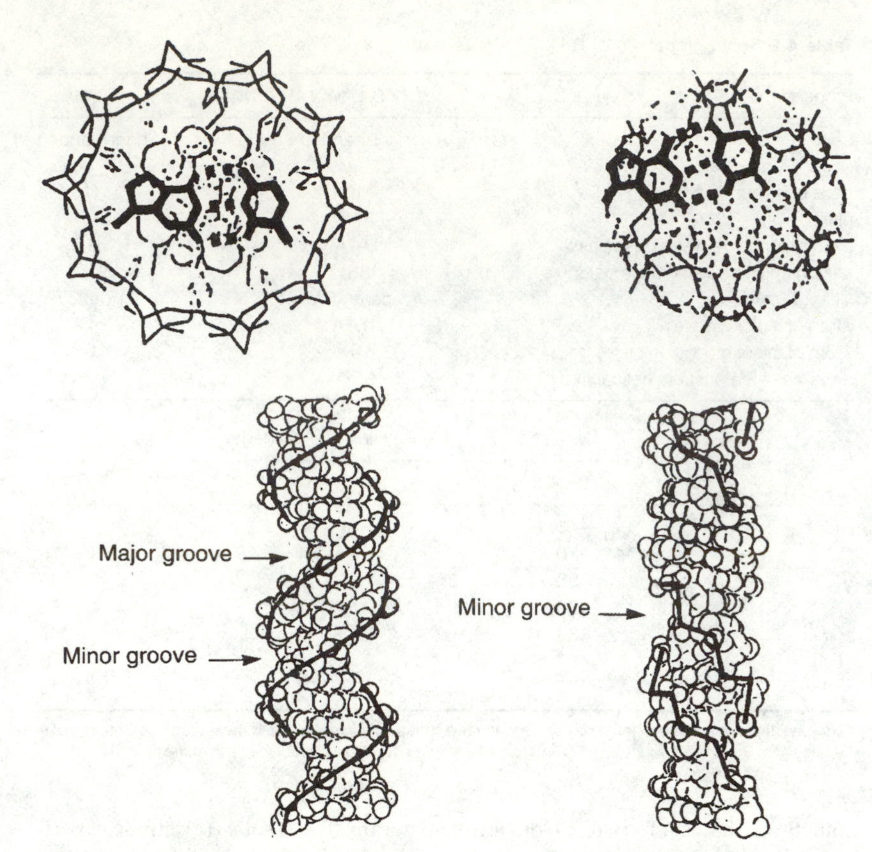

Figure 4.17 The B and Z forms of DNA. The upper figures show the location of a GC pair in both. (From Wang *et al.*, 1979. Reprinted with permission from *Nature*, vol. 282, pp. 680–6 and with the authors' permission. © 1979 Macmillan Magazines Ltd.)

Table 4.3 Properties of Z-DNA

Property	Z-DNA
helix	left-handed
sugar	C2′-*endo*/C3′-*endo**
base pairs per turn	12
pitch	44.6 Å
axial distance between base pairs	3.7 Å
torsion angle between base pairs	−30° (−10°, −50°)†
tilt angle	−7°

	Rotation angles (deg)	
	C	G
α	138	100
β	−94	−108
γ	80	−70
δ	48	−130
ε	180	−140
ζ	−170	56
χ	20	−100

*The first value is relative to C, the second to G.
†The values in parentheses refer to CpG and GpC respectively.

The B and Z forms are more directly compared in Fig. 4.17, while Fig. 4.18 shows the three types of structure, A, B and Z, as they appear respectively in crystals of GGGGCCCC (form A), CGCGAATTCGCG (form B) and CGCGCG (form Z).

4.4.3 Distortion of the helix and new parameters

Local distortions can be seen very clearly in Fig. 4.18, and it is obvious that the geometrical parameters used previously to describe 'ideal' double helices will be

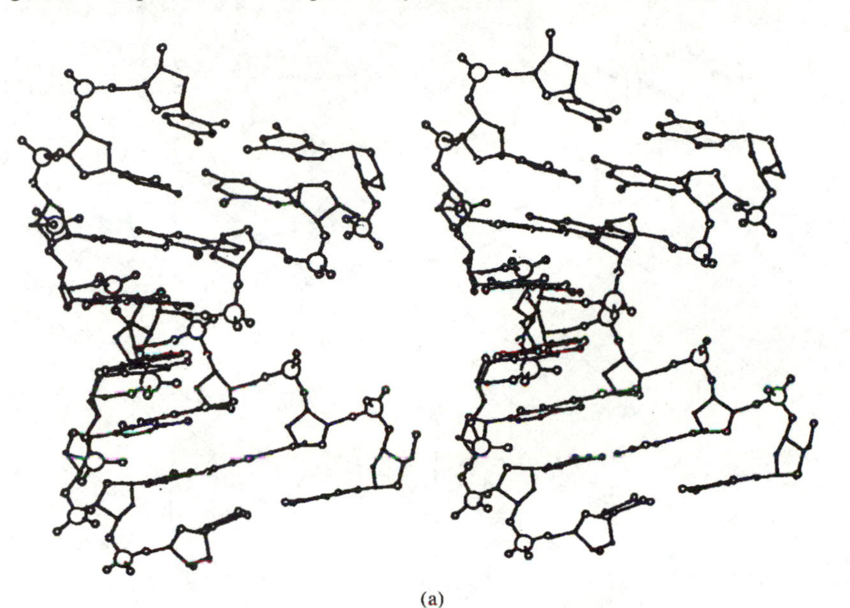

(a)

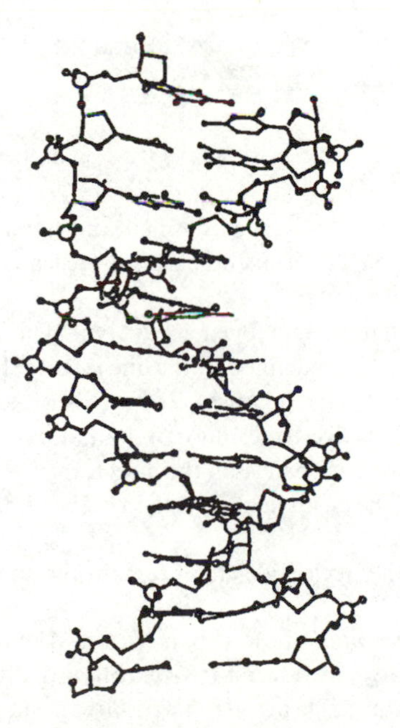

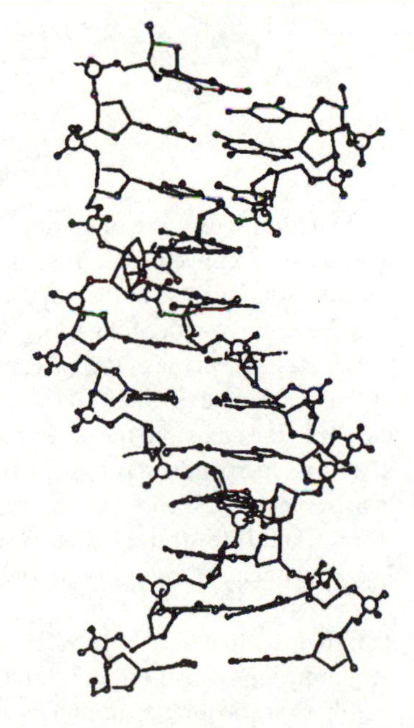

(b)

Figure 4.18(a),(b)

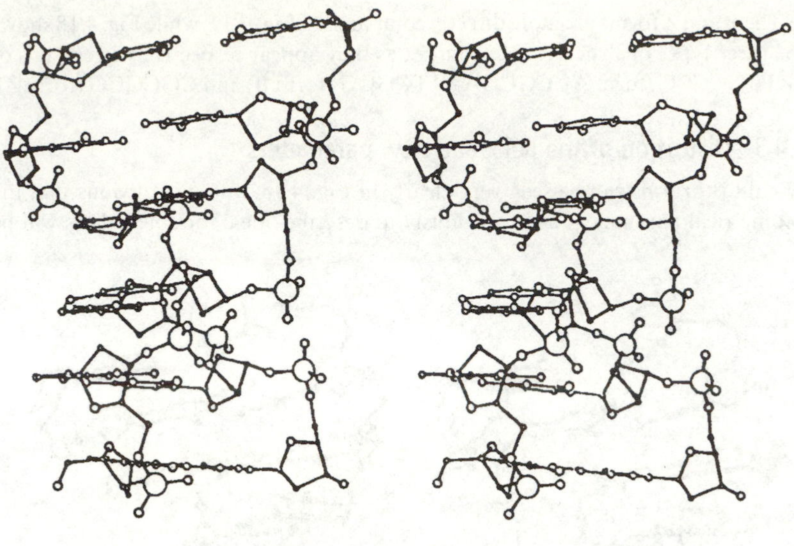

(c)

Figure 4.18 (a) The A structure: the octamer d(GGGGCCCC). (From McCall *et al.*, 1985. Reprinted from *Journal of Molecular Biology* with the permission of Academic Press.) (b) The B structure: the dodecamer d(CGCGAATTCGCG). (From Drew and Dickerson, 1981. Reprinted from *Journal of Molecular Biology* with the permission of Academic Press.) (c) The Z structure: the hexamer d(CGCGCG). (From Wang *et al.*, 1979. Reprinted with permission from *Nature*, vol. 282, pp. 680–6 and with the authors' permission. © 1979 Macmillan Magazines Ltd.)

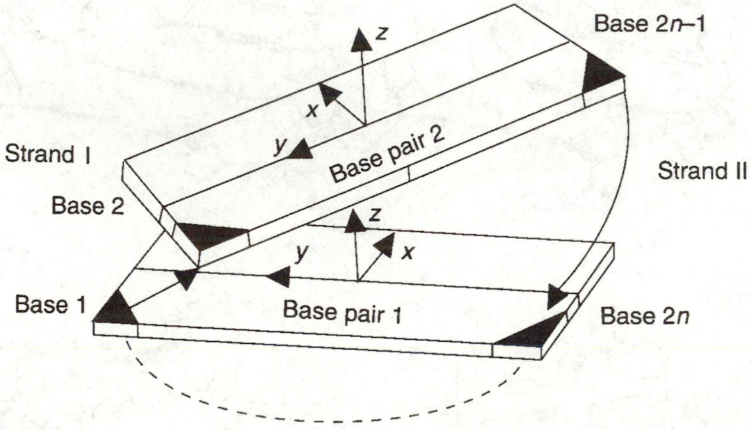

Figure 4.19 Definition of the reference frame: the double helix is viewed from the minor groove. The shaded corners of the plates representing the bases correspond to the bond between C1′ and the sugar. The arrows on each curve are pointing in the 5′ → 3′ direction. On strand I, bases are numbered from 1 to n, on strand II from $n+1$ to $2n$.

inadequate. A new set of parameters has recently been defined and agreed by international convention. As usual, greater detail in the description leads to a considerable increase in the number of parameters.

1. We begin by redefining a local coordinate system for each base pair. The x and y axes are in the plane of the base, the y axis being defined by the line joining the C6 of the pyrimidine (T or C) to the C8 of the purine (G or A). The x axis is perpendicular to the y axis and the positive direction is from the minor to the major groove. The z axis, perpendicular to the plane of the base pair, is in the direction 5′ → 3′ of whichever strand is chosen to be strand I. The positive y axis is then in the direction from strand II to strand I (Fig. 4.19).

2. The angles α, β, γ, δ, ε, ζ and χ in the nucleotide chain remain the same as before.

3. In order to describe the *relative movement* of one base pair with respect to the neighbouring one, the three angles already introduced (twist, roll and tilt, see Fig. 4.16) are no longer enough. Note that the angle of *roll* (ρ) is positive if the angle

between the two base pairs opens out towards the small groove (Fig. 4.20a). Similarly, the angle of *tilt* τ (rotation around the x axis) is positive if it opens out towards strand I (Fig. 4.20b). There is also *translation* defined along the same axes: *rise* D_z along the z axis, *slide* D_y along the y axis, and *shift* D_x along the x axis. All six parameters (illustrated in Fig. 4.21) may be defined either with reference to a local coordinate system or with reference to a single coordinate system in which the z axis is that of the double helix. The values in the two systems are in general different.

However, another set of parameters is needed to define the relative motion of one base with respect to the other within a base pair (Fig. 4.22). This motion may be *coordinated* or *opposed*. Coordinated rotation involves the angles of *tip* θ and *inclination* η. Opposed rotation involves an angle of *opening* σ around Oz, an angle of *propeller twist* ω around Oy and an angle of *buckle* λ around Ox. Coordinated and opposed translations are also possible: two *coordinated displacements* d_x and d_y; and three *opposed translations*: a *stagger* S_z along the z axis, a *stretch* S_y along the y axis and a *shear* S_x along the x axis.

In all, and in addition to the seven angles of rotation of the chain, 16 parameters are in principle required for a complete description of the geometry of nucleic acid helices. Such a list of angles and distances only has a meaning:

(1) if all these parameters can be determined experimentally, which is only possible from accurate measurements obtained by X-ray diffraction from oligonucleotide crystals or from interproton distances obtained by NMR in solution;

(2) if the parameters are involved in nucleic acid functions and more especially in describing the deformations induced by specific complexes with various small molecules or proteins.

To complete the description of structures differing from the Watson–Crick model, we must consider different base pairings. Instead of establishing hydrogen bonds between N6 and O4, N1 and N3 in the AT pair or between O6 and N4, N1

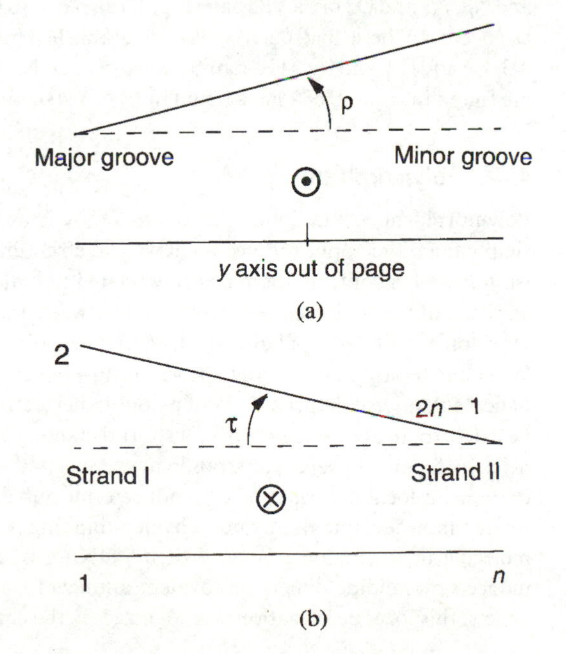

Figure 4.20 (a) A positive angle of roll; (b) a positive angle of tilt.

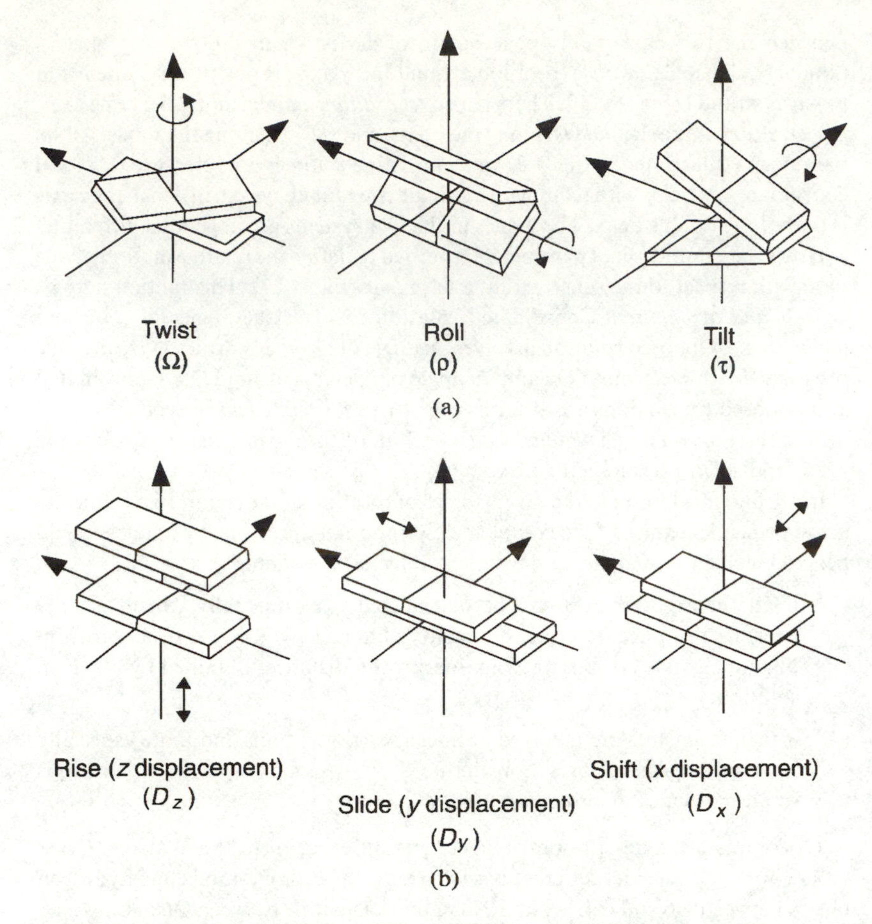

Figure 4.21 (a) Definition of angles between two successive base pairs. The rotations about the z, y and x axes are anticlockwise. (b) Definition of translations of one base pair with respect to another along the x, y and z axes.

and N3, N2 and O2 in the GC pair (Fig. 4.14), pairings known as *Hoogsteen pairings* can occur. In these, hydrogen bonds are established for an AT pair between N7 and N3, N6 and O4, and for a GC pair between N7 and N3, O6 and N4 (Fig. 4.23). These are found in triple DNA helices and in tRNA structures.

4.4.4 **Polymorphism**

Polymorphism of the double helices in DNA is an intrinsic property resulting simply from the application of the laws and structural schemes of organic chemistry. Is such a conformational variety related to biological mechanisms? In other words, can we find precise correlations between the mode of replication, transcription, repair and regulation of DNA and the polymorphism of the double helix? We can still only give a partial answer to this, but some experimental data clearly indicate that local deformations of a double-helix structure, the passage from one helical form to another, even over a short distance, or the opening out of a double helix to form two separate strands are associated with specific processes. For example, a local deformation is produced not only by the covalent binding of a foreign molecule but also by non-covalent binding (intercalation, positioning of a molecule in the major groove, protein–DNA association). In the case of UV-induced pyrimidine dimers or covalent adducts from mutagenic or carcinogenic agents, this local deformation is recognized by the enzymatic repair system. DNA

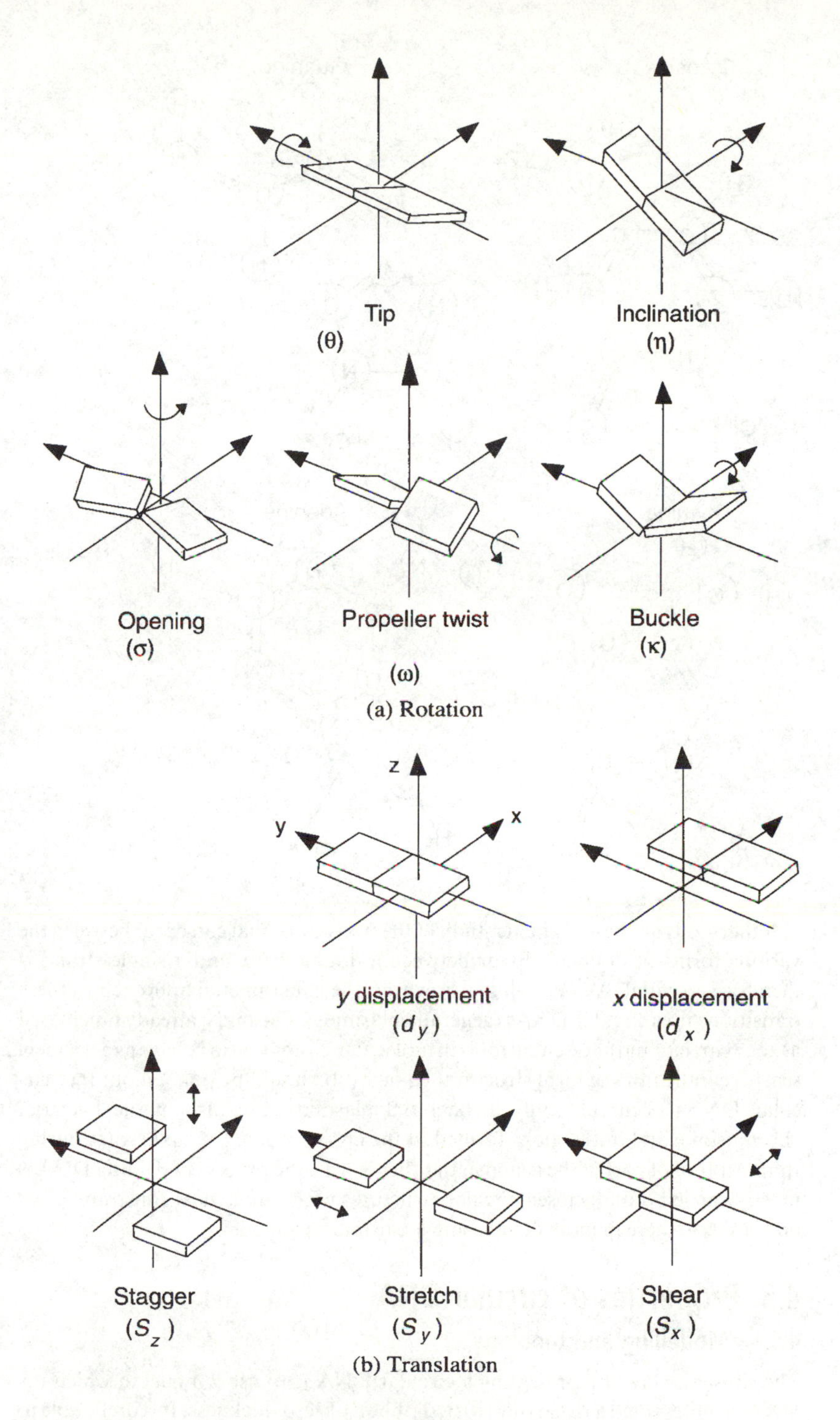

Figure 4.22 (a) Definition of rotations of one base with respect to the other within a base pair. The two movements are *coordinated* in the upper row and *opposed* in the lower row. (b) Definition of translations of one base with respect to the other within a base pair. The movements are *coordinated* in the upper row and *opposed* in the lower row.

replication or transcription requires an opening up of the double helix. Similarly, transient Z forms may appear during transcription (upstream from the polymerase) or through the presence of a covalent adduct on a base occurring in a given sequence.

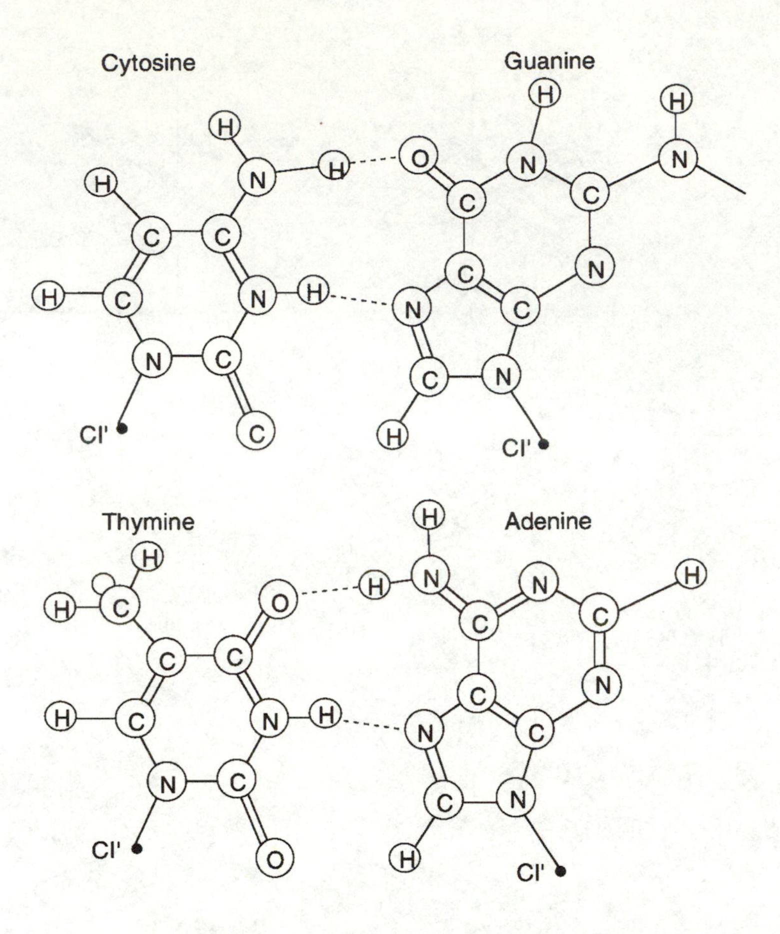

Figure 4.23 Examples of two Hoogsteen pairings between CG and AT. Note the protonation of cytosine at N1.

A thermodynamic and kinetic study of the transitions that can occur between the various forms of double helix or between a double helix and a single strand is therefore essential. As we shall see, however, the experimental approach to these transitions uses circular DNAs (generally plasmids). Plasmids, already much used as gene carriers and as essential tools in molecular biology, also behave as extremely sensitive indicators of local structural changes. It should be pointed out that circular DNAs occur not only in bacterial plasmids, but also in the bacterial chromosome and in the loops formed in the chromosomes of eukaryotes during transcription of part of the genome. In other words, the process of reading DNA is mostly carried out on closed circular structures with topological constraints. We now analyse these in more detail using a plasmid as a model.

4.5 Properties of circular DNA

4.5.1 Modelling and topology

The simplest way of representing a circular DNA is to use a model in which the DNA is represented as a narrow twisted ribbon of zero thickness. It is preferable to take the axis of the double helix as that of the ribbon and also to specify that the surface of the ribbon is always perpendicular to the local dyad axis. To remind ourselves of the two antiparallel helices of DNA, the two edges of the ribbon are given directions represented by arrows with opposite senses. If the ribbon is closed on itself (each edge being joined to itself without forming a Möbius strip), each edge describes a closed curve in space and the two curves thus described are *topologically linked*, i.e. they cannot be separated without cutting one of them. Three numbers characterize the ribbon in space (Crick, 1976): the linking number L, the twist number T and the writhing number W.

The *linking number* L is defined as the number of times the closed curve forming one of the edges of the ribbon spatially cuts the other curve forming the other edge of the ribbon. It is therefore an integer. To calculate L, consider a projection of the ribbon on to a plane and find all the points where a segment of one of the curves passes above (or below) the other. A value $+1$ is allocated to a point at which the segment of the upper edge (represented by the arrow) has to be rotated clockwise to be oriented along the direction of the arrow representing the segment of the lower edge. A value -1 is allocated to a point where the same operation involves an anticlockwise rotation. Or again, if the observer – like Ampère's (backstroke) swimmer – is traversed from foot to head by the arrow of the upper strand and looks at the lower strand, L will be $+1$ if the lower strand is on the right-hand side and -1 if it is on the left-hand side. Remember that this convention is the result of the two opposite directions given arbitrarily to the two edges of the ribbon.

The linking number L will be obtained by summing over all the crossings and dividing by 2, and it therefore describes the linking of the two curves in space: it must be 0 if the two curves can be separated. It remains constant during all deformations of the ribbon that do not tear it: this is a *topological* property. The image of a ribbon of given L in a mirror is a ribbon with the same magnitude of L but of opposite sign. A *relaxed* circular DNA molecule, i.e. one with the axis of the double helix lying in a plane, has value of L equal to the number of turns in the helix.

To define the *twist* T, consider a coordinate system $\boldsymbol{t}, \boldsymbol{u}, \boldsymbol{v}$ on the ribbon defined as follows: at a point M on edge A (see Fig. 4.24), draw the tangent vector $\boldsymbol{t}$ and the vector $\boldsymbol{u}$ perpendicular to $\boldsymbol{t}$ at M and resting on the second edge B. The vector $\boldsymbol{v}$ is defined by $\boldsymbol{v} = \boldsymbol{t} \times \boldsymbol{u}$. When M moves on A, the $(\boldsymbol{t}, \boldsymbol{u}, \boldsymbol{v})$ system rotates around $\boldsymbol{t}$ and, after moving round the whole of the closed curve A, the system has rotated through an angle ϕ. T is defined as $\phi/2\pi$.

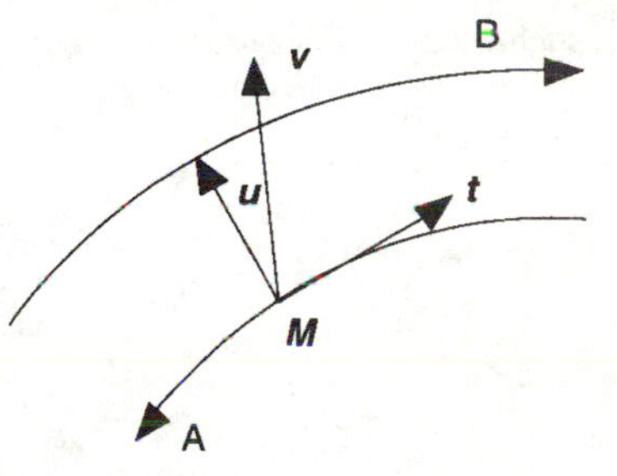

Figure 4.24 Coordinate system for definition of T.

The sign of T depends on the direction of rotation, i.e. that of the helix described by each edge. It is a number forming an intrinsic property of the ribbon, but is not necessarily an integer. It has become a *metric* property instead of a topological property. T may be changed by any deformation of the ribbon, and can be calculated per section of the closed ribbon, so that we shall then have $T_{AB} + T_{BC} + \ldots$. Note also that T is independent of the direction of the path since the sense of a helix is an intrinsic property.

Comparing L and T:

1. For a single closed curve such as the axis of the double helix, L and T are meaningless.
2. For a relaxed circular DNA, i.e. one in which the axis of the ribbon describes a plane curve, with 5000 base pairs and 10 base pairs per turn, we have $L = +500$, $T = +500$, so that $L - T = 0$. If the closed curve representing the axis of the helix does not lie in a plane, $L - T$ is not zero.

The *writhing* number W is defined as $L - T$. It is a geometrical property rather than a topological one. Consider a closed curve with a given value for W. If, around this curve taken as axis, we construct a helical ribbon, it cannot take on just any value for T since L must be an integer. If, for example, $W = 1.7$, then $T = -2.7$ ($L = -1$) or $T = -1.7$ ($L = 0$) or $T = 0.7$ ($L = +1$).

A few examples will make the meaning of L, T and W clearer.

1. If we form, with a closed circular ribbon, two superimposed loops, the twist is very small ($T = 0$). Since $L = +2$, we have $W = 2$ (the axis of the ribbon describes a helix).
2. Consider the telephone cable that connects the handset to the base. The attachment of this cable at its two ends is equivalent to a certain topological constraint defined by a number $L = T + W$. At each instant the cable will take on values of T and W compatible with the elasticity (modulus of rigidity and bending modulus) of the material from which it is made. If the cable is not extended (handset on the base) T is small and W large: the axis of the cable describes a helix. When the handset is in use the cable is stretched: the twist T increases and W decreases.
3. More precisely, for a closed ribbon wound round a cylinder and describing a right-handed helix, we have

$$T = N \sin \alpha$$

where N is the number of turns of the helix and α the angle between the tangent to the helix and a plane perpendicular to the axis of the helix. In this case:

$$W = N - N \sin \alpha = N(1 - \sin \alpha)$$

If α tends to zero (telephone cable unextended), W tends to N. If α increases and tends to $\pi/2$ (extended cable), W tends to zero.

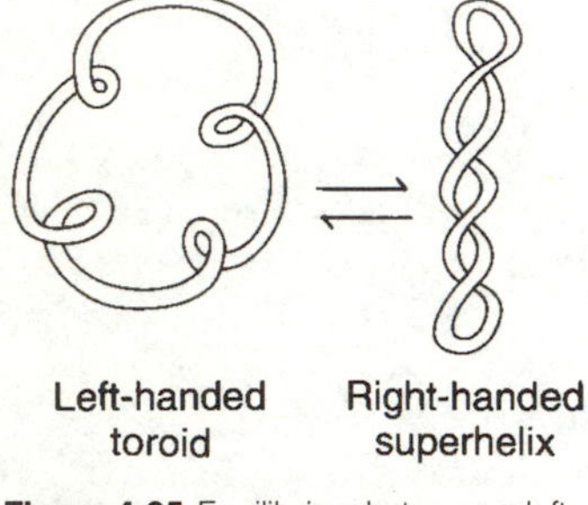

Figure 4.25 Equilibrium between a left-handed toroid and a right-handed superhelix.

For most of the supercoiled circular DNAs found in cells (plasmids or viral DNAs such as that of SV40), the number L is less than a certain value L_0 that would characterize a molecule having the same total number of base pairs but relaxed, i.e. where no twisting constraint remained. Since $L < L_0$, $\Delta L = L - L_0$ is a negative number: there is a larger number of base pairs per turn and hence fewer turns of the helix in the closed circular form than in the relaxed form. Despite this constraint, in a given medium the circular form will tend to recover the value of the angle of torsion Ω between two base pairs by writhing on itself. The writhing number W in this case will therefore be negative and we can also say that circular DNA forms negative 'supercoils'. These may either occur as toroidal shapes or as superhelices (Fig. 4.25), the two forms being in equilibrium.

Writhing reduces the torsional energy but introduces a curvature that increases with the bending energy. Equilibrium is established between the two deformations so as to minimize the total mechanical energy. The relationship between $\Delta L = L - L_0$, the W number and the change in twist ΔT can be written

$$\Delta L = W + \Delta T \qquad (4.1)$$

Since the twist energy is generally much greater than the bending energy, ΔT is small and W large. The sum of the two energies forms the free energy $G(W, T)$, which may be considered as *potential energy* stored in the molecular structure.

Note that L_0 is an *operational* quantity (all constraints have been removed) and is no longer a topological invariant. Depending on the external conditions (temperature, medium, ionic strength, etc.), the value of L_0 may vary. In addition, the quantities T and W cannot at present be directly measured experimentally. Because of this, the quantity $\Delta L = L - L_0$, called the *number of supercoils* τ, is the one generally used.

A specific variation in L is also defined by

$$\sigma = (L - L_0)/L_0 \tag{4.2}$$

known as the *supercoil density*. For most natural circular DNAs, σ lies between -0.03 and -0.09. Once the circle is closed, ΔL remains constant as long as no break (in single or double strands) occurs.

In vivo, the function of enzymes known as *topoisomerases* is to alter the value of the linking number L. They can cut either one DNA strand (*topoisomerases I*) or two (*topoisomerases II*). They then allow a part of the DNA to pass through the cut, and re-establish the bond between the free ends. ATP is not required in any of these processes. Topoisomerases I change L by $+1$ and topoisomerases II change it by $+2$. In both cases, there is a reduction in the number of supercoils and if the process continues we end up with the relaxed form. On the other hand, to introduce negative supercoils, i.e. to reduce L in comparison with L_0, there is another enzyme, *gyrase*, which consumes ATP. In energy terms, the creation of negative supercoils will favour further unwinding of DNA and hence the *replication*, *transcription* and *recombination* processes. Any variation in ΔT will be reflected by a change in W, i.e. by a change in the tertiary structure of the DNA.

4.5.2 Physical properties of circular DNA

The presence of supercoils in a circular DNA is easily detected by electron microscopy but no quantitative determination is possible because of the very method itself. The projection of a three-dimensional object on to a plane does not allow supercoils to be counted. On the other hand, the compact form resulting from a high value of W has a hydrodynamic behaviour different from that of the relaxed form. The latter has a coefficient of friction f (ratio of the frictional force to the speed of the particle) greater than that of the supercoiled form. The *electrophoretic mobility* $U = qE/f$ (where q is a quantity proportional to the charge and E is the electric field) and the *sedimentation constant* $s = m^*/f$, where m^* is the effective mass of the molecule, will therefore vary with W as shown in the curves of Fig. 4.26.

In both cases, the mobility and the sedimentation constant have minimum values when $W = 0$. On the other hand, the *intrinsic viscosity* (η), which varies as a power of the mean size of the molecule, is maximized when $W = 0$. These methods were

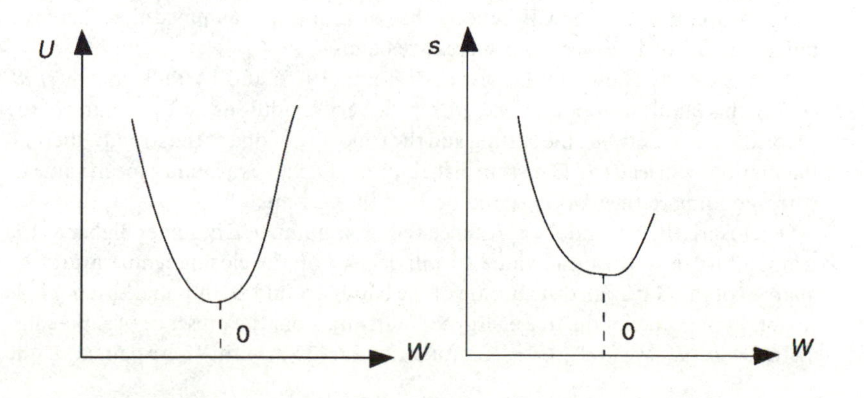

Figure 4.26 Variations of electrophoretic mobility U and sedimentation constant s with W.

used, for example, to follow the unwinding produced by an intercalating molecule such as ethidium bromide (ETB). The plane aromatic ring is sandwiched between two adjacent base pairs so that the distance from each of the base pairs to the intercalated aromatic ring is almost identical to the initial distance between the two base pairs (3.4 Å in the B form). This elongation of the double helix is accompanied by an unwinding ($\Delta\Omega < 0$) whose value depends on the type of aromatic molecule. For ethidium bromide, $\Delta\Omega = -26°$. Hence, if ETB is progressively intercalated into a circular DNA initially having τ negative supercoils, a relaxed form is obtained for a number N of intercalated molecules given by

$$N = 2\pi\tau/\Delta\Omega$$

For $\tau = -25$ and $\Delta\Omega = -26$, we find $N = 346$. The unwinding produced by the intercalation has removed the constraint by taking the DNA back to its relaxed state under experimental conditions.

It is precisely this intercalation of ETB that was the origin of the first work on circular DNA. Beyond 346 molecules, we can continue to intercalate ETB and hence to unwind DNA, this time by creating a positive supercoiling ($W > 0$). In accordance with theoretical predictions, it can be shown experimentally that the first part of the process ($W \to 0$) is easier to realize than the second ($W > 0$) since in the latter we progressively increase the total free energy of the DNA.

4.5.3 Gel electrophoresis

The first hydrodynamic methods were soon replaced by a simpler method requiring only small quantities of material: *agarose gel electrophoresis*. It has become the first choice of methods for studying covalently closed circular DNAs.

One-dimensional electrophoresis

Firstly, it becomes easy to separate a closed supercoiled DNA from a relaxed DNA (with a cut single strand): because of its compact form, the former migrates more rapidly than the latter. The two species appear as two bands several centimetres apart. If electrophoresis is then carried out after reaction with topoisomerase I, several bands appear, corresponding to a family of *topoisomers*, so called because they differ from each other only by their L value (Fig. 4.27a). Their migration depends on the value of W. This is because the resistance offered by the gel to motion in the electric field is smaller when W is large, although there is of course a limiting value. However, since T is constant in a given medium and at a given temperature, the change in W reflects that of L and it is assumed that there is a difference ΔL of 1 between two successive bands.

As was clearly shown by Depew and Wang (1975) and by Pulleyblank *et al.* (1975), this family of topoisomers reflects the very conditions of functioning of the topoisomerase. Between the cutting and the reclosure of one of the strands, thermal fluctuations will lead to a Gaussian distribution of L values around a mean value L_0 corresponding to the working conditions of the enzyme.

On closure, $W = 0$ and $L_0 = T$. Since the twist number T becomes higher as the temperature falls, the mean value L_0 will depend on the closing temperature. An analysis of the Gaussian distribution of the bands around L_0 thus makes it possible to obtain the value of the free energy ΔG_τ arising from the presence of supercoils. With $\tau = L - L_0$, we find $\Delta G_\tau = K\tau^2$ for a circular DNA with N base pairs. For one

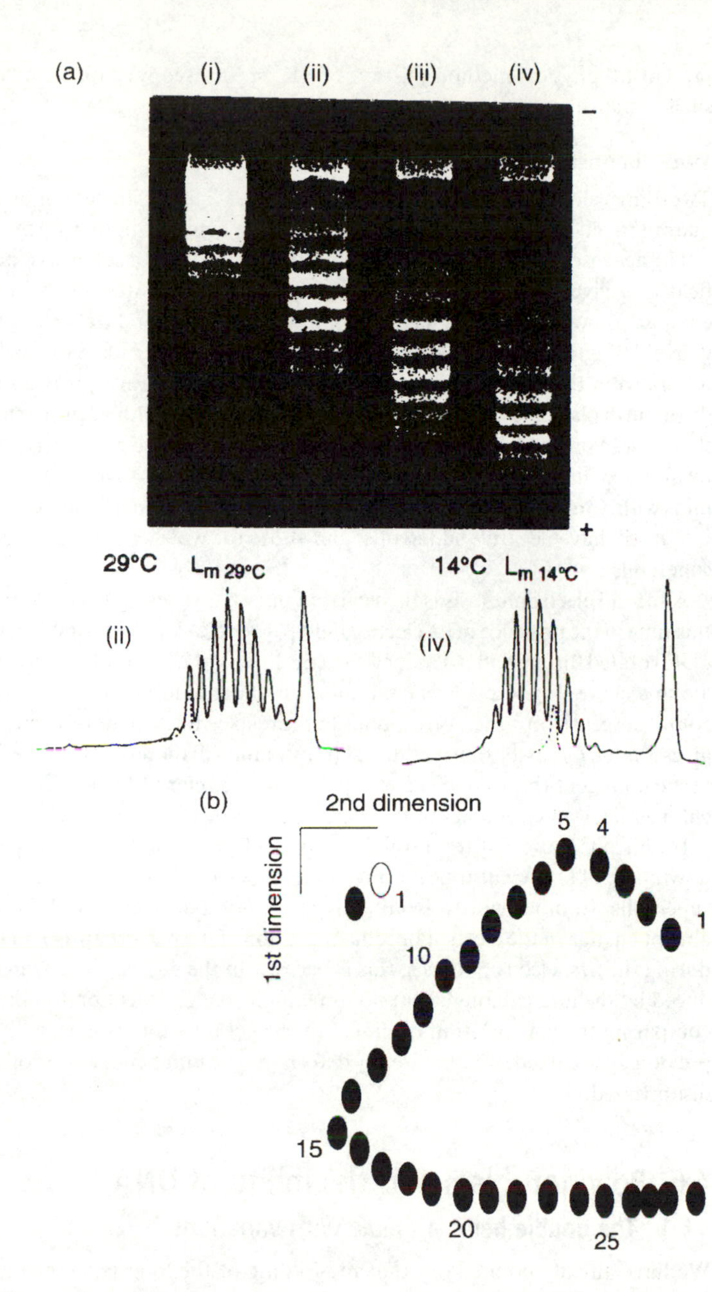

Figure 4.27 (a) Electrophoresis of topoisomers of PM2 DNA as revealed by ethidium bromide fluorescence. The four runs (i), (ii), (iii) and (iv) are for different temperatures of ligation (37, 29, 21 and 14°C respectively). The densitometric patterns (ii) and (iv) show Gaussian distributions. The dotted curve indicates the presence of a small amount of linearized PM2. (b) Two-dimensional electrophoretic pattern for topoisomers. (From Depew and Wang, 1975. Reprinted from *Proceedings of the National Academy of Sciences, USA* with the authors' permission.)

base pair, we can then write

$$\Delta G_\tau / N = NK(\tau/N)^2 \tag{4.3}$$

Since both τ/N and $\Delta G_\tau/N$ are almost constant, NK must be constant, as confirmed by experiment. We find $NK \approx 1100RT$.

To obtain some idea of the sensitivity of the 'tool' provided by a plasmid having, for example, 4000 base pairs, consider a rise in temperature of 10°C. This is accompanied by an unwinding of only 0.1° per base pair, but of 400° for the whole plasmid, i.e. greater than one supercoil, and this is easily detected. There is at present no physical method (X-ray, NMR, spectroscopy) capable of measuring such a small angular variation.

Two-dimensional electrophoresis

Two-dimensional electrophoresis was developed a few years ago in order to deal quantitatively with the transition from a double helix to another form.

The agarose gel plate on which the sample is placed is first subjected to an electric field in a given buffer solution (e.g. 100 mM of trisborate-EDTA at pH 8). In comparison with a plasmid having a cut in the single strand (relaxed form), other plasmids migrate more quickly as the number of supercoils τ increases and each spot differs from the next by a change in L of one unit. However, it is impossible to distinguish plasmids with positive supercoils from those with negative supercoils in this way. In order to do this, a certain number n of negative supercoils are removed by an intercalating agent such as chloroquine added in sufficient quantities. Plasmids with a number of negative supercoils $\tau = n$ are in relaxed form; those for which $\tau > n$ still have negative supercoils; and those for which $\tau < n$ now have positive supercoils.

A second electrophoresis is then carried out on this new plasmid population but this time in the presence of an electric field perpendicular to the first (or with the gel plate rotated through 90°). Under these conditions, the topoisomer spots are spread along a curve (Fig. 4.27b). The minimum in the mobility at the apex with a horizontal tangent (top right) corresponds to the relaxed form in the first buffer and thus gives a value $L_{0,1}$ in the first medium. The minimum in the mobility at the apex with a vertical tangent (bottom left) corresponds to the relaxed form in the presence of chloroquine, which enables us to determine a value of $L_{0,2}$ in the second medium.

In this population of topoisomers, any conformational change resulting in an unwinding of DNA will appear above a certain value of the free energy stored in the supercoils. In other words, from a given topoisomer, we shall often observe an abrupt change in the electrophoretic migration. This perturbation can only occur during the first electrophoresis; this is because in the second the unwinding introduced by the intercalating agent no longer allows the transition to take place. By comparing the new position occupied in the gel by a topoisomer with the one it previously occupied, we can directly determine the number of supercoils that have disappeared.

4.6 Polymorphism and flexibility of DNA

4.6.1 The double helix: a theme with variations

We have already pointed out that the folding of the long DNA chain involving several tens or hundreds of thousands of nucleotides, either in a phage head or in a chromosome, implies that the molecule is extremely flexible.

The three principal forms of double helix (A, B, Z) actually represent three families since the distances and angles characteristic of a given type are only mean values. Recent crystallographic studies on small helical fragments have highlighted the variation in the parameters from one base to another. The sequence of bases, which constitutes the information carried by the genome, is reflected in the structure. This idea enables us to understand more clearly the mechanisms of specific recognition and the concept of 'consensus' sequence. Such sequences, proposed for prokaryote or eukaryote promoters, for the repressor–operator interaction in the

SOS system, for the interaction between messenger RNA and ribosomal RNA, etc., must be understood as generating special local structures and dynamics.

The following sections list some of the particular structural features frequently encountered.

The kink

First conceived as a model (Crick and Klug, 1975; Sobell *et al.*, 1977), this is assumed to be a fairly sudden bend in the axis of the double helix. We shall see in section 25.2 dealing with the structure of the nucleosome (Luger *et al.*, 1997) that such bends occur at several points but that the angle through which the DNA is bent is much smaller than that found in the first structure determined at low resolution (Richmond *et al.*, 1984). In addition, many proteins interact with DNA and produce a localized curvature in it (see section 24.2 and Fig. 24.16).

The loop

The rupture of hydrogen bonds over several base pairs, and the separation of the two nucleotide chains, produces loops of various sizes. *In vitro* denaturation experiments have already revealed the existence of such structures and their migration along the helix. Similar models have been proposed for the initiation of replication and transcription. In the case of replication, the combined action of the unwinding proteins and the proteins binding themselves specifically to the single strand triggers the creation and progression of the replication fork. In transcription, the RNA-polymerase bound to the −35 region of DNA creates a loop in the −10 region.

Breathing

The temporary breaking of hydrogen bonds is caused by the rapid partial rotation of one base in a pair, making the hydrogen atoms in the NH groups (G and T) accessible and enabling them to be exchanged with neighbouring protons in the presence of a catalyst. This exchange mechanism, which can be studied either by fluorescence-monitored stopped flow or by proton NMR, shows that the opening occurs on average every millisecond or tens of milliseconds at room temperature (see Chapter 9).

This plasticity of the double helix is also evident in the phenomenon of *premelting*. Well below the temperature T_m marking the transition from the double helix to the single strand, a reversible conformational change can be observed by various spectroscopic methods.

The cruciform structure

In the presence of self-complementary palindromic sequences (A . . . GA′ . . . G′) separated by several base pairs, a ‘cross’ may be formed by an exchange of complementarity (Fig. 4.28). The loops closing the two arms consist of a few bases not participating in the palindromic sequences. These single-stranded regions, resembling the loops in tRNA, will be sensitive to specific nucleases (such as the S_1 endonuclease) or to certain reagents (e.g. OsO_4 with thymine).

Appearance of a Z form

For alternating sequences of the GC type or more generally of the pU−pY type, a left-handed double helix can be formed locally, provided there are enough base pairs capable of adopting this conformation. This $B \rightarrow Z$ transition is facilitated,

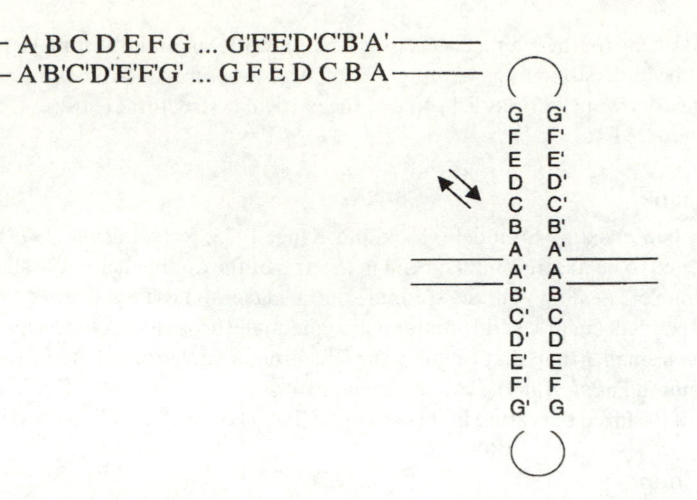

Figure 4.28 The cruciform structure.

for example, by the torsional energy stored in the supercoiled circular DNAs. The transition zone between the left-handed and right-handed helix is also sensitive to the S_1 endonuclease. The conformation of such a junction is not known precisely.

The conformational changes (kink, loop, cruciform, Z form) are generally caused by constraints due to the medium (pH, ionic strength) or by the molecule itself (supercoiling energy), but they can also be triggered by the binding of specific ligands to the DNA. In *intercalation*, the change in conformation is caused solely by ligand interaction.

Intercalation

A molecule with a plane aromatic ring (fluorene, acridine, phenanthridine, etc.) may be intercalated in DNA, i.e. slipped between two base pairs. Since the distance between aromatic molecules is determined by van der Waals radii (of the order of 3.4 Å), two adjacent base pairs are about 6.8 Å from each other. This elongation of the DNA molecule is accompanied by a local unwinding and a change in the sugar puckering. All such local changes in conformation naturally require accurate structural studies.

Three- and four-stranded helices

Alternating sequences of the $(dTdC)_n(dAdG)_n$ type located in negative supercoiled plasmids placed in a medium of pH < 7 take on an unusual tertiary structure. Triple helices and a kink are formed in a DNA conformation called H-DNA, whose existence has been demonstrated *in vitro*. Such alternating sequences are found for example in the human genome about every 150 kb. The supercoils may be induced during transcription but at the moment there is no proof that triple helices exist *in vivo*.

Four-stranded structures have recently been discovered in oligonucleotide crystals. They provide a good model for telomeric DNA located at the end of chromosomes containing a G-rich and a C-rich strand. These specialized sequences could play a major role in maintaining the stability and integrity of chromosomes.

4.6.2 Flexibility

Conformational studies of DNAs and their double-helix structure show that there is only a limited range of angles in the nucleotide chain: in other words, there is some rigidity in the structure. If we add to this the Coulomb repulsion between the charged phosphate groups, we should describe DNA to begin with as a fairly rigid molecule, at least over small lengths of the chain. Yet DNA exhibits polymorphism around the three A, B and Z families and the change from one form to another does require a certain 'plasticity' in the molecule. However, the real problem is in fact that DNA chains fold in so many ways. We have already pointed out how much a long DNA molecule must be folded inside a phage head or a chromosome. In both cases, positively charged molecules (polyamines, histones, protamines) play an important role. More recently, however, without the help of any other molecule, it has proved possible to build closed DNA circles containing not many more than a hundred base pairs. The small radius of curvature (of the order of 50–60 Å) raises the problem of the flexibility of the chain: it is difficult to understand such a high flexibility unless, at certain points in the molecule, there are regions already curved because of their sequence or at least some regions that are more flexible than others. The general term *flexibility* thus embraces both static aspects (curvature) and dynamic aspects related to the existence of a double helix departing from the ideal linear rigid model.

The study of flexibility must therefore begin with an overall look at the whole DNA molecule, which will involve the general principles of polymer mechanics in a given medium. A more detailed analysis on a local scale will then be undertaken in order to take into account the sequence and any irregularities in the geometry of the double helix.

The worm-like chain (continuous curvature)

An interesting model has been proposed by Kratky and Porod (1949) to describe all states between the two extreme models of the perfectly flexible chain with free rotation (or Gaussian chain) and the perfectly rigid rod-shaped chain.

Consider N segments of length a each making a small angle θ with the previous one (lying on a cone of vertical semi-angle θ around the previous segment – see Fig. 4.29). The mean value $\langle h \rangle$ of the projection on the first segment of the

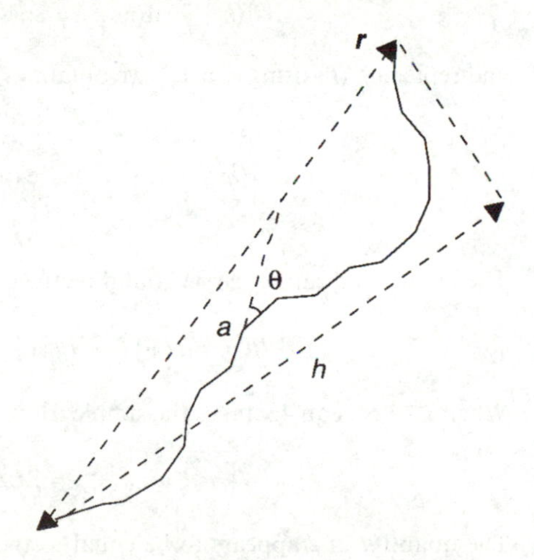

Figure 4.29 Planar model of Kratky and Porod (1949) (worm-like chain) consisting of N segments of length a making an angle θ with adjacent segments. In this example, h is the projection of the vector r joining the two ends on to the first segment.

end-to-end distance is given by

$$\langle h \rangle = a \sum_{k=0}^{N} x^k = a(1 - x^N)/(1 - x) \tag{4.4}$$

where $x = \cos\theta$.

The *persistence length* L_P is defined as the limiting value of h as $N \to \infty$. We then have:

$$L_P = a/(1 - x) \tag{4.5}$$

(As $\theta \to 0$, the chain can be viewed as one for which L_P remains finite, i.e. in which $a \to 0$, which amounts to introducing a continuous curvature). Since θ is small, $\cos\theta \approx 1 - \theta^2/2$ and

$$L_P = 2a/\theta^2 \tag{4.6}$$

Note that the persistence length does not depend on the length L along the curve, i.e. Na, but is an intrinsic property of the polymer in a given medium. It is related to the structure and the interactions and not to the molecular mass. In eqn 4.4, we can replace $x^N = (1 - \theta^2/2)^N$ by $\exp(-N\theta^2/2)$ and putting the total length of the chain $L = Na$ and using eqn 4.6, we obtain

$$x^N = \exp(-L/L_P)$$

and hence

$$\langle h \rangle = L_P[1 - \exp(-L/L_P)] \tag{4.7}$$

As $L \to \infty$, $\langle h \rangle \to L_P$, which agrees with the definition. If, on the other hand, the chain is small enough for $L \ll L_P$, we obtain $\langle h \rangle \approx L$. This short chain behaves like a rigid rod of length L. With this *worm-like* model, we can thus account for both local rigidity and the flexibility of a sufficiently long chain.

Another comparison between this model and those describing the behaviour of conventional polymers can be obtained by calculating $\langle h^2 \rangle$. If $\boldsymbol{r}$ is the vector joining the two ends, we have $\langle h^2 \rangle = \langle \boldsymbol{r} \cdot \boldsymbol{r} \rangle$. Hence

$$\mathrm{d}\langle h^2 \rangle = \mathrm{d}\langle \boldsymbol{r} \cdot \boldsymbol{r} \rangle = 2\langle \boldsymbol{r} \cdot \mathrm{d}\boldsymbol{r} \rangle = 2\langle h \rangle\, \mathrm{d}L \tag{4.8}$$

and replacing $\langle h \rangle$ using eqn 4.7, we obtain $\langle h^2 \rangle$ by integration:

$$\langle h^2 \rangle = 2L_P \int_0^L [1 - \exp(-L/L_P)]\, \mathrm{d}L$$

The integration can be carried out directly and leads to

$$\langle h^2 \rangle = 2L_P\{L - L_P[1 - \exp(-L/L_P)]\} \tag{4.9}$$

When $L \to \infty$, eqn 4.9 takes the simple form

$$\langle h^2 \rangle = 2LL_P$$

The quantity $2L_P$ appears to be equal to the *statistical link* in a Gaussian chain equivalent to this infinite *worm-like* chain. When $L \to 0$, the rigid behaviour

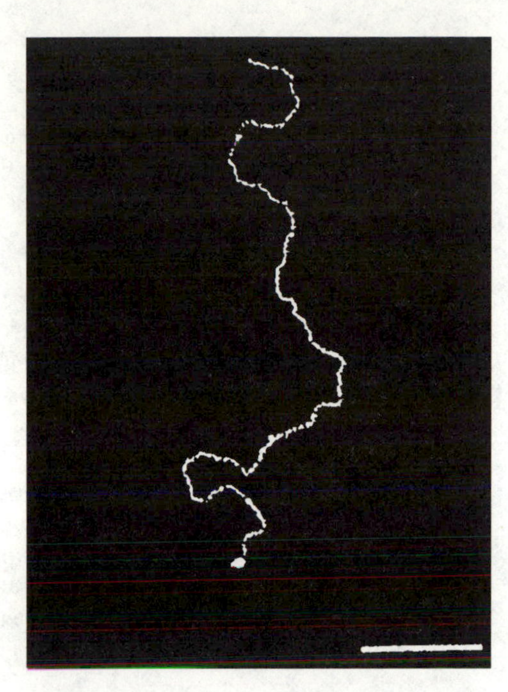

(a)

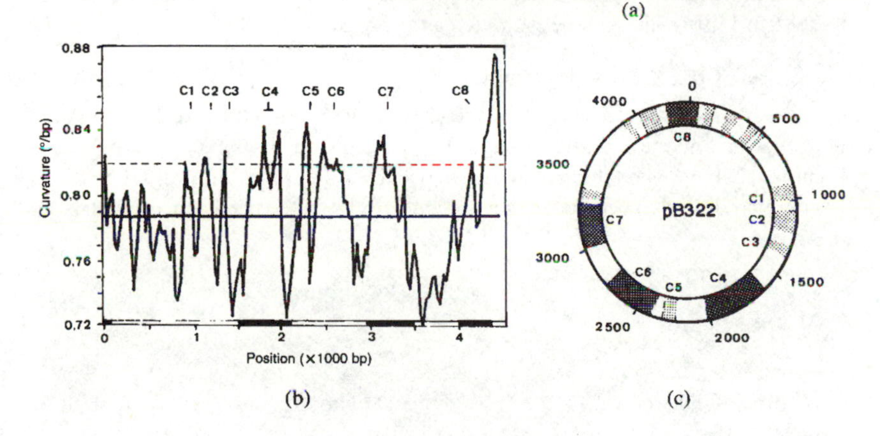

Figure 4.30 Curvature of a pBR322 DNA molecule determined by electron microscopy. (a) View of one molecule from 29 to 4361 labelled at the 4361 end. The bar corresponds to 200 nm. (b) Variation of the curvature with position along the molecule. The horizontal line gives the mean value of the curvature. (c) The curvature represented on the pBR circle: the darker the zone, the greater the curvature. (From Muzard *et al.*, 1990. Reprinted from *The EMBO Journal* by permission of Oxford University Press and with the authors' permission.)

previously described by eqn 4.7 is given precisely by

$$\langle h^2 \rangle_L \rightarrow L^2[1 - (1/3)(L/L_P) + (1/12)(L/L_P)^2 - \cdots] \tag{4.10}$$

If, for example, we consider a B-DNA with a persistence length $L_P = 450$ Å, we find for $L = 680$ Å (length of DNA in the nucleosome) that $\langle h^2 \rangle^{1/2} = 544$ Å. There is already considerable flexibility.

The chain model with continuous curvature is only of interest if both quantities L and L_P can be measured. Two groups of methods are available for this.

Electro-optical methods

Measurements are made of the linear birefringence or dichroism produced by an electric field applied to a solution of DNA. The orientation of the molecules by the electric field will depend on their rigidity, with rods being much more easily oriented than chains. The magnitude of the optical phenomenon depends both on the orientation and on the intrinsic physical properties of the molecule. However, interpretation of the data is difficult because of the approximations made in hydrodynamic models. Moreover, in order to use pulsed electric fields that are strong enough, measurements have to be made at low ionic strength.

Electron microscopy

Another approach is to measure the lengths L and L_P directly for a molecule deposited on the grid of an electron microscope (Fig. 4.30). Although measuring L poses no problem, that of L_P requires measurements of the angle θ between two segments separated by a distance s counted along the chain by using the relationship (only valid in 3D):

$$\cos\theta \approx \exp(-s/L_P)$$

The method can be used at any ionic strength and is independent both of the polydispersity and of molecular aggregation, since each molecule is dealt with separately. On the other hand, artefacts may be present depending on the method used for depositing the DNA on the grid. If insufficient time is made available to re-establish conformational equilibrium (sudden freezing for example), we then observe the projection on to a plane of a three-dimensional structure in solution.

From all the results obtained by either of these two methods, the persistence length of DNA in 1 mM NaCl can be estimated as approximately 600 Å.

Flexibility of the worm-like chain

Once the geometrical parameters of the chain have been defined (essentially the total length L and the persistence length L_P), we can study its dynamics. How does this model behave when thermal agitation occurs? As our intuition would predict, the energy required to bend the chain is proportional to the persistence length (see Box 4.1).

Box 4.1 **Calculation of worm-like chain flexibility**

It will be assumed to begin with that the chain bends but remains in a plane. We can define a force per unit length of the chain dG/ds required to produce a curvature $d\theta/ds$, where dG is the elementary change in free energy, and for small values of the curvature we can write:

$$dG/ds = (dG/d\theta)(d\theta/ds) + \tfrac{1}{2}(d^2G/ds^2)^2(d\theta/ds)^2 + \cdots \qquad (1)$$

In the absence of any permanent bending moment, $dG/d\theta = 0$ and

$$dG/ds = g(d\theta/ds)^2/2 \qquad (2)$$

putting $d^2G/d\theta^2 = g$ (with dimensions ML^3T^{-2}).

The total energy required to bend a chain of length L is therefore

$$\Delta G = (g/2)\int_0^L (d\theta/ds)^2\, ds$$

If we assume for small displacements that $\theta = ks$, where $k = \mathrm{d}\theta/\mathrm{d}s$ is a constant (circular bending), then

$$\Delta G = gk^2 \int_0^L \mathrm{d}s/2 \tag{3}$$

If θ_L is the angle between the ends of the chain,

$$\theta_L = \int_0^L k\,\mathrm{d}s = kL \tag{4}$$

so that

$$k = \theta_L/L \tag{5}$$

and

$$\Delta G = g\theta_L^2/2L \tag{6}$$

The bending coefficient g is therefore equal to twice the energy required to bend a unit length of the polymer chain through one radian. In fact the bending of the chain is a phenomenon related to the thermal agitation, and the probability of bending characterized by an energy ΔG is given by the Boltzmann factor $\exp(-\Delta G/RT)$. We thus have:

$$\langle\theta_L^2\rangle = \int_0^\pi \exp(-\Delta G/RT)\theta_L^2\,\mathrm{d}\theta_L \Big/ \int_0^\pi \exp(-\Delta G/RT)\,\mathrm{d}\theta_L$$

$\langle\theta_L^2\rangle$ being the variance of the distribution, which is

$$\langle\theta_L^2\rangle = \int_0^\pi \exp(-g\theta_L^2/2LRT)\theta_L^2\,\mathrm{d}\theta_L \Big/ \int_0^\pi \exp(-g\theta_L^2/2LRT)\,\mathrm{d}\theta_L = LRT/g \tag{7}$$

The bending was assumed to be in a plane, i.e. around an axis in the plane and perpendicular to the chain. For an axis perpendicular to both the plane and the chain, we should find the same result, and since the two modes of deformation are assumed independent, we have finally

$$\langle\theta_L^2\rangle = 2LRT/g \tag{8}$$

The mean distance between the ends projected on to the first link is

$$\langle h\rangle = \int_0^L \langle\cos\theta_s\rangle\,\mathrm{d}s \tag{9}$$

with $\langle\cos\theta_s\rangle = 1 - \langle\theta_s^2\rangle/2 + \cdots$. Since eqn 8 above is satisfied more accurately for a segment $s \ll L$, we have

$$\langle\cos\theta_s\rangle = 1 - RTs/g \approx \exp(-RTs/g) \tag{10}$$

and hence, substituting this in the expression for $\langle h\rangle$ and integrating:

$$\langle h\rangle = (g/RT)[1 - \exp(-RTL/g)] \tag{11}$$

Comparing eqn 4.7 and eqn 11 above gives:

$$L_\mathrm{P} = g/RT \tag{12}$$

The persistence length L_P varies as $1/T$ for a constant bending coefficient, and substituting the value for g in eqn 6 in Box 4.1 gives

$$\Delta G = RTL_P\theta_L^2/2L \qquad (4.11)$$

(an expression first obtained by Landau and Lifschitz).

Measurements of flexibility

The DNA chain is rigid over a short distance (comparable with a rod) and becomes flexible over large distances. The persistence length L_P introduced by the *worm-like chain* model determines an approximate boundary between the two types of behaviour. Several methods are available for the measurement of L_P, all of which depend on a determination of the radius of gyration R_G, since it can be shown for the worm-like model that

$$R_G^2 = 2L_PL\{1/6 - L_P/2L + L_P^2/L^2 - (L_P^3/L^3)[1 - \exp(-L/L_P)]\} \qquad (4.12)$$

For large molecules like DNA, L_P/L is very small and we can use the approximation $R_G^2 \approx L_PL/3$.

The radius of gyration can be determined directly by light scattering or small-angle X-ray scattering, or more indirectly from hydrodynamic quantities like the translational Brownian diffusion coefficient (determined by inelastic scattering of light), the intrinsic viscosity or the decay of linear electric dichroism.

4.6.3 The double-strand to single-strand transition

Thermodynamic aspects

Following the model of Zimm and Bragg (1959) (see section 3.5.3), we denote paired bases by '1' (AT or AU and GC) and unpaired ones by '0'. There are two contributions to the energy of formation of a base pair: one is from the hydrogen bonds (ΔG_1) and other is from the increase in the stacking energy when changing from two bases to two stacked base pairs (ΔG_2). The latter term is absent for the first base pair created after a sequence of unpaired bases. Keeping the notation s and σs defined in Chapter 3, we then have

$$s = \exp[-(\Delta G_1 + \Delta G_2)/RT] \qquad (4.13)$$

$$\sigma s = \exp(-\Delta G_1/RT) \qquad (4.14)$$

The factor σ is thus equal to $\exp(\Delta G_2/RT)$. Since ΔG_2 generally has a high negative value, $\sigma \ll 1$ and can then be considered as a nucleation parameter. However, a special feature of the double helix must be taken into account: the formation of 'loops' when a disorganized region is located between two double-stranded regions.

If j base pairs are disrupted, the loop is formed from $m = 2(j+1)$ nucleotides. It can be shown that the difference in entropy between this loop and a free chain with the same number of links is $(3R/2)\ln(j+1)$ provided the loops are large enough and any effect from excluded volume can be neglected. Since $S = R\ln W$ (on a molar scale), the probability of closing the loop varies as $(j+1)^{-3/2}$. A more rigorous expression for small loops is of the form:

$$(1 - e^{Aj})/(j+1)^{3/2}(1 - e^{Bj}) \qquad (4.15)$$

where A and B are constants. In both expressions, the closure of a loop is accompanied by a $\Delta S < 0$. This additional entropy change must be included in σ, which will then be given by

$$\sigma = (j+1)^{-3/2} \exp(\Delta G_2/RT) \tag{4.16}$$

The results for very long chains are identical to those already given.

For long chains, it is also possible to apply this simple modelling to the calculation of the mean number $\langle h \rangle$ of helical regions at a given temperature. Since each helical region must be 'primed', it contributes to the partition function by a factor σ. We can therefore write

$$\langle h \rangle = \partial \ln Z / \partial \ln \sigma = n \, \partial \ln \lambda_1 / \partial \ln \sigma = n\sigma \, \partial \lambda_1 / \lambda_1 \, \partial \sigma \tag{4.17}$$

Hence, using the value of λ_1:

$$\langle h \rangle = n\sigma s / \lambda_1 [(1-s)^2 + 4\sigma s]^{1/2} \tag{4.18}$$

$\langle h \rangle$ is a maximum for $s = 1$ and has the value

$$\langle h \rangle_{\max} = n\sigma^{1/2}/2(1+\sigma^{1/2}) \approx n\sigma^{1/2}/2 \tag{4.19}$$

If, for example, $\sigma = 10^{-4}$, then $\langle h \rangle = n/200$, which means that a helical region exists on average every 200 base pairs when $T = T_m$ $(s = 1)$. Since $\langle k \rangle = n/2$ at this point, the ratio $\langle k \rangle / \langle h \rangle$ defines a kind of mean length. In this example, $\langle k \rangle / \langle h \rangle = \sigma^{1/2} \approx 100$.

In other words, the molecule can be viewed during the 'melting' process as being formed on average from sequences of 100 base pairs separated by denatured regions of the same length.

Heterogeneity of base pairs

The calculation can be made more realistic by taking into account the difference between AT and GC, the finite length of sequences (in the case of RNAs, for example), the charges carried by the phosphate groups, etc. A description of the various methods used lies outside the scope of this book, but the influence of some parameters on the helix–coil transition can be calculated much more simply. If, for example, we wish to take into account the two types of pairing AT and GC, we can introduce two parameters s_A and s_B instead of one, with the definitions

$$s_A = \exp(-\Delta G_{AA}/RT) \quad \text{and} \quad s_B = \exp(-\Delta G_{BB}/RT) \tag{4.20}$$

$\pi \Delta G_{XY}$ being the free-energy change when a base pair type X is added to a double helix terminated by a Y pair. Only two values are used by assuming that

$$\Delta G_{AB} + \Delta G_{BA} = \Delta G_{AA} + \Delta G_{BB} \tag{4.21}$$

If the DNA consists of ν pairs of type B (GC) and $1-\nu$ pairs of type A (AT), we have for the transition temperature

$$\nu \Delta G_{BB} + (1-\nu)\Delta G_{AA} = 0 \tag{4.22}$$

or, putting $k = s_A/s_B$:

$$\nu \ln k + \ln s_A = 0 \tag{4.23}$$

Taking as a standard a polymer formed solely from AT pairs that 'melts' at a temperature T_A, eqn 3.22 can be integrated, and using eqn 4.23 gives:

$$(\Delta H/R)(1/T_A - 1/T_m) = \nu \ln k \tag{4.24}$$

which implies a linear relationship between $1/T$ and ν. We may then write:

$$\Delta T_m = T_m - T_A = RT_A T_m \nu \ln k/\Delta H \tag{4.25}$$

and since $T_A T_m$ varies only a little, there is in practice a linear relationship between T_m and the fraction ν of GC pairs.

Experimentally, ΔT_m is found to be approximately 0.43°C per %GC pairs under normal conditions (0.1 M Na^+ and pH 7).

Influence of charges and ions

The breaking or formation of a double helix implies an interaction between two negatively charged polyelectrolytes. If G_{ds} denotes the free energy of the double strand and G_{ss} that of a single strand in the absence of any Coulomb interaction, the quantity

$$\Delta G = G_{ss} - G_{ds} \tag{4.26}$$

represents the change in free energy accompanying the melting (all quantities refer to one nucleotide). Since $\Delta G = \Delta H - T\Delta S$, and since the change is favoured by a rise in temperature, ΔH is positive. The number of degrees of freedom of a single strand is greater than that of a double helix and so ΔS is also positive. The ratio $\Delta H/\Delta S$ defines the transition temperature T_m for which $\Delta G = 0$.

In the double helix, however, each phosphate group experiences the electric potential created by the other phosphate groups situated both in the same strand and in the other strand. Only the first type of Coulomb interaction occurs with the single strand. A term ΔG_e representing an electrostatic repulsive energy must therefore be added to the free energy of the double helix. The transition is thus characterized by a change in free energy $\Delta G'$ given by

$$\Delta G' = \Delta G - \Delta G_e \tag{4.27}$$

The new transition temperature $T_m{}'$ is thus

$$T'_m = \Delta H - \Delta G_e/\Delta S = T_m - \Delta G_e/\Delta S \tag{4.28}$$

Since ΔG_e and ΔS are both positive, this energy due to repulsion between the phosphate groups destabilizes the double helix.

At very low ionic strengths ($\approx 10^{-4}$ M), ΔG_e is large enough for the DNA to be partly denatured at ordinary temperatures. Since ΔG_e is a linear function of $\log I$ (I being the ionic strength), the same applies to T_m, and the variation of T_m with base content and with the ionic strength can be represented as:

$$T_m = 16.6 \log I + 0.43(\%\text{GC}) + 81.5 \tag{4.29}$$

which agrees quite well with experimental results. It is interesting to note the orders of magnitude of ΔH and ΔS. As an example, we find for a certain DNA, per mole of base pairs, that $\Delta H \approx 8$ kcal and $\Delta S \approx 24$ e.u. (entropy units), which gives $T_m \approx 91$°C. At room temperature (20°C) and under the same conditions, we have $\Delta G \approx 1.6$ kcal. This is a relatively small stabilizing energy for the double helix but nevertheless the probability of finding denatured DNA at room temperature is negligible because of the cooperative effect. Far from T_m (e.g. at room temperature), the probability of finding an opened base pair at a given instant is therefore very small.

A more detailed analysis of the phenomenon shows that the energies (other than electrostatic) vary considerably, depending on the relative arrangement of the four nucleotides, and that some sequences (of the AT/TA type) are less stable than others. Such a local heterogeneity in stability is taken into account in more elaborate theories of the 'melting' of a double helix. These predict that there are regions or domains of different stability throughout the length of the DNA molecule. A rise in temperature is accompanied by the successive 'melting' of these domains in the order of their stability. Such a succession of organized regions and open looped zones has in fact been observed by electron microscopy.

Moreover, by combining a temperature rise in very small jumps ΔT with a very precise detection of changes in absorbance ΔA, a curve of dA/dT against T can be obtained and forms a kind of detailed map of the stability of a DNA. In some cases, it can be compared with functional domains (see Fig. 4.31).

It is also possible to detect regions of lower stability inside plasmids. By increasing the density of superhelices, local 'melting' occurs at constant temperature. The stored energy is used locally to unwind the double helix. Unwinding ten base pairs causes approximately one negative supercoil to disappear.

4.6.4 B–Z transition

It is only recently that Z forms of DNA have been discovered in a bacterial plasmid in which an alternating GC sequence had been inserted. This sequence changes into

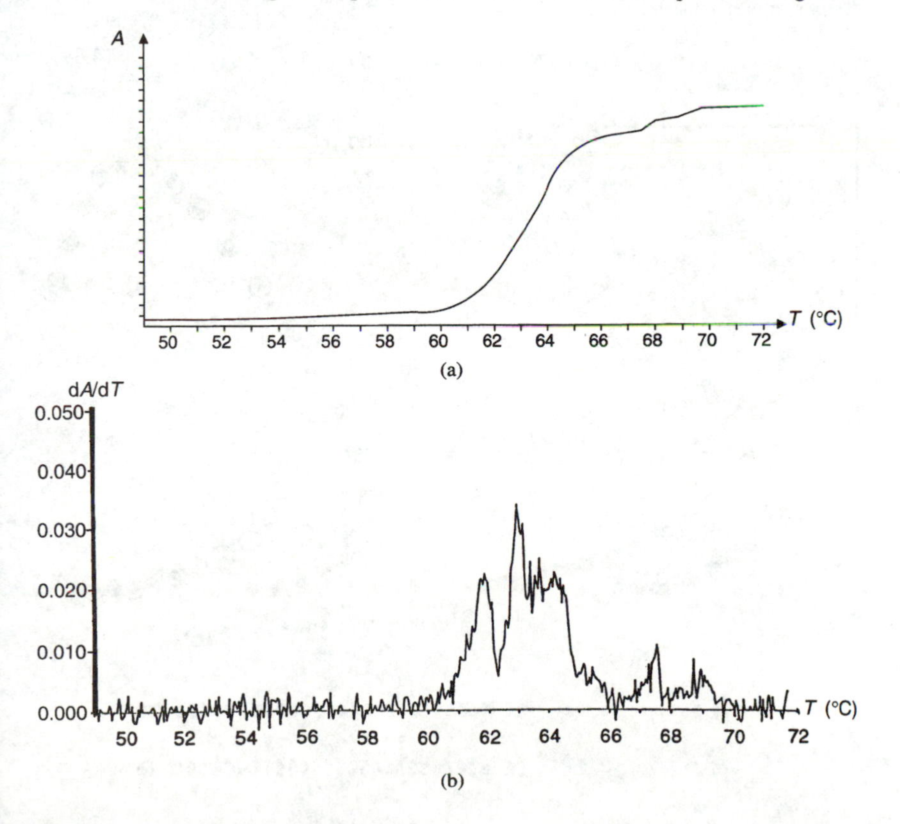

Figure 4.31 DNA melting profiles (linearized SV40 after ECOR1 cutting): (a) integral curve – variation of absorbance at 270 nm with temperature; (b) derivative curve – plot of dA/dT against T. In both cases the temperature is raised in steps of 0.05°C and the error in the absorbance is less than 10^{-4}. Curve (b) can be regarded as a superposition of Gaussian curves representing melting of domains with different GC contents. (From Gabarro, 1978. Reprinted from *Analytical Biochemistry* with the permission of Academic Press.)

a Z form with a sufficient density of supercoils. From a thermodynamic point of view, the transition occurs when it is energetically preferable to change from B to Z rather than to increase the twisting and writhing energy of covalently closed circular DNA. Detection relies on the fact that the change from a right-handed to a left-handed helix introduces a mean twist of about 66° per base pair. Comparing the curve of electrophoresis spots corresponding to the different topoisomers of two plasmids, it can be seen that for the one containing the $(CG)_n$ sequence there is a sudden break at a certain number of negative supercoils (Fig. 4.32). The electrophoretic mobility of these circular DNAs depends solely on W, two spots of the same mobility having equal values of $W = L - T$. A reduction ΔT in the twist is compensated at constant W by a corresponding reduction in ΔL. However, one spot in a two-dimensional electrophoretic pattern differs from the next by $\Delta L = 1$. If, for example, the nth spot after the transition has migrated a distance equal to that of the $(n-5)$th spot, it means that $\Delta L = \Delta T = -5$.

Two quantities characterize the B → Z transition, similar to those introduced into the Zimm and Bragg theory of the helix–coil transition.

1. If ΔG_{BZ} is the free-energy change corresponding to the change of one base pair from the B form to the Z form, the parameter s is defined by

$$s = \exp(-2\Delta G_{BZ}/RT) \tag{4.30}$$

   the factor 2 arising from the fact that the unit in the Z form consists of two base pairs.

2. If ΔG_j is the free energy required to *create a BZ junction*, the nucleation parameter σ is defined by

$$\sigma = \exp(-2\Delta G_j/RT) \tag{4.31}$$

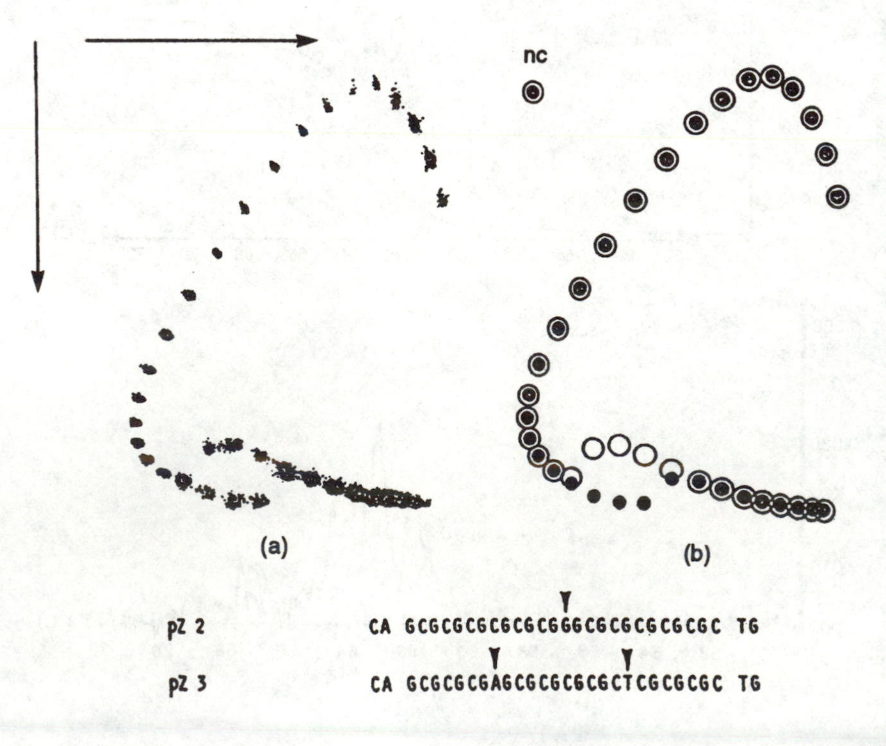

Figure 4.32 B–Z transition in a plasmid with an inserted GC sequence. Comparison of two plasmids, pZ2 with one 'error' and pZ3 with two 'errors'. (a) The two-dimensional electrophoretic pattern produced on the gel by the two plasmids simultaneously.
(b) Interpretation of (a): open circles represent pZ2, full circles represent pZ3, and nc is a plasmid with one single-strand break. It can be seen that pZ3 needs a greater number of supercoils than pZ2 to adopt a Z structure. The transition occurs at −16.7 in pZ2 and at −19.7 in pZ3. (From Ellison *et al.*, 1985. Reprinted from *Proceedings of the National Academy of Sciences, USA* with the authors' permission.)

the factor 2 this time arising from the existence of two junctions with a B form at each end of the sequence that changes into the Z form.

It is then possible (see Box 4.2) to calculate ΔT as a sum depending on the two parameters s and σ and a parameter b corresponding to the unwinding at the junction. Experimentally, n and the number of supercoils are varied (Fig. 4.33). Curves showing the variation of ΔT with $L - L_0$ are obtained and can be used, with a best-fit procedure, to calculate the three quantities ΔG_{BZ}, ΔG_j and b. It is found (Peck and Wang, 1983) that $\Delta G_{BZ} = 0.33$ kcal mol^{-1}, $\Delta G_j = 5.0$ kcal mol^{-1} of junction (or $\sigma = 2.4 \times 10^{-4}$ at room temperature) and $b = -0.4$ turn per junction.

Unexpectedly, it is found that the increase in ionic strength is accompanied by an *increase* in ΔG_{BZ}, whereas for an alternating poly-d(CG) the B → Z transition is

Box 4.2 Calculation of ΔT for the B → Z transition

Suppose that, out of n (GC) units, i of them adjacent to each other change into the Z form ($i < n$). There are $n - i + 1$ ways of placing these i units in the whole sequence. The partition function describing the transition over the whole sequence is

$$P_L = 1 + \sum_{i=1}^{n} (n - i + 1)\sigma s^i$$

This is only true for a linear DNA or a circular DNA with a single strand break. For a supercoiled DNA, the change in L contributing to the free energy through a term in L^2 must be taken into account. If the relaxed form (L_0) is taken as reference, we obtain

$$\Delta G = K(L - L_0)^2$$

with $K = 1100RT/N$, where N is the total number of base pairs (bp) in the plasmid.

When the length i changes from 0 to i_0, there is a total change in twist of $ai_0 + 2b$, where b is a parameter to be determined that corresponds to the unwinding at the junction. Assuming that the B form has 10.5 bp/turn and the Z form has 12 bp/turn, the value of a can be calculated:

$$a \approx -2(1/10.5 + 1/12) \approx -0.357$$

The new partition function P_S then takes the form

$$P_S = \exp[-K(L - L_0)^2/RT] + \sum_{i=1}^{n} (n - i + 1)\sigma s^i \exp\{-[K(L - L_0) - ai - 2b]^2\}/RT$$

and the mean value of ΔT will be given by

$$\langle \Delta T \rangle = \sum_{i=1}^{n} (ai + 2b)(n - i + 1)\sigma s^i \exp\{-[K(L - L_0) - ai - 2b]^2/RT\}/P_S$$

A comparison of this value for $\langle \Delta T \rangle$ and the experimental value enables the B → Z transition parameters to be calculated.

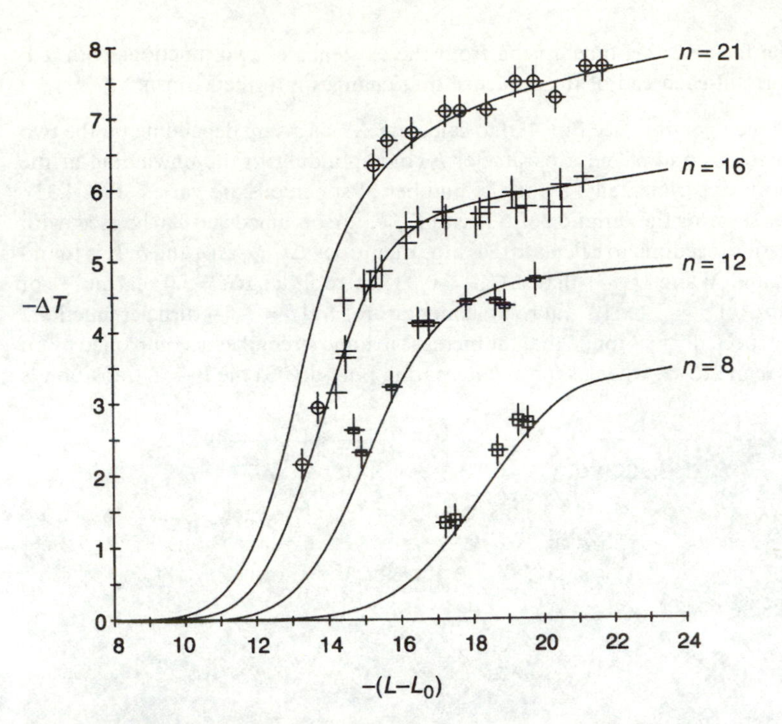

Figure 4.33 Variation of twist number with the number of supercoils during the B → Z transition. (From Peck and Wang, 1983. Reprinted from *Proceedings of the National Academy of Sciences, USA* with the authors' permission.)

only possible at a high ionic strength. Moreover, the process is very slow. In 0.1 M NaCl at 20°C, the relaxation time is of the order of an hour.

4.6.5 Formation of a cruciform structure

A cruciform structure appearing in a plasmid produces a delay in the one-dimensional electrophoretic migration. In two dimensions, a sudden shift of the topoisomers is observed, as in the case of the B → Z transition. This is because the change from the B form to a cruciform structure occurs above a certain density of negative supercoils. It is accompanied by unwinding, i.e. by $\Delta T < 0$ and hence by $\Delta\tau < 0$. If the palindromic sequence is long enough, the cruciform structure is the stable one for $s > 0.03$. However, since the rate of formation is very slow, preheating at 65°C is necessary to accelerate the change from the B form to the cruciform structure. The latter is therefore present during the first electrophoresis. The introduction of an intercalating agent before the second electrophoresis unwinds the DNA, reduces s and destabilizes the cruciform structure. Thus, for a topoisomer of given $L - L_0$, there are two forms (with and without cruciform), which are separated in the first electrophoresis but migrate together during the second. The transition occurs over a small interval of $\tau = L - L_0$ and thus appears to be highly cooperative.

An analysis of the results leads to a free-energy change of 17 kcal mol^{-1} unfavourable to the formation of a cruciform structure.

4.6.6 Three-stranded and four-stranded helices

At an acidic pH (pH 4), a mixture of equal proportions of oligonucleotides $(dT.dC)_n$ and $(dG.dA)_n$ leads to the formation of a *triple helix*. One strand $(dT.dC^+)$, where C^+ is the protonated cytosine, becomes detached from a part of the double helix and is accommodated in the major groove of the other part of the double helix (Fig. 4.34). The TAT and C^+GC associations of the triple helix thus formed cause a Watson–Crick and Hoogsteen pairing to occur simultaneously (Fig. 4.35). Two single-stranded regions appear with a flexible kink acting as a hinge between the double and triple helix conformations.

In a crystallographic study of a complex between a deoxyoligopurine and two peptide nucleic acids (PNA) (Betts *et al.*, 1995), a triple helix was obtained at 2.5 Å resolution. Deoxyribose is C3′-*endo* inside an A helix but the tilt is that of a B form. The PNA_2-DNA_1 triplex in fact forms a previously unknown helix with 16 groups of three bases per turn and a large cavity around the axis.

If the favourable alternating sequence is inserted into a plasmid, the triple helix may be formed at a neutral pH as long as a sufficient number of negative supercoils are present. As in transitions studied previously, the stress energy stored in the plasmid is used to create a new structure incorporating some unwinding of the double helix (appearance of a single strand) and reducing the number of negative supercoils. The formation of a triple helix can therefore be detected and measured using two-dimensional electrophoresis.

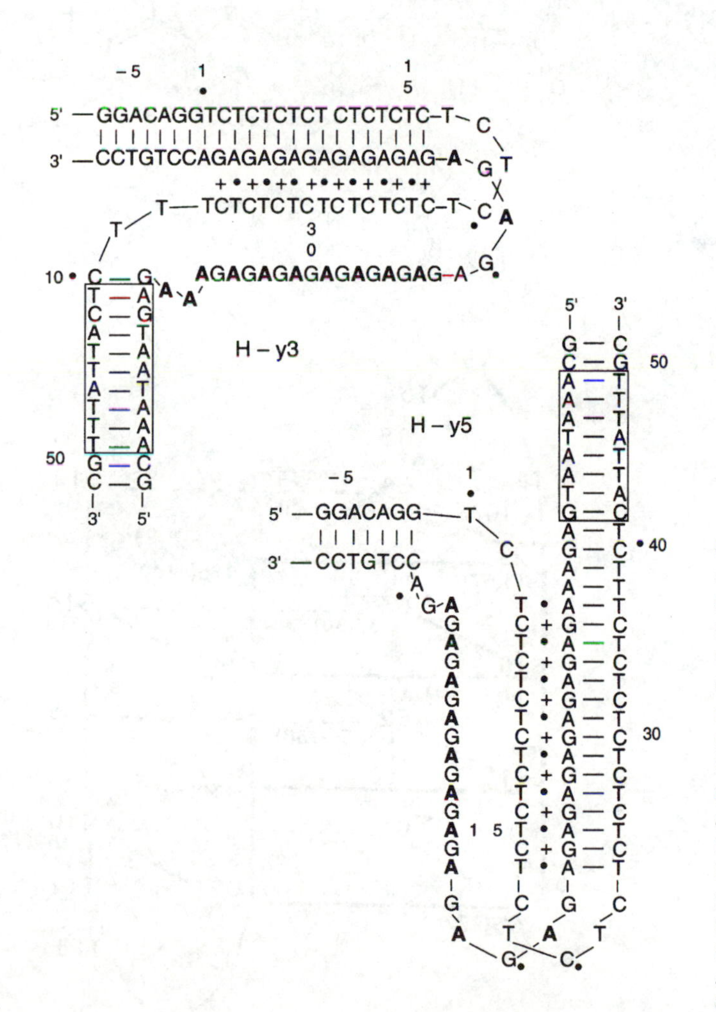

Figure 4.34 Two types of triple helix depending on whether the 3′(Hy3) or the 5′(Hy5) moiety of the TC strand is used. Hoogsteen pairs are indicated by dots when they are between A and T and by crosses when they are between G and protonated C. Bases in bold type indicate preferred points of attack in triple-stranded H-DNA by chemical probes such as diethylpyrocarbonate for A, osmium tetroxide for T and methoxyaniline for C. (From Htun and Dahlberg, 1989. Reprinted with permission from *Science*, vol. 243, pp. 1571–6 and with the authors' permission. © 1989 American Association for the Advancement of Science.)

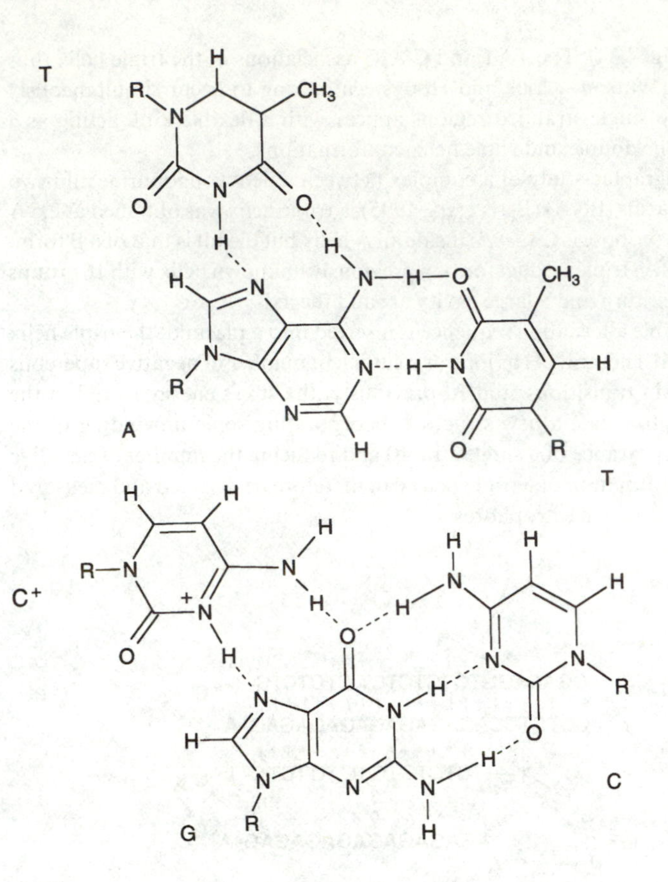

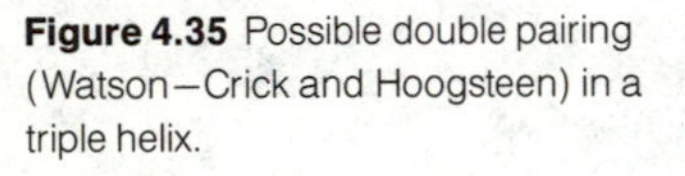
Figure 4.35 Possible double pairing (Watson–Crick and Hoogsteen) in a triple helix.

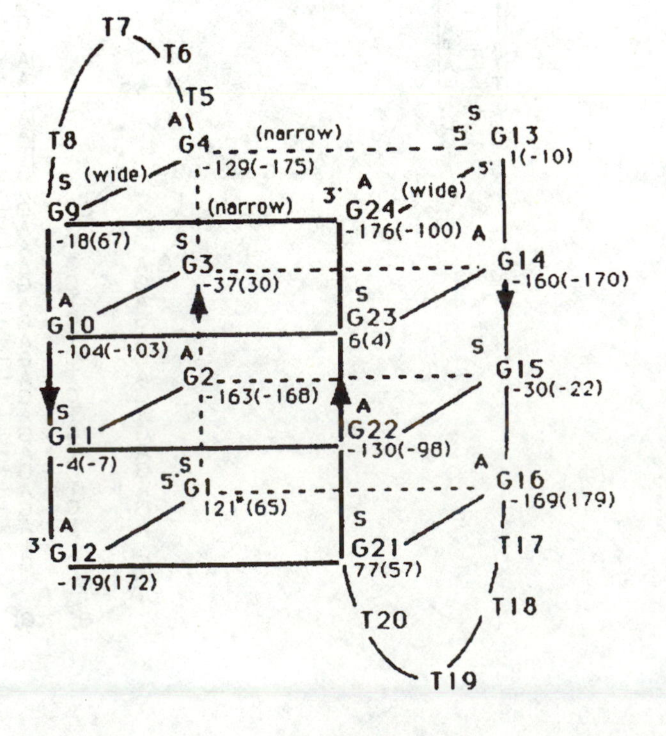

Figure 4.36 Spatial distribution of G and T tracts in the crystal. The four T groups form loops joining the antiparallel G helices. In molecule A, G1–G12 is one strand, G13–G24 is the other. In molecule B (not shown), the corresponding numbering is G30–G42 and G43–G54. The letters S and A near each guanine indicate *syn* and *anti* conformations respectively; the numbers are values of χ, those in brackets referring to the B form. (From Kang *et al.*, 1992. Reprinted with permission from *Nature*, vol. 356, pp. 126–31 and with the authors' permission. © 1992 Macmillan Magazines Ltd.)

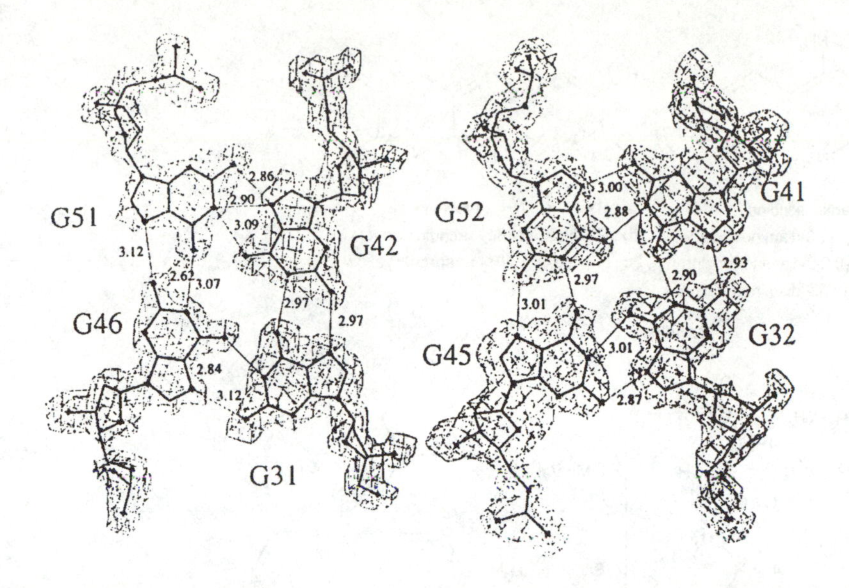

Figure 4.37 Organization of guanine bases in the quadruplex. The skeleton and electron density of four G bases are drawn inside the first two planes of the B molecule. Note the cyclic system of hydrogen bonds with their respective lengths. (From Kang *et al.*, 1992. Reprinted with permission from *Nature*, vol. 356, pp. 126–31 and with the authors' permission. © 1992 Macmillan Magazines Ltd.)

Evidence for four-stranded DNA structures has recently been revealed in a crystallographic study (Kang *et al.*, 1992) of the synthetic oligonucleotide d($G_4T_4G_4$). The crystal contains a complex of two $G_4T_4G_4$ units held together by cyclic hydrogen bonding of four Gs (Figs 4.36 and 4.37). The four-G pairing was indeed found about 30 years ago in guanosine gels and in assemblies of poly(G) or poly(I) (Guschlbauer *et al.*, 1990). The sugar–phosphate backbone has an alternating *syn–anti* conformation. Apart from modelling a telomeric structure, such a four-stranded DNA could play a role as transient structures in sister chromatid exchange.

4.6.7 Intercalation

Lerman (1961) proposed an entirely new model for the interaction between DNA and certain molecules with more than one aromatic ring (Fig. 4.38). The planar molecule is intercalated between two base pairs. The distance between the molecule and each of the adjacent pairs is roughly the same as that between double-helix base pairs, i.e. 3.4 Å for the B form. The intercalation process thus causes an elongation of the molecule and an untwisting of the helix.

This general picture is confirmed by a series of measurements in solution. The increase in length L of the molecule can be measured directly by electron microscopy and indirectly by the scattering of light and the increase in the intrinsic viscosity (η) of small rigid DNA fragments (since in this case η varies as L^3). The measurements are roughly in agreement with a relationship of the type $L = L_0(1 + 2r)$, where r is the ratio of the molar concentration of the intercalated molecules to that of the nucleotides. Measurement of the unwinding requires the use of supercoiled closed circular DNA. The unwinding angle depends on the nature of the intercalating agent, i.e. the stacking energy between the planar aromatic molecule and the adjacent base pairs.

The initial model enabled a whole series of problems to be tackled. These were:

- the calculation of the binding isotherm;
- the intercalation energy and the role of charges, hydration and supercoiling;

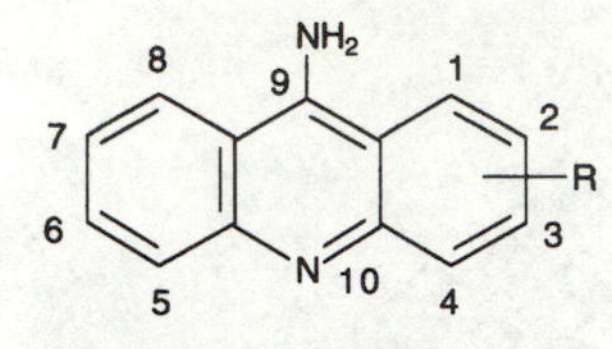

1a : R = H : 9-aminoacridine
1b : R = 1-NH$_2$: 1,9-diaminoacridine
1c : R = 2-NH$_2$: 2,9-diaminoacridine
1d : R = 3-NH$_2$: 3,9-diaminoacridine

2a : R = H, X = NH$_2$: proflavin
2b : R = H, X = NMe$_2$: acridine orange
2c : R = Me, X = NH$_2$: acridine yellow

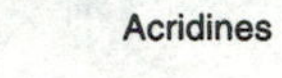

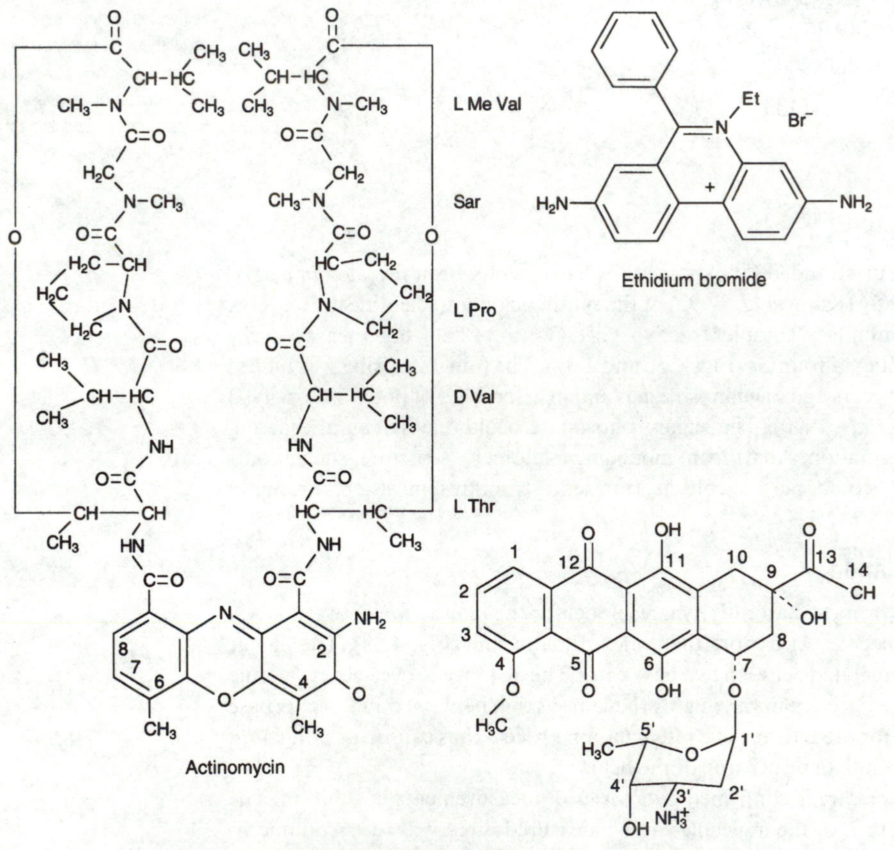

Figure 4.38 Examples of intercalating molecules.

- the exact geometry of the complex;
- the kinetics of the process.

We now examine each of these aspects, and we shall find that many structural and chemical properties of DNA are involved.

Binding isotherm

The calculation of the binding isotherm for an intercalating molecule will depend on an assumption: intercalation between the ith and $(i + 1)$th base pairs introduces such a structural distortion that the probability of intercalation between the $(i-1)$th and ith and between the $(i+1)$th and $(i+2)$th becomes negligible. This binding with *exclusion* of neighbouring sites is easily calculated from the matrix model developed previously in connection with the helix–coil transition. Instead of the '1's and '0's denoting monomeric units that are *organized* into helices or *not organized*, here they will denote an *occupied* or *vacant* site respectively. The accommodation of the intercalating ligand by the polymer is thus represented by a sequence of '0's and '1's.

We then have to define the statistical weight of an occupied site with respect to that of a vacant site, in the same way as we defined the statistical weight s of a monomer organized into a helix with respect to that included in a statistical chain. The change in free energy ΔG_1 accompanying the intercalation can be written

$$\Delta G_1 = \Delta G_0 - RT \ln A \tag{4.32}$$

where ΔG_0, the standard free-energy change, is $-RT \ln K$, K being the intrinsic binding constant and A being the concentration of the free ligand. Hence:

$$\Delta G_1 = -RT \ln(KA) \tag{4.33}$$

The isotherm is calculated in Box 4.3 and gives the Scatchard isotherm (graph of r/A against r) as

$$r/A = K(1-2r)^2/(1-r) \tag{4.34}$$

where r is the ratio of the mean number of sites occupied by the intercalating agent to the total number of sites, i.e. to the total number of base pairs. Equation 4.34 shows that $r/A = 0$ when $r = 0.5$, in agreement with the main assumption of one intercalating molecule every two base pairs.

However, the isotherm is no longer a straight line but is a curve that is concave upwards (Fig. 4.39). Moreover, the slope at the origin is no longer $-K$ but $-3K$, as can be seen by finding the value of $\mathrm{d}(r/A)/\mathrm{d}r$ for $r = 0$. It is therefore possible, when experiments are limited to small values of r, to obtain a quasi-linear isotherm that, for $r/A = 0$, gives an intersection with the horizontal axis at $r = 1/3$, which is meaningless.

Energetics

Finding the energy balance favourable to intercalation is a complex matter.

From the *DNA viewpoint*, there is a loss of stacking energy between adjacent base pairs and the crossing of several potential barriers during the structural changes of the sugars and the phosphodiester chain. Moreover, the chains of water molecules that contribute to the stability of the DNA are highly disturbed at the position of the intercalation site. The increased separation of the charged phosphate groups around the intercalation site locally reduces the charge parameter (see section 18.2.2) by altering the distribution of counter-ions.

Box 4.3 **Calculation of the binding isotherm**

In the partition function Z describing the state of the polymer, the quantity $\exp(-\Delta G_1/RT) = KA$ is the statistical weight of a site with a ligand bound to it. Of the four possibilities for the situation in two neighbouring sites (00, 01, 10, 11), only the first three are retained because of the exclusion hypothesis. The matrix describing the process is constructed by seeking the statistical weight of the $(i+1)$, $(i+2)$ sequence knowing that of the sequence i, $(i+1)$ according to the following:

		Sequence i, $(i+1)$ 00	01	10
Sequence $(i+1)$,$(i+2)$	00	1	0	1
	01	KA	0	KA
	10	0	1	0

and hence the transfer matrix is:

$$M = \begin{pmatrix} 1 & 0 & 1 \\ KA & 0 & KA \\ 0 & 0 & 0 \end{pmatrix}$$

which multiplies a column vector $\boldsymbol{a}_i$ to give the column vector $\boldsymbol{a}_{i+1}$.

As we saw in section 3.5.3, the diagonalization of the matrix simplifies the calculation of Z. The so-called 'secular' equation is obtained by subtracting a parameter λ from each element of the principal diagonal in $\boldsymbol{M}$ and putting the determinant of $\boldsymbol{M}$ equal to zero. This gives

$$\lambda(\lambda - 1) = KA$$

Since only the largest root, λ_1, remains in Z, we have

$$\lambda_1 = [1 + (1 + 4KA)^{1/2}]/2$$

Referring again to section 3.5.3, we have

$$Z = \lambda_1^n$$

and the mean number ν of occupied sites is given by

$$\nu = \partial \ln Z/\partial \ln A$$

or

$$\nu = n[(1 + 4KA)^{1/2} - 1]/2(1 + 4KA)^{1/2}$$

As in any process of binding a ligand to a macromolecule, it is convenient to define the ratio $r = \nu/n$ of the mean number of sites occupied by the intercalating molecule to the total number n of sites, i.e. to the total number of *base pairs*. We have

$$r = \nu/n = [(1 + 4KA)^{1/2} - 1]/2(1 + 4KA)^{1/2}$$

or again

$$r = [1 - (1 + 4KA)^{-1/2}]/2$$

Extracting A from this equation and forming r/A, we obtain the new expression for the Scatchard isotherm (graph of r/A against r):

$$r/A = K(1 - 2r)^2/(1 - r)$$

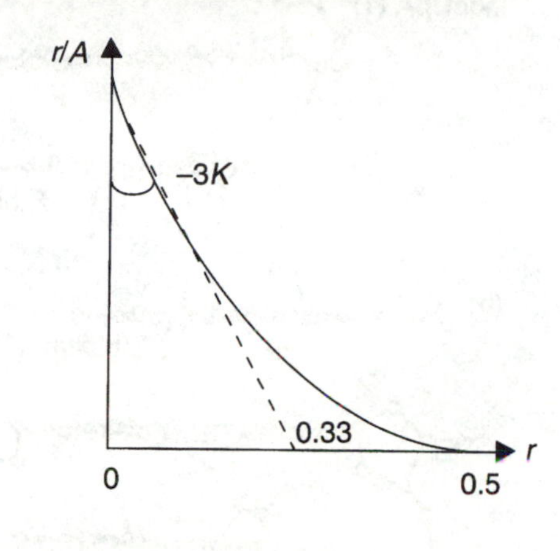

Figure 4.39 The binding isotherm.

For covalently closed circular DNAs or more complex organizations like the supercoiling of DNA around a histone core in a nucleosome, the unwinding induced by the intercalation modifies the topological constraint. In the presence of negative supercoils (see section 3.5.2), intercalation is favoured, but it is disfavoured if it creates positive supercoils. In both cases, the stored twisting and bending energy will be respectively added to or subtracted from the intrinsic binding energy.

From the *viewpoint of the intercalating molecule*, there is van der Waals interaction with the base pairs on both sides of it. The energy of this interaction will depend on the geometry of the intercalating molecule and on the various atomic groups substituted in the aromatic rings. It will also depend, for a given nucleotide sequence XpY, on whether X is in the 3′ or the 5′ position (Fig. 4.40).

Most intercalating molecules have a positively charged group (generally a nitrogen atom). Such a cation will be attracted to the immediate neighbourhood of the DNA, where its concentration may be much higher than in the rest of the solution. Moreover, there is competition between the intercalating cation and the usual counter-ions (Na^+, K^+ or Mg^{2+}). The binding constant will therefore generally be a decreasing function of the ionic strength.

Lastly, the hydration of the free intercalating molecule is considerably modified, because the volume in which the van der Waals envelopes of the various aromatic molecules (bases and intercalating molecule) are stacked is largely hydrophobic.

These general reflections on the energetics of the process show how difficult it is to include all the factors and to derive general laws. The large number of parameters involved also explains some of the contradictory results on recognition specificity, i.e. the preferential binding of the intercalating molecule to certain sequences. For example, 9-aminoacridine binds more strongly to CpG than to GpG.

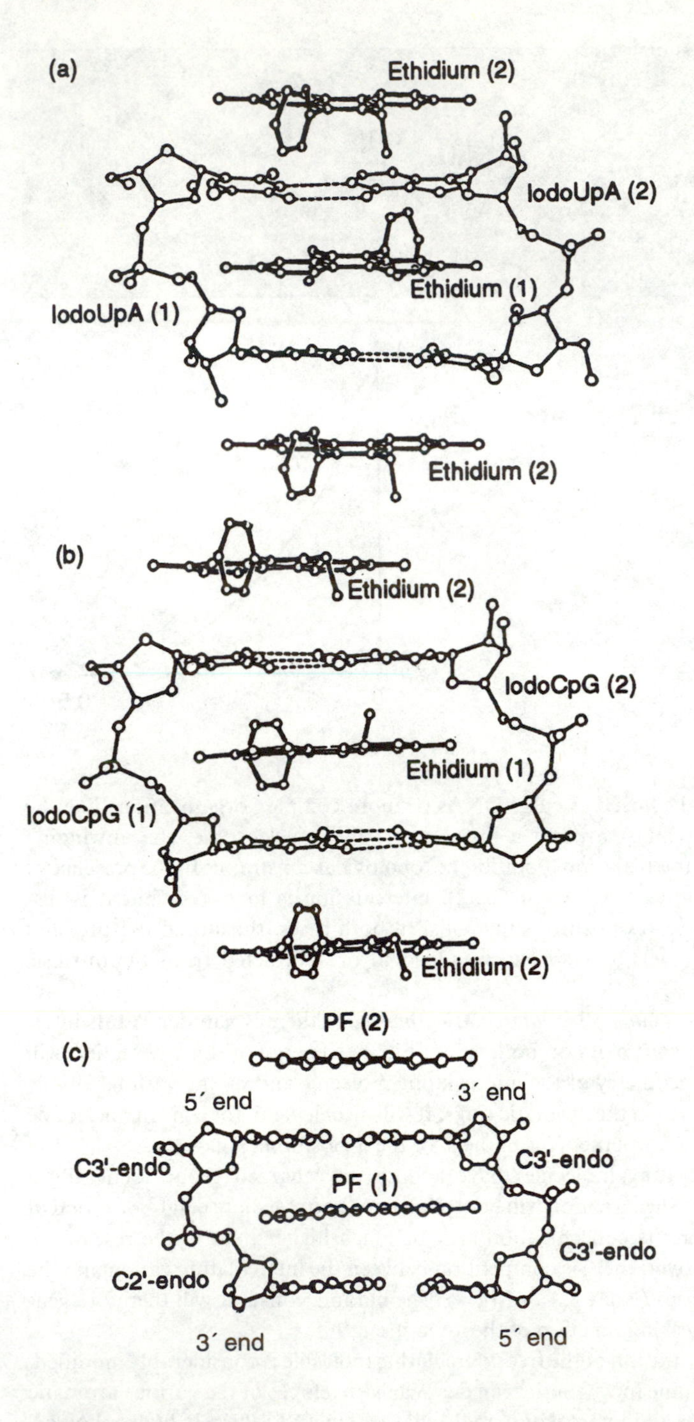

Figure 4.40 Position of the ethidium bromide molecule between two base pairs: (a) iodoUpA and (b) iodoCpG. (From Tsai *et al.*, 1977. Reprinted from *Journal of Molecular Biology* with the permission of Academic Press.) (c) Proflavin molecule between two CpG pairs. (From Shieh *et al.*, 1980. Reprinted from *Nucleic Acids Research* by permission of Oxford University Press and with the authors' permission.)

Geometry

Crystallization of complexes between intercalating molecules and dinucleotides had to be achieved before the local changes of structure in DNA after intercalation could be described in detail (Figs 4.41 and 4.42). In most cases, there is an alternation of both types of sugar puckering, each characteristic of the B form and the A form.

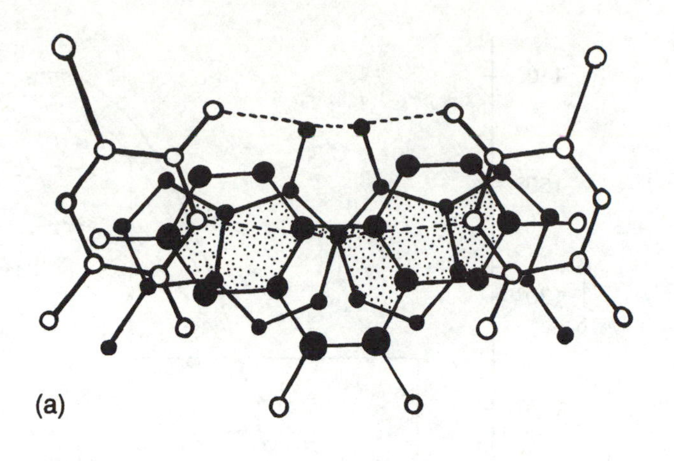

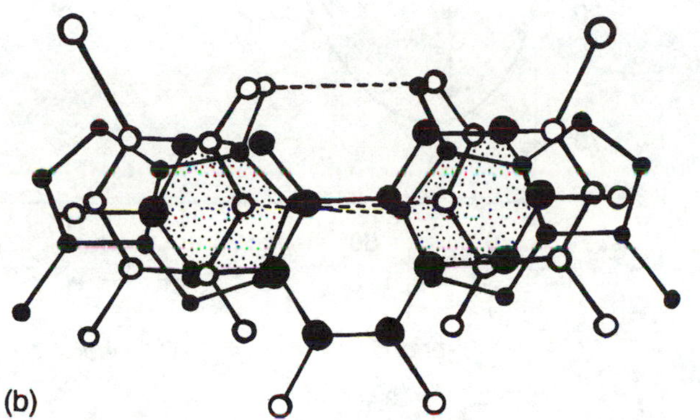

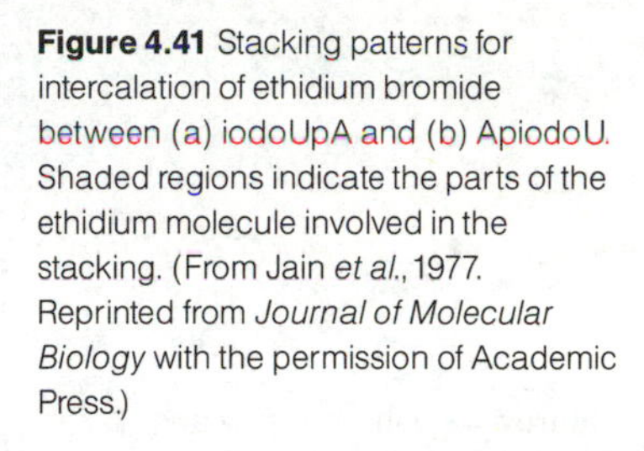

Figure 4.41 Stacking patterns for intercalation of ethidium bromide between (a) iodoUpA and (b) ApiodoU. Shaded regions indicate the parts of the ethidium molecule involved in the stacking. (From Jain *et al.*, 1977. Reprinted from *Journal of Molecular Biology* with the permission of Academic Press.)

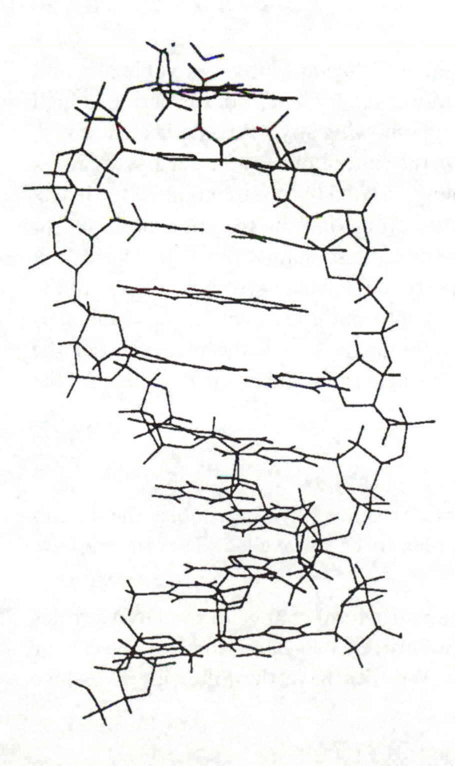

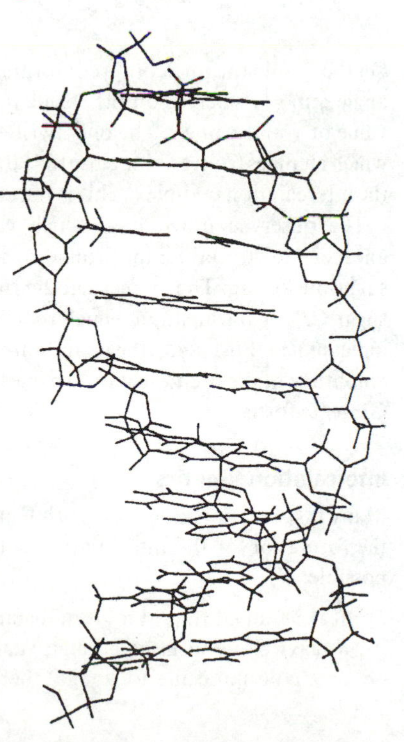

Figure 4.42 Stereomodel of 9-aminoacridine intercalated into an octamer between a TA pair and a GC pair. The 9-amino group is turned towards the minor groove. (From Woodson and Crothers, 1988. Reprinted with permission from *Biochemistry*. © 1988 American Chemical Society.)

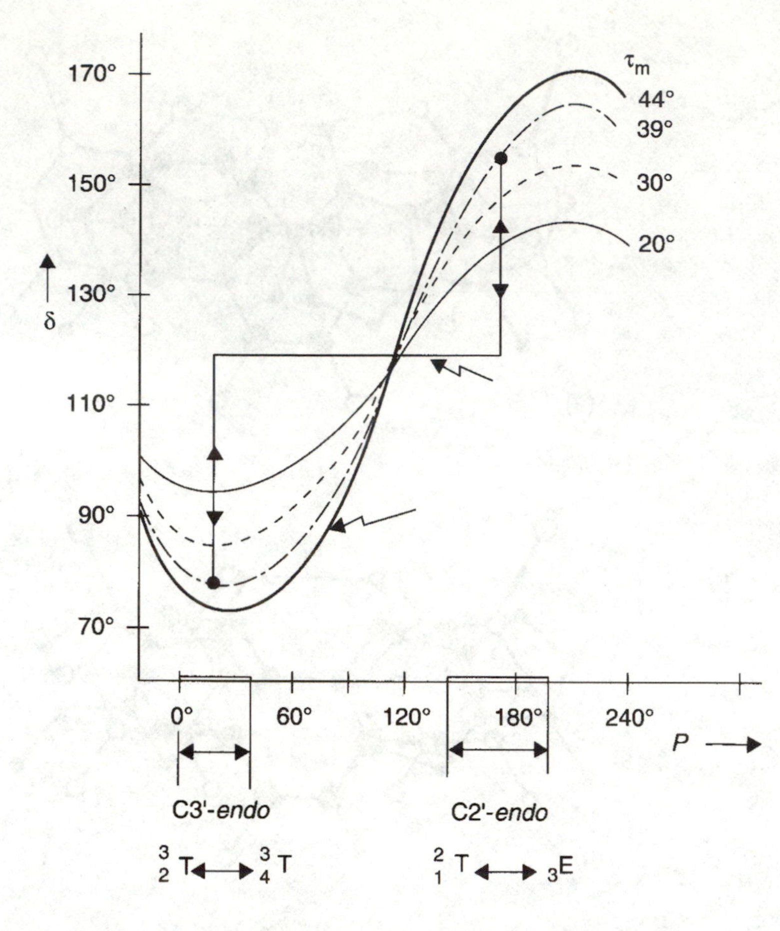

Figure 4.43 Variation of the angle δ with the pseudorotation phase P and amplitude τ_m.

On the 5′ side the nucleotide conformation is C3′-*endo*, with a low value of χ and angle γ in g^+; whereas on the 3′ side it adopts the C2′-*endo* conformation, a high value of χ and γ in g^+. The only twist angle showing any appreciable change is β, which changes from a value close to 180° in the dimer alone to a value close to 220° in the intercalation complex. This is the change related to the stretching of the helix.

The observed unwinding, which varies from one intercalating molecule to another (26° for ethidium bromide), seems to arise mainly from the alternating sugar puckering. The increase in the rigidity of the helix may be partly due to the sugar C3′-*endo* transition, which reduces conformational fluctuations. In fact, as indicated in Fig. 4.43, the variation of the angle δ with the phase P and the amplitude of the pseudorotation passes through a minimum near $P = 0$, i.e. for the C3′-*endo* form.

Intercalation kinetics

Many fast-reaction kinetics experiments (see section 9.8) have enabled the steps in the formation of the intercalation complex to be unravelled. Two theories are possible:

1. It is assumed that at a given instant a structural change in the DNA creates a 'cavity', i.e. a large enough space between two base pairs for the planar aromatic molecule to 'lodge' there. We then have the following two-stage

kinetic process:

$$P \underset{k_{-1}}{\overset{k_1}{\rightleftarrows}} P^*$$

$$P^* + L \underset{k_{-2}}{\overset{k_2}{\rightleftarrows}} PL$$

in which P denotes DNA, P* the new open conformation and L the intercalating molecule. A local transformation (which will depend on the medium and the temperature but also on the sequence) is followed by the intercalation step in which the aromatic molecule takes its place in the preformed cavity (Fig. 4.44).

2. The positively charged molecule is first positioned near the phosphate by a diffusion-controlled mechanism. The aromatic part is temporarily lodged in one of the grooves while waiting for the conformational fluctuation, creating an intercalation site. In this case, we have the following kinetic process:

$$P + L \underset{k_{-1}}{\overset{k_1}{\rightleftarrows}} PL_1 \underset{k_{-2}}{\overset{k_2}{\rightleftarrows}} PL_2$$

in which PL_1 is the 'external' complex and PL_2 the intercalation complex.

Note that the rate constants k_1 and k_{-1} in the first process relate to the conformational change of the DNA. In the second process, they characterize a

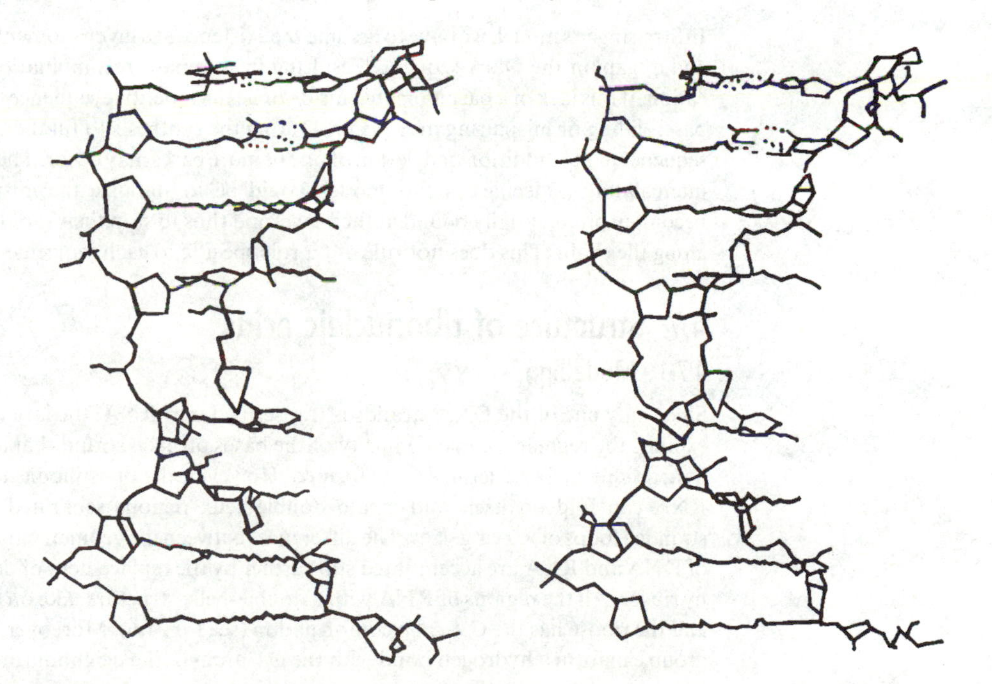

Figure 4.44 Stereomodel of the opening between two base pairs (CpG) to intercalate a planar aromatic molecule.

diffusional encounter, while the conformational change is implicated in the transition from PL_1 to PL_2. While the first step takes place on a time scale of a microsecond, the second is much slower and the lifetime of the complex ($1/k_{-2}$) is of the order of 10 ms. A study of the variation of reaction rates with temperature gives an activation energy of about 15–20 kcal mol^{-1}.

These two values are approximately those found by NMR for the exchange of imino protons, i.e. those involved in the hydrogen bonds between base pairs. In both cases, there is a rotation of the bases but, for the intercalation, we must assume a coupling between that motion and the variation in β.

Biological consequences of intercalation

Crick and co-workers (Brenner *et al.*, 1961) proposed a model for indirect mutagenesis by intercalation of acridine molecules. At each intercalation site there is a *frameshift* generally leading to a non-functional protein. However, if three intercalation sites succeed each other, the reading frame is recovered. In these triple mutants a functional protein can again be obtained as long as the modified part does not play an important role. We can thus say that the study of the mutagenesis induced by intercalation was the first indication of the existence of a triplet code.

Streisinger (1966) proposed a model for this frameshift mutagenesis based on several experimental facts:

1. There are regions in a gene much more susceptible to modifications than others; these are called *hotspots* and correspond to repetitive sequences.
2. Mutations by addition are more frequent than those by deletion.
3. In all cases, DNA replication must take place, indicating that there is a link between mutation and the replication fork.

In Streisinger's model, we have to assume the existence at a given moment of a break and a 'gap' in the DNA sequence resulting from repair, recombination or replication. If this lack of a base is located inside or near a repetitive sequence of the same base, sliding or mispairing may occur. During the synthesis to fill the 'gap' in the sequence, either addition or deletion of one or more bases may occur. The role of the intercalating molecules in this model would be to stabilize the mispairing by becoming preferentially bound at their level and thus to stop their rapid migration along the chain. This does not rule out a role specific to each molecule.

4.7 Structure of ribonucleic acids

4.7.1 Modelling

Since only one of the DNA strands is transcribed into RNA, the latter no longer exhibits the regular complementarity of the bases on each strand that allows very long double-helix structures to be formed. However, the phosphodiester chain of RNA can fold on itself and create double-helix regions separated by single-stranded loops of varying sizes. The differences between the geometry and structure of DNA and RNA are accentuated still further by the replacement of deoxyribose by ribose. All the regions of RNA with a double-helix structure take on the A form and the ribose has the C3′-*endo* conformation (see Fig. 4.6). Moreover, the 2′-OH group can form a hydrogen bond with the O4′ atom of the neighbouring 3′ ribose, which stabilizes and stiffens the structure.

With any A, G, U and C sequence in the chain, there is *a priori* a tremendous variety of conformations with the Watson–Crick type pairings A–U, G–C and

G–U. Rules for constructing a secondary structure therefore had to be sought from the base sequence. The problem is immediately simpler than that of proteins: there are only four bases forming three types of complementary pairs and the double helix is the only possible secondary structure.

Since the formation of a loop is discouraged because of the entropy change that would occur ($\Delta S < 0$), while each new base pairing reduces the free energy, there must exist a structure for which the free energy is a minimum. The free-energy difference ΔG between the structured RNA and a random chain must have as negative a value as possible. Both the initiation and propagation processes already studied for DNA have to be present to build a double helix. For RNA, the initiation by creating the first base pair also gives rise to a fairly large loop. The stability of the double helix, and hence the elongation mechanism, will depend on a comparison of the melting points of the AU and GC regions with the temperature of the experiment. In general, it can be shown (Tinoco *et al.*, 1971; see Box 4.4) that, at 25°C, $\Delta G_{\text{elongation}} = -1.2\,\text{kcal}$ for AU, $-2.4\,\text{kcal}$ for GC and 0 for GU. Moreover, $\Delta G_{\text{initiation}} = -2.3RT[B - 1.5\ln(m+1)]$, where B is a constant of the order of -3 and m is the number of unpaired bases in the loop.

These simple rules were improved later (see the review by Turner *et al.*, 1987) when more structures had been determined, making it possible to increase the number of parameters and to take into account the immediate neighbourhood of each base pair. 'Flat' models of RNA can then be built in which the double helix is represented by a ladder and the loop by a flat curve. For tRNA, this is the so-called *clover-leaf* model (Fig. 4.45). A maximum number of GC or AU pairings occurs in A-form double-helix regions (*stems*), and unusual types of nucleotide (ψU, 5MeG,

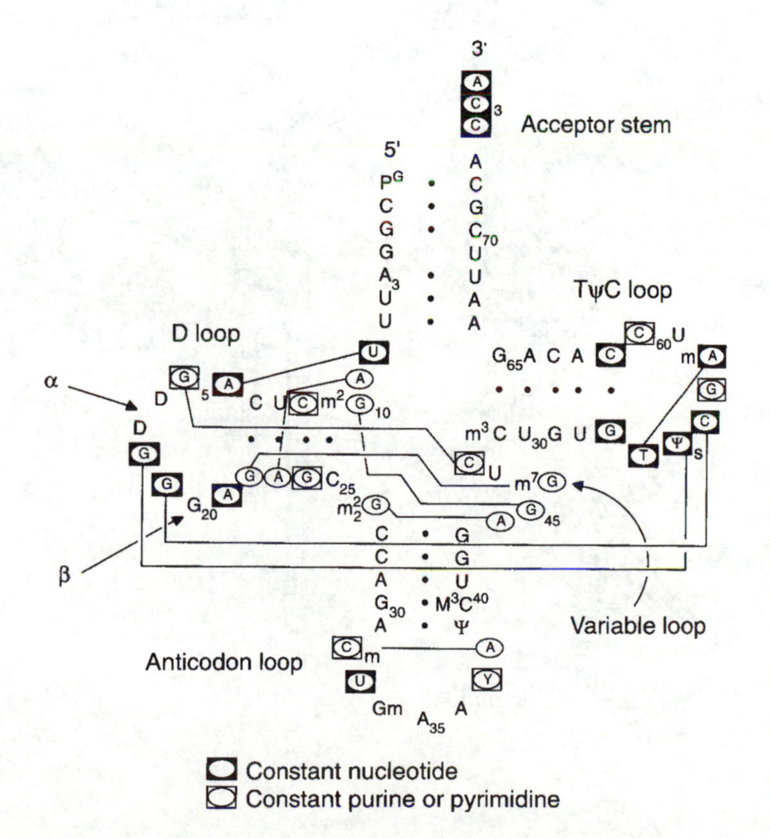

Figure 4.45 Clover-leaf structure of tRNA^{Phe}. The solid lines connecting circled bases indicate hydrogen bonding in the tertiary structure. Dots indicate standard Watson–Crick pairing between complementary bases. Numbering is from the 5′ end to the 3′ end where the acceptor triplet CCA is located. (From Quigley and Rich, 1976. Reprinted with permission from *Science*, vol. 194, pp. 796–806 and with the authors' permission. © 1976 American Association for the Advancement of Science.)

DU, etc.) are located in single-stranded regions forming hairpin structures. Some base pairs are always found in the same position when all the known tRNA structures are compared. In 16S ribosomal RNA, we find a complex mixture of loops and stems (Fig. 4.46).

4.7.2 Three-dimensional structure

It is obvious that plane constructions like those above do not represent the true spatial structure of RNA. Just as in the assembly of a 3D model from a flat cardboard blank cut to shape, the planar arrangement has to be folded, stuck together and thus built up into a three-dimensional structure. It is immediately obvious that such a structure plays a decisive role in the function of RNAs if we simply consider the process of translation.

Several RNAs play absolutely vital roles in biological processes. Messenger RNA, for instance, carries the nucleotide sequence to be translated. Its precursor, which migrates from the nucleus into the cytoplasm inside nucleoprotein complexes, must undergo a prior *splicing* involving particular tertiary structures.

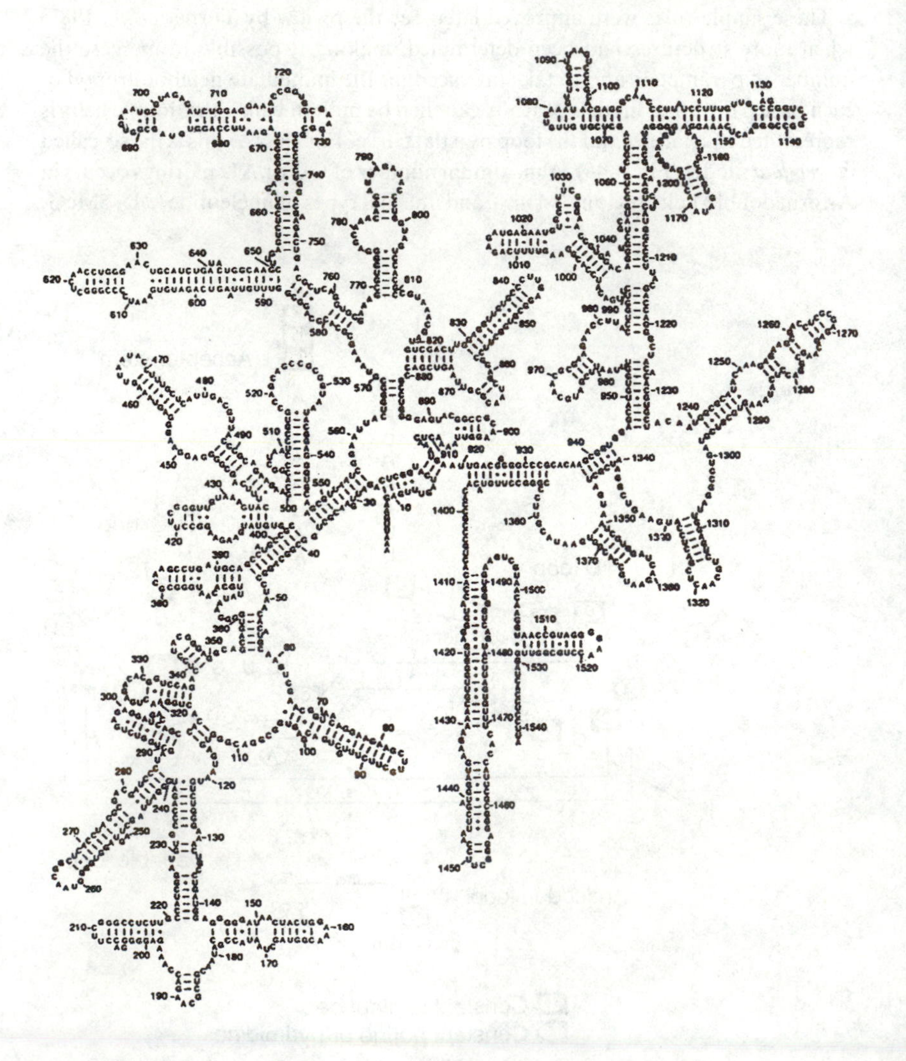

Figure 4.46 Secondary structure of *E. coli* 16S RNA. (From Stern *et al.*, 1988. Reprinted from *Journal of Molecular Biology* with the permission of Academic Press.)

Box 4.4 **The energetics of RNA**

We saw that, for the helix–coil transition temperature T_m, $\Delta G = 0$ and hence $\Delta S = \Delta H / T_m$. We can therefore write the variation in propagation free energy (elongation of the double helix) at any temperature as

$$\Delta G_{prop} = \Delta H(1 - T/T_m)$$

If T_A is the melting point of the AU pairs and T_G that of the GC pairs, it has been shown that

$$T_m^{-1} = T_A^{-1} + f_{GC}(T_G^{-1} - T_A^{-1})$$

where f_{GC} is the fraction of GC pairs in the double-helix structure. In 1 M NaCl, $T_A = 78°C$ and $T_G = 152°C$. With $\Delta H = -8$ kcal, we have for $T = 25°C$

$$\Delta G_{AU} = \Delta H(1 - T/T_A) = -1.2\,\text{kcal}$$

$$\Delta G_{GC} = \Delta H(1 - T/T_G) = -2.4\,\text{kcal}$$

If we take −1.2 kcal as a 'unit', +1 involves the formation of an AU pair and +2 that of a GC pair. By convention, 0 is attributed to the formation of a GU pair. If there are N_{AU} and N_{GC} AU and GC pairs respectively, the propagation energy involves only $(N-1)$ pairs.

Using the relationship between T_m, T_A and T_G, and by an approximate calculation, we obtain

$$\Delta G_{prop} = N_{AU}\Delta H(1 - T/T_A) + N_{GC}\Delta H(1 - T/T_G) - \Delta H(1 - T/T_m)$$

For initiation, a loop of m unpaired bases must be created. The corresponding change in free energy has the form

$$\Delta G_{init} = -2.3RT[B - 1.5\ln(m+1)]$$

with $B = -3$ (mean of experimental values).

The total change in free energy ΔG is the sum

$$\Delta G = \Delta G_{prop} + \Delta G_{init}$$

Since $\Delta H(1 - T/T_m) = -1.8$ kcal and $-2.3RTB = 4.14$, the total ΔG is given by

$$\Delta G = -1.2N_{AU} - 2.4N_{GC} + 5.9 + 2.1\ln(m+1)$$

Transfer RNA, through its anticodon loop, makes a selection from various codons and introduces the amino acids it carries into the peptide chain under construction. Ribosomal RNAs, with bound ribosomal proteins, define a precise tertiary structure of the ribosome and the positioning of the other two types of RNA.

Apart from their role in translation, we should mention the importance of RNA structures in the morphology of viruses and in that of *viroids*, the short fragments of RNA responsible for several tropical plant diseases. Finally, at a later stage we shall be looking at the role of this three-dimensional structure in the enzyme-like behaviour of certain RNAs (ribozymes). In all these examples, we need to know the spatial structure of the RNA. So far, however, the only structures to have been obtained are those of a few tRNAs and viral RNAs together with that of a ribozyme.

The tertiary tRNA structure was initially determined from X-ray diffraction patterns of $tRNA^{Phe}$ crystals (Kim *et al.*, 1974; Robertus *et al.*, 1974) and more recently on $tRNA^{Asp}$ crystals (Moras *et al.*, 1980). A comparison of this three-dimensional structure (Fig. 4.47) with the clover-leaf model enables us to understand what forces are involved in the folding of a nucleotide chain containing between 74 and 91 nucleotides. Hydrogen bonds play a decisive role in many cases:

1. In double-helix regions (stems), the standard Watson–Crick pairing is observed except for the existence of G–U pairs (Fig. 4.48a).
2. In other regions, hydrogen bonds are found in inverse Hoogsten pairings between A and U (Fig. 4.48b) in triple helices and in junctions between bases and backbone (sugars or phosphates) or between two backbone atoms.
3. Some hydrogen bonds enable the chain to form a kink. For example, at the end of the TψC loop of $tRNA^{Phe}$, a sudden kink is produced due to a hydrogen bond between N3 of uracil 55 (ψU) and phosphate 58, and another between O2′ of the ribose in residue 55 and nitrogen N7 of guanine 57. A similar situation occurs at the beginning of the anticodon loop, with a hydrogen bond between N3 of uracil 33 and phosphate 36, and another between ribose O2′ of residue 33 and N7 of adenine 35 (Fig. 4.48c). In both cases, the chain rotates through 180° in the space of three nucleotides.
4. Hydrogen bonds between the ribose 2′-OH group and a neighbouring phosphate may also create a bulge, i.e. push out one or more bases, like those forming the variable loop. This bond is generally associated with a ribose transition from the C3′-*endo* form to the C2′-*endo* form.

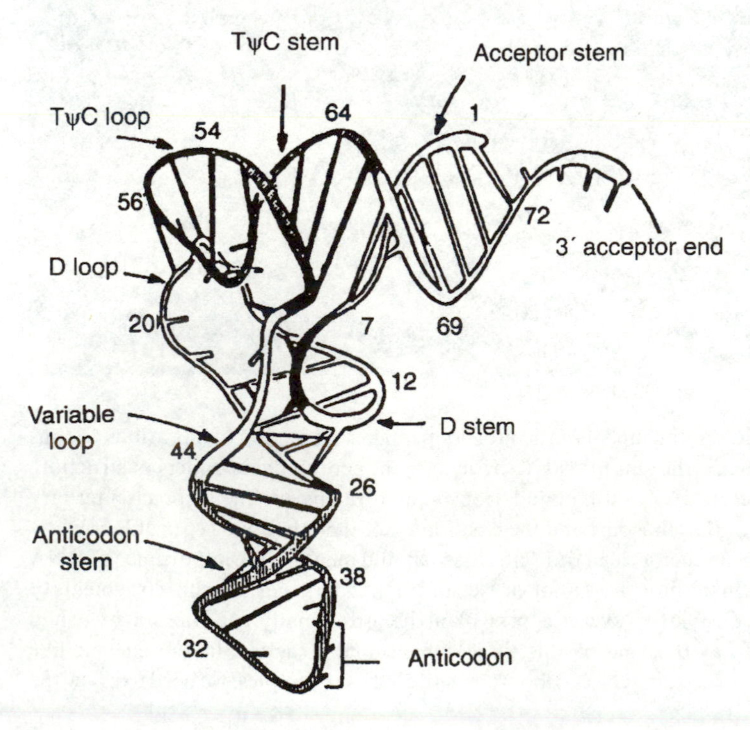

Figure 4.47 Tertiary structure corresponding to the clover-leaf model of Fig. 4.45. Hydrogen bonds between bases are shown as cross-rungs joining the coiled ribbons of the phosphodiester backbone. The black rungs represent hydrogen bonds stabilizing the tertiary structure. (From Quigley and Rich, 1976. Reprinted with permission from *Science*, vol. 194, pp. 796–806 and with the authors' permission. © 1976 American Association for the Advancement of Science.)

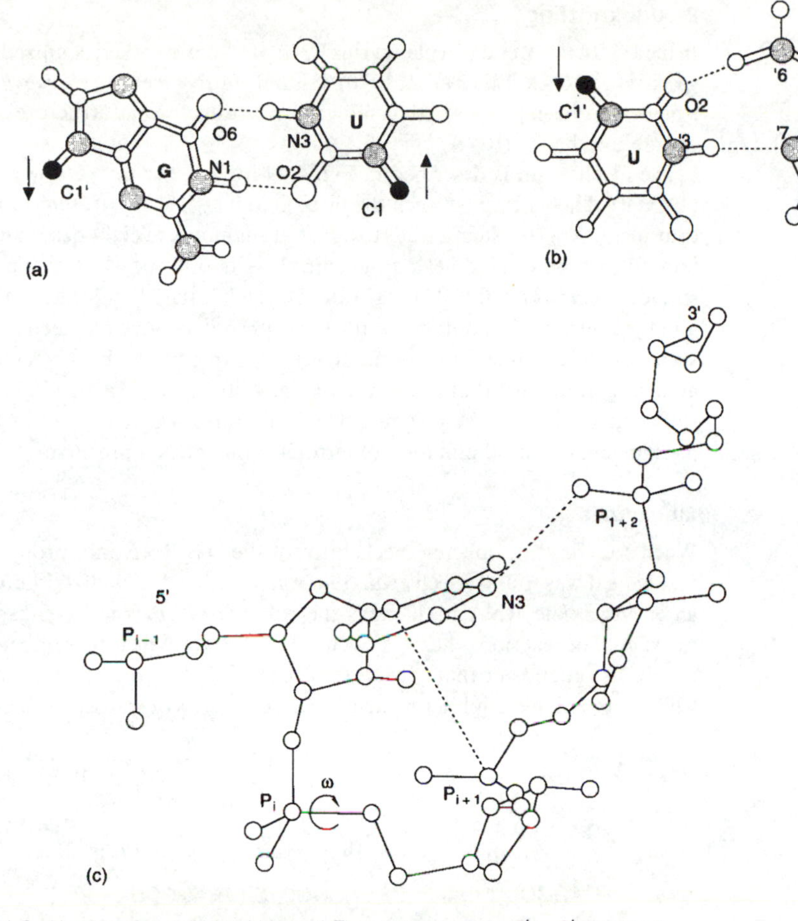

Figure 4.48 (a) G–U pairing (so-called wobble pair). The arrows show the $5' \rightarrow 3'$ direction in each sugar–phosphate strand. (b) Reversed Hoogsteen pairing between U and A. (c) Formation of a sudden kink. The angle $\alpha(\omega)$ of the phosphate P_i differs from its usual value, thus ensuring the rotation of the following bonds. There is a van der Waals bond between the phosphate P_{i+1} and the uracil, and a hydrogen bond between P_{i+2} and N3 of uracil.

We might have hoped that some general rules governing the folding could be deduced from other tertiary tRNA structures, enabling us to set up prediction algorithms applicable to other types of RNA such as ribosomal RNA.

Using a few such rules, but mostly making use of many topological constraints, a three-dimensional model of 16S RNA as it might be inside the 30S ribosome has recently been constructed. The choice of the folding in the model was mainly guided by the location of ribosomal proteins determined by neutron diffraction after selective deuteration and by a determination of their binding sites deduced from accessibility measurements. When we remember all the tRNA models proposed before 1974, none of which corresponded to the structure determined by X-ray diffraction, it is surely advisable to await the true structure of a ribosome now being determined by various physical methods.

Pseudoknotting

In the 1970s, the 3′ end of a plant virus RNA was found to be recognized by enzymes specific for tRNA (Pinck *et al.*, 1970). It is only fairly recently (Pleij *et al.*, 1985) that a model has been proposed that shows a three-dimensional structure of this 3′ end resembling that of tRNA.

The mechanism is described in Fig. 4.49 for the Turnip Yellow Mosaic Virus (TYMV). The pairing between the three guanines (13–15) of stem 1 and the three cytosines (25–27) of stem 2 leads to a spatial alignment of eight base pairs organized into an A-type double-helix fragment. This is only possible if the two single-stranded regions A(10)C(11)U(12) and U(21)C(22)U(23)C(24) cross the helix and give the illusion of a knot: hence the term *pseudoknot* given to such a structure. If such a modified three-dimensional structure of this end of the RNA is constructed and compared with that of a tRNA (Fig. 4.50), there is a striking resemblance sufficient to 'baffle' the enzyme. Such structures are envisaged in the splicing mechanism, i.e. the elimination of introns in the mRNA precursor.

Ribozymes

When studying the splicing mechanism of the 26S RNA in a protozoan (*Tetrahymena*), it was found (Cech *et al.*, 1981) that the RNA was cut without the help of an enzyme. The RNA itself plays the role of the enzyme by engaging in self-cleavage. The reaction scheme is given in Fig. 4.51 for type I (there are four different types). The guanosine that triggers the process is only involved through the 3′-OH, which acts as a nucleophilic group and causes a series of transesterifications, i.e. an

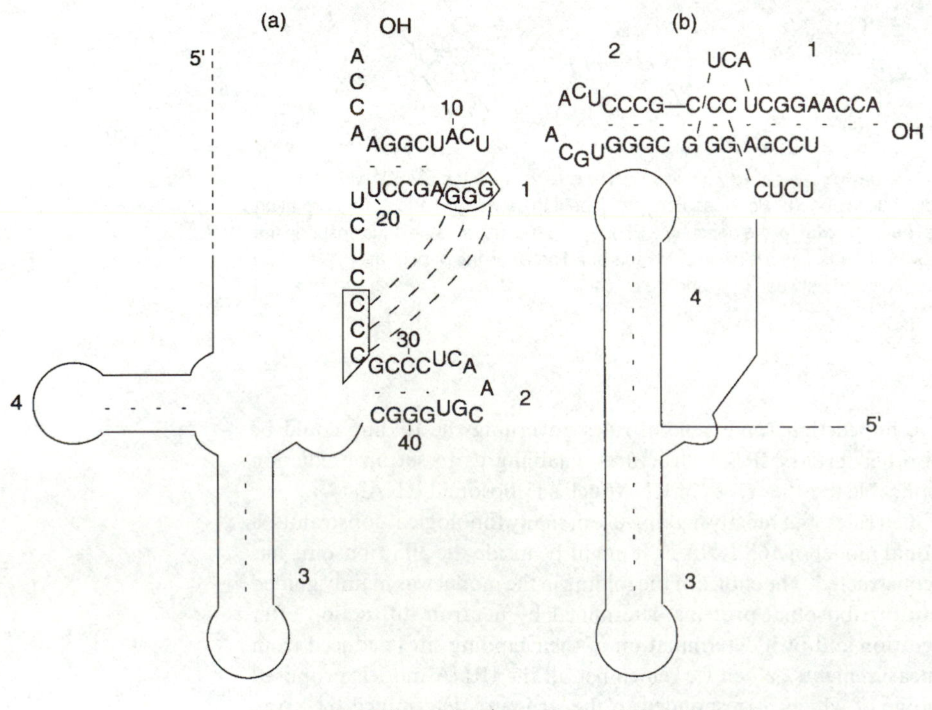

Figure 4.49 Folding of the 3′ end of TYMV RNA: (a) indication of pairings to be made: (b) type of folding to achieve the pairings. (From Pleij *et al.*, 1985. Reprinted from *Nucleic Acids Research* by permission of Oxford University Press and with the authors' permission.)

exchange between phosphodiester bonds without changing their number. The released intron is then subjected to a succession of cleavages and cyclizations. This catalytic activity is related to the tertiary structure of RNA.

The first crystal structure of a catalytic RNA molecule has recently been determined (Pley *et al.*, 1994) in the case of a hammerhead ribozyme. The crystal was produced with a complex between a 34-nucleotide RNA and a 13-nucleotide single-stranded DNA inhibitor. The latter becomes a substrate for the ribozyme when a

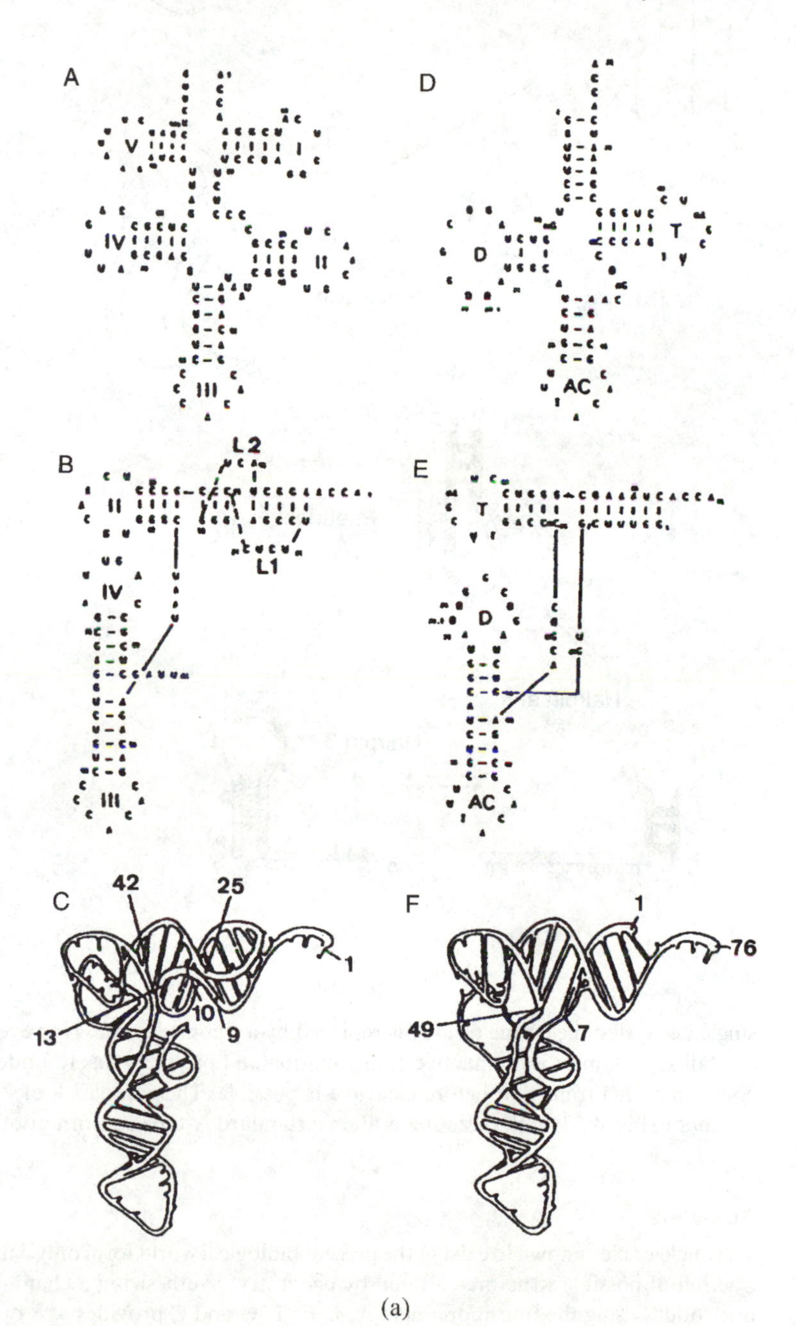

(a)

Figure 4.50(a)

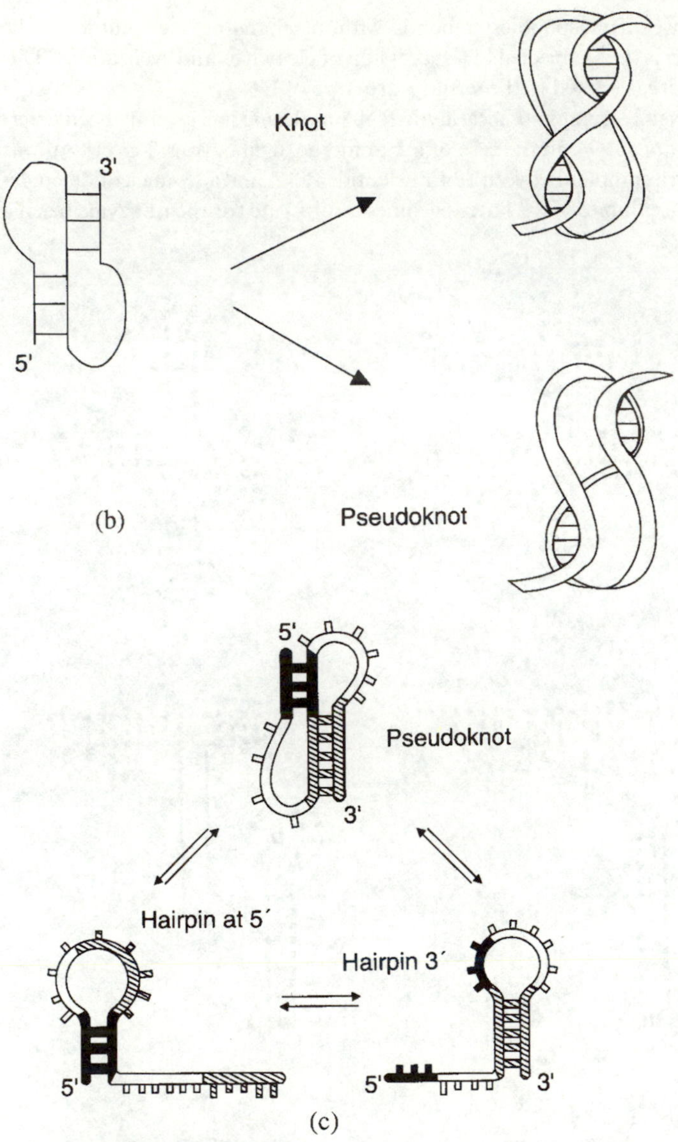

Figure 4.50 (a) Comparison between the pseudoknot structure of TYMV and that of tRNA. A, B, C: secondary structure of the 3′ end of TYMV RNA, model of folding and diagram of tertiary structure. D, E, F: the same for yeast $tRNA^{Val}$, except that F is a canonical three-dimensional tRNA structure. (From Dumas *et al.*, 1987. Reprinted from *Journal of Biomolecular Structure and Dynamics* with the permission of Adenine Press and the authors.) (b) Folded structures can form a knot or a pseudoknot. Only the latter is observed but it is difficult to predict this from the energy balance. (c) Possible equilibria between a pseudoknot and two hairpin structures depend on temperature, ionic strength and RNA sequence. (From Puglisi *et al.*, 1991. Reprinted with permission from *Accounts of Chemical Research.* © 1991 American Chemical Society.)

single deoxyribonucleotide residue is replaced by a ribonucleotide. However, the crystallized complex is an inactive conformation and probably has to undergo a conformational transition before cleavage is possible. The complex looks like a wishbone (Fig. 4.52) with three stems all in a standard A-type conformation.

Aptamers

The nucleic acids known to exist in the present biological world form only a minute fraction of possible structures offered by chemistry. Synthesizing a chain of 200 nucleotides using the four monomers A, U or T, G and C provides 4^{200} or 10^{120}

different molecules! It is therefore tempting to try to see whether random combinations of nucleotides might lead to molecules with new properties, such as the specific recognition of small or large molecules or novel catalytic activities. This is the principle underlying the production of *aptamers*. Clever screening of a population consisting of no more than 10^{15} molecules (25 positions with the four

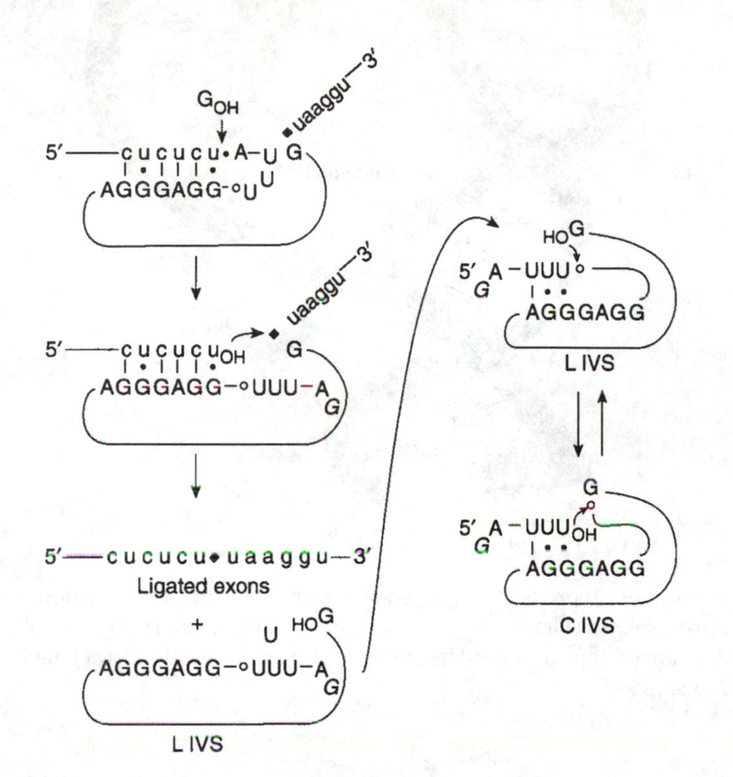

Figure 4.51 Self-splicing mechanism of a group I RNA (left column) and autocyclization (right column). After nucleophilic attack of G, the splicing leads to two linked exons and to the release of one intron, which is then cyclized. L IVS and C IVS are linear and circular introns respectively. (From Cech, 1990. Reprinted with permission from *Annual Review of Biochemistry*, vol. 59 and with the author's permission. © 1990 Annual Reviews Inc.)

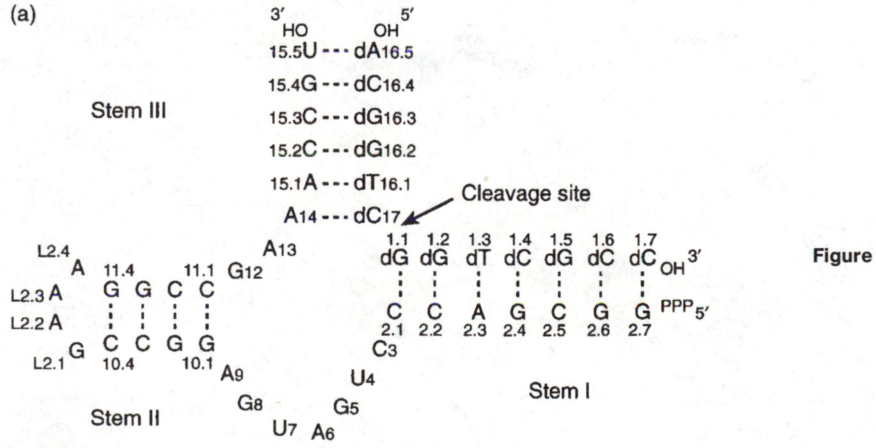

Figure 4.52(a)

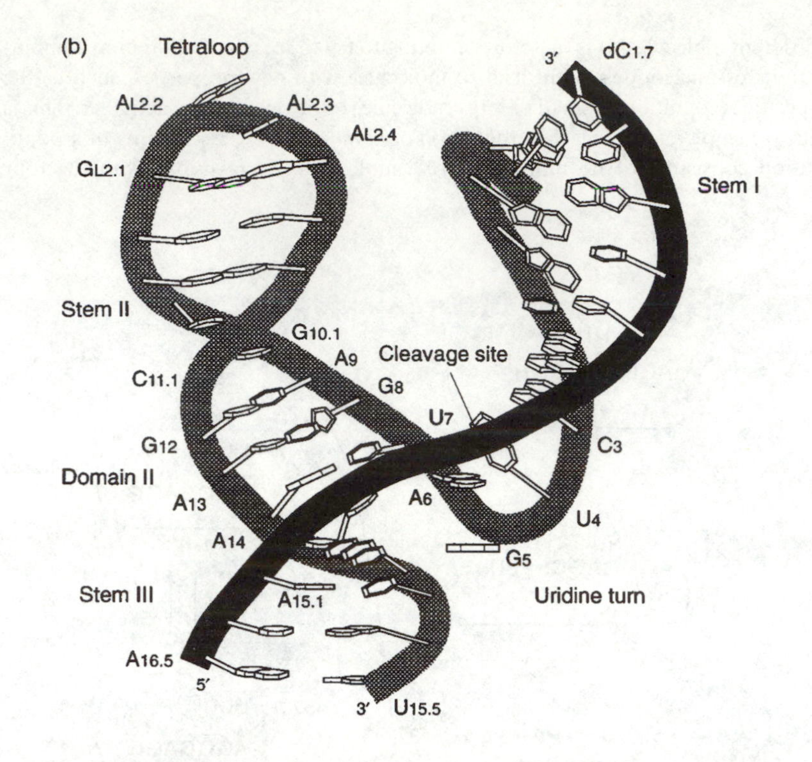

Figure 4.52 (a) 'Flat' representation of the complex in which the hammerhead shape clearly appears. Watson–Crick pairings are indicated by dotted lines. The stem numbering scheme is indicated. (b) The ribonucleotide strand is grey and deoxyribonucleotide strand is black. The three stems are clearly identified. The C3–A6 loop is reminiscent of the $tRNA^{Phe}$ pseudouridine loop and the two are easily superimposed. (From Pley *et al.*, 1994. Reprinted with permission from *Nature*, vol. 372, pp. 68–74 and with the authors' permission. © 1994 Macmillan Magazines Ltd.)

nucleotides) using precisely the targeted substrate enables new functional molecules to be isolated. The almost limitless possibilities here open up an immense field to the creative imagination and to a radically new pharmacology (for a review, see Gold *et al.*, 1995).

5

Conformation of proteins

Proteins are involved in all stages of cellular activity: some, embedded wholly or partly in membranes, regulate inflow and outflow of molecules and ions to allow signals to be exchanged with the external medium; others provide for the packing and shaping of certain subcellular complexes, like the histones in chromosomes or the ribosomal proteins in ribosomes. Most of them will behave as specific enzymes with one (or sometimes several) well-defined functions either alone or in enzymatic complexes.

5.1 Sequence

All proteins are formed by sequences of the motif:

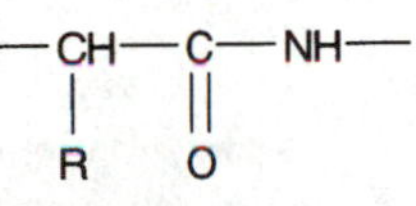

where the chemical group R, called the *side chain* and bound to a carbon called C^{α}, is one of a list of 20 possibilities (Table 5.1). Corresponding to the motif is the monomeric unit called an amino acid:

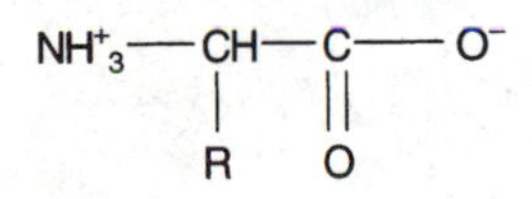

in which the two endgroups have been written in their ionized form. Amino acids form chains by losing a water molecule from the carboxy and amino groups at the ends to produce the peptide bond –CO–NH–. Ribosomes provide the sites for protein synthesis: information brought to the genome by messenger RNA is transcribed according to the *genetic code*, which, for each nucleotide triplet, specifies a corresponding amino acid. A given amino acid can be coded by more than one triplet (degeneracy of the code) (Table 5.2).

Except for that in glycine, the C^{α} of amino acids is an *asymmetric carbon*. It is a remarkable fact that all the amino acids present in proteins are of the L type. One way of representing the spatial orientation of the chain is to look at the C^{α} from the direction of the hydrogen atom bound to it: in a clockwise sense we should 'read' the sequence CO, R, N for an L-amino acid (Fig. 5.1).

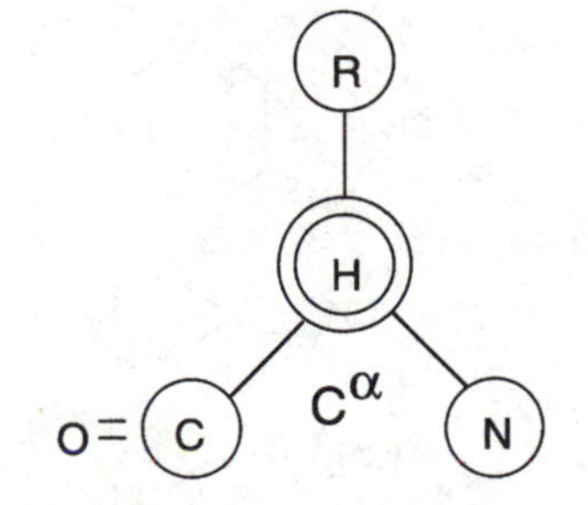

Figure 5.1 Orientation of atoms in an L-amino acid. The diagram shows the group in a view towards C^{α} from the hydrogen attached to it.

There are of course enantiomeric D-amino acids (mirror images of the L-amino acids). The chiral choice of the L form is one of the puzzles that arises when studying the origin of life on the Earth.

Table 5.1 Amino acids

Amino acid	Three-letter symbol	One-letter symbol	Side chain R
Glycine	Gly	G	$-H$
Aliphatics			
alanine	Ala	A	$-CH_3$
valine	Val	V	$-CH-(CH_3)_2$
leucine	Leu	L	$-CH_2-CH-(CH_3)_2$
isoleucine	Ile	I	$-CH(CH_3)-CH_2-CH_3$
proline*	Pro	P	C(=O), N, C–H, CH_2, CH_2, CH_2 (ring)
cysteine	Cys	C	$-CH_2-SH$
methionine	Met	M	$-(CH_2)-S-CH_3$
Aromatics			
histidine	His	H	$-CH_2-C=CH$, N, NH, CH (ring)
phenylalanine	Phe	F	$-CH_2-$ (benzene ring)
tyrosine	Tyr	Y	$-CH_2-$ (benzene ring) $-OH$
tryptophan	Trp	W	CH_2, C, CH, NH (indole ring)
Polar			
asparagine	Asn	N	$-CH_2-C(=O)NH_2$
glutamine	Gln	Q	$-(CH_2)_2-C(=O)NH_2$
serine	Ser	S	$-CH_2\,OH$
threonine	Thr	T	$-CH\,OH-CH_3$
Charged			
lysine	Lys	K	$-(CH_2)_4-NH_3^+$

Table 5.1 (*Continued*)

Amino acid	Three-letter symbol	One-letter symbol	Side chain R
arginine	Arg	R	$-(CH_2)_3-NH-C(=NH_2^+)-NH_2$
aspartate	Asp	D	CH_2-COO^-
glutamate	Glu	E	$(CH_2)-COO^-$

*Note that in the case of proline the ring is closed with the NH group before C^α:

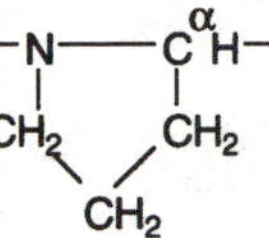

Table 5.2 Genetic code

First base	Second base				Third base
	U	C	A	G	
U	Phe	Ser	Tyr	Cys	U
	Phe	Ser	Tyr	Cys	C
	Leu	Ser	Stop	Stop	A
	Leu	Ser	Stop	Trp	G
C	Leu	Pro	His	Arg	U
	Leu	Pro	His	Arg	C
	Leu	Pro	Gln	Arg	A
	Leu	Pro	Gln	Arg	G
A	Ile	Thr	Asn	Ser	U
	Ile	Thr	Asn	Ser	C
	Ile	Thr	Lys	Arg	A
	Met	Thr	Lys	Arg	G
G	Val	Ala	Asp	Gly	U
	Val	Ala	Asp	Gly	C
	Val	Ala	Glu	Gly	A
	Val	Ala	Glu	Gly	G

The sequence of amino acids in the polypeptide chain defines the *primary structure* or *sequence* of the protein. By convention, the sequence is written from the N-terminal to the C-terminal. At each end, the protein macromolecule generally has one of the charged groups of an amino acid. Thus, the N-terminal is of the type

$$NH_3^+-HC^\alpha R-CO-NH-$$

and the C-terminal is

$$-NH-HC^\alpha R-CO-O^-$$

The protein sequence has for many years been determined by biochemical methods. Automatic sequencers are now available to speed up the process and can operate with very small amounts of protein. However, they cannot at present deal with sequences longer than 50 amino acids, so that the protein has to be cut into a

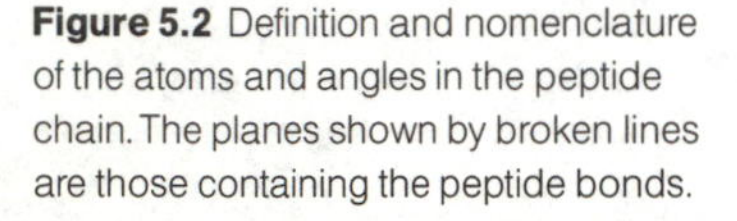

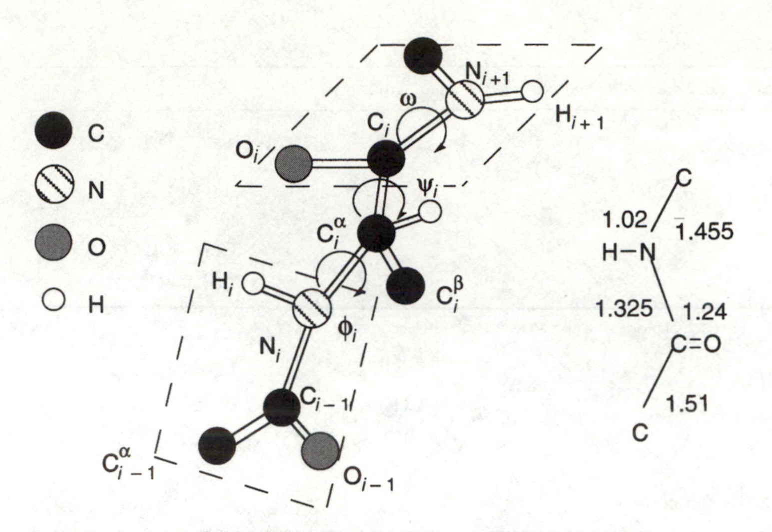

Figure 5.2 Definition and nomenclature of the atoms and angles in the peptide chain. The planes shown by broken lines are those containing the peptide bonds.

sufficient number of fragments (with proteases) while ensuring that there is enough overlap to align the sequences on each fragment correctly. If the protein gene is isolated, the sequencing of the corresponding section of DNA can generally be carried out more quickly, and in principle this allows the protein sequence to be determined by applying the rules of the genetic code.

5.2 Conformational parameters of the peptide bond

Quantum-mechanical calculations and X-ray diffraction patterns from dipeptide and oligopeptide crystals both give the C–N distance in the peptide bond as 1.325 Å, a value much closer to the length of the double C=N bond (1.25 Å) than to that of a single bond. The double-bond character of the C–N bond and the fact that the *trans* position of the O and H atoms is more stable by 4 kcal mol^{-1} than the *cis* position, together lead to the C, O, N and H atoms lying in the same plane. To describe the slight deformation that sometimes occurs, an angle of rotation about the C–N peptide bond ω is introduced but never differs very much from 180°.

Successive planes are linked as indicated in Fig. 5.2. The junction of the $(i-1)$th and ith amino acids creates a plane containing the O_{i-1}, C_{i-1}, N_i and H_i atoms. Similarly, the junction of the ith and $(i+1)$th amino acids creates another plane containing the O_i, C_i, N_{i+1} and H_{i+1} atoms. The only remaining degrees of freedom correspond to rotations about the N_i–C_i^α and C_i^α–C_i bonds. These are defined by the angles ϕ_i and Ψ_i respectively. The conformation adopted by a chain of n amino acids thus involves n pairs (ϕ, Ψ), i.e. $2n$ parameters (Fig. 5.2).

5.3 Spatial organization and related problems

When we consider the way in which proteins are synthesized in the cell, we see that a peptide chain growing at the rate of about 50 amino acids per second must progressively organize itself if it is to have a well-defined structure when it leaves the ribosome.

Moreover, a number of chemical changes described as post-translational (acetylation, phosphorylation, binding of sugars or lipids, specific cleavage of a small peptide to activate an enzyme, etc.) will only be able to occur on a protein that already possesses its three-dimensional structure. In general, therefore, there must be sufficient sequence-governed structural information to give the protein a unique and reproducible three-dimensional structure in the medium in which it occurs.

The first determination of a three-dimensional protein structure was that of myoglobin crystals in 1961, obtained by X-ray diffraction. Since then, many thousands of protein structures have been determined at various resolutions ranging between 1.8 and 3.5 Å. More recently, with the technological and methodological advances in NMR, complete structures of small proteins ($M < 15\,000$) have been determined in solution. However, we should remember that the number of proteins in living matter is now known to number 20 000 or more. At the same time, the difficulties of crystallization when using X-ray diffraction and of size and solubility when using NMR are often serious obstacles to the use of these methods. They are also expensive and time-consuming when a full three-dimensional structure is to be established. It is therefore important to find other methods, especially reliable ones enabling us to predict a structure, i.e. to specify the spatial distribution of all the atoms in the protein from its amino acid sequence.

The following general points need to be considered in connection with the search for protein structures:

1. Consider a protein with a sequence of N residues. The selection of the sequence cannot have been made by trial and error, i.e. using only point mutations of the genome. Even with extremely high mutation rates, the age of the Universe (about 5×10^{17} s) would be a long way short of the time needed to explore the 20^N possibilities (e.g. $20^{300} = 10^{400}$ for a protein with 300 residues). As Ptitsyn and Volkenstein (1986) have pointed out, we have to assume that evolution can only 'edit' a randomly selected sequence insofar as the sequence fulfils a desired function. Hence there is no one-to-one relationship between sequence and function: the structure is not determined by the function, while the reverse is generally true. If a structural 'code' does exist, it must be highly degenerate since different amino acid patterns can lead to the same spatial structure. It follows that the frequently used test of structural homology is probably a sufficient but not a necessary condition for function homology.

2. A protein with a new function can result from the assembly in messenger RNA of several shorter nucleotide sequences, each already having been selected for a simpler function. Such fusion of functional domains shortens the time for evolution and it is common to find enzymes in upper eukaryotes that result from the assembly of more primitive prokaryotic enzymes.

3. From a thermodynamic point of view, we find the following contributions to the total free energy of a protein:

(a) conformational entropy corresponding to the loss of degrees of freedom due to the bonding of amino acids and the restricted motion of side chains;

(b) energy of intramolecular hydrogen bonds and of hydrogen bonds between the protein and external water molecules;

(c) energy of van der Waals bonds or, more generally, of hydrophobic 'bonds' (see Chapter 13);

(d) Coulomb energy of electrostatic bonds (salt bridges, internal chelates) and coupling energy between dipoles formed by α-helices;

(e) valence bond energy and energy from topological stress in disulphide bridges.

From this list (which is not exhaustive) we can try to determine the total free energy of the protein and look for the conformation that minimizes the free energy. However, because there is such a large number of variables, there are also many minima and it is difficult to find which of these would correspond to the lowest energy and thus to the stable state of the protein. Moreover, it is also possible for the protein to be not in a stable state but in a metastable state with a lifetime long enough to give it the same characteristics as that of a stable state.

Instead of seeking to approach a three-dimensional structure directly from the sequence, it seems preferable to proceed one step at a time, attempting to build up the protein structure gradually from 'prefabricated' components and deriving certain rules for assembly. We shall see later that this is not a very realistic approach to protein morphogenesis.

5.4 Analysis of some secondary structures

5.4.1 α-helices

The above-mentioned prefabricated components are also called *secondary structures* since they correspond to the second stage of organization, the first being the

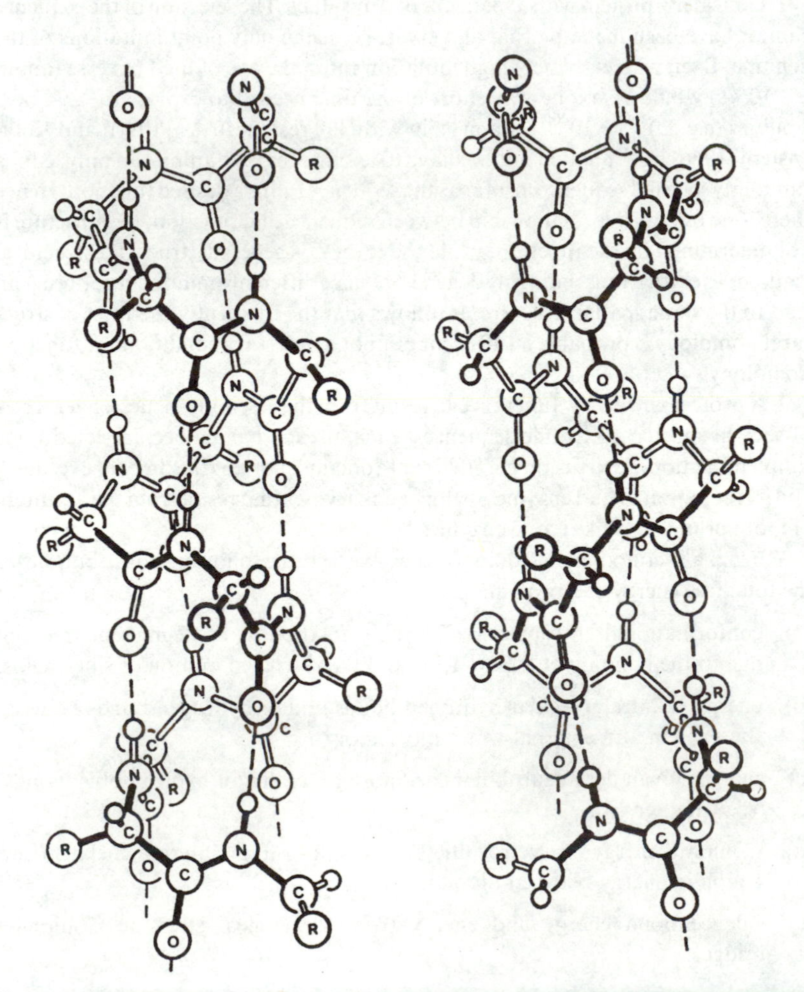

Figure 5.3 Left-handed and right-handed α-helices.

ear sequence of amino acids or *primary structure* (see Section 3.4). In peptide ains, the *right-handed α-helix* is the one most frequently found (Fig. 5.3). The ial distance p between two successive monomeric units is 1.5 Å and the ratio tween the pitch P of the helix and p is not an integer ($P/p = 3.6$). Five turns and monomers are thus required before two exactly superposed monomers are countered again. The rotation angle between two successive monomeric units 00° and the values of ϕ and Ψ (see Table 5.3) are the same for all residues forming rt of the helix. The helical structure is stabilized by hydrogen bonds parallel to the is between the C=O group of the ith monomer and the N–H group of the $(i+4)$th onomer (Fig. 5.3). In Fig. 5.2, we isolated the planes containing the C, O, H and N oms for the sake of clarity, but the true division of the peptide chain into residues shown in Fig. 5.4.

The theoretical prediction made in 1951 by Pauling *et al.* has been largely rne out by all known protein structures. Systematic studies of synthetic poly-ptides, both theoretical and experimental, have extended the family of helical ms (Table 5.3).

.2 3_{10} helices and β-turns

e 3_{10} *helix* occurring in some proteins is often located at the C-terminal end of α-helix (Fig. 5.5). The stabilizing hydrogen bonds are no longer parallel to the s and connect atoms i and $i+3$. The subscript 10 describing the helix is the mber of atoms involved in the 'cycle' joining the C=O and N–H groups (Fig. 5.6), th of which participate in the hydrogen bond. In fact, the 3_{10} helix is only one of possible forms of peptide structures in which hydrogen bonds occur between the and $(i+3)$th residue. They are defined by the six intervening ϕ and Ψ values. is applies equally to *β-turns*, two examples of which are shown in Fig. 5.7 pe I and type II). They are found between two helical structures (α-helices or

le 5.3 Parameters of helical peptide structures

ıcture	ϕ (deg)	ψ (deg)	Number of residues per turn	Distance between monomers (Å)
t-handed α-helix	−57	−47	3.6	1.5
helix	−49	−26	3.0	2.0
t-handed polyglycine II	−80	150	3	3.1
handed polyproline I	−83	158	3.33	1.9
handed polyproline II	−78	149	3	3.12
allel β-sheet	−119	113	(2)	(3.25)
parallel β-sheet	−139	135	(2)	(3.47)

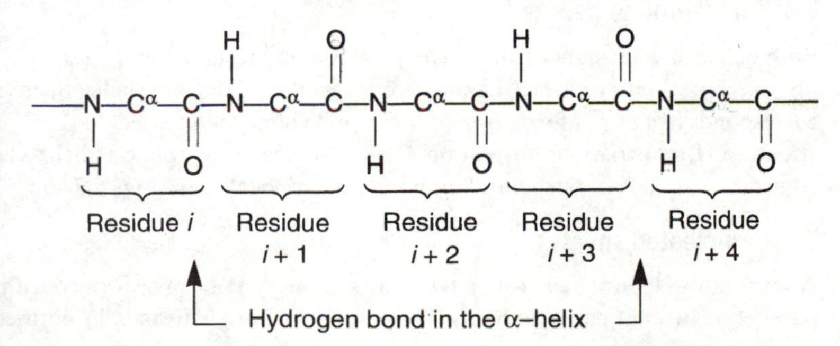

Figure 5.4 Hydrogen bonds stabilizing the helical structure.

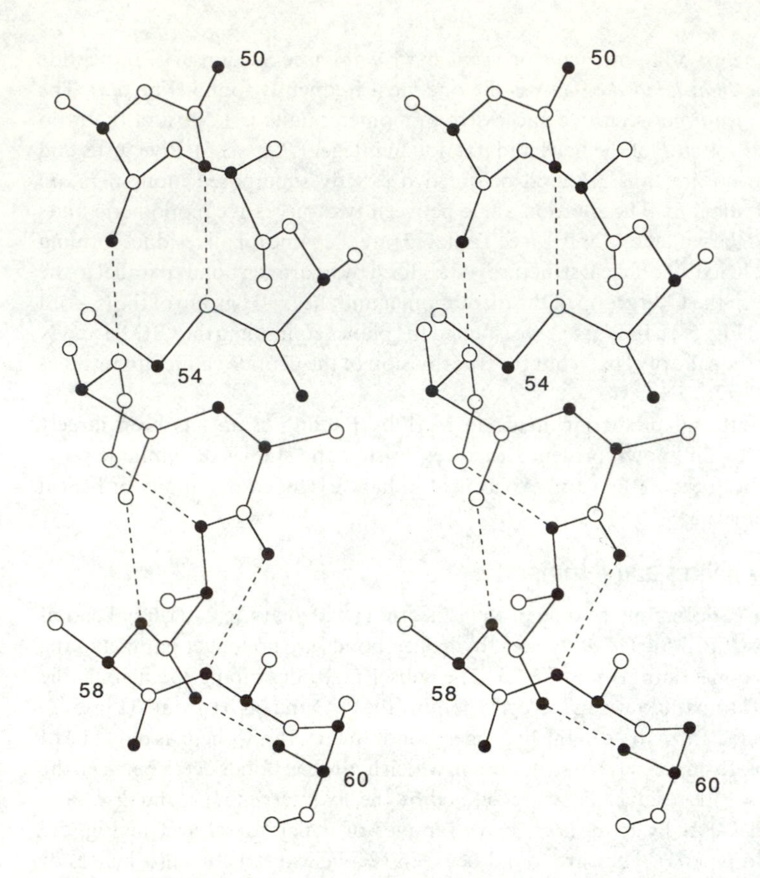

Figure 5.5 Stereoscopic view of a 3_{10} helix (from residue 54 to residue 60) located after an α-helix in ribonuclease. Hydrogen bonds between i and $i+3$ are not as well aligned as in an α-helix and C=O groups are tilted more towards the outside. (From Baker and Hubbard, 1984. Reprinted from *Progress in Biophysics and Molecular Biology*, vol. 44, pp. 97–179, © 1984, with kind permission from Elsevier Science Ltd., The Boulevard, Langford Lane, Kidlington, Oxford OX5 1GB, UK.)

antiparallel β-sheets). The N–H and C=O groups located in the bend are pre rential interaction sites, either making hydrogen bonds with other polar grou (side chains of other parts of the protein, water molecules) or establishing coval bonds with glycan chains in glycoproteins.

A statistical study of a large number of proteins shows that the commonest ami acids found in a β-turn are as follows:

- in position 1 – **Asn**, Cys, Asp, His
- in position 2 – **Pro**, Ser
- in position 3 – **Asn**, **Gly**, Asp
- in position 4 – Trp, Gly.

5.4.3 Polyproline helices

Both helices are left-handed but their special shapes result from the *cis* and *tr* proline conformations, which cannot occur together in the same helix. Structu corresponds to a *cis* conformation of the peptide bond with respect to the C^{α}, a structure II to a *trans* conformation (Fig. 5.8). The polyproline II helix can considered as the basic structural monomer of tropocollagen (Figs 5.9 and 5.1

5.4.4 β-pleated sheets

A very curious type of helix with a two-fold symmetry axis can be formed with t parallel or two antiparallel chains. A β-sheet structure is formed in both ca

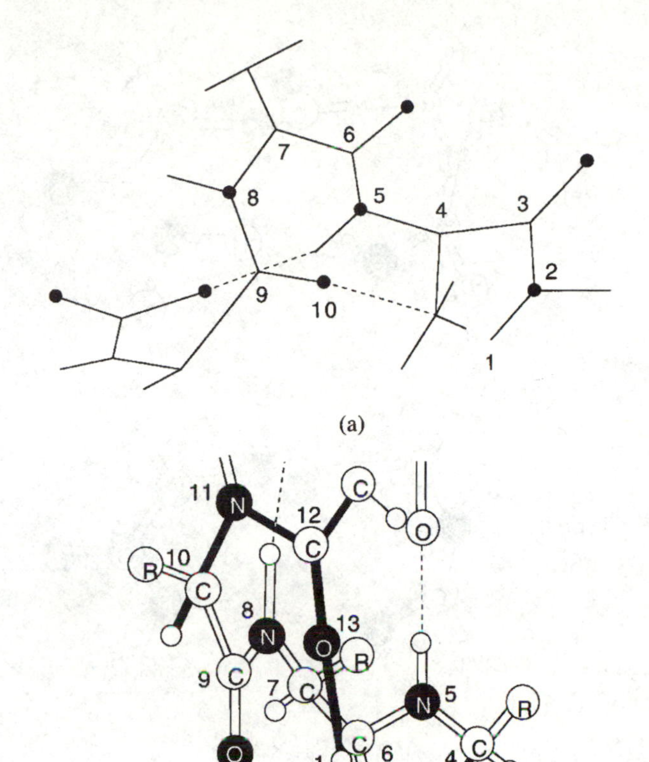

Figure 5.6 (a) A diagram of a short 3_{10} helix in carbonic anhydrase (from residue 159 to residue 164) indicating the numbering of atoms, from the H of NH to the O of C=O, participating in the 'cycle'. (b) For comparison, a diagram indicating the 13 atoms participating in the same type of 'cycle' in part of an α-helix.

pleated like an accordion with its side chains R located alternately on each side of the strip (Fig. 5.11). We shall see later the important role played by such β structures in protein organization. The situation is different in α-helices, causing the average length of the hydrogen bond in α-helices to be 3 Å and in β-sheets to be 2.91 Å. The contribution of the hydrogen bond towards the stability of the secondary structure is therefore higher in the latter.

5.5 Prediction of secondary structures

Is it possible to predict from the amino acid sequence what type of secondary structure (α-helix, β-sheet, β-turn) the peptide chain will adopt? In principle, two angles ϕ and ψ are sufficient to define the conformation of a dipeptide. If the chain consists of a regular sequence of the same amino acid (a homopolypeptide), two methods are available for the conformational calculation.

5.5.1 Ramachandran's method

In this method, the various interaction potentials (studied in section 2.5) can be replaced by a potential given by the hard-sphere model (treating atoms as impenetrable spheres like those used in building molecular models). The starting point is a

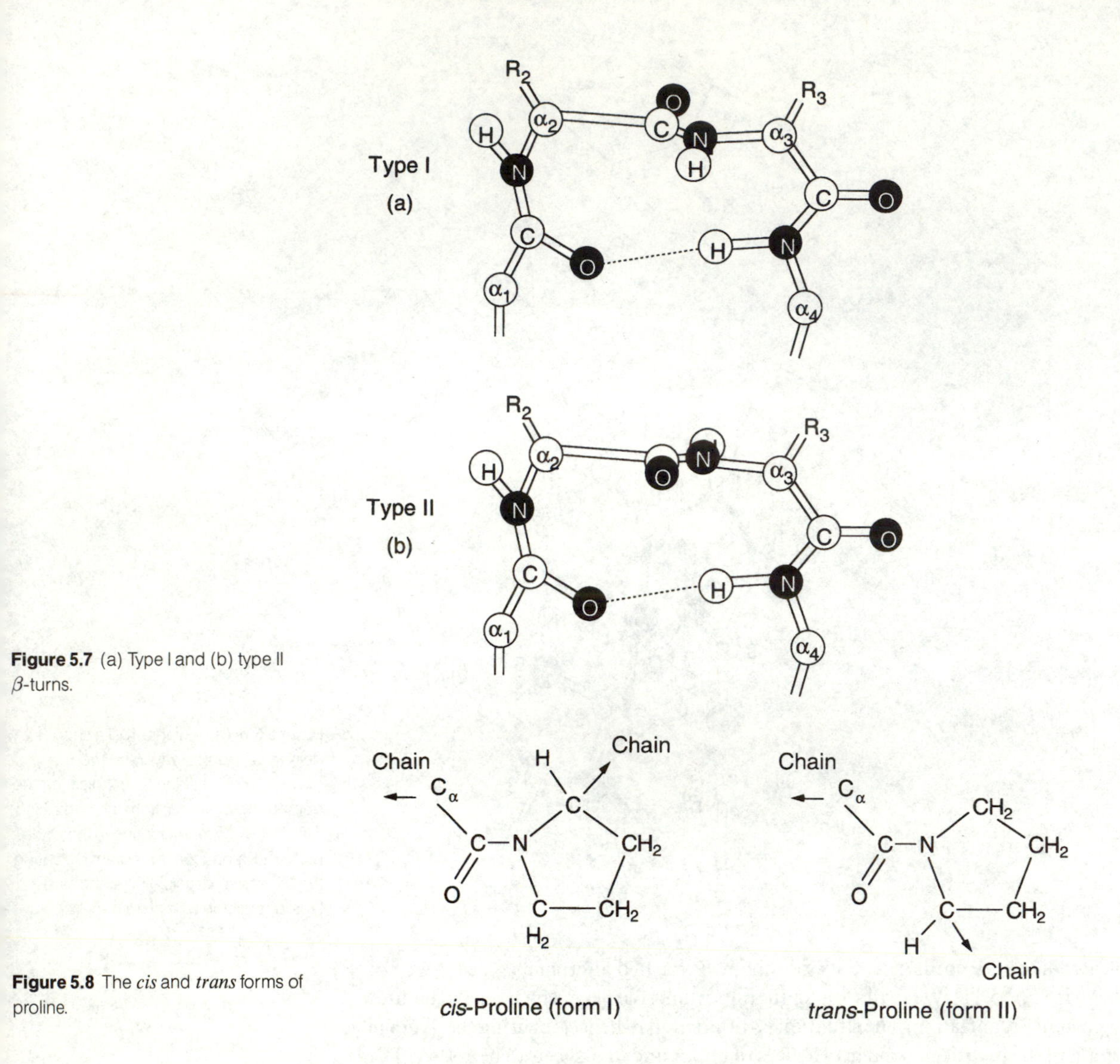

Figure 5.7 (a) Type I and (b) type II β-turns.

Figure 5.8 The *cis* and *trans* forms of proline.

Table 5.4 Contact distances (Å); values in parentheses give the absolute minima (maximum deformation of electron shells)

	C	N	O	H
C	3.20 (3.00)	2.90 (2.80)	2.80 (2.70)	2.40 (2.20)
N		2.70 (2.60)	2.40 (2.60)	2.40 (2.20)
O			2.70 (2.60)	2.40 (2.20)
H				2.00 (1.90)

table giving the acceptable distances in ångströms between the centres of atoms before any 'collision' occurs (Table 5.4).

Referring to Fig. 5.2, we can see for example that varying Ψ_i, the rotation angle about the $C_i^\alpha - C_i$ bond, the 'spheres' representing O_i or $N_i + 1$ 'collide' with those representing C_i^β or H_i. Some values of Ψ_i are therefore excluded. A similar argument can be applied to ϕ_i and all other pairs of atoms. In a map of ϕ against Ψ, we can thus delimit areas within which a certain type of secondary structure can exist for a given residue (Fig. 5.12).

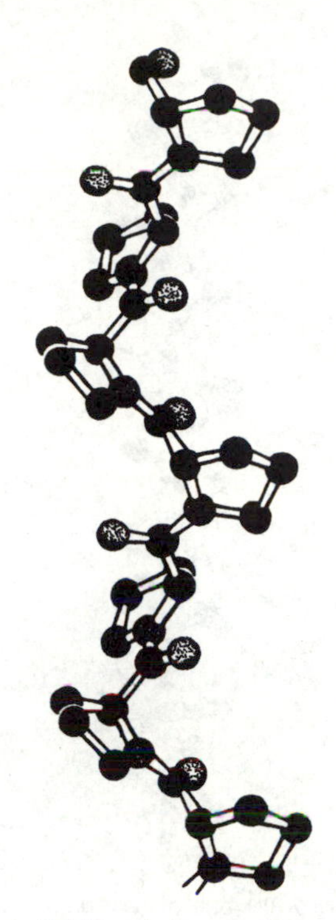
Figure 5.9 Model of tropocollagen.

5.5.2 Use of potential functions

Using analytical expressions for the potential functions (torsion, Lennard-Jones, hydrogen bond, etc.), and with an appropriate choice of parameters, the homopolypeptide potential energy can be determined for each pair (ϕ, Ψ) and equipotential lines can be drawn in the (ϕ, Ψ) plane. Regions with a potential minimum (Fig. 5.13) correspond to stable structures.

We can see by comparing Figs 5.12 and 5.13 that the results obtained by the more sophisticated method differ only slightly from those obtained using the hard-sphere model.

However, this deals only with homopolypeptides. If we attempt to extend the process to a sequence with 20 different amino acids, the problem is unsolvable by either method. To explain this, we only have to plot on a (ϕ, Ψ) graph (or on a Ramachandran plot) the observed values for 1000 residues (except glycine) belonging to proteins whose tertiary structure is well known (Fig. 5.14). Is it possible to predict this result with a hard-sphere model? We only have to look at the collision of the β-carbon of residue n with the CO of residue $n - 1$ or with the NH of residue $n + 1$. The former will limit the angle ϕ, the latter the angle Ψ. We thus obtain a graph (Fig. 5.15) in which the allowed cross-hatched areas fit experimental points quite well. There is however a large number of points between the α-helix and β-sheet regions that are not in allowed areas. In any case, the method only gives allowed areas and not a precise structure. We thus have to find another approach that is based on a statistical study of proteins with known structures.

5.5.3 Statistical predictions

An idea common to all the various methods is to attribute to each amino acid in a protein a specific *structural information content*. In other words, we allocate to each amino acid in a peptide chain a probability of forming an α-helix, a β-sheet or some other structure. The probability is obtained from the complete structural analysis of many proteins by assuming that, over a large number, the *frequency* with which we find, for example, an amino acid in an α-helix is equal to the propensity of the same amino acid to form an α-helix. If several of these 'α-helix formers' are found following each other in the sequence, this part of the chain will have a high probability of being organized into an α-helix.

In this way, we find numerical values for the frequency, showing that the presence of E, A and L, and to a lesser extent of H, M, Q, W, V and F, favours the formation of an α-helix, whereas that of G, P, Y and N disrupts the helix or stops it if it has already been formed. Similarly, the presence of M, V or I strongly favours the formation of a β-sheet. By calculating the mean value of probabilities as we move along the sequence, we can compare it with a threshold value to determine those regions more likely to become α-helices and those more likely to become β-sheets.

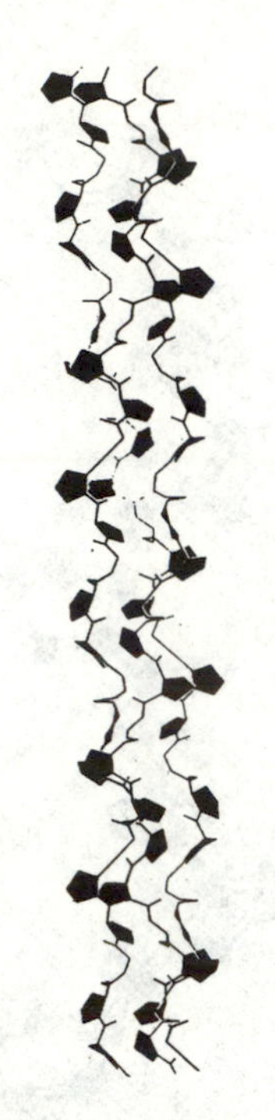

Figure 5.10 Model of the collagen triple helix with about 3.3 residues per turn in the case of a repeated Gly–Pro–Pro sequence. Each strand is a helix with a structure similar to that of polyproline. The three strands are connected by hydrogen bonds between the NH of glycine and the C=O of residues in other strands. The whole molecule forms a rod-shaped structure particularly resistant to stretching. (From *Biochemistry*, 3rd edn, by L. Stryer. © 1988 by Lubert Stryer. Used with permission of W.H. Freeman and Company.)

All these empirical methods rely on the apparently obvious assumption that the conformation adopted by each residue in a peptide chain depends not only on the type of amino acid but on its neighbourhood as well. Recent studies of 17-mer peptides in which different amino acids are substituted by chemical synthesis actually show that the tendency of a given amino acid to form a helix depends on the sequence in which it occurs. The structural information consists of the two types of data: the nature of the amino acid and the environment. Lacking any knowledge of the steric information code, the secondary structure of the peptide chain appears to be random while being completely defined in principle. It is thus tempting to deal with this problem by using information theory, treating the amino acid sequence as an *input message* whose correct interpretation leads to an exact corresponding conformation, i.e. the *output message*. We therefore have to evaluate the information carried by an event y (a given amino acid sequence) in order to obtain an event x, i.e. the conformation adopted by one of the amino acids in the sequence. This procedure is based on the definition of *conditional probabilities*.

Let $P(x \mid y)$ be the conditional probability of x, i.e. the probability that x occurs when y is known for certain to occur. $P(x)$ would be the probability of x without any information. *Information* $I(x:y)$ can thus be defined (Fano, 1961) by

$$I(x:y) = \ln[P(x \mid y)/P(x)] \tag{5.1}$$

If $P(x, y)$ is the probability that the two events occur simultaneously, then

$$P(x, y) = P(x \mid y)P(y). \tag{5.2}$$

In the case where two events are strictly independent, i.e. y has no effect on x, then

$$P(x, y) = P(x)P(y) \tag{5.3}$$

In this case, comparing eqns 5.2 and 5.3 gives $P(x) = P(x \mid y)$, which, after putting $I(x:y) = 0$ in eqn 5.1, means the following: the fact that y occurs gives no information about the probability of x. In all other cases, there are two possibilities. Either $P(x \mid y) > P(x)$, so the information is positive and y favours x, or $P(x \mid y) < P(x)$ and y disfavours x. The event y is to be selected from 20 possibilities but the event x is arbitrarily reduced to three types of structure: the α-helix (H), β-sheet (E) and unorganized structure (C). Note that the last-named group does not mean that the structure is statistically like that of a chain with free rotation (a Gaussian chain) but that it is not included in the first two categories.

What can in fact be observed in all the many known protein structures is the *frequency* with which a message y (sequence) gives rise to one of the structures H, E or C. For large numbers, frequency and probability can be equated. This is true to a first-order approximation when only one given residue is considered. The *contingency table* contains $3 \times 20 = 60$ inputs and if we operate from a databank on 12 000 residues for example, the mean value of 200 residues per input validates the approximation.

If we now wish to include information per *pair* of residues, there are $3 \times 20 \times 20 = 1200$ inputs and the average of only 10 requires a more sophisticated treatment involving information theory. A much larger databank would be needed for *triplets* of residues. Predicted percentages are already distinctly improved when pairs of residues are taken into account. Using data from 68 proteins, 65 per cent of the three types of structure are correctly located in the sequence.

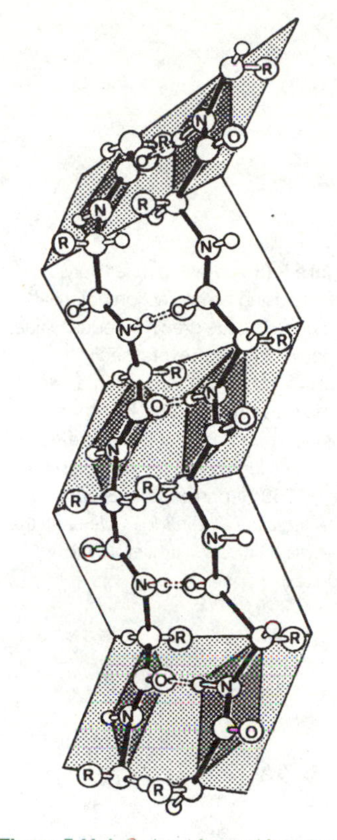

Figure 5.11 A β-sheet formed from two antiparallel peptide chains. Note the accordion-like pleated structure and the position of the CONH planes in successive pleats.

\nother approach was recently suggested, involving the use of a neuronal model. ere is in fact some formal parallelism between the prediction of a secondary ıcture and reading a text aloud. The input message consists of a series of letters embled into words and the output message is a series of phonemes either pro- ınced by the phonation system or produced by a speech synthesizer. Just as this ding requires an apprenticeship using reference words, a neurone circuit can be ght to make structures from a series of amino acids by taking into account the tionships already existing in 100 proteins.

Each unit i of the neural network is reached by 'outputs' from all the j units nected to it. The total input of the ith unit is characterized by a unit 'activation' $= \sum_j w_{ij}s_j + b_i$, where w_{ij} is the weight allocated to the connection (i, j) and b_i is the arization state of the ith unit. The output state of the ith unit is therefore given by

$$s_i = (1 + e^{-Ei})^{-1}$$

hat s_i varies from 0 to 1 as E_i varies from $-\infty$ to $+\infty$.

'he input sequence is examined through a 'window' of 13 amino acids (six on h side of the one in question). The simplest version of the output consists of three ıctural types: α-helix, β-sheet and unorganized chain. The *apprenticeship* ins with giving random values to the w_{ij} lying between -0.1 and $+0.1$, fol- ed by presenting each of the proteins of the selected series, i.e. those with wn sequence and structure. At the end of a series of cycles, the value of E_i st remain constant to within 10^{-4} for each protein. Two results stand out:

Despite the large number of structures used, the successful percentage is no better (about 65 per cent) than previous statistical methods.

On the other hand, the prediction is greatly improved if we examine it within a given class of proteins, e.g. those having only α-helices as their secondary structure or having only β-sheets.

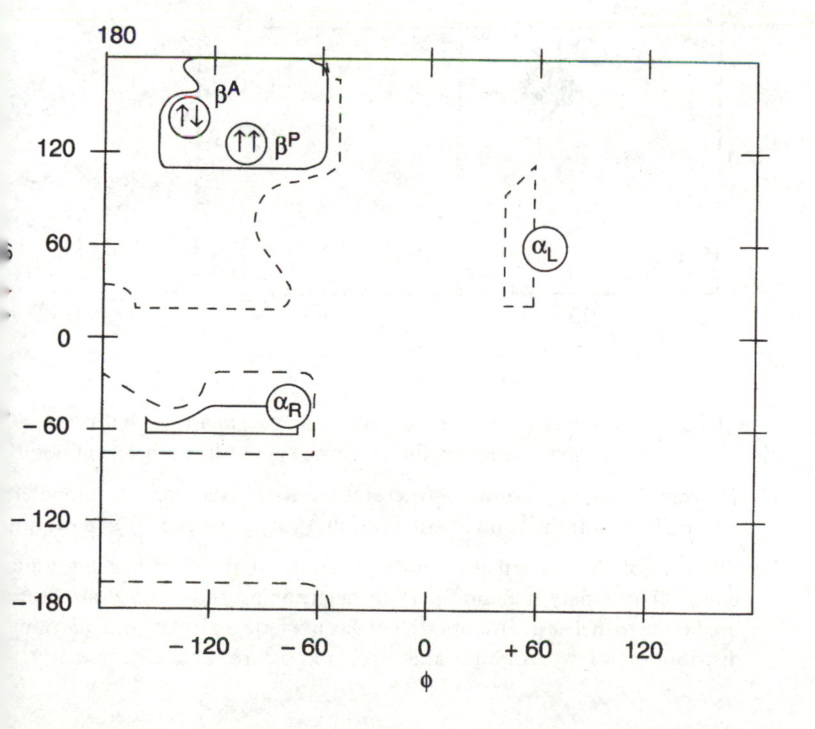

Figure 5.12 A $\Psi - \phi$ map for poly-L-alanine in Ramachandran's method. The contours drawn with solid lines give the allowed distances in Table 5.4. The contours in broken lines correspond to the minimum distances in Table 5.4 (values in parentheses). Note the symmetry about the diagonal of the contours corresponding to α_R and α_L. (From Flory, 1969. Reprinted from *Statistical Mechanics of Chain Molecules* with the permission of Carl Hanser Verlag, Munich.)

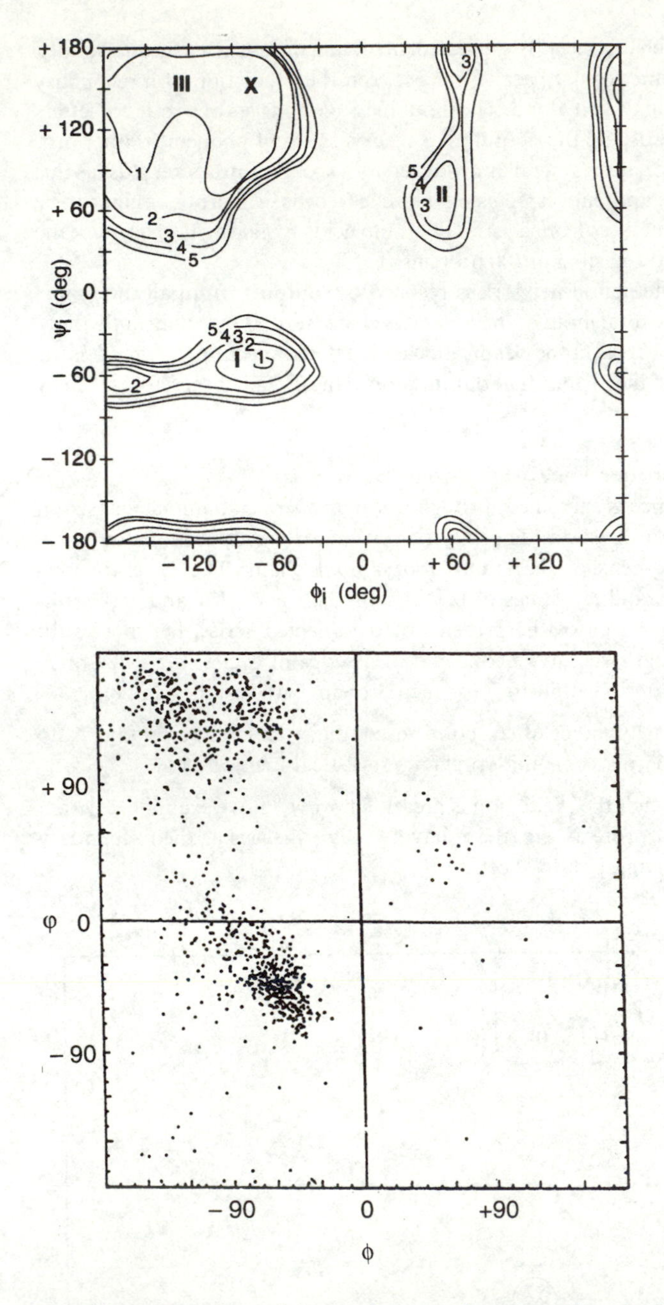

Figure 5.13 A ϕ–Ψ map for poly-L-alanine using an interaction potential. The contours are drawn for equal values of potential energy expressed in kcal mol^{-1}, with the zero taken arbitrarily at the cross in region III. The other potential minima in I and II define stable structures for a given (Ψ, ϕ) pair. (From Flory, 1969. Reprinted from *Statistical Mechanics of Chain Molecules* with the permission of Carl Hanser Verlag, Munich.)

Figure 5.14 A ϕ–Ψ map determined from 1000 non-glycine residues in eight proteins whose structures have been refined at high resolution. (From Richardson, 1981. Reprinted from *Advances in Protein Chemistry* with the permission of Academic Press and the author.)

In the present state of the art, it seems clear that all methods that propose dicting a secondary structure from the sequence neglect two important points

1. The role of packing secondary structures into a relatively compact globule. introduces constraints and interactions that vary from one protein to anot
2. The role of the solvent and more generally of the so-called hydroph bonds. These play a major part in determining the tertiary structure predicting α-helices, attempts have been made to take into account distribution of hydrophilic and hydrophobic residues. Particularly g

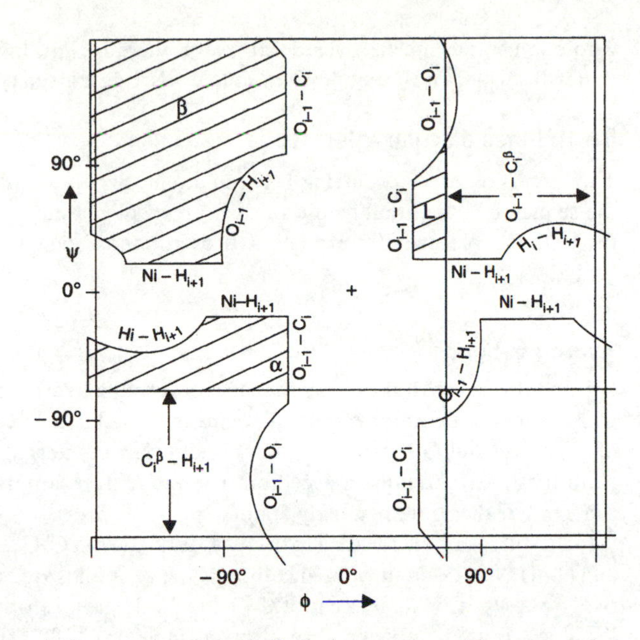

Figure 5.15 A $\Psi-\phi$ map similar to those in Ramachandran's method (hard-sphere approximation). Allowed regions are cross-hatched. Other conformations are ruled out by collisions of the atoms indicated on the borders of each region. (From Richardson, 1981. Reprinted from *Advances in Protein Chemistry* with the permission of Academic Press and the author.)

results are obtained if one side of the helix is in contact with a non-polar medium and the other with the aqueous solvent. A limiting case is that of transmembrane α-helices in ion channels.

The influence of the two parameters, packing and solvent, is automatically averaged in any statistical analysis. As a result, the structural specificity of each protein disappears.

5.6 Tertiary structure

When the tertiary structure of a protein is determined, either from X-rays or NMR, we find in each case a precise spatial distribution of atoms characteristic of the protein and its function, whose uniqueness is often the very condition for its crystallization. It is becoming increasingly clear that this *unique* three-dimensional protein structure does not result from a succession of secondary structures themselves dictated by the sequence. On the contrary, these secondary structures (α-helix, β-sheet, β-turn), while being favoured by a given amino acid sequence, are only stable if they form part of the globular structure of the protein. Two observations confirm this:

1. Short α-helices (12 residues on average) are found in proteins but the same helices isolated in water are not stable (see section 3.5). On the other hand, a statistical study of helix lengths in proteins with known structures indicates that helices are found less frequently as their length increases.
2. The secondary structures of a pentapeptide of defined sequence can be examined from databanks of protein structures. It is found that the secondary structure is preserved in only 20 per cent of cases (Argos, 1987). Since several amino acids are almost equally likely to form an α-helix or a β-sheet, only a small amount of energy is needed to change from one structure to the other.

Generally speaking, the structure adopted by the protein can be considered as an equilibrium state under the influence of compensating energy differences, whose source must be discovered. Moreover, there are intramolecular interactions that will strengthen the equilibrium state by locking a given conformation.

5.6.1 Intramolecular interactions

Three types of bond, occurring between atomic groups a great distance apart in the sequence but spatially close to each other, play a major role in establishing the tertiary structure of a protein: hydrogen bonds, ionic bonds and disulphide bridges.

Hydrogen bonds

Figure 5.16 shows all the groups capable of forming hydrogen bonds in proteins, with the maximum number of bonds being indicated in each case. From this point of view, Asn and Gln can play an important part in internal protein bridges. At neutral pH, Asp and Glu are acceptors, Lys and Arg are donors, while Ser, Thr and Tyr can be either but are mostly donors.

Let us first examine the case of the peptide bond with C=O as the acceptor group and N–H as the donor group, since most hydrogen bonds are established with these two groups (nearly 90 per cent of C=O and N–H groups form hydrogen bonds). While N–H almost always forms only a single bond (80 per cent), C=O can be involved in two or more bonds (35 per cent).

If we also make a comparison between the other groups in the peptide chain and water, they are shared almost equally in the case of C=O, whereas almost 70 per cent of N–H groups are bound to chain atoms and only 20 per cent to water. Other hydrogen bonds are established between the C=O and N–H groups and atoms of side chains belonging to remote residues in the chain. For example, if Ser or Thr are in position i, the majority of hydrogen bonds with C=O will be made with positions $i-4$ and $i-3$ of the latter, 80 per cent of them corresponding to bonds between C=O and N–H in an α-helix. For Ser, Gln or Glu in position i, on the other hand, most bonds are made with N–H if N–H is in position $i+2$ or $i+3$. Not all the potential sites of side chains are equally involved in hydrogen bonds. The proportion varies from 100 per cent for His, 85 per cent for Asp and Trp, 77 per cent for Glu, to 68 per cent for Tyr, and about 50 per cent for the other side chains.

Unlike Ser and Thr, Asn and Gln often share 'long-distance' hydrogen bonds between their NH_2 groups and a C=O of the peptide chain. These are true bridging agents inside the proteins.

Finally, it is more difficult to form hydrogen bonds between groups of amino acids when the groups are *mobile* (importance of molecular dynamics), when the hydrogen bond is suffering too much *distortion* or when there is *competition with water molecules*. In the last case, however, the water molecules are an integral part of the structure. Their dual function as donor and acceptor (see section 13.5.2) and their small size makes them a 'keystone' of the structure.

Ionic bonds

These will occur between side chains with opposite charges forming a true ion pair, i.e. a system of two charges whose attractive Coulomb energy is much greater than the thermal energy. Several ion pairs are illustrated in Fig. 5.17.

In spite of the length and flexibility of its side chain, Arg forms only a few hydrogen bonds with external water molecules and mainly establishes them with

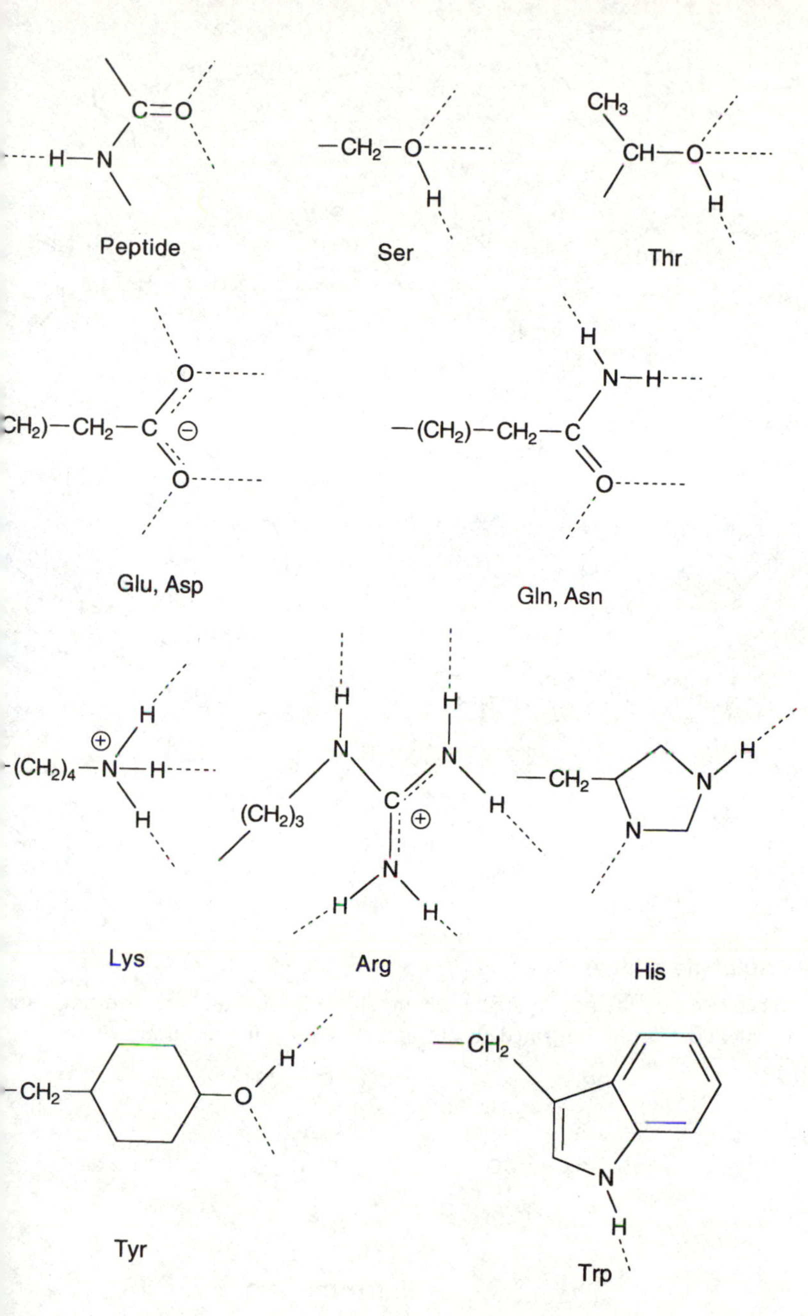

Figure 5.16 Groups capable of forming hydrogen bonds in proteins. (From Baker and Hubbard, 1984. Reprinted from *Progress in Biophysics and Molecular Biology*, vol. 44, pp. 97–179, © 1984, with kind permission from Elsevier Science Ltd., The Boulevard, Langford Lane, Kidlington, Oxford OX5 1GB, UK.)

ıer atoms at the surface of the protein. In connection with these electrostatic eractions, it should be pointed out that a majority of Asp and Glu form hydrogen nds with the α-helix N-terminal, while a majority of Lys and Arg do so with its :erminal. This seems to be due to the electric dipole field associated with the ıelix (see Chapter 15).

onic bonds (ion pairs) are formed at distances of the order of 2.7–3.0 Å, i.e. ch less than the Debye length in the medium in which the protein is dissolved. ch ion pair contributes about 2–3 kcal mol^{-1} to the total energy of the lecule.

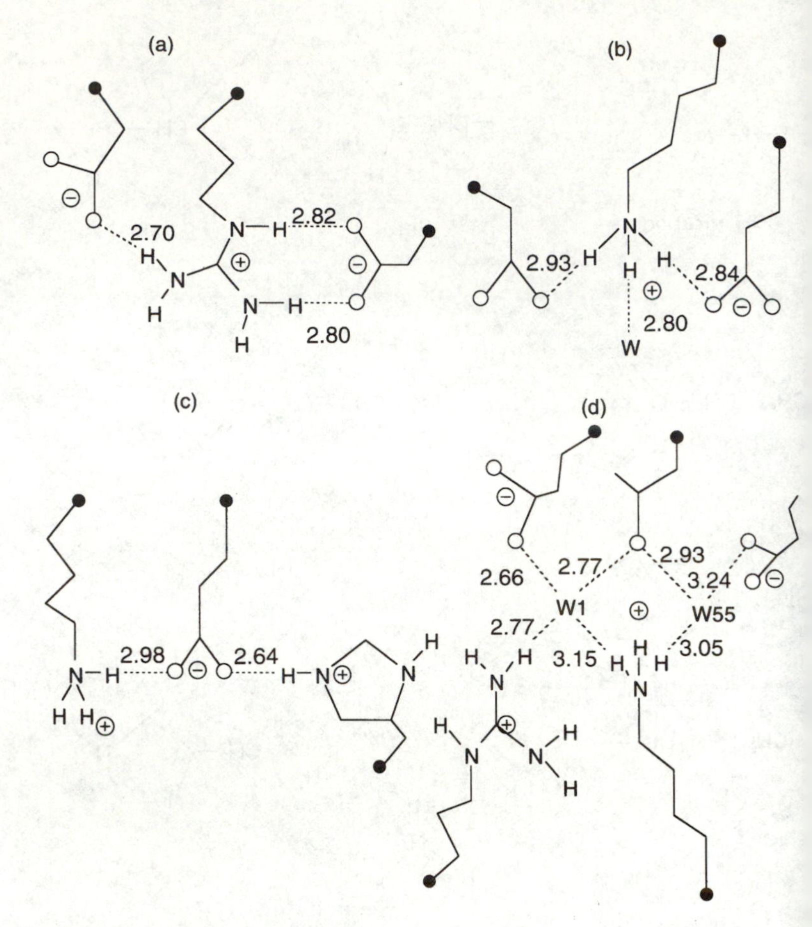

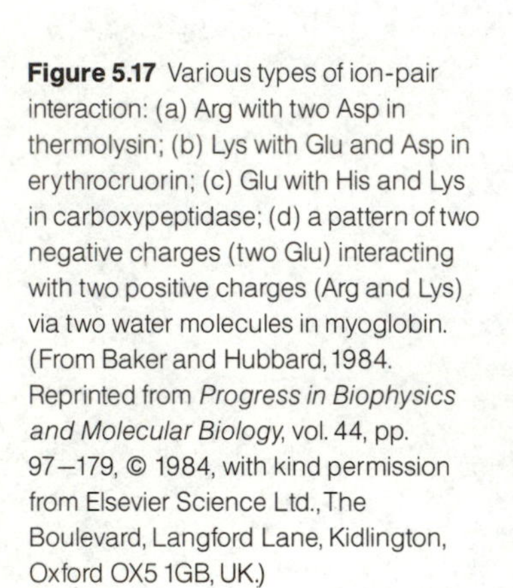
Figure 5.17 Various types of ion-pair interaction: (a) Arg with two Asp in thermolysin; (b) Lys with Glu and Asp in erythrocruorin; (c) Glu with His and Lys in carboxypeptidase; (d) a pattern of two negative charges (two Glu) interacting with two positive charges (Arg and Lys) via two water molecules in myoglobin. (From Baker and Hubbard, 1984. Reprinted from *Progress in Biophysics and Molecular Biology,* vol. 44, pp. 97–179, © 1984, with kind permission from Elsevier Science Ltd., The Boulevard, Langford Lane, Kidlington, Oxford OX5 1GB, UK.)

Disulphide bridges

When two side chains of cysteine become spatially close and are oxidized, a sin covalent S–S bond is formed according to the following reaction:

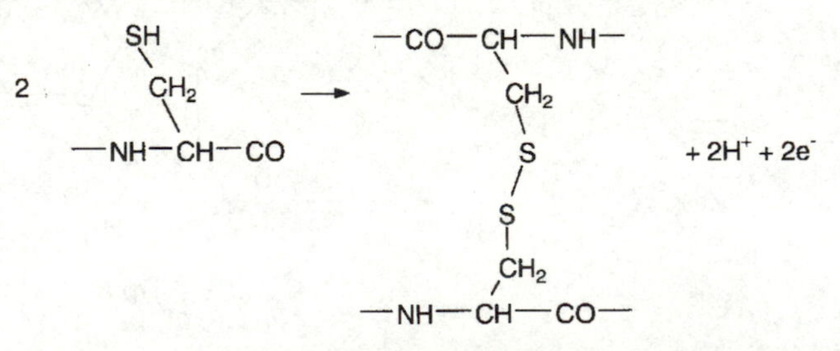

The presence of an electron acceptor such as O_2 favours the formation of an S bond, while its rupture is favoured by the presence of an electron donor such dithiothreitol ($SH{-}CH_2{-}CHOH{-}CHOH{-}CH_2{-}SH$), which is cyclized to g the oxidized form:

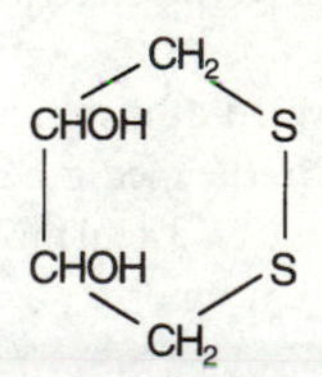

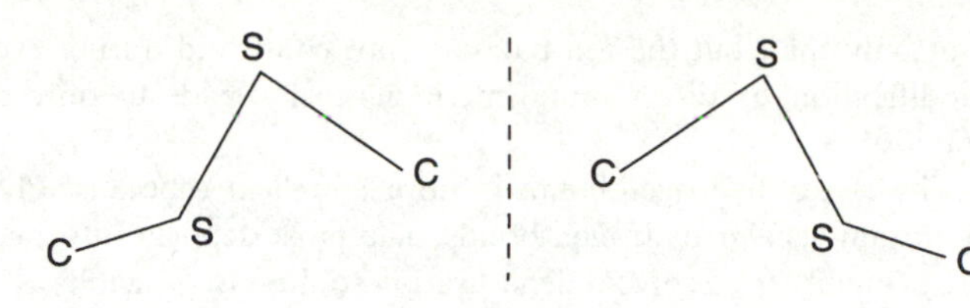

Figure 5.18 A disulphide bond and its mirror image.

he distance between the two sulphur atoms is about 2.07 Å compared with 1.55 Å ›r C–C, 1.82 Å for C–S and 1.47 Å for O–O. In the four-atom group C–S–S–C, ıe dihedral angle between the two S–C bonds is about 90°, which implies a otential barrier to rotation about the S–S bond in the range 10–20 kcal mol^{-1}.

A disulphide bridge and its mirror image are illustrated in Fig. 5.18. There is ıus a strong *topological constraint* on the peptide chain, but at the same time the ability of the three-dimensional structure (particularly its thermal stability) is creased. Many denaturation processes in proteins containing free SH groups egin with the formation of S–S bonds during their handling in solution. Such onds give the protein a structure completely different from the native one.

The role played by disulphide bridges in protein folding can be examined as llows. The kinetics of their formation (by exchange of the reduced form with a agent of the R–S–S–R type) can be followed by blocking the intermediate stages ith a reagent such as iodoacetamide ICH_2–$CONH_2$, which binds any SH groups ıat are still free according to the reaction:

he process is illustrated in Fig. 5.19 for a hypothetical protein with six SH groups Creighton, 1978, 1990).

$$-CH_2—SH + ICH_2\,CONH_2 \rightarrow —CH_2—S—CH_2—CONH_2 + HI$$

general, all three types of bond (hydrogen bond, ionic bond, disulphide bridge) ay an important part in the *maintenance* of the compact three-dimensional otein structure. It is obvious that S–S bonds cannot explain the tertiary structure proteins without SH groups. Ionic bonds are few in number: on average there e five ion pairs for 150 residues, i.e. a stabilization energy of 10 ± 5 kcal mol^{-1}.

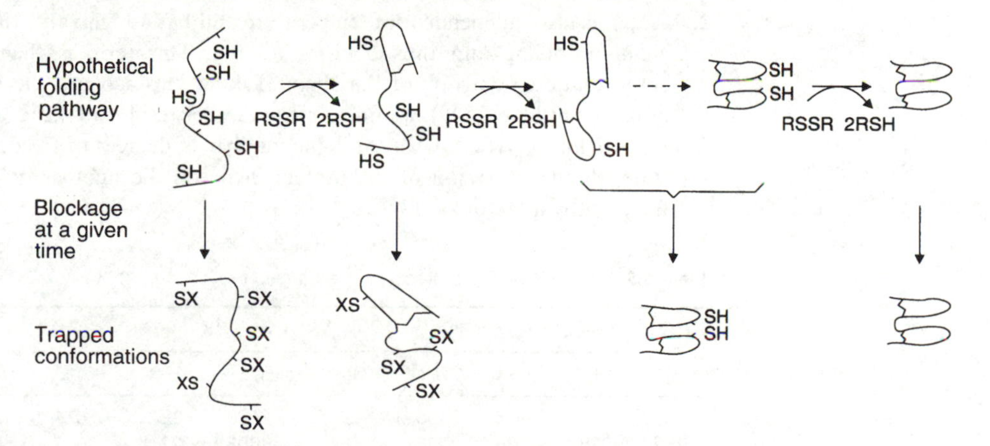

gure 5.19 Procedure for finding intermediate stages in folding in the case of a hypothetical protein h six cysteines. From the unfolded state, disulphide bridges are created by the action of -S–S–R. The folding process can be stopped at any time by adding a reagent such as doacetamide IX which replaces the SH proton with an X group. (From Creighton, 1990. Reprinted m *The Biochemical Journal* with the permission of the Biochemical Society.)

Not only that, but the ion pairs are not conserved during evolution and the modification by direct mutagenesis generally leads to only slight changes i stability.

The case of hydrogen bonds is more complicated because of the large numbe of intramolecular hydrogen bonds. The most detailed information about thes arises mainly from solvent denaturation studies. In principle, all solvents that ca form strong hydrogen bonds with the peptide chain should favour the opening an hence the denaturation of the protein. This is not so for quaternary ammoniu salts, which denature proteins without disrupting α-helices. Dioxan, on the oth hand, an acceptor of hydrogen bonds, should not compete with C=O groups y denatures proteins. Lastly, if we compare the denaturing action of ethanol an glycol, we find that the latter, which is capable of forming more hydrogen bond is less denaturing.

In short, none of the three types of bond can on its own explain the formation a compact structure.

5.6.2 Folding forces

Thermodynamics of unfolding

Studies on protein denaturation (Privalov, 1979; Baldwin, 1986; Privalov and Gi 1988) have provided some important pieces of information. Measurements c lysozyme (Sturtevant, 1977) have given us standard $\Delta H°$, $\Delta S°$ and $\Delta G°$ valu characterizing the unfolding of proteins (Table 5.5). These values can then b interpreted as the sum of two contributions. The first comes from *hydration* of a the groups previously embedded in the globular part of the protein: ΔH_{hyd} $\Delta S_{hyd}°$ and $\Delta G_{hyd}°$ values can be determined from experiments on the transfer liquid hydrocarbons to water. The second contribution, described as *residual*, simply the difference between the two sets of values. The results in Table 5.5 brin out the following points:

1. $\Delta H_{res}°$ is almost independent of temperature and favours chain folding. It the major contribution around 25°C. Above 60°C, however, ΔH_{hy} becomes the dominant term and is always favourable to folding.

2. $\Delta S_{res}°$ is also independent of temperature and has a high value that favou chain unfolding and thus opposes folding. This term is the change conformational entropy of the chain. Taking into account the number lysozyme residues (129), we find a mean ΔS per residue of 4.2 cal K^{-1}. Sin $\Delta S = R\ln(\Omega_{den}/\Omega_{nat})$, where Ω is the number of degrees of freedom in eac state, the denaturation of the protein increases the number of degrees freedom by a factor of 8.

Table 5.5 Thermodynamics of lysozyme denaturation

T (°C)	$\Delta H°$ (kJ mol^{-1})			$\Delta S°$ (kJ mol^{-1}K^{-1})			$\Delta G°$ (kJ mol^{-1})		
	Obs	Hyd	Res	Obs	Hyd	Res	Obs	Hyd	Re
10	137	−78	215	247	−2026	2273	67.4	495	−4
25	236	20	216	586	−1688	2274	60.7	523	−4
60	469	248	221	1318	−964	2282	27.2	569	−5
100	732	509	223	2067	−224	2291	−41.4	593	−6

Source: from Baldwin (1986).

3. There are also two significant temperatures: T_h and T_s. For the first, around 22°C, $\Delta H_{hyd}{}^\circ = 0$. At this temperature, the change in free energy is entirely entropic as regards the hydration mechanism. This is the entropy required to organize the water molecules around the hydrophobic groups. For the second, around 113°C, $\Delta S_{hyd}{}^\circ = 0$. At this (hypothetical) temperature, the free-energy change would be entirely enthalpic.

Hydrophobic interactions

One aspect of hydrophobic interactions is the contact between aromatic side chains (V, F, W). From an analysis of 34 protein structures with 580 pairs of aromatic side chains (Burley and Petsko, 1985), we can make the following comments:

1. The average distance between two aromatic rings lies in the range 4.5–7.0 Å, but in general there are no stacking interactions similar to those between nucleic acid base pairs. More than two-thirds of the dihedral angles between two aromatic rings are between 50° and 90°.
2. These interactions make a major contribution to the stability of the tertiary structure and mainly occur in regions not accessible to the solvent. These aromatic pairs could form nucleation sites during the folding of the peptide chain.

Helix formation

A recent experimental study (Chakrabartty *et al.*, 1994) on 58 peptides in solution showed that, in the absence of helix-stabilizing side-chain interactions, the propensity for amino acids to form a helix in solution is much less than that indicated by the statistical studies described in section 5.5.3. Ala appears to be the only helix former, Leu and Arg are neutral and all the other amino acids, in terms of free energy, have a tendency to be helix breakers. Helix formation in solution is therefore very unlikely. In predictions of secondary structure, on the other hand, the statistical analysis implicitly (but partially) takes into account the local and non-local interactions within proteins.

Lattice models

The recent development of *lattice models* has thrown new light on protein folding forces. In such models, amino acids are represented by spatially connected beads distributed over a two-dimensional square lattice or a three-dimensional cubic lattice with sites either empty or filled with one bead. *Excluded volume* is taken into account by forbidding two different residues from occupying the same lattice site. Bond angles are restricted to $\pm 90°$ and 180°, and bond length is normalized to 1. The coordinates of the ith residue are denoted by $\boldsymbol{r}_i$ and *contact* occurs between non-adjacent residues when $|\boldsymbol{r}_i - \boldsymbol{r}_j| = 1$. The number of accessible conformations depends on the number of contacts; when the latter number is a maximum, the protein is said to be in a *compact state*.

Details of atomic structure are lost, but such simplified models offer advantages over more realistic ones. The lattice model allows conformations to be counted directly and the entropy and free energy to be calculated from the partition function. To mimic protein behaviour, so-called *HP models* are often used in which there are only two types of bead: hydrophobic (H) and polar (P) (Fig. 5.20). If a protein is considered as a block copolymer with polar and non-polar sequences

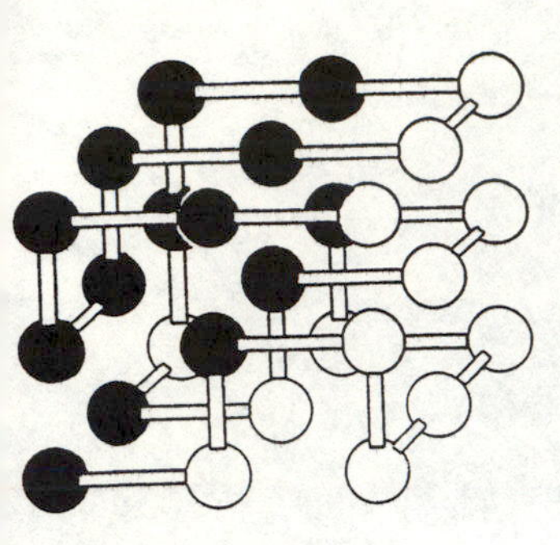

Figure 5.20 Lattice model of an HP sequence of 27 monomers. Black circles: hydrophobic groups (H); white circles: polar groups (P). (From Shakhnovich and Gutin, 1993. Reprinted with permission from *Proceedings of the National Academy of Sciences, USA*, vol. 90, pp. 7195–9. © 1993 National Academy of Sciences, USA.)

randomly distributed, the average length of α-helices and β-chains present can be estimated from simple assumptions. Eleven residues per α-helix and six residues per β-chain are found on average, and these are close to experimental values.

Two major features of proteins are the heterogeneity and the long-range nature of amino acid interactions. Both can be included in HP models and lead to two fundamental properties of proteins: a single ground state and a cooperative folding transition. An increase in secondary structures with compactness is observed. Locally compact and non-articulated structures like α-helices and β-sheets are favoured because they help the rest of the chain to configure into a compact conformation.

5.6.3 Folding theory

According to a simple model (Baldwin, 1986; Dill, 1990; Dill *et al.*, 1995), the final spatial structure of the peptide chain is the result of an equilibrium between two opposing forces.

The first is associated with hydrophobic interactions. To minimize the solvation energy, the chain, independently of locally created secondary structures, will adopt a spatial configuration that tends to minimize contacts of hydrophobic groups with water and to favour contacts of hydrophilic groups with water. In lattice models, the HH attraction minimizes the exposure of H to solvents and causes P monomers to move to the surface, and the compactness of the whole structure induces local secondary structures. However, the compactness is not maximized, so that pockets and cavities occur and the general shape is not spherical.

London–van der Waals interactions are created between aliphatic and aromatic side chains but their packing does not seem to play a dominant role in the folding. A random packing (nuts-and-bolts model) occurs rather than jigsaw-puzzle packing.

The second force is associated with the loss of conformational entropy. A peptide chain folded into a unique configuration has lost most of its degrees of freedom. The side chains contribute a large entropy due to excluded volume and their degrees of freedom are strongly coupled to those of the backbone. Fluorescence measurements confirm the sudden decrease in their mobility. The only degrees of freedom left are those corresponding to internal rotation and vibration about the average conformation. This decrease in entropy associated with chain folding corresponds to a positive ΔG unfavourable to the compact structure. As can be seen in Table 5.5 for lysozyme, the observed ΔG° is small (60 kJ or 14 kcal at 25°C) compared with both antagonistic energies. By 'locking' the system, the three bonds previously studied (ionic, hydrogen, disulphide) contribute a stabilizing free energy obviously found in ΔG_{res}.

Even if this scheme is valid, there is still the difficult problem of going from a low-degeneracy to a non-degenerate state. On the other hand, all NMR studies on small proteins reveal not a unique native state but the presence of many equivalent structures in equilibrium.

5.6.4 Modes of representation

Even when we have found the conditions for building a tertiary structure, the general problem of three-dimensional prediction has not been solved. The only thing clearly established is that a structural code is not to be found in the amino acid sequence. Structural information now seems to be non-local rather than local. As a result, information derived from structures determined by X-ray rystallography can be used to refine the description because it offers new ways f representing the protein.

he 'inside' of a protein

'he thermodynamics of protein denaturation has been compared to that of the ransition from an organic phase to water. This amounts to assuming that the inside f a protein is similar to a liquid organic solvent. Using this, liquid dioxan is often onsidered to be one of the best models for the interior of a protein. However, if the *acking density*, i.e. ratio of the van der Waals volume to the volume actually ccupied, is evaluated (Richards, 1974), we find for most globular proteins a value f between 0.7 and 0.75. This is approximately equal to the value of 0.74 for a ompact packing of spheres and values of between 0.7 and 0.8 for crystals of organic olecules. In terms of the occupied volume, therefore, it would seem that the inside f a protein is close to being in a crystalline state.

However, the internal dynamics and the apparent viscosity of protein interiors re not those of a crystal (see sections 9.3 and 9.4). Since it has a compressibility only ne-tenth of that of a liquid, a protein could be better described as a *dense liquid*.

omains

)ne way of representing the atomic distribution in a protein is to build a *distance ap*. Let r_{ij} be the distance between the C^α atoms of residues i and j. In a plot of j gainst i, a numerical value r_{ij} can be attributed to each point (i, j) and lines of equal alues of r can be drawn for given distances (4, 8, 12 Å for example). Since $r_{ij} = r_{ji}$, e need only consider the isosceles right-angled triangle of side n, where n is the otal number of residues (Fig. 5.21). In this type of diagram, α-helices appear as egments parallel to the diagonal (slope -1) separated by a distance of three resi- ues, according to the properties of the hydrogen bonds in the α-helix. An anti- arallel β-sheet, on the other hand, will be a segment of slope $+1$.

In the α-helix, for each hydrogen bond there is a kind of spatial loop formed by a roup of 13 atoms:

$$O_i—C_i—N_{i+1}—C^\alpha_{i+1}—C_{i+1}—N_{i+2}—C^\alpha_{i+2}—C_{i+2}—N_{i+3}—C^\alpha_{i+3}—C_{i+3}—N_{i+4}—H$$

'hose two ends are involved in a hydrogen bond parallel to the axis of the helix. ince the distances between atoms are small and the values of i and j are close to each ther, the corresponding points (i, j) are near the diagonal. In the antiparallel β- ructure, on the other hand, the pairing of two chains brings atoms with very ifferent values of i and j close together. Moreover, equal distance are found for e pairs $(j+1, i-1)$, (j, i) and $(j-1, i+1)$, which places the points perpendicular the diagonal. Points and contours appearing at very different values of i and j veal 'long-range' interactions. Near the diagonal, we observe clusters of points t short distances that belong to *domains*, i.e. regions with compact packing and n overall globular shape.

Another way of detecting the position of these domains is to use the concept of *ccessible area*. Suppose that a peptide chain of n residues folded into its tertiary ructure is divided arbitrarily into two regions, one including the residues 0 to i nd the other including the residues $i+1$ to n. If A_1 is the accessible area of the rst region and A_2 that of the second, the quantity $B = A_1 + A_2 - A$, where A is e total accessible area, will define an *interface* area between the two regions. If e protein consists practically of only one domain, the curve $B = f(i)$ will be

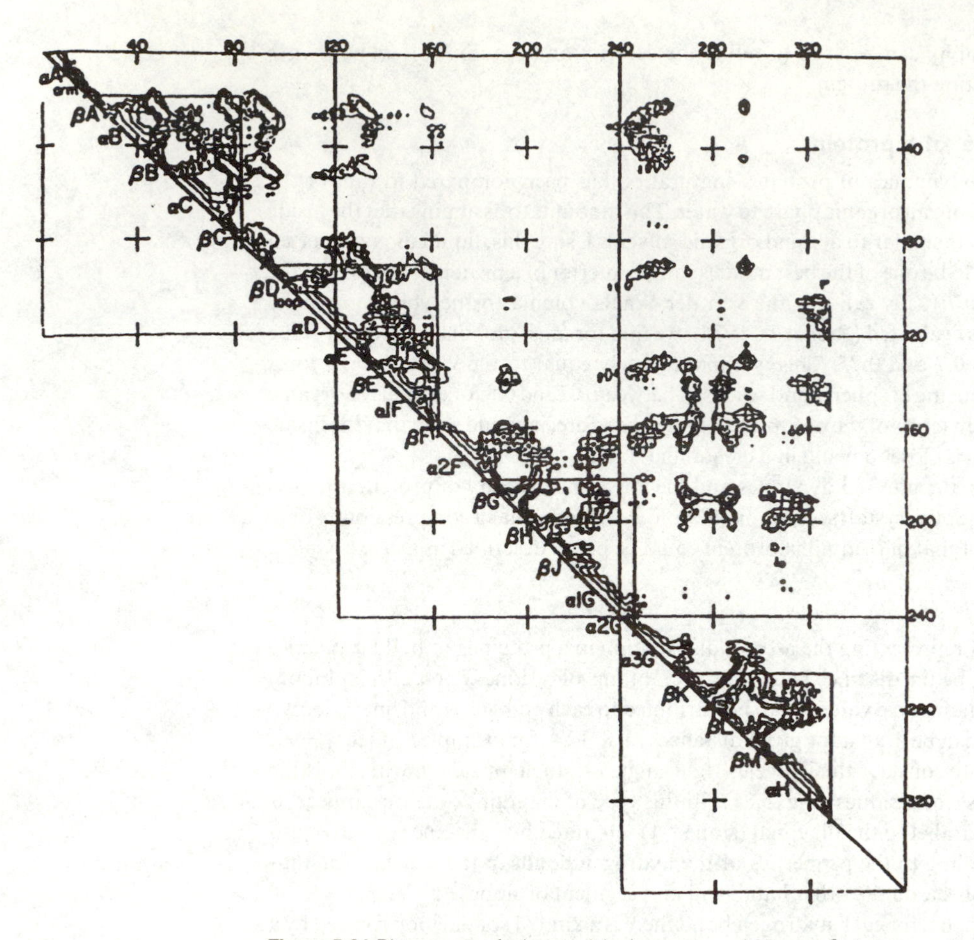

Figure 5.21 Distance map for lactate dehydrogenase. α-helix and β-sheet secondary structures are indicated along the diagonal. Contours are plotted for distances of 0, 4, 8,.... 16 Å. The four triangles near the diagonal define four protein domains. (From Rossmann and Liljas, 1974. Reprinted from *Journal of Molecular Biology* with the permission of Academic Press.)

continuous and bell-shaped because the interface area increases and the decreases as the point i goes from one end of the protein to the other. If, on th contrary, there are two or more different domains, the B interface must pas through a minimum when crossing the domain boundary. For this value of i th point separates two sub-units, which, having a minimum contact area, behave a two distinct domains. An examination of the curves $B(i)$ for a whole series c proteins enables the domains to be identified more systematically than by a simpl visual inspection of a three-dimensional protein model (Fig. 5.22).

The ratio A/A_G is calculated for each domain, A being the accessible area of th domain and A_G being the area of a 'globular' protein with the same mass M as tha of the domain. In fact:

$$A_G = 11.1M^{2/3} \qquad (A_G \text{ in Å}^2;\ M \text{ in daltons})$$

As soon as the domain differs from a globular shape, the ratio A/A_G become greater than 1. In fact, the ratio is generally found to be less than 1, indicating compact globular shape for the domain.

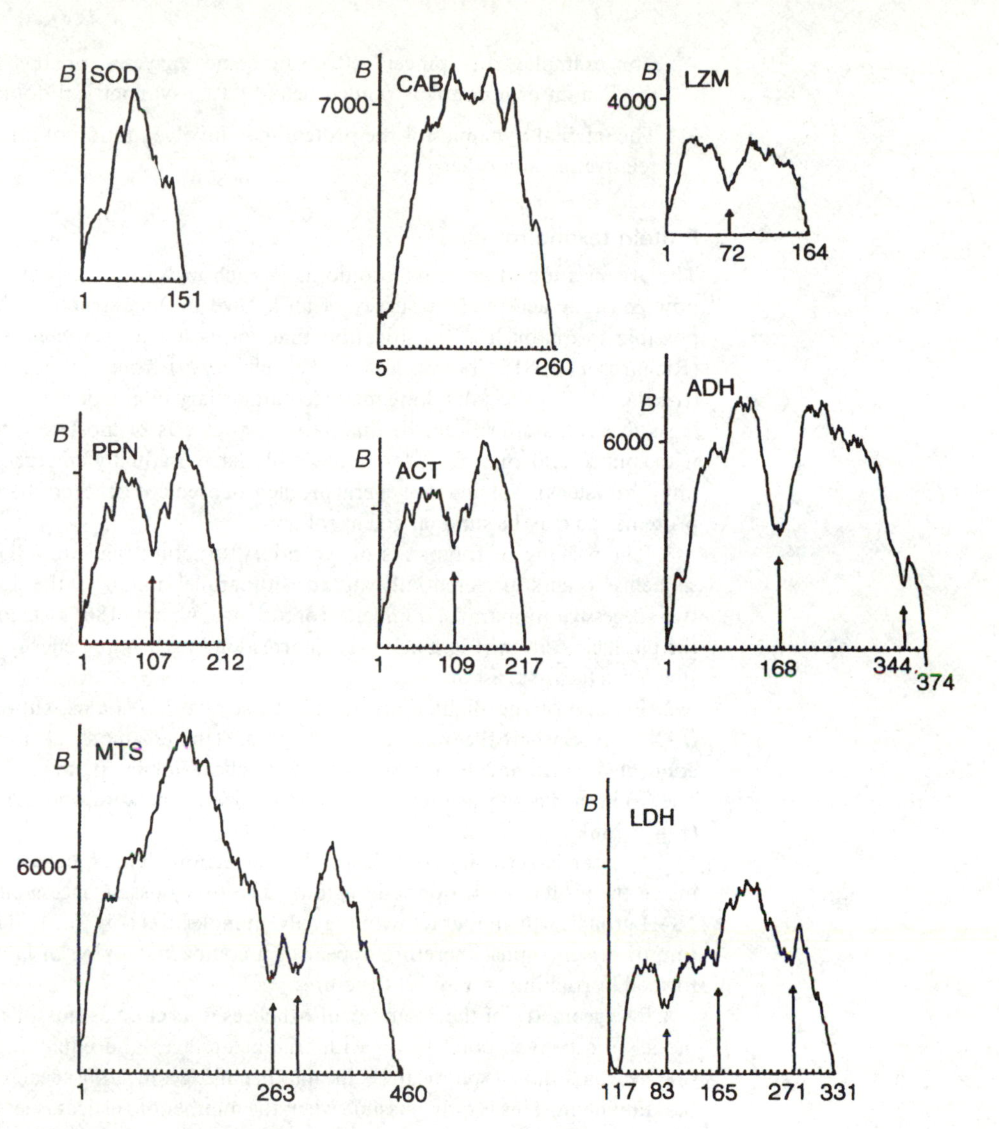

Figure 5.22 Variations in the *B* area of the (1, *i*) (*i*, *n*) interface with *i*: SOD, superoxide dismutase; CAB, carbonic anhydrase; LZM, hen lysozyme; PPN, papain; ACT, actinidin; ADH, alcohol dehydrogenase; MTS, methionyl-tRNA synthetase; LDH, lactate dehydrogenase. (From Janin and Wodak, 1983. Reprinted from *Progress in Biophysics and Molecular Biology*, vol. 42, pp. 21 – 78, © 1983, with kind permission from Elsevier Science Ltd., The Boulevard, Langford Lane, Kidlington, Oxford OX5 1GB, UK.)

The domain concept is important for three reasons:

1. The domain must be a stage in the folding of the peptide chain as soon as it is synthesized. In some cases, there may be a correspondence between gene splicing into exons and the splitting of the tertiary structure into domains.
2. A domain often defines a functional region of the protein. Two identical domains in two different proteins must have very similar if not identical functions. Two or more domains are sometimes necessary for the function

(for example, the 'pincer' effect in some enzymes or the recognition mechanism of a double-stranded helix by two symmetrical domains).

3. The internal dynamics of the protein may involve motion of all its domains relative to each other.

Protein taxonomy

The organization of proteins into domains each with its own special function is now generally accepted. A survey of all known protein structures has made it possible to establish a classification that forms a true *taxonomy* of proteins (Richardson, 1981; Chothia and Finkelstein, 1990) from secondary structures (α-helices and β-sheets), taking into account tertiary interactions.

As in any classification, the qualitative aspect was of most concern to begin with, but several rules for protein assembly have gradually emerged. These are important steps in solving the general problem of predicting the tertiary structure of proteins and may be summarized as follows:

1. The packing of fragments of secondary structure that are adjacent in the sequence occurs preferentially in an antiparallel manner: the loop joining two successive antiparallel fragments rotates through only 180° instead of 360° for the parallel position (Fig. 5.23a). The corresponding bending energy is therefore smaller in the first case.

2. Because of the slight right-handed twist of the β-chains, sub-units of the β–X–β type, where the two β-chains are parallel in the same sheet, are more easily connected by fragments of a right-handed helical linkage (i) than a left-handed one (ii) in such a way as once again to minimize the bending energy of the loops (Fig. 5.23b).

3. There are no crossings or knots in the connections. The connections cannot be inside the protein (hydrophobic region) since they possess accessible C=O and N–H groups, which interact with the solvent molecules (Fig. 5.24). The architecture of proteins must therefore appear as a compact stacking of layers that are formed by packing of α and β structures.

4. The geometry of the assembly of α-helices is as close as possible to an ideal model of a convex polyhedron with triangular faces and equal edges. Such a polyhedron is almost spherical, i.e. the number of edges meeting at each vertex must be a minimum. This is only possible when the number of vertices is less than 12. A polyhedron with $2n$ vertices can represent the packing of n helices. For a given polyhedron, each vertex can correspond to the end of a helix and each edge to the axis of a helix. As a result, several architectures are possible, but only a limited number. An octahedron will accommodate three helices, a dodecahedron four, a hexadecahedron five and an icosahedron six (Fig. 5.25). In all these assemblies, one 'face' of the helix turned towards the inside is hydrophobic, the other turned to the outside has enough polar groups to make it soluble in the aqueous medium.

Clearly, this classification cannot be applied to transmembrane proteins of ion channels where a bundle of α-helices passes right through the membrane. As we shall see later, however, the amphiphilic character of the α-helices is still present. The polar part may be turned towards the inside of the channel and involved in the mechanism of ionic transport.

5. β structures can be packed into sheets in a large number of ways because of their great flexibility. Again, because of the twist in the β chains, their assembly into sheets leads to a skewed surface. Instead of being assembled in twisted sheets, the

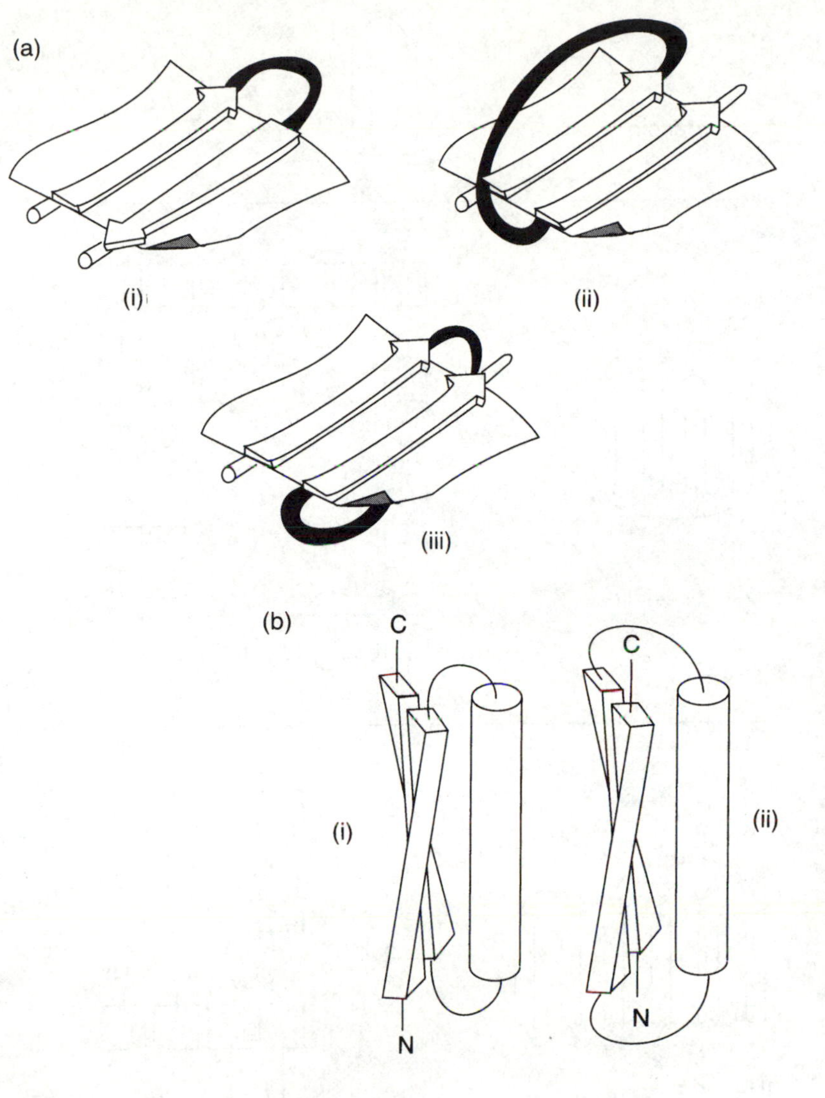

Figure 5.23 (a) The three main types of connection between β-strands: (i) hairpin or same end; (ii) cross-over or opposite end; and (iii) left-handed cross-over. (From Richardson, 1981. Reprinted from *Advances in Protein Chemistry* with the permission of Academic Press and the author)
(b) Connections between sub-units of type β–X–β where two β-chains are parallel in the same sheet. These are made more easily by fragments of right-handed helical linkage (i) than by fragments of left-handed linkage (ii). (From Chothia and Finkelstein, 1990. Reprinted with permission from *Annual Review of Biochemistry*, Vol. 59 and with the authors' permission © 1990 Annual Reviews Inc.)

β chains might also form a kind of 'barrel' incorporating 5–13 chains. Like elastic bandages, the inner diameter of the barrel will depend on the general twist of the surface. Figures give a better idea of this than a lengthy description (Figs 5.26 and 5.27).

6. Assembling α and β structures together leads to an apparently wide variety of spatial organizations (Figs 5.28–5.33) because the possible combinations of n_α α-chains and n_β β-chains differ from each other by only a small amount of free energy. The following comments are relevant to this.

(a) As the size of the protein increases, the difference between the surface area of the protein and that of a close-packed cubic stack of the same number of residues tends to zero. If V is the mean volume per residue (about 150 Å^3), the cube would have a side L_0 given by $(NV)^{1/3}$, where N is the number of residues. The specific surface area per residue of such a cubic packing would be $6L_0^2/N$ or about $170N^{1/3}$ in Å^3.

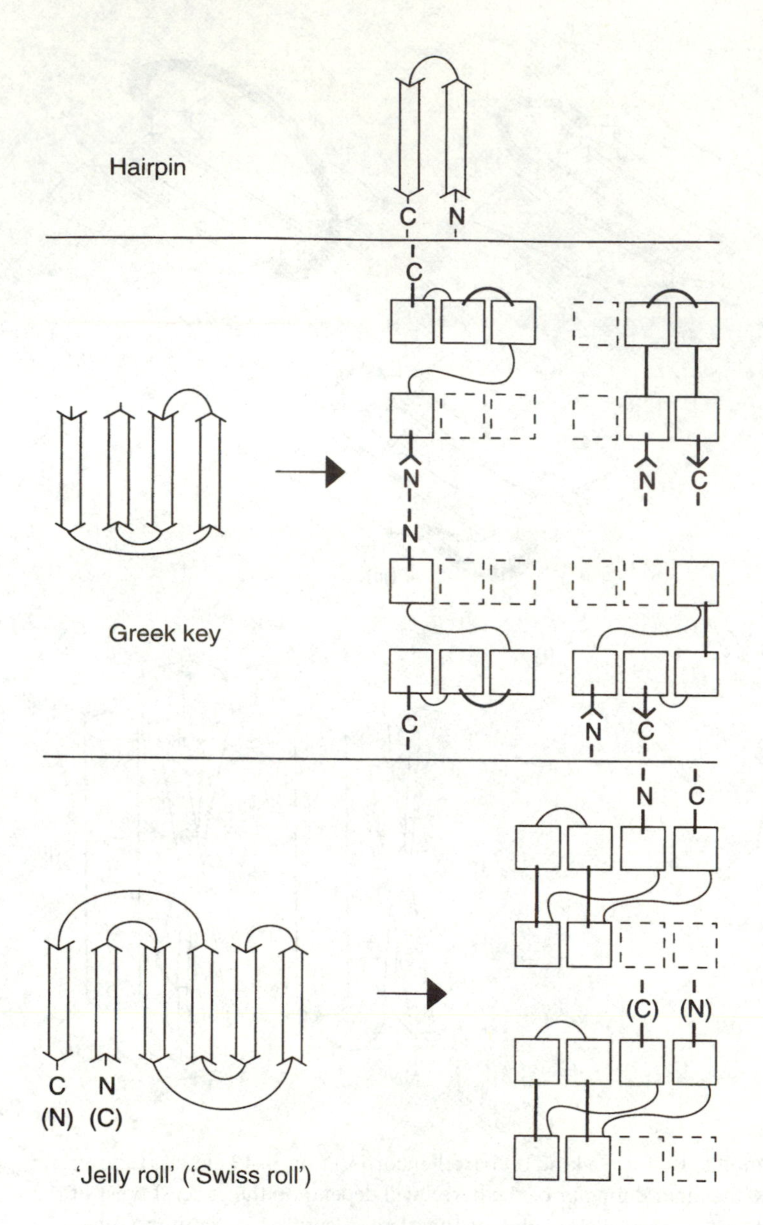

Figure 5.24 Chain topology in β-sheet proteins. (From Chothia and Finkelstein, 1990. Reprinted with permission from *Annual Review of Biochemistry*, vol. 59 and with the authors' permission. © 1990 Annual Reviews Inc.)

(b) Since the difference in free energy between several types of packing (including the cubic close-packed) is very small, the exact secondary structure is largely determined by local interactions inside structural domains themselves induced by the collapse of the tertiary structure. This explains the relative success of the predictions of secondary structure we examined above.

(c) Do the tertiary structures obtained from protein crystals after analysis of their X-ray diffraction pattern give a true or distorted image of the protein? We might say that the relatively loose packing of the molecules in the crystal leaves a large volume for the solvent medium, which may not be significantly different from the cellular medium in which the protein is working. On the other hand, it has been shown for some enzymes that

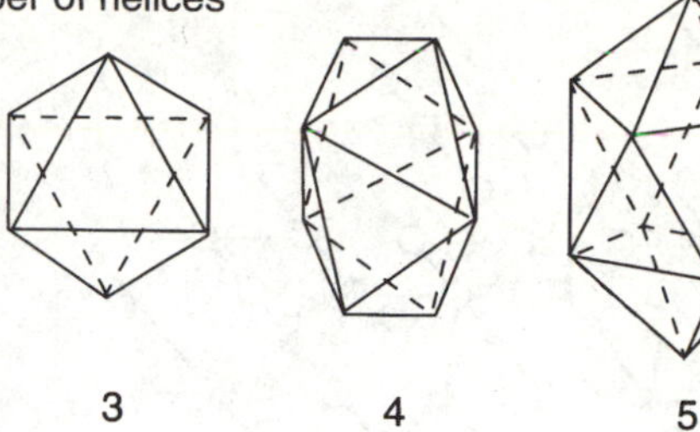

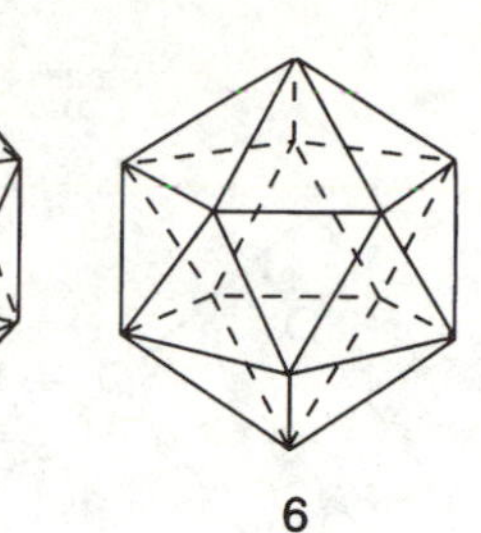

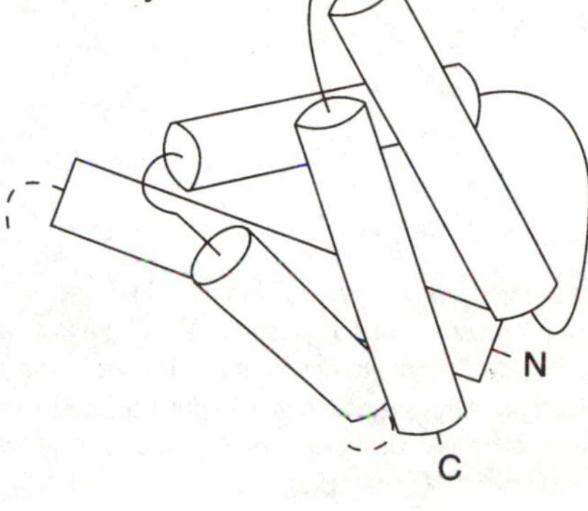

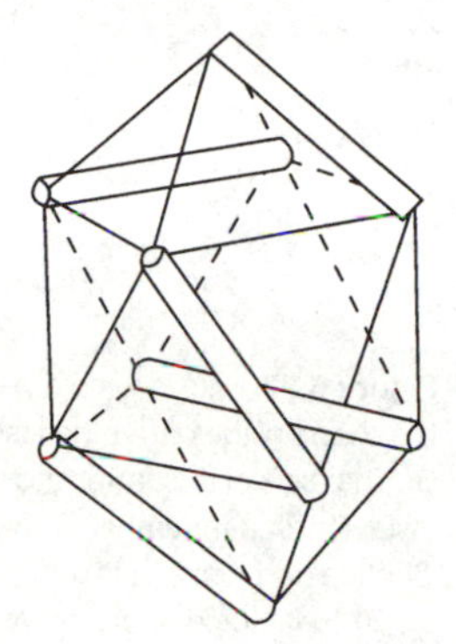

Figure 5.25 Polyhedra for ideal packing of α-helices. (From Chothia and Finkelstein, 1990. Reprinted with permission from *Annual Review of Biochemistry*, vol. 59 and with the authors' permission. © 1990 Annual Reviews Inc.)

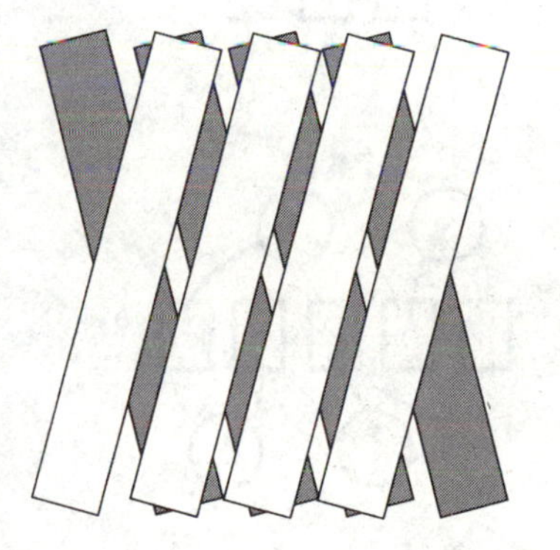

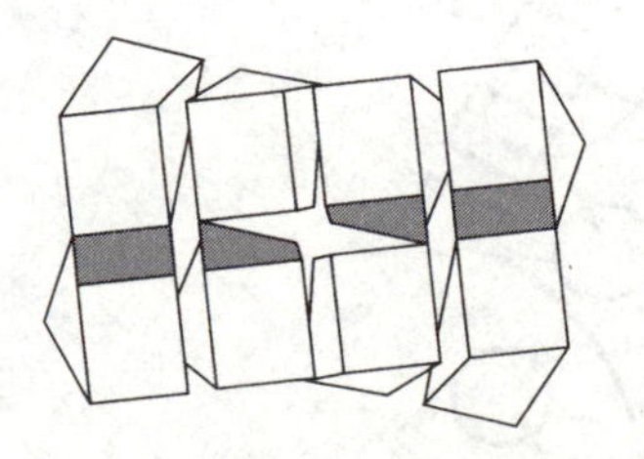

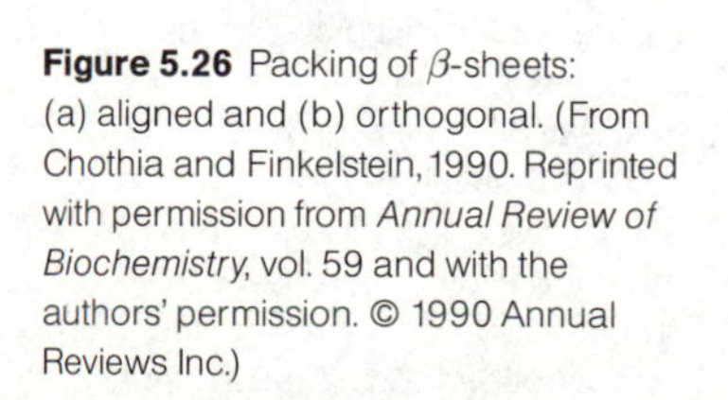

Figure 5.26 Packing of β-sheets: (a) aligned and (b) orthogonal. (From Chothia and Finkelstein, 1990. Reprinted with permission from *Annual Review of Biochemistry*, vol. 59 and with the authors' permission. © 1990 Annual Reviews Inc.)

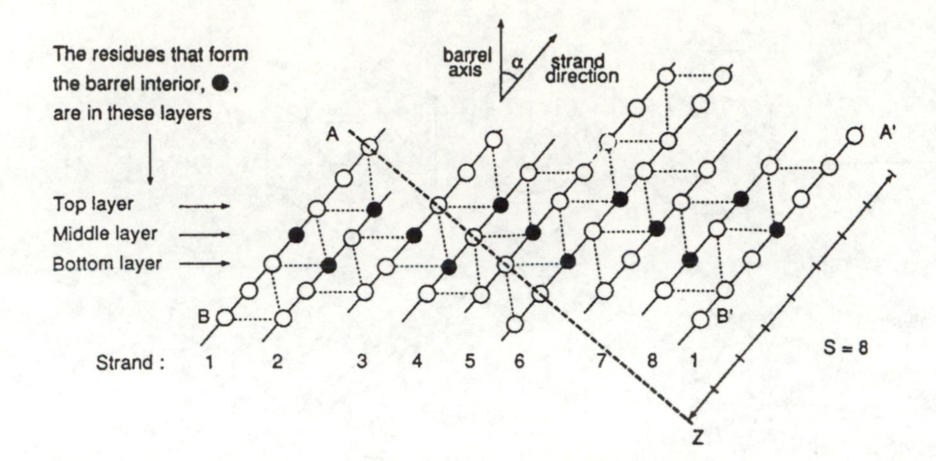

Figure 5.27 Packing into a β-barrel. Eight β-sheet strands are involved here. To get some idea the shape of the barrel, the first and last strands must be superimposed by sticking A to A′ and B′. The axes of the barrel and strand are inclined to each other at an angle α as indicated in the inset. The shear number S measures the stagger of the strands: here of eight residues. (From Chothia and Finkelstein, 1990. Reprinted with permission from *Annual Review of Biochemistry*, 59 and with the authors' permission. © 1990 Annual Reviews Inc.)

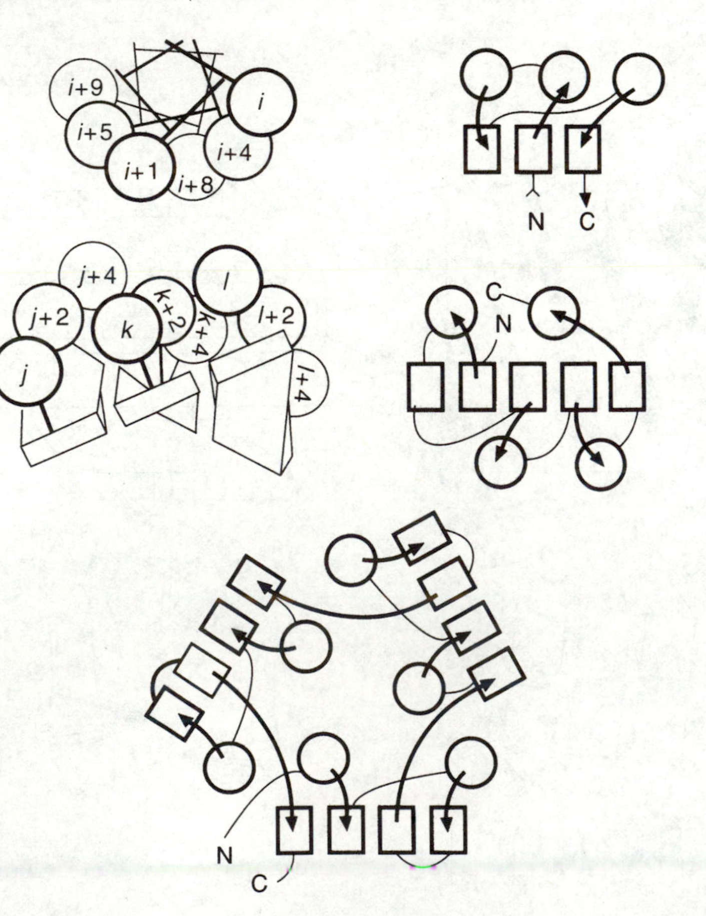

Figure 5.28 Packing of secondary structures in α/β-type proteins. Side chains of α-helices are denoted by i; those of β-sheets by j, k, l. (From Chothia and Finkelstein, 1990. Reprinted with permission from *Annual Review of Biochemistry*, vol. 59 and with the authors' permission. © 1990 Annual Reviews Inc.)

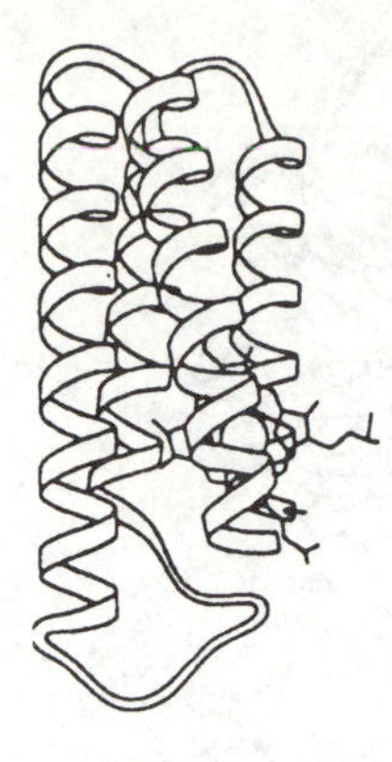

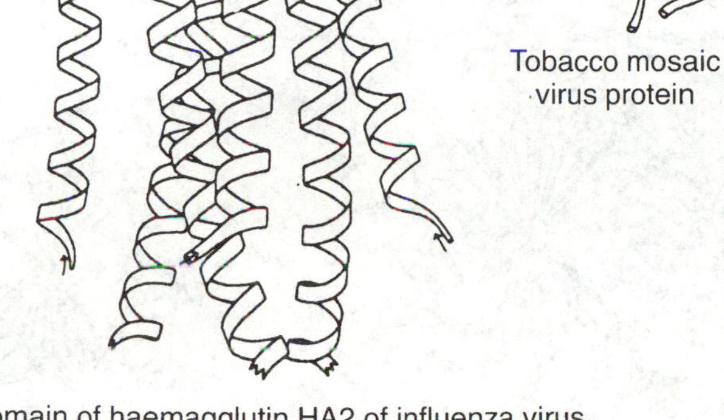

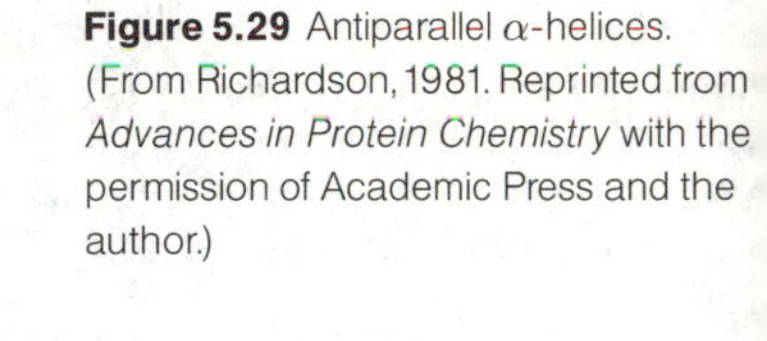

Figure 5.29 Antiparallel α-helices. (From Richardson, 1981. Reprinted from *Advances in Protein Chemistry* with the permission of Academic Press and the author.)

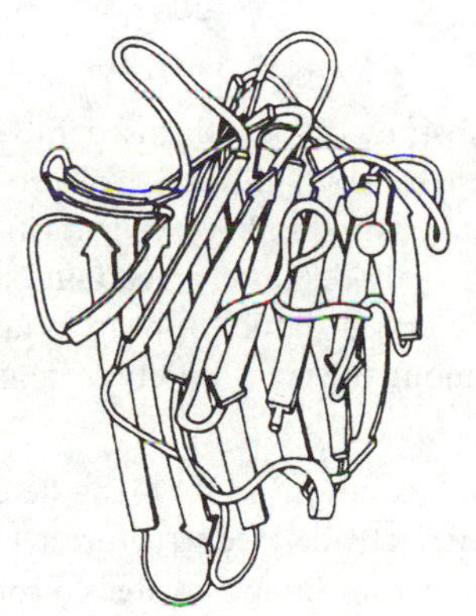

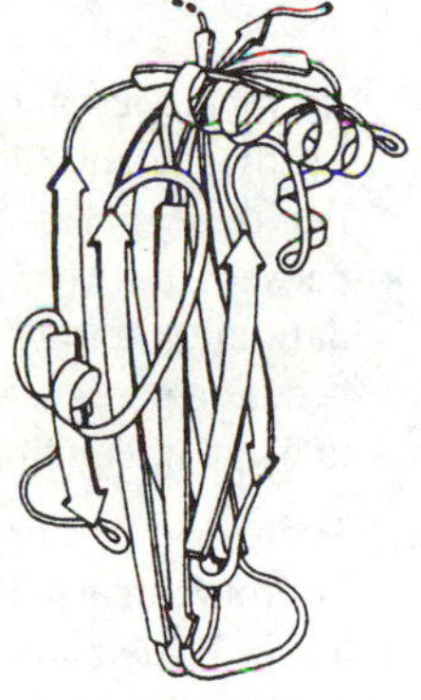

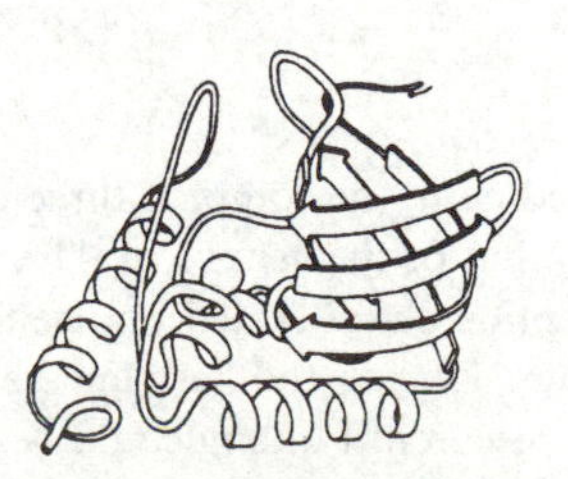

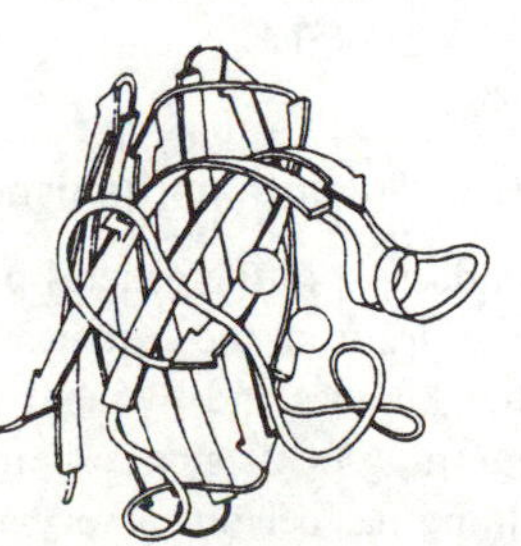

Figure 5.30 Antiparallel β structures. (From Richardson, 1981. Reprinted from *Advances in Protein Chemistry* with the permission of Academic Press and the author.)

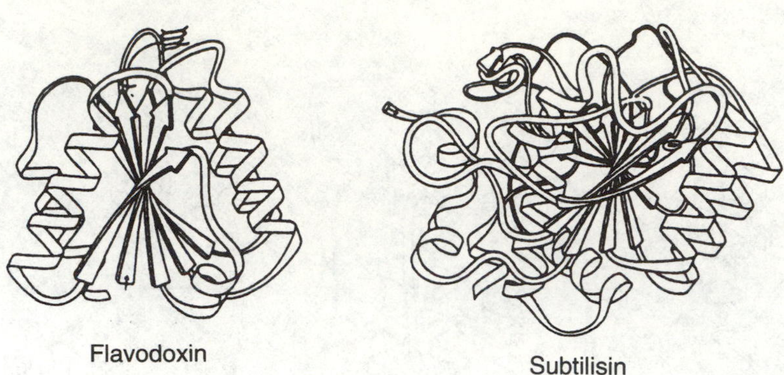

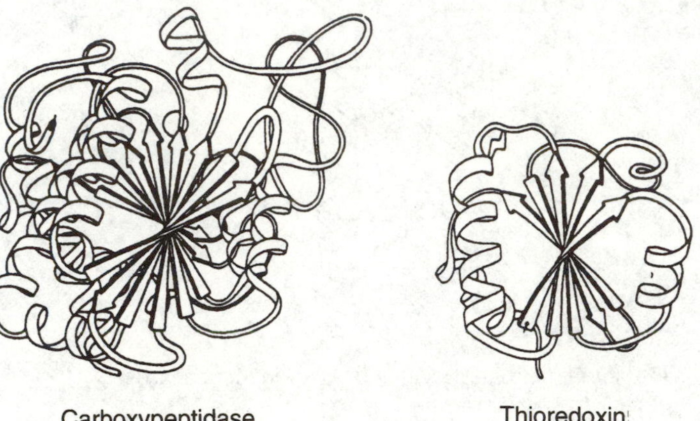

Figure 5.31 Parallel α and β structures. (From Richardson, 1981. Reprinted from *Advances in Protein Chemistry* with the permission of Academic Press and the author.)

enzymatic action takes place on a substrate that has diffused into the crystal. Lastly, a recent study of a small protein by X-ray diffraction of a crystal and by two-dimensional NMR in solution confirms the excellent structural fit between the two models built by each method. However, certain experimental data indicate that constraints of the crystal lattice appear to favour or to 'freeze' a particular conformation among the great variety of conformations in dynamic equilibrium in solution.

(d) Proteins giving crystals with a high resolution (< 2 Å) may be exceptions. A protein may not have a rigid and perfectly defined structure at every point. In the Fc domain of immunoglobulins, for example, there is disorder due to the existence of several possible configurations. Can such a disorder be related to the general idea that, in any organization of the living world, there is already in a macromolecule a margin of uncertainty that increases with the size of the system?

5.6.5 **Folding mechanisms**

In addition to the general problem of predicting the compact three-dimensional structure from the sequence, there is also that of the genesis of the structure. By what kinetics and by what stages is a peptide chain folded on itself at the very beginning of its synthesis in the ribosome? During the past few years, protein folding has become a very active field of research. Folding could be similar to a

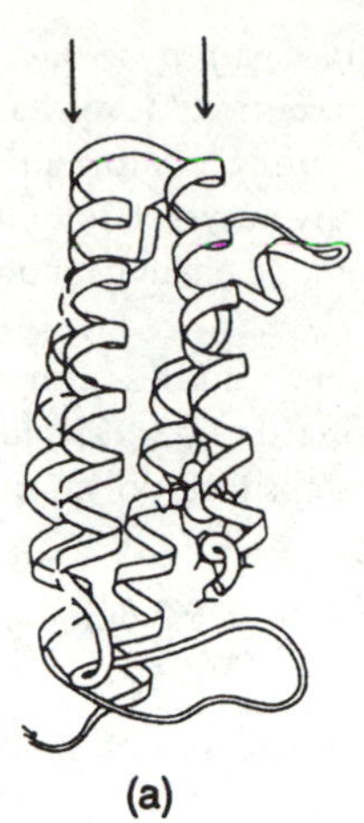

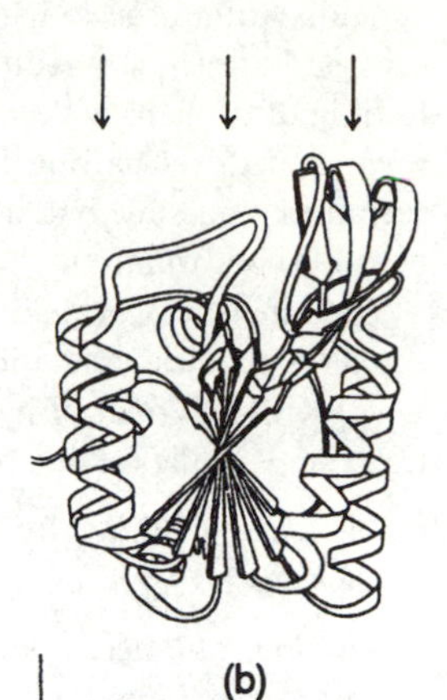

(c)

Figure 5.32 Examples of protein domains with different numbers of layers of backbone structure: (a) two-layer cytochrome c′; (b) three-layer phosphoglycerate kinase; and (c) four-layer triosephosphate isomerase. The arrows point to the backbone layers. (From Richardson, 1981. Reprinted from *Advances in Protein Chemistry* with the permission of Academic Press and the author.)

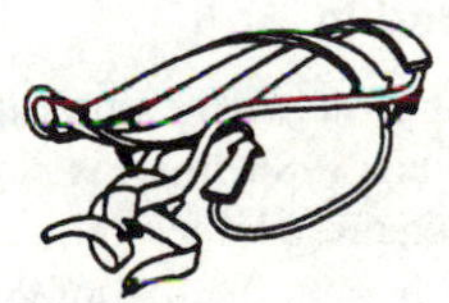

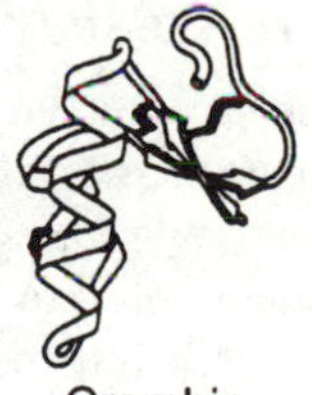

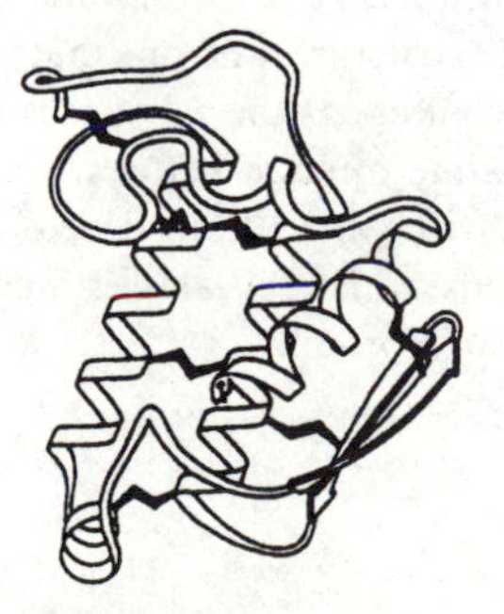

Figure 5.33 Disulphide-rich domains. (From Richardson, 1981. Reprinted from *Advances in Protein Chemistry* with the permission of Academic Press and the author.)

crystallization process with *nucleation centres* and a fast *propagation* from the centres. In practice, it is difficult to locate nucleation centres. Moreover, during t folding of proteins with disulphide bridges, many 'wrong' conformations may formed with mispairing between SH groups. Many ways of folding therefo seem to be possible but, as discussed above, the peptide chain cannot explore possible conformations.

The kinetics of protein folding is a problem recently addressed by using m ecular dynamics simulations on a three-dimensional lattice (Karplus and Sa 1995) with a 27-mer of hydrophobic and polar groups (HP polymer). There a three steps in the folding process:

- a rapid collapse from the random coil state to a random semicompa globule;
- a slow rate-determining search through the semi-compact states to find transition state;
- a rapid folding from this transition state to the native one.

Unexpectedly, only a small number of configurations would be capable of c lapsing into compact states. They correspond to a large enough energy g between the initial unfolded state and the final folded one and do not reflec particular HP sequence.

This new approach, although at present limited to small proteins, looks ve promising and could be decisive in finding a satisfactory explanation for assembly of tertiary structures.

Chaperone proteins have recently been discovered and isolated. Through stro interactions with 'nascent' peptide chains, they can favour the assembly of stable final tertiary structure by keeping the peptide chain inside a kind of 'bask and thus greatly reducing the number of possible configurations. In a rece structure determination, chaperonin appears in fact to be a hollow cylinder (Ant *et al.*, 1996).

5.6.6 The structure–function relationship

When evolutionary trees are drawn up for proteins, they appear to indicate alm no natural selection (Ptitsyn and Volkenstein, 1986); there is simply an 'imprinti of random processes. Why is the peptide chain so large compared to its active si Its main role is to 'protect' this site from thermal fluctuations. Once incorpora into a structure that can only be modified at the price of a large free-energy chan the active site preserves a well-defined geometry essential for its functioning. other words, a particular three-dimensional structure is not essential for t function but simply *a* three-dimensional structure, and one that is not determin by the enzymatic function. Such a concept opens the way to the creation of co pletely new proteins and at the same time offers a better understanding of origin of life. A parallel can be drawn with calculations developed from a blo copolymer of polar and non-polar regions, which reveal the similarity between architecture of the copolymer and that of a protein.

Part I Conformation of biopolymers

Conclusion

In this first Part dealing with the conformation of biopolymers, we have placed the emphasis on proteins and nucleic acids, which are now considered as macromolecules essential to cell life and, as such, are still widely studied. If we had decided to cover the polysaccharides as well, and more generally all molecules carrying a glycan group, an additional chapter would have been necessary. The wide variety of three-dimensional structures adopted by glycans shows their major role in many recognition mechanisms. For example, specific interactions between particular proteins known as *lectins* capable of specific recognition of sugars are important processes in intercellular recognition. At present, however, there are only *descriptions* of interactions like these between sugars and macromolecules. Complete structures are still only poorly known and modelling is still in its infancy. That is the reason why we have for the moment omitted discussion of this new field of research. It may nevertheless soon be possible to unravel a recognition code relying on structural complementarities.

As was mentioned in the Preface, we have deliberately avoided any systematic discussion of techniques or physical methods. In the special case of conformational studies, X-ray diffraction by biopolymer crystals, two- or three-dimensional NMR of macromolecule solutions, electron microscopy and electron diffraction are the main experimental methods, and these have been the basis of a large number of results presented and analysed. It is not impossible that more recent approaches, such as atomic force microscopy or near-field optical microscopy, might also become methods of choice in some cases.

Apart from that, however, a general remark should be made about the first three methods mentioned above. The 'appearance' of a macromolecule is always regarded as an essential source of information for the biochemist, and modelling it on a computer screen to program the building of new molecules is, in a similar way, becoming invaluable.

We should remember, however, that the molecular 'appearance' is always a reconstruction from experimental data, which never allow us to have a 'direct' view of the macromolecule. With X-ray diffraction, NMR and microscopy, the observable quantity is related to the observed molecular quantity through a Fourier transform. With X-rays and microscopy, this mathematical transform occurs implicitly in the diffraction equation, which establishes a general relationship between a vector $\boldsymbol{r}$ in molecular space and a vector $\boldsymbol{s}$ in the observed space.

The phase of the scattered wave is given by the scalar product $2\pi \boldsymbol{r} \cdot \boldsymbol{s}$. The vecto has a magnitude $2\pi\theta/\lambda$, where 2θ is the angle between the direction of observati and the incident radiation and λ is the wavelength of the radiation, so that s has t dimensions of L^{-1}. For a given phase, as the modulus of $\boldsymbol{r}$ decreases, so that o increases. An examination of large values of $\boldsymbol{s}$ thus gives information about t atomic structure of macromolecules. The 'resolution' of a structure from a d fraction pattern in fact depends on the quality of the diffraction at the wid possible angle from the track of the incident beam.

If, in the near future, X-ray lasers are developed, we can anticipate that hol graphy will restore a three-dimensional view of the crystal structure being studie In this case, the Fourier transform would be given directly by the optical dev without any of the constraints due to the loss of phases inherent in current metho

With NMR, it is the measurement of the resonance frequencies v that allow given proton to be located in the molecule. In pulsed NMR, the observable i function of time t, and similarly the two functions v and t are related by a Four transform in which this time the phase is given by $2\pi vt = \omega t$.

A final comment concerns the general problem of the relationship betwe structure (or conformation) and biological function. The three-dimensio structure mentioned above as needed by the biologist and the model-maker is in fa only a first approximation. Such a static description is no longer sufficient. Polym dynamics in all its aspects is essential for understanding the role of the polymer a predicting its function. In almost all specific interactions, one conformation chosen by the partners that are present from amongst a large collection of equa probable conformations in the energy range defined by thermal movement.

Part I Conformation of biopolymers

References and further reading

Biopolymers generally

Cantor, C.R. and Schimmel, P.R. (1980). *Biophysical chemistry*, part III. Freeman, San Francisco.

Chu, B. (1967). *Molecular forces*. Wiley, New York.

Flory, P.J. (1969). *Statistical mechanics of chain molecules*, 1st edn. Carl Hanser Verlag, Munich.

Flory, P.J. (1989). *Statistical mechanics of chain molecules*, 3rd edn. Wiley, New York.

Hirschfelder, J.O., Curtiss, C.F. and Bird, R.B. (1967). *Molecular theory of gases and liquids*. Wiley, New York.

Kollman, P.A. and Allen, L.C. (1972). The theory of the hydrogen bond. *Chemical Review*, **22**, 283–303.

Pimental, G.C. and McClellan, A.L. (1971). Hydrogen bonding. *American Review of Physical Chemistry*, **22**, 347–85.

Schellman, J.A. (1958). The factors affecting the stability of hydrogen-bonded polypeptide structures in solution. *Journal of Physical Chemistry*, **62**, 1485.

Volkenstein, M.V. (1963). *Configurational statistics of polymer chains*. Wiley, New York.

Weiner, S.J., Kollman, P.A., Case, D.A., Singh, U.C., Ghio, C., Alagona, C. *et al.* (1984). A new force field for molecular mechanical simulation of nucleic acids and proteins. *Journal of the American Chemical Society*, **108**, 765–84.

Zimm, B.H. and Bragg, J.K. (1959). Theory of the phase transition between helix and random coil in polypeptide chains. *Journal of Chemical Physics*, **31**, 526.

Nucleic acids

Altona, C. and Sundaralingam, M. (1972). Conformational analysis of the sugar ring in nucleosides and nucleotides. A new description using the concept of pseudorotation. *Journal of the American Chemical Society*, **94**, 8205–12.

Arber, W. and Linn, S. (1969). DNA modification and restriction. *Annual Review of Biochemistry*, **38**, 467–500.

Betts, J.A., Josey, J.A., Veal, J.M. and Jordan, S.R. (1995). A nucleic acid triple helix formed by a peptide nucleic acid–DNA complex. *Science*, **270**, 1838–41.

Brenner, S., Barnett, L., Crick, F.H.C. and Orgel, A. (1961). The theory of mutagenesis. *Journal of Molecular Biology*, **3**, 121–4.

Cech, T.R. (1990). Self-splicing of group I introns. *Annual Review of Biochemistry*, **59**, 543–68.

Cech, T.R. and Bass, B.L. (1986). Biological catalysis by RNA. *Annual Review of Biochemistry*, **55**, 599–629.

Cech, T.R., Zaug, A.J. and Grabowski, P.J. (1981). In vitro splicing of the ribosomal RNA precursor of *Tetrahymena*; involvement of a guanosine nucleotide in the excision of the intervening sequence. *Cell*, **27**, 487–96.

Courey, E.J. and Wang, J.C. (1983). Cruciform formation in a negatively supercoiled DNA may be kinetically forbidden under physiological conditions. *Cell*, **33**, 817–29.

Crick, F.H.C. (1976). Linking numbers and nucleosomes. *Proceedings of the National Academy of Sciences, USA*, **73**, 2369–2643.

Crick, F.H.C. and Klug, A. (1975). Kinky helix. *Nature*, **255**, 530–3.

Crothers, D.M. (1968). Calculation of binding isotherms for heterogeneous polymers. *Biopolymers*, **6**, 575–84.

Crothers, D.M., Kallenbach, N. and Zimm, B.H. (1965). The melting transition of low molecular weight DNA: theory and experiments. *Journal of Molecular Biology*, **11**, 802–20.

Delarue, M. and Moras, D. (1980). RNA structure. *Nucleic Acids and Molecular Biology*, **3**, 182–96.

Depew, R.E. and Wang, J.C. (1975). Conformational fluctuations of DNA helix. *Proceedings of the National Academy of Sciences, USA*, **72**, 4275–9.

Dickerson, R.E. (1989). Definition and nomenclature of nucleic acids structure components. *Nucleic Acids Research*, **17**, 1797–1803.

Dickerson, R.E. and Drew, H.R. (1981). Structure of a B-DNA dodecamer: influence of base sequence on helix structure. *Journal of Molecular Biology*, **149**, 761–86.

Drew, H.R. and Dickerson, R.E. (1981). Structure of a B-DNA dodecamer: geometry of hydration. *Journal of Molecular Biology*, **151**, 535–56.

Dumas, P., Moras, D., Florentz, C., Giegé, R., Verlaan, P., van Belkum, A. *et al.* (1987). 3D graphics modelling of the tRNA-like 3′ end of TMV RNA: structural and functional implications. *Journal of Biomolecular Structure and Dynamics*, **4**, 707–27.

Ellison, M.J., Kelleher, R.J., Wang, A.H.J., Habener, J.F. and Rich, A. (1985). Sequence-dependent energetics of the B–Z transition in supercoiled DNA containing nonalternating purine–pyrimidine sequences. *Proceedings of the National Academy of Sciences, USA*, **82**, 8320–4.

Gabarro, J. (1978). Numerical analysis of thermal denaturation of nucleic acids. *Analytical Biochemistry*, **91**, 309–22.

Gold, L., Polisky, B., Uhlenbeck, O. and Yarus, M. (1995). Diversity of oligonucleotide function. *Annual Review of Biochemistry*, **64**, 763–97.

Guschlbauer, W., Chantot, J.F. and Thiele, D. (1990). Four-stranded nucleic acid structures 25 years later: from guanosine gels to telomer DNA. *Journal of Biomolecular Structure and Dynamics*, **8**, 491–511.

Htun, H. and Dahlberg, J.E. (1989). Topology and formation of triple-stranded H-DNA. *Science*, **243**, 1571–6.

Jain, S.C., Tsai, C.-C. and Sobell, H.M. (1977). Visualization of drug–nucleic acid interactions at atomic resolution. II: Structure of an ethidium/dinucleoside monophosphate crystalline complex, ethidium:5-iodocytidylyl(3′–5′)guanosine. *Journal of Molecular Biology*, **114**, 317–31.

Kang, C.H., Zhang, X., Ratliff, R., Moyzis, R. and Rich, A. (1992). Crystal structure of four-stranded *Oxytricha* telomeric DNA. *Nature*, **356**, 126–31.

Kennard, O. and Hunter, W.N. (1989). Oligonucleotide structure: a decade of results from single crystal X-ray diffraction studies. *Quarterly Review of Biophysics.*, **22**, 327–79.

Kim, S.H., Sussmann, J.L., Suddath, F.L., Quigley, G.J., McPherson, A., Wang, A.H.J. *et al.* (1974). The general structures of tRNA molecules. *Proceedings of the National Academy of Sciences, USA*, **71**, 4970–4.

Kratky, O. and Porod, G. (1949). Röntgenuntersuchung gelöster Fadenmolekule. *Recueil des Travaux Chimiques*, **68**, 1106.

Lerman, L.S. (1961). Structural considerations in the interaction of DNA and acridines. *Journal of Molecular Biology*, **3**, 18–30.

Luger, K., Mäder, A.W., Richmond, R.K., Sargent, D.F. and Richmond, T.J. (1997). Crystal structure of the nucleosome core particle at 2.8 Å resolution. *Nature*, **389**, 251–60.

McCall, M., Brown, T. and Kennard, O. (1985). The crystal structure of d(GGGGCCCC). A model for poly(dG)poly(dC). *Journal of Molecular Biology*, **183**, 385–96.

Maxam, A.M. and Gilbert, W. (1977). A new method for sequencing DNA. *Proceedings of the National Academy of Sciences, USA*, **74**, 560–64.

Moras, D., Comarmond, M.B., Fischer, J., Thierry, J.C., Ebel, J.P. and Giegé, R. (1980). Crystal structure of $tRNA^{Asp}$. *Nature*, **288**, 669–74.

Muzard, G., Théveny, B. and Révet, B. (1990). Electron microscopy mapping of pBR322 DNA curvature. Comparison with theoretical models. *The EMBO Journal*, **9**, 1289–98.

Peck, L.J. and Wang, J.C. (1983). Energetics of B–Z transition in DNA. *Proceedings of the National Academy of Sciences, USA*, **80**, 6206–10.

Pinck, M., Yot, P., Chapeville, F. and Duranton, H. (1970). Enzymatic binding to the 3′ end of TYMV RNA. *Nature*, **226**, 954–6.

Pleij, C.W.A., Rietwald, K. and Bosch, L. (1985). A new principle of RNA folding based on pseudoknotting. *Nucleic Acids Research*, **13**, 1717–23.

Pley, H.W., Flaherty, K.M. and McKay, D.B. (1994). Three-dimensional structure of a hammerhead ribozyme. *Nature*, **372**, 68–74.

Pohl, F.M. and Jovin, T.M. (1972). Salt-induced cooperative conformational change of a synthetic DNA: equilibrium and kinetic studies with poly(dG-dC). *Journal of Molecular Biology*, **67**, 375–96.

Puglisi, J.D., Wyatt, J.R. and Tinoco, I. (1991). RNA pseudoknots. *Accounts of Chemical Research*, **24**, 152–8.

Pulleyblank, D.E., Schure, M., Tang, D., Vinograd, J. and Vosberg, J. (1975). Action of nicking–closing enzyme on super-coiled and nonsupercoiled closed circular DNA: formation of a Boltzmann distribution of topological isomers. *Proceedings of the National Academy of Sciences, USA*, **72**, 4280–4.

Quigley, G.J. and Rich, A. (1976). Structural domains of transfer RNA molecules. *Science*, **194**, 796–806.

Reiss, C. and Arpa-Gabarro, T. (1977). Thermal transition spectroscopy: a new tool for submolecular investigation of biological macromolecules. *Progress in Molecular and Subcellular Biology*, **5**, 1–30.

Richmond, T.J., Finch, J.T., Rushton, B., Rhodes, D. and Klug, A. (1984). Structure of the nucleosome core particle at 7 Å resolution. *Nature*, **311**, 532–7.

Robertus, J.D., Ladner, J.E., Finch, J.T., Rhodes, D., Clark, B.F.C. and Klug, A. (1974). Structure of yeast $tRNA^{Phe}$ at 3 Å resolution. *Nature*, **250**, 546–51.

Sanger, F., Nicklen, S. and Coulson, A.R. (1977). DNA sequencing with chain-terminating inhibitors. *Proceedings of the National Academy of Sciences, USA*, **74**, 560–4.

Shieh, H.S., Berman, H.M., Dabrow, M. and Neidle, S. (1980). The structure of drug–deoxydinucleoside phosphate complex: generalized conformational behaviour of interaction complexes with RNA and DNA fragments. *Nucleic Acids Research*, **8**, 85–97.

Sobell, H.M., Tsai, C.-C., Jain, S.C. and Gilbert, S.G. (1977). Visualization of drug–nucleic acid interactions at atomic resolution. III: Unifying structural concepts in understanding drug–DNA interactions and their broader implications in understanding protein–DNA interactions. *Journal of Molecular Biology*, **114**, 333–65.

Stern, S., Weiser, B. and Noller, H.F. (1988). Model for the three-dimensional folding of 16S ribosomal RNA. *Journal of Molecular Biology*, **204**, 447–81.

Streisinger, G., Okada, Y., Emrich, J., Newton, J., Tsugita, A., Terghazi, E. *et al.* (1966). Frameshift mutation and the genetic code. *Cold Spring Harbor Symposia on Quantitative Biology*, **31**, 77–84.

Sutcliffe, J.G. (1979). Complete nucleotide sequence of the *E. coli* pBR322. *Cold Spring Harbor Symposia on Quantitative Biology*, **43**, 77–90.

Tinoco, I., Uhlenbeck, O.C. and Levine, M.D. (1971). Estimation of secondary structure in RNA. *Nature*, **230**, 362–7.

Tsai, C.-C., Jain, S.C. and Sobell, H.M. (1977). Visualization of drug–nucleic acid interactions at atomic resolution. I: Structure of an ethidium/dinucleoside monophosphate crystalline complex, ethidium:5-iodouridylyl(3′–5′)adenosine. *Journal of Molecular Biology*, **114**, 301–15.

Turner, D.H., Sugimoto, N., Jaeger, J.A., Longfellow, G.E., Freier, S.M. and Kierzer, R. (1987). Improved parameters for prediction of RNA structure. *Cold Spring Harbor Symposia on Quantitative Biology*, **52**, 123–33.

Von Kitzing, E., Lilley, D.M.J. and Diekmann, S. (1990). The stereochemistry of a four-way DNA junction: a theoretical study. *Nucleic Acids Research*, **18**, 2671–83.

Wada, A., Yabuki, S. and Husimi, Y. (1980). Fine structure in the thermal denaturation of DNA: high temperature-resolution spectrophotometric studies. *CRC Critical Review of Biochemistry*, **9**, 87–144.

Wang, A.H.J., Quigley, G.J., Kolpak, F.J., Rawford, J.L.C., van Boom, J.H., van der Marel, G. and Rich, A. (1979). Molecular structure of a left-handed double helical DNA fragment at atomic resolution. *Nature*, **282**, 680–6.

Wang, J.C., Peck, L.J. and Becherer, K. (1983). DNA supercoiling and its effects on DNA structure and function. *Cold Spring Harbor Symposia on Quantitative Biology*, **47**, 85–91.

Watson, J.D. and Crick, F.H.C. (1953a). The structure of DNA. *Cold Spring Harbor Symposia on Quantitative Biology*, **18**, 123–31.

Watson, J.D. and Crick, F.H.C. (1953b). A structure for deoxyribonucleic acid. *Nature*, **171**, 737–8.

Woodson, S.A. and Crothers, D.M. (1988). Binding of 9-aminoacridine to bulged-base DNA oligomers from a frameshift hotspot. *Biochemistry*, **27**, 8904–14.

Proteins

Anton, A.A., Dogson, E.J. and Dogson, G.G. (1996). Circular assemblies. *Current Opinion in Structural Biology*, **6**, 142–50.

Argos, P. (1987). Analysis of sequence-similar pentapeptides in unrelated protein tertiary structures. *Journal of Molecular Biology*, **197**, 331–48.

Baker, E.N. and Hubbard, R.E. (1984). Hydrogen bonding in globular proteins. *Progress in Biophysics and Molecular Biology*, **44**, 97–179.

Baldwin, R.L. (1986). Temperature dependence of the hydrophobic interaction in protein folding. *Proceedings of the National Academy of Sciences, USA*, **83**, 8069–72.

Burley, S.K. and Petsko, G.A. (1985). Aromatic–aromatic interactions: a mechanism of protein structure stabilization. *Science*, **229**, 23–8.

Burley, S.K. and Petsko, G.A. (1988). Weakly polar interactions in proteins. *Advances in Protein Chemistry*, **39**, 125–89.

Chakrabartty, A., Kortemme, T. and Baldwin, R.L. (1994). Helix propensities of the amino acids measured in alanine-based peptides without helix-stabilizing side-chain interactions. *Protein Science*, **3**, 843–52.

Chothia, C. and Finkelstein, A.V. (1990). The classification and origins of protein folding patterns. *Annual Review of Biochemistry*, **59**, 1007–39.

Chou, P.Y. and Fasman, G. (1974). (1) Conformational parameters for amino acids in helical, β-sheet and random coil regions calculated from proteins. (2) Prediction of protein conformation. *Biochemistry*, **13**, 211–45.

Creighton, T.E. (1978). Experimental studies of protein folding and unfolding. *Progress in Biophysics and Molecular Biology*, **33**, 231–97.

Creighton, T.E. (1990). Protein folding. *Biochemical Journal*, **270**, 1–16.

Dill, K.A. (1990). Dominant forces in protein folding. *Biochemistry*, **29**, 7155–8133.

Dill, K.A., Bromberg, S., Yue, K., Fiebig, K.M., Yee, D.P., Thomas, P.D. *et al.* (1995). Principles of protein folding. A perspective from simple exact models. *Protein Science*, **4**, 561–602.

Fano, R. (1961). *Transmission of information*. Wiley, New York.

Gaboriaud, C., Bissery, V., Benchetrit, T. and Mornon, J.P. (1987). Hydrophobic cluster analysis: an efficient new way to compare and analyse amino acid sequences. *FEBS Letters*, **224**, 149–55.

Gibrat, J.F., Garnier, J. and Robson, B. (1987). Further developments of protein secondary structure prediction using information theory. *Journal of Molecular Biology*, **198**, 425–43.

Holley, L.H. and Karplus, M. (1989). Protein secondary structure prediction with a neural network. *Proceedings of the National Academy of Sciences, USA*, **86**, 152–6.

Janin, J. and Wodak, S.J. (1983). Structural domains in proteins and their role in the dynamics of protein function. *Progress in Biophysics and Molecular Biology*, **42**, 21–78.

Karplus, M. and Sali, A. (1995). Theoretical studies of protein folding and unfolding. *Current Opinion in Structural Biology*, **5**, 58–73.

Kim, P.S. and Baldwin, R.L. (1990). Intermediates in the folding reaction of small proteins. *Annual Review of Biochemistry*, **59**, 631–60.

Pauling, L., Corey, R.B. and Brandon, H.R. (1951). The structure of proteins: two hydrogen-bonded helical configurations of the polypeptide chain. *Proceedings of the National Academy of Sciences, USA*, **37**, 205–11.

Privalov, P.L. (1979). Stability of proteins. *Advances in Protein Chemistry*, **33**, 167–241.

Privalov, P.L. and Gill, S.J. (1988). Stability of protein structure and hydrophobic interaction. *Advances in Protein Chemistry*, **39**, 191–234.

Ptitsyn, O.B. and Volkenstein, M.V. (1986). Protein structures and neutral theory of evolution. *Journal of Biomolecular Structure and Dynamics*, **4**, 137–56.

Qian, N. and Sejnowski, T.J. (1988). Predicting the secondary structure of globular proteins using neural network models. *Journal of Molecular Biology*, **202**, 865–84.

Richards, F.M. (1974). The interpretation of protein structures: total volume, group volume, distribution and packing density. *Journal of Molecular Biology*, **82**, 1–14.

Richardson, J.S. (1981). The anatomy and taxonomy of protein structure. *Advances in Protein Chemistry*, **34**, 167–339.

Robson, B. and Pain, R.H. (1971). Analysis of the code relating sequence to conformation in proteins: possible implication for the mechanism of formation of helical regions. *Journal of Molecular Biology*, **58**, 237–59.

Rossmann, M.G. and Liljas, A. (1974). Recognition of structural domains in globular proteins. *Journal of Molecular Biology*, **85**, 177–181.

Scheraga, H.A. (1984). Protein structure and function: from a colloidal to a molecular view. *Carlsberg Research Communications*, **49**, 1–55.

Schulz, G.E. (1988). A critical evaluation of methods for prediction of protein secondary structure. *Annual Review of Biophysics and Biophysical Chemistry*, **17**, 1–21.

Shakhnovich, E.I. and Gutin, A.M. (1993). Engineering of stable and fast-folding sequences of model proteins. *Proceedings of the National Academy of Sciences, USA*, **90**, 7195–9.

Stryer, L. (1988). *Biochemistry*, 3rd edn. Freeman, San Francisco.

Sturtevant, J.M. (1977). Heat capacity and entropy changes in processes involving proteins. *Proceedings of the National Academy of Sciences, USA*, **74**, 2236–40.

Part II

Dynamics of biopolymers

Introduction

The form adopted by biological entities, the subject of Part I, is one of the basic features of the living world; another feature, independent of the first and immediately perceptible to any observer, is that of *movement*. The immobility of death (at least on the scale of the individual being) is in marked contrast to the continual movement characteristic of life. Motion and life are intimately bound together in all aspects of the world around us.

Clearly, not all motion is an indication of life: the 'dialogue between wind and sea' is merely a very complex phenomenon combining aerodynamics and hydrodynamics, one that is difficult to predict because of the large number of parameters involved and the need to know the initial conditions (which may be more or less random). Similarly, evidence of life is not contained in the ruffling of plants by a breeze or a gale, but rather in the slow movements associated with pod-shedding or phototropism, whose dynamics can be studied using time-lapse photography.

The movements made by *living organisms* range from the rapid fluttering of insect wings to the slow rotation of sunflower heads. Motion at the level of *individual organs or cells* is exemplified by heartbeats, electrical brain waves and the undulations of cilia or flagella. At a microscopic level, there are the movements involved in cell division with its well-orchestrated chromosome ballet, in DNA transcription and replication with their impressive retinues of spatially and temporally coordinated enzymes, in metabolic cycles, in membrane channels opening to let through a flow of ions, and in the processing of substrates by enzymes: all these provide examples of *macromolecular dynamics or kinetics*. On an even smaller scale, that of *atomic groups within macromolecules*, we can still detect very rapid movements: of amino acids in proteins, base pair opening in nucleic acids, conformational change in rhodopsin upon the arrival of a photon, and so on.

The time scales required to describe all these phenomena span 15 orders of magnitude, but there are often functional connections between phenomena taking place at widely different rates. At each level in the description of a living organism a *functional order* (Careri *et al.*, 1975) appears to exist, bringing together several structures working to a well-defined aim. The function associated with each structure can only be defined through its interactions with the others. Functional order thus becomes a set of correlations between a series of biochemical events, which can then be described by observable or computable physical quantities. Their time variations correspond to the various dynamic levels described above.

The intramolecular level

The classical static description at this level consists either of an average value of atomic positions or of a configuration fixed preferentially by a set of constraints such as those found in a crystal. As with any atomic assembly, a biopolymer is in fact a system fluctuating with time. Such fluctuations arise from continuous thermal motion, i.e. an exchange of energy with the external medium through never-ending molecular collisions. The mean thermal energy is $k_B T$, which, under standard conditions ($T = 298$ K or 25°C) with $k_B = 1.385 \times 10^{-23}$ J K^{-1}, is equal to 4.04×10^{-21} J or about 2.4 kJ mol^{-1}. The amplitudes of the thermal fluctuations follow a Maxwellian distribution like the distribution of molecular velocities in a gas. Fluctuations of the order of $2k_B T$, $3k_B T$, $4k_B T$, etc., can thus occur with decreasing but significant probabilities according to the normal distribution. The energy exchanged in this way at a given instant allows a biopolymer to undergo conformational changes, subject to all the geometrical and energetic constraints of the atomic assembly forming the biopolymer. If we know all the parameters of the interaction laws described in Part I, we can calculate the thermally induced motion and model the molecular dynamics. Conversely, the internal molecular dynamics can be studied and measured using many physical techniques described later in Chapter 9 (fluorescence, electron or nuclear magnetic resonance, proton exchange, inelastic Rayleigh scattering, etc.), allowing the interaction parameters to be determined with precision.

For rapid movements, atoms or atomic groups can be considered as moving with small amplitudes in a viscous pseudo-fluid (e.g. the rotation of a tyrosine aromatic ring about the valence bond, or the puckering of a deoxyribose). For slower movements, with greater amplitudes, all the forces holding together the average macromolecular structure have to be taken into account. In this case, parts of the molecule may move as a whole (for example, motion about bonds acting as 'hinges') and the viscosity of the medium in which the macromolecule is dissolved then introduces a significant damping factor.

Investigating the internal dynamics of biopolymers therefore entails a comparison of experimental data with models that simulate molecular movements as simply as possible.

The molecular and intermolecular levels

The thermal motion of the solvent molecules surrounding the macromolecule causes random displacements of the latter's centre of gravity and random rotations of the whole molecule. Both the translational and rotational movements are characteristic of *Brownian motion*. This is the normal way in which a molecule moves in the cellular medium in the absence of any other external force. If there is an external force capable of being described as a chemical or electrical potential gradient, the motion of the molecule obeys Newton's laws but the underlying Brownian motion has to be regarded as a random force added to the others (drag force and frictional force).

The equation of classical mechanics derived from Newton's laws is transformed into a Langevin equation capable of describing molecular transport in a fluid medium. In fact, as we shall see, it is often better to abandon the microscopic description provided by the Langevin equation and to seek solutions of a partial differential equation. Such solutions require a knowledge of initial conditions and boundary conditions. Coupling with other types of transport or with chemical reactions can also be introduced. The functions of space and time forming solutions of this equation enable us to describe the whole of a cellular domain and its evolution with time. As demonstrated by Prigogine and his school, stable periodic structures can persist far from equilibrium. As functions of space, they may provide a first explanation of morphogenetic processes. As functions of time, they might explain the mechanisms underlying many biological clocks.

6

Fluctuations around an equilibrium conformation

6.1 The harmonic oscillator model

In order to describe the rapid movements of atoms about a valence bond as well as the oscillatory motion of macromolecular domains (which is slower and has a larger amplitude), it is natural to choose the *harmonic oscillator* as a model. This is a simple mechanical system frequently used as a model in a large number of physical problems. It consists of a particle of mass m capable of frictionless motion along the x axis and experiencing a restoring force directly proportional to the distance x between the current position of the mass and its equilibrium position taken as the origin O (Fig. 6.1).

Putting the restoring force $F = -kx$, where k is the *spring constant* (SI unit $\mathrm{N\,m^{-1}}$), the equation of motion takes the form

$$m\,\mathrm{d}^2x/\mathrm{d}t^2 + kx = 0 \tag{6.1}$$

Solutions to this equation are known to be sinusoidal functions such as

$$x = x_0 \sin(\omega t + \phi) \tag{6.2}$$

or, in complex form,

$$x = x_0 \exp[\mathrm{i}(\omega t + \phi)] \tag{6.3}$$

where ω is the *pulsatance* or *angular frequency* in rad $\mathrm{s^{-1}}$, related to the frequency ν in Hz by $\omega = 2\pi\nu$.

Substituting eqn 6.2 or 6.3 into eqn 6.1 gives

$$k = m\omega^2 = 4\pi^2\nu^2 m \tag{6.4}$$

so that the angular frequency ω is directly related to the mechanical constants m and k of the oscillator. The quantities x_0 and ϕ in eqn 6.2 or 6.3 depend on the initial conditions: the velocity and position at $t = 0$.

Figure 6.1 The harmonic oscillator of mass m has a restoring force proportional to its displacement from its equilibrium position O.

The kinetic energy of the harmonic oscillator $E_k = \frac{1}{2}mv^2$ can be obtained from the velocity

$$v = dx/dt = x_0\omega\cos(\omega t + \phi)$$

Since

$$v^2 = x_0^2\omega^2\cos^2(\omega t + \phi) = x_0^2\omega^2(1 - x^2/x_0^2)$$

we obtain

$$E_k = \tfrac{1}{2}m\omega^2(x_0^2 - x^2) \tag{6.5}$$

The potential energy under the restoring force is clearly $E_p = \frac{1}{2}kx^2$ and from eqn 6.4:

$$E_p = \tfrac{1}{2}m\omega^2 x^2 \tag{6.6}$$

The total energy of the oscillator is therefore

$$E_t = E_k + E_p = \tfrac{1}{2}m\omega^2 x_0^2 = \tfrac{1}{2}kx_0^2 \tag{6.7}$$

During the oscillations, the total energy remains constant but there is a continuous exchange between kinetic and potential energy. This process can be illustrated by plotting the energy $E_p = \frac{1}{2}kx^2$ against x as in Fig. 6.2, where the horizontal lines represent the total energy E_t. The maximum amplitude x_0 is defined by the points where the straight line representing E_t intersects the parabola. The kinetic energy at these points is zero, and at the origin O it is a maximum. At any point M, the potential energy is given by the value of E and the kinetic energy by the segment MP.

Before going any further, three important features of the harmonic oscillator need to be brought out.

6.1.1 The shape of the potential well

For a one-dimensional harmonic oscillator, the variation of potential energy with x is represented by a parabola that becomes 'flatter' as the constant $k = m\omega^2$ becomes smaller. More precisely, with a potential energy of the order of $k_B T$, the maximum amplitude x_0 is given by

$$x_0 = \omega^{-1}(2k_B T/m)^{1/2} \tag{6.8}$$

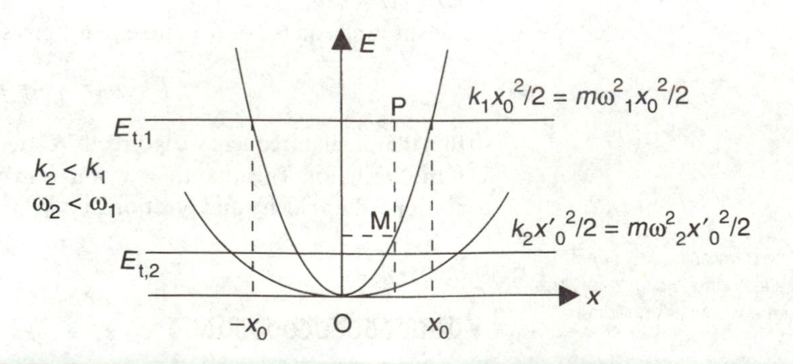

Figure 6.2 Energy diagram for harmonic oscillators.

Given an oscillator with a mass of 100 daltons or $(1/6) \times 10^{-24}$ kg and an angular frequency $\omega = 10^{12}$, and taking $k_B T = 4 \times 10^{-21}$ J, we obtain $x_0 = 0.22$ nm. Equation 6.8 shows that the deviation x_0 from the equilibrium position becomes large with smaller values of ω.

6.1.2 **Strong damping**

This is usually the case with internal oscillations of groups inside the macromolecule, the damping resulting from various interactions, particularly from the viscous coupling with the solvent. It is represented by a frictional constant f in such a way that the frictional force F_f is given by

$$F_f = fv = f \, dx/dt \tag{6.9}$$

The solution of the damped oscillator equation reveals two types of damping conditions characterized by two decay functions $\exp(-\beta t)$ and $\exp(-\gamma t)$ describing the return to equilibrium, where

$$\beta = [f + (f^2 - 4km)^{1/2}]/2m \tag{6.10a}$$

$$\gamma = [f - (f^2 - 4km)^{1/2}]/2m \tag{6.10b}$$

The exponential functions can be written in the forms $\exp(-t/\tau_1)$ and $\exp(-t/\tau_2)$, defining two relaxation times

$$\tau_1 = \beta^{-1} \qquad \text{and} \qquad \tau_2 = \gamma^{-1} \tag{6.11}$$

When $f^2 \gg 4km$, β and γ tend to the limiting values

$$\beta = f/m \qquad \text{with} \qquad \tau_1 = m/f \tag{6.12}$$

$$\gamma = k/f \qquad \text{with} \qquad \tau_2 = f/k \tag{6.13}$$

The quantity τ_1 is the limiting time occurring in the solution of the Langevin equation (see Chapter 7). Its physical significance will emerge when studying Brownian motion in the presence of an external force: as soon as $t > \tau_1$, the particle moves uniformly.

Returning to the previous example of a mass of 100 daltons, f can be replaced to a first approximation by $k_B T/D_T$, where D_T is the coefficient of diffusion for an ion in solution, say about $10^{-9}\,\text{m}^2\,\text{s}^{-1}$. This gives $f^2 = 16 \times 10^{-24}$ and $km = m^2\omega^2 = (1/36) \times 10^{-24}$, so in fact $f^2 \gg 4km$. We thus obtain $\tau_1 = 4 \times 10^{-13}$ s and $\tau_2 = 24 \times 10^{-12}$ s. The angular frequency ω would have to be 10 times as great for τ_1 and τ_2 to have the same order of magnitude. In other words, the random movements of the oscillator are 10 times as rapid as its natural frequency. Such a strongly damped harmonic oscillator could thus also be described as a Brownian particle obeying Langevin's equation. We shall encounter this process later in molecular dynamics.

6.1.3 **The concept of an attractor**

The energy of a harmonic oscillator can also be expressed in terms of two variables: p, the linear momentum; and x, the distance from the equilibrium position. In

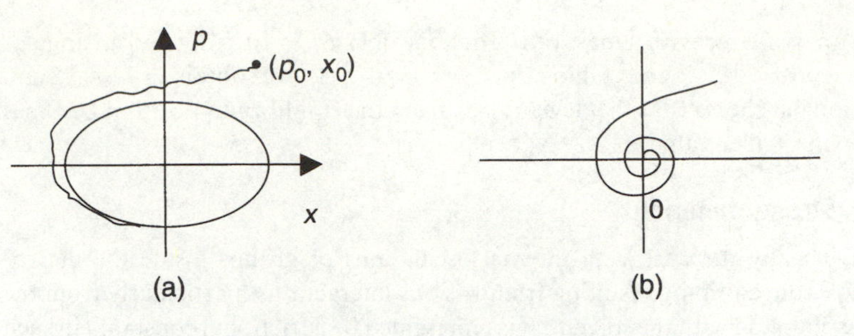

Figure 6.3 Motion of an oscillator plotted in two-dimensional phase space: (a) sustained oscillations; (b) damped oscillations.

that case,

$$E_k = p^2/2m \qquad \text{and} \qquad E_p = kx^2/2 \tag{6.14}$$

If the motion is represented by a point moving in a *two-dimensional phase space* defined by the orthogonal coordinates (p, x), eqn 6.7 for the total energy is replaced by

$$p^2/2m + kx^2/2 = \text{constant} \tag{6.15}$$

which is the equation of an ellipse with an axial ratio of k/m. The smaller the value of k/m, the greater the elongation of the ellipse along the x axis.

Any point on the ellipse corresponds to the state of the oscillator at a given time, assuming that E_t is constant, i.e. that the oscillations are *sustained*. Whatever the initial conditions of the oscillator may be (represented by the point (p_0, x_0) in Fig. 6.3a), the point (p, x) giving the state of the oscillator will describe a curve in phase space that approaches the ellipse of the sustained oscillator with a constant energy E_t as $t \rightarrow \infty$. The ellipse is defined as an *attractor* in phase space (Fig. 6.3a).

For a non-sustained damped oscillator, on the other hand, the final state is at the origin (0, 0) and the diagram in phase space will be a spiral ending at the origin, now regarded as the attractor point (Fig. 6.3b). The concept of an attractor can be extended to much more complex systems, leading to a classification of certain periodic motions such as the electric waves emitted by the brain.

6.2 Coupled oscillators and normal modes

In many cases, particularly those involving the oscillatory motion of electrons or atoms inside a molecule, isolated oscillators of the kind examined above do not exist. Instead, there are many oscillators with interactions or *coupling* between them. In the limiting case of very strong coupling, the motion taking place is collective. For example, the vibrations of an atom on both sides of its equilibrium position are the result of coupling with many other atoms through the relevant valence bonds. Because of the coupling, the frequencies differ markedly from those of isolated oscillators and must be calculated from sets of coupled equations.

Two coupled oscillators

We consider first the simplest problem of two identical coupled oscillators, a problem whose results reveal all the properties of a more complex system. Let each oscillator have a mass m and a spring constant k and let them be coupled through a third spring with a constant c. The system parameters are m, k and c. We show in

Box 6.1 Study of two coupled harmonic oscillators

The system is illustrated in Fig. 6.4, where the masses are displaced from their equilibrium positions by x_1 and x_2 and where the damping is assumed to be negligible. The equations of motion for each mass are

$$m\,\mathrm{d}^2x_1/\mathrm{d}t^2 = -kx_1 + c(x_2 - x_1) \tag{1}$$

$$m\,\mathrm{d}^2x_2/\mathrm{d}t^2 = -kx_2 - c(x_2 - x_1) \tag{2}$$

giving the following pair of equations:

$$m\,\mathrm{d}^2x_1/\,\mathrm{d}t^2 + (k + c)x_1 = cx_2 \tag{3a}$$

$$m\,\mathrm{d}^2x_2/\,\mathrm{d}t^2 + (k + c)x_2 = cx_1 \tag{3b}$$

We assume solutions of the form $x = \exp[\mathrm{i}(\omega t + \phi)]$ since $e^{\mathrm{i}\omega t}$ will be a factor in every term and will cancel throughout. We then obtain

$$-m\omega^2 \exp(\mathrm{i}\phi_1) + (k + c)\exp(\mathrm{i}\phi_1) = c\exp(\mathrm{i}\phi_2) \tag{4a}$$

$$-m\omega^2 \exp(\mathrm{i}\phi_2) + (k + c)\exp(\mathrm{i}\phi_2) = c\exp(\mathrm{i}\phi_1) \tag{4b}$$

so that

$$k + c - m\omega^2 = c\exp[\mathrm{i}(\phi_2 - \phi_1)] \tag{5a}$$

$$k + c - m\omega^2 = c\exp[\mathrm{i}(\phi_1 - \phi_2)] \tag{5b}$$

Multiplying the last two equations, we obtain

$$(k + c - m\omega^2)^2 = c^2$$

which gives the following two equations to determine ω:

$$k + c - m\omega^2 = \pm c \tag{6}$$

from which we obtain the two solutions

$$\omega_{\mathrm{A}}^2 = k/m \qquad \text{and} \qquad \omega_{\mathrm{B}}^2 = (k + 2c)/m \tag{7}$$

The first solution corresponds to the natural frequency of each spring ($\omega_{\mathrm{A}} = \omega_0$). Substituting eqn 7 into eqn 5 gives the corresponding values of ϕ. For ω_{A}, $\phi_1 = \phi_2$ and the oscillators are in phase. For ω_{B}, $\phi_1 = \phi_2 + \pi$ (remembering that $e^{\mathrm{i}\pi} = -1$) and the oscillators are in phase opposition.

Box 6.1 that there are two distinct solutions for the angular frequency given by

$$\omega_{\mathrm{A}} = (k/m)^{1/2} \qquad \text{and} \qquad \omega_{\mathrm{B}} = [(k + 2c)/m]^{1/2} \tag{6.16}$$

In the first case (ω_{A}), the oscillators are in phase: the centre spring is not deformed and exerts no force on the masses, which move as if they were uncoupled. In the second case (ω_{B}), the oscillators are in phase opposition and the centre spring is alternately stretched and compressed. These two solutions are known as the two *vibrational normal modes* of the system. Any linear combination of them is also a solution because the differential equations are linear.

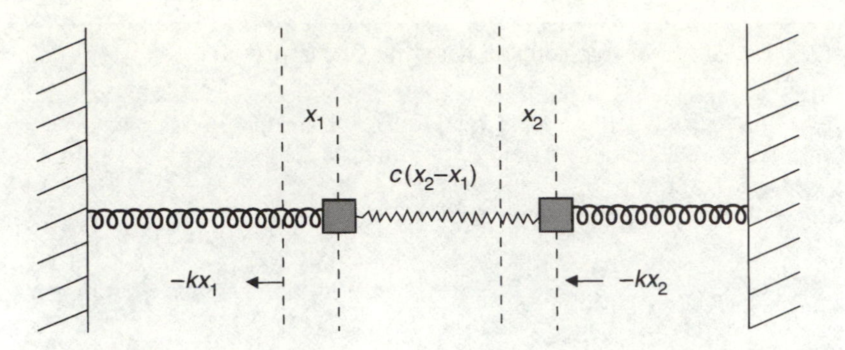

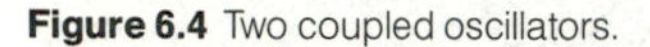

Figure 6.4 Two coupled oscillators.

Many coupled oscillators

Extending the treatment to n coupled oscillators (see Box 6.2) shows that there are then n normal modes. As examples, consider the CO_2 and H_2O molecules. There are three normal modes in both cases, in all of which the centre of gravity of the molecule must remain at rest.

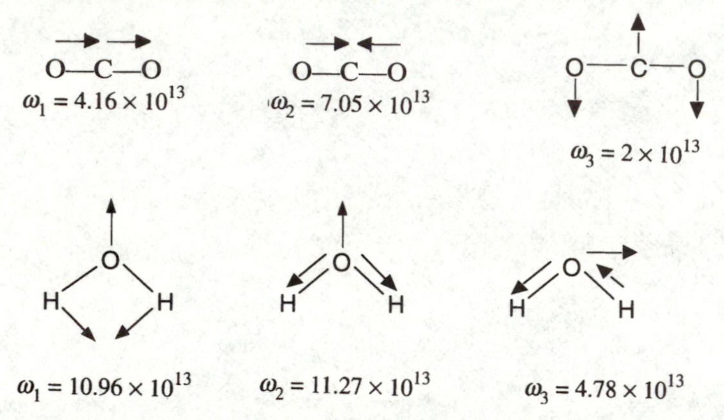

When these molecules are subjected to a forced harmonic oscillation, resonance will occur when the applied frequency equals one of the normal mode frequencies (known as eigenfrequencies). Absorption of energy occurs at these frequencies, giving rise to an *absorption spectrum*. This is quite simply the same as the vibrational spectrum of the molecule and it can be detected by infrared or Raman spectroscopy.

Three-dimensional oscillators

In order to extend the treatment of the set of one-dimensional coupled oscillators to a three-dimensional system, there are two problems to be faced.

The first is whether the internal dynamics of a molecule can be studied in the framework of classical mechanics rather than that of quantum mechanics. To examine this, consider one of Heisenberg's uncertainty relations:

$$\Delta E \Delta \tau \approx \hbar \qquad (\hbar = h/2\pi) \tag{6.17}$$

where ΔE is an increment of energy and $\Delta\tau$ a time interval. In macromolecules, $\Delta E \approx k_B T \approx 4 \times 10^{-21}$ J at room temperature, which leads to $\Delta\tau \approx 2 \times 10^{-14}$ s ($h = 1.054 \times 10^{-34}$ J s). Because atomic velocities inside molecules are much lower than those in the gaseous state ($< 10^3$ m s^{-1}), the displacement during a time

Box 6.2 **Many coupled oscillators**

When there is a large number of coupled oscillators, all with their movements in the same direction, eqns 3 in Box 6.1 can easily be generalized by noting that the equations are of the form

$$\mathrm{d}^2x_i/\mathrm{d}t^2 + \sum_j f_{ij}x_j = 0 \tag{1}$$

(For two oscillators $i = 1$ or 2, $f_{11} = (k+c)/m = f_{22}$, $f_{12} = f_{21} = -c/m$.)

The n linear equations in eqn 1 can be written in matrix form as

$$(\mathrm{d}^2\boldsymbol{x}_i/\mathrm{d}t^2) = (-\boldsymbol{f}_{ij})(\boldsymbol{x}_j) \tag{2}$$

where the left-hand side and the second factor on the right are column vectors and $(\boldsymbol{f}_{ij})$ is a square matrix with the number of rows and columns equal to the number of oscillators. If the matrix $(\boldsymbol{f}_{ij})$ had only diagonal elements $-\lambda_i$, eqn 2 would reduce to a set of simple differential equations:

$$\mathrm{d}^2x_i/\mathrm{d}t^2 = -\lambda_i x_i \tag{3}$$

whose solutions would be oscillatory motions with angular frequencies $\omega_i = \lambda_i^{1/2}$. Substitution of eqn 3 into eqn 1 would give equations of the type

$$(f_{11} + \lambda_1)x_1 + f_{12}x_2 + \cdots = 0$$

Thus, to diagonalize $\boldsymbol{f}_{ij}$ we only have to equate to zero its determinant with λ added to each diagonal element:

$$\begin{vmatrix} f_{11}+\lambda & f_{12} & \cdots & f_{1i} & \cdots & f_{1n} \\ f_{21} & f_{22}+\lambda & & & & \\ \vdots & & \ddots & & & \vdots \\ f_{i1} & & & f_{ii}+\lambda & & \\ \vdots & & & & \ddots & \vdots \\ f_{n1} & & \cdots & & \cdots & f_{nn}+\lambda \end{vmatrix} = 0$$

This is an equation of the nth degree whose n solutions $\lambda_1, \lambda_2, \ldots$ define all the possible modes of vibration of the system of n oscillators. These are the *normal modes*, equal in number to the number of oscillators.

We can check the calculation by applying it to the case already studied in Box 6.1. Eqns 3 in Box 6.1 can be written in matrix form as

$$\begin{pmatrix} \dfrac{\mathrm{d}^2x_1}{\mathrm{d}t^2} \\ \dfrac{\mathrm{d}^2x_2}{\mathrm{d}t^2} \end{pmatrix} = \begin{pmatrix} -(k+c)/m & c/m \\ c/m & -(k+c)/m \end{pmatrix} \begin{pmatrix} x_1 \\ x_2 \end{pmatrix}$$

Diagonalization leads to a quadratic equation:

$$[-(k+c)/m + \lambda]^2 - c^2/m^2 = 0$$

with roots $\lambda_\mathrm{A} = k/m$ and $\lambda_\mathrm{B} = (k+2c)/m$, in agreement with eqn 7 in Box 6.1 when we put each $\lambda = \omega^2$.

interval $\Delta\tau$ is much less than 0.2 Å. The classical approach is therefore valid for processes taking longer than 10^{-13} s and when displacements are of the order of a fraction of 1 Å.

The second problem concerns the spatial distribution of coupled oscillators for which a calculation of the normal modes requires new equations independent of any particular coordinate system. This is achieved by using Lagrange's equations with generalized coordinates q_i and their equilibrium values q_i°. The potential energy E_p and kinetic energy E_k are given by

$$E_\mathrm{p} = \tfrac{1}{2}\sum_{ij} K_{ij}(q_i - q_i^\circ)(q_j - q_\mathrm{j}^\circ)$$

$$E_\mathrm{k} = \tfrac{1}{2}\sum_{ij} M_{ij}(\partial q_i/\partial t)(\partial q_j/\partial t)$$

where K_{ij} and M_{ij} are the elements of $n \times n$ matrices equal to the second derivative of the potential and kinetic energies with respect to the generalized coordinates and velocities taken at the equilibrium positions q_i° and q_j°: in other words, where

$$K_{ij} = \partial^2 E_\mathrm{p}/\partial q_i \partial q_j \qquad \text{and} \qquad M_{ij} = \partial^2 E_\mathrm{k}/\partial(\partial q_i/\partial t)\partial(\partial q_j/\partial t)$$

Each of the q_I coordinates is assumed to execute harmonic oscillations about the equilibrium position q_i°, so that we can write

$$q_i = q_i^\circ + \sum_k A_{ik} Q_k$$

where $Q_k = \alpha_k \cos(\omega_k t + \phi_k)$ is a normal coordinate of amplitude α_k and phase ϕ_k and where A_{ik} is the element of the matrix that transforms the generalized coordinates q into normal coordinates Q. The potential and kinetic energies then take the simple forms

$$E_\mathrm{p} = \tfrac{1}{2}\sum_k^n \omega_k^2 Q_k^2$$

$$E_\mathrm{k} = \tfrac{1}{2}\sum_k^n \partial Q_k^2/\partial t$$

7

Brownian motion

The molecules in a gas or liquid undergo rapid and irregular movements, which can be detected throughout the medium by microscopic particles or macromolecules dispersed in it. These particles experience rapidly repeated collisions (10^{14} per second) from the molecules of the medium and as a result apparently move around in a completely random manner. In fact they receive a series of short impulses that are virtually uncorrelated with each other so that it is impossible to examine the dynamics of each particle in detail. However, we shall see later how such random motion can be modelled, at least tentatively.

The characteristic parameters of this *Brownian motion* (from its discovery by Brown in 1827) can be measured by a number of experimental methods. These range from the plotting against time of the position of a given particle as observed under an optical microscope, and the optical examination of molecu-lar diffusion in a liquid, to inelastic Rayleigh scattering or fluorescence recovery after photo-bleaching (FRAP). The theoretical study in this chapter, however, involves concepts and assumptions that are explained and discussed at a simple level rather than being dealt with by detailed and elaborate mathematical treatments, which can be found in specialized books.

7.1 Random variables and the autocorrelation function

A satisfactory yet simple model of Brownian motion must involve the concept of a *random variable* and some of the properties of functions of a random variable.

7.1.1 Random variables

A random variable x can take on a series of values $(x_1, x_2, \ldots, x_n)$ with respective probabilities $p(x_1), p(x_2), \ldots, p(x_n)$, which must clearly satisfy

$$\sum_i p(x_i) = 1$$

The function $p(x)$ is also called a *probability function*. The spatial coordinates of a particle undergoing Brownian motion provide a good example of a random variable. With such a variable, two quantities can be defined that are important because they are related directly to measurements. They are:

- the *mean value*, defined by

$$\langle x \rangle = \sum_{i=1}^{n} x_i p(x_i) \tag{7.1}$$

- the *variance*, defined by

$$\mathrm{var}(x) = \langle (x - \langle x \rangle)^2 \rangle = \langle x^2 \rangle - \langle x \rangle^2 \tag{7.2}$$

With a very large number of x_i values, the discrete distribution can be replaced by a continuous distribution or *probability density* $f(x)$ defined as

$$f(x) = \mathrm{Lim}_{\Delta x \to 0}\, p(x < X < x + \Delta x)/\Delta x$$

with the normalization condition

$$\int_0^\infty f(x)\,\mathrm{d}x = 1$$

and the definition

$$p(a < X < b) = \int_a^b f(x)\,\mathrm{d}x$$

For example, the function

$$f(x) = (2\pi\sigma^2)^{-1/2} \exp(x - \langle x \rangle)^2 / 2\sigma^2 \tag{7.3}$$

is the *normal distribution* function or *Gaussian distribution* about the mean value $\langle x \rangle$ with a variance σ^2. It can be shown (see Box 7.1) that such a function describes Brownian motion along the x direction.

7.1.2 Stochastic processes

This term embraces all phenomena in which the random variable (or variables) is (are) an undefined function of time, $y(t)$: observations made at a given time t give different values of $y(t)$. We only consider probability distributions, and two of these are particularly important:

(1) $W_1(y, t)\,\mathrm{d}y$ is the probability of finding the quantity y in the interval $\{y, y + \mathrm{d}y\}$ at time t;

(2) $W_2(y_1, t_1, y_2, t_2)\,\mathrm{d}y_1\,\mathrm{d}y_2$ is the probability of *simultaneously* finding y_1 in the interval $\{y_1, y_1 + \mathrm{d}y_1\}$ at time t_1 and y_2 in the interval $\{y_2, y_2 + \mathrm{d}y_2\}$ at time t_2.

A process is said to be *purely random* if

$$W_2(y_1, t_1, y_2, t_2) = W_1(y_1, t_1) W_2(y_2, t_2) \tag{7.4}$$

which is equivalent to assuming that there is no correlation between positions y_1 and y_2 whatever the times t_1 and t_2. Equation 7.4 in fact describes an ideal limiting case, since there is always a correlation between y_1 and y_2 when $t_2 - t_1 = 0$.

Box 7.1 **Random walk in one dimension**

A particle makes a series of jumps from the origin O ($x = 0$) either forwards ($x > 0$) or backwards ($x < 0$). To simplify the problem and avoid pointless calculations, it will be assumed (i) that the jumps are of equal length a and (ii) that the probability of jumping forwards or backwards is the same and is equal to $\frac{1}{2}$. The latter assumption is equivalent to tossing a coin to see which direction the particle jumps each time. The problem is to find the probability $W(m, N)$ that the particle starting from the origin will reach the point $x = ma$ after N jumps.

If N_1 jumps are made on one side (e.g. the positive side) and N_2 on the other, we have $N = N_1 + N_2$ and $m = N_1 - N_2$ or

$$N_1 = (N+m)/2 \qquad \text{and} \qquad N_2 = (N-m)/2 \tag{1}$$

The number of ways in which the N jumps can be made is given by the binomial coefficient $N!/N_1!N_2!$. Moreover, the probability of a given sequence of N_1 positive jumps and N_2 negative jumps is $(\frac{1}{2})^{N_1}(\frac{1}{2})^{N_2}$ or $(\frac{1}{2})^N$. Finally, therefore

$$W(m, N) = (\tfrac{1}{2})^N N!/(N/2 + m/2)!(N/2 - m/2)! \tag{2}$$

This probability function is also called the *binomial distribution*. As an example, if $N = 20$, the graph plotted in Fig. 7.1 gives the variation of $W(m, N)$ with m. More generally, if $m \ll N$ the Stirling approximation $\ln x! \approx n \ln x - x + \ln(2\pi x)/2$ can be used to calculate the factorials. Taking the natural logarithm of both sides of eqn 2:

$$\ln W = -N \ln 2 + \ln N! - \ln(N/2 + m/2)! - \ln(N/2 - m/2)! \tag{3}$$

and using Stirling's approximation and simplifying, we obtain

$$\ln W = N \ln 2 + N \ln N - (N/2)\ln(N^2 - m^2) + N \ln 2 - m^2/N + \tfrac{1}{2}\ln(2/\pi N)$$

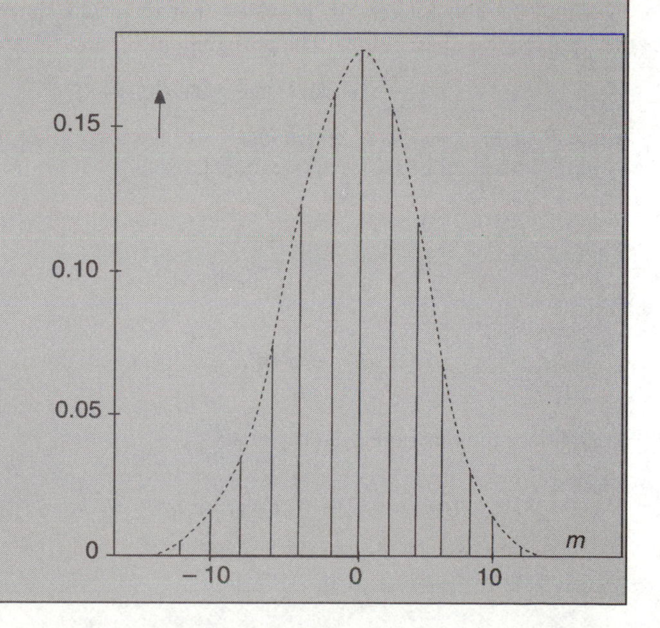

Figure 7.1 The distribution $W(m, N)$ for $N = 20$ and for m varying between -12 and $+12$. Note that the distribution is already very nearly a Gaussian one centred on $m = 0$ (indicated by the dotted curve).

Since $(N/2)\ln[N^2/(N^2-m^2)] = -(N/2)\ln(1-m^2/N^2) \approx m^2/2N$, we have finally:

$$W = (2/\pi N)^{1/2} \exp(-m^2/2N) \tag{4}$$

$W(m, N)$ is a Gaussian distribution centred on $m = 0$ with a variance $\sigma^2 = N$. Until now, we have been assuming a discrete series of jumps of length a. If N is large enough, the distribution in Fig. 7.1 becomes so dense that $W(m, N)$ can be considered as a function of a continuous variable $x = ma$, leading to

$$W(x, N)\,\mathrm{d}x = (2\pi\sigma)^{-1/2} \exp(-x^2/\sigma^2)\,\mathrm{d}x \tag{5}$$

with $\sigma^2 = Na^2$. Note, however, that $\mathrm{d}x$ is a small enough quantity to justify the mathematical treatment but a large one compared with the jumps made by the particle.

In Brownian motion, the random process is in a *steady state*, which means that $W_1(y, t)\,\mathrm{d}y$ is simply the probability of finding y in the interval $\{y, y + \mathrm{d}y\}$ without the need to specify the time t. In other words, the probability distribution function defining the process over a given time interval depends only on the length of this interval and not the time at which it occurs. More precisely, the integral

$$\langle y(t_0)\rangle = \lim_{T\to\infty} (1/T) \int_{t_0}^{t_0+T} y(t)\,\mathrm{d}t \tag{7.5}$$

becomes independent of t_0 and can therefore be written

$$\langle y\rangle = \lim_{T\to\infty} (1/T) \int_0^T y(t)\,\mathrm{d}t \tag{7.6}$$

Similarly, $W_2(y_1, y_2, t)$ becomes the probability of finding a pair of values y_1 and y_2 within the above intervals, provided that there is a time interval $t = t_2 - t_1$ between the two observations. The probabilities W_1 and W_2 are related by

$$W_2(y_1, y_2, t) = W_1(y_1)P_2(y_1 \mid y_2, t) \tag{7.7}$$

where P_2 is the *conditional probability* that, given y_1, y lies in the interval $\{y_2, y_2 + \mathrm{d}y_2\}$ after a time t. Clearly, we must have

$$P_2(y_1 \mid y_2, t) > 0$$

$$\int_0^\infty P_2\,\mathrm{d}y_2 = 1$$

$$W_1(y_2) = \int_0^\infty W_1(y_1)P_2(y_1 \mid y_2, t)\,\mathrm{d}y_1$$

In addition, the function P_2 must satisfy

$$P_2(y_1 \mid y_2, t) = \int_0^\infty P_2(y_1 \mid y, t_0)P_2(y \mid y_2, t-t_0)\,\mathrm{d}y$$

for any t_0 in the interval $\{0, t\}$.

This set of properties defines a *first-order Markov process*, a mathematical scheme that exactly fits Brownian motion, whether *inter*molecular or *intra* molecular, the latter involving internal movements of macromolecular chains.

7.1.3 The autocorrelation function

A random quantity $y(t)$ generally has two different values $y(t)$ and $y(t + \tau)$ when measured at times t and $t + \tau$. However, as τ tends to zero, the value at $t + \tau$ becomes less independent of that at t. Conversely, as τ tends to infinity, the value at $t + \tau$ tends to become absolutely independent of that at t. There is thus a correlation between successive events, which depends on the time interval τ separating them. This is quantified by an *autocorrelation function* $G(\tau)$, which is the time average of the product of two random variables measured at times t and $t + \tau$:

$$G(\tau) = \langle y(t)y(t+\tau)\rangle = \lim_{T\to\infty} (1/T) \int_0^T y(t)y(t+\tau)\,\mathrm{d}t \tag{7.8}$$

In order to calculate $G(\tau)$ in practice, the time over which the fluctuating quantity $y(t)$ is being observed is divided into short intervals Δt during which the variation in $y(t)$ is negligible (Δt must be small compared with the correlation time τ_c of the process, which is defined later). It can then be shown (see Box 7.2) that the function $G(\tau)$ decreases from $\langle y^2\rangle$ at $\tau = 0$ to $\langle y\rangle^2$ when $\tau \to \infty$. We can write

$$G(\tau) = \langle y\rangle^2 + (\langle y^2\rangle - \langle y\rangle^2)f(\tau) \tag{7.9}$$

where $f(\tau)$ is a decreasing function of τ.

If we now consider the fluctuations with time δy of $y(\tau)$, it can be shown (see Box 7.2) that

$$\langle \delta y(0)\delta y(\tau)\rangle = \langle \delta y^2\rangle f(\tau) \tag{7.10}$$

Box 7.2 Calculation of the autocorrelation function

Putting $\tau = n\Delta t$ and $t = j\Delta t$, where n and j are integers, the integrals in eqns 7.6 and 7.8 are replaced by the sums

$$\langle y\rangle = \lim_{T\to\infty} (1/N) \sum_{j=1}^{N} y_j \tag{1}$$

$$G(\tau) = \langle y(t)y(t+\tau)\rangle = \lim_{T\to\infty} (1/N) \sum_{j=1}^{N} y_j y_{j+n} \tag{2}$$

If $\tau = 0$, then

$$G(0) = \lim_{T\to\infty} (1/N) \sum_{j=1}^{N} y_j^2$$

The inequality

$$\sum_{j=1}^{N} y_j^2 > \sum_{j=1}^{N} y_j y_{j+n}$$

is always valid since the products $y_j y_{j+n}$ could be negative, so that $G(\tau)$ is a decreasing function of τ. When $\tau = 0$, $G(0) = \langle y^2 \rangle$; as $\tau \to \infty$, the two quantities become no longer correlated and the correlation function is simply the product of two probabilities:

$$\lim_{\tau \to \infty} G(\tau) = \langle y(t)y(t+\tau) \rangle = \langle y \rangle^2$$

The autocorrelation function $G(\tau)$ therefore decreases from $\langle y^2 \rangle$ to $\langle y \rangle^2$ as τ changes from 0 to ∞ and can be written in the form

$$G(\tau) = G(0)f(\tau) = \langle y \rangle^2 + (\langle y^2 \rangle - \langle y \rangle^2)f(\tau) \tag{3}$$

where $f(\tau)$ is a decreasing function of τ with $f(\tau) = 1$ when $\tau = 0$. The fluctuation of $y(t)$ with time, δy, is defined by

$$\delta y(t) = y(t) - \langle y \rangle \tag{4}$$

Using this to replace $y(t)$ in eqn 7.8, we obtain for a steady state

$$\langle y(0)y(\tau) \rangle = \lim_{T \to \infty} (1/T) \int_0^T [\langle y \rangle + \delta y(0)][\langle y \rangle + \delta y(\tau)]\mathrm{d}t \tag{5}$$

Since $\langle \delta y(0) \rangle = \langle \delta y(\tau) \rangle = 0$, eqn 5 becomes

$$\langle y(0)y(\tau) \rangle = \langle y \rangle^2 + \langle \delta y(0)\delta y(\tau) \rangle \tag{6}$$

Since we also have that the classical relationship defining the variance of the distribution is

$$\langle \delta y^2 \rangle = \langle y^2 \rangle - \langle y \rangle^2 \tag{7}$$

combining eqns 3, 5 and 7 gives

$$\langle \delta y(0)\delta y(\tau) \rangle = \langle \delta y^2 \rangle f(\tau) \tag{8}$$

The function $f(\tau)$ is often an exponential of the form $\exp(-\tau/\tau_c)$, where τ_c is known as the *correlation time* of the physical property in question. Hence

$$\langle \delta y(0)\delta y(\tau) \rangle = \langle \delta y^2 \rangle \exp(-\tau/\tau_c) \tag{7.11}$$

which is the autocorrelation function of the fluctuations.

All the above definitions and calculations become significant when the internal movements of a macromolecule are studied by correlation spectroscopy or inelastic scattering of light. In both cases, the experimental measurements are related directly to the autocorrelation function (see Chapter 9).

7.2 Brownian motion and coefficients of diffusion

As pointed out by Wang and Uhlenbeck (1945), two theoretical approaches to Brownian motion are possible. Firstly, we can emphasize the time variation of the displacement (or any other physical variable) by using the *correlation function* and its Fourier transform: the *power spectrum*. On the other hand, we can use the *diffusion equation*, i.e. the partial differential equation that the distribution function of the random variable or variables of the system must satisfy. We shall use the second approach in the analysis that follows. In other cases, such as those of inelastic

Rayleigh scattering or NMR, the first approach has to be used. In any given problem, of course, the two approaches lead to the same result.

7.2.1 Mathematical analysis

Every 'jump' of a molecule from one position to another due to collisions with neighbouring molecules is a random event. Its probability depends only on that of the preceding event and on the probability function defining the transition from one position to another. A set of events governed by such rules constitutes, as we have seen, a first-order Markov process.

One-dimensional Brownian motion can be treated in an elementary way as a random walk process along the x axis. However, a more elaborate but still simple treatment is also possible (Chandrasekhar, 1943).

Consider the simple case of one-dimensional Brownian motion along the x axis with jumps of mean length a each taking a time τ, which is in fact large compared with the mean time between successive collisions (of the order of 10^{-14} s). Similarly, a is not the distance covered between two collisions, but the algebraic sum of all the displacements during the time τ. Such a definition means that we can assert that there is no correlation between the motion of a particle observed at time t and that observed at time $t + \tau$. Only a small number of collisions is necessary for this assumption to be valid.

As a result, τ is still small enough to be considered mathematically as a first-order differential to which the rules of differential calculus can be applied. Under these conditions (see Box 7.3), the function $n(x, t)$ giving the number or concentration of molecules at time t and position x is found to obey a partial differential equation called the *diffusion equation*:

$$\partial n/\partial t = D_{\mathrm{T}}\, \partial^2 n/\partial x^2 \tag{7.12}$$

The translational coefficient of diffusion D_{T} is defined by

$$D_{\mathrm{T}} = \langle \Delta x^2 \rangle / 2\tau \tag{7.13}$$

where $\langle \Delta x^2 \rangle$ is the variance of the distribution of jump lengths a.

The treatment can be extended as follows to three dimensions and to rotational motion of the particles about themselves.

In extending the treatment to three dimensions, the coefficient of diffusion is the same for each dimension. In addition, since there is assumed to be no correlation between the motions along the three axes, the mean square of the displacement in space $\langle a^2 \rangle$ will be

$$\langle a^2 \rangle = \langle a^2 \rangle_x + \langle a^2 \rangle_y + \langle a^2 \rangle_z = 3\langle a^2 \rangle_x = 6 D_{\mathrm{T}} \tau \tag{7.14}$$

and the diffusion equation will be

$$\partial n/\partial t = D_{\mathrm{T}}(\partial^2 n/\partial x^2 + \partial^2 n/\partial y^2 + \partial^2 n/\partial z^2) = D_{\mathrm{T}} \nabla^2 n \tag{7.15}$$

where ∇^2 is the Laplacian operator.

In extending the treatment to the rotational motion of the particles, a rotational coefficient of diffusion is defined from the mean square of elementary rotations $\Delta\alpha$ about one of the molecular axes by

$$D_{\mathrm{R}} = \langle \Delta\alpha^2 \rangle / 2\tau \tag{7.16}$$

Box 7.3 **Calculation of the translational coefficient of diffusion**

The derivation in this box follows the treatment of Chandrasekhar (1943). Let $n(x, t)$ be the number of molecules at position x at time t. At time $t + \tau$, the number *at the same point* is $n(x, t + \tau)$. The latter can be considered as resulting from that present at the previous jump (i.e. at position $x - a$) at time t by writing

$$n(x, t+\tau) = \int_{-\infty}^{+\infty} n(x-a, t) f(a)\, \mathrm{d}a \tag{1}$$

where $f(a)$ is the distribution function of the jump a satisfying the following normalization and symmetry conditions:

$$\int_{-\infty}^{+\infty} f(a)\, \mathrm{d}a = 1 \qquad \text{and} \qquad \int_{-\infty}^{+\infty} af(a)\, \mathrm{d}a = 0 \tag{2}$$

where the latter arises from the microscopic reversibility of the process.

A Taylor expansion of the expressions for n in eqn 1 gives

$$n(x, t+\tau) = n(x, t) + \tau(\partial n/\partial t)_{\tau=0} + (\tau^2/2)(\partial^2 n/\partial t^2)_{\tau=0} + \cdots$$

$$n(x-a, t) = n(x, t) - a(\partial n/\partial t)_{a=0} + (a^2/2)(\partial^2 n/\partial t^2)_{a=0} + \cdots$$

and hence

$$n(x, t) + \tau(\partial n/\partial t) + (\tau^2/2)(\partial^2 n/\partial t^2) + \cdots$$
$$= \int_{-\infty}^{+\infty} n(x, t) f(a)\, \mathrm{d}a - (\partial n/\partial x) \int_{-\infty}^{+\infty} af(a)\, \mathrm{d}a + (\partial^2 n/\partial x^2) \int_{-\infty}^{+\infty} (a^2/2) f(a)\, \mathrm{d}a$$

The conditions in eqn 2 mean that the first term on the right is equal to $n(x, t)$ and the second term is zero, so that we are left with

$$\tau(\partial n/\partial t)_{a=0} = (\partial^2 n/\partial x^2)_{a=0} \frac{1}{2} \int_{-\infty}^{+\infty} a^2 f(a)\, \mathrm{d}a$$

The function $n(x, t)$ thus obeys the partial differential equation

$$\partial n/\partial t = D_{\mathrm{T}} \partial^2 n/\partial x^2 \tag{3}$$

provided that

$$D_{\mathrm{T}} = (1/2\tau) \int_{-\infty}^{+\infty} a^2 f(a)\, \mathrm{d}a \tag{4}$$

The integral in this equation defines $\langle a^2 \rangle$ (the variance of a) and the translational coefficient of diffusion is therefore

$$D_{\mathrm{T}} = \langle a^2 \rangle / 2\tau \tag{5}$$

Any small rotation $\Delta\theta$ about the centre of gravity is the superposition of two elementary rotations about perpendicular diameters:

$$\langle \Delta\alpha^2 \rangle = \langle \Delta\beta^2 \rangle = 2D_{\mathrm{R}}\tau \qquad \text{and} \qquad \langle \Delta\theta^2 \rangle = 2\langle \Delta\alpha^2 \rangle = 4D_{\mathrm{R}}\tau \qquad (7.17)$$

The orientation function will give the spatial distribution of the molecular orientations, normally described by two angles, θ made with the z axis and ϕ made with the x axis. Such a function, $n(\theta, \phi, t)$, will later be shown to obey a partial differential equation somewhat more complex than eqn 7.15.

7.2.2 Correspondence with macroscopic laws

Translational diffusion

Brownian motion corresponds to the macroscopic property of translational diffusion. This is described by Fick's equations, which are based on a purely phenomenological approach.

Translational diffusion, i.e. a *transport of matter*, occurs when unequal concentrations appear at two points in a gaseous, liquid or solid medium. It therefore seems reasonable to assume that the *flux of matter* J (amount crossing unit area per unit time, of dimensions $\mathrm{ML^{-2}T^{-1}}$) is proportional to the *concentration gradient* to a first approximation, and to put

$$J = -A\,\partial c/\partial x \qquad (7.18)$$

where A is a constant and the minus sign indicates that the transport of matter always takes place from a region of high concentration to one of low concentration. Since the dimensions of $\partial c/\partial x$ are $\mathrm{ML^{-4}}$, those of A are $\mathrm{L^2T^{-1}}$, the same as those of a translational coefficient of diffusion.

Assuming, again for simplicity, that the diffusion is one-dimensional, we calculate the flux of matter through a small volume element $S\,\mathrm{d}x$ defined by two planes of area S perpendicular to the x axis at x and $x + \mathrm{d}x$:

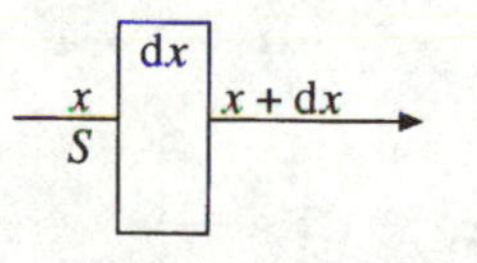

During a time $\mathrm{d}t$, a quantity of matter $J_x S\,\mathrm{d}t$ enters the volume and a quantity $J_{x+\mathrm{d}x} S\,\mathrm{d}t$ leaves it. Since $J_{x+\mathrm{d}x} = J_x + (\partial J/\partial x)\,\mathrm{d}x$, the resultant quantity entering the volume is

$$\mathrm{d}m = -(\partial J/\partial x) S\,\mathrm{d}x\,\mathrm{d}t$$

If c is the concentration in the volume element, its change $\mathrm{d}c$ is such that $\mathrm{d}m = S\,\mathrm{d}x\,\mathrm{d}c$, and equating the two values for $\mathrm{d}m$, we have the equation of continuity:

$$\partial c/\partial t = -\partial J/\partial x \qquad (7.19)$$

Combining this with eqn 7.18 gives *Fick's second law*, or the partial differential equation governing diffusion:

$$\partial c/\partial t = A\,\partial^2 c/\partial x^2 \qquad (7.20)$$

Because n and c are proportional, eqns 7.12 and 7.20 are identical if $A = D_{\mathrm{T}}$ and if D_{T} is assumed to be independent of c and thus of x in the partial derivative of J.

Such an assumption is implicit in the mathematical treatment of the Markov process since the function $f(a)$ is assumed to be symmetrical, i.e. independent of whether the values of $n(x, t)$ are to the right or the left.

When the diffusion takes place in three dimensions x, y, z, then instead of eqn 7.20 we obtain

$$\partial c/\partial t = D_{\mathrm{T}}(\partial^2 c/\partial x^2 + \partial^2 c/\partial y^2 + \partial^2 c/\partial z^2) \tag{7.21}$$

Rotational diffusion

The distribution of molecular orientations is described by a function $f(\theta, \phi, t)$ representing the density of the ends of vectors bound to the molecules over a sphere of unit radius. The angles θ and ϕ are the usual angles in spherical coordinates: θ is the angle between the vector and the z axis, and ϕ is the angle between the projection of the vector on to the xy plane and the x axis. The function f in rotational diffusion is equivalent to the concentration c in translational diffusion. Only translational diffusion corresponds to a transport of matter. The function f satisfies a partial differential equation similar to eqn 7.12 but a little more complicated since it involves the angles θ and ϕ. The equation is derived in Box 7.4, where the final result is

$$\partial f/\partial t = D_{\mathrm{R}}[(\sin\theta)^{-1}\partial(\partial f/\sin\theta\,\partial\phi)/\partial\phi + (\sin\theta)^{-1}\partial(\sin\theta\,\partial f/\partial\theta)\partial\theta] \tag{7.22}$$

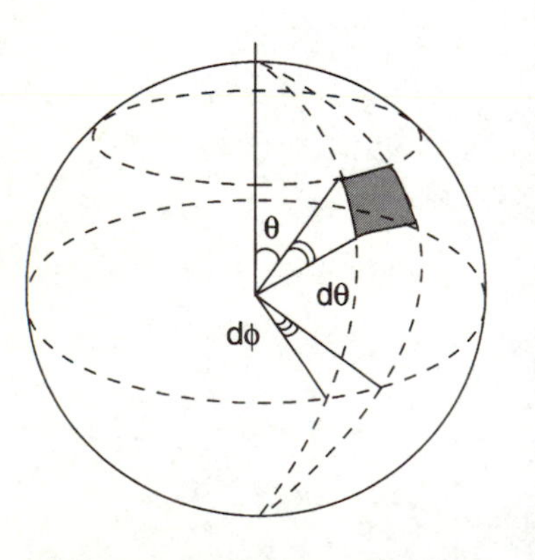

Figure 7.2 Spherical coordinates for the components of grad f.

Box 7.4 Establishment of the rotational diffusion equation

A rotational flux can also be defined by a vector $\boldsymbol{J}_{\mathrm{R}}$ related to a vector gradient grad by

$$\boldsymbol{J}_{\mathrm{R}} = -D_{\mathrm{R}}\,\mathrm{grad}\,f \tag{1}$$

This is similar to eqn 7.18. Note that D_{R}, which is the ratio of the mean square of an angle (in radians) and a time, is measured in s^{-1} and has the dimensions T^{-1}. The quantity f, which is a surface distribution of a density of points, has the dimensions L^{-2}, as has the gradient (the ratio of f and an angle). It follows that the dimensions of J_{R} are $L^{-2}T^{-1}$, so it is indeed a flux, i.e. the number of vector ends on the surface of the sphere that traverse a unit area per unit time.

The components of the vector grad are the partial derivatives of f with respect to θ and ϕ (Fig. 7.2): the first is $\partial f/\partial\theta$ and the second is $(1/\sin\theta)\,\partial f/\partial\phi$. The quantity f must also satisfy the normalization condition

$$\int_0^{\pi} \mathrm{d}\theta \sin\theta \int_0^{2\pi} f(\theta, \phi, t)\,\mathrm{d}\phi = 1 \tag{2}$$

The components of $\boldsymbol{J}_{\mathrm{R}}$ are thus:

$$J_{\mathrm{R},\theta} = -D_{\mathrm{R}}\partial f/\partial\theta \qquad \text{and} \qquad J_{\mathrm{R},\phi} = (-D_{\mathrm{R}}/\sin\theta)\partial f/\partial\phi \tag{3}$$

To obtain a relationship between the spatial derivatives of f and the time variation $\partial f/\partial t$, we examine the resultant fluxes across a surface element ABCD (Fig. 7.3). The resultant flux across the elementary segments AB and CD is $(\partial J_{\mathrm{R},\phi}/\partial\phi)\,\mathrm{d}\phi\,\mathrm{d}\theta\,\mathrm{d}t$, while that across AD and BC is $[\partial(J_{\mathrm{R},\theta}\sin\theta)/\partial\theta]\,\mathrm{d}\phi\,\mathrm{d}\theta\,\mathrm{d}t$. In addition, during a time interval $\mathrm{d}t$, there is a change $(\partial f/\partial t)\,\mathrm{d}t$ in the number of points contained in a surface element, i.e. a resultant change of $(\partial f/\partial t)\sin\theta\,d\phi\,d\theta\,dt$. The sum of the three changes must be zero since there is no other cause of a change in the density function f. This leads to the *equation of continuity*:

$$\sin\theta\,\partial f/\partial t + \partial(J_{R,\phi})/\partial\phi + \partial(J_{R,\theta}\sin\theta)/\partial\theta = 0 \tag{4}$$

Replacing $J_{R,\phi}$ and $J_{R,\theta}$ from eqn 3 gives the partial differential equation governing rotational diffusion

$$\partial f/\partial t = D_R\{(\sin\theta)^{-1}\partial[(\sin\theta)^{-1}\partial f/\partial\phi]/\partial\phi + (\sin\theta)^{-1}\partial(\sin\theta\,\partial f/\partial\theta)/\partial\theta\} \tag{5}$$

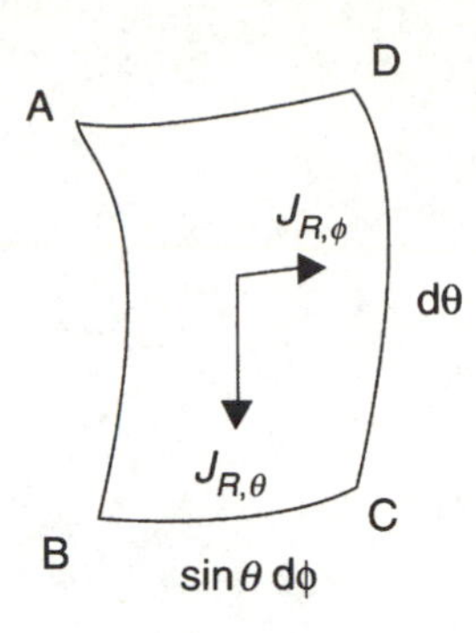

Figure 7.3 Surface element of Fig. 7.2 for calculation of flux.

We shall see later how this equation can be used to interpret fluorescence polarization phenomena.

7.3 Langevin's equation

Having established the mathematical background for dealing with Brownian motion, we now need to look further into the physical aspects, and to do this we must take into account the medium in which the particles are moving. This involves finding an equation describing the dynamics of a particle of mass m, taking into account all the forces acting on it.

7.3.1 Establishing the equation

Returning to the simple one-dimensional case with translation along the x axis, there are two forces acting on the particle.

The first is a frictional force F_1 proportional to the velocity $v = dx/dt$, i.e.

$$F_1 = f\eta_0\,dx/dt \tag{7.23}$$

where η_0 is the viscosity of the medium through which the particle is moving, related to the transfer of linear momentum between elements of the moving fluid, and where f is a geometrical factor depending only on the shape of the particle. Equation 7.23 assumes that the medium around the particle is *continuous*, yet it is actually experiencing repeated collisions with molecules of solvent, which means that the latter is regarded as *discontinuous*. The contradiction can only be resolved by using a molecular model of the viscous forces, but despite this eqn 7.23 is valid when the molecule or particle moving through the medium is significantly larger than the molecules of the solvent and when it is moving with a low velocity.

The second force acting on the particle is a time-dependent random force $F(t)$. The only component of this to be considered is that along the direction of displacement. Two important assumptions are made about this force: (i) it is independent of the velocity v, and (ii) its variations are much more rapid than those of v.

The equation of motion for the particle then becomes

$$m\,d^2x/dt^2 = -f\eta_0\,dx/dt + F(t) \tag{7.24}$$

known as *Langevin's equation*. Here, the mean work done by the random force $F(t)$ is assumed to be zero over a time long compared with that between two successive collisions but sufficiently short to be considered mathematically as equivalent to a very short time interval δt.

Note that giving the work a mean value of zero is entirely in agreement with the laws of thermodynamics. In the medium at constant temperature where the Brownian motion is taking place, which is at equilibrium on a scale larger than that of the individual particles, the *mean work* is zero according to the second law. In this case, it is found (see Box 7.5) that, after a very short time of the order of $m/f\eta_0$, i.e. between 2 and 20 ps for macromolecules, the mean square distance travelled during a time t is given by

$$\langle x^2 \rangle = 2tk_B T/f\eta_0$$

Box 7.5 **Solution of Langevin's equation**

Multiplying eqn 7.24 by x and averaging the various quantities over a time δt eliminates the random term and gives

$$m\langle x\,\mathrm{d}^2x/\mathrm{d}t^2\rangle + f\eta_0\langle x\,\mathrm{d}x/\mathrm{d}t\rangle = 0 \quad (1)$$

At temperature T, the particle has a mean kinetic energy according to the equipartition of energy given by

$$m(\langle \mathrm{d}x/\mathrm{d}t\rangle)^2 = k_B T/2 \quad (2)$$

Since $x\,\mathrm{d}x/\mathrm{d}t = \frac{1}{2}\mathrm{d}(x^2)/\mathrm{d}t$, and hence

$$x\,\mathrm{d}^2x/\mathrm{d}t^2 = \tfrac{1}{2}\mathrm{d}^2(x^2)/\mathrm{d}t^2 - (\mathrm{d}x/\mathrm{d}t)^2 \quad (3)$$

we have

$$\tfrac{1}{2}m\,\mathrm{d}^2(\langle x\rangle^2)/\mathrm{d}t^2 - (\langle \mathrm{d}x/\mathrm{d}t\rangle)^2 + \tfrac{1}{2}\eta_0\,\mathrm{d}(\langle x\rangle^2)/\mathrm{d}t = 0 \quad (4)$$

Putting $\mathrm{d}(\langle x\rangle^2)/\mathrm{d}t = z$ gives

$$\tfrac{1}{2}m\,\mathrm{d}z/\mathrm{d}t - k_B T + \eta_0 f_z/2 = 0 \quad (5)$$

whose general solution is

$$z = 2k_B T/f\eta_0 + A\exp(-\eta_0 ft/m) \quad (6)$$

where A is a constant of integration. The exponential term rapidly vanishes since the *relaxation time* $\tau = m/f\eta_0$ is of the order of 10^{-13} s (compare this with the formula for the sedimentation coefficient defined in section 7.4.2). A steady state is quickly established in which

$$z = 2k_B T/f\eta_0 = \mathrm{d}(\langle x\rangle^2)/\mathrm{d}t$$

and hence

$$\langle x\rangle^2 = (2k_B T/f\eta_0)t \quad (7)$$

Comparing eqn 7.13 and eqn 7, we deduce that

$$D_T = k_B T/f\eta_0 \quad (8)$$

and comparing this with the definition of eqn 7.18,

$$D_T = k_B T / f \eta_0 \qquad (7.25)$$

The same approach can be applied to rotational motion by replacing forces with torques. In particular, a frictional couple is exerted, given by $\Gamma = -C\eta_0\, \mathrm{d}\alpha/\mathrm{d}t$, where C is the friction constant in rotation. Hence

$$D_R = k_B T / C\eta_0 \qquad (7.26)$$

It is clear from their definitions that D_T and D_R have dimensions L^2T^{-1} and T^{-1} respectively. Equation 8 in Box 7.5 and eqn 7.26 then show that f has the dimensions of length and C those of volume. In the translational case, f corresponds to a *molecular length* to within a multiplying factor, while C corresponds to a *molecular volume*. For a sphere of radius R and volume V, for example, $f = 6\pi R$ and $C = 8\pi R^3 = 6V$.

As examples, we can calculate the rotational Brownian motion for certain atomic groups. Using the calculated dimensions of several groups found in protein side chains, the rotational coefficient of diffusion can be determined from

$$D_R = k_B T / 6\eta_0 V \qquad (7.27)$$

where V is the volume of the chemical group, assumed to be spherical, and η_0 is the viscosity of the medium in which the group is moving. At 20°C and taking $\eta_0 = 10^{-3}\,\mathrm{N\,s\,m^{-2}}$ for water, the relaxation time $\tau_c = 1/6D_R = V\eta_0/k_B T = 0.246V$ is given in picoseconds when V is in Å^3. Table 7.1 gives the values of V and τ_c for various residues.

It is also possible to calculate τ_c for a set of spheres of increasing radius a, giving a good approximation of the tumbling motion of globular proteins as in Table 7.2.

This type of solution to Langevin's equation corresponds to the simple Einstein result that the mean square of the displacement from the origin, $\langle (x - x_0)^2 \rangle$, is given by $2D_T t$. This result is no longer valid as $t \to 0$ since it leads to infinite velocities. The calculation was repeated by Chandrasekhar (1943) and by Uhlenbeck and Ornstein (1930) with two additional conditions: a finite initial velocity and a final velocity distribution ($t \to \infty$) identical to a Maxwellian distribution. This leads to far more elaborate expressions. However, for times $t \gg \beta^{-1}$ where $\beta = f\eta_0/m$, the new method gives the Einstein–Langevin solution again. At any time t:

$$(x - x_0)^2 = (2k_B T/m\beta^2)(\beta t - 1 + \mathrm{e}^{-\beta t}) \qquad (7.28)$$

Table 7.1 Values of V and τ_c for various atomic groups

Group	V (Å3)	τ_c (ps)
CH_3	26	6.4
OH	13.6	3.3
NH_3^+	21	5.2
phenyl ring (Phe)	75.2	18.5
phenylene ring (Tyr)	68.1	16.8
imidazole ring (His)	58.6	14.4
guanidinium (Arg)	52	12.8
indole ring (Trp)	103	25.3

Table 7.2 Values of τ_c for spheres of radius a

a (Å)	τ_c (ns)	a (Å)	τ_c (ns)
5	0.13	30	28
10	1.03	40	66
15	3.5	50	129
20	8.3	75	435
25	16	100	1032

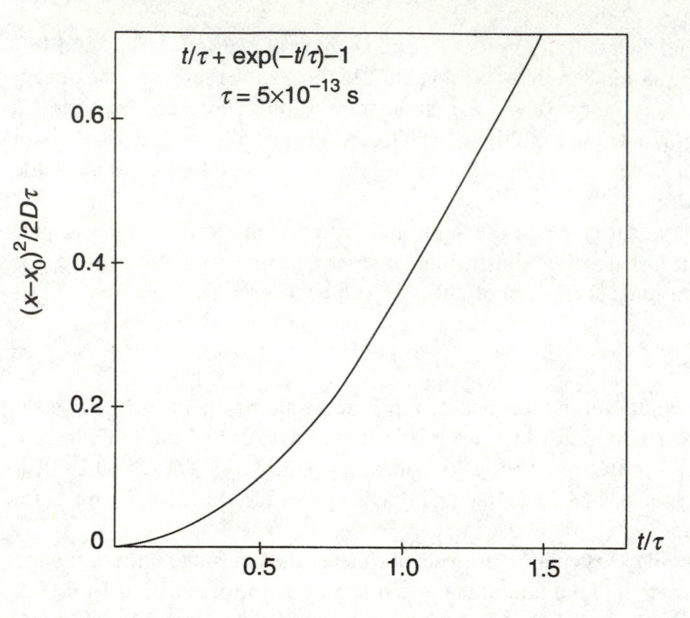

Figure 7.4 Variation of $(x - x_0)^2/2D\tau$ with t/τ (eqn 7.29). For small t/τ, the curve obeys a square law and the Brownian particle behaves as a free particle. For very long times, the variation becomes linear and obeys Einstein's diffusion equation: $(x - x_0)^2 = 2D_T t$.

which, as $t \to 0$, behaves as the product of the square of a velocity and the square of a time. Using the definition of D_T, eqn 7.28 can be written

$$(x - x_0)^2/2D_T\tau = t/\tau - 1 + e^{-t/\tau} \tag{7.29}$$

At very short times, the Brownian motion of the particle is the same as that of a free particle with a mean square velocity equal to $k_B T$. In the neighbourhood of $t = \tau$, a transition to the diffusive process occurs, and for $t \gg \tau$, $(x - x_0)^2$ becomes proportional to t and no longer to t^2 (Fig. 7.4).

7.3.2 Harmonic oscillator with Brownian motion

In this case we have to add a restoring force to Langevin's equation. Again using a simplified one-dimensional model, the restoring force will be $F = -kx = -m\omega^2 x$ (see Chapter 6), giving

$$d^2x(t)/dt^2 = -\tau^{-1}\, dx(t)/dt + A(t) - \omega^2 x \tag{7.30}$$

The solution of this equation involves very complicated expressions and we shall not attempt it here. We can, however, determine the mean value $\langle x^2 \rangle$ (see Box 7.6) with the simple result:

$$\langle x^2 \rangle = k_B T/m\omega^2 \tag{7.31}$$

Box 7.6 **Mean square amplitude of a harmonic oscillator**

Since the potential energy $E_p = m\omega^2 x^2/2$, we have that

$$\langle x^2 \rangle = \int_0^\infty x^2 \exp(-m\omega^2 x^2/2k_B T)\,dx \bigg/ \int_0^\infty \exp(-m\omega^2 x^2/2k_B T)\,dx$$

The integrals are of the type

$$\int_0^\infty x^n \exp(-ax^2)\,dx$$

dealt with in the Mathematical Appendix. For $n = 0$, the value is $\frac{1}{2}(\pi/a)^{1/2}$ and for $n = 2$, it is $\frac{1}{4}(\pi/a^3)^{1/2}$, so that putting $a = -m\omega^2/2k_B T$ we have the result

$$\langle x^2 \rangle = k_B T/m\omega^2$$

This expression gives a satisfactory order of magnitude for the mean displacement as a function of the mass and frequency of the oscillator. Assuming that $\langle x^2 \rangle^{1/2}$ is of the order of 1 Å, $m\omega^2$ is approximately unity. Large atomic groups, such as the two parts of lysozyme that pivot about a 'hinge' in the molecule, then have periods of the order of a few nanoseconds. With amplitudes 10 times greater, the periods are still only a few tens of nanoseconds.

The solution of Langevin's equation examined above involved a few simple examples only, but it can be generalized to a whole macromolecular chain, making it possible in principle to calculate internal fluctuations by including all interactions between the atoms in the molecule and the role of the solvent (see Box 7.7).

7.4 Brownian motion with external forces

We have seen so far that Brownian motion can be described either on a molecular scale by means of Langevin's equation or on a macroscopic scale using the diffusion equations and the coefficients of diffusion, D_T and D_R. The latter are also molecular parameters. We shall see later that modelling the internal stochastic dynamics of a macromolecule is based mainly on the use of Langevin's equation.

The situation is different when we consider the dynamics of biopolymers as describing their motion inside the cell where the interactions they experience originate. Here, we must take into account forces acting on the molecules apart from thermal agitation, the great majority of which are of either electrical or chemical origin.

Electrical forces

These forces arise through the existence in the medium carrying the molecules of an electric potential Ψ, a function of the spatial coordinates. It is usually possible to calculate this from the distribution of fixed and mobile charges. In particular, there is a potential distribution in the neighbourhood of a macromolecule that reflects the distribution of charged groups and hence the structure of the polyelectrolyte or polyampholyte (see Chapter 17). A similar situation occurs at the level of a

Box 7.7 Conformational fluctuations of a macromolecular chain

The derivation in this box follows the treatment of Hopfinger (1973). If ϕ_k is the angle of rotation about the kth bond, the Brownian motion of the chain can be described by a Langevin-type equation:

$$I\,\mathrm{d}^2\phi/\,\mathrm{d}t^2 + [f]\,\mathrm{d}\phi/\,\mathrm{d}t = \sum_{i,j}^{N} T(i,j) + U(t)$$

On the left, the first term is an *inertial* term, I being the moment of inertia of the chain about the kth bond; $[f]$ is a tensor representing the viscous forces affecting the motion due to interactions with all the other segments of the chain. On the right, $T(i, j)$ is the resultant torque due to bound and unbound atoms; and $U(t)$ is a torque from random forces.

Two approximations can be made. The first is that the inertial term can be neglected in comparison with the frictional term, and the second is that the tensor $[f]$ can be replaced by a scalar f. The latter assumption means that the friction exerted on the kth segment is taken to be independent of both its position and its orientation.

Integrating over a short time τ gives

$$\phi_k(t+\tau) - \phi_k(t) = \tau/f \sum T(i,j) + U(t)$$

In this form, the calculation can be carried out numerically using a random selection of examples to simulate $U(t)$, replacing f by a Stokes' equation and calculating $T(i, j)$ from various interaction potentials.

membrane or any subcellular organelle (nucleus, ribosomes, mitochondria, etc.). The spatial variation of Ψ is described by a vector gradient (grad Ψ), whose components are the partial derivatives of Ψ with respect to the spatial coordinates (such as the Cartesian coordinates x, y, z or the spherical coordinates r, θ, ϕ). A molecule with a charge q placed in such a gradient experiences a force given by

$$F_\mathrm{e} = -q \,\mathrm{grad}\, \Psi = q\boldsymbol{E} \tag{7.32}$$

where $\boldsymbol{E}$ is the electric field.

Chemical forces

These arise through variations in the species concentrations. Here, we define a *chemical potential* μ_i for a species i as the *partial molar free energy*, i.e. the partial derivative $\partial G/\partial N_i$ of the free energy G with respect to the number N_i of moles of the ith species, the number of moles of other species being held constant.

The chemical potential μ_i (SI unit: $\mathrm{J\,mol^{-1}}$) depends on the concentration c_i of each species, or more exactly on its activity a_i according to the classic relationship

$$\mu_i = \mu_i^\circ + RT \ln a_i \tag{7.33}$$

where μ_i° is a standard potential depending on the units chosen for the concentration. If the concentration of H^+ ions is used, the chemical potential of H^+ is measured by the pH except for a numerical constant.

A generalized force $\boldsymbol{F}_\mathrm{c}$ is associated with the potential μ_i by

$$F_\mathrm{c} = -\mathrm{grad}\, \mu_i \tag{7.34}$$

where, as in the electrical case, the components of the vector gradient are the partial derivatives of μ_i with respect to the spatial coordinates.

7.4.1 General equation for a transport process

Molecular transport will take place under the influence of both the electrical and chemical forces, and our next task is to determine its characteristics. When a molecule is subjected to an external force $\boldsymbol{F}$, Langevin's equation (eqn 7.24) can be written in the general form

$$m\boldsymbol{a} + f\eta_0 \boldsymbol{v} - \boldsymbol{F} - \boldsymbol{A}(t) = 0 \tag{7.35}$$

where $\boldsymbol{a}$ is the acceleration and $\boldsymbol{v}$ the velocity.

When $\boldsymbol{F} = 0$, we have seen that this stochastic differential equation corresponds to a partial differential equation obtained from the definition of a *diffusion flux*

$$J_{\mathrm{D}} = D_{\mathrm{T}}\, \partial c/\partial x$$

and from the equation of continuity.

When $\boldsymbol{F} \neq 0$, if we retain the usual simplified one-dimensional model, the molecule is accelerated and its speed increases until the frictional force $f\eta_0\, \mathrm{d}x/\mathrm{d}t$ is exactly equal and opposite to X, the x component of F. The resultant external force is then zero and the molecule therefore moves with a uniform velocity

$$\mathrm{d}x/\mathrm{d}t = F/f\eta_0 \tag{7.36}$$

As we saw in the solution of Langevin's equation, this velocity is reached in a very short time of the order of 10^{-12} s. The resultant uniform molecular motion through the solvent is a transport of matter characterized by a *convection flux* J_{C} defined by

$$J_{\mathrm{C}} = c\, \mathrm{d}x/\mathrm{d}t = cF/f\eta_0 \tag{7.37}$$

Because Brownian motion is still present, the total flux of matter J_{T} is the sum of the convection flux and the diffusion flux:

$$J_{\mathrm{T}} = -D_{\mathrm{T}}\, \partial c/\partial x + cF/f\eta_0 \tag{7.38}$$

Applying the same equation of continuity to J_{T} as used previously to obtain the partial differential equation for diffusion, we obtain *Smoluchowski's equation*

$$\partial c/\partial t = \partial (D_{\mathrm{T}}\, \partial c/\partial x - cF/f\eta_0)/\partial x \tag{7.39}$$

The solution of this equation, a function $c(x, t)$, describes how the system evolves. It can be viewed as the migration at uniform velocity of a *mobile boundary* between solvent and solution with Brownian motion occurring on both sides. In other words, if we define a new variable x' representing the position of the molecule relative to a coordinate system moving at the uniform speed $F/f\eta_0$, the function $c(x', t)$ is similar to the solutions found for the partial differential equation describing translational Brownian diffusion with boundary conditions and initial conditions imposed on the transport process. The transport of matter will still be accompanied by a Brownian motion tending to create a uniform distribution of molecules, i.e. to enlarge the mobile boundaries created by the applied external force.

As examples, we now look at two methods used to separate molecules: sedimentation and electrophoresis.

7.4.2 Sedimentation

Sedimentation is defined as the action on a molecule of a gravitational force, whether natural or artificial. Natural gravity operates on particles suspended at rest in a fluid. This is the process occurring in so-called 'sedimentation basins' in which mineral salts are deposited slowly on geological time scales under the action of gravity. Artificial gravity is created when the molecular solution is placed in a rapidly rotating rotor. F is then the *centrifugal force* experienced by a particle of mass m rotating with an angular speed ω at a distance x from the rotation axis:

$$F = m\omega^2 x$$

In practice, when a particle or macromolecule of mass m is in a liquid medium of density ρ_0, m must be replaced by an apparent mass m^* that takes into account the Archimedean upthrust. In a two-component medium (solvent and solute), we have

$$m^* = m(1-\bar{V}\rho_0) \tag{7.40}$$

where $\bar{V}$ is the partial specific volume of the solute. In a more complex medium such as a biopolymer in a salt solution, the value of m^* has to be modified to take preferential solvation into account. The velocity is then, from eqn 7.36:

$$v = \mathrm{d}x/\mathrm{d}t = m^*\omega^2 x/f\eta_0 \tag{7.41}$$

This relationship seems to indicate that v is an increasing function of x and thus to contradict the assumption of uniform motion. In fact, at any instant there is equilibrium between the forces and the motion can be considered sufficiently uniform in spite of the slow increase of v with x.

The speed of sedimentation is characterized by an experimental quantity s, the *sedimentation constant*, defined by

$$s = (\omega^2 x)^{-1}\,\mathrm{d}x/\mathrm{d}t \tag{7.42}$$

Here s is a measurable quantity with the dimensions of time. Because the velocities are generally very low (of the order of a few millimetres per hour) and the angular velocities very high (modern centrifuges can run at up to 75 000 rpm), s is quoted in units of 10^{-13} s, called a *svedberg* after Theodor Svedberg, a pioneer in ultra-centrifugation techniques.

The equation for the transport of matter depends on the conditions under which the measurement is made. If migration is taking place in a cylindrical tube in a radial direction x, the sedimentation flux $J_\mathrm{C} = c\,\mathrm{d}x/\mathrm{d}t$ is given by

$$J_\mathrm{C} = cs\omega^2 x \tag{7.43}$$

so that the transport equation is

$$\partial c/\partial t = \partial(D_\mathrm{T}\,\partial c/\partial x - cs\omega^2 x)/\partial x \tag{7.44}$$

whose solution depends on the initial conditions.

It is quite difficult to obtain a general solution to eqn 7.44, so we shall consider a special case in which the diffusion process can be neglected ($D_\mathrm{T} = 0$). This means that the molecule is large enough for diffusion to be negligible during the time the sedimentation is being observed. However, the results obtained from such a special case enable us to understand the main features of the sedimentation process.

Equation 7.44 now becomes simply

$$\partial c/\partial t = -s\omega^2\, \partial(cx)/\partial x = -s\omega^2(c + x\,\partial c/\partial x)$$

with, as initial conditions, a uniform concentration between the initial position x_0 (meniscus) and x_1 (bottom of cell). Since there is no diffusion, the mobile boundary is represented by a step function and the coordinate x_f of this boundary satisfies the equation defining s. After integration, this gives

$$x_f = x_0 \exp(s\omega^2 t) \tag{7.45}$$

since at time $t = 0$, the boundary is at $x_f = x_0$. We now change the variable by putting $u = x_0 \exp(s\omega^2 t)$, giving $u\,\mathrm{d}c/\mathrm{d}u = -c$ and hence $c = A/u$. A is a constant of integration obtained by using the fact that at $t = 0$, $c = c_0$ and $u = x_0$, so that $A = c_0 x_0^2$. Replacing u, we obtain:

- for $x_1 > x > x_f$:

$$c(x, t) = c_0 \exp(-s\omega^2 t) \tag{7.46a}$$

- for $x_0 < x < x_f$:

$$c(x, t) = 0 \tag{7.46b}$$

An examination of these solutions enables us to understand the concept of a mobile boundary, which, in this ideal case, marks the sharp separation between a region of zero concentration and one with a finite concentration.

7.4.3 Electrophoresis

A molecule with a charge q placed in an electric field $\boldsymbol{E}$ experiences a force $q\boldsymbol{E}$. Since it is also in a medium of viscosity η_0, there is also a viscous force $\boldsymbol{F}_f = f\eta_0\boldsymbol{v}$ opposing the motion. When the two forces are equal, the particle moves with a uniform velocity given by

$$\boldsymbol{v} = q\boldsymbol{E}/f\eta_0 \tag{7.47}$$

The *mobility* of the molecule u is defined by $\boldsymbol{v} = \mathrm{u}\boldsymbol{E}$, so that

$$u = q/f\eta_0 \tag{7.48}$$

These expressions are similar to those defining the sedimentation constant s of the same molecule: $v = s\omega^2 x$ with $s = m^*/f\eta_0$. There are, however, clear differences in that s has the dimensions of time while u has those of $\mathrm{M^2V^{-1}T^{-1}}$, and in that u on the molecular scale depends on the charge and shape of the molecule.

Experimentally, an electric current I is passed through an electrophoresis cell of cross-sectional area S giving a current density $\boldsymbol{i}$ of magnitude I/S. This is related to the electric field through Ohm's law in the form $\boldsymbol{i} = \sigma\boldsymbol{E}$, where σ is the conductivity of the medium.

Assuming a uniform current density and migration in one dimension, eqn 7.48 gives

$$u = v\sigma S/I \tag{7.49}$$

for the resultant transport of matter. This is characterized by a convective flux due to the electric field given by

$$J_E = cv = cuE \tag{7.50}$$

to which is added the flux due to diffusion, $J_D = -D_T\,\partial c/\partial x$. The general equation governing the whole process is obtained by adding the two contributions and using the equation of continuity, eqn 7.19:

$$\partial c/\partial t = \partial(D_T\,\partial c/\partial x - cuE)/\partial x \tag{7.51}$$

This equation is formally identical to the differential equation for sedimentation (in a rectangular cell), so that the transport of matter due to electrophoresis also gives rise to a mobile boundary at a value of x given by

$$x_f = uEt \tag{7.52}$$

around which there is a distribution of concentration whose partial derivative $\partial c/\partial x$ is a Gaussian curve centred on $x = x_f$.

The examples of sedimentation and electrophoresis have been treated in an elementary way but one that gives us some understanding of Smoluchowski's equation (eqn 7.39), and how its solution gives us the distribution of the concentration in simple geometrical cases. However, the two processes of sedimentation and electrophoresis are also used to separate molecules. In this case, the discrimination is improved if the random displacements due to Brownian motion can be made negligible in comparison with the overall drift movement. Since the thermal agitation is almost constant, there are only two possible ways of achieving this.

The first is to increase the external force considerably so that the Brownian motion becomes negligible while the experiment is being conducted: this will involve a higher speed of rotation in the ultracentrifuge or a high voltage applied in electrophoresis. The other way is to increase the resistance to motion by using a medium with a high viscosity (sucrose solution, agarose or acrylamide gel). In practice, both are used together.

7.4.4 **Biopolymer in an electrochemical potential**

We now return to the general case considered at the beginning of section 7.4: that of a biopolymer placed in an *electrochemical potential* μ_{ec}, the sum of the chemical and electric potentials. The latter is given by $NZe\Psi = FZ\Psi$, the energy of a mole of solute with a charge Ze at a potential Ψ (F being the Faraday, the product of Avogadro's number N and the elementary charge e, i.e. about 96 500 C). Thus:

$$\mu_{ec} = \mu_0 + RT\ln a + ZF\Psi \tag{7.53}$$

The generalized force along the x direction, X_c, is given by $-\partial\mu/\partial c = -(RT/a)\,\partial a/\partial x$. It is easy to show that, at low concentrations, $a \approx c$ and $X_c \approx (-RT/c)\,\partial c/\partial x$. On a molecular scale, the force is

$$X = X_c/N = (-k_B T/c)\partial c/\partial x$$

and the flux corresponding to this is

$$J = cX/f\eta_0 = (-k_B T/f\eta_0)\partial c/\partial x = -D_T\,\partial c/\partial x$$

i.e. the flux due to diffusion.

At infinitely dilute concentrations, therefore, the diffusion process can be considered as resulting from a generalized force applied to the molecule given by the negative gradient of the chemical potential. Of course, such a purely thermodynamic interpretation reveals nothing about the mechanism but does lead to the same macroscopic laws for diffusion.

At finite concentrations, on the other hand, a and c cannot be equated and in this case:

$$X = (-k_{\mathrm{B}}T/a)\partial a/\partial x = (-k_{\mathrm{B}}T/\gamma c)\partial(\gamma c)/\partial x$$

leading to

$$J = (-k_{\mathrm{B}}T/\gamma f\eta_0)\partial(\gamma c)/\partial x$$

which is equivalent to defining a translational coefficient of diffusion D_{T}' that now depends on the concentration and is usually written

$$D_{\mathrm{T}}' = D_{\mathrm{T}}(1 + \partial \ln \gamma/\partial \ln c) \tag{7.54}$$

The laws governing Brownian motion are modified in such a way that random jumps to the left and right about a given position on the x axis (see Box 7.1) no longer have the same probability. In the presence of an electrochemical potential, Smoluchowski's equation takes the form

$$\partial c/\partial t = \partial[D_{\mathrm{T}}'(c)\,\partial c/\partial x - cuE]/\partial x \tag{7.55}$$

and its solution can only be obtained numerically.

8

Conformational changes

Many biological processes are correlated with a conformational change in a biopolymer either as a whole or in a limited molecular domain. One example is the change from a double-stranded helix to a single-stranded one during the transcription or replication of DNA. Another is the transition from the right-handed B-DNA to the left-handed Z-DNA recently demonstrated *in vivo* and related to the processing of RNA polymerase during transcription. Also well known are local conformational changes in the double helix induced by the intercalation of a planar aromatic molecule between two base pairs, and we should also note the existence of transient forms such as cruciform structures or regions with triple helices.

It has long been known that the binding of an oxygen molecule to the haem group of proteins such as myoglobin, haemoglobin or the cytochromes is accompanied by quite a large structural perturbation of the protein. In haemoglobin, such a conformational change explains the cooperative effect between the four haem groups (Hill phenomenon). More generally, the binding of a substrate to an enzyme is accompanied by a conformational change. Finally, allosteric behaviour, flip-flop mechanisms, etc., are amongst a number of models studied either by biochemical methods or, in some favourable cases, by physico-chemical methods (structure determination or the study of the transition from one structure to another).

8.1 Kinetics of a conformational change

8.1.1 Simple change of conformation

Analysis of the reaction

The reaction is represented by the following simple kinetic scheme of a transition from a state A to a state B:

$$\mathrm{A} \underset{k_{-1}}{\overset{k_1}{\rightleftharpoons}} \mathrm{B}$$

where k_1 and k_{-1} are the reaction rates (s^{-1}) of the two transitions. We have

$$\mathrm{d}[\mathrm{A}]/\,\mathrm{d}t = -k_1[\mathrm{A}] + k_{-1}[\mathrm{B}] \tag{8.1}$$

If [A′] and [B′] are the equilibrium concentrations of the two states, and if we put

$$x_A = [A] - [A'] \qquad \text{and} \qquad x_B = [B] - [B'] \tag{8.2}$$

(noting that $[A'] + [B'] = [A] + [B]$), then eqn 8.1 becomes

$$dx_A/dt = -k_1[A'] + k_{-1}[B'] - k_1 x_A + k_{-1} x_B \tag{8.3}$$

The first two terms on the right cancel each other, and since $x_A + x_B = 0$, we obtain

$$dx_A/dt = -(k_1 + k_{-1})x_A \tag{8.4}$$

which can be directly integrated to give

$$x_A = x_A^0 e^{-\lambda t}$$

with

$$\lambda = k_1 + k_{-1} \tag{8.5}$$

This exponential decrease of x_A with time from an initial value x_A^0 can therefore be characterized by a *relaxation time* τ given by

$$1/\tau = k_1 + k_{-1} \tag{8.6}$$

Note also that $\ln x_A$ is a linear function of time.

Coupled reactions

The transition from a state A of all or part of a macromolecule to a state B very often takes place through a series of intermediate states, some of which may be difficult to detect because of their very short life-time. Consider, for example, the simplest case:

$$A_0 \underset{k_{-1}}{\overset{k_1}{\rightleftarrows}} A_1 \underset{k_{-2}}{\overset{k_2}{\rightleftarrows}} A_2$$

Using variables defined in a similar way to those in eqn 8.2, i.e. x_0, x_1 and x_2, we have

$$dx_1/dt = k_1 x_0 - (k_{-1} + k_2)x_1 + k_{-2}x_2 \tag{8.7}$$

$$dx_2/dt = k_2 x_1 - k_{-2}x_2 \tag{8.8}$$

This system of two coupled differential equations can be solved by the method developed in Box 8.1. In general, there are two values λ_1 and λ_2 that satisfy eqn 8.8, so that

$$x_1 = C_{11}\exp(-\lambda_1 t) + C_{12}\exp(-\lambda_2 t) \tag{8.9a}$$

$$x_2 = C_{21}\exp(-\lambda_1 t) + C_{22}\exp(-\lambda_2 t) \tag{8.9b}$$

the coefficients C_{ij} depending on the initial conditions. The quantities x_1 and x_2 are no longer single-exponential functions of time and $\ln x$ no longer varies linearly with t.

Box 8.1 Solution of two coupled kinetic equations

Because $x_0 + x_1 + x_2 = 0$, eqns 8.7 and 8.8 can be written in the form

$$\mathrm{d}x_1/\mathrm{d}t = -(k_1 + k_{-1} + k_2)x_1 + (k_{-2} - k_1)x_2 \quad (1a)$$

$$\mathrm{d}x_2/\mathrm{d}t = k_2 x_1 - k_{-2}x_2 \quad (1b)$$

or

$$-\mathrm{d}x_1/\mathrm{d}t = a_{11}x_1 + a_{12}x_2 \quad \text{and} \quad -\mathrm{d}x_2/\mathrm{d}t = a_{21}x_1 + a_{22}x_2 \quad (2)$$

where the coefficients a_{11}, a_{12}, a_{21} and a_{22} are related to the rate constants by

$$a_{11} = k_1 + k_{-1} + k_2 \qquad a_{12} = -(k_1 - k_{-2}) \qquad a_{21} = -k_2 \qquad a_{22} = k_{-2} \quad (3)$$

The solutions of eqns 2 are of the form

$$x_1 = C_1 \exp(-\lambda t) \quad \text{and} \quad x_2 = C_2 \exp(-\lambda t)$$

so that

$$\mathrm{d}x_1/\mathrm{d}t = -\lambda C_1 \exp(-\lambda t) = -\lambda x_1$$

and

$$\mathrm{d}x_2/\mathrm{d}t = -\lambda C_2 \exp(-\lambda t) = -\lambda x_2 \quad (4)$$

Substituting eqns 4 into eqns 2 gives a system of two linear algebraic equations

$$(a_{11} - \lambda)x_1 + a_{12}x_2 = 0 \quad \text{and} \quad a_{12}x_1 + (a_{22} - \lambda)x_2 = 0 \quad (5)$$

whose solution is obtained by equating the determinant of coefficients to zero: when λ is one of the roots of the quadratic equation

$$\begin{vmatrix} a_{11} - \lambda & a_{12} \\ a_{21} & a_{22} - \lambda \end{vmatrix} = 0$$

$$\lambda^2 - (a_{11} + a_{22})\lambda + a_{11}a_{22} - a_{12}a_{21} = 0$$

More generally, when there are n possible states A with the sequence A_0, A_1, $A_2, \ldots, A_n$, the $(n+1)$ variables x_i must obey the conservation equation

$$\sum_{i=0}^{n} x_i = 0$$

As a result there are only n independent equations and the determinant is formed by subtracting λ from each of the diagonal elements a_{ii}. The nth-order equation obtained by equating the determinant to zero (the secular equation) generally has n roots, leading to solutions of the type

$$x_i = \sum_{j=1}^{n} C_{ij} \exp(\lambda_j t)$$

The main problem in practice is to split the experimental $x(t)$ curve into a *unique* sum of exponential functions. The required experimental accuracy quickly becomes so demanding that the numerous intermediate steps are inaccessible and their very existence remains questionable. In some cases, the rates of transition can be considerably slowed down by lowering the temperature of experiments. Any intermediate steps with long enough life-times to be measured can then be detected.

8.1.2 Diffusion-controlled bimolecular reactions

In many cases, the conformational change of a molecule or macromolecule is triggered by its encounter with a *ligand* or an *effector*, or by the binding of the macromolecule itself to another free macromolecule or to a membrane receptor. We therefore now examine the bimolecular reaction consisting of the encounter of a molecule A with a molecule B leading to a complex C. As we saw above, the motion of molecules both *in vitro* and *in vivo* obeys the laws of Brownian motion, so we have to examine the coupling of these laws with kinetic ones.

In the bimolecular reaction $A + B \rightarrow C$, a necessary condition for the formation of the complex C is the prior encounter of two molecules A and B freely diffusing in the medium. Such a process imposes a limit on the rate constant that we shall now attempt to determine. Formally, we can write:

$$A + B \underset{k_{-1}}{\overset{k_1}{\rightleftharpoons}} (A \ldots B) \underset{k_{-2}}{\overset{k_2}{\rightleftharpoons}} C$$

where $(A \cdots B)$ is a transient state of small concentration resulting from the temporary association of two molecules A and B because of their collision. It can be assumed that $(A \cdots B)$ is in a steady state and hence that $d[A \cdots B]/dt = 0$. We then have:

$$d[A]\cdots[B]/dt = k_1[A][B] - k_{-1}[A]\cdots[B] - k_2[A]\cdots[B] + k_{-2}[C] = 0 \qquad (8.11)$$

Hence

$$[A]\cdots[B] = (k_1[A][B] + k_{-2}[C])/(k_{-1} + k_2) \qquad (8.12)$$

Since

$$d[C]/dt = k_2[A]\cdots[B] - k_{-2}[C] \qquad (8.13)$$

using eqn 8.12 gives

$$d[C]/dt = k_1k_2[A][B]/(k_{-1} + k_2) - k_{-1}k_{-2}[C]/(k_{-1} + k_2) = k_{12}[A][B] - k_{21}[C] \qquad (8.14)$$

where

$$k_{12} = k_1k_2/(k_{-1} + k_2) \quad \text{and} \quad k_{21} = k_{-1}k_{-2}/(k_{-1} + k_2) \qquad (8.15)$$

The reaction is said to be diffusion-controlled when $k_2 \gg k_{-1}$, i.e. when the complex or the product is produced as soon as the complex $(A \cdots B)$ is present and before any dissociation occurs. Equations 8.15 then become

$$k_{12} = k_1 \quad \text{and} \quad k_{21} = k_{-1}k_{-2}/k_2 \qquad (8.16)$$

To calculate k_1 (see Box 8.2), it will be assumed that each of the molecules A and B is spherical and that the efficiency of the collision is independent of where the area of encounter is located on the spherical surface. If this is not the case, a steric factor

Box 8.2 **Calculation of a diffusional collision**

As a first approach, we assume that the only interaction between A and B is collisional. Fick's law gives us the flux of matter due to molecule A (with spherical symmetry):

$$J_A = -D_A\, \partial C_A/\partial r \tag{1}$$

Assuming firstly that B is at rest, a spherical region precluding any approach by A can be drawn around B of radius $r_0 = r_A + r_B$. Hence, for the centre of gravity of A we have

- $r < r_0$

$$C_A = 0 \tag{2a}$$

- $r \to \infty$

$$C_A \to C_A^o \tag{2b}$$

where C_A^o is the mean concentration of A in solution expressed as the number of molecules per unit volume. The amount of A per second flowing towards B through a sphere of radius r with centre at the centre of gravity of B is given by

$$\mathrm{d}q_A/\mathrm{d}t = -4\pi r^2 J_A = 4\pi r^2 D_A\, \partial C_A/\partial r \tag{3}$$

This is easily integrated to give

$$(\mathrm{d}q_A/\mathrm{d}t) \int_{r_0}^{\infty} \mathrm{d}r/r^2 = 4\pi D_A \int_0^{C_A^o} \mathrm{d}C_A \tag{4}$$

and hence

$$\mathrm{d}q_A/\mathrm{d}t = 4\pi D_A r_0 C_A^o \tag{5}$$

In this expression, r_0 is in m, D_A in $m^2\,s^{-1}$ and C_A in molecules m^{-3}, so that $\mathrm{d}q_A/\mathrm{d}t$ is in molecules s^{-1} and gives the number of molecules of A colliding with B per second.

If we now take into account the movement of B, assuming that the movements of A and B are uncorrelated, D_A is replaced by $D_A + D_B$. Moreover, the probability of collision is proportional to the number of B molecules per unit volume C_B^o. The total amount of the complex formed per second is therefore

$$\mathrm{d}q_{AB}/\mathrm{d}t = 4\pi r_0 (D_A + D_B) C_A^o C_B^o \tag{6}$$

To express this in moles of AB, we use the fact that the number of molecules per m^3 is $10^3 N$ where N is Avogadro's number, and that the molar concentration is defined per litre of solution. Thus:

$$10^3 N \mathrm{d}(AB)/\mathrm{d}t = 4\pi r_0 (D_A + D_B) N^2 (A)(B) 10^6$$

The rate constant k_1 in $mol^{-1}\,s^{-1}$ is therefore

$$k_1 = 4\pi r_0 (D_A + D_B) 10^3 N \tag{7}$$

If there is an interaction potential U between A and B, eqn 1 must be amended to include an additional flux. U is assumed to be a potential with spherical symmetry, which is only roughly true because of domains of preferential interaction on the molecular surface. The molecule A experiences a force $X = -\partial U/\partial r$ and therefore moves with a velocity $v = X/f\eta_0 = XD_A/k_BT$, giving an additional flux

$$J_C = -C_A D_A(\partial U/\partial r)/k_BT \quad (8)$$

Equation 1 thus becomes

$$J_A = -D_A[dC_A/dr + (dU/dr)C_A/k_BT] \quad (9)$$

The factor in square brackets on the right can be written in the form

$$\exp(-U/k_BT)d[C_A \exp(U/k_BT)]/dr \quad (10)$$

In addition, the boundary conditions are

- $r < r_0$

$$C_A = 0 \quad \text{and} \quad U = U_0 \quad (11a)$$

- $r \to \infty$

$$C_A \to C_A^0 \quad \text{and} \quad U \to 0 \quad (11b)$$

From eqns 3, 9, 10 and 11, we obtain

$$(dq_A/dt)\int_{r_0}^{\infty} e^{U/k_BT} dr/r^2 = 4\pi D_A \int_0^{C_A^0} d(C_A e^{U/k_BT}) \quad (12)$$

The integral on the left-hand side can only be evaluated if an explicit expression for $U(r)$ is known. Otherwise, we can put

$$\int_{r_0}^{\infty} \exp(U/k_BT)dr/r^2 = (fr_0)^{-1} \quad (13)$$

where f is a numerical factor. If, for example, U is an electrostatic potential of the form $z_A z_B e^2/4\pi\varepsilon_r\varepsilon_0 r$, z_A and z_B being the charges on A and B, the integration is simple since $1/r^2$ is the derivative of U except for a numerical coefficient. If we put

$$y = z_A z_B e^2/4\pi\varepsilon_r\varepsilon_0 k_BT \quad (14)$$

then

$$(fr_0)^{-1} = y^{-1}[\exp(y/r_0) - 1]$$

from which

- if $y \ll 1$, $f \approx 1$;
- if $y > 0$ (repulsive potential), $f < 1$;
- if $y < 0$ (attractive potential), $f > 1$.

Inserting eqn 14 into eqn 12 and treating the mutual approach in a similar way, we finally obtain

$$k_1 = 4\pi f(D_A + D_B)10^3 Nr_0 \quad (15)$$

generally less than 1 must be introduced into the expression for k_1. It is found (eqn 7 in Box 8.2) that

$$k_1 = 4\pi r_0 f(D_A + D_B)10^3 N \quad (8.17)$$

where r_0 is the sum of the radii r_A and r_B of the reacting molecules, and D_A and D_B are their respective translational diffusion coefficients. The numerical factor f takes into account interactions between the molecules other than collisions. Thus, with $D_A = 5 \times 10^{-10}$ (small molecule) and $D_B = 5 \times 10^{-11}$ (protein), and with $r_0 = 2 \times 10^{-9}$ m, we find with $f = 1$ that

$$k_1 = 8.3 \times 10^9 \text{mol}^{-1}\text{s}^{-1}$$

We shall see later in Part V (Chapter 23) that experimental values of k_1 for interactions between two macromolecules are far too high and that this leads us to propose another mechanism for diffusional collision.

8.2 Chemical relaxation

The study of coupling between several reactions using classical kinetics leads to a set of differential equations whose solution rapidly becomes more complicated as the number of reactions increases. Another experimental method described later is to displace the system abruptly from the equilibrium state and follow its subsequent evolution.

8.2.1 Basic principle

Consider a system characterized by a measurable quantity S in an equilibrium state where $S = S_0$. If the system is displaced from equilibrium by an abrupt change in an *intensive* parameter (pressure, temperature, electric field, etc.), the system evolves towards a new equilibrium state S'. A readjustment of the perturbed system of this kind is known as *chemical relaxation*. Assuming that the changes are always small, the rate of change can be considered as proportional to the difference $S - S'$, so that

$$\mathrm{d}S/\mathrm{d}t = -k(S - S') \quad (8.18)$$

Putting

$$x = S - S_0 \quad (x \ll S) \qquad \text{and} \qquad x' = S' - S_0 \quad (8.19)$$

we have

$$\mathrm{d}x/\mathrm{d}t = -k(x - x') \quad (8.20)$$

When the difference in the value of S between the two equilibrium states (i.e. x') is independent of time, or when $x(t)$ is a step function (i.e. $x = 0$ for $t < 0$ and $x =$ constant for $t > 0$), eqn 8.20 can be integrated directly to give

$$x = x'(1 - \mathrm{e}^{-kt}) \quad (8.21)$$

or, putting $1/k = \tau$:

$$x = x'(1 - \mathrm{e}^{-t/\tau}) \quad (8.22)$$

Here τ is a relaxation time, and the variation of x with t is shown in Fig. 8.1. Equation 8.20 can also be written in the form

$$\tau \mathrm{d}x/\mathrm{d}t = x' - x \quad (8.23)$$

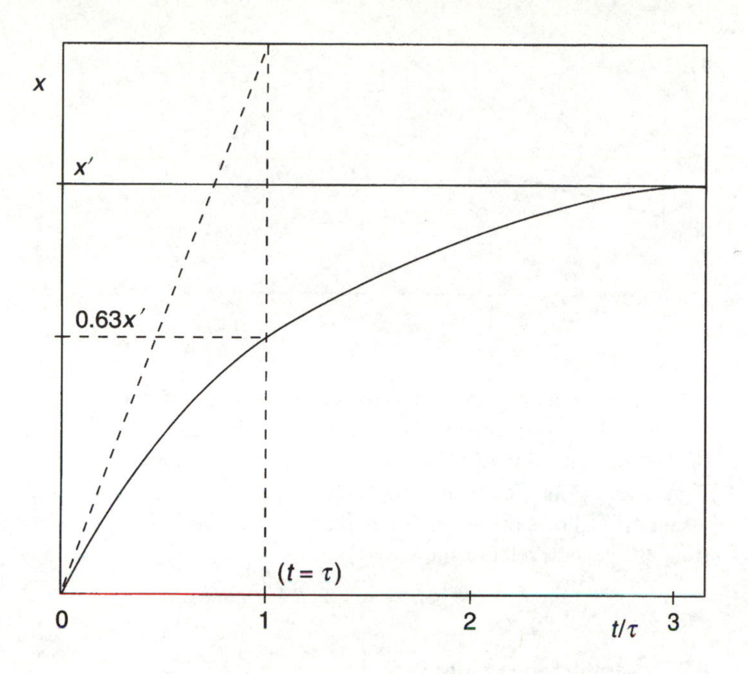

Figure 8.1 Variation of x with t for a chemical reaction in which x' is independent of time.

8.2.2 Application to a bimolecular reaction

Consider the reaction defined by the scheme

$$\mathrm{A + B} \underset{k_{-1}}{\overset{k_1}{\rightleftharpoons}} \mathrm{C}$$

Note that the two rate constants are not expressed in the same units: k_1 is in $\mathrm{mol^{-1}\,s^{-1}}$ and k_{-1} is in $\mathrm{s^{-1}}$. We have

$$\mathrm{d[C]/d}t = k_1[\mathrm{A}][\mathrm{B}] - k_{-1}[\mathrm{C}] \tag{8.24}$$

With the definitions of x and x' in eqns 8.19 and using the relationships $-x_\mathrm{C} = x_\mathrm{A} = x_\mathrm{B}$ and $-x'_\mathrm{C} = x'_\mathrm{A} = x'_\mathrm{B}$, we obtain

$$[\mathrm{A}] = [\mathrm{A}'] + x_\mathrm{A} - x'_\mathrm{A} \qquad [\mathrm{B}] = [\mathrm{B}'] + x_\mathrm{A} - x'_\mathrm{A} \qquad [\mathrm{C}] = [\mathrm{C}'] - (x_\mathrm{A} - x'_\mathrm{A})$$

and introducing these into eqn 8.24 gives us

$$\begin{aligned}-\mathrm{d}x_\mathrm{A}/\mathrm{d}t = {} & k_1[\mathrm{A}'][\mathrm{B}'] - k_1[\mathrm{C}'] + (x_\mathrm{A} - x'_\mathrm{A})k_1[\mathrm{A}' + \mathrm{B}'] \\ & + k_1(x_\mathrm{A} - x'_\mathrm{A})^2 + k_{-1}(x_\mathrm{A} - x'_\mathrm{A})\end{aligned} \tag{8.25}$$

The first two terms on the right cancel because $[\mathrm{A}'][\mathrm{B}']$ and $[\mathrm{C}']$ correspond to a new equilibrium state. The fourth term is negligible since we are still assuming a small-amplitude perturbation, so eqn 8.25 becomes

$$-\mathrm{d}x_\mathrm{A}/\mathrm{d}t = (x_\mathrm{A} - x'_\mathrm{A})(k_1[\mathrm{A}' + \mathrm{B}'] + k_{-1}) \tag{8.26}$$

Comparing eqns 8.26 and 8.23 gives us:

$$\tau^{-1} = k_1[\mathrm{A}' + \mathrm{B}'] + k_{-1} \tag{8.27}$$

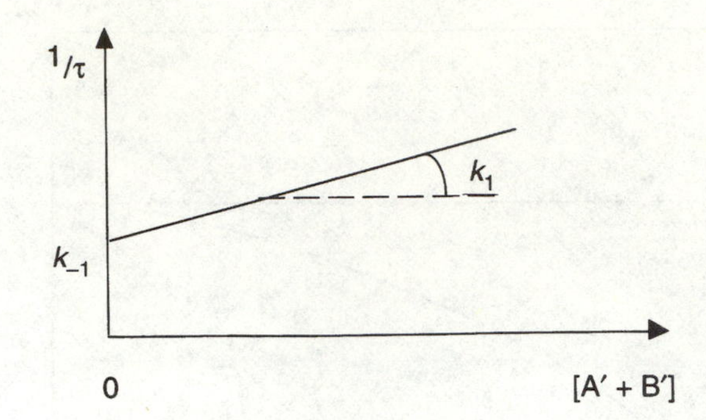

Figure 8.2 Variation of $1/\tau$ with $[A'] + [B']$.

Plotting $1/\tau$ against $[A' + B']$ yields a straight line of slope k_1 and intercept k_{-1} (Fig. 8.2). The two rate constants can therefore be determined from measurements of the concentrations $[A']$ and $[B']$ at equilibrium and of the relaxation time. If measurements are made at various temperatures, a graph of $\ln k_1$ and $\ln k_{-1}$ against $1/T$ allows the activation enthalpies ΔH_1 and ΔH_{-1} to be determined by using the general relationship

$$\ln k = -\Delta H/RT + \text{constant} \tag{8.28}$$

8.2.3 Multi-step reactions

In many interactions between a molecule and a macromolecule, a first collisional, and generally non-specific, step is followed by a specific interaction such as intercalation in DNA, binding of the substrate to the enzymatic site, recognition of a receptor, and so on. The simplest way to represent the reaction is to consider two steps:

$$\mathrm{A + B} \underset{k_{-1}}{\overset{k_1}{\rightleftharpoons}} \mathrm{C_1} \underset{k_{-2}}{\overset{k_2}{\rightleftharpoons}} \mathrm{C_2}$$

where the complex C_1 formed at the beginning is transformed into a complex C_2. Details of the calculation are given in Box 8.3, where the notation used in eqn 5 leads to the final expression for $1/\tau$:

$$1/\tau = \{a_{11} + a_{22} \pm [(a_{11} + a_{22})^2 - 4(a_{11}a_{22} - a_{12}a_{21})]^{1/2}\}/2 \tag{8.29}$$

as in eqn 6 in Box 8.3.

Equation 8.29 can often be simplified, for example if

$$a_{11} = k_1[\mathrm{A}' + \mathrm{B}'] + k_{-1} \gg a_{22} = k_2 + k_{-2} \tag{8.30}$$

i.e. the first step is much faster than the second, so that the bimolecular process has time to reach equilibrium before the conformational change $C_1 \rightleftharpoons C_2$ has started. In this case, we find two relaxation times:

$$1/\tau_1 \approx a_{11} = k_1[\mathrm{A}' + \mathrm{B}'] + k_{-1} \tag{8.31}$$

$$1/\tau_2 \approx k_{-2} + k_2/\{1 + (k_{-1}/k_1[\mathrm{A}' + \mathrm{B}'])\} \tag{8.32}$$

We again find a linear variation of $1/\tau_1$ with $[A' + B']$, thus enabling both k_1 and k_{-1} to be determined. The variation of $1/\tau_2$ with $[A' + B']$, on the other hand, is hyperbolic.

Box 8.3 **Two-step kinetics**

Denoting concentrations by letters and equilibrium values by primes, we define x_A, etc., as before by

$$[A] = [A'] + x_A \qquad \ldots \qquad [C_2] = [C_2'] + x_2 \tag{1}$$

with $x_A = x_B = -(x_1 + x_2)$. We then have

$$-d[A]/dt = k_1[A][B] - k_{-1}[C_1] \tag{2a}$$

$$d[C_2]/dt = k_2[C_1] - k_{-2}[C_2] \tag{2b}$$

which, replacing x_1 by $-(x_A + x_2)$, leads to

$$-dx_A/dt = k_1[A' + B'] + k_{-1}x_A + k_{-1}x_2 \tag{3}$$

and similarly

$$dx_2/dt = k_2([C_1'] + x_1) - k_{-2}([C_2'] + x_2) = -k_2(x_A + x_2) - k_{-2}x_2$$

since $k_2[C_1'] - k_{-2}[C_2'] = 0$, and finally

$$-dx_2/dt = k_2x_A + (k_2 + k_{-2})x_2 \tag{4}$$

Equations 3 and 4 can be put in matrix form

$$\begin{pmatrix} dx_A/dt \\ dx_2/dt \end{pmatrix} = \begin{pmatrix} a_{11} & a_{12} \\ a_{21} & a_{22} \end{pmatrix} \begin{pmatrix} x_A \\ x_2 \end{pmatrix}$$

by putting

$$a_{11} = k_1[A' + B'] + k_{-1} \qquad a_{12} = k_{-1} \qquad a_{21} = k_2 \qquad a_{22} = k_2 + k_{-2} \tag{5}$$

The solutions are obtained by adding $-1/\tau$ to each element of the principal diagonal and equating the determinant of the matrix to zero. The roots of the resultant quadratic equation give the values of $1/\tau$:

$$1/\tau = \{a_{11} + a_{22} \pm [(a_{11} + a_{22})^2 - 4(a_{11}a_{22} - a_{12}a_{21})]^{1/2}\}/2 \tag{6}$$

as in eqn 8.29.

From eqn 8.32, given that $k_1/k_{-1} = K_1$, the equilibrium constant, we can determine k_2 and k_{-2}. This type of behaviour is illustrated in Fig. 8.3 for the intercalation of a planar aromatic molecule into a double helix of DNA. The first step is the non-specific electrostatic binding to phosphate groups; the second is the intercalation proper, with the large conformational changes accompanying it.

Reactions often need the presence of a catalyst C and in this case the reaction scheme takes the form:

$$A + C \underset{k_{21}}{\overset{k_{12}}{\rightleftharpoons}} B + C$$

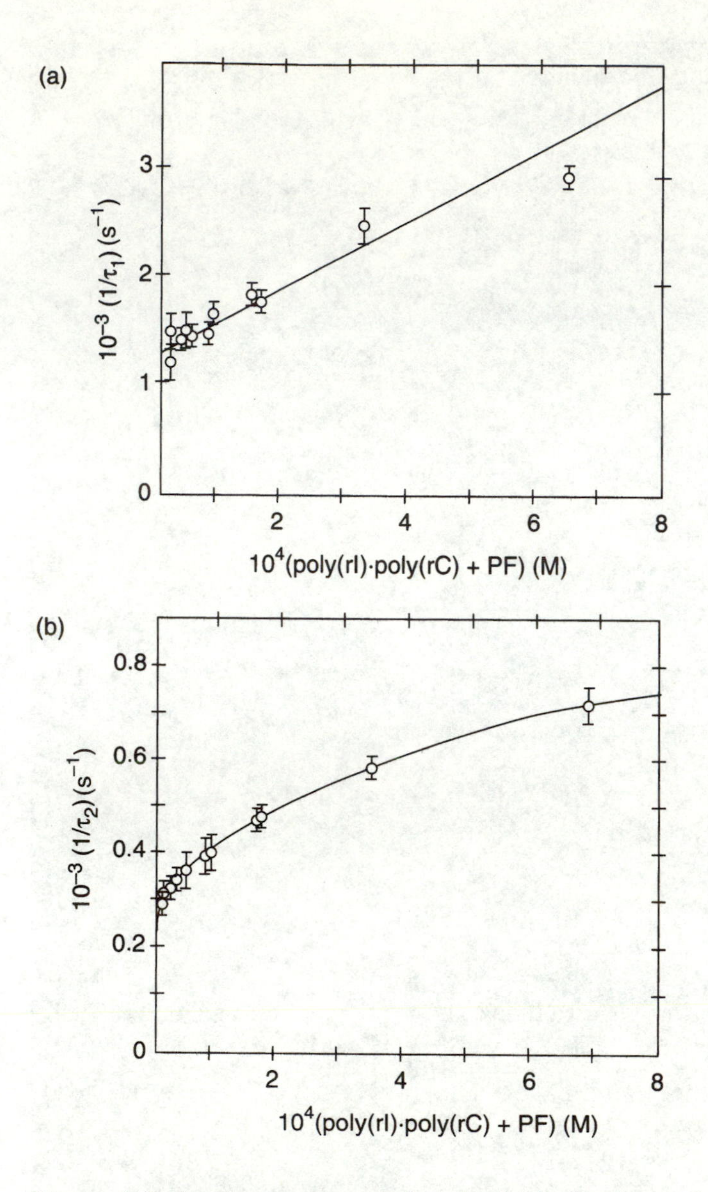

Figure 8.3 Temperature-jump kinetic study of the intercalation of proflavin (PF) in poly(rI).poly(rC): (a) variation of $1/\tau_1$ with $[A' + B']$ (i.e. poly(rI).poly(rC) + PF); (b) variation of I/τ_2 with $[A' + B']$. (From Steenbergen and Mohr, 1973. Reprinted from *Biopolymers* with the permission of John Wiley & Sons. © 1973 John Wiley & Sons.)

Since by definition the same amount of catalyst is left after the reaction as before it, the same conventions as above lead to

$$x_A = -x_B \qquad x'_A = -x'_B$$

$$[A] = [A'] + x_A - x'_A \qquad [B] = [B'] - (x_A - x'_A)$$

Hence

$$-dx_A/dt = k_{12}[A'][C] - k_{21}[B'][C] + k_{12}[C](x_A - x'_A) + k_{21}[C](x_A - x'_A)$$

The first two terms on the right cancel and again comparing what is left with eqn 8.23, we obtain:

$$1/\tau = (k_{12} + k_{21})[C] \tag{8.33}$$

This is a precise quantitative relationship embracing the intuitive idea that the reaction proceeds more rapidly when the catalyst concentration increases. This idea will be used later when studying proton exchange.

8.2.4 **Transition from double helix to single strand**

An interesting example of transition kinetics between two structures is the conformational change during the renaturation or denaturation of a self-complementary oligonucleotide, i.e. the transition from a double-helical structure to a single-stranded one.

Since such an equilibrium depends on temperature, the relaxation mechanism can easily be studied by measuring the absorbance at 260 nm of the oligonucleotide solution after a jump in temperature. The experiment was carried out with AAA ... UUU (A_nU_n). The reaction scheme is:

$$2\text{M} \underset{k_{-1}}{\overset{k_1}{\rightleftharpoons}} \text{D}$$

where M denotes the single-stranded form and D the double-helical form. We have

$$\text{d[D]}/\text{d}t = k_1[\text{M}]^2 - k_{-1}[\text{D}]$$

and since $[\text{D}]-[\text{D}'] = -([\text{M}]-[\text{M}'])/2$, we also have $x_\text{M} = -2x_\text{D}$ and $x'_M = -2x'_D$. Replacing [M] with $[\text{M}'] + x_\text{M} - x'_M$ as before, using the relationship that $k_1[\text{M}']^2 - k_{-1}[\text{D}'] = 0$, and neglecting terms in $(x_\text{M} - x'_\text{M})^2$, we obtain

$$-\text{d}x_\text{M}/\text{d}t = (4k_1[\text{M}'] + k_{-1})(x_\text{M} - x'_\text{M})$$

which, again by comparison with eqn 8.23, gives

$$1/\tau = 4k_1[\text{M}'] + k_{-1}$$

This result has to be expressed in terms of the total concentration of the single strand $C_\text{T} = 2[\text{D}] + [\text{M}]$. Because $k_1/k_{-1} = [\text{D}']/[\text{M}']^2$, we have

$$[\text{M}']^2 = k_{-1}(C_\text{T} - [\text{M}'])/2k_1$$

However,

$$1/\tau^2 = 16k_1^2[\text{M}']^2 + 8k_1k_{-1}[\text{M}'] + k_{-1}^2$$

and replacing $[\text{M}']^2$ by its value finally gives

$$1/\tau^2 = 8k_1k_{-1}C_\text{T} + k_{-1}^2$$

A graph of $1/\tau^2$ against C_T therefore gives a straight line from which k_1 and k_{-1} can be obtained. A_4U_4, for example, gives $k_1 \approx 10^6\ \text{mol}^{-1}\ \text{s}^{-1}$ and $k_{-1} \approx 10^4\ \text{s}^{-1}$.

When the same problem is approached with DNA, it becomes much more complex because we then have to take account of the great length of the molecule and the mechanism of base-pair opening and unwinding. In addition, there is the fact that the stability of various domains in DNA depends on their base composition. Finally, the presence of loops and supercoils would complicate the process even further. In fact, such a problem is never encountered inside the cell: whether for replication or transcription, the transition of DNA from one form to another always depends on its interaction with a protein.

8.3 Kinetics of the helix–coil transition

The kinetics of the helix–coil transition in proteins retains its significance *in vivo* since it occurs mainly during the folding of the peptide chain just formed on the ribosome.

We use the same notation as used previously in the Zimm and Bragg theory (section 3.5.3) to describe this transition. Statistical weights 1, s and σs are given respectively to an element of the coil, an element of the helix and an element at the end of the helix. The *nucleation parameter* σ, which is much less than 1, is a measure of the difficulty of initiating a helix starting from a coil.

In addition, we assign '0' to an element of the coil and '1' to an element of the helix. Each polymer molecule can thus be represented by a succession of '0's and '1's. The kinetic description can be restricted to the study of the transition from one triplet to another by envisaging a nucleation stage and a growth stage for the helix, both stages being reversible. The eight possible triplets take part in the following reactions, where k_1 and k_0 are the rate constants for the addition and subtraction of an element respectively:

- Growth stage

$$001 \underset{k_0}{\overset{k_1}{\rightleftharpoons}} 011 \qquad 100 \underset{k_0}{\overset{k_1}{\rightleftharpoons}} 110$$

The parameter s is in fact an equilibrium constant equal to k_1/k_0.

- Nucleation stage

$$000 \underset{k_0}{\overset{\sigma k_1}{\rightleftharpoons}} 010 \qquad 101 \underset{\sigma k_0}{\overset{k_1}{\rightleftharpoons}} 111$$

Helix formation **Coil formation**

Having set up the reaction scheme, the calculation is carried out as in Box 8.4. The relaxation time for the helix–coil transition is found to be about 10^{-7} s. Because this is such a short time (less than a microsecond), the creation of an α-helix is undoubtedly not a limiting factor in the folding of the peptide chain.

8.4 Crossing a potential barrier

Until now, we have only considered formal aspects of the transition from one conformation to another. We have shown how experimental parameters can be determined and related to kinetic quantities characterizing the reaction. To go any further we must investigate the reaction mechanism itself and in particular we must analyse the energies involved in conformational changes.

Such changes only occur after crossing a *potential barrier* between the initial and final conformations. The transition from one state to another using only thermal

Box 8.4 **Kinetic study of the helix–coil transition**

Let f be the fraction of molecules in a given state. The fraction of helices θ will then be

$$\theta = f_{011} + f_{110} + f_{010} + f_{111} \tag{1}$$

The kinetic equation giving the time variation of θ is given by

$$\begin{aligned} \mathrm{d}\theta/\mathrm{d}t = {} & k_1(f_{001} + f_{100}) - k_0(f_{011} + f_{110}) \\ & + \sigma k_1 f_{000} - k_0 f_{010} + k_1 f_{101} - \sigma k_0 f_{111} \end{aligned} \tag{2}$$

Using eqn 1 in eqn 2 and the relationship $\sum f_i = 1$, we obtain

$$\mathrm{d}\theta/\mathrm{d}t = k_1 - (k_0 + k_1)\theta + (1 - \sigma)(k_0 f_{111} - k_1 f_{000}) \tag{3}$$

or, after dividing both sides by k_0:

$$(1/k_0)\mathrm{d}\theta/\mathrm{d}t = s - (1 + s)\theta + (1 - \sigma)(f_{111} - s f_{000}) \tag{4}$$

To go any further, we have to go over the process again and consider it as the sum of a series of elementary processes. Each of them corresponds to an element of conformational change and can therefore be described by first-order kinetics with a relaxation time τ_i that depends on the state of the neighbouring elements, as in any cooperative phenomenon. The overall process will then exhibit a spectrum of relaxation times, which amounts to writing

$$\theta = \theta_\infty + (\theta_0 - \theta_\infty)\sum_i \beta_i \exp(-t/\tau_i) \tag{5}$$

where θ_0 is the helical fraction at $t = 0$, θ_∞ is the fraction as $t \to \infty$, and β_i is the statistical weight of elements with a relaxation time τ_i. Differentiating eqn 5 with respect to time gives

$$\mathrm{d}\theta/\mathrm{d}t = -(\theta_0 - \theta_\infty)\sum_i \beta_i \tau_i^{-1} \exp(-t/\tau_i) = \Delta\theta \sum_i \beta_i \tau_i^{-1} \exp(-t/\tau_i) \tag{6}$$

where $\Delta\theta = \theta_\infty - \theta_0$.

We define a mean relaxation time τ^* as the value obtained experimentally from the initial slope of the curve represented by eqn 5:

$$1/\tau^* = \Delta\theta^{-1}(\mathrm{d}\theta/\mathrm{d}t)_{t=0} \tag{7}$$

since $\exp(-t/\tau_i) = 1$ when t = 0. Now choose an equilibrium value of θ and consider a variation of s rather than of t. In other words, the quantity $\theta(s)$ is given an initial value θ_0 and s is changed by Δs so that the new value $\theta(s + \Delta s)$ is equal to the final value θ_∞:

$$\Delta\theta = \theta_\infty(s + \Delta s) - \theta(s)$$

The rate of transition from s to $s + \Delta s$ is obtained by differentiating eqn 4 with respect to s, i.e.

$$(1/k_0)(\mathrm{d}\theta/\mathrm{d}t)_{t=0} = [1 - \theta - (1 - \sigma)f'_{000}]\Delta s \tag{8}$$

with the equilibrium value f'_{000} replacing f_{000}. Combining eqns 7 and 8 gives

$$1/\tau^* = k_0[1 - \theta - (1 - \sigma)f'_{000}]\,\Delta s/\Delta\theta \quad (9)$$

An interesting point is the transition temperature at which $\theta = \frac{1}{2}$ and $s = 1$ ($k_1 = k_0$). In this case, we have $\Delta s/\Delta\theta = 4\sigma^{1/2}$ (see eqn 3.21) and $f'_{000} = (1 - 2\sigma^{1/2})/2$. Substituting these values in eqn 9 and remembering that σ is very small, we obtain

$$1/\tau^* \approx 4\sigma k_0 = 4\sigma k_1$$

and since k_1 is of the same order as the diffusional rate constant (10^{10}) and $\sigma \approx 10^{-4}$, we obtain $\tau^* \approx 10^{-7}$ s.

energy has a probability determined by the height of the barrier. An enzyme, like any catalyst, will lower the barrier by a suitable spatial arrangement of the atoms taking part in the reaction.

8.4.1 Activated complex theory

This theory was first proposed by Eyring and co-workers (see the book by Glasstone *et al.*, 1941). In a simple chemical reaction like $A + B \rightarrow C$, the transition from the initial state (A and B separate) to the final state (C) is assumed to take place after crossing a potential barrier of height equal to the activation energy of the reaction. At the top of the barrier there is a configuration called the *activated complex* $(AB)^*$, which must be adopted by the reactants before the transformation into C can take place.

To calculate the concentration of the activated complex, it is assumed to be in thermodynamic equilibrium even though it is in fact at a maximum potential energy. This is equivalent to saying that the species $(AB)^*$ could be dissociated again to produce A and B. In the reaction scheme

$$A + B \rightarrow (AB)^* \rightarrow C$$

we can write

$$[AB^*] = K^*[A][B] \quad (8.34)$$

The reaction rate is given by the number of activated complex molecules crossing the potential barrier per second, which can be written

$$-dA/dt = [AB^*] \times \text{crossing frequency} \quad (8.35)$$

The crossing frequency is the frequency ν with which the activated complex is transformed into C, which happens when one of its vibrational modes is replaced by a translational mode. Since the vibrational energy is $k_B T$, we have that

$$\nu = k_B T/h \quad (8.36)$$

The reaction rate can also be written in standard form as

$$-\mathrm{d}[\mathrm{A}]/\mathrm{d}t = k_1[\mathrm{A}][\mathrm{B}] \quad (8.37)$$

where k_1 is a rate constant in $\mathrm{mol}^{-1}\,\mathrm{s}^{-1}$. From eqns 8.35 and 8.37 we obtain

$$k_1 = K^* k_\mathrm{B} T/h \quad (8.38)$$

In practice, a factor called the *transmission coefficient* x has to be introduced on the right representing the generally unknown probability that the activated complex $(\mathrm{AB})^*$ does indeed give the final product C and not the original reactants A and B. Since the rate constant K^* can be expressed in terms of the standard change in free energy by

$$\Delta G_0^* = -RT \ln K^* \quad (8.39)$$

eqn 8.38 with the transmission coefficient gives

$$k_1 = (x k_\mathrm{B} T/h) \exp(-\Delta G_0^*/RT)$$

Replacing the free energy G by $H - TS$, we can separate the entropy term from the enthalpy term to give

$$k_1 = A(k_\mathrm{B} T/h) \exp(-\Delta H_0^*/RT) \quad (8.40)$$

where $A = x \exp(-\Delta S_0^*/R)$. The volume change in solids or liquids is generally negligible so that $\Delta H_0^* = \Delta E_0$, where E_0 is the *activation energy*.

The model of an activated complex used to describe a transient state in Eyring's theory has been given a considerable boost by the discovery of *abzymes*, i.e. antibody molecules capable of acting like enzymes. To begin with we take Pauling's idea that the function of an enzyme is to stabilize a transition state. If therefore we prepare an antibody directed against an atomic assembly mimicking the structure of a transition state, we ought to obtain an enzyme capable of arranging the reactants correctly and making them react. Before this idea could be exploited and before abzymes could be produced (Lerner *et al.*, 1991), several advances were necessary: a better knowledge of the structure of the transition state, improvements in organic synthesis to build a stable molecule having a structure identical to that of the transition state and, finally, the discovery of monoclonal antibodies.

Eyring's theory involves only the height of the potential barrier at the top of which there is a short-lived species, the activated complex. The barrier is crossed only when the vibrational energy of the activated complex is exchanged with the kinetic energy of translation. No mechanism for the crossing is proposed since the theory is based on thermodynamics and quantum mechanics.

In fact, a potential barrier may exist in a number of macromolecular structures. One occurs, for example, in the movement of a small molecule within a protein matrix, in the movement of an ion in an ion channel, in the transition from *cis*- to *trans*-proline, in the transition from C2′-*endo* to C3′-*endo* puckering in deoxyribose, etc.

Barrier crossing by a diffusional process must take into account the medium and its viscosity: this is the starting point of Kramers' theory.

8.4.2 Kramers' theory

This theory is due to Kramers (1940). Our model here is a particle moving in a potential $V(x)$ of the form:

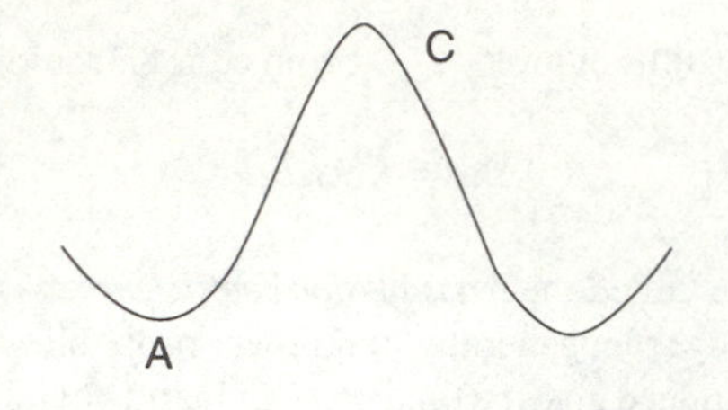

What is the velocity of a particle starting at the potential well A when it crosses the potential barrier at C by diffusion?

As a first approach, assume that the barrier height is much greater than the thermal energy, i.e. $mV_C \gg k_B T$. The problem can then be treated as one of a quasi-stationary system with an almost Maxwellian distribution around A. Since the density of particles is assumed to be very low beyond C compared with its equilibrium value, the small diffusion of particles through C will tend to restore equilibrium everywhere. When the calculation has been carried through as in Box 8.5, we obtain the frequency N with which the particle trapped in the potential well at A

Box 8.5 Crossing a potential barrier according to Kramers' theory

Following Kramers (1940), we introduce the following variables:

- $\beta = f\eta_0/m = 1/s$ (except for the factor $1-\bar{V}\rho$) where s is the sedimentation constant; β^{-1} is thus of the order of 10^{-13} s
- $q = \beta k_B T/m$; hence

$$D_T = k_B T/f\eta_0 = k_B T/\beta m = q/\beta^2 \quad (1)$$

- $K = X/m$, where X is the external force acting on the particle
- $w(x, t)$, the probability distribution function.

The total flux of particles due to Brownian motion and the force X is given by eqn 7.38:

$$J_T = -D_T\, \partial w/\partial x + wX/f\eta_0 = -q/\beta^2\, \partial w/\partial x + Kw/\beta \quad (2)$$

The continuity equation $\partial w/\partial x = -\partial J_T/\partial x$ leads to Smoluchowski's equation (see eqn 7.39):

$$\partial w/\partial t = \partial(q\beta^{-2}\, \partial w/\partial x - Kw/\beta)/\partial x \quad (3)$$

Assuming that the force X is the derivative of a potential, i.e. that

$$K = -\partial U/\partial x \quad (4)$$

it is then possible to express J_T in terms of $\exp(\beta U/q)$. Using eqn 4, we thus have

$$\mathrm{d}[\exp(\beta U/q)]/\mathrm{d}x = -K\beta q^{-1} \exp(\beta U/q) \quad (5)$$

so that

$$\mathrm{d}[w\exp(\beta U/q)]/\mathrm{d}x = [\exp(\beta U/q)](\mathrm{d}w/\mathrm{d}x - w\beta K/q) \tag{6}$$

Combining eqns 2 and 6 gives the flux J_T in the form

$$J_\mathrm{T} = -q\beta^{-2}\exp(-\beta U/q)\mathrm{d}[w\exp(\beta U/q)]/\mathrm{d}x \tag{7}$$

But J_T is assumed to be a steady-state flux, so that eqn 7 can be integrated between A and B to give

$$J_\mathrm{T}\int_\mathrm{A}^\mathrm{B}\exp(\beta U/q)\mathrm{d}x = [-q\beta^2 w\exp(\beta U/q)]_\mathrm{A}^\mathrm{B}$$

or

$$J_\mathrm{T}\int_\mathrm{A}^\mathrm{B}\beta\exp(\beta U/q)\mathrm{d}x = [(k_\mathrm{B}T/m)w\exp(\beta U/q)]_\mathrm{A}^\mathrm{B} \tag{8}$$

Replacing β/q in this by $m/k_\mathrm{B}T$, gives

$$J_\mathrm{T} = [(k_\mathrm{B}T/m\beta w)\exp(mU/k_\mathrm{B}T)]_\mathrm{A}^\mathrm{B}\Big/\int_\mathrm{A}^\mathrm{B}\exp(mU/k_\mathrm{B}T)\mathrm{d}x \tag{9}$$

As assumed at the beginning, the number ν_A of particles near A obeys the Maxwell–Boltzmann distribution:

$$d\nu_\mathrm{A} = w_\mathrm{A}\exp(-mU/k_\mathrm{B}T)\mathrm{d}x \tag{10}$$

For small x, the potential U can be taken as that of a harmonic oscillator with x giving the distance from the equilibrium position:

$$U = \omega_\mathrm{A}^2 x^2/2 \tag{11}$$

and so, integrating eqn 10:

$$\nu_\mathrm{A} = w_\mathrm{A}\int_0^\infty\exp(-m\omega_\mathrm{A}^2x^2/2k_\mathrm{B}T)\mathrm{d}x = w_\mathrm{A}\omega_\mathrm{A}^{-1}(2\pi k_\mathrm{B}T/m)^{1/2} \tag{12}$$

If we now look at the numerator of eqn 9 and substitute the value of U given by eqn 11, we see that

$$[w\exp(m\omega_\mathrm{A}^2x^2/2k_\mathrm{B}T)]_\mathrm{A}^\mathrm{B} \approx w_\mathrm{A}$$

since $x = 0$ at A (x is the distance from A) and since w is almost zero at B. Using this in eqn 9 gives

$$J_\mathrm{T} = k_\mathrm{B}T(m\beta)^{-1}w_\mathrm{A}\Big/\int_\mathrm{A}^\mathrm{B}\exp(mU/k_\mathrm{B}T)\mathrm{d}x \tag{13}$$

The ratio J_T/ν_A is the flux expressed as the number of particles divided by the number of particles at A. This is therefore the escape frequency N (in s^{-1}) for a particle trapped in the potential well at A to cross the potential barrier at C. Replacing J_T and ν_A using eqns 12 and 13, we obtain

$$N = \omega_A \beta^{-1} (k_B T / 2\pi m)^{1/2} \Big/ \int_A^B \exp(mU/k_B T) \mathrm{d}x \tag{14}$$

Since $mV_C/k_BT \gg 1$, the main contribution to the integral is from the region immediately near C where the potential U can be assumed to have the form

$$U \approx V_C - \omega_C^2 (x - x_C)^2 / 2$$

Under these conditions

$$\int_A^B \exp(mU/k_B T) \mathrm{d}x \approx \exp(mV_C/k_B T) \int_A^B \exp[-m\omega_C^2 (x - x_C)^2]/2k_B T \mathrm{d}x$$

$$= (2k_B T/m\omega_C^2)^{1/2} \exp(mV_C/k_B T) \tag{15}$$

Substituting eqn 15 into eqn 14 gives finally:

$$N = \omega_A \omega_C \exp(-mV_C/k_B T)/2\pi\beta \tag{16}$$

can escape through the potential barrier:

$$N = \omega_A \omega_C \exp(-mV_C/k_B T)/2\pi\beta \tag{8.41}$$

where ω_A and ω_C are the angular frequencies of harmonic oscillators at points A and C and where $\beta = f\eta_0/m$. Equation 8.41 was first proposed by Kramers for the velocity with which the potential barrier is crossed in a purely diffusional manner. Both the viscosity of the medium η_0 and the coefficient of friction f are involved in β. Equation 8.41 is valid only as long as it is limited to what happens over times greater than β^{-1}. It thus applies where the frictional factor $f\eta_0$ is large.

9

Experimental methods

In the first three chapters of Part II we have been studying various conceptual and theoretical aspects of the dynamics of biopolymers. In the last two, we look at the experimental side of the subject: in this chapter we give a brief survey of some widely used physical methods, while Chapter 10 will be devoted to a new approach to biopolymer dynamics using modelling methods.

9.1 Raman spectroscopy

9.1.1 Introduction

The interaction between light and molecules can only be rigorously studied within the framework of quantum mechanics, in which the incident light of frequency ν consists of photons each with energy $h\nu$. The photons interact with electrons of the molecule causing them to be raised from the ground state of energy E_0, say, to an excited state of energy E_1. The photon energy can only be absorbed if

$$h\nu \geq E_1 - E_0$$

What happens if $h\nu$ is less than $E_1 - E_0$? There are two possibilities:

1. The first is that the photons may simply be scattered, each molecule acting as a scattering centre sending photons in all directions. The energy of the scattered photon is the same as that of the incident photon. The scattered intensity varies as λ^{-4}. This is *elastic Rayleigh scattering*.

2. The second possibility is that a small number of photons exchange energy with molecular electrons. The energy involved is much less than that exchanged in electronic transitions (which is of the order of an electronvolt). It originates from coupling between photons and the motion of electrons dragged along by atoms to which they are bound in the molecule. In cases where this motion is translational or rotational Brownian motion of the whole molecule, *inelastic Rayleigh scattering* occurs and causes a broadening of the scattered frequency band (see section 9.6). When the energy corresponds to intramolecular vibrations, *Raman scattering* occurs. The frequency of the scattered photons is the sum of or difference between the initial frequency ν and the natural frequencies of vibration of the molecule ν_v. Instead of a broadening, a Raman spectrum is obtained corresponding to the discrete frequency values of the electronic ground state (not the excited state). Because there is only a small probability of such an energy transfer and because the

scattered photons are only a very small proportion (about 10^{-5}) of the incident photons, Raman spectra have very low intensities. Figure 9.1 illustrates the principle underlying the Raman effect.

An analysis of Raman spectra gives information about the various vibrational frequencies characteristic of atomic groups in the molecule. Since these frequencies can also be calculated from a model of coupled harmonic oscillators, the Raman spectrum can be used directly to investigate molecular dynamics.

9.1.2 An elementary treatment

Up to a certain point, the absorption of light can be described using a classical theory in which the electron is considered as a harmonic oscillator of natural frequency ν_0 and the electric field E of frequency ν is assumed to exert a periodic force on the electron. It can then be shown classically that the periodic displacement of the electron produced by the force creates an induced electric dipole moment

$$p(t) = \alpha(\omega)E\cos(\omega t) \tag{9.1}$$

where α is the polarizability, assumed for the time being to be a scalar quantity dependent on the angular frequency $\omega = 2\pi\nu$, and E is the electric field in the

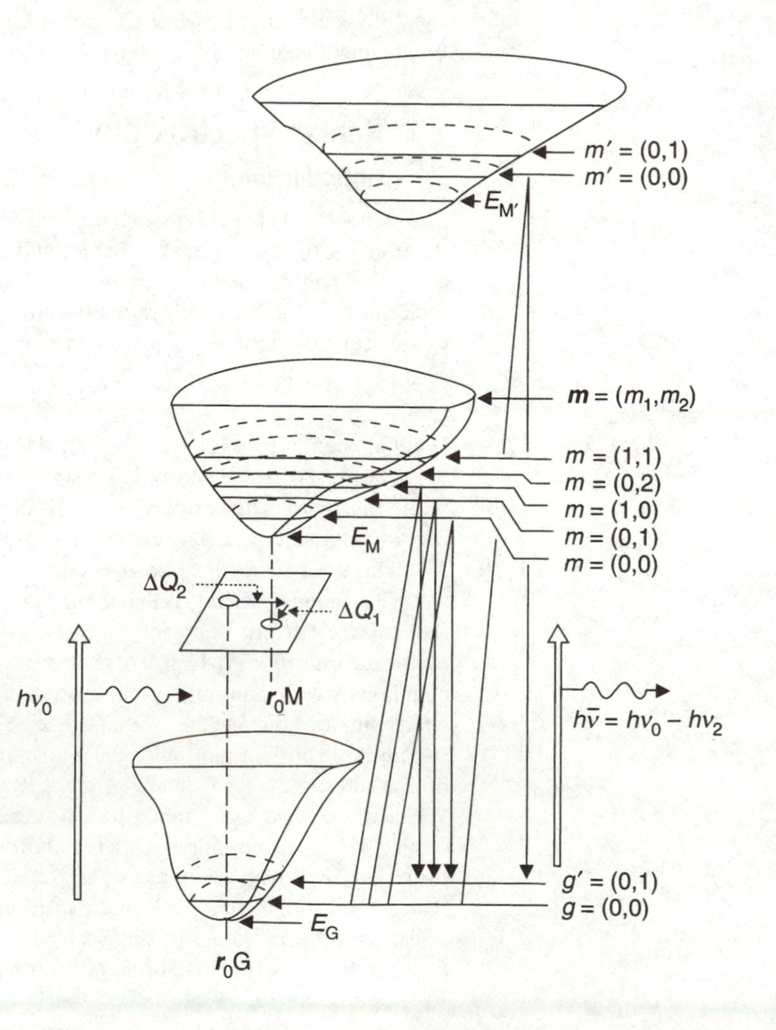

Figure 9.1 Energy levels giving rise to Raman spectra with transitions between the ground state (G) and excited states (M, M') represented in two dimensions Q_1 and Q_2. The vectors $\boldsymbol{r}_0^G$ and $\boldsymbol{r}_0^M$ correspond to the equilibrium positions in the ground and excited states, respectively. The components of the vibronic vector $\boldsymbol{m}$ are the vibrational quantum numbers $m_1, m_2, \ldots, m_n$ of the normal modes 1, 2, ..., n in the excited state M. (From Warshel, 1977. Reprinted with permission from *Annual Review of Biophysics and Bioengineering*, vol. 6 and with the author's permission. © 1977 Annual Reviews Inc.)

incident wave. To include the effect of the motion of the atoms to which the electrons are bound, the polarizability is also assumed to depend on this motion. For a given vibrational frequency ν_1, we can write

$$\alpha(\omega) = \alpha_0(\omega) + \delta\alpha(\omega)\cos(\omega_1 t) \tag{9.2}$$

where α_0 is the polarizability when the various atoms are in their equilibrium positions, $\delta\alpha$ is the magnitude of the change in polarizability and $\omega_1 = 2\pi\nu_1$. Hence

$$p(t) = E_0[\alpha_0(\omega) + \delta\alpha(\omega)\cos(\omega_1 t)]\cos(\omega t) = p_0(t) + \delta p(t)$$

where

$$p_0(t) = E_0\alpha_0(\omega)\cos(\omega t)$$

and where

$$\delta p(t) = E_0\delta\alpha(\omega)\cos(\omega_1 t)\cos(\omega t) = \tfrac{1}{2}E_0\delta\alpha(\omega)\{\cos[(\omega - \omega_1)t] + \cos[(\omega + \omega_1)t]\} \tag{9.3}$$

The first dipole moment $p_0(t)$ corresponds to elastic Rayleigh scattering at the frequency ν of the incident light. The second dipole moment $\delta p(t)$ corresponds to the Raman spectrum in two frequency bands $\omega - \omega_1$ and $\omega + \omega_1$. The first is called the *Stokes band* and extends from a few cm^{-1} to about 3500 cm^{-1}. The second, the *anti-Stokes band*, is limited to a few hundred cm^{-1}. It is characteristic of molecules already in an excited state when the incident photon arrives. The molecule does of course have a whole spectrum of vibrational frequencies ν_i $(i = 1, 2, \ldots, N)$ where N is the number of normal modes of vibration. These are obtained (see Chapter 6) by solving the differential equations describing the coupling between all the harmonic oscillators representing the motion of atoms. The magnitude $\delta\alpha$, on the other hand, depends on the normal coordinates through the relationships

$$\delta\alpha = (\partial\alpha/\partial q_i)_0 \delta q_i \tag{9.4}$$

the zero subscript indicating that the partial derivative is calculated at the equilibrium value $q_{i,0}$. This relationship assumes that the harmonic oscillator is the correct model for all the vibrations.

9.1.3 Comments on the problem

The above elementary treatment, although capable of producing relationships between measured quantities (intensity and frequency of Raman lines) and molecular parameters (amplitude and frequency of atomic motion), cannot be considered as satisfactory for the following reasons:

1. The anisotropy of the polarizability must be included. Each component of the induced dipole moment depends on all components of the electric field: in Cartesian coordinates this property is expressed by

$$\begin{aligned} p_x &= \alpha_{xx}E_x + \alpha_{xy}E_y + \alpha_{xz}E_z \\ p_y &= \alpha_{yx}E_x + \alpha_{yy}E_y + \alpha_{yz}E_z \\ p_z &= \alpha_{zx}E_x + \alpha_{zy}E_y + \alpha_{zz}E_z \end{aligned}$$

in which the nine coefficients α_{ij} are the components of the *polarizability tensor*.

2. The quantum approach involves the orthonormal wavefunctions Ψ describing the vibrational states of the molecule. For example, the transition from a vibrational state described by Ψ_1 to another described by Ψ_2 depends on integrals of the type

$$A_{ij} = \int_0^\infty \Psi_i \alpha_{ij} \Psi_j \, d\tau$$

   where the integration extends over the whole molecular volume and α_{ij} is a matrix element representing the polarizability tensor. In the same way, A_{ij} can be considered as one element of a matrix representing a vector quantity $\boldsymbol{A}$, such that the square of the modulus of $\boldsymbol{A}$ gives the intensity of the Raman line.

3. The relationship between the magnitude of the change in polarizability and the normal coordinates assumes that the harmonic oscillator model is valid. In fact, anharmonic terms have to be introduced in many cases. Coupling between rotation and vibration must also be taken into account.

9.1.4 Resonance Raman scattering

In this method, the molecule is excited with photons of sufficient energy to produce a transition from the ground state to an excited state involving an electronic transition of a group of atoms (a *chromophore*), such as an aromatic ring of the molecule. Under these conditions, a preferential energy exchange will take place with the vibrational oscillators coupled directly to the chromophore. The intensity of the corresponding Raman line can then be amplified several thousand times, with two consequences:

1. It is now possible to operate with low concentrations of the biopolymer (in the range 10^{-4} to 10^{-5} M).
2. It is also possible to pinpoint the vibrational spectrum related to the excited chromophore. For example, by selecting a substrate having absorption bands (in the visible or near-ultraviolet) clearly distinct from those of the enzyme, the dynamics of the substrate can be followed during an enzymatic reaction.

9.2 Fluorescence depolarization

9.2.1 Definitions and calculations

Fluorescence emission is largely characterized by the following two parameters:

1. The *lifetime of the excited state* τ_0. The molecule, after capturing a photon $h\nu$, is in the lowest vibrational level of the excited state. When a photon $h\nu'$ is emitted ($\nu' < \nu$), the molecule falls back to one of the vibrational levels of the ground state. The time between these two events is the lifetime of the excited state.
2. The *fluorescence quantum yield* Φ_0. This is the ratio of the number of quanta emitted per molecule to the number absorbed.

Both parameters are modified by several processes (internal conversion, falling back to the triplet state, presence of an excited-state inhibitor) that reduce τ and Φ to their experimentally determined values.

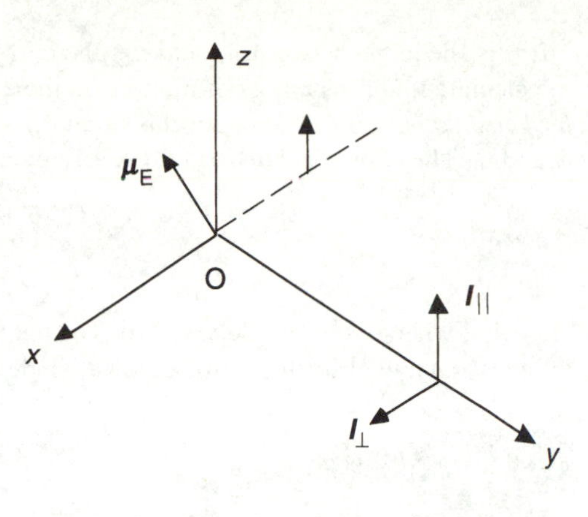

Figure 9.2 Polarization of fluorescence. The vector $\boldsymbol{\mu}_E$ is the emission dipole moment. The electric field $\boldsymbol{E}$ of the incident wave defines the z axis, while observations are made along the y axis. The emitted light has two components, $I_{\parallel}$ and $I_{\perp}$, parallel and perpendicular respectively to $\boldsymbol{E}$.

Polarization of fluorescence

A third parameter is the state of polarization of the emitted light. The transition from the ground state to the excited state (absorption) can be associated with a transition electric dipole or *absorption dipole* of moment $\boldsymbol{\mu}_{\mathrm{A}}$. Similarly, the transition from the excited state to the ground state can be associated with an *emission dipole* of moment $\boldsymbol{\mu}_{\mathrm{E}}$. The fluorescent light produced by all the emitting molecules (hence for any orientation of $\boldsymbol{\mu}_{\mathrm{E}}$) is conventionally observed along a y axis perpendicular to the $z\mathrm{O}x$ plane defined by the electric field E_z along z and the direction of propagation of the light along x (Fig. 9.2).

To characterize the partially polarized state of this light, we define two intensities $I_{\parallel}$ and $I_{\perp}$, the polarization of the first being parallel to E_z and that of the second perpendicular to E_z. If $\boldsymbol{u}_x$ and $\boldsymbol{u}_z$ are unit vectors along the x and z axes respectively, then

$$I_{\parallel} \approx (\boldsymbol{\mu}_{\mathrm{E}} \cdot \boldsymbol{u}_z)^2 = \mu_z^2 \qquad \text{and} \qquad I_{\perp} \approx (\boldsymbol{\mu}_{\mathrm{E}} \cdot \boldsymbol{u}_x)^2 = \mu_x^2 \tag{9.5}$$

The state of polarization is defined either by the *polarization p*, where

$$p = (I_{\parallel} - I_{\perp})/(I_{\parallel} + I_{\perp}) \tag{9.6}$$

or by the *polarization anisotropy r*, where

$$r = (I_{\parallel} - I_{\perp})/(I_{\parallel} + 2I_{\perp}) \tag{9.7}$$

The quantity $I_{\parallel} + 2I_{\perp}$ is the total intensity of the fluorescence light that would be observed with unpolarized incident light. There is clearly a relationship between p and r given by

$$p = 3r/(2 + r) \qquad \text{or} \qquad r = 2p/(3 - p) \tag{9.8}$$

Fundamental polarization

If the fluorescent molecules are fixed in a highly viscous or solid medium, the emission dipole μ_{E} will have the same orientation with respect to the absorption dipole μ_{A} for all molecules, but their respective directions will be different since they depend on the shape of the wavefunctions defining the states '0' and '1'. No molecular rotation is possible but, since the medium is isotropic, the molecular orientation is randomly distributed.

If α is the angle between μ_A and μ_E, the corresponding values of p and r can be calculated. Our assembly of motionless molecules gives rise to what is called *fundamental polarization*, and for this situation we denote the values of p and r by p_0 and r_0. They are calculated as in Box 9.1, giving

$$p_0 = (3\cos^2\alpha - 1)/(3 + \cos^2\alpha) \tag{9.9}$$

$$r_0 = (3\cos^2\alpha - 1)/5 \tag{9.10}$$

Note that when $\cos^2\alpha = 1/3$, $p = r = 0$. The light is completely depolarized when the emission and absorption dipoles are at angle of about 55°.

Box 9.1 Calculation of p and r

For equal μ_E and μ_A

Assume first that μ_E and μ_A are both equal to the same vector μ. In Cartesian coordinates, the direction of μ is defined by angles θ (between μ and the z axis) and ϕ (between the projection of μ on to the xy plane and the x axis), with $0 < \theta < \pi$ and $0 < \phi < 2\pi$:

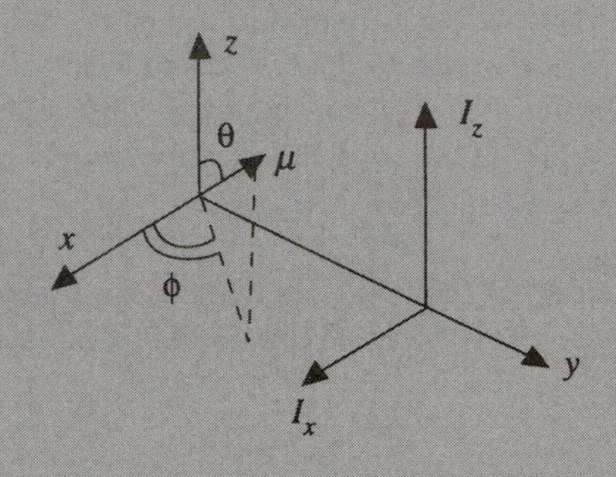

We have

$$\mu_x = \mu \sin\theta\cos\phi \qquad \mu_y = \mu\sin\theta\sin\phi \qquad \mu_z = \mu\cos\theta \tag{1}$$

In a given direction (θ, ϕ), the number of oscillators is equal to the product of the surface element $d\sigma = \sin\theta\, d\theta\, d\phi$ of a sphere of unit radius and the probability of excitation $\cos^2\theta$:

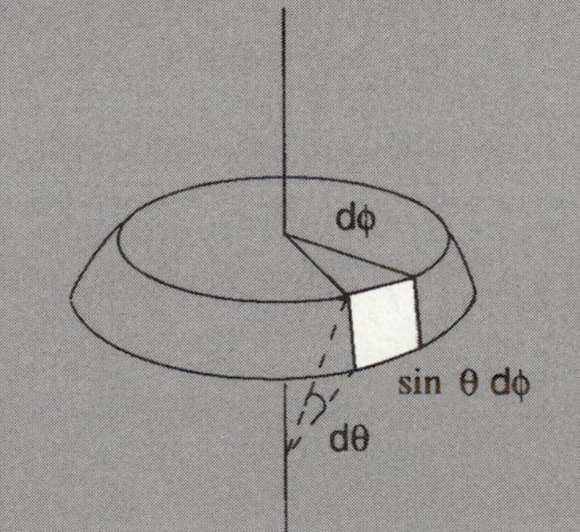

Along the y axis there are only two intensity components $I_x = \mu_x^2$ and $I_z = \mu_z^2$ corresponding to the two components of the electric field in the transverse wave emitted along the y axis. For all possible molecular orientations, we must calculate the mean values $\langle I_x \rangle$ and $\langle I_z \rangle$:

$$\langle I_x \rangle = (1/4\pi) \int\int \mu_x^2 \cos^2\theta \sin\theta \, d\theta \, d\phi \tag{2}$$

$$\langle I_z \rangle = (1/4\pi) \int\int \mu_z^2 \cos^2\theta \sin\theta \, d\theta \, d\phi \tag{3}$$

Replacing μ_x^2 and μ_z^2 using eqn 1, we see that the following integrals have to be evaluated:

$$\int_0^\pi \cos^4\theta \sin\theta \, d\theta = (1/5)[\cos^5\theta]_0^\pi = 2/5 \tag{4}$$

and

$$\int_0^\pi \cos^2\theta \sin\theta \, d\theta = -(1/3)[\cos^3\theta]_0^\pi = 2/3 \tag{5}$$

so that

$$\int_0^\pi \cos^2\theta \sin^2\theta \sin\theta \, d\theta = 2/3 - 2/5 = 4/15 \tag{6}$$

We also have that

$$(1/2\pi) \int_0^{2\pi} \cos^2\phi \, d\phi = 1/2 \tag{7}$$

Using eqns 4, 5, 6 and 7 in $\langle I_x \rangle$ and $\langle I_z \rangle$ gives

$$\langle I_x \rangle = \mu^2/15 \qquad \langle I_z \rangle = \mu^2/5 \tag{8}$$

The vertical component is three times as great as the horizontal component. Hence, using the definitions of p and r:

$$p_{(\alpha=0)} = p_{00} = (\langle I_z \rangle - \langle I_x \rangle)/(\langle I_z \rangle + \langle I_x \rangle) = 0.5 \tag{9}$$

$$r_{(\alpha=0)} = r_{00} = (\langle I_z \rangle - \langle I_x \rangle)/(\langle I_z \rangle + 2\langle I_x \rangle) = 0.4 \tag{10}$$

For any μ_E and μ_A

In general, μ_E and μ_A are not equal and are at an angle α to each other. Let the directions of the two vectors be defined by the angles θ, ϕ and θ', ϕ' respectively. Spherical trigonometry gives the following relationship between θ, θ' and α:

$$\cos\theta' = \cos\theta\cos\alpha + \sin\theta\sin\alpha\cos(\phi - \phi') \tag{11}$$

In addition, the expressions for I_x and I_z become

$$I_x = \mu_E^2 \cos^2\theta \sin^2\theta' \cos^2\phi' \tag{12}$$

$$I_z = \mu_E^2 \cos^2\theta \cos^2\theta' \tag{13}$$

Using

$$\cos^2\theta' = \cos^2\theta\cos^2\alpha + \sin^2\theta\sin^2\alpha\cos^2(\phi - \phi') + 2\sin\theta\sin\alpha + \cos\theta\cos\alpha\cos(\phi - \phi') \tag{14}$$

and $\sin^2\theta' = 1 - \cos^2\theta'$, and substituting these into eqns 12 and 13, the calculation of $\langle I_x\rangle$ and $\langle I_z\rangle$ once again involves the integrals in eqns 4 and 6, with, of course, the conditions $\langle\cos^2(\phi - \phi')\rangle = \langle\cos^2\phi\rangle = 1/2$ and $\langle\cos(\phi-\phi')\rangle = 0$. Finally, replacing $\sin^2\alpha$ with $1 - \cos^2\alpha$ and rearranging terms, we obtain

$$\langle I_x\rangle = (\mu_E^2/15)(2 - \cos^2\alpha) \quad (15)$$

$$\langle I_z\rangle = (\mu_E^2/15)(1 + 2\cos^2\alpha) \quad (16)$$

When $\alpha = 0$, we find the result already calculated in the first part of this box. From the general values for $\langle I_x\rangle$ and $\langle I_z\rangle$ and from the definitions of p and r, we arrive at:

$$p_0 = (3\cos^2\alpha - 1)/(3 + \cos^2\alpha) \quad (17)$$

$$r_0 = (3\cos^2\alpha - 1)/5 \quad (18)$$

Finally, comparing these with the values for $\alpha = 0$ (denoted by subscripts 00):

$$1/p_0 - 1/3 = (1/p_{00} - 1/3)[2/(3\cos^2\alpha - 1)] \quad (19)$$

$$r_0/r_{00} = (3\cos^2\alpha - 1)/2 \quad (20)$$

Brownian motion

If the fluorescent molecules are free to move in a solution, they are subject to Brownian motion and, in particular, they rotate about an axis as time progresses, making a random angle θ with the axis such that $\langle\theta^2\rangle = 2D_R t$. During the lifetime τ of the excited state, the molecule rotates through an angle that is larger with longer lifetimes and with greater values of the rotational coefficient of diffusion D_R. This is equivalent to introducing a random angular variable $\theta(t)$ between μ_A and μ_E instead of the constant angle α. The state of polarization of the fluorescence light will be altered and D_R can then be determined from a study of p or r. In other words, the vector μ_E bound to the molecule becomes an indicator of its angular position. The expressions for p and r are unchanged, but their mean values must be calculated, and hence the mean value $\langle\cos^2\theta(t)\rangle$. For this, we use the partial differential equation governing rotational diffusion (see eqn 5 in Box 7.4):

$$\sin\theta\,\partial f/\partial t = D_R\{\partial[(\sin\theta)^{-1}\partial f/\partial\phi]/\partial\phi + \partial(\sin\theta\,\partial f/\partial\theta)/\partial\theta\}$$

where the function $f(\theta, \phi, t)$ describes the distribution of molecular orientations and represents the density, over the surface of a sphere of unit radius, of the ends of the vectors bound to the molecule. In our case, only θ undergoes random variations, the function f depends only on θ and t so that the terms in $\partial/\partial\phi$ disappear. Hence

$$\sin\theta\,\partial f/\partial t = D_R\partial[(\sin\theta)\partial f/\partial\theta]/\partial\theta \quad (9.11)$$

For similar reasons, the normalization condition is simply

$$2\pi\int_0^T f(\theta, t)\sin\theta\,d\theta = 1$$

Putting $u(t) = \langle\cos^2\theta(t)\rangle$, Box 9.2 shows that

$$u(t) = (1/3)[1 + 2\exp(-6D_R t)] \tag{9.12}$$

This can be used to calculate an expression for $r(t)$ defined by eqn 9.7, giving, as in eqn 9.10:

$$r(t) = (3\langle\cos^2\theta(t)\rangle - 1)/5$$

Box 9.2 **Calculation of $u(t)$**

From the diffusion equation (eqn 9.11), the aim is to calculate

$$u(t) = 2\pi \int_0^\pi \langle\cos^2\theta\rangle f(\theta, t) \sin\theta \, d\theta$$

In order to obtain $\partial f/\partial t$, we differentiate $u(t)$:

$$du/dt = 2\pi \int_0^\pi \cos^2\theta \, \partial f/\partial t \sin\theta \, d\theta$$

and, replacing $\partial f/\partial t$ by its value from eqn 9.11, we have

$$du/dt = 2\pi D_R \int_0^\pi \cos^2\theta \, \partial/\partial\theta(\sin\theta \, \partial f/\partial\theta) \, d\theta$$

A first integration by parts leads to

$$du/dt = 2\pi D_R[\cos^2\theta \, (\sin\theta \, \partial f/\partial\theta)]_0^\pi + 2\pi D_R \int_0^\pi 2\cos\theta \sin^2\theta \, \partial f/\partial\theta \, d\theta$$

The first term on the right is zero, and carrying out a second integration by parts on the second term gives

$$2\pi[2\cos\theta \sin^2\theta \, f(\theta, t)]_0^\pi - 2\pi D_R \int_0^\pi 2f(\theta, t)(2\sin\theta\cos^2\theta - \sin^3\theta) \, d\theta$$

Once again, the first term of this is zero and after expanding the second term using $\sin^3\theta = \sin\theta(1 - \cos^2\theta)$ we obtain

$$du/dt = 2D_R - 6D_R u$$

To integrate this, we put $u(t) = A(t)\exp(-6D_R t)$. After replacement and elimination, we get

$$dA(t)/dt = 2D_R \exp(-6D_R t)$$

so that

$$A(t) = (1/3)\exp(-6D_R t) + A_0$$

and hence

$$u(t) = 1/3 + A_0 \exp(-6D_R t)$$

Since $u = 1$ when $t = 0$, $A_0 = 2/3$, and finally:

$$u(t) = (1/3)[1 + 2\exp(-6D_R t)]$$

and, replacing $\langle \cos^2 \theta(t) \rangle$ by $u(t)$,

$$r(t) = 2 \exp(-6D_R t)/5 \tag{9.13}$$

Note that the last calculation assumes that the angle α between μ_A and μ_E is zero, as shown by calculating $r(t)$ for $t = 0$, i.e. 0.4. In the general case of a non-zero angle α, eqn 9.13 becomes

$$r(t) = 0.2 \exp(-6D_R t)(3 \cos^2 \alpha - 1) \tag{9.14}$$

as in eqn 18 in Box 9.1, or

$$r(t) = r_0 \exp(-6D_R t) \tag{9.15}$$

where r_0 is the fundamental anisotropy given by eqn 9.10.

9.2.2 **Measurements**

Lifetime

In pulse fluorescence measurements, a solution of fluorescent molecules is irradiated by a short intense light pulse and the emitted fluorescence is analysed over the time domain. Even though the pulse is very short, it does extend over a finite time during which the intensity varies according to a function of time $g(t')$ characteristic of the light source. For each value of this intensity, the fluorescence intensity $I_F(t)$ takes the form of a *convolution integral* (see Mathematical Appendix):

$$I_F(t) = \int_0^t g(t') i_F(t - t')\, dt' \tag{9.16}$$

where $i_F(t-t')$ is the intensity emitted at each instant by the fluorescent species. When only one lifetime is present, $i_F(t-t')$ is a simple exponential function that can be obtained by a *deconvolution* process (Fig. 9.3). It can be shown (see Mathematical Appendix) that, if $J(\nu)$, $A(\nu)$ and $B(\nu)$ are the Fourier transforms of $I_F(t)$, $f(t')$ and $i_F(t-t')$ respectively, then

$$J(\nu) = A(\nu)B(\nu) \tag{9.17}$$

where ν is a frequency or the reciprocal of a time. In other words, a convolution integral in time space is transformed into a simple product of two functions in frequency space. From $J(\nu)$ and $A(\nu)$, we can obtain $B(\nu)$ and then $i_F(t)$ after an inverse Fourier transform.

Measurements of D_R

From the definition of r in eqn 9.7, and replacing I by $I_0 \exp(-t/\tau)$, we obtain

$$I_{\parallel} - I_{\perp} = r_0 I_0 \exp[-(6D_R + 1/\tau)t]$$

or

$$I_{\parallel} - I_{\perp} = r_0 I_0 \exp(-t/\tau') \tag{9.18}$$

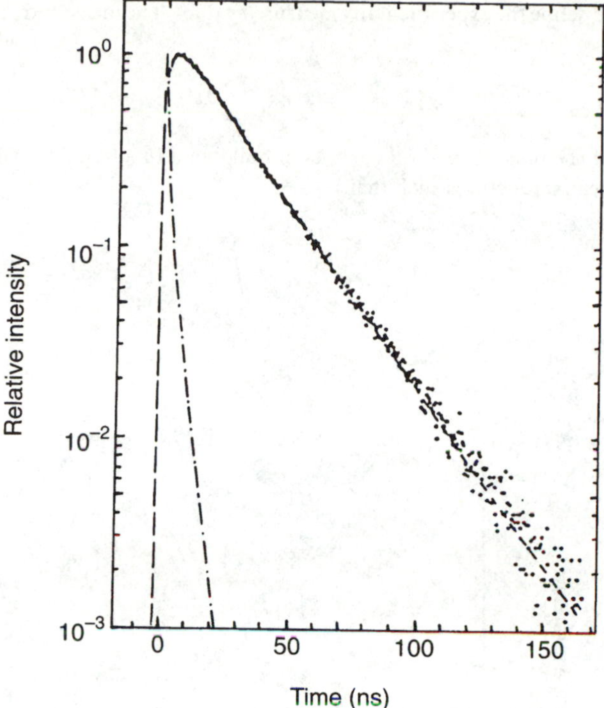

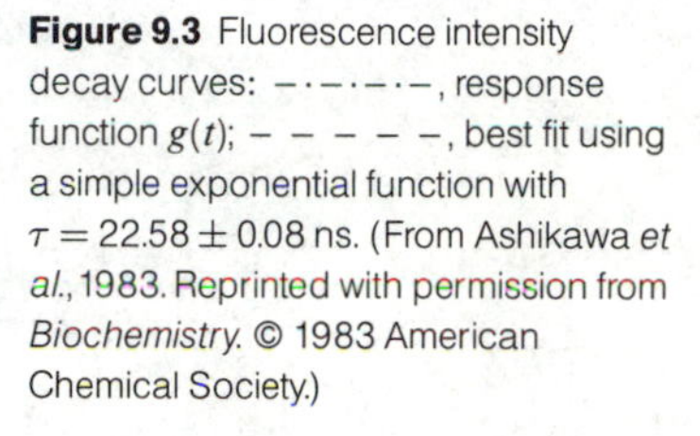

Figure 9.3 Fluorescence intensity decay curves: –·–·–·–, response function $g(t)$; – – – – –, best fit using a simple exponential function with $\tau = 22.58 \pm 0.08$ ns. (From Ashikawa *et al.*, 1983. Reprinted with permission from *Biochemistry*. © 1983 American Chemical Society.)

with

$$1/\tau' = 1/\tau + 6D_R \tag{9.19}$$

From the exponential fluorescence decay curve, τ' and hence D_R can be determined. The measurement of D_R will clearly depend on the respective values of τ and the *correlation time* τ_c defined by

$$\tau_c = 1/6D_R \tag{9.20}$$

In the special case of a large molecule, D_R is small and $\tau_c \gg \tau$. The situation is similar to that of a rigid medium in which p is almost equal to p_0 and hence $1/\tau \approx 1/\tau'$. The use of fluorescence depolarization to measure the rotational coefficient of diffusion is therefore limited to the study of macromolecules with sizes approximately equal to that of proteins. For example, a spherical protein of radius 3 nm has $D_R = 2 \times 10^6\,\text{s}^{-1}$ at 20°C in water ($\eta_0 = 10^{-3}\,\text{N s m}^{-2}$). If $\tau = 10^{-8}$ s, $6D_R\tau = 0.12$, which can easily be measured but with an uncertainty in D_R that is becoming significant. To study larger molecules, fluorescent groups with longer lifetimes would have to be available. Some molecules with a triplet state could be used.

Cases with more than one D_R

This is clearly the most common situation. The motion of an ellipsoidal molecule already involves two rotational coefficients of diffusion. We then put

$$S_{\text{exp}}(t) = I_{\parallel}(t) + 2I_{\perp}(t) \qquad \text{and} \qquad D_{\text{exp}}(t) = I_{\parallel}(t) - I_{\perp}(t) \tag{9.21}$$

to define the experimentally measured values. The measured anisotropy r_{exp} is then (see eqn 9.7)

$$r_{\text{exp}} = D_{\text{exp}}/S_{\text{exp}} \tag{9.22}$$

The present method consists in building a function $S(t-t')$ by summing exponential functions such that

$$S(t)_{\text{calc}} = \int_0^t g(t')S(t-t')\,\mathrm{d}t' \tag{9.23}$$

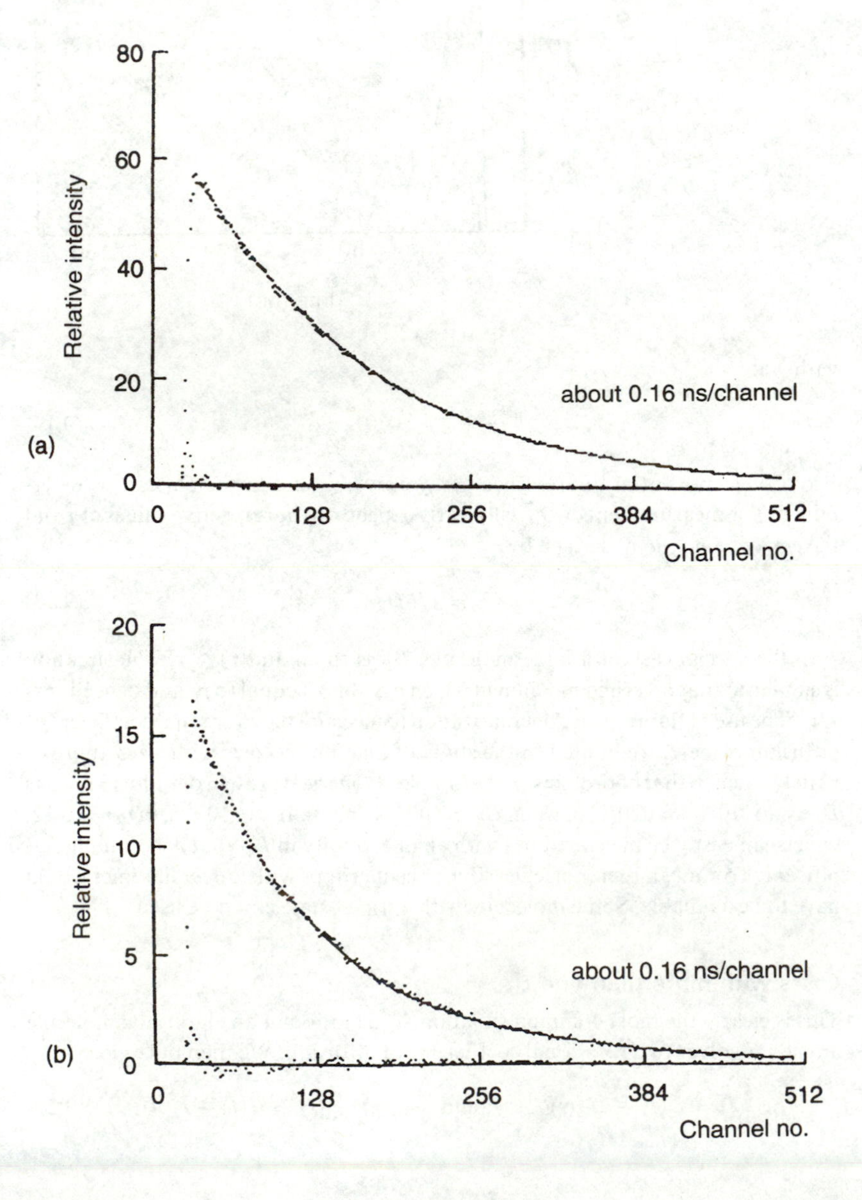

Figure 9.4 Fluorescence decay curves. In both diagrams, the points scattered around the time axis show the differences between the calculated curve and the observed values. The model used for $r(t)$ was a sum of two exponentials. (a) Comparison between $S_{\text{calc}}(t)$ and experiment; (b) comparison between $D_{\text{calc}}(t)$ and experiment. (From Thomas *et al.*, 1980. Reprinted from *Biophysical Chemistry*, vol. 12, pp. 177–88, © 1980 with kind permission from Elsevier Science NL, Sara Burgerhartstraat 25, 1055 KV Amsterdam, The Netherlands.)

is the best fit of the experimental curve $S_{exp}(t)$. Similarly, a function $D(t-t')$ can be defined as

$$D(t)_{calc} = \int_0^t g(t')D(t-t')\,dt' \qquad (9.24)$$

In this expression, D is replaced by the product $r(t)S(t)$ where $r(t)$ is constructed by using an *a priori* model. The value of $r(t)$ is adjusted to obtain the best fit between D_{calc} and D_{exp}. Figure 9.4 shows an example of this method.

9.3 Proton exchange

9.3.1 Introduction

All reactions of the acid–base type involve a transfer of protons. We shall see the particular importance of the coupling of such reactions with redox systems (electron-type proton pumps) to account for membrane energetics. Proton transfer is also involved in many enzymatic reactions. The respective conformations of the enzyme and the substrate favour proton transfer by creating local ionization conditions differing considerably from those of the medium.

Apart from these processes, proton exchange is also a widely used method for detecting the static and dynamic accessibility inside biopolymers of groups capable of exchanging a proton. In every case, a *donor–acceptor pair* is involved.

9.3.2 Exchange mechanism

Each member A and B of the pair may exist in a protonated or non-protonated form (generalized concept of acid and base), and for each the following equilibria can be written:

$$AH \rightleftharpoons A^- + H^+$$

$$BH \rightleftharpoons B^- + H^+$$

for which two dissociation equilibrium constants can be defined, namely

$$K_A = [A^-][H^+]/[AH] \quad \text{and} \quad K_B = [B^-][H^+]/[BH] \qquad (9.25)$$

The pK of both members of the pair can be defined from these constants by $pK_A = -\log K_A$ and $pK_B = -\log K_B$, so that

$$K_A = 10^{-pK_A} \quad \text{and} \quad K_B = 10^{-pK_B}$$

The transfer from donor to acceptor is described globally by the equilibrium reaction

$$AH + B^- \rightleftharpoons BH + A^-$$

If K_e is the corresponding equilibrium constant, then

$$K_e = [BH][A^-]/[AH][B^-] = [BH][A^-][H^+]/[AH][B^-][H^+] = K_A/K_B$$

or

$$K_e = 10^{\Delta pK} \tag{9.26}$$

where

$$\Delta pK = pK_B - pK_A \tag{9.27}$$

It is agreed to define ΔpK by

$$\Delta pK = pK_{acceptor} - pK_{donor} \tag{9.28}$$

and eqn 9.26 then shows that K_e will be greater than or less than 1 depending on the sign of ΔpK.

Eigen (1964) showed that the transfer mechanism will include intermediate steps and that the transfer must be written

$$AH + B \rightleftharpoons AH\cdots B \rightleftharpoons A\cdots HB \rightleftharpoons A + HB$$

The height of the potential barrier depends on the donor–acceptor pair. It varies with the length of the hydrogen bond but also depends on the presence of charges and on bond distortion. After a diffusion-controlled collision between the two members of the pair, a transient complex is formed in which protons are redistributed and which is then dissociated to give the final species. Such an exchange is described in general by second-order kinetics. The total rate constant is the product of the rate constant k_D of the transient ternary complex formation and the fraction of the complex in which the exchange may take place. As in any equilibrium, this fraction is equal to

$$K_e/(1 + K_e) = 10^{\Delta pK}/(1 + 10^{\Delta pK}) = (1 + 10^{-\Delta pK})^{-1} \tag{9.29}$$

In practice, proton exchange always takes place in the presence of an excess of acceptor B, whose concentration can be assumed to be constant. We are therefore dealing with a pseudo-first-order reaction and we have

$$k_{exchange} = k_D[B]K_e/(1 + K_e) = k_D[B]/(1 + 10^{-\Delta pK}) \tag{9.30}$$

If, for example, B is the OH^- group of water, $[B] = 10^{-7}$ at pH 7 with $pK_B = 14$. If $pK_A = 14$, $\Delta pK = 0$ and $k_{exchange} = k_D \times 10^{-7}/2$. Since k_D is of the order of $10^{10}\,M^{-1}\,s^{-1}$, $k_{exchange} \approx 10^3$ and the exchange time is of the order of a millisecond. It becomes longer when pK_B (donor) increases. It becomes shorter when, instead of pure water, a high concentration of a basic catalyst is introduced into the medium.

Note that, when $\Delta pK \ll 0$, $k_{exchange} \approx k_D[Cat] \times 10^{-\Delta pK}$ (s^{-1}), where [Cat] is the concentration of the catalyst. If, on the other hand, $\Delta pK \gg 0$, $k_{exchange} \approx k_D[Cat]$ (s^{-1}). For example, if the catalyst is an imidazole group ($pK = 7.7$) with a 10 mM concentration and if the donor has a pK of 12, $\Delta pK = -4.3$ and $k_{exchange} \approx 10^{3.7}$.

The exchange mechanism has been examined in this section for an atom completely exposed to external protons. The exchange rate constant is therefore an intrinsic quantity. In practice, proton exchange must take into account the accessibility of the atom and the group. The accessibility is not only defined in static terms but also depends on the conformational dynamics of the atomic group containing the atom in question or of the whole molecule. It is in this way that measurements of the proton exchange rate give access to molecular dynamics. We demonstrate this with two widely studied examples: nucleic acids and proteins.

9.3.3 Nucleic acids: coupling with an opening–closing mechanism

In the case of DNA, imino proton exchange can only occur when base pairing is disrupted. The exchange thus reflects the internal dynamics of the double helix. Conversely, measurement of the exchange rate will give information about the dynamics of DNA. With NMR, proton exchange can be followed individually for each base pair, so that a genuine map of the dynamics can be drawn up as a function of the sequence. The exchange scheme is of the following type:

$$\text{Closed (C)} \underset{k_C}{\overset{k_O}{\rightleftharpoons}} \text{Open (O)} \xrightarrow[k_1]{} \text{Exchange with catalyst (Cat)}$$

This is a very general exchange scheme, not only valid for protons, and depends on the following assumptions:

- the reaction takes place only with the open form
- there is only one open form to be considered
- during the opening time, the exchange reaction takes place as if sites were open all the time (e.g. as in single strands or mononucleotides).

We know that the fraction of open base pairs in DNA cannot be large, which implies that $k_O \ll k_C$, so that the open state (O) can be assumed to be a steady state, giving

$$d[O]/dt = k_O[C] - (k_C + k_1[Cat])[O] = 0$$

and hence

$$[O] = k_O[C]/(k_C + k_1[Cat]) \tag{9.31}$$

Because the catalyst is always in excess with respect to DNA, the observed rate constant k_{obs} is for a pseudo-first-order reaction:

$$d[C]/dt = -k_{obs}[C]$$

However, $d[C]/dt = k_C[O] - k_O[C]$ and using eqn 9.31 we have

$$k_{obs} = k_O k_1[Cat]/(k_C + k_1[Cat]) \tag{9.32}$$

(an expression similar to Michaelis' equation for the enzyme–substrate reaction), or

$$1/k_{obs} = 1/k_O + 1/Kk_1[Cat] \tag{9.33}$$

where $K = k_O/k_C$ is the open/closed equilibrium constant.

Since the reaction is pseudo-first-order, $1/k_{obs}$ is equivalent to a relaxation time τ_{obs} characterizing the exchange experimentally. The quantity $\tau_O = 1/k_O$ is the base-pair opening time, so that

$$\tau_{obs} = \tau_O + 1/Kk_1[Cat] \tag{9.34}$$

The graph of τ_{obs} against 1/[Cat] is clearly a straight line (Fig. 9.5). The value of τ_O is obtained by extrapolating [Cat] to ∞. In other words, with an infinite catalyst concentration, the kinetics will be dominated by the base-pair opening rate.

Proton exchange between the catalyst and the imino group of T or G involved in the hydrogen bond is reflected experimentally by a broadening of the proton resonance line. If Δx is the mid-height increase in line width, it can be shown that $\tau_{obs} = (\pi \Delta x)^{-1}$. The relaxation time T_1 can also be measured with ($T_{1,Cat}$) and without ($T_{1,0}$) the catalyst. It can be shown that $(\tau_{obs})^{-1} = T_{1,Cat} - T_{1,0}$.

Catalysts must be basic, so that transfer takes place from the base (G or T) as a donor to the catalyst as acceptor. Acceptor pK values vary from 7.7 (imidazole) to 8.4 (triethanolamine), 8.9 (Tris) and 9.2 (NH_3).

The pK values of DNA donor groups are around 3 for N1 of A, N3 of C and N7 of G, but approximately 12 for N1 of G and N3 of T (imino protons). In the latter case, NH_3 appears to be the best catalyst ($\Delta pK = -2.7$). With NH_3 or imidazole as catalyst, τ_{obs} indeed varies linearly with 1/[Cat]. Extrapolation to 1/[Cat] = 0 gives measured lifetimes τ_O of the order of milliseconds (1–7 ms for AT, 7–40 ms for GC). These lifetimes differ from one base pair to another, indicating that the opening process involves only one base pair at a time and not a whole set of base pairs as suggested in some models. The lifetime τ_O depends on the local structure and hence indirectly on the base sequence.

Information can be obtained not only about τ but about K as well. If we put $k_1[\text{Cat}] = k_{tr}$, eqn 9.34 becomes

$$\tau_{obs} = \tau_O(1 + k_C/k_{tr}) \qquad (9.35)$$

The rate constant k_{tr} (proportional to [Cat]) can also be written as αk_i where $k_i = 1/\tau_i$ is the proton transfer rate constant for a single nucleotide. The numerical factor α characterizes the accessibility of the imino group in the open pair. Hence:

$$\tau_{obs} = \tau_O(1 + k_C \tau_i/\alpha) = \tau_O + \tau_i/K\alpha$$

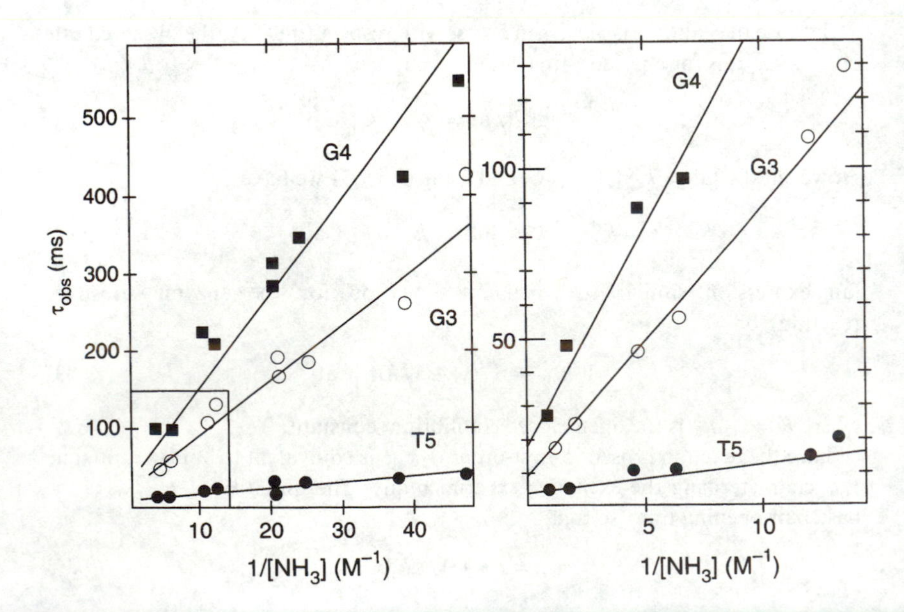

Figure 9.5 Relaxation time τ_{obs}, the experimental manifestation of proton exchange, plotted against 1/[NH_3] for a double-stranded self-complementary oligodeoxynucleotide d-CGCGATCGCG with ammonia acting as a catalyst. The diagram on the right is an enlargement of the framed area in the left-hand diagram. Note the difference between the behaviour of guanines and thymines. Extrapolation of [NH_3] to ∞ shows that τ_O(GC) is 2–3 times as great as τ_O(AT). (From Leroy *et al.*, 1988. Reprinted from *Journal of Molecular Biology* with the permission of Academic Press.)

Measuring τ_{obs} both with the nucleotide and with the base pair at the same catalyst concentration enables $K\alpha$ to be determined. We find a value of about 10^{-5} for AT pairs and of about 10^{-6} for GC pairs. If we assume that $\alpha = 1$, this means that one AT pair out of 10^5 and one GC pairut of 10^6 is open at any time. Since $K = k_O/k_C$ and $k_O \approx 10^3$, k_C is of the order of 10^8 to 10^9.

The relaxation time corresponding to the closing process is therefore very short, of the order of 1–10 ns. This order of magnitude agrees very well with the time scale (determined by other methods) of base-pair movement in DNA.

It only remains, therefore, to explain why the opening time τ_O is so long in comparison. It can be shown that, among all the rotational fluctuations experienced by base pairs on the scale of a nanosecond, there is a small probability (10^{-6}) that a fluctuation will occur large enough to trigger a base-pair opening (even a partial one) and proton exchange. The mean time between two such events is thus of the order of $10^{-9}/10^{-6}$ or 10^{-3} s.

9.3.4 **Proteins**

The exchange kinetics of the native protein are generally compared with those of the denatured protein. The differences in this case must be explained in terms of conformation and accessibility. The denatured protein will then be compared with model compounds to explain the exchange process in terms of a chemical mechanism.

The only exchange rate that can be measured is that of the amide NH group. For other proteins (with COOH, imidazole, NH_3^+ or NH_2 side chains), the exchange process is too fast.

Exchange with the NH group

The exchange is made with either a basic or acidic catalyst and the exchange rate constant can be expressed simply by (Woodward and Hilton, 1979):

$$k_{obs} = k_H[H^+]^x + k_{OH}[OH]^y + k_{H_2O} \tag{9.36}$$

where diffusional collisions and ΔpK are taken into account in k_H (acid catalysis) and k_{OH} (base catalysis). Direct exchange with water is represented by k_{H_2O}, which is generally negligible. The order of the reaction is defined by x and y. If k_{obs} is plotted against the pH, the curve passes through a minimum when the first two terms on the right of eqn 9.36 are equal (Fig. 9.6). If homopolypeptides and proteins (native or denatured) are compared, several differences appear.

1. Whereas we find $x = y = 1$ for polypeptides (i.e. first-order kinetics), x and y for proteins vary in a range between 0.4 and 0.65, reflecting the effect of side chains (steric hindrance and perturbation of the local pK).
2. Two orders of magnitude are sufficient to describe the variation in k_{obs} at 25°C for polypeptides, but eight orders of magnitude are necessary for proteins.
3. The mean activation energy is about 20 kcal mol^{-1}, but there is a large range of variation. A difference of eight orders of magnitude (i.e. a factor of 10^8 covering the range of values) corresponds to an energy difference of 11 kcal mol^{-1} ($e^{11/0.6}$ being $\approx 10^8$). For example, for bovine pancreatic trypsin inhibitor (BPTI) at pH 5.6 and 22°C, k_{obs} changes from 2×10^{-6} for Arg 20 to 1.1×10^{-2} for Arg 17.

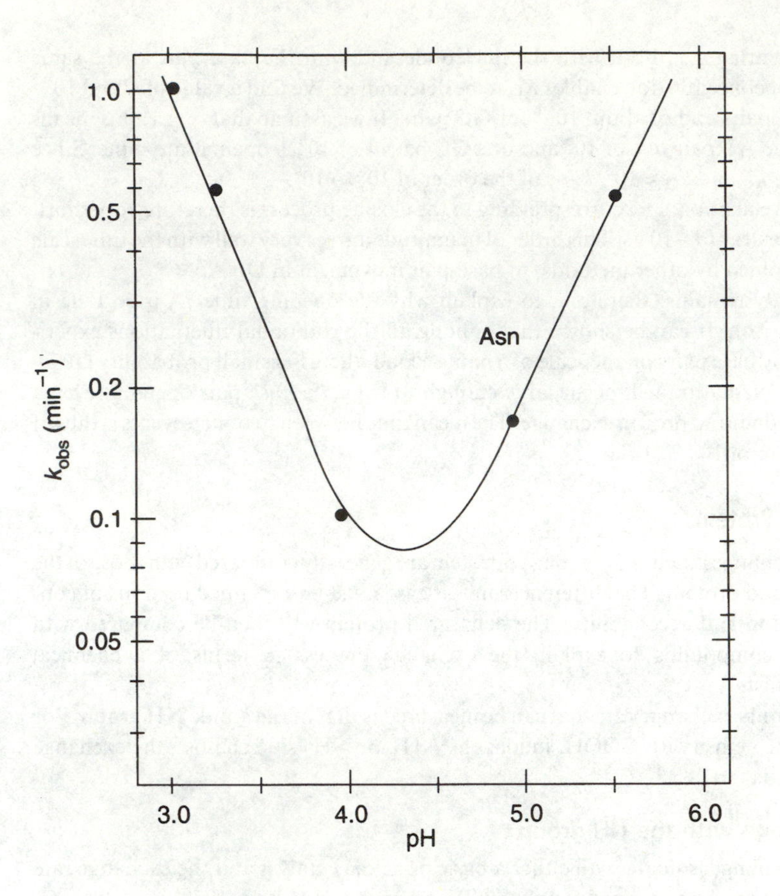

Figure 9.6 H–T exchange as a function of pH at 0°C for poly D–L-asparagine (Asn), showing that the exchange rate constant passes through a minimum. (From Molday *et al.*, 1972. Reprinted with permission from *Biochemistry*. © 1972 American Chemical Society.)

Accessibility

Proton exchange depends on the ability of H_3O^+, OH^- or H_2O to penetrate the protein, which might take place by two processes: temporary channels through the protein matrix resulting from internal conformational fluctuations; or a local uncoiling exposing the peptide chain to the solvent. There are two reasons why the second process has to be abandoned.

1. Uncoiling cannot be envisaged on the molecular scale since the exchange rates differ considerably from one proton to another. Cooperative uncoiling, of either the α-helix or the β-sheet, is not observed experimentally. Moreover, although classes of exchangeable protons corresponding to locally denatured regions can be defined inside a protein, we cannot assert that all the protons in a given class belong to the same region of the molecule.
2. Observed changes in activation energy should correspond to local denaturation, which must be the limiting step in the model. However, we have to assume fast denaturation, and in that case the exchange rate should reflect only the intrinsic value for each residue. It would then vary by only two orders of magnitude, in complete disagreement with the experimental results (eight orders of magnitude). This means that a mechanism for solvent penetration must be found.

In one model, the molecule or ion would reach the exchange site after a diffusional process similar to that occurring in a solid, i.e. along a channel of complicated shape. The observed variation in k would be explained by the variation in the number of steps between the protein surface and the site. There are, however, the following good reasons for ruling out this process:

1. The properties of the solvent inside the protein should depend mainly on the local environment, and the exchange should therefore be independent of the pH.
2. The experimentally measured activation enthalpy is much greater than the few kcal mol^{-1} characteristic of the potential barrier against solvent penetration and it once again becomes impossible to explain the range of eight orders of magnitude.
3. In some proteins, internal water molecules can be rapidly exchanged with solvent molecules. Hydrogen atoms forming hydrogen bonds with the water molecules, on the other hand, are only slowly exchanged.

In a second model, the solvent can only diffuse inside the protein through rapid fluctuations of the matrix. The process here is similar to diffusion in a liquid. The rate decreases slowly as the molecular size increases. The limiting step reflects a fluctuation in the protein structure, disrupting some of the hydrogen bonds or creating a cavity. The large variation in activation energy is explained by the detailed local structure. There is some similarity between this model and that of local denaturation and any difference arises from the magnitude of the fluctuation.

Exchange scheme

The model developed for the imino protons of DNA can be used to study exchange in proteins: the proton must be in an open and accessible conformation in order to be exchanged. Thus, for a H $\rightleftharpoons$ D (proton–deuteron) exchange, we have

$$\mathrm{D} + \mathrm{Closed(H)} \underset{k_C}{\overset{k_O}{\rightleftharpoons}} \mathrm{D} + \mathrm{Open(H)} \overset{k_{\mathrm{exchange}}}{\rightleftharpoons} \mathrm{Open(D)} + \mathrm{H}$$

The rate constant k_{exchange} is what would be measured in a completely denatured state of the protein. As usual, we distinguish two limiting mechanisms.

- First mechanism:

$$k_O + k_C \ll k_{\mathrm{exchange}} \qquad \text{and} \qquad k_{\mathrm{obs}} = k_O \tag{9.37}$$

  The exchange mechanism is so fast that, as soon as a proton is accessible, it is instantly replaced by D. The transition to the open form is thus the limiting step. In this case, the dependence on pH is the same as that of k_O.

- Second mechanism:

$$k_O + k_C \gg k_{\mathrm{exchange}}$$

  which leads to

$$k_{\mathrm{obs}} = k_O k_{\mathrm{exchange}}/(k_O + k_C) \tag{9.38}$$

  The term $k_O/(k_O + k_C)$ is the probability of finding an open form at equilibrium. The measurement of k_{obs} in this case leads to the determination

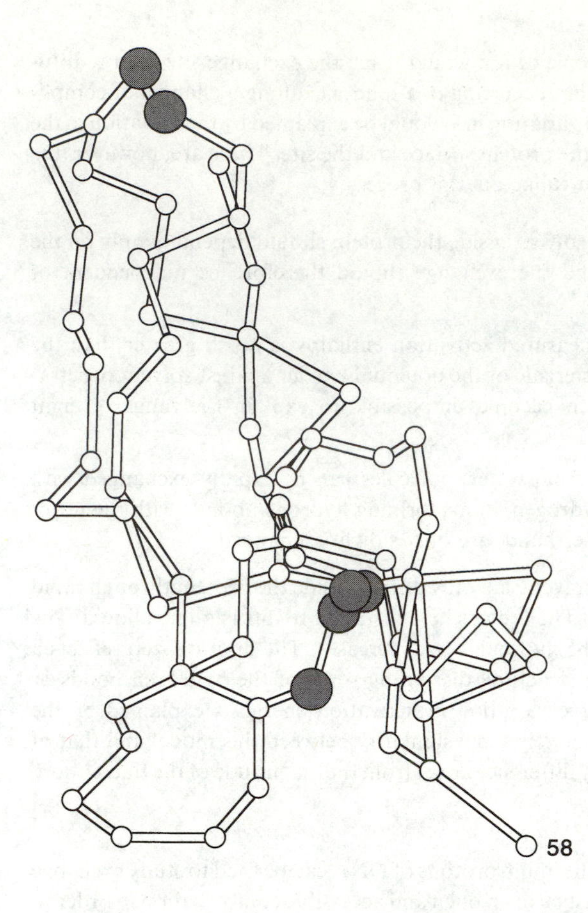

Figure 9.7 Peptide chain of BPTI from X-ray diffraction.

of the ratio $K = k_O/k_C$ and then to the free-energy change ΔG_O corresponding to site opening. Note that the variation of k_{obs} from one proton to another depends both on that of $k_{exchange}$ and on that of K.

If k_{obs} is plotted against $k_{exchange}$ with the second mechanism, all the protons that are exchanged because of the same local internal motion must lie on the same straight line of slope $k_O/(k_O + k_C)$.

The case of BPTI

This is a small protein with 58 amino acids and containing 53 peptide bonds with labile protons on the NH group. Since the tertiary structure is known, the accessible area for each atom can be calculated. By comparing the accessibility of protons and their exchange capability, three classes of exchangeable protons have been distinguished by NMR experiments (Figs 9.7 and 9.8):

1. For 42 protons, there is agreement between the two parameters (the proton on an accessible nitrogen is also one that is more rapidly exchangeable).
2. The proton of Glu49 is not accessible but is readily exchanged.
3. For 10 protons, on the other hand, there is high accessibility correlated with slow exchange.

A variation in temperature produces a change from the second to the first mechanism at 50°C. At that point, fluctuations with a high activation enthalpy

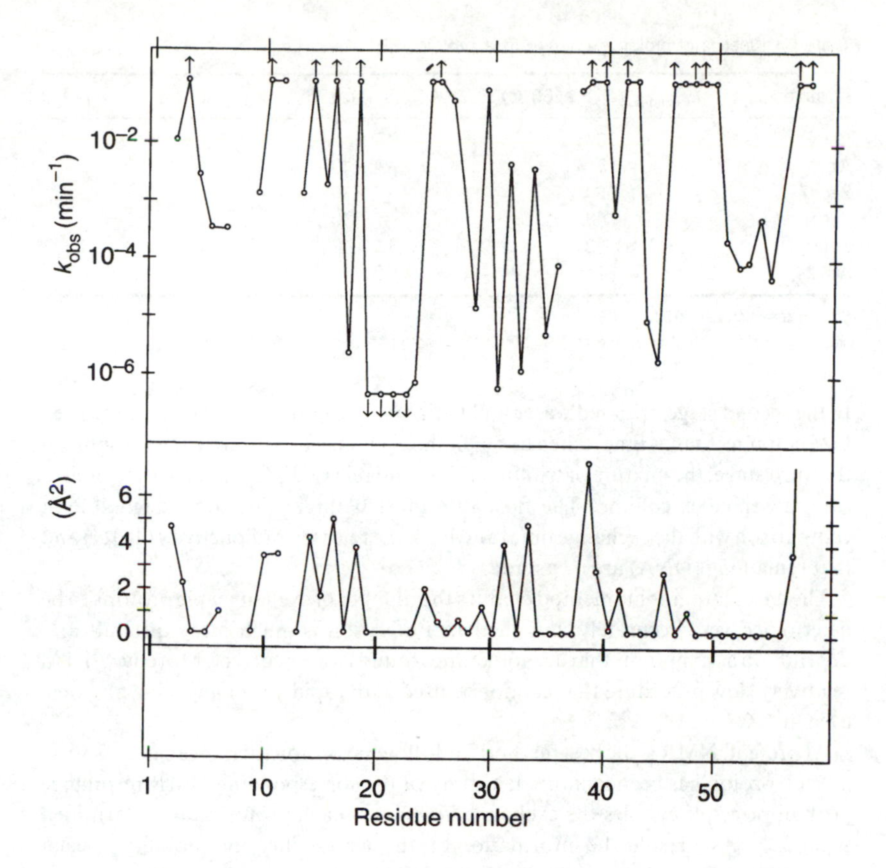

Figure 9.8 Proton exchange in amide groups of BPTI at pD 3.5 and 36°C. The upper diagram shows log(k_{obs}) for each proton, whose position is given by the numbering of the amino acid in the lower diagram. The arrows pointing upwards or downwards indicate that the exchange is respectively too fast or too slow to be measured. The exchange rate is compared with the accessibility (in Å^2) of the N atoms in the peptide chain. (From Wagner and Wüthrich, 1982. Reprinted from *Journal of Molecular Biology* with the permission of Academic Press.)

increase, with larger amplitudes and longer lifetimes than those at lower temperatures: a longer time is required for large-amplitude fluctuations to 'discover' the return pathway. Changing the pH is equivalent to modifying the charge distribution in the protein. Typical curves plotting k_{obs} against pH show two alterations:

(1) minima are flattened because of the deprotonation of COOH groups and the formation of salt bridges with COO^-, which make the conformations more rigid;

(2) minima are shifted to higher pH values.

9.3.5 Technical aspects

Exchange in proteins was first measured by using infrared absorption around 1500 cm^{-1} to follow the replacement of H by D in the peptide bond. In nucleic acids, the H–D replacement in bases produces a spectral shift to shorter wavelengths. This property has been used to study exchange kinetics in which the change in absorbance at 290 nm is detected by stopped-flow experiments.

A chromatographic method has also been developed for nucleic acids in which the exchange occurs between H and T (^{3}H or tritium). There are three stages in the method. In the first, exchange with tritiated water is produced after incubation of the polymer with TOH:

$$\text{DNA-H} + \text{TOH} \rightarrow \text{DNA-T} + \text{HOH}$$

Table 9.1 Kinetics of proton exchange in a protein (BPTI)*

Amino acid	$k_{exchange}$ (s^{-1}) (at 68°C)	k_{obs} (s^{-1}) (at 36°C)	Accessible area ($Å^2$)
Gly 12	0.67	1.7×10^{-3}	3.3
Ala 25	1.3	1.7×10^{-3}	1.8
Ser 47	2.9	1.7×10^{-3}	0
Cys 5	1.83	3.17×10^{-4}	0
Ile 18	0.93	3.3×10^{-8}	0
Asn 24	1.47	8.3×10^{-9}	0

Source: from Wagner (1983).

*$k_{exchange}$ and k_{obs} are defined in eqns 9.37 and 9.38; all data obtained at pD 3.5.

In the second stage, tritiated water and tritiated DNA are separated in a Sephadex G25 column. At that time, taken as $t = 0$, the exchange of tritium with H begins. In the third stage, the mixture of tritiated DNA and released TOH is passed through a second Sephadex column. The flow rate must be high (flow time negligible in comparison with the exchange time) and both the released radioactivity (TOH) and that remaining (DNA) are measured.

The chromatographic method enables the number of exchangeable protons to be determined very accurately, but their identification is much more difficult and depends on assumptions that are sometimes found to be incorrect. Moreover, it is a relatively slow procedure that cannot be used with rapidly exchangeable protons, even at 4°C.

At present, NMR is the best method for following proton exchange since, as soon as each proton has been identified, a study of the corresponding NMR resonance peak in principle enables the exchange kinetics of each proton to be determined separately. As a result, the information on the accessibility and mobility of each proton is much more complete (see Table 9.1).

There are two components of the longitudinal relaxation time T_1: one arising from the overall motion of the molecule and the other from proton exchange. We can write:

$$T_1^{-1} = (T_1)_o^{-1} + k_{obs} \tag{9.39}$$

For large enough molecules

$$(T_1)_o^{-1} \approx 0.1\gamma^4 h^2 \tau_c \sum_i r_i^{-6}$$

where γ is the gyromagnetic ratio and τ_c is the correlation time corresponding to the rotational coefficient of diffusion for the molecule, given by $\tau_c = \eta V / k_B T$, η being the viscosity of the medium and V the molecular volume. Assuming that proton exchange is negligible at 273 K, the value of T_1 measured at this temperature enables $(T_1)_o$ to be determined at temperature T. Hence

$$(T_1)_{o(T)} = (T_1)_{o(273)}(273\eta_T / T\eta_{273}) \tag{9.40}$$

Subtracting this calculated value from the experimental value of T_1 gives k_{obs}.

9.4 X-rays and temperature factors

The probability of finding an atom at a distance x from its equilibrium position obeys a Gaussian law provided that each atom is assumed to be undergoing isotropic harmonic vibrations. It is therefore determined solely by the mean square $\langle x^2 \rangle$ of its displacement. It can be shown (see Box 9.3) that the structure factor of each atom must be multiplied by $\exp(-B\sin^2\theta/\lambda^2)$, the *Debye–Waller factor*. In this expression, $B = 8\pi^2\langle x_i^2 \rangle$ applies to each component x_i ($\langle x_1^2 \rangle = \langle x_2^2 \rangle = \langle x_3^2 \rangle$).

Box 9.3 **Calculation of the temperature factor**

This treatment follows that of Guinier (1964). The *ergodic principle* applies to the thermal motion of the atoms in a crystal: the ensemble of displacements of all the atoms at a given time is statistically identical to the ensemble of displacements with time of a given atom.

Let $\boldsymbol{x}$ be the displacement of an atom with respect to one of the lattice points. It is assumed that there is only one atom per unit cell whose mean position can be taken as a lattice point and which has a scattering factor f. X-ray crystallographic theory shows that a *structure factor* F can be defined for a unit cell expressed in the form

$$F = f\exp(-2\pi i\boldsymbol{s}\cdot\boldsymbol{x})$$

The vector $\boldsymbol{s}$ describing the distribution of the diffracted X-rays has modulus $s = (2\sin\theta/\lambda)$, where 2θ is the angle between the incident and diffracted beam and λ is the wavelength of the X-rays.

The mean value $\langle F \rangle$, for small values of $\boldsymbol{x}$ (a small fraction of 1 Å), is given approximately by

$$\langle F \rangle \approx \langle f(1 - 2\pi i\boldsymbol{s}\cdot\boldsymbol{x} - 2\pi^2(\boldsymbol{s}\cdot\boldsymbol{x})^2 + \ldots)\rangle$$

Since the lattice point is the equilibrium position of the atom, $\langle \boldsymbol{s}\cdot\boldsymbol{x} \rangle = 0$ and hence

$$F \approx f(1 - 2\pi^2 s^2\langle x_1^2 \rangle)$$

where x_1 is the projection of $\boldsymbol{x}$ on to $\boldsymbol{s}$. If x_2 and x_3 are the projections of $\boldsymbol{x}$ on the other two axes of the reference frame, then $x^2 = x_1^2 + x_2^2 + x_3^2$. Assuming that the displacements are isotropic, $x_1^2 = x_2^2 = x_3^2$ and therefore $x_1^2 = x^2/3$. Since $s = 2\sin\theta/\lambda$, we obtain

$$\langle F \rangle = f(1 - 8\pi^2 x^2 \sin^2\theta/3\lambda^2)$$

or

$$\langle F \rangle \approx f\exp(-8\pi^2 x^2 \sin^2\theta/3\lambda^2)$$

The diffracted intensity, which varies as $|F|^2$ is therefore multiplied by

$$D = \exp(-16\pi^2 x^2 \sin^2\theta/3\lambda^2)$$

Because D increases with θ, the intensity becomes weaker the greater the Bragg angle in the diffraction pattern. For each component x_i of the displacement, $\langle F \rangle$ is usually written in the form

$$\langle F \rangle = f\exp(-B\sin^2\theta/\lambda^2)$$

where $B = 8\pi^2\langle x_i^2 \rangle$ is known as the *Debye–Waller factor*.

The temperature coefficient can also be considered as corresponding to a Gaussian distribution in the electronic cloud around each atom.

9.4.1 Determination and significance of B

The value of B can only be determined if a high-resolution structure is available. The Debye–Waller factor is introduced into each atomic structure factor and B then becomes an additional parameter in the refinement of the structure. Generally speaking, however, B can only be determined if additional constraints are imposed (bond lengths and angles, flatness of rings, chirality, etc.). In addition, two bonded atoms must have values of B that differ by no more than 10 Å^2.

9.4.2 Interpretation of B

It should be pointed out that the Gaussian 'noise' around any atom may arise from sources other than thermal motion. Firstly, some disorder may be produced near the atom by the existence of several protein configurations in the crystal, and this may still be present at the temperature of the experiment. Secondly, there may be defects in the crystal lattice: a molecule may not have exactly the same arrangement in one unit cell as it does in the adjacent one. Such 'static' disorder is independent of temperature and has both rotational and translational components.

Moreover, the value of $\langle x^2 \rangle$ determined from B is the sum of at least three contributions, from vibrations, from the conformation and from the lattice:

$$\langle x^2 \rangle = \langle x^2 \rangle_{\text{vibr}} + \langle x^2 \rangle_{\text{conf}} + \langle x^2 \rangle_{\text{latt}}$$

The last term is estimated in two different ways:

1. If the molecule incorporates a metal atom, we can measure the Mössbauer effect, which is independent of any lattice defects. Subtracting the $\langle x^2 \rangle$ obtained from Mössbauer measurements from the $\langle x^2 \rangle$ determined by X-ray measurements on the same atom gives $\langle x^2 \rangle_{\text{latt}} = 0.025$ Å^2 (Frauenfelder *et al.*, 1979; Hartmann *et al.*, 1982), which accounts for less than 15 per cent of the total mean square displacement.
2. If the molecule crystallizes in two different forms, as in the case of lysozyme (Artymiuk *et al.*, 1979), the two corresponding values of $\langle x^2 \rangle$ can be compared and it becomes clear that the contribution of $\langle x^2 \rangle_{\text{latt}}$ is small enough to be negligible except when atoms forming part of amino acids are in contact with two molecules in the crystal.

We can make an attempt to distinguish between $\langle x^2 \rangle_{\text{vibr}}$ and $\langle x^2 \rangle_{\text{conf}}$ by varying the temperature. The harmonic vibrations in a potential well decrease with T until the Debye temperature (around 100 K) is reached, below which $\langle x^2 \rangle_{\text{vibr}}$ remains constant. On the other hand, the conformational states of the protein trapped in the crystal have a corresponding number of fairly deep potential wells (*sub-states*). If the temperature is not high enough to allow the potential barriers to be crossed, then only variations in $\langle x^2 \rangle_{\text{vibr}}$ will be observed. In general, we must not expect very high accuracy in the absolute value of B since several noise factors occur in its experimental determination (uncertainty in atomic coordinates, radiation damage, absorption of X-rays, etc.). It can nevertheless be assumed that relative values of B enable significant comparisons to be made between atoms.

9.4.3 Results and molecular dynamics

There are two useful ways of presenting the results. Either B can be plotted against the amino acid position in the chain, or the movement of chains can be represented by a 'blurred' region superimposed on the image of the tertiary structure of the protein (myoglobin and lysozyme).

Several points can be raised about the observed differences. Firstly, the haem 'pocket' is rigid, whereas the lysozyme 'cleft' is the most mobile region. Secondly, the atoms involved in rigid structures (α-helices, β-sheets) are less 'restless' than the others. Thirdly, the atomic mobility generally increases with distance r from the centre of the protein. In lysozyme, the $\langle x^2 \rangle$ distribution changes suddenly above $r = 14$ Å, a value roughly equal to the 'radius' of the $30 \times 30 \times 45$ (Å^3) ellipsoid representing a particle equivalent to the protein (Fig. 9.9).

A fourth point is that the side chains are more mobile than the main chain. A regular variation in mobility may occur along a given side chain. In myoglobin, for example, the following values of $\langle x^2 \rangle$ in Å^2 are found for K98:

C1	C2	C3	C4	C5	N6
0.195	0.21	0.23	0.25	0.26	0.275

This is a case of a lysine that protrudes from the protein and makes contact with the solvent: the motion of the N-terminal is greater than that of the α-carbon. In the same protein, on the other hand, K77, which forms a salt bridge with E18, has almost the same value of $\langle x^2 \rangle$ (0.11 ± 0.01 Å^2) for all the atoms in the side chain.

The final point is that, when the temperature is reduced (e.g. from 300 K to 80 K), a large decrease in $\langle x^2 \rangle$ is observed for some atoms, pointing to a value of zero extrapolated to 0 K. The greatest temperature effects are generally observed with atoms having the largest value of x.

If the atoms are moving inside a protein, they have to find a place for themselves in small cavities. Their motion generates new cavities, which become available for other atoms, and so on. Such a process implies some correlation between all atomic movement in the protein matrix. This concept of concerted fluctuations has already been invoked in explaining proton exchange (see section 9.3) and is also involved in the diffusion-controlled movement of a molecular ligand towards its specific site (as in the case of O_2 or CO, which bind to the haem group of haemoglobin, or that of Xe entering internal cavities in the same protein).

The observed $\langle x^2 \rangle$ can thus be regarded as the sum of two types of movement:

- fast local fluctuations that are very similar from one atom to another
- collective motion with a lower frequency and a structure-dependent amplitude.

In studying these movements, the X-ray and spectroscopic methods (optical, Mössbauer and NMR) are complementary experimental approaches.

9.5 Flash photolysis

9.5.1 General aspects

This is a pulse technique in which a short but intense light flash is used to disturb a system in equilibrium. This induces a series of unstable or metastable states, which are then generally studied by spectroscopic methods. Figure 9.10 shows the main components of the experimental arrangement: the source of the light flash, the cell in which the reaction occurs and the systems for observation and recording.

The intensity of the light flash is calibrated by measuring the photochemical degradation of standard dye solutions. Some idea of the number of photons emitted can be obtained from the following example: a 0.5 J flash at a wavelength of 600 nm produces 1.5×10^{18} photons over a period of a few microseconds.

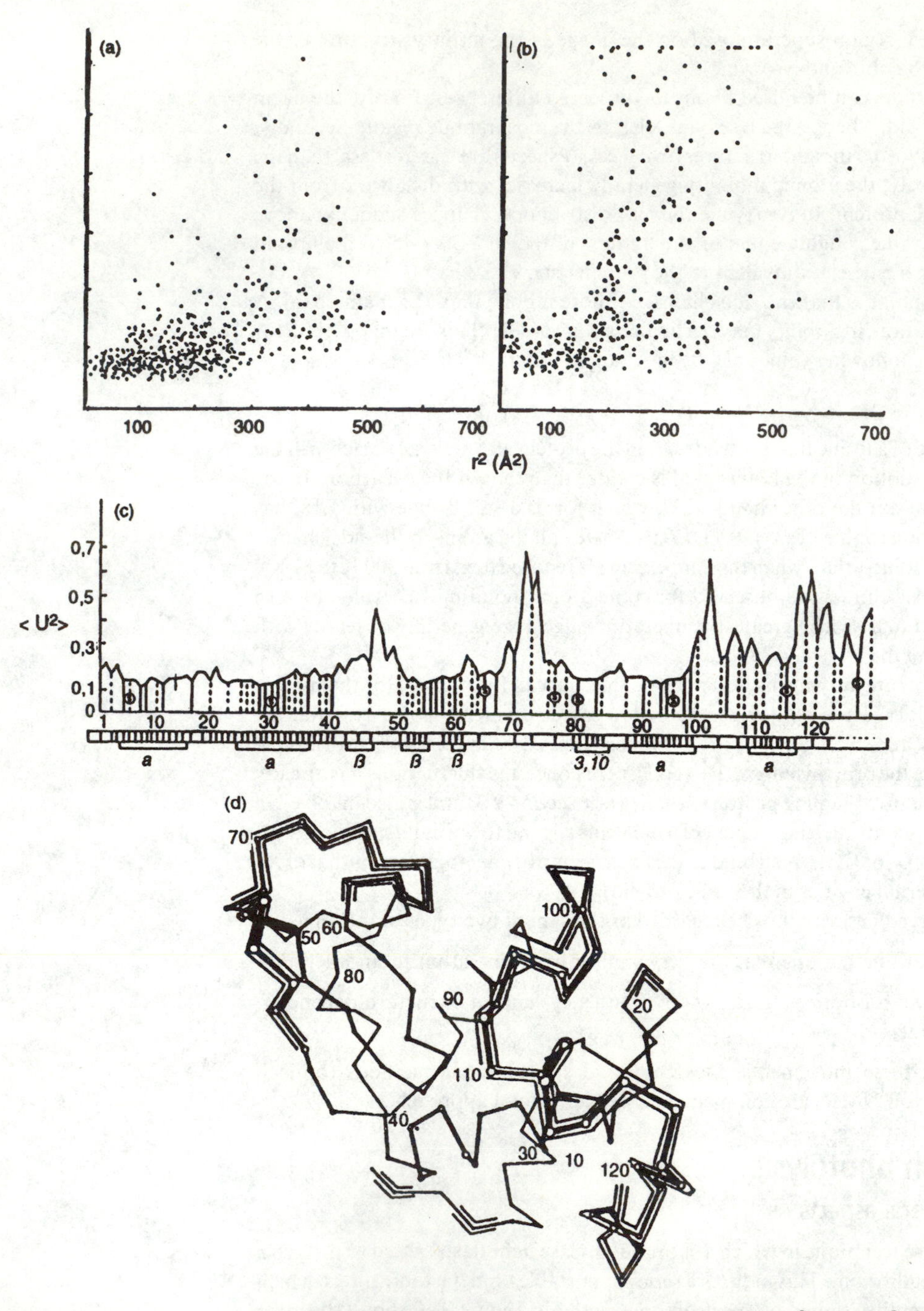

Figure 9.9 Temperature factors for human lysozyme. (a) A plot of $\langle x^2 \rangle$ against r^2, where r is the distance between the atom and the centroid of the peptide chain. (b) A similar diagram for atoms of the side chains. (c) Values of $\langle x^2 \rangle$ ($\langle U^2 \rangle$ here) along the peptide chain. Solid vertical lines are for internal residues; dotted lines are for residues lying in the surface. The bottom line indicates residues involved in hydrogen bonds. (d) The widening of the peptide backbone corresponds to an $\langle x^2 \rangle$ greater than 0.2 Å^2. Note the characteristic vertical cleft in the lysozyme molecule. (From Artymiuk *et al.*, 1979. Reprinted with permission from *Nature*, vol. 280, pp. 563–8 and with the authors' permission. © 1979 Macmillan Magazines Ltd.)

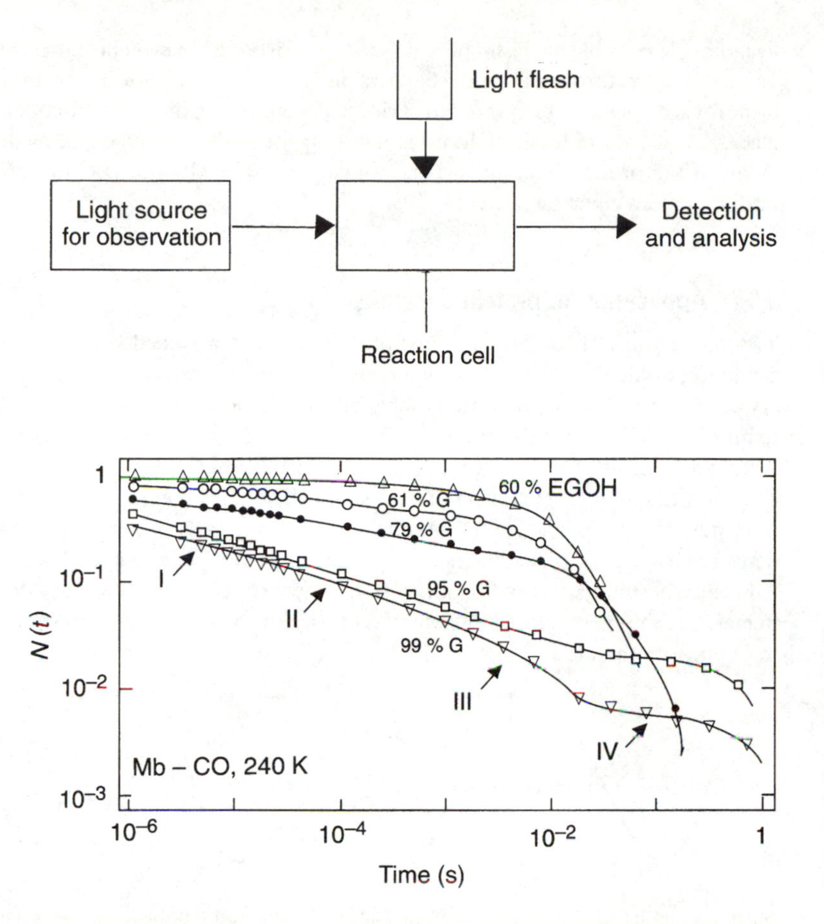

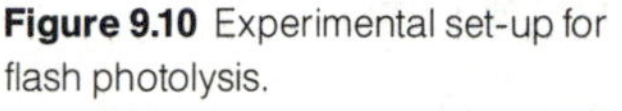
Figure 9.10 Experimental set-up for flash photolysis.

Figure 9.11 Combination of CO with the myoglobin molecule at 240 K. $N(t)$ is the fraction of haem groups not re-bound to a CO molecule at time t after photodissociation. EGOH = ethylene glycol; %G = percentage of glycol by weight in water–glycol mixtures. (From Beece *et al.*, 1980. Reprinted with permission from *Biochemistry*. © 1980 American Chemical Society.)

To study protein dynamics, a respiratory protein like myoglobin, which binds CO or O_2 on its haem group, may be chosen. The photolysis flash releases a gas molecule from the active site and the kinetics of the rebinding of the ligand is followed by monitoring the variation in the absorption of the haem group in the Soret band. The rate constants will depend on the potential barriers crossed by the gas molecule during its displacement, but the barrier heights in their turn will depend on the internal dynamics of the protein. In other words, the evolution of the system reflects the internal movements in the protein.

Measurements made by Beece *et al.* (1980) are shown in Fig. 9.11 and cover the range from a microsecond to a second. The rate constants k are a function of the temperature T and the viscosity of the medium η. The temperature is varied between 200 and 300 K, while the viscosity can be made to cover a range of five orders of magnitude if water–methanol, water–glycerol and water–sucrose mixtures are used. It is found that $\ln k$ varies linearly with $1/T$ for a given viscosity of the medium and that it also varies linearly with $\ln \eta$ (i.e. $k \approx \eta^{-a}$).

If the protein were a rigid matrix, however, the viscosity of the external medium would have no effect on any crossing of potential barriers inside the protein. Moreover, in the range of viscosities from 10 to 10^3 cP, k varies as η^{-1}, which is reminiscent of Kramers' equation for crossing a potential barrier (see section 8.4). This leads to the idea that what is in fact occurring is a Brownian diffusion of the ligand made possible by the motion of the atoms forming the protein matrix. Such diffusion occurs through potential barriers that are no longer static but are present in the form of 'gates', which are open or closed depending on the fluctuations taking place. Because the motions of the protein tertiary structure depend on the medium in which the protein molecule occurs, we can now see why the viscosity of the medium has an effect on rate constants.

9.5.2 Application to protein dynamics

Two models have been proposed for proteins. It was assumed from the initial crystallographic studies that a protein could be considered as a rigid aperiodic crystal. On the other hand, proton exchange and NMR suggest a constantly fluctuating system in which the protein molecule 'tries' all the structures allowed at a given temperature. As is often the case, reality lies halfway between the two extremes. In fact, the complexity of a three-dimensional protein structure and the large number of possible conformational states mean that a protein is preferably described as a 'glass' or, as suggested by more recent models, as a neural network.

From a thermodynamic point of view, the free energy G can be viewed as offering numerous minima separated by potential barriers, each minimum corresponding to a conformational *sub-state*:

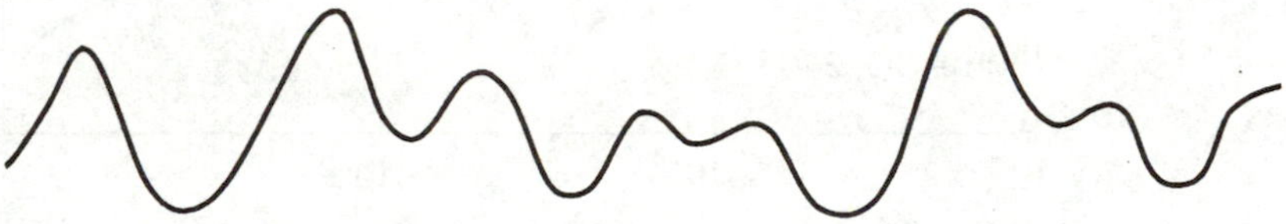

Such a scheme only takes into account one conformational coordinate, whereas in fact G must be envisaged in a multi-dimensional space. In other words, the protein can be represented as a highly degenerate energy state. We might even assume that each sub-state is in its turn split into other sub-states, etc. Under these conditions, the state of the protein may very well depend on the experimental method used. We can imagine, for example, that crystalline constraints lead to a very small number of sub-states in a crystal. The structure determined by X-ray diffraction would then be only one possibility among many. It might also be the case that each of these sub-states carries out the given biological function (an enzymatic one, for example) with different kinetic parameters.

Low-temperature studies of the rebinding of CO or O_2 to various respiratory proteins clearly show that the number $N(t)$ of molecules not yet bound is not a simple exponential function of time, as would be predicted if it were a question of crossing a potential barrier. There are at least two relaxation times, if not more. Two interpretations of this are possible:

(1) several reaction pathways are available in each of the protein molecules, but the molecules are identical;

(2) only one reaction pathway exists, but there is a heterogeneous population of molecules.

To distinguish between these two models, the system is illuminated not with a single flash but with a series of flashes separated by a time τ_m having a value between the limiting values of the experimentally observed relaxation times. If the population is heterogeneous, we should expect the same population of molecules to contribute to the decay of $N(t)$ after each flash, so that the decay curve would remain the same. On the other hand, with a homogeneous population but with several pathways per molecule, the most rapid rebinding processes gradually vanish and $N(t)$ decreases less and less as time passes.

Experiment shows unambiguously that the first interpretation is the correct one. We are therefore dealing with a non-homogeneous system at low temperature, each protein molecule being in a different sub-state from any other.

9.6 Inelastic scattering of light

9.6.1 General principle

In studying Rayleigh scattering of light, a relationship is established between the scattered intensity and the optical properties of the medium. If measurements are made with a pure liquid or, even better, with a crystal, the scattering almost disappears.

The distance between the scattering centres in a crystal is very small compared with the wavelength of the light: 0.15 nm compared with 500 nm. Consider a division of the crystal into a large number of elementary cubes of side $\lambda/2$: in a 1 mm^3 crystal there will be 64×10^9 such cubes. Any scattering centre emitting radiation in a given direction has a corresponding equivalent scattering centre in one of the elementary cubes that will emit radiation with an opposite phase and will thus destructively interfere with the first. It follows that *a perfect crystal does not scatter light*. The same argument could be used with a liquid, but now each elementary cube would be a volume element with N scattering centres *on average*. If N were the same for each elementary volume, there would be no scattered light as in a crystal. However, because of fluctuations in the liquid density, N varies continuously in two ways: with time for a given elementary volume, and with different elementary volumes at a given time. The destructive interference of the emitted radiation is never quite complete and light is scattered *because of the fluctuations*.

It follows that, if the scattered light could be analysed, information could be obtained about the fluctuations, i.e. about the Brownian motion of molecules in solution. This analysis depends on the fact that any random function of time $y(t)$ can be represented by an infinite Fourier series with coefficients a_n and b_n and an angular frequency ω_n for values of n from 1 to ∞ (see Mathematical Appendix). This method will be applied to the function describing the time variation of the amplitude of fluctuations of concentration in the solution. To do this, we consider such fluctuations as the superposition of an infinite number of *concentration waves* with suitable amplitudes and wavelengths.

Consider the propagation of such a plane wave of wavelength Λ (Fig. 9.12), which creates in the solution a sinusoidal distribution of concentration with wavefronts Λ apart. These plane waves can be thought of as equivalent to the lattice planes in crystals that diffract X-rays. Using this analogy, the scattered radiation can be regarded as reflection from a concentration plane and will only be observed when Bragg's law is satisfied, i.e. when

$$\Lambda = (\lambda/2) \sin \theta \tag{9.41}$$

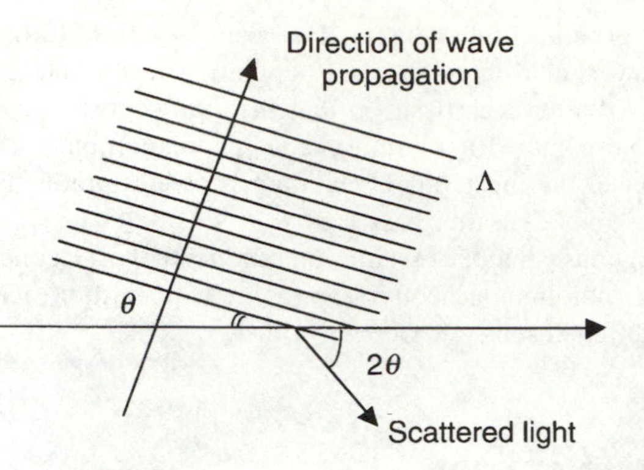

Figure 9.12 Propagation of a plane concentration wave.

where 2θ is the scattering angle and λ is the wavelength of the light in the medium (given by $\lambda = \lambda_0/n$, where λ_0 is the wavelength of the incident light *in vacuo* and n is the refractive index of the medium). The calculation then involves finding the correlation function $G(\tau)$ associated with concentration fluctuations and thus with the corresponding fluctuations of refractive index. The calculation is carried out in Box 9.4 and shows that

$$G(\tau) = G(0)\exp(-K^2 D_T \tau) \tag{9.42}$$

where $K = 4\pi\lambda^{-1}\sin\theta$.

9.6.2 Calculation of the scattered intensity

Since the scattered intensity varies with the square of the polarizability (or of the refractive index – see Box 9.4), it can be assumed without a rigorous proof that the correlation function for the scattered intensity varies as G^2, i.e. as $\exp(-2K^2 D_T \tau)$.

According to the Wiener–Khinchin theorem (see Mathematical Appendix), the scattered intensity is simply the Fourier transform of $G^2(\tau)\exp(i\omega_0\tau)$, where ω_0 is the angular frequency $2\pi\nu_0$ of the incident wave. This is a function of ω defined by:

$$I(\omega) = (2/\pi)\int_0^\infty G^2(\tau)\exp[i(\omega_0 - \omega)\tau]\,d\tau \tag{9.43}$$

Replacing G using eqn 9.42 and keeping only the real part of $I(\omega)$, we obtain

$$\Re(I(\omega)) = 4G^2(0)K^2 D_T/[\pi(2K^2 D_T)^2 + (\omega_0 - \omega)^2] \tag{9.44}$$

or, putting $1/\tau_c = 2K^2 D_T$:

$$\Re(I(\omega)) = 2G^2(0)\tau_c/\pi[1 + \tau_c^2(\omega_0 - \omega)^2] \tag{9.45}$$

In this form, the intensity distribution appears as a *Lorentzian curve* centred on ω_0 (Fig. 9.13). It passes through a maximum when $\omega = \omega_0$ and the bandwidth is defined by the value of ω at half the maximum height, i.e. for

$$\tau_c^2(\omega_0 - \omega)^2 = 1$$

or

$$\omega_0 - \omega = \Delta\omega = \tau_c^{-1} = 2K^2 D_T \tag{9.46}$$

which represents half the bandwidth.

Box 9.4 **Calculation of the correlation function**

Let $\delta c(\Lambda, x, t)$ be the x component of the amplitude of a 'concentration' wave with wavelength Λ. The quantity $\delta c(\Lambda, x, t)$ obeys Fick's law of diffusion:

$$\partial[\delta c(\Lambda, x, t)]/\partial t = D_T \partial^2[\delta c(\Lambda, x, t)]/\partial x^2 \tag{1}$$

where D_T is the translational coefficient of diffusion for the diffusing molecule. The variation along x is sinusoidal so that $\delta c = \sin(2\pi x/\Lambda)$ and hence

$$\delta c(\Lambda, x\, t) = \delta c(\Lambda, t) \sin(2\pi x/\Lambda) \tag{2}$$

Substituting eqn 2 into eqn 1 gives

$$\partial\delta c(\Lambda, t)/\partial t = -4\pi^2 D_T \delta c(\Lambda, t)/\Lambda^2$$

and hence

$$\delta c(\Lambda, t) = \delta c(\Lambda, 0) \exp(-4\pi^2 D_T \tau/\Lambda^2) \tag{3}$$

From this, we obtain the correlation function associated with the random variable $\delta c(\Lambda, t)$:

$$G(\tau) = \langle \delta c(\Lambda, t)\delta c(\Lambda, 0)\rangle = \langle \delta c^2(\Lambda, 0)\rangle \exp(-4\pi^2 D_T \tau/\Lambda^2) \tag{4}$$

The quantity

$$K = 2\pi/\Lambda = 4\pi(\sin\theta)/\lambda = 4\pi n(\sin\theta)/\lambda_0 \tag{5}$$

is simply the difference between the modulus of the incident wavevector and that of the scattered wavevector. The two wavevectors of magnitude $2\pi/\lambda$ make an angle 2θ with each other:

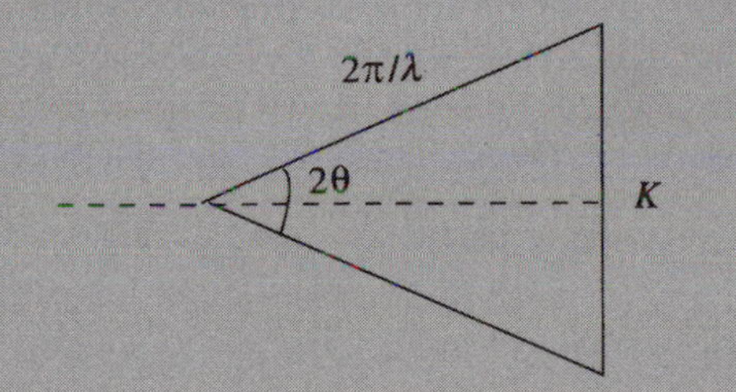

Hence $K = (2\pi/\lambda)(2\sin\theta)$, and finally, putting $\langle \delta c^2(\Lambda, 0)\rangle = G(0)$,

$$G(\tau) = G(0) \exp(-K^2 D_T \tau) \tag{6}$$

To obtain an order of magnitude, assume that the scattered light produced by monochromatic radiation with $\lambda_0 = 500$ nm in a liquid of refractive index 1.33 (a dilute aqueous solution) is observed at 90° ($\theta = 45°$). With $D_T = 5 \times 10^{-11}\,\mathrm{m^2\,s^{-1}}$, we find that

$$\Delta\nu = \Delta\omega/2\pi \approx 9\text{kHz}$$

Since the frequency $\nu_0 = \omega_0/2\pi = 6 \times 10^{14}$ Hz, a relative variation of about 10^{-11} in frequency has to be detected. This entails the use of a highly monochromatic source, i.e. one having a natural bandwidth of the order of a few tens of hertz. Inelastic Rayleigh scattering has only become a possible method with the advent of lasers as light sources.

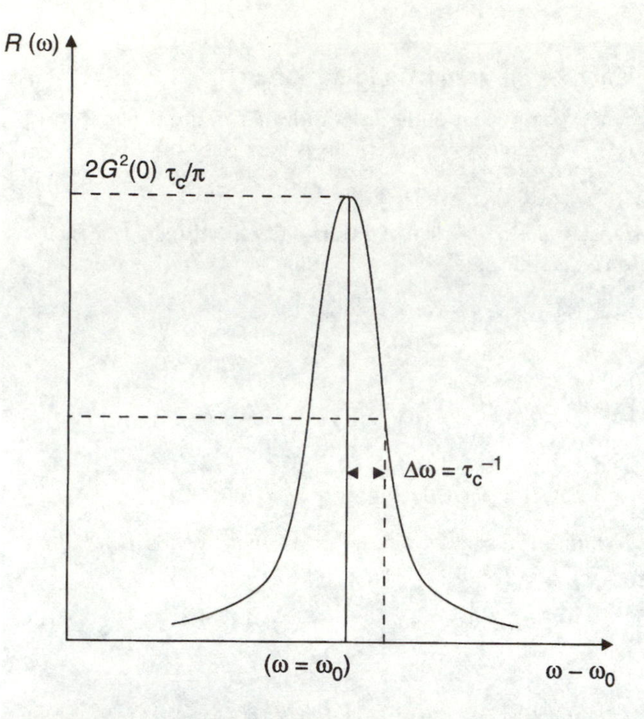

Figure 9.13 Shape of the Lorentzian curve centred on ω_0. The half-width τ_c^{-1} is $2K^2 D_T$.

However, a resolution of 10^{-11} is outside the range of classical dispersion devices. The broadening of the incident line can only be measured by using a beat frequency method similar to that used for the reception of frequency-modulated radio waves. Superimposed on the electric field of the scattered wave with frequency ω is a very small fraction of the incident electric field with frequency ω_0, the fraction being obtained for example by reflection on a small screen. The intensity resulting from the superposition of the two fields is measured. Since the total electric field is

$$E_T = E_0 \cos(\omega_0 t) + E_1 \cos(\omega t)$$

the intensity $I = AE^2$ takes the form

$$I = A\{E_0^2 \cos^2(\omega_0 t) + E_1^2 \cos^2(\omega t) + E_0 E_1 [\cos((\omega - \omega_0)t) + \cos((\omega + \omega_0)t)]\} \tag{9.47}$$

At high frequencies, the detector only gives an average value, i.e. 1/2 for $\cos^2(\omega t)$ and 0 for $\cos[(\omega + \omega_0)t]$, and hence

$$I = A(E_0^2 + E_1^2)/2 + AE_0 E_1 \cos[(\omega - \omega_0)t] \tag{9.48}$$

The first term on the right is a steady component and the second is a low-frequency alternating component, which can easily be detected and amplified. Values of $\Delta\omega = \omega - \omega_0$ are measured for various values of θ and plotted against $\sin^2\theta$. According to eqn 9.46, this should give a straight line of slope proportional to D_T. This is an absolute, fast and reasonably accurate method requiring only a small amount of material. The method can be extended and adapted to other types of investigation into the dynamics of macromolecular solutions.

9.7 Inelastic neutron scattering

Another method by which the internal movements in a macromolecule can be detected is inelastic neutron scattering. Neutrons are produced by nuclear fission in a nuclear reactor: they are uncharged particles with a mass of 1.675×10^{-24} g, very nearly equal to that of a proton. When they travel at a velocity v, they can also be represented by a wave of wavelength given by de Broglie's relation

$$\lambda = h/mv \qquad \text{or} \qquad \lambda = 3.956 \times 10^{-7}/v$$

where λ is in m and v in m s^{-1}. Wavelengths of the order of 0.5–1 nm (interatomic spacing) can only be obtained with velocities of around 500 m s^{-1}. The source is a sphere of liquid deuterium immersed in the reactor. A velocity selector is used to give a 'monochromatic' or monoenergetic beam. Neutrons are scattered by atomic nuclei and the amplitude of the *coherent* scattered wave is proportional to a *diffusion length* b_N representing a phase change in the wave due to the action of the nuclear potential. The length b_N can also be regarded as the radius of a rigid sphere from which the neutron would bounce, although such a model cannot explain the negative value of b_N for the hydrogen atom.

The following gives values of b_N in units of 10^{-12} cm for most of the atoms encountered in biopolymers:

	H	D	C	N	O	P	S
b_N	−0.374	0.667	0.665	0.94	0.58	0.5	0.285

There are two points to be made about these values. The first is that the diffusion length is independent of the scattering angle because the size of the nucleus (10^{-4} Å) is negligible compared with the equivalent neutron wavelength. The second is that, although X-ray scattering increases with electron density (i.e. with atomic number), b_N is of the same order of magnitude for all atoms. The contribution from hydrogen atoms is therefore much greater relative to that from others than it is with X-rays.

The coherent diffusion length for a molecule is obtained by adding the b_N values of all its atoms. For a biopolymer, therefore, we define a mean diffusion length $\sum b_N/V$ where V is the molecular volume, a quantity that takes the place of the electron density in X-ray scattering.

The determination of molecular dimensions is based on the use of coherent scattered radiation (i.e. without change of wavelength). However, as with the scattering of light, there is some inelastic scattering because of the thermal motion of the atoms. This effect is particularly large with neutrons, since the energy of thermal neutrons is close to that of the thermal motion in molecules. The *incoherent* diffusion length is abnormally high for protons, so that it is mainly their movement that will be detected. Both linear momentum and energy are transferred between the macromolecular protons and neutrons. The scattered intensity will therefore depend on the temperature and its value can be compared with that deduced from calculations using molecular dynamics. There is, however, almost no angular variation of the inelastic scattering that would depend on the protein structure. It is thus impossible to distinguish the vibrational modes of each atom.

9.8 Fast-reaction kinetics

Conformational changes triggered by an external agent (change of temperature, pH or ionic strength, collision with a molecule) can be followed using kinetic studies. These require (i) as accurate a definition as possible of zero time, (ii) detection that is fast and sensitive but does not perturb the process and (iii) the ability to make measurements over a large range of rate constants. A number of methods have been developed since the first experiments carried out by Chance.

9.8.1 Stopped flow

The reaction is initiated in this method by mixing two solutions. These may have different pH values or different ionic strengths, and may also contain two different molecules, A and B. The mixing process must always be as rapid as possible (definition of the time zero) and the flow must be stopped after a very short time.

The first type of apparatus (Fig. 9.14) consists of two syringes whose plungers are suddenly depressed by means of compressed gas. The stop of the stopping syringe is adjusted so as to reduce the volume to a minimum while ensuring that the measuring cell is completely filled with a homogeneous solution. The main part of the system is the mixing cell formed by several regulating nozzles, which create the turbulent flow required for the rapid mixing of solutions A and B.

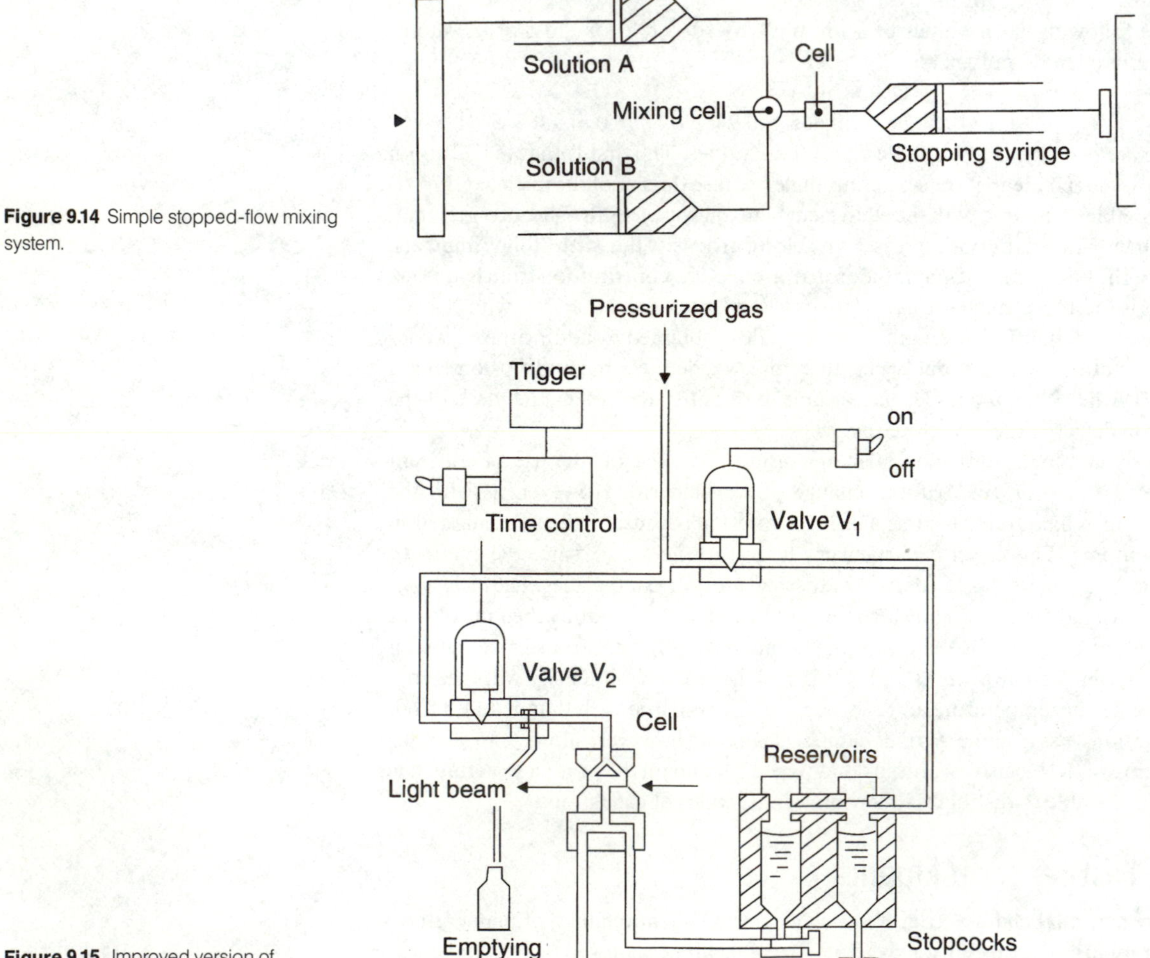

Figure 9.14 Simple stopped-flow mixing system.

Figure 9.15 Improved version of stopped-flow mixing system.

By using a minimum volume and a suitable machining of the mixing cell, the *dead time* can be reduced to less than a millisecond in the best apparatus available. To improve the system, particularly by reducing the dead time, a new generation of apparatus (Fig. 9.15) uses a closed system with pressurized nitrogen. The start and stop are triggered by solenoid valves controlled by a microprocessor program. The dead time can be reduced with this system to about 0.25 ms.

The detection of a conformational change generally involves the use of an optical property such as variations in absorbance, circular dichroism or fluorescence. The possibilities of this method are illustrated by two examples: one with a DNA–daunomycin interaction (Fig. 9.16) and one with an enzyme–substrate reaction (Fig. 9.17). It is also possible to couple a stopped-flow apparatus with detection using small-angle X-ray scattering.

(a)

(b)

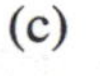

(c)

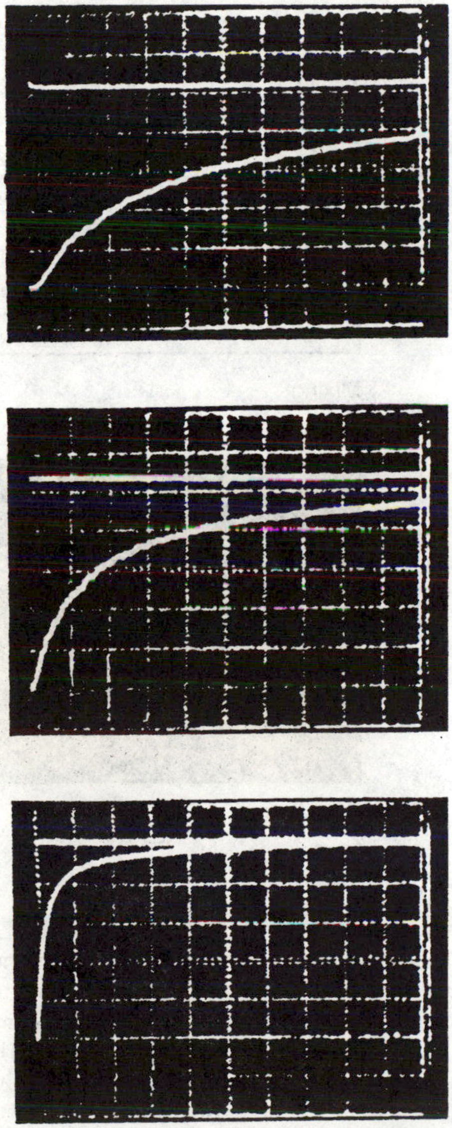

Figure 9.16 Kinetics of the daunomycin–DNA interaction at 21 °C with a final molar daunomycin/base-pair ratio of 0.125. The change in absorbance at 480 nm is recorded as a function of time. The horizontal scales are (a) 2 ms per division, (b) 5 ms per division, and (c) 50 ms per division. (From Chaires *et al.*, 1985. Reprinted with permission from *Biochemistry.* © 1985 American Chemical Society.)

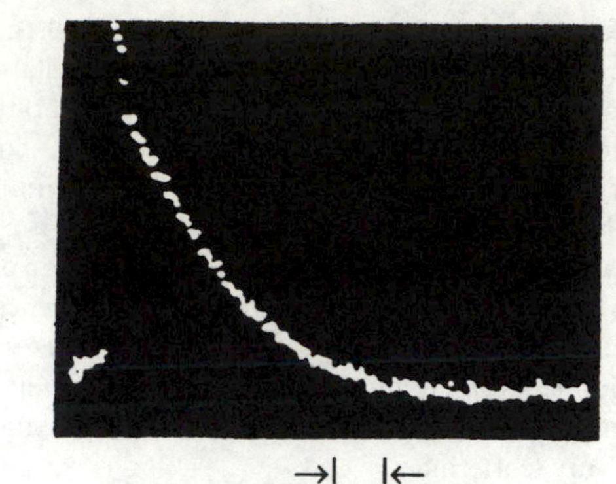

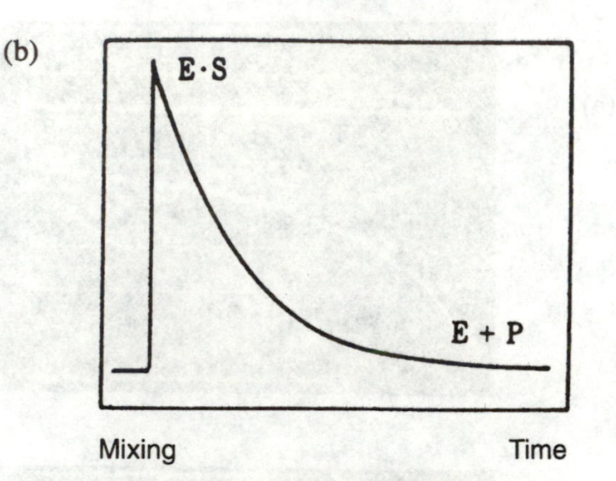

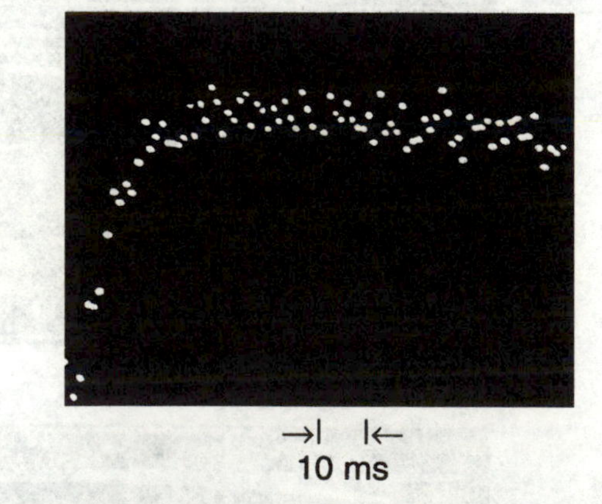

Figure 9.17 Kinetics of an enzyme – substrate reaction in 0.1 M KCl / 10^{-3} M $MgCl_2$ / 10 mM Tris at pH 9 and 23°C. The enzyme is a porcine leucine aminopeptidase and the substrate is a dansylated derivative of Leu-Gly, whose fluorescence is observed at 515 nm. Diagram (b) is a schematic drawing of what is observed in (a), i.e. a sudden increase in fluorescence immediately after the mixing ($E + S \rightarrow ES$) followed by an exponential decrease corresponding to the transformation $ES \rightarrow E + P$. Diagram (c) is an enlargement of the beginning of the reaction. (From van Wart and Lin, 1983. Reprinted from *Proceedings of the National Academy of Sciences, USA* with the authors' permission.)

9.8.2 Temperature jump

To achieve shorter reaction times, i.e. conformational changes occurring in the microsecond range, relaxation methods have to be used. The commonest of these is the temperature-jump method using the system illustrated in Fig. 9.18. The solution is suddenly heated by the Joule heat generated by discharging a capacitor C through the resistance of the conducting solution contained in the Perspex measuring cell (which may be thermostatically controlled). Measurement is initiated by the

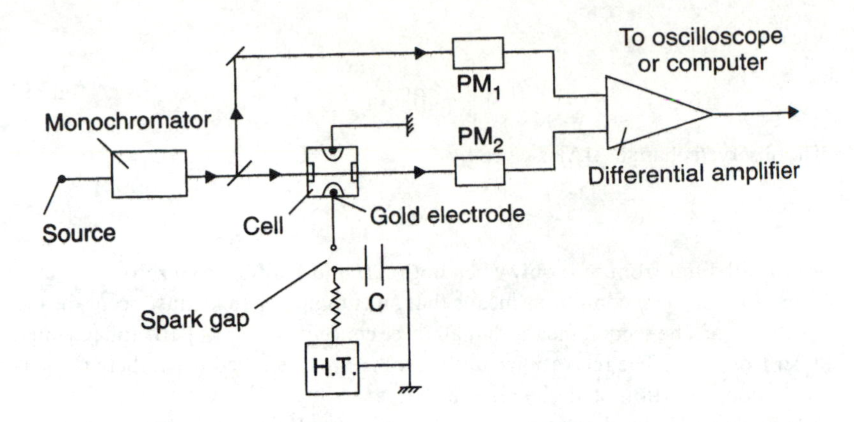

Figure 9.18 A temperature-jump system.

voltage induced in a small coil near the spark gap. The reference photomultiplier (PM1) corrects for the fluctuations of the lamp.

If the resistance of the cell contents is R and the capacitance of the capacitor is C, the equation governing the discharge is

$$R(dQ/dt) + Q/C = 0$$

whose solution is

$$Q = Q_0 \exp(-t/RC)$$

where Q_0 is the initial charge on the capacitor, so that the current is

$$I = (-Q_0/RC)\exp(-t/RC)$$

The amount of heat released is proportional to I^2 and hence to $\exp(-2t/RC)$, thus defining a characteristic time for the system of $\tau_0 = RC/2$. If $R = 100\,\Omega$ and $C = 0.05\,\mu F$, this gives $\tau_0 = 2.5\,\mu s$. Although the time τ_0 can be reduced by using smaller capacitors, this also reduces the energy released, $W = CV^2/2$. With $V = 30\,000$ V, this represents $W = 22.5$ J or about 5 cal, sufficient to raise the temperature of a 1 ml cell by 5°C in 2.5 μs.

Irrespective of the sensitivity of the method used for detection, the very nature of this technique imposes conditions that must be satisfied if the perturbation of the system is to be observed. Consider the reaction $A + B \rightleftharpoons C$ with $K = [C]/[A][B]$, so that

$$d \ln K = d[C]/[C] - d[A]/[A] - d[B]/[B]$$

However, $d[C] = -d[A] = -d[B]$, so that

$$d \ln K = -d[A]([A]^{-1} + [B]^{-1} + [C]^{-1}) \quad (9.49)$$

But K varies with temperature according to the standard thermodynamic law

$$d \ln K/dT = \Delta H/RT^2 \quad (9.50)$$

and using this in eqn 9.49 gives us

$$d[A] = -\Delta H([A]^{-1} + [B]^{-1} + [C]^{-1})^{-1}\, dT/RT^2 = \Delta H \Gamma\, dT/RT^2$$

where

$$\Gamma = ([A]^{-1} + [B]^{-1} + [C]^{-1})^{-1} \tag{9.51}$$

The observed change $\Delta[A]$ is given by

$$\Delta[A] = \Gamma \Delta H \Delta T / RT^2 \tag{9.52}$$

which will differ from zero only when both ΔH and Γ differ from zero.

The first of these conditions means that an enthalpy change must occur during the reaction. The second means that all three concentrations [A], [B] and [C] must differ from zero. The relaxation can therefore only be studied when there is a sufficient concentration of all the reagents.

In the case of a simple conformational change $A \rightleftharpoons B$, we denote the fraction of A that is unchanged by $1-\alpha$, i.e.

$$[A] = [A]_0(1-\alpha) \qquad [B] = [A]_0\alpha \qquad K = [B]/[A] = \alpha/(1-\alpha)$$

so that

$$\Delta K/K = \Delta\alpha/\alpha + \Delta\alpha/(1-\alpha) = \Delta\alpha/\alpha(1-\alpha) \tag{9.53}$$

The change in concentration $\Delta[A]$ due to the temperature jump is therefore

$$\Delta[A] = [A]_0\Delta\alpha \qquad \text{or} \qquad \Delta[A] = [A]_0\alpha(1-\alpha)\Delta K/K \tag{9.54}$$

and $\Gamma = \alpha(1-\alpha)$ is therefore a maximum when $\alpha = 0.5$, i.e. when $K = 1$ and $[A] = [B]$.

The experimental conditions must be such that K is sufficiently close to 1 to produce an observable change in concentration, taking into account the experimental uncertainty. Figure 8.3 in the previous chapter showed two curves of the relaxation times τ_1 and τ_2 plotted against $[A' + B']$ (see eqns 8.31 and 8.32).

In another type of temperature-jump system, the heating pulse is provided by a near-infrared power laser. This radiation is absorbed by the water and the heating time is consequently reduced to a few tenths of a microsecond.

More recently, a particularly rapid temperature-jump apparatus has been devised (Ballew *et al.*, 1996). Two pulsed (1 ns) infrared beams of wavelength 1.54 μm are sent in opposite directions through the cell, producing uniform and very rapid heating (at $5 \times 10^9\,K\,s^{-1}$). Changes in the solution contained in the cell are followed by absorption, circular dichroism or fluorescence. With a dead time of less than 20 ns, very fast conformational changes can be observed.

9.8.3 Other experimental aspects

Thermodynamics

Intensive variables other than temperature, such as pressure P or electric field E, can be used. The system is displaced from equilibrium by a sudden change in either parameter, e.g. by using a *shock wave* generated in a tube by the explosion of a thin membrane. In either case, enthalpy changes are no longer involved: with P, for example, we have

$$\partial \ln K/\partial P = -\Delta V/RT \tag{9.55}$$

where ΔV is the change in volume of the system. This can be applied to several associations between proteins and nucleic acids for which the reaction is almost athermal ($\Delta H \approx 0$) but in which there is a considerable volume change.

If the electric field E is varied, we have

$$\partial \ln K / \partial E = \Delta M / RT \tag{9.56}$$

where M is the electric dipole moment.

Periodic changes

Instead of using a sudden change in a physical quantity $x(t)$, where x is a step function, we can subject the system to a periodic variation of x. If the frequency is high enough, the system in equilibrium will only be able to follow such a continuous perturbation with some delay or *phase lag*, which can be related to values of the rate constants.

Equation 8.23 in the previous chapter can be written as

$$\tau \, \mathrm{d}x/\mathrm{d}t + x = A \exp(\mathrm{i}\omega t) \tag{9.57}$$

where A is a real quantity. The solution of this equation is a periodic function $x = B\exp(\mathrm{i}\omega t)$, which, substituted in eqn 9.57, gives

$$B = A/(1 + \mathrm{i}\omega t) \tag{9.58}$$

The amplitude B is therefore a complex quantity, which can be written in the form:

$$B = A(1 - \mathrm{i}\omega\tau)/(1 + \omega^2\tau^2)$$

or

$$B = A\mathrm{e}^{-\mathrm{i}\phi}/(1 + \omega^2\tau^2)^{1/2} \tag{9.59}$$

where $\tan\phi = \omega t$. The response of the system therefore has a real part

$$\mathfrak{R}(\omega\tau) = (1 + \omega^2\tau^2)^{-1} \tag{9.60}$$

and an imaginary part

$$\mathfrak{I}(\omega\tau) = \omega\tau(1 + \omega^2\tau^2)^{-1} \tag{9.61}$$

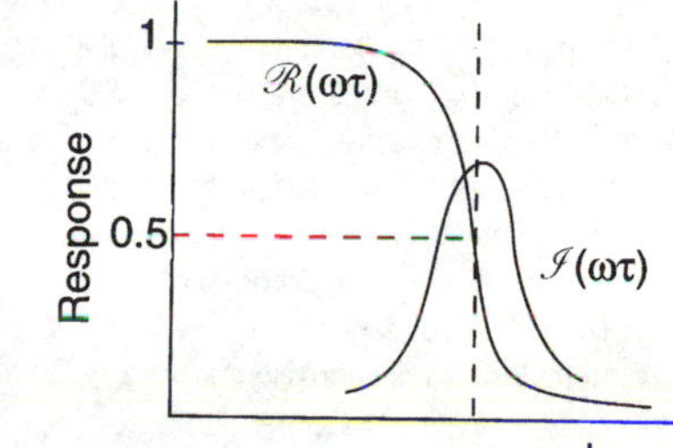

Figure 9.19 Real and imaginary parts of the response to a periodic stimulus.

whose variations with $x = \ln(\omega\tau)$ are shown in Fig. 9.19 (i.e. $\mathfrak{R} = (1 + \mathrm{e}^{2x})^{-1}$ and $\mathfrak{I} = \mathrm{e}^x(1 + \mathrm{e}^{2x})^{-1}$ · $\mathfrak{I}(\omega\tau)$ goes through a maximum when $\omega\tau = 1$ and $\mathfrak{R}(\omega\tau)$ has a point of inflection at the same value.

Another way of looking at the result is to say that the response of the system, while at the same frequency as the perturbation, lags behind it by a phase angle ϕ and with a measurable amplitude given by

$$A' = A/(1 + \omega^2\tau^2)^{1/2} \tag{9.62}$$

The curves of Fig. 9.19 show that, when ω becomes small, the chemical system is in phase ($\tan\phi \approx 0$) with the incident perturbation and remains permanently close to its initial equilibrium value. There is no energy loss. When $\omega \to \infty$, on the other hand, the chemical system at equilibrium can no longer follow the external perturbation and $\mathfrak{I}(\omega\tau)$ becomes constant, while the dissipative part becomes zero.

The response of the chemical system is usually studied in the range where τ and ω^{-1} are the same order of magnitude. The imaginary component is the one normally investigated through the phase shift, since it corresponds to an energy loss

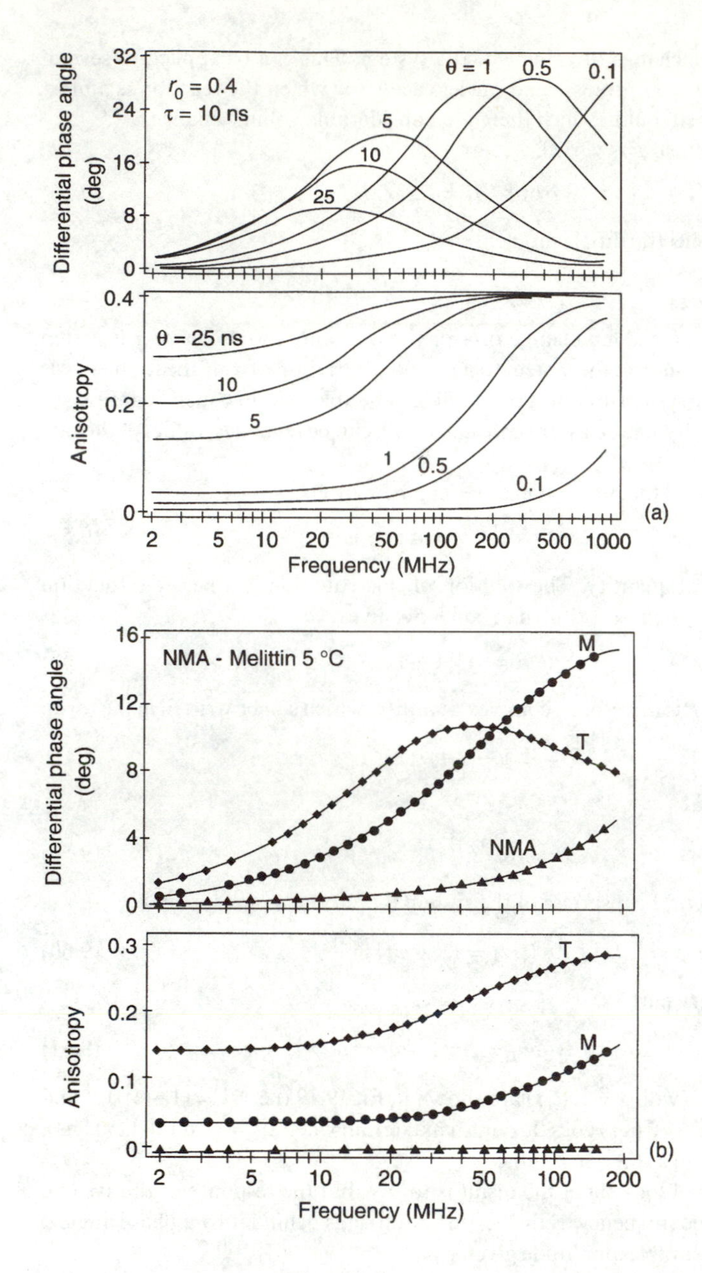

Figure 9.20 (a) Simulation of phase fluorimetry. The differential phase angle $\Delta(\omega) = \phi_{\parallel} - \phi_{\perp}$ and the anisotropy $r(\omega)$ are plotted against frequency for correlation times θ ranging from 100 ps to 25 ns. The fluorescence lifetime τ is 10 ns and $r_0 = 0.4$. (b) Differential phase angle and anisotropy for *N*-methylanthraniloylamide (NMA; ▲), an NMA–melittin monomer complex (M; •) and an NMA–melittin tetramer complex (T; ◆). All experiments were carried out at 5°C. (From Maliwal *et al.*, 1986. Reprinted from *Biochimica et Biophysica Acta*, vol. 873, pp. 173–81, © 1986 with kind permission from Elsevier Science NL, Sara Burgerhartstraat 25, 1055 KV Amsterdam, The Netherlands.)

that passes through a maximum when $\omega\tau = 1$. This is a phenomenon very similar to resonance in a harmonic oscillator subject to an external periodic force.

When the reaction is characterized by several relaxation times, a range of ω values must be explored. For each value of ω, we obtain a pair of values $A'(\omega)$ and $\phi(\omega)$, which are then compared using a least-squares method with values calculated from a model. Thus, after the covalent binding of a fluorescent group to a protein (melittin), the dynamics of the protein alone or of its complexes with lipids can be followed by analysing the fluorescence decay curves with a phase spectrofluorimeter (Fig. 9.20).

10

Methods of modelling intramolecular dynamics

Our treatment of the dynamics of biopolymers in Part II began with a description, in the first three chapters, of some simple physical models representing the internal movements and conformational changes in macromolecules, which allowed us to analyse their Brownian motion with or without an applied force field. The next chapter then briefly examined a whole group of experimental methods yielding macroscopic parameters related to molecular motion. We now see how a closer connection can be made between these two approaches by using physical models (harmonic oscillator and Brownian particle) to construct a much more detailed description of polymer dynamics that takes its structure into account. There are three possible ways of doing this:

(1) describing the molecule as a set of coupled harmonic oscillators – this leads to *dynamic analysis in terms of normal modes*;

(2) generalizing Newton's equation by calculating the trajectory of each atom as a result of its interaction with all neighbouring atoms – this is known as *molecular dynamics*;

(3) using Langevin's equation in which each atom is assumed to move in a viscous medium under the influence of external forces that are either random or derived from a potential – this leads to *stochastic dynamics*.

Each of these methods will be described and discussed at an elementary level, illustrating them by examples and showing their advantages and drawbacks. Apart from its importance in providing a picture of the dynamic aspect of a molecule, modelling now plays an essential part in refining molecular structures obtained by X-ray diffraction or NMR. This means that the static and dynamic aspects are becoming increasingly difficult to separate. This is a new molecular viewpoint, calling into question the specific recognition processes between molecules and hence raising doubts about a purely structural interpretation of biological functions.

10.1 Dynamics in terms of normal modes

The study of coupled harmonic oscillators in section 6.2 led to the calculation of normal modes for the system. These form a set of vibrations characterized by their angular frequency ω_k, their amplitude α_k and their phase at the origin ϕ_k. Such a calculation can be applied to biopolymers provided the generalized equilibrium coordinates q_i° are known (i.e. we know the three-dimensional structure in terms of variables such as the rotation angles about the valence bonds) and provided the potential energy E_p is expressed in terms of the coordinates q_i.

The K_{ij} term, i.e. the second derivative of E_p with respect to the coordinates, can be calculated by numerical differentiation of E_p, assuming small changes in q_i. The M_{ij} term, i.e. the second derivative of E_p with respect to the atomic velocities, can also be obtained numerically. In this way, a small protein may be characterized by hundreds of normal modes.

The model also allows the following data to be obtained.

Fluctuations

The fluctuations in each atomic position can be calculated from values of α_k and A_{jk}, giving information about local dynamics, which can be compared with that deduced, where possible, from temperature factors.

Effect of temperature

The amplitude α_k of a normal mode of angular frequency ω_k depends on temperature. Introducing the normal coordinate $Q_k = \alpha_k \cos(\omega_k t + \phi_k)$ into the equation for the harmonic oscillator gives

$$\langle \omega_k^2 Q_k^2 / 2 \rangle = k_B T/2 \tag{10.1}$$

since the mass is implicitly included in the value of Q_k. Using the expression for Q_k given above, the left-hand side of eqn 10.1 is

$$\langle \omega_k^2 \alpha_k^2 \cos^2(\omega_k t + \phi_k)/2 \rangle = \omega_k^2 \alpha_k^2 / 4 \tag{10.2}$$

so that

$$\alpha_k^2 = 2k_B T/\omega_k^2 \tag{10.3}$$

Correlations

To find the correlated movements between two atoms with positions r_i and r_j, we have to calculate the quantity

$$\langle \Delta r_i \Delta r_j \rangle = A_{ij} A_{jk} \alpha_k^2 \langle \cos^2(\omega_k t + \phi_k) \rangle = \tfrac{1}{2} A_{ij} A_{jk} \alpha_k^2 \tag{10.4}$$

Using eqn 10.3 then gives us

$$(\Delta r_i \Delta r_j) = k_B T A_{ij} A_{jk} / \omega_k^2 \tag{10.5}$$

Overall, the dynamics of normal modes describes the atomic movements quite satisfactorily, particularly the correlations between those of two atoms. However, using the quadratic approximation for the potential restricts the motion to very small amplitudes about the equilibrium position. Moreover, even the lowest frequencies (a few cm^{-1}) correspond in proteins or nucleic acids to times much shorter than those correlated with biological properties. For example, if the frequency is 1 cm^{-1}, its value in Hz is equal to the velocity of light in cm s^{-1} (3×10^{10}), which corresponds to a period of 33 ps. Figure 10.1 illustrates two slow modes for a small protein.

Two improvements have recently been introduced: the inclusion of anharmonic terms (i.e. correlations between movements) and the introduction of a frictional

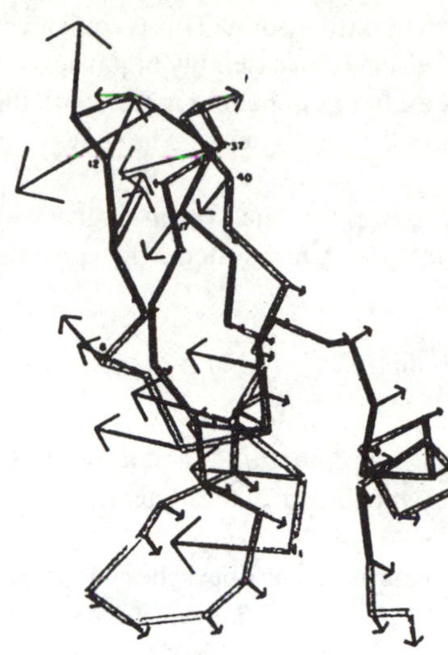

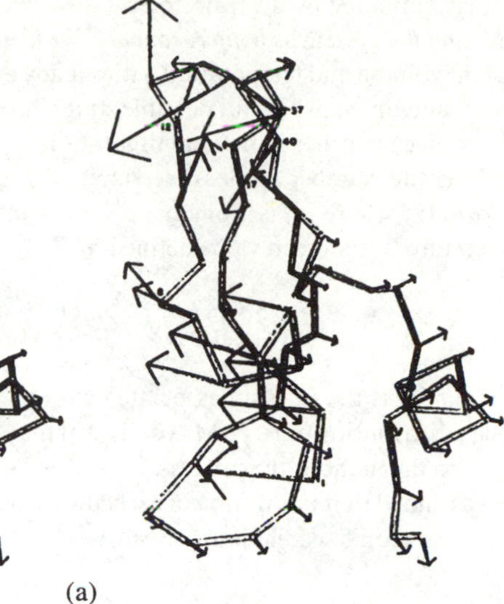

(a)

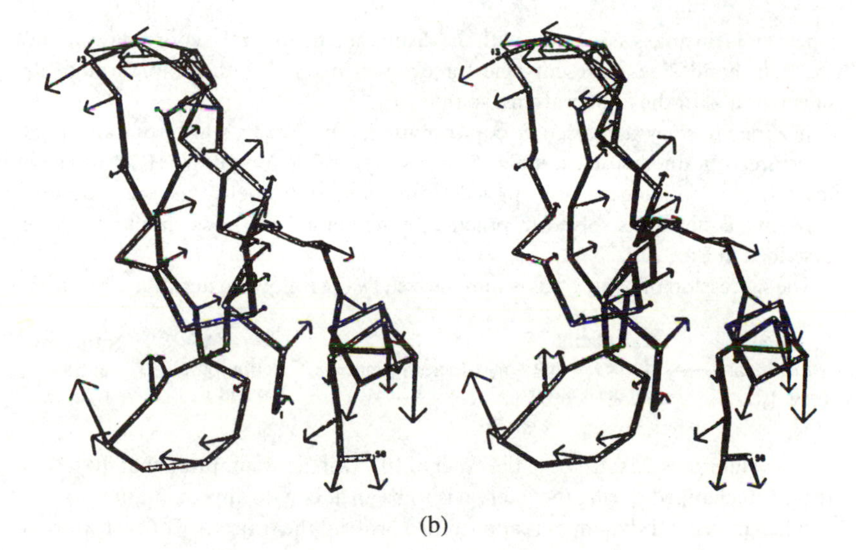

(b)

Figure 10.1 Normal-mode dynamics for BPTI. The amplitudes and directions of the displacements are indicated by the arrows, whose length has been drawn on a larger scale for clarity. The atomic motion is, of course, symmetrical about the equilibrium position. The two modes in (a) and (b) are for the two lowest frequencies, i.e. 4.56 and 5.40 cm^{-1}. (From Levitt *et al.*, 1985. Reprinted from *Journal of Molecular Biology* with the permission of Academic Press.)

term into the harmonic equation to obtain a form similar to that of Langevin's equation. The latter modification means that the normal modes become damped, so that

$$Q_k = \alpha_k \cos(\omega_k t + \phi_k) \exp(-b_k t)$$

where b_k is the damping coefficient.

10.2 Molecular dynamics

In this method, the motion of the N atoms comprising the molecule is traced out in a configuration space of $3N$ dimensions with coordinates q_k. A fluctuation in the system is simulated by the trajectory of the representative point. This is equivalent to treating the system as a *microcanonical ensemble*, i.e. an assembly of particles at constant volume and temperature without any exchange of heat or matter with the outside, and one in which all possible states have the same energy. The states only differ from each other in their configuration.

If W is the number of possible states, the entropy of the system is given by $S = k_B \ln W$, where k_B is Boltzmann's constant. In such a system, the constant temperature T is a mean value defined by

$$3Nk_BT/2 = \sum_k \tfrac{1}{2} m_k \langle (\mathrm{d}q_k/\mathrm{d}t)^2 \rangle \tag{10.6}$$

where $\mathrm{d}q_k/\mathrm{d}t$ is the velocity associated with the coordinate q_k. At time $t = 0$, the velocity distribution must be Maxwellian at the temperature of the system and must remain so throughout the process.

It is assumed that the motion of each atom i in the molecule obeys the generalized form of Newton's second law $F = ma$, i.e.

$$m_k\,\mathrm{d}^2q_k/\mathrm{d}t^2 = -\partial/\partial q_k \left[\sum_{\text{pairs}} E_p(r_{ij}) \right] \tag{10.7}$$

where m_k is the mass associated with the displacement along the coordinate q_k and the right-hand side represents the forces exerted on the ith atom through its interaction with the j other atoms, so that $j \neq i$.

In order to vary a molecular conformation, we have to start from an initial structure (obtained, for example, from X-ray diffraction or NMR), to which methods for refinement are applied using minimum energy as the criterion. Newton's equation is solved in practice by numerical methods similar to those described in Box 10.1.

The succession of operations is summarized by the following iterative algorithm:

Initial structure at $t = 0, v_0, r_0$	$\longrightarrow$	Calculation of forces acting on each atom	Incrementation $\longrightarrow$	$t = t + \Delta t$ Calculation of v and r	$\rightarrow$	Structure at time $t + \Delta t, v, r$

The time intervals Δt must be of the order of 10^{-15} s if the assumption that the forces remain unchanged during the interval is to be an adequate approximation.

When molecular dynamics is applied to a protein, the trajectory of each atom is described in detail and the real molecular deformations can then be followed. The flexibility can be characterized by a fluctuation σ such that

$$\sigma^2 = \langle x^2 \rangle - \langle x \rangle^2$$

The time and spatial correlations can also be calculated by forming averages such as $\langle x(0)x(t) \rangle$ or $\langle \boldsymbol{x}_i \cdot \boldsymbol{x}_j \rangle$.

Several problems arise in the method. The first is that because Δt is so small, the simulation of the dynamics can only cover a small period of time lest the computational time and cost become prohibitive. A period of 100 ps involves 100 000 iterations and is currently the practical limit. It is therefore impossible to envisage studying slow movements of large amplitude. If these do occur, the mean structure obtained will not necessarily be the true one. The following sketch shows how it

Box 10.1 Solution of Newton's equations

The practical solution of Newton's equations uses an approximate method due to Verlet. For a very small increment of time δt, q_k is expanded around t:

$$q_k(t+\delta t) = q_k(t) + (\delta t)^2 q_k/2! + (\delta t)^3 q_k/3! + \ldots \tag{1}$$

and similarly for $q_k(t-\delta t)$. The two expressions are added so that the odd terms cancel each other. Keeping only the first even term in $(\delta t)^2$ and replacing q_k by F_k/m_k, we obtain

$$q_k(t+\delta t) = 2q_k(t) - q_k(t-\delta t) + (\delta t)^2 F_k/m_k \tag{2}$$

which is Verlet's algorithm. If terms in $(\delta t)^4$ and above are to be neglected, δt must be small enough. This forces us to use time increments of the order of 10^{-15} s and means that we have to follow 10 000 movements to describe the molecule for 10 ps.

Moreover, the algorithm only guarantees that the kinetic energy is constant for sufficiently small value of δt, so that the kinetic energy has to be calculated repeatedly to check that the method remains consistent. For this purpose, the mean velocity

$$\delta q_k(t)/\delta t = [q_k(t+\delta t) - q_k(t-\delta t)]/2\delta t \tag{3}$$

is identified with the instantaneous velocity and the kinetic energy is calculated using eqn 10.6.

The calculation with the algorithm in eqn 2 requires knowledge both of the initial configuration at $t=0$ and of that at time $-\delta t$. This entails finding the velocity distribution at time $-\delta t$ that will ensure that the N atoms at $t=0$ obey a Maxwell–Boltzmann distribution at a temperature T. In this velocity distribution, a random choice of values is made to define the position at time $-\delta t$ using

$$q(-\delta t) = q(0) - (\delta t)\,\mathrm{d}q_k/\mathrm{d}t \tag{4}$$

At each step, all the forces exerted on the ith atom must clearly be calculated by using the $N(N-1)/2$ distances r_{ij}: this explains the time required for the calculation and why it is at present impossible to describe molecular motion by this method for a period of time longer than 100 ps.

may depend on the choice of initial structure.

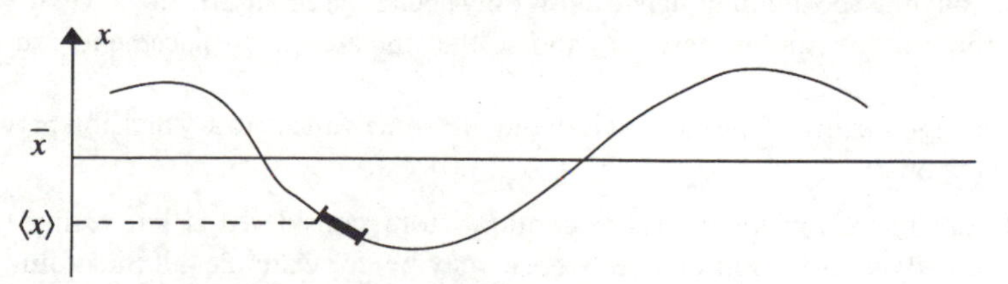

The sinusoidal curve represents a large-amplitude variation taking place, for example, over tens of microseconds. The range covered by molecular dynamics is indicated by the short thickened segment, and it is clear that the corresponding mean value $\langle x \rangle$ may differ considerably from the true mean value.

Molecular dynamics normally uses a system of $3N$ variables: the Cartesian coordinates of the N atoms in the molecule. All these coordinates are in fact undergoing high-frequency vibrations, a situation that forces us to choose a 'step' Δt of the order of 10^{-15} s. The majority of these movements are of small amplitude and are insignificant, so that to a first approximation we can 'freeze' them, i.e. keep the bond lengths and valence angles constant. The molecule is replaced by a collection of rigid atomic groups connected by fictitious bonds around which the rotation angles form the only variables (Mazur *et al.*, 1991). Lagrange's equation can be written for this new set of variables, thus greatly reducing the number of variables in the calculation. At the same time, Δt can be increased 100-fold to a value of the order of 0.1 ps. With 100 000 steps as before, we can now follow the dynamics over 10 ns, a period comparable with that for fluorescence and NMR.

Two other problems arise with charged biopolymers: the inclusion in the analysis of a specific solvent environment (water and ions), and the treatment of the long-range forces present in electrostatic interactions. The complete solvent environment has recently been included in the simulation of protein crystals or solvated protein systems using molecular dynamics (see the review by Brooks, 1995). The reaction field (see section 15.3.3) has also been used recently to mimic long-range forces: the effect of this electric field can be included in the molecular dynamics calculations. Such extensions are costly in terms of computer time, but computer power is increasing all the time and makes the processes more feasible.

10.3 Stochastic dynamics

To avoid some of the constraints of molecular dynamics, particularly the restriction on the length of time covered by any calculation, another form of dynamics can be used. This no longer depends on solutions of Newton's equation but on that of Langevin's equation (see section 7.3). For a particle moving with velocity $\boldsymbol{v}_i$ we can write

$$m\,\mathrm{d}\boldsymbol{v}_i/\mathrm{d}t = -m\beta_i\boldsymbol{v}_i - \operatorname{grad} E_\mathrm{p}(r_i) + \boldsymbol{R}_i \tag{10.8}$$

where the term $-m\beta_i\boldsymbol{v}_i$ is a frictional force. The external forces represented by the term $-\operatorname{grad} E_\mathrm{p}(r_i)$ include those contributed by the solvent. The force $\boldsymbol{R}_i$ is that due to collisions from the solvent molecules giving energy to the particle, unlike the frictional force, which is a damping force returning energy to the solvent from the particle.

The choice of the elementary time increment Δt is governed by two apparently contradictory requirements: it must be large enough to allow time for many collisions but must be small enough to allow only negligible changes in the forces (except for those in the random force $\boldsymbol{R}_i$) and so that the atomic displacements are very small.

The significance of the method is that it introduces an arbitrary division between two regions:

(1) the first region, known as $\{\beta\}$, contains atoms considered as 'interesting' from the dynamics point of view – these may be, for example, all the atoms in a protein molecule, but the $\{\beta\}$ ensemble could also describe a protein domain;

(2) the second region, known as $\{\alpha\}$, contains atoms considered as 'uninteresting' and forming part of the 'solvent' in a generalized sense of the term.

If, for example, we consider 400 β atoms 'immersed' in 3600 α atoms, stochastic dynamics only has to deal with 1200 coordinates and hence with 1200 corresponding equations for the motion of the β atoms. In molecular dynamics, we should have to deal instead with 4000 atoms and solve 12 000 equations.

However, the β_i values have to be calculated and then we can use molecular dynamics. When $\beta \gg (\Delta t)^{-1}$, there is strong damping and this leads to diffusional conditions with only one parameter, the coefficient of diffusion, D_T or D_R.

Part II Dynamics of biopolymers

Conclusion

In Part II, we have tried to present almost all the basic concepts needed to study the dynamics of a macromolecule. If this is compared with Part I, it is clear that the dynamic behaviour of biopolymers is much less well known than their structure.

Although powerful general methods are available for determining precise three-dimensional structures over a wide range of molecular weights, nothing equivalent exists in dynamics. Each of the methods briefly described enables us to study a particular range of time, but overlapping of the methods is rarely possible. There is, it is true, some similarity between the time scales of NMR and fluorescence (10^{-8} s to 10^{-9} s), but can we really compare mechanisms detected by proton exchange in DNA with those detected by intercalation, even if the time scales are similar?

Moreover, there are still many unexplored regions of the ranges of time relevant to the functions of biopolymers. What method is at present available, for instance, to provide information about the dynamics of the proteins and nucleic acids involved in replication, transcription or translation?

Faced with this shortage of experimental data, it is tempting to resort to increasingly sophisticated modelling because of the availability of more powerful computers and more efficient algorithms. However, any modelling unsupported by experiment must be distrusted, for it only needs small changes in the parameters to pass from acceptable to totally absurd values for the amplitudes of molecular movements. In addition, as we saw above, molecular dynamics cannot explore times longer than a nanosecond.

It is for these reasons that the main results in biopolymer dynamics have been obtained in processes involving the refinement of structures yielded by X-ray diffraction or NMR spectroscopy. Interaction between these two methods appears to be an essential factor in detecting a series of conformations or structures in dynamic equilibrium with each other.

Both theoretical and experimental advances are therefore needed before biopolymer dynamics can provide a complete description of the functioning of cellular machinery at the molecular level.

Part II Dynamics of biopolymers

References and further reading

Artymiuk, P.J., Blake, C.C.F., Grace, D.E.P., Oatley, S.J., Phillips, D.C. and Sternberg, M.J.E. (1979). Crystallographic studies of the dynamic properties of lysozyme. *Nature*, **280**, 563–68.

Ashikawa, I., Kinosita, K., Ikegami, A., Nishimura, Y., Tsuboi, M., Watanabe, K., *et al.* (1983). Internal motion of DNA in chromatin: nanosecond fluorescence studies of intercalated ethidium. *Biochemistry*, **22**, 6018–26.

Ballew, R.M., Sabelko, J. and Gruebele, M. (1996). Direct observation of fast protein folding: the initial collapse of apomyoglobin. *Proceedings of the National Academy of Sciences, USA*, **93**, 5759–64.

Beece, D., Eisenstein, L., Frauenfelder, H., Good, D., Marden, M.C., Reinish, L., *et al.* (1980). Solvent viscosity and protein dynamics. *Biochemistry*, **19**, 5147–57.

Brooks, C.L. (1995). Methodological advances in molecular dynamics simulations of biological systems. *Current Opinion in Structural Biology*, **5**, 211–5.

Careri, P., Fasella, P. and Gratton, E. (1975). Statistical time events in enzymes: a physical assessment. *CRC Critical Review of Biochemistry*, **3**, 141–64.

Chaires, J.B., Dattagupta, N. and Crothers, D.M. (1985). Kinetics of the daunomycin–DNA interaction. *Biochemistry*, **24**, 260–7.

Chandrasekhar, S. (1943). Stochastic problems in physics and astronomy. *Reviews of Modern Physics*, **15**, 1–89.

Coffey, W.T. (1980). Rotational and translational Brownian motion. *Advances in Molecular Relaxation and Interaction Processes*, **17**, 169–337.

Doi, M. and Edwards, S.F. (1986). *The theory of polymer dynamics*. Oxford University Press, Oxford.

Eigen, M. (1964). *Angewandte Chemie: International Edition in English*, **3**, 1–19.

Eigen, M. and de Maeyer, L. (1972). *Techniques of organic chemistry*, vol. VIII, part 2, pp. 895–1054. Interscience, New York.

Englander, S.W. and Englander, J.J. (1972). Hydrogen–tritium exchange. *Methods in Enzymology*, **26c**, 406–13.

Frauenfelder, H., Petsko, G.A. and Tsnernoglou, D. (1979). Temperature-dependent X-ray diffraction as a probe of protein structural dynamics. *Nature*, **280**, 558–63.

Friedman, H.L. (1985). *A course in statistical mechanics*. Prentice-Hall, Englewood Cliffs, NJ.

Glasstone, S., Laidler, K.J. and Eyring, H. (1941). *The theory of rate processes*. McGraw-Hill, New York.

Guinier, A. (1964). Théorie et technique de la radiocristallographie, 3rd edn. Dunod, Paris.

Hartmann, H., Parak, F., Steigemann, W., Petsko, G.A., Ringe-Ponzi, D. and Frauenfelder, H. (1982). Conformational substates in a protein: structure and dynamics of metmyoglobin at 80 K. *Proceedings of the National Academy of Sciences, USA*, **79**, 4967–71.

Hopfinger, A.J. (1973). *Conformational properties of macromolecules*. Academic Press, New York.

Karplus, M. and McCammon, J.A. (1981). The internal dynamics of globular proteins. *CRC Critical Review of Biochemistry*, **9**, 293–349.

Karplus, M. and McCammon, J.A. (1983). Dynamics of proteins: elements and function. *Annual Review of Biochemistry*, **53**, 263–300.

Kramers, H.A. (1940). Brownian motion in a field of force and the diffusion model of chemical reactions. *Physica*, **7**, 284.

Lerner, R.A., Benkovic, S.J. and Schultz, P.G. (1991). At the cross-roads of chemistry and immunology: catalytic antibodies. *Science*, **252**, 659–67.

Leroy, J.L., Kochoyan, M., Huynh-Dinh, T. and Guéron, M. (1988). Characterization of base-pair opening in deoxynucleotide duplexes using catalyzed exchange of the imino proton. *Journal of Molecular Biology*, **200**, 223–38.

Levitt, M. (1982). Protein conformation, dynamics and folding by computer simulation. *Annual Review of Biophysics and Bioengineering*, **11**, 251–71.

Levitt, M., Sander, C. and Stern, P.S. (1985). Protein normal-mode dynamics: trypsin inhibitor, crambin, ribonuclease and lysozyme. *Journal of Molecular Biology*, **181**, 423–47.

McCammon, J.A. and Harvey, S. (1987). *Dynamics of proteins and nucleic acids*. Cambridge University Press, Cambridge.

McCammon, J.A. and Karplus, M. (1988). The dynamics picture of protein structure. *Accounts of Chemical Research*, **16**, 187–93.

McConnell, B. and von Hippel, P.H. (1970). Hydrogen exchange as a probe of the dynamic structure of DNA. Effect of base composition and destabilizing salt. *Journal of Molecular Biology*, **50**, 297–316 and 317–32.

Malcolm, A.D.B. (1975). Biochemical applications of relaxation kinetics. *Progress in Biophysics and Molecular Biology*, **30**, 205–25.

Maliwal, B.P. and Lakowicz, J.R. (1986). Resolution of complex anisotropy decays by variable frequency phase-modulation fluorometry: a simulation study. *Biochimica et Biophysica Acta*, **873**, 161–72.

Maliwal, B.P., Hermetter, A. and Lakowicz, J.R. (1986). A study of protein dynamics from anisotropy decays obtained by variable frequency phase-modulation fluorometry: internal motions of *N*-methylanthranoyl melittin. *Biochimica et Biophysica Acta*, **873**, 173–81.

Mazur, A.K., Dorofeev, V.E. and Abagyan, R.A. (1991). Derivation and testing of explicit equations of motion for molecules described by internal coordinates. *Journal of Computational Physics*, **92**, 261–72.

Molday, R.S., Englander, S.W. and Kallen, R.G. (1972). Primary structure effects on peptide group hydrogen exchange. *Biochemistry*, **11**, 150–8.

Pecht, I. and Rigler, R. (1977). *Chemical relaxation in molecular biology*. Springer-Verlag, Berlin.

Petsko, G.A. and Ringe, D. (1984). Fluctuations in protein structure from X-ray diffraction. *Annual Review of Biophysics and Bioengineering*, **13**, 331–71.

Ringe, D. and Petsko, G.A. (1985). Mapping protein dynamics by X-ray diffraction. *Progress in Biophysics and Molecular Biology*, **45**, 197–235.

Steenbergen, C. and Mohr, S.G. (1973). A T-jump kinetic study of the binding of proflavin to polyIpolyC. *Biopolymers*, **12**, 791–8.

Teitelbaum, H. and Englander, S.W. (1975). Open states in native polynucleides: (1) Hydrogen-exchange study of adenine-containing double helices. (2) Hydrogen-exchange study of cytosine-containing double helices. *Journal of Molecular Biology*, **92**, 55–78.

Thomas, J.C., Allison, S.A., Appelof, C.J. and Schurr, J.M. (1980). Torsion dynamics and depolarization of fluorescence of linear macromolecules. Fluorescence polarization anisotropy measurements on a clean viral 029 DNA. *Biophysical Chemistry*, **12**, 177–88.

Uhlenbeck, G.E. and Ornstein, L.S. (1930). On the theory of the Brownian motion. *Physical Review*, **36**, 823–41.

van Wart, H.E. and Lin, S.H. (1983). Stopped-flow cryoenzymology: detection of catalytic intermediates not observable at ambient temperatures. *Proceedings of the National Academy of Sciences, USA*, **80**, 7506–9.

Wagner, G. (1983). Characterization of the distribution of internal motions in the basic pancreatic trypsin inhibitor using a large number of internal NMR probes. *Quarterly Review of Biophysics*, **16**, 1–57.

Wagner, G. and Wüthrich, K. (1982). Amide proton exchange and surface conformation of the basic pancreatic trypsin inhibitor in solution. *Journal of Molecular Biology*, **160**, 343–61.

Wang, M.C. and Uhlenbeck, G.E. (1945). On the theory of the Brownian motion (II). *Reviews of Modern Physics*, **17**, 323–42.

Warshel, A. (1977). Interpretation of resonance Raman spectra of biological macromolecules. *Annual Review of Biophysics and Bioengineering*, **6**, 273–300.

Woodward, C.K. and Hilton, B.D. (1979). Hydrogen exchange kinetics and internal motions in proteins and nucleic acids. *Annual Review of Biophysics and Bioengineering*, **8**, 99–127.

Part III

Hydration of biopolymers

Introduction

Water is a major constituent in all known forms of life. It forms 55 per cent by weight of an adult human body, although, as indicated in the table below, it is unevenly distributed. The water content of the body is kept constant by an average daily intake of 2.5 litres for an adult as part of solid or liquid food. With increasing age, the proportion decreases: from 55 per cent at age 40 to 47 per cent at 70.

Proportions of water in various tissues and tissue fluids (wt%)

muscle	79	myelin	40
heart	78–81	epidermis	64
liver	71	placenta	87
kidney	81	blood plasma	93
spleen	79	pleural fluid	98
brain	70–84	cerebrospinal fluid	96–99
nerve	56	saliva	99.4

Until the advent of NMR, there was no way of carrying out an *in vivo* study of the water present in tissues. Even with NMR, the water detected is essentially 'free' water, in which the movements of the water molecules are sufficiently rapid to allow proton resonance lines to be observed. The broadening of these lines when produced by strongly bound water makes it undetectable.

In imaging by NMR, parameters such as nuclear spin density or relaxation time T_1 are measured. As with most such techniques involving instrumentation, the image is not obtained directly but only after processing the collected data. Here, the production of a two- or three-dimensional image requires a magnetic field gradient to be superimposed on the steady magnetic induction B_0. As a result of this, each point in the sample has a different precession frequency from its neighbours and spatial displacements are then translated into changes of resonant frequency. By combining pulse sequences of both the radiofrequency field and the gradient field, and by using three-dimensional Fourier analysis, the nuclear spin density can be reconstructed over a plane or in a volume element of the sample.

Images of the free water in a given tissue enable normal and pathological samples to be compared. Moreover, with the computational speeds now available, it is even possible to monitor the movement of fluids such as the blood flow in the heart in real time.

As regards the bound water, there are few experimental methods at present available enabling us to specify its role in biological processes and to propose any models for phenomena such as effector–receptor binding, the integrity of membrane structures, the folding of macromolecular chains, charge transfer (proton or electron) between donor–acceptor pairs, etc. Difficulties arise simply because no comparison can be made with the same process in the absence of water: we cannot, for example, examine the spatial folding of a completely dehydrated peptide chain.

In dealing with hydration in this Part, therefore, we can only offer a series of experimental and often complementary approaches, coupled with static and dynamic models. This will enable us to gain at least some insight into the role played by water in the structure and functioning of biological systems.

The macromolecules occurring in the cellular medium or in various body fluids (blood, lymph, cerebrospinal fluid, etc.) must be soluble in water. This implies that there are many mechanisms governing interactions between water molecules and the atoms located at the solvent–macromolecule interface. The term *hydration* covers all such phenomena, which are not only involved in solubility but also play a role in the morphogenesis, the stability and the dynamics of biopolymers.

When hydration is defined in this way, it corresponds to non-covalent associations of water molecules. Moreover, such molecules are not site-bound but are continually changing places with external water molecules. The *residence time*, i.e. the average time spent by a water molecule in a hydration site, might at least qualitatively give us some idea of the hydration energy. The residence time is never longer than a nanosecond.

It should now be clear that hydration as defined above differs considerably from chemical hydration, which is the covalent addition of a water molecule to another molecule. This process generally involves both H^+ and OH^- ions and leads to another compound with different chemical properties (the hydration of ethylene, $CH_2{=}CH_2$, for example, gives ethanol, CH_3CH_2OH).

11

Properties of water

It is paradoxical that we should still know so little about a liquid that is universally in daily use and of such importance biologically. This chapter discusses what we do know of the physical properties of water, using the classical distinction between its three phases (vapour, liquid and solid) with well-defined transition temperatures when a change of phase occurs.

11.1 The water molecule

We should expect an sp bond between the s orbital of H and the two 2p orbitals of O in H_2O. The angle HOH should be 90° but is actually 104.5°. Hybridization occurs between the 2s and 2p orbitals of O to form sp^3 bonds in the molecule. Two O orbitals remain available, corresponding to the lone-pair electrons and thus bringing about the almost tetrahedral symmetry of the water molecule (Fig. 11.1) with the following parameters: OH distance = 0.957 Å, HOH angle = 104.5°, angle between free orbitals of O ≈ 109°.

There are two important consequences of the geometrical and electronic properties of H_2O.

1. The charge distribution can be represented diagrammatically by two positive point charges at the positions of the H atoms and two negative point charges in a plane perpendicular to that of the molecule. Water therefore has a high dipole moment $p = 1.834$ D (debye) $= 6.11 \times 10^{-30}$ C m, and a quadrupole moment of about 1.87×10^{-39} C m^2. The mean polarizability is 1.444×10^{-30} m^3; the distribution of polarizability is not known but is very nearly isotropic (see Box 11.1).

2. The presence of two free orbitals on the O atom and the polarity of the OH bonds account for the ease with which the tetrahedral coordination of H_2O occurs, especially with other water molecules. The energy of the association of water molecules into dimers, trimers, etc., can be calculated, and out of all the possible associations, the trimer

```
H       H       H
|       |       |
O—H····O—H····O—H
```

is the most stable. Next come the tetramer with S_4 symmetry, the asymmetric pentamer and the hexamer with S_6 symmetry.

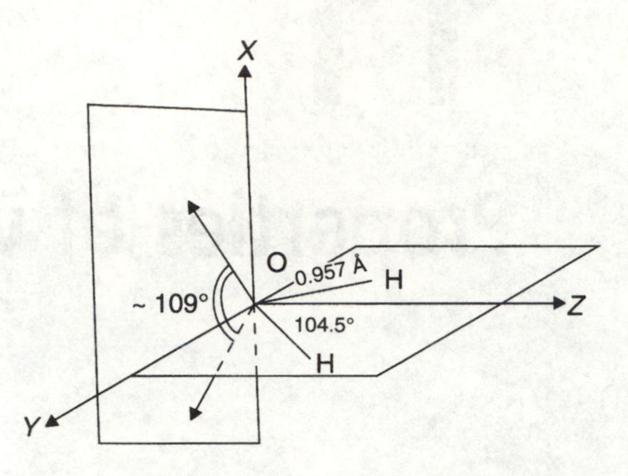

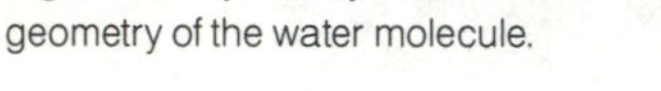
Figure 11.1 Symmetry elements and geometry of the water molecule.

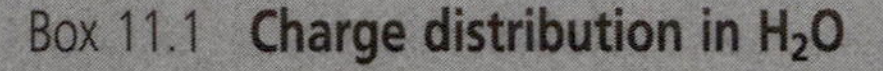
Box 11.1 Charge distribution in H_2O

The water molecule can be represented diagrammatically by

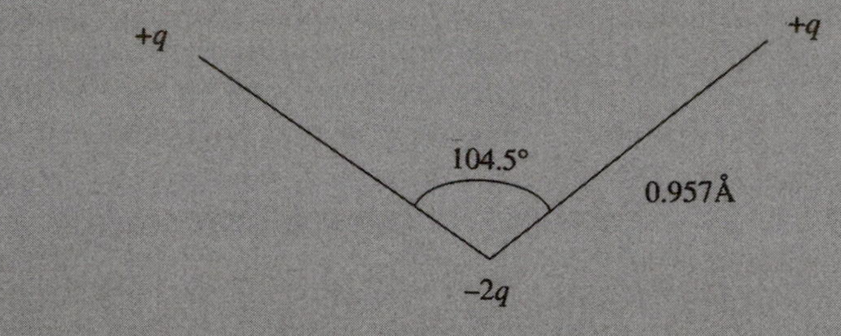

The distance d between the centroids of the positive and negative charges is given by

$$d = 0.957\cos(52.25°) = 0.586\,\text{Å}$$

so that to obtain a dipole moment of 6.11×10^{-30} C m the charge $2q$ must be equal to 1.043×10^{-19} C, i.e. $0.65e$, where e is the elementary charge. The charge q is thus $0.325e$.

The three vibrational normal modes of H_2O, D_2O and HOD are shown in Fig. 11.2: the fundamental and harmonic frequencies of these yield the infrared spectrum of a dilute water vapour. When we compare the frequencies for H_2O, D_2O and HOD, we see that the HOD molecule appears to have almost uncoupled OH and OD stretching bands, since their frequencies are very close to the corresponding ones in H_2O and D_2O. As a result, HOD is often used as an infrared probe of the structure of water in its different states.

11.2 Ice

11.2.1 Structure

Water does not have one unique crystalline solid state, but a large number existing over a wide range of temperatures and pressures. In this chapter, we shall only look at ice I and ice II, since the conditions under which they form and are stable are the closest to those of liquid water. Ice I exists at 0°C under normal pressures. It consists

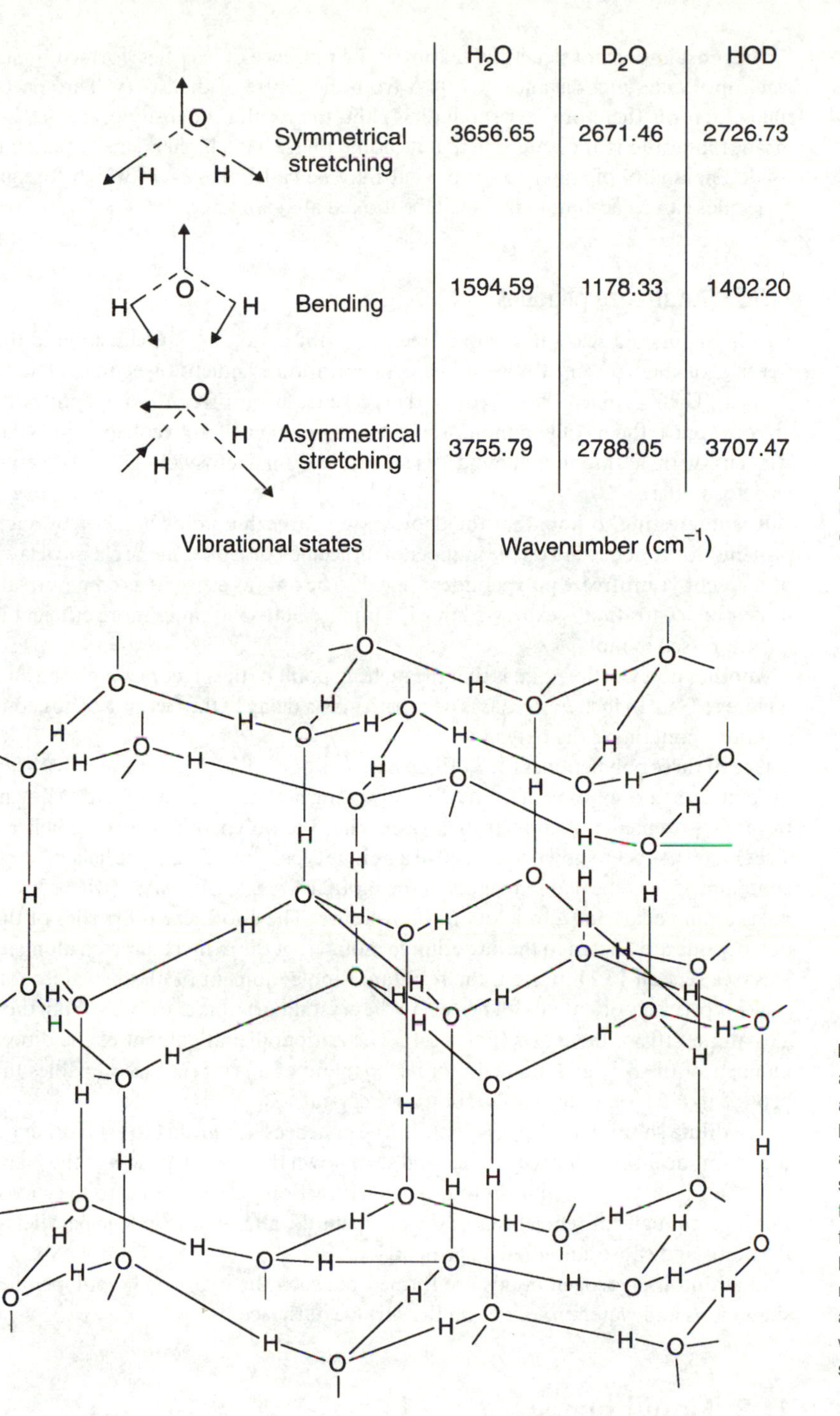

Figure 11.2 Vibrational ground states of H_2O, D_2O and HOD with their corresponding wavenumbers.

Figure 11.3 Structure of ice I. The oxygen atoms are in successive layers formed by a network of chair-shaped hexagons. Each layer is the mirror image of the adjacent one. Each oxygen atom is surrounded by four hydrogen atoms: the two belonging to the molecule are 1.01 Å from the oxygen atom; the other two belonging to two neighbouring molecules are at 1.75 Å. Each oxygen atom is at the centre of a tetrahedron, which means that all four angles have the same value of $\cos^{-1}(1/3) = 109° 47'$.

of chair-shaped hexagonal layers (Fig. 11.3) having opposite puckering in two superimposed rings. Each oxygen atom lies at the centre of a tetrahedron formed by four other oxygen atoms separated by 2.76 Å. The hydrogen bonds between OH and O stretch the OH distance slightly to 1.01 Å.

The most important structural feature is the presence of 'cavities' formed by six water molecules at a distance of 2.94 Å from the centre of the cavity. The space-filling factor is therefore very small (0.34) and the resultant density of ice (0.924) means that water is anomalous in that its liquid phase has a higher density than the solid. The ability of water to form what may be called 'cages' in which foreign molecules can be accommodated will be looked at again later.

11.2.2 **Antifreeze peptides**

Fish living in cold sea-water with a freezing point close to $-2°C$ must avoid the freezing of their serum, whose saline concentration is much lower than that of the sea. Their problem is solved by the presence of antifreeze polypeptides or glycoproteins, the most common of which is a polymer of a glycotripeptide Ala-Ala-Thr with a threonine-bound disaccharide (β-galactosyl($1\rightarrow3$)-α-*N*-acetyl-galactosamine).

It is interesting to note that the depression of freezing point induced by such proteins does not depend on the number of molecules, as it does in the classical laws of cryoscopy: antifreeze polypeptides appear to be only as efficient as common salt if their concentration is expressed in $g\,l^{-1}$ but are 300–500 times more efficient if it is expressed in $mol\,l^{-1}$.

Another noteworthy point is that the melting point of the glycoprotein solutions is close to 0°C. In fact, *hysteresis* is occurring, i.e. a delay in the freezing of aqueous solutions containing the polymers.

One of these polypeptides ($M = 1650$) has been crystallized and studied by X-ray diffraction at a resolution of 2.5 Å, thus providing new data. It was already known, from measurements of circular dichroism, that the polypeptides form α-helices. The crystal has been shown to consist of a unique type of packing of the helices, each containing four repeated sequences of the motif TAAXAAAAAA (where X is a polar amino acid), i.e. a total of 36 Ala residues. The antifreeze properties of the polypeptide are related to the large dipole moment of the α-helix directed along its axis (see section 15.1). In ice I, the resultant dipole moment of all the water molecules is precisely oriented with respect to the crystallographic axes: 50.8° from the *c* axis and 60° from the *a* axis (Fig. 11.4). The antiparallel alignment of the dipole moment of the α-helix (and hence of its axis) and of a crystal of ice I enables the peptide to latch on to the surface of the ice crystal.

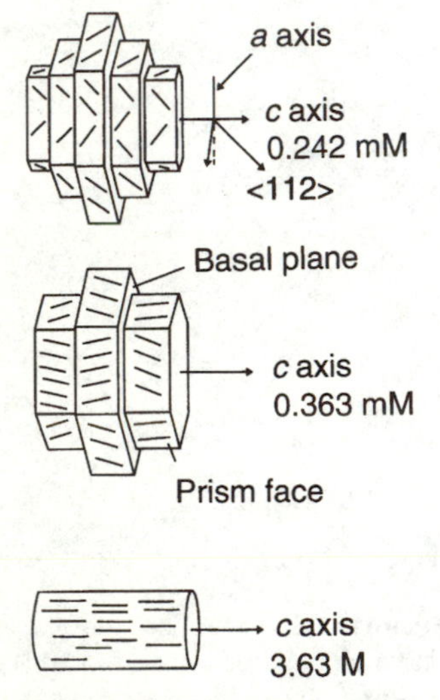

Figure 11.4 Schematic diagram of ice crystals formed in antifreeze peptide solutions with increasing concentrations. (From Yang *et al.*, 1988. Reprinted with permission from *Nature*, vol. 333, pp. 232–7 and with the authors' permission © 1988 Macmillan Magazines Ltd.)

In a dilute solution of the polypeptide, the α-helices lie parallel to each other on the prismatic faces of the seeds of ice and slow down the crystal growth in the plane of its basal surface, but allow it to grow in a direction perpendicular to the *c* axis. For high enough concentrations of the polypeptide, all the α-helices lie parallel to the *c* axis and cause a needle-like growth.

In addition, hydrogen bonds are formed between the hydrophilic polypeptide side chains and water molecules on the surface of the ice.

11.3 **Liquid water**

Liquid water is well known for its unusual physical properties given its low molecular weight. Unlike its direct homologue, H_2S, and unlike molecules with analogous structures such as CH_4, water is a liquid from 0°C to 100°C under atmospheric pressure. To maintain itself in a liquid state in this way, there must be strong interactions between its molecules, and this leads to speculation as to whether some form of order might be present.

Table 11.1 lists some of the properties of both H_2O and D_2O.

11.3.1 Structural study

Bernal and Fowler (1933) were the first to provide evidence of a local structure in liquid water from their classical X-ray studies. From the pattern yielded by small-angle X-ray scattering, it is possible to obtain the *distribution function* $g(r)$ of distances r between one water molecule and another (Fig. 11.5) at 4°C. The quantity $4\pi r^2 g(r)\,dr$ is the number of molecules at distances between r and $r + dr$ from the origin. A perfectly isotropic distribution has $g(r) = 1$ for any r. The fact that $g(r)$ goes through several maxima shows that some short-range order exists, defined by 'neighbours' at distances of 2.8, 3.7, 4.5 and 7 Å, but with none beyond 8 Å. A sphere of radius 8 Å contains about 70 water molecules, giving some idea of the

Table 11.1 Comparison between H_2O and D_2O

Property	H_2O	D_2O
molecular weight	18.015	20.028
melting point (°C)	0.00	3.81
triple point (°C)	0.01	3.82
boiling point (°C)	100.00	101.42
temperature of maximum density (°C)	3.98	11.23
critical temperature (°C)	374.15	371.2
ΔH melting (kcal mol^{-1})	1.436	1.501
ΔH vaporization (kcal mol^{-1})	10.70	11.11
ΔS melting (cal K^{-1})	5.26	5.42
ΔS vaporization (cal K^{-1})	38.63	40.11
molar volume of liquid at melting point	18.018	18.118

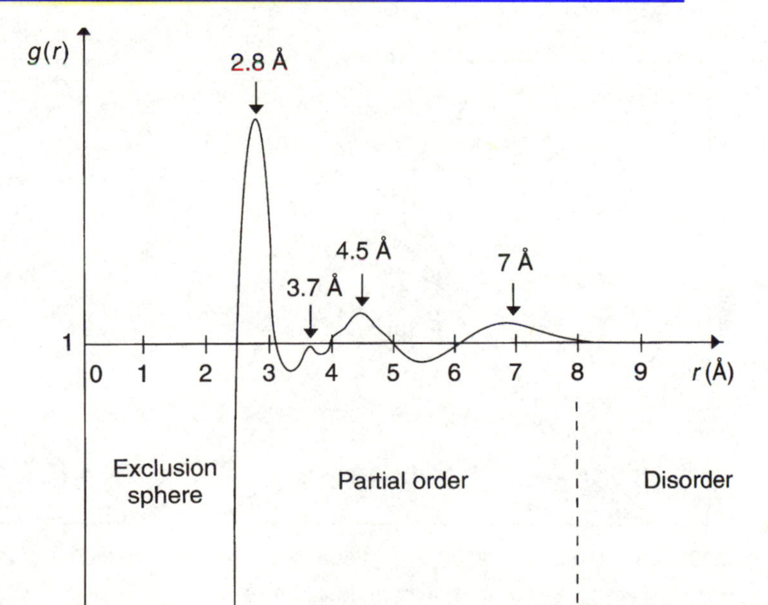

Figure 11.5 Distribution function $g(r)$ for liquid water at 4°C.

size of the ordered domains. The first peak, at 2.8 Å at 4°C, shifts to 2.94 Å at 200°C under pressure.

If a graph of $4\pi r^2 g(r)$ against r is plotted, the area under the peak at 2.8 Å shows that each molecule is surrounded on average by 4–5 molecules. This figure is 12 for close-packing of spheres (with a space-filling factor of 0.74), while it is found to be 8–10 for many simple liquids. Liquid water therefore has a very open structure and an almost tetrahedral arrangement of molecules. However, the first peak cannot be identified with a single Gaussian distribution, which seems to indicate the existence of a mixture of structures even at short range. In addition, while the peaks at 2.8, 4.5 and 7 Å are also found in ice, the peak at 3.7 Å is unique to liquid water.

Another method of studying the structure of water is to analyse its infrared and Raman spectra. To avoid the coupling between the vibrational modes of two molecules, measurements are made either of the frequency of the OD band (by dissolving HOD in H_2O) or of that of the OH band (by dissolving in D_2O). In the latter case, the wavenumber (3439 cm^{-1}) lies between that of ice I (3277 cm^{-1}) and water vapour (3707 cm^{-1}). Although the bandwidth is smaller than that of an isolated water molecule, it is much larger (160 cm^{-1}) than in ice (5 cm^{-1}). Such a width can only be explained by a structural disorder with a wide variety of local environments. A variation of the position of the Raman band with temperature occurs between 20 and 90°C (Fig. 11.6) together with the existence of an isosbestic point characterizing an equilibrium between two species clearly visible in the figure. Nevertheless, each of the species retains a bandwidth that is still very large.

By studying HOD in water at 400°C up to a pressure of 5000 bar, any density between 0.016 and 0.9 can be obtained, thus decoupling the variations in temperature and density. It is then found that the characteristic frequency of OD tends to a limit of 2620 cm^{-1} and not 2720 cm^{-1} as in the dilute vapour. The bandwidth must therefore be due mainly to rotational interactions between a given molecule and its neighbours, and does not involve hydrogen bonding (see Box 11.2).

The above two methods of investigating the structure of water are complementary and cover different structural ranges. In fact, liquid water is examined

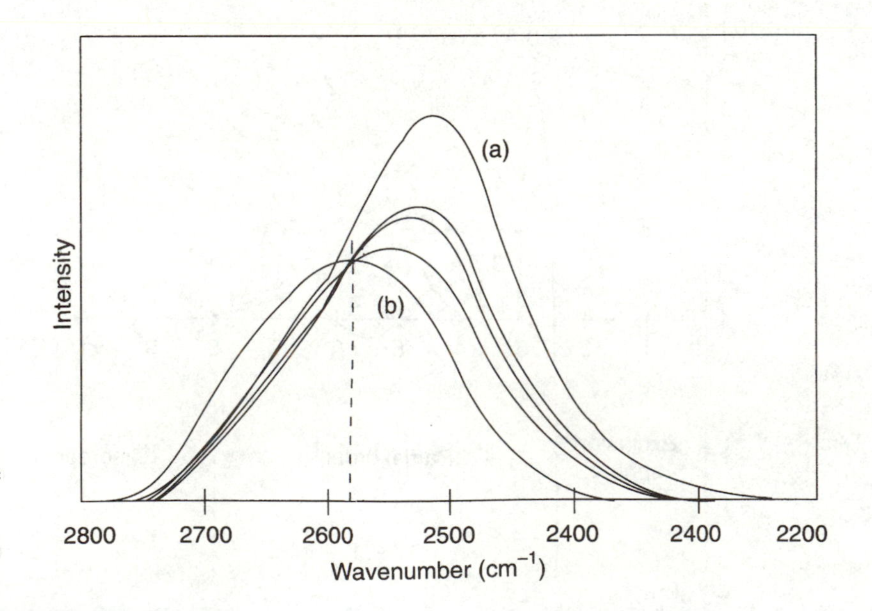

Figure 11.6 Raman spectra of OD: (a) asymmetric band at 25°C; (b) variations with temperature. The dashed vertical line shows the isosbestic point. (From Walrafen, 1968. Reprinted from *Journal of Chemical Physics* with the permission of the American Institute of Physics and the author.)

Box 11.2 Relationship between the heat capacity C_v and energy fluctuations

Another way of representing the instantaneous local disorder in water is to estimate the amplitudes of energy fluctuations. In a closed system (constant N, V, T) in thermal equilibrium with the external medium (a canonical ensemble), the energy fluctuation can be defined as usual by

$$\sigma_E^2 = \langle E^2\rangle - \langle E\rangle^2$$

By definition:

$$\langle E\rangle = \sum E_j \exp(-E_j/k_B T) / \sum \exp(-E_j/k_B T)$$

and hence

$$\langle E\rangle \sum \exp(-E_j/k_B T) = \sum E_j \exp(-E_j/k_B T)$$

Differentiating both sides with respect to T at constant N and V gives

$$(\partial\langle E\rangle/\partial T)_{V,N} \sum \exp(-E_j/k_B T) + (\langle E\rangle/k_B T^2) \times \sum E_j \exp(-E_j/k_B T) = (1/k_B T^2) \sum E_j^2 \exp(-E_j/k_B T)$$

so that

$$\begin{aligned} k_B T^2(\partial\langle E\rangle/\partial T) &= \sum E_j^2 \exp(-E_j/k_B T) / \sum \exp(-E_j/k_B T) \\ &\quad - \left[\sum E_j \exp(-E_j/k_B T) / \sum \exp(-E_j/k_B T)\right]^2 \\ &= \langle E^2\rangle - \langle E\rangle^2 = \sigma_E^2 \end{aligned}$$

However, when V and N are held constant, the internal energy of a canonical ensemble depends only on the temperature and the partial derivative $(\partial E/\partial T)_{V,N}$ is simply the heat capacity at constant volume C_v. Finally, therefore

$$\sigma_E^2 = k_B T^2 C_v$$

For water C_v is 1 cal g^{-1} and the molar specific heat is 18 cal. Since R is about 2 cal K^{-1}, $\sigma_E = 6T$ cal, giving an amplitude of fluctuations of 1.64 kcal mol^{-1} at 0°C and 2.24 kcal mol^{-1} at 100°C. The latter is almost a quarter of the molar latent heat of vaporization at this temperature, which is about 9.7 kcal mol^{-1}.

on very different time scales depending on the physical method used (Eisenberg and Kauzmann, 1969) and because of this the actual structures detected will be very different. Thus (Fig. 11.7), we can distinguish three types of structure: instantaneous (I), vibrational (V) and diffusional (D). The first is a 'snapshot' of water taken over a very short time of less than 10^{-13} s. Molecules are caught in a given oscillatory and vibrational state, which generally differs from their average state. If the observation time is extended to between 10^{-11} s and 10^{-13} s, the image will be 'fuzzy' because of the oscillations and vibrations, but if any order is present in the liquid it must be detected from the instantaneous positions of the molecules. Inelastic scattering of neutrons allows the molecular motion to be studied in more detail. If the interaction time τ_i of the neutrons with water molecules is long enough

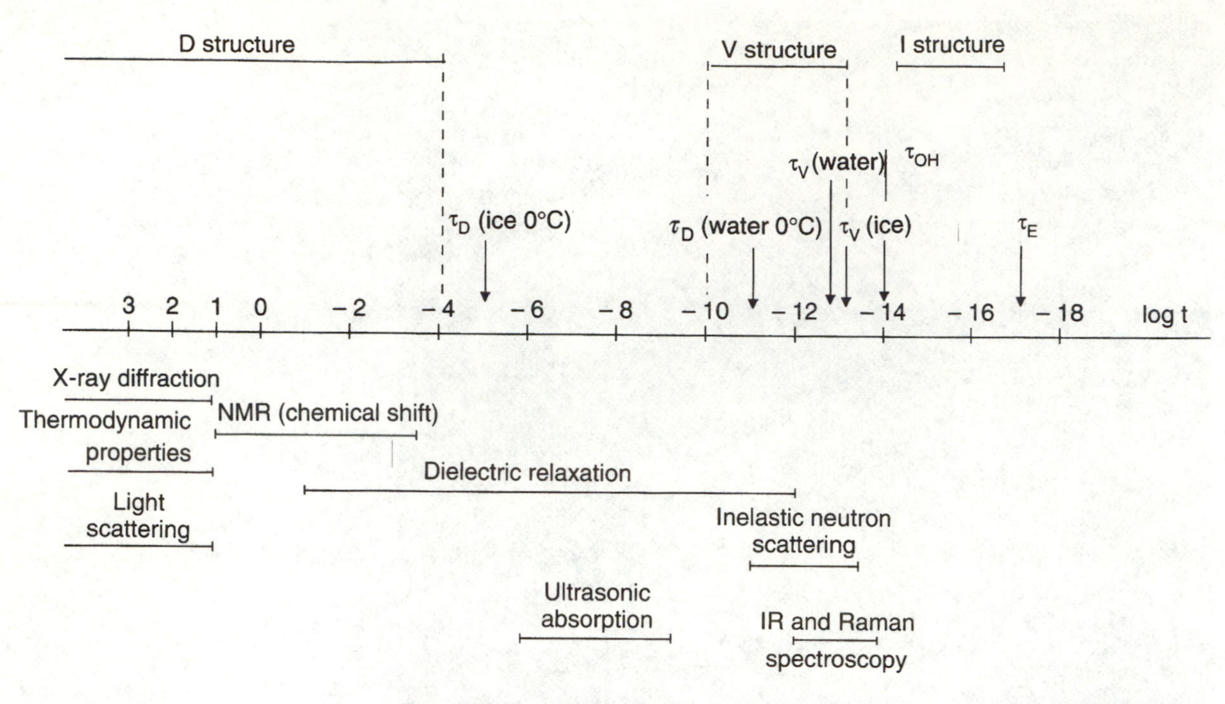

Figure 11.7 Time scales for the lifetimes of water structures: τ_E = period of electron in a Bohr orbit; τ_{OH} = mean period of the ν_T mode; τ_D = average lifetime between two successive reorientations. (From Eisenberg and Kauzmann, 1969. Reprinted from *The Structure and Properties of Water*, 1969, by permission of Oxford University Press.)

(slow neutrons), a process of diffusion by abrupt jumps is observed, the mean time between two successive jumps being of the order of 2×10^{-12} s. No long-range order can be detected, but at short range there is a tetrahedral-like coordination different from that of ice I. There are no free water molecules present, as there are in water vapour: this is the V structure.

If, finally, the observation time exceeds 10^{-11} s, the image is completely blurred by the various molecular movements (vibration, rotation, translation). In this case, therefore, it is preferable to define a new structure (D) by considering the spatial distribution of molecules with respect to one of them taken as origin. This is equivalent to putting a camera on a given water molecule and taking a snapshot as it moves. The photograph would not be completely blurred because some local order persists in the mutual arrangement of molecules, and it will correspond to the information provided by the distribution function $g(r)$. There are many experimental techniques for obtaining information about the V and D structures, but there is at present no way of discovering anything about the I structure, which remains purely theoretical.

Another point should be made about the lifetime of the V structure. As the temperature is gradually lowered, τ_D (see Fig. 11.7) increases and the V structure persists over longer times. At a very low temperature (−80°C), τ_D is of the order of several days or weeks, and this new structure is a *glass*. By rapidly freezing a thin pellet of water, it might be possible to fix a structure similar to the V structure, though it will of course differ from one point to another in the glass. When a macromolecule is present, the observed structure will be an instantaneous picture of the dynamic equilibrium between the macromolecule and its aqueous environment. This forms the basis for new methods of preparing specimens for deposition on the grid of an electron microscope.

11.3.2 Models of liquid water

Many models of liquid water have been proposed as a result of all these observations. However, these must not only have an explanatory value but a predictive one as well. They must be able to provide calculated values for physical properties such as phase change temperatures, latent heat capacities and specific heat capacities that agree with experimental values.

The models put forward can be divided into two groups, depending on whether the water molecule is assumed to have a single type of environment or to have more than one. In the latter case, water is viewed as a mixture or an interstitial compound.

The mixture model

In the mixture model of Frank and Wen (1957), liquid water is assumed to be in the form of molecular clusters arranged as in the ice I lattice, with a very short lifetime (10^{-11} s) and with disordered regions. A large number of water molecules would be present in the clusters because of the cooperative nature of hydrogen bond formation.

The advantage of this model is that it lends itself to a quantitative treatment (Nemethy and Scheraga, 1962). Several energy levels accessible to the water molecule are defined and a partition function Z calculated from the population in the various levels. The other thermodynamic functions can be calculated from Z with as few adjustable parameters as possible. The values for these parameters are chosen to give the best fit between calculated and observed values of the physical properties. The agreement for some quantities is quite good, but the $g(r)$ function and the Raman and infrared bandwidths are more difficult to predict.

A crucial factor, however, is that the model depends on the assumption that the hydrogen bonds between water molecules are highly directional, whereas quantum-mechanical calculations show that such bonds are deformable and that rotation about them is easy. One model (Grunwald, 1986) assumes a mixture of a tetra-coordinated structure A and a penta-coordinated structure B and is able to explain many of the properties. The relative amount of states A and B depends on the type of physical property (see Box 11.3).

The interstitial model

It is assumed in the interstitial model that non-coordinated water molecules can be accommodated in the cavities of a limited ice I lattice. There are, however, serious criticisms of this:

1. It is impossible to see how molecules of liquid water could exist in these 'cavities' without interacting strongly with the lattice molecules and thus destroying the lattice.
2. The accommodation of water molecules would necessarily increase the lattice volume and such a process is improbable at normal temperatures and pressures.
3. The interstitial model is too close to a solid-state model to be capable of interpreting the thermodynamic properties of liquid water.

Box 11.3 A two-state model

This is the model of Grunwald (1986). Water can be considered as a mixture of two states A and B, the first having a low energy, low entropy and large volume, and the second having a high energy, high entropy and small volume. Such a model has always been regarded with reservations by theoreticians, but it does provide a correct interpretation of spectral data, both Raman and infrared, by using an equilibrium between the two states according to a law of mass action for which

$$\mathrm{d}/\mathrm{d}T[\ln(\mathrm{B})/(\mathrm{A})] = (1/RT^2)\Delta H_{\mathrm{AB}}$$

This is an internal isomerization of the lattice and not a dissociation of water 'polymers'. We cannot assume that the A state is an 'iceberg' structure with directional hydrogen bonds and the B state is one in which hydrogen bonds are broken, for calculation shows that hydrogen bonds can be bent and twisted and hence suffer considerable deformation but cannot be broken. It is preferable to view A and B as two types of organization around a molecule. State A has tetrahedral coordination (Bernal and Fowler model) and state B has five-fold coordination with bifurcated hydrogen bonds.

If α is the fraction in state A, then $K = (1-\alpha)/\alpha$ and

$$\Delta G_{\mathrm{AB}} = -RT \ln K = \Delta H_{\mathrm{AB}} - T\Delta S_{\mathrm{AB}}$$

Measurements of heat capacity give $\Delta H_{\mathrm{AB}} = 2.5\ \mathrm{kcal\,mol^{-1}}$ and $\Delta S_{\mathrm{AB}} = 6.7\ \mathrm{kcal\,mol^{-1}\,K^{-1}}$. This gives the following variation of α with temperature:

t (°C)	0	25	62	83
α	0.76	0.694	0.587	0.532

The tetrahedral model predominates at low temperatures.

Coordination number

Each of the two states is characterized by a coordination number m defined as the number of hydrogen bonds between a water molecule and its nearest neighbours. The mean coordination number $\langle m \rangle$ at any temperature can then be defined by

$$\langle m \rangle = \alpha m_{\mathrm{A}} + (1-\alpha) m_{\mathrm{B}}$$

with $m_{\mathrm{A}} = 4$ and $m_{\mathrm{B}} = 5$. Using the values in the above 'table' gives $\langle m \rangle = 4.4$ at 0°C and 4.9 at 83°C. However, these values can also be derived from the area under the 2.9 Å peak in $g(r)$. The agreement is satisfactory, with the most recent measurements giving $\langle m \rangle = 4.4$ at 25°C.

Polarizability

In Kirkwood's theory of the dielectric constant of water, a coordination parameter g is introduced to take into account the field produced by the dipole moments of neighbouring water molecules. If μ is the intrinsic dipole moment of a water molecule and μ' the dipole moment due to nearest neighbours, we have

$$\mu\mu' = (g-1)\mu^2$$

Similarly, we can attribute values g_A and g_B respectively to the two states to account for the variation in dielectric constant with temperature. In this way, we find $g_A = 3.0 \pm 0.1$ and $g_B = 1.9 \pm 0.2$, indicating that the B state is about half as polar as the A state.

Volume

The molar volumes of states A and B are in the ratio 6 : 5, which corresponds to the ratio of the numbers of water molecules contained in a given total volume. If we adopt for the A state the value $v_A = 20\,\text{ml mol}^{-1}$ (corresponding to the density of ice), then we deduce that $v_B \approx 16.7\,\text{ml mol}^{-1}$, giving a volume change of $-3.3\,\text{ml mol}^{-1}$ when going from A to B. This lies within the considerable range of values (from -3 to -7) obtained from several different experimental methods interpreted by a two-state model.

An analysis of the physical properties of water can be taken further within the framework of this model. However, it should be pointed out that it relies on two configurations A and B that are time averages, which would no longer be valid over very short times, for example in the infrared domain. In such cases, a multi-state or almost continuous model would be better. The comments made previously about the time scale for observation of the structure of water (Fig. 11.7) should be kept in mind.

There is at present no entirely satisfactory model, but what is certain is that liquid water is indeed a mixture of several types of molecule depending on the degree of hydrogen bonding. For each type, the physical properties vary because of local changes in molecular interactions. The rapid interconversion of the different types means that liquid water appears to be a homogeneous pure liquid at sufficiently long time scales. In any case, it is clearly impossible to attribute to water a so-called 'memory'!

As we shall see later, problems related to the hydration of macromolecules are paradoxically simpler than those related to water alone, insofar as the structure and dynamics of bound water molecules can be analysed at all.

12

Hydrophobic and hydrophilic molecules

Molecules can be divided into two main categories as regards their hydration:

(1) those without charged groups or atoms (O, N, etc.) capable of forming hydrogen bonds – these are described as *hydrophobic* and are mostly molecules with aliphatic and aromatic hydrocarbon chains;

(2) those which, because they are charged or contain nucleophilic groups, can easily bind water molecules – these are *hydrophilic* substances.

The mechanisms by which the two groups become hydrated differ markedly from each other and need to be analysed in detail.

12.1 Hydration of hydrophobic groups

When a hydrocarbon is transported from an organic to an aqueous phase, a thermodynamic study shows that the corresponding ΔG is positive (unfavourable reaction). When the two phases are in equilibrium, their chemical potentials are equal. If X_o is the molar fraction of the hydrocarbon in the organic phase and X_w the same in the aqueous phase, we have (see Box 12.1):

$$\mu_o^\circ + RT \ln X_o = \mu_w^\circ + RT \ln X_w \qquad (12.1)$$

or

$$RT \ln(X_w/X_o) = \mu_o^\circ - \mu_w^\circ \qquad (12.2)$$

where μ_o° and μ_w° are the standard chemical potentials of the hydrocarbon in the two phases.

Since ΔG is positive, $\mu_o^\circ < \mu_w^\circ$ and hence $X_w \ll X_o$, so that only a very small fraction is dissolved in the water. If, for example, $\Delta G = 4\,\text{kcal mol}^{-1}$, $X_w/X_o = 1.3 \times 10^{-3}$ at room temperature. The positive value of ΔG is a result of the high value of ΔS, even though ΔH is often negative. Note that the measurements of the thermodynamic quantities must take account of the way in which the solute is transported from one phase to the other.

Box 12.1 Transition from an organic to an aqueous phase

The chemical potential μ_w of the hydrocarbon solution should be written in its exact form as

$$\mu_w = \mu_w^\circ + RT\ln X_w + RT\ln f_w$$

where f_w is the activity coefficient.

Because of the very low solubility of the hydrocarbon in water, we can take $f_w = 1$ and $RT\ln f_w = 0$, since the term represents that part of the chemical potential resulting from interactions of the solute molecules with themselves. Moreover, the term $RT\ln X_w$ is a purely statistical contribution to μ_w arising from the entropy of mixing two components assumed to have the same size of molecule. The term μ_w° is therefore the only one representing the free energy of the interaction between the dissolved molecule and water.

Similarly, for the same hydrocarbon in an organic solvent:

$$\mu_o = \mu_o^\circ + RT\ln X_o + RT\ln f_o \tag{12.3}$$

Assuming that μ_o° is almost independent of the type of solvent, eqn 12.3 can still be applied to a pure liquid hydrocarbon. We then have $\ln X_o = 0$, $\ln f_o = 0$ and $\mu_o = \mu_o^\circ$. The difference $\mu_o^\circ - \mu_w^\circ$ can then be measured by the transfer of liquid hydrocarbon from the pure liquid to the aqueous solution. For n-alkanes, it is found that

$$\mu_o^\circ - \mu_w^\circ = -2436 - 884n_c \tag{12.4}$$

where μ is in cal mol^{-1} and n_c is the number of carbon atoms.

It is interesting to compare the values of $\Delta\mu$ obtained for $n_c = 5$ (pentane) and $n_c = 6$ (hexane) with those for non-linear or substituted hydrocarbons with the same value of n_c:

$n_c = 5$		$n_c = 6$	
pentane	−6856	hexane	−7740
cyclopentane	−6000	cyclohexane	−6730
methylcyclopentane	−6880	methylcyclohexane	−7730

Saturated cyclic hydrocarbons are less hydrophobic than linear ones, but the simple addition of a methyl group increases $\Delta\mu$ by between 0.8 and 0.9 kcal mol^{-1} and reaches a value close to that for linear hydrocarbons.

Another comparison can be made with aromatic hydrocarbons having $n_c = 6$. For benzene, $\Delta\mu = -4620$ and for toluene $\Delta\mu = -5430$, so that these are less hydrophobic than the saturated linear compounds, probably because of the van der Waals interactions between the π electrons of the carbon ring and the water molecules. The difference of 0.8 kcal mol^{-1} between benzene and toluene reflects that found above for the addition of a methyl group.

In a mixture with i components, extensive quantities such as entropy and volume can be associated with intensive quantities known as *partial molar quantities*. For example, the partial molar entropy of the ith component is defined by

$$\langle S_i \rangle = (\partial S/\partial N_i)_{T,P,N_j}$$

Since $S = (-\partial G/\partial T)_{P,N}$, we have that

$$\langle S_i \rangle = -\partial/\partial N_i(\partial G/\partial T)_{P,N_j} = -\partial/\partial T(\partial G/\partial N_i)_{P,N_j} = -\partial\mu_i/\partial T$$

With $\mu_i = \mu_i^\circ + RT\ln X_i$, where X_i is the molar fraction of the ith component, we obtain

$$\langle S_i \rangle = -R\ln X_i - (\partial\mu_i^\circ/\partial T)_{P,N_j} \tag{12.5}$$

The $-R\ln X_i$ term, known as the *cratic* contribution, arises from the mixing entropy. The term

$$S_i^\circ = -(\partial\mu_i^\circ/\partial T)_{P,N_j} \tag{12.6}$$

is the unitary entropy.

Applying these relationships to the bimolecular reaction $A + B \rightleftharpoons AB$ and noting that one AB molecule is obtained for each pair of A and B molecules, we have for a molar solution ($c_i = 1$):

$$\Delta S_i = \Delta S_i^\circ + R\ln X_i \tag{12.7}$$

For water, $c_i = 55.6$, so $X_i = 1/55.6$ and hence $RT\ln X_i = 8\,\text{cal mol}^{-1}\,\text{K}^{-1}$. At 300 K, this cratic term contributes $-2.4\,\text{kcal mol}^{-1}$ to the free energy. Moreover, the change in unitary entropy is

$$\Delta S_i^\circ = \Delta S_i - R\ln X_i \tag{12.8}$$

This entropy change is a true representation of the hydration process and is the quantity used in what follows. For example, for n-butane $\Delta H = -0.85\,\text{kcal mol}^{-1}$, $\Delta S = -22.3\,\text{cal mol}^{-1}\,\text{K}^{-1}$ and $\Delta G = 5.9\,\text{kcal mol}^{-1}$. More generally, for all aromatic and aliphatic hydrocarbons, $-2.9 < \Delta H < 0.7\,\text{kcal mol}^{-1}$ and $-23 < \Delta S < -18\,\text{cal mol}^{-1}\,\text{K}^{-1}$ and hence $2.6 < \Delta G < 6\,\text{kcal mol}^{-1}$. The large negative value of ΔS reflects the organization of water molecules around hydrophilic groups to form 'cages'. In other words, the contact between water and the hydrophobic group is much less favoured than contact between two water molecules. Moreover, this solvation of a hydrophobic group is accompanied by a decrease in volume due to a better use of the available space. It is therefore impossible to conceive of the water molecules arranged around the hydrophobic group as they are in ice I (iceberg model). When water is frozen at 273 K, the latent heat is $80\,\text{cal g}^{-1}$, giving $\Delta H = -1.44\,\text{kcal mol}^{-1}$ and $\Delta S = -5.3\,\text{cal mol}^{-1}\,\text{K}^{-1}$. It would therefore need four water molecules to be 'frozen' to give a value for ΔS of around $-20\,\text{cal mol}^{-1}\,\text{K}^{-1}$. The corresponding value of ΔH would then be about -5.8 kcal, which is much too high.

In the recent two-state model (tetra-coordinated A and penta-coordinated B), the introduction of a hydrophobic compound would tend to increase the amount of A (the state with the lowest energy and entropy and the largest volume).

With most solvents, the solute takes the place of solvent molecules in the liquid network, but in water a hydrophobic solute cannot replace a water molecule. Instead of the substitution, therefore, an *interstitial* compound is formed and this is why steric and spatial characteristics of the molecules are so important. Similarly, the 'loosest' structure will be the one that accommodates the solute molecules most easily, leading to the high negative value of ΔS. The favourable change in enthalpy ($\Delta H < 0$) is another way of expressing the structural rearrangement in water.

12.2 Hydration of ions

12.2.1 Structural aspect

In the strong electric field (of the order of 10^{10} V m^{-1}) near an ion, the permanent dipoles of the water molecules take up a direction pointing towards the central ion. The orientation energy $W = -\boldsymbol{p} \cdot \boldsymbol{E}$, where $\boldsymbol{p}$ is the dipole moment of the water molecule, is between $10k_BT$ and $20k_BT$. In the first hydration shell around the ion, the water molecules can no longer rotate freely and because of this they form a structure very different from that in ice I, called the first hydration shell or ordered *region I* (Frank and Wen, 1957). Between this ordered layer and the bulk of the water there must be a second and somewhat disordered *region II*, since local order similar to that revealed by X-rays cannot be established. Ions can be classified according to the relative importance of layers I and II.

With a large ratio of charge to ionic radius, the number of ordered water molecules in region I is greater than in II and the salt solution will be more ordered than pure water: this is so for Li^+, F^- and Mg^{2+}. For Cs^+, Rb^+ and ClO_4^-, on the other hand, the salt solution will be more 'liquid' than pure water, i.e. it has less local order. Such a qualitative description is just as valid as the previous one based on parameters like the radius of the hydrated ion and the hydration number, whose values depend mainly on the method used.

Another approach has more recently been proposed: the introduction of a *correlation time* τ_c between the reorientating motions of water molecules, which can be measured by NMR. It is then possible to calculate τ_c in the hydration sphere (layer I) and to compare it with the value $(\tau_c)_0$ in pure water. In fact, we find that $\tau_c > (\tau_c)_0$ for structure-making ions and $\tau_c < (\tau_c)_0$ for structure-breaking ions. In other words, experiment shows that $1/T_1$ increases with concentration in the first case and decreases in the second case, T_1 being the longitudinal relaxation time of the proton.

12.2.2 Thermodynamic aspect

The free energy of hydration ΔG_H for an ion can be defined as the work required to move the ion from vacuum ($\varepsilon_1 = 1$) to a medium of dielectric constant ε_2 approximately equal to 80. An expression for ΔG_H has been given by Born (see eqn 15.37):

$$\Delta G_H = (q^2/8\pi\varepsilon_0 r)(1-\varepsilon_2^{-1}) \tag{12.9}$$

where q is the ionic charge and r the ionic radius. Born used the following process to calculate ΔG:

(1) remove the charge from the ion in vacuum (work ΔW_1);

(2) transfer the uncharged ion from vacuum to water (work ΔW_2 assumed to be zero);

(3) recharge the ion in water (work ΔW_3).

However, we have to decide what we mean by the *ionic radius*. If we take it as equal to that determined by crystallographic measurements, the calculated values for ΔG are too high compared with the experimental ones. The discrepancy can be accounted for by several factors: (i) r should be increased to the value given by its van der Waals radius, thus taking better account of the ionic hindrance in solution;

(ii) the value of ΔW_2 should be different from zero; and (iii) ε_2 ought to be reduced because near the ion the water molecules no longer have the rotational freedom they have in free water, thus lowering the dielectric constant.

However, it appears preferable (Rashin and Honig, 1985) to reconsider the Born process. Take, for example, the discharging of a cation with charge $+e$ *in vacuo*, for which an electron must be added to obtain a neutral atom. Transporting this atom into water is equivalent to creating a spherical cavity of radius R inside which the electron density of the solvent is negligible. To evaluate R, we can call on the electron density distribution inside an ionic crystal of the type X^-M^+, where X^- is a halogen and M^+ an alkali (Fig. 12.1). The electron density due to a cation M^+ located at E becomes significant at a point D such that AD (with X^- at A) is equal to the anionic radius. On the other hand, because the electron cloud of X^- cannot penetrate the free valence orbital of M^+, the electron density of X^- becomes significant only at a distance from the centre of M^+ equal to the radius EB of this orbital.

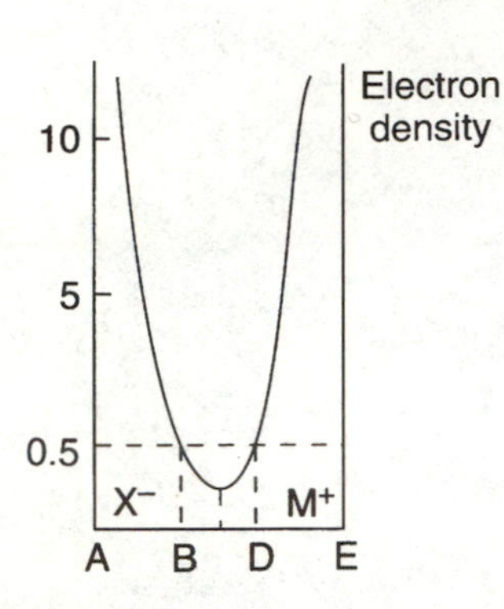

Figure 12.1 Electron density distribution in an alkali halide.

Because of this asymmetry between X^- and M^+, different radii are proposed for the cavities of anions and cations. For the cation the radius is equal to that of the orbital, i.e. almost the van der Waals radius. For the anion, the radius is equal to that of the ion. The validity of these assumptions can be tested by comparing calculated and observed values of the hydration enthalpies of each ion, obtained from measurements on salts.

Since $\Delta G = \Delta H - T\Delta S$, $\Delta H = \Delta G - T\,\mathrm{d}(\Delta G)/\mathrm{d}T$, so that from eqn 12.9:

$$\Delta H_{\mathrm{calc}} = (q^2/8\pi\varepsilon_0 r)[1 - 1/\varepsilon - (T/\varepsilon^2)\,\mathrm{d}\varepsilon/\mathrm{d}T] \qquad (12.10)$$

ΔH_{calc} varies as r^{-1}, but the calculated values are systematically 6–7 per cent higher than those observed, leading to a similar correction factor of the same order for both cavity radii to obtain agreement between ΔH_{calc} and ΔH_{exp}. The results are listed in Table 12.1, and the experimental points now lie very close to a straight line given by $\Delta H = Kr^{-1}$ where K is a constant (Fig. 12.2).

We can also consider the entropy change taking place when an ion dissolves in water (change in partial molar entropy), taking the value zero for a proton.

Table 12.1 Corrected ionic radii and hydration enthalpies

Ion	Corrected radius (Å)	$-\Delta H_{\mathrm{exp}}$	Ionic radius (Å)
Li^+	1.32	126	0.61
Na^+	1.68	98.5	0.96
K^+	2.17	78.4	1.33
Rb^+	2.31	73.4	1.48
Cs^+	2.51	67.6	1.66
Mg^{2+}	1.45	462.4	0.65
Ca^{2+}	1.86	384	0.99
Zn^{2+}	1.34	492	0.74
Cd^{2+}	1.51	435	0.97
Hg^{2+}	1.54	440	
NH_4^+	2.1	77.2	
F^-	1.42	119.3	1.34
Cl^-	1.94	85.3	1.81
Br^-	2.09	78.8	1.95
I^-	2.34	69	2.17
OH^-	1.50	109	

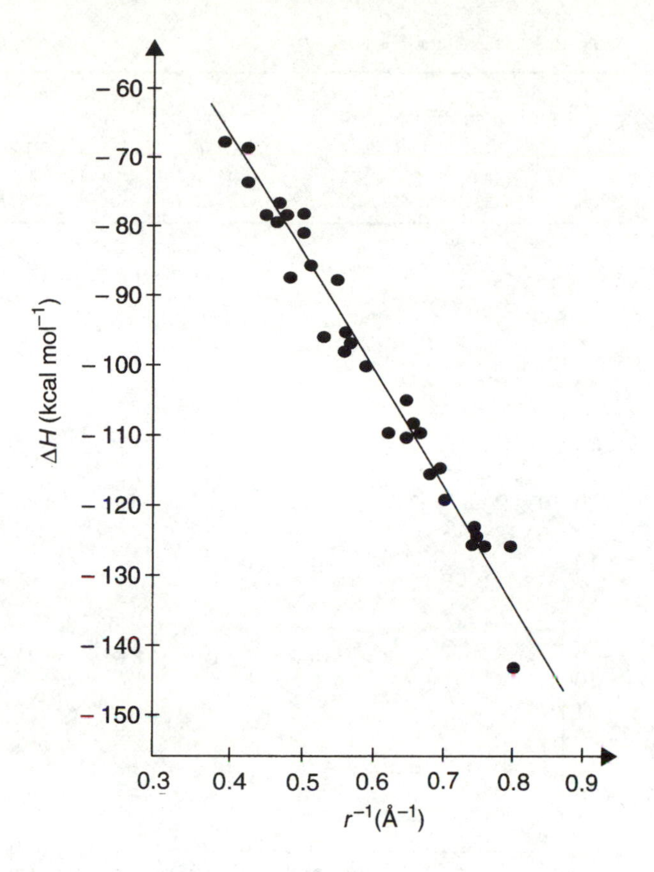

Figure 12.2 Variation of the hydration enthalpy of ions with their corrected radii. The values are related to one elementary charge. (From Rashin and Honig, 1985. Reprinted with permission from *Journal of Physical Chemistry*. © 1985 American Chemical Society.)

Table 12.2 Unitary entropy of ions (cal $mol^{-1}K^{-1}$)

Li^+	−4.6	Ba^{2+}	−5	F^-	−10.3
Na^+	6.4	Sr^{2+}	−17.4	Cl^-	5.2
K^+	16.5	Ca^{2+}	−21.2	Br^-	11.3
NH_4^+	19	Mg^{2+}	−36.2	NO_3^-	27
Rb^+	20.7	H^+	0	SO_4^{2-}	−3.6
Cs^+	23.8				

Table 12.2 gives values of ΔS for various series of cations and anions. The sign of ΔS is directly related to the 'structure-making' or 'structure-breaking' character ofthe ions.

Another experimental method of studying the structural changes in liquid water produced by ions in solution is to measure the viscosity η of salt solutions and to compare this with the viscosity η_0 of pure water at the same temperature. A relationship

$$\eta/\eta_0 = 1 + Ac^{1/2} + Bc$$

is obtained, where c is the molar concentration of the salt, and A is a positive coefficient reflecting the electrostatic interactions between ions that contribute to the energy of fluid flow. The coefficient B, on the other hand, may be positive or negative and can be interpreted in structural terms. If, for example, B is negative (which occurs only in the presence of certain ions), the viscosity of the solution may

Table 12.3 Values of the coefficient *B*

Li	Na	K	NH_4	Rb	Cs	Ba
0.149	0.086	−0.007	−0.007	−0.03	0.045	0.22
Sr	Mg	Cl	Br	I	NO_3	SO_4
0.265	0.385	−0.007	−0.042	−0.068	−0.046	−0.209

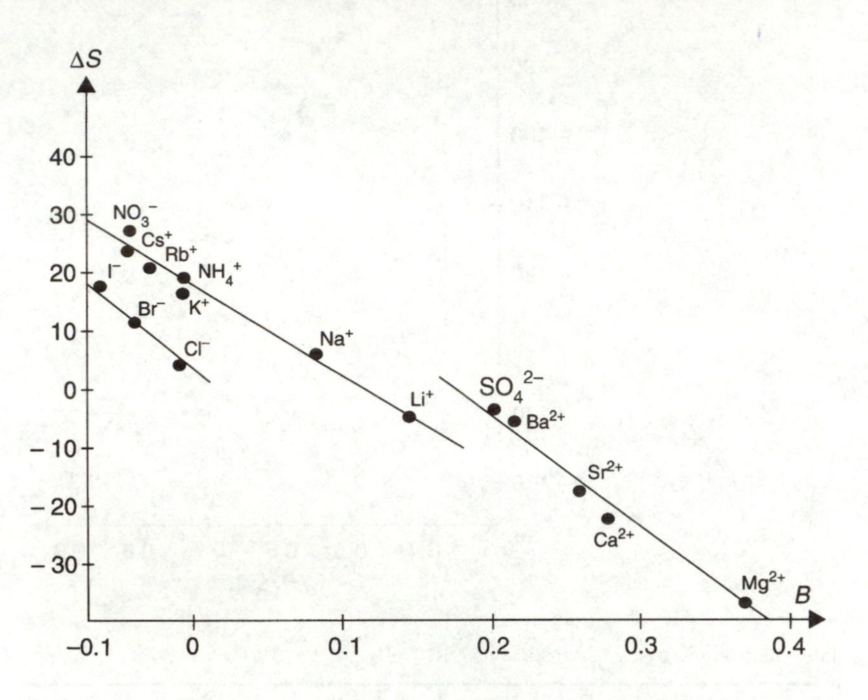

Figure 12.3 Variation of the unitary ionic entropy *S* with the coefficient of viscosity *B*.

be lower than that of water at a sufficiently high salt concentration. Although this behaviour is dynamic rather than static, a negative *B* can be interpreted as corresponding to the disappearance of the short-range order present in pure water. *B* can only be determined experimentally for salt solutions, i.e. when there is a cation–anion pair. However, a value can be attributed to each ion if a reference value is chosen by assuming that K^+ and Cl^- make equal contributions to the *B* of KCl solutions (Gurney, 1953). The results for a number of ions are given in Table 12.3.

If the unitary entropy is plotted against *B* for each ion (Fig. 12.3), we find linear relationships defining the various families (alkalis, alkaline earths, halogens, etc.), thus implying that both parameters reflect the structural changes in liquid water. However, the distinction between ions shows that parameters other than the electric field around the ion must be involved.

12.2.3 Lyotropic series

Hofmeister (1888) established a sequence of salts according to their ability to precipitate euglobulins. The salts are classified in *increasing* order of the concentrations needed for precipitation and hence in *decreasing* order of their efficiency:

$$Na_3(\text{citrate}) > Li_2SO_4, Na_2SO_4, K_2HPO_4, Na_2HPO_4 > (NH_4)_2SO_4 > MgSO_4$$
$$> KCH_3COO, NaCH_3COO > NaCl > NaNO_3$$

From this it is possible to deduce an ionic order, which has since been found in many other processes such as protein stabilization, enzyme inhibition and DNA stabilization. In denaturing native conformations or double-helical structures, the following sequence is found, in order of *increasing* efficiency:

- for anions

$$SO_4^{2-} < CH_3COO^- < Cl^- < Br^- < NO_3^- < ClO_4^- < I^- < CNS^-$$

- for cations

$$(CH_3)_4N^+ < NH_4^+ < Rb^+, K^+, Na^+, Cs^+ < Li^+ < Mg^{2+} < Ca^{2+} < Sr^{2+}$$

Clearly, the effect of an ion is not related solely to its charge, and other factors must be involved. These include the ionic polarizability, which increases with size (e.g. Cs^+ has a higher polarizability than water), the presence of a dipole moment, the presence of hydrophobic groups such as $(CH_3)_4N^+$ and the possibility of forming additional hydrogen bonds (ClO_4^-, NO_3^-) with water molecules. A detailed study of the local properties of water around each of the ions has still to be carried out.

We now turn to a phenomenon that is easier to analyse and is of great practical importance: the salting-out mechanism. We follow the treatment of Edsall and Wyman (1958). The majority of water-soluble organic molecules become less soluble if water is replaced with a salt solution. Conversely, the addition of an organic compound to a salt solution lowers the solubility of the salt. In some cases (e.g. salts of divalent cations) phase separation may occur. A denser layer rich in water and salt separates from a lighter layer rich in the organic compound.

This reciprocal effect on solubility has long been used for the selective precipitation of proteins by the addition of a salt such as ammonium sulphate. It is also involved in processes for growing biopolymer crystals, so that a detailed analysis is worth undertaking.

The basic idea is that the introduction of ions into a mixture of water and organic molecules produces a rearrangement of water molecules. In the language of classical thermodynamics, the addition of ions is said to reduce the activity of the water and increase that of the organic compound. The fact that charged ions are involved clearly points to the phenomenon being mainly electrical. In a medium of dielectric constant ε subjected to an electric field E, the stored energy per unit volume is $\varepsilon\varepsilon_0 E^2/2$. The electric field produced by the ion itself decreases by a factor $1/\varepsilon$. Hence, at a given distance r from an ion of charge Ze, the energy of the system is lowered as the dielectric constant increases. The dielectric constant of water (about 80) is generally higher than that of an organic medium, so that water molecules will be preferentially bound around ions to lower the energy of the system. This leads to an exclusion of the organic molecules (see Box 12.2).

The presence of an organic solvent such as ethanol thus increases the activity of the salt, and conversely a high salt concentration will increase the activity of the organic compound and hence reduce its solubility. The reciprocity of this effect can easily be deduced by expressing the variation in free energy with the number of moles, for we have

$$\mathrm{d}G = \sum_i \mu_i \, \mathrm{d}n_i$$

Box 12.2 Salting out

The derivation in this box follows Edsall and Wyman (1958). A calculation taking all the parameters into account would be very complex. Instead, we restrict the treatment to a simple model, which, although apparently unrealistic, gives a reasonably satisfactory explanation of the observed phenomena and allows us to predict other salting-out mechanisms, at least qualitatively.

The medium contains n_1 moles of water, n_2 moles of the organic compound and n_3 moles of a monovalent salt dissociated into n_3 moles of anions and cations with ionic charges $-e$ and $+e$ respectively. In a medium of dielectric constant ε, the electrical energy of the cation and anion are $n_3e^2/8\pi\varepsilon\varepsilon_0 r_c$ and $n_3e^2/8\pi\varepsilon\varepsilon_0 r_a$, where r_c and r_a are the respective ionic radii, assuming the ions are spherical charges. The total electrical energy can be expressed as $n_3e^2/8\pi\varepsilon\varepsilon_0 b$ where $1/b = 1/r_a + 1/r_c$. In water the same ion pair would have an energy given by $n_3e^2/8\pi\varepsilon_w\varepsilon_0 b$ where ε_w is the dielectric constant of water.

The difference between the two terms

$$\Delta W = (e^2/8\pi\varepsilon_0 b)(1/\varepsilon - 1/\varepsilon_w)n_3$$

is the difference between the electrical energy of the ion dissolved in water and dissolved in solution.

The chemical potential of the salt is the sum of a chemical term μ_3 and an electrical term μ_3' given by

$$\mu_3' = \partial(\Delta W)/\partial n_3 = A(1/\varepsilon - 1/\varepsilon_w)$$

where $A = n_3e^2/8\pi\varepsilon_0 b$, a quantity characteristic of the salt.

The chemical term is given by the classical expression in terms of the molar fraction X by

$$\mu_3 = \mu_3^\circ + RT\ln(X_3/X_3^\circ)$$

Since n_3 salt molecules give n_3 cations and n_3 anions, we have

$$X_3 = 2n_3/(n_1 + n_2 + 2n_3)$$

The total chemical potential with respect to its standard value is

$$(\mu_3 - \mu_3^\circ)_{\text{total}} = A(1/\varepsilon - 1/\varepsilon_w) + RT\ln(X_3/X_3^\circ)$$

In this expression, however, it is only the first term on the right that plays any part in the salting-out process. An electrical activity coefficient f_3 can be defined from μ_3' such that

$$RT\ln f_3 = A(1/\varepsilon - 1/\varepsilon_w)$$

If, for example, we take $b = 10^{-10}$ m, $T = 300$ K, $e = 1.6 \times 10^{-19}$ C, $\varepsilon = 25$ (ethanol) and $\varepsilon_w = 80$, we find that $f_3 \approx 2000$. The salt is therefore 2000 times less soluble in ethanol than it is in water.

where the sum is over all components of the system and where μ_i is the chemical potential (or partial molar free energy) of the ith component. Hence

$$\partial\mu_i/\partial n_j = \partial^2 G/\partial n_i\,\partial n_j = \partial\mu_j/\partial n_i$$

or in terms of activities, a_i and a_j:

$$\partial \ln a_i/\partial n_j = \partial \ln a_j/\partial n_i$$

which gives us a direct relationship between the change in activity of one component (e.g. the salt) and the change in activity of the other (e.g. the organic compound).

Such an elementary treatment can only give us a general idea of the salting-out mechanism. In the special case of proteins, where high salt concentrations have to be used, there is no valid theory leading to a quantitative prediction of the process. There is an empirical law of the type

$$\log S = \log S_0 - KI$$

where I is the ionic strength, S is the solubility, S_0 its hypothetical value for $I = 0$ and K is a constant generally proportional to the size of the protein but also depending on the kind of ion used. We then obtain the Hofmeister series again, as in many other phenomena.

12.3 Hydration of amphiphilic molecules

12.3.1 Definition and properties of amphiphilic molecules

A molecule that has both a polar group and an aliphatic chain will have new solvation properties. Such a molecule is called *amphiphilic*: the *polar head* will undergo solvation like an ion, while the *hydrocarbon chain* will do so like a hydrophobic molecule. The idea of an amphiphilic molecule becomes much more important whenever an interface is created between a polar and non-polar medium. An example is the interface appearing in *emulsions* of oil in water or of water in oil.

The energy required to form such interfaces is generally too great for the system to be stable. This is reflected by the changes occurring in the emulsion as time passes: first a *coalescence* of droplets of a given medium, followed by a separation into two phases, as with oil and water where one phase forms a layer over the other, depending on their relative densities. The emulsion can be stabilized by using an amphiphilic compound known as an *emulsifying agent* or *surfactant*. By attaching itself to each of the two phases, the amphiphilic molecules establish a connection between the two media which reduces or eliminates the surface tension at the interface and allows the emulsion to remain stable, possibly indefinitely. The industrial importance of amphiphilic compounds can be appreciated by quoting one figure: their annual production amounts to about 1 kg for every human being on the planet.

12.3.2 Quaternary ammonium ions

Quaternary ammonium ions $(NR_4)^+$, where R is a short aliphatic chain, are good examples of compounds providing another way of studying models of hydration: by varying the length of the aliphatic chain and then measuring the change in ionic mobility with temperature. If short chains are bound, as in Me_4N^+, the ionic structure-breaking effect predominates and the ionic mobility (the product of conductivity and viscosity of the solvent) must decrease with T. For ions like Bu_4N^+ or Pr_4N^+, on the other hand, the cages formed around the aliphatic chains, which greatly reduce the ionic mobility at room temperature, 'melt' as T increases, thus increasing the ionic mobility (Fig. 12.4). If, now, an OH group is introduced

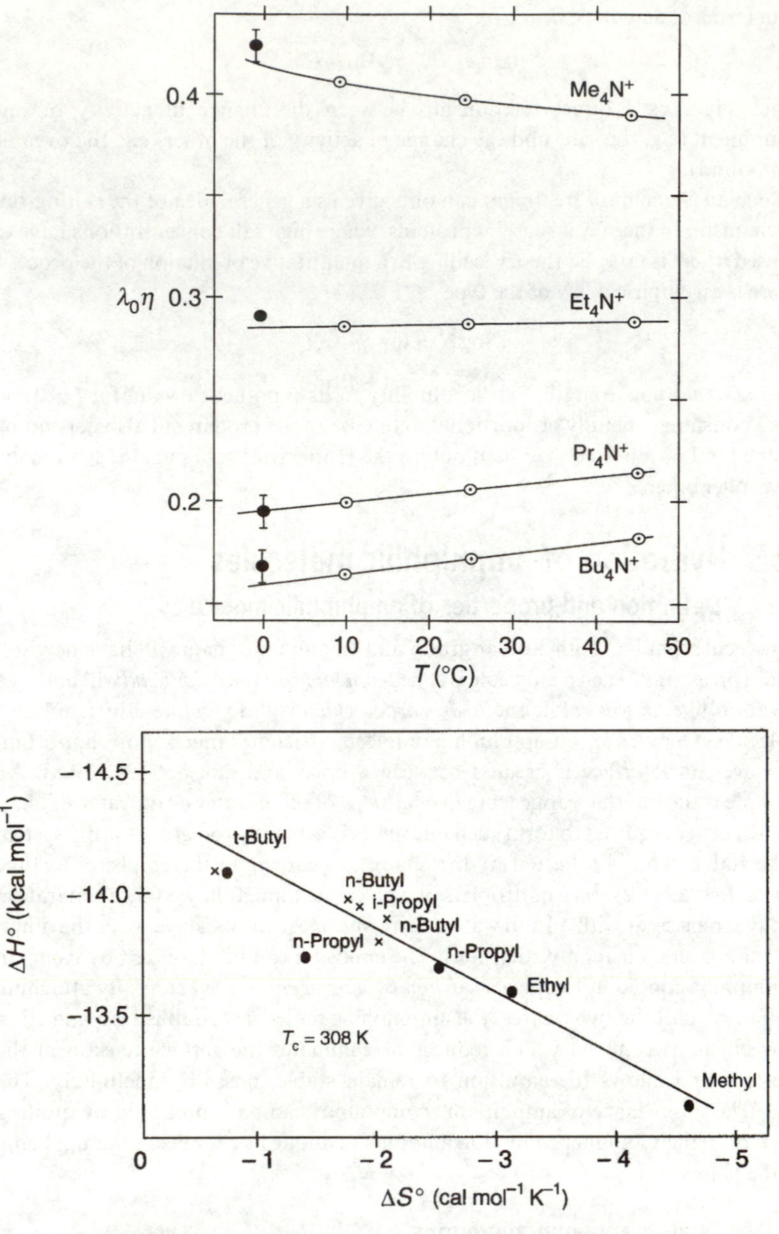

Figure 12.4 Variation of mobility with temperature for quaternary ammonium ions. (From Kay and Evans, 1966. Reprinted with permission from *Journal of Physical Chemistry*. © 1966 American Chemical Society.)

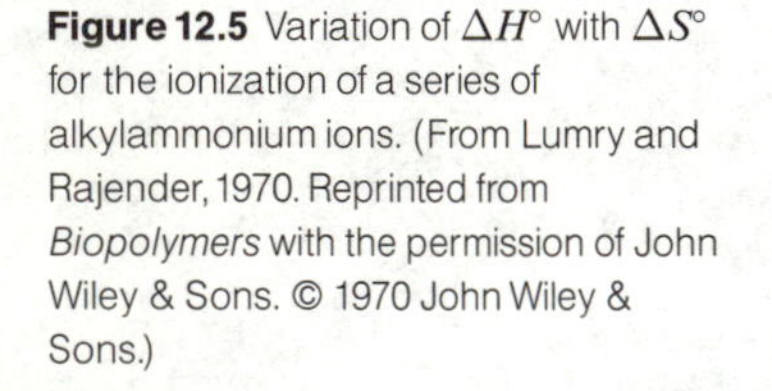

Figure 12.5 Variation of $\Delta H°$ with $\Delta S°$ for the ionization of a series of alkylammonium ions. (From Lumry and Rajender, 1970. Reprinted from *Biopolymers* with the permission of John Wiley & Sons. © 1970 John Wiley & Sons.)

into the aliphatic chain, the mobility increases because partial solvation can occur without a cage effect: thus, $Me_3(EtOH)N^+$ has a higher mobility than Me_3PrN^+. This type of compound also gives rise to *compensation phenomena* in which the change in standard free energy of solution $\Delta G°$ is very small and often close to zero because of compensation between the enthalpy and entropy terms $\Delta H°$ and $T\Delta S°$. Moreover, in many isothermal processes, there is a linear relationship between ΔH and ΔS (see Box 12.3). This occurs in a family of compounds with a variable chemical structure, e.g. the series of alkylammonium ions when their ionization energy is concerned (Fig. 12.5). It also occurs for a given molecule whose solubility is measured in a mixture of water and organic solvent.

Box 12.3 Study of compensation phenomena

In all cases, the slope of the straight line obtained when ΔH is plotted against ΔS defines a *compensation temperature* T_c. Classically, we have

$$\Delta H^\circ = \Delta H_0^\circ + \int_0^\infty \Delta C_p \, dT$$

$$\Delta S^\circ = \Delta S_0^\circ + \int_0^\infty (\Delta C_p / T) \, dT$$

and we then see that a linear relationship between the two quantities is only possible if $\Delta C_p = 0$. This could occur by chance, but has no general significance. We are therefore led to distinguish between the chemical part of ΔH° and ΔS° denoted conventionally by the subscript 'a' and the solvation part denoted by the subscript 'b', at the same time realizing that an experimental determination of the two parts is not easy. A linear relationship is assumed between ΔH_b° and ΔS_b° (both related to the state of solvation):

$$\Delta H_b^\circ = \alpha + T_c \Delta S_b^\circ$$

From the definition of G and replacing ΔH° by $\Delta H_a^\circ + \Delta H_b^\circ$ and ΔS° by $\Delta S_a^\circ + \Delta S_b^\circ$, we obtain:

$$\Delta G^\circ = \alpha + \Delta H_a^\circ - T\Delta S_a^\circ - (T - T_c)\Delta S_b^\circ$$

The quantity $\alpha' = \alpha + \Delta H_a^\circ - T_c \Delta S_a^\circ$ can be introduced since ΔH_a° and ΔS_a° are assumed to be almost constant in a given series, and hence

$$\begin{aligned}\Delta G^\circ &= \alpha' - (T - T_c)\Delta S_a^\circ - (T - T_c)\Delta S_b^\circ \\ &= \alpha' + T_c(\Delta S_a^\circ + \Delta S_b^\circ) - T(\Delta S_a^\circ + \Delta S_b^\circ)\end{aligned}$$

It follows that

$$\Delta H^\circ = \alpha' + T_c \Delta S^\circ$$

i.e. the experimentally observed relationship.

Underlying this *ad hoc* explanation of compensation phenomena, there must be an interpretation using an equilibrium between two states of water W_1 and W_2 (e.g. two different types of order for the water molecules around a solvated molecule). We should then have

$$T_c = (\Delta H / \Delta S) \qquad \text{for the transition } W_1 \rightarrow W_2$$

The observed compensation temperature would essentially measure a type of transition between two structures of bound water.

12.3.3 Micellar organization

As soon as the hydrophobic chain is long enough, new properties appear. Beyond a certain concentration of amphiphiles in water, a new type of order is established in the solution in the form of *micelles*, consisting of fairly well-defined aggregates of a certain number of amphiphilic molecules. The limiting concentration is known as the *critical micelle concentration* (CMC) (Fig. 12.6). Two opposing forces come

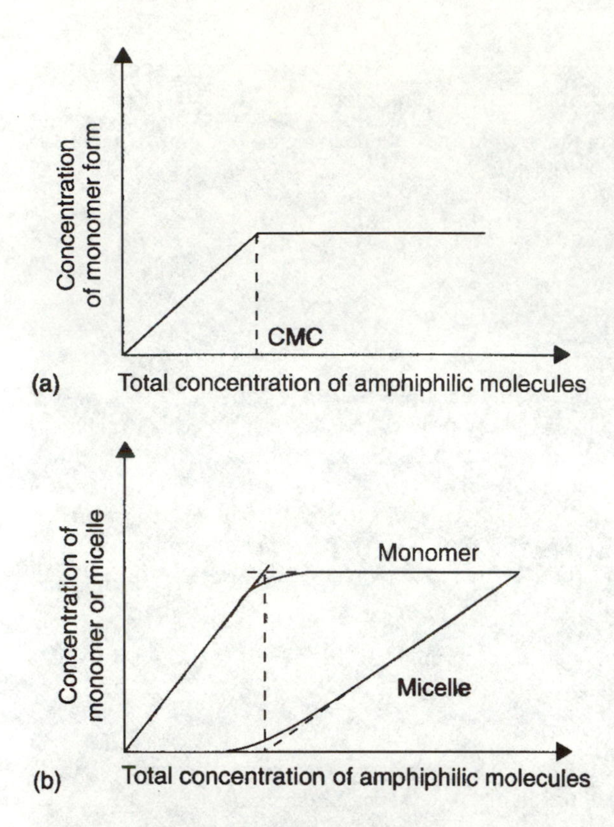

Figure 12.6 The emergence of the micellar form with increasing concentration: (a) definition of the CMC; (b) true variation of monomer and micelle concentrations.

into play in building the micelle (Tanford, 1973) (see Box 12.4):

(1) an *attractive* van der Waals force between aliphatic chains tending to draw them together within a volume to prevent them dissolving individually;

(2) a *repulsive* force between polar heads, which are outside the volume and create a hydrophilic interface with the water.

The shape and size of a micelle will depend on the area covered by the polar heads, i.e. on the ratio between the total surface area of the micelle and the number N_p^+ of polar heads. There are in general three types of micellar geometry: spherical, cylindrical and double-layer (Fig. 12.7).

The morphology and stability of a micellar structure depend on the number of carbon atoms in the hydrophobic chain and on the number of chains inside the micelle. In broad terms, the chain will be almost entirely within the 'internal medium' of the micelle, which can be compared with a liquid hydrocarbon, i.e. a medium with a low dielectric constant.

A more quantitative treatment as in Box 12.4 enables us to define an area for the polar head depending on the shape of the micelle.

The polar head area decreases as we pass from the spherical to the planar double layer. Moreover, Box 12.4 shows that the maximum number of chains N_{max} packed in a spherical micelle varies as the square of the number of carbon atoms, and for $n' = 12$ for example $N_{max} = 56$. In order to pack in the greatest number of chains, the micelle has to become a prolate ellipsoid with its minor axis still equal to L. After that, there is a sudden change to one of the other two shapes, cylindrical or double layer, which in principle can accommodate an infinite number of chains. The choice

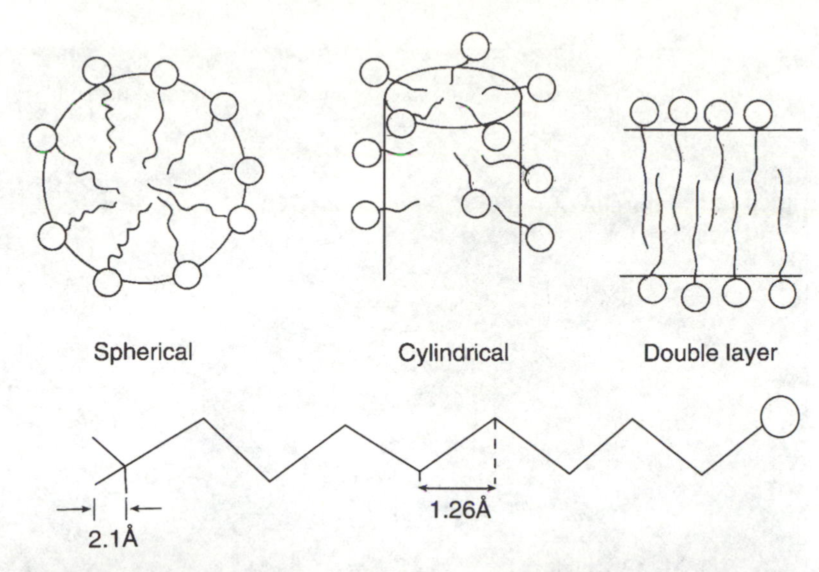

Figure 12.7 The three forms of micellar organization.

Figure 12.8 Geometry of a micellar chain.

Box 12.4 **Micellar structure**

This argument follows Cantor and Schimmel (1980). Let n' be the number of carbon atoms inside the micelle. The maximum length of the chain L_m in Fig. 12.8 is given by

$$L_m(\text{Å}) = 2.1 + 0.63 + 1.26(n'-1)$$

where, on the right-hand side, the first term (2.1) represents hindrance of the terminal CH_3 and the second term (0.63) is half the C–C distance projected on to the axis. This equation simplifies to

$$L_m(\text{Å}) = 1.5 + 1.26n' \quad (1)$$

For a large enough n', eqn 1 can be replaced by a relationship of the form $L_m \approx \alpha n'$. Since the stability of a micelle depends on an equilibrium between two opposing forces, the area allocated to each polar head must be calculated separately for each overall shape. However, there is also a relationship between the number N of chains in the micelle each containing n' carbon atoms and the volume V of the micelle. This takes the form

$$V = \beta n' N \quad (2)$$

since N is directly proportional to the mass of the micelle.

Spherical micelle of radius L (assumed $\approx \alpha n'$)

In this case:

$$\text{total surface area} \qquad A = 4\pi L^2 = 4\pi\alpha^2 n'^2 \quad (3)$$

$$\text{volume} \qquad V = 4\pi L^3/3 = 4\pi\alpha^3 n'^3/3 \quad (4)$$

From eqns 2 and 4:

$$N = 4\pi\alpha^3 n'^2/3\beta \quad (5)$$

and the area per polar head (assuming only one chain is bound to the head) is

$$A/N = 3\beta/\alpha \tag{6}$$

Cylindrical micelle of radius L (assumed $\approx \alpha n'$) and height h (assumed $\gg L$)

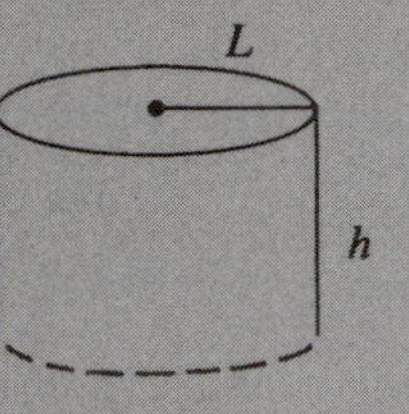

Here:

$$\text{surface area} \quad A \approx 2\pi Lh = 2\pi h\alpha n'$$
$$\text{volume} \quad V = \pi L^2 h = \pi\alpha^2 n'^2 h = \beta n' N$$

so that

$$N = \pi\alpha^2 hn'/\beta$$

Hence

$$A/N = 2\beta/\alpha \tag{7}$$

Planar double layer

In this simple case:

$$\text{volume} \quad V = 2LA/2 = LA = \alpha n' A$$

and also from eqn 2

$$V = \beta n' N$$

so that

$$A/N = \beta/\alpha \tag{8}$$

between these two depends on the repulsive forces between the charged polar heads, which increase as we go from the cylinder to the double layer since the area per polar head is halved.

When two hydrophobic chains are bound to the polar head, the double-layer form is favoured. This is what happens with natural or artificial membranes containing phospholipid chains.

The material in Boxes 12.5 and 12.6 will help in understanding the thermodynamics of micelles.

Box 12.5 **Transfer of a hydrocarbon from water to the micelle**

The structure of aliphatic chains inside the micelle is very similar to that of a liquid hydrocarbon, with only a small extra constraint near the polar heads. To see this, we simply consider the transfer of a hydrophobic compound between water and the micelle. To a first approximation, the interior of the micelle is assumed to be an ideal solution of the compound, for which

$$\mu^\circ_{\text{micelle}} - \mu^\circ_{\text{w}} = RT \ln(X_{\text{w}}/X_{\text{micelle}})$$

where $\mu^\circ_{\text{micelle}}$ is the chemical potential of the amphiphile molecule in the micellar aggregate.

With alkanes, we obtain a linear relationship

$$\mu^\circ_{\text{micelle}} - \mu^\circ_{\text{w}} = -1934 - 771 n_c$$

which is very close to that found for the transition from the liquid alkane to its aqueous solution.

The entropy and enthalpy changes are of the same order of magnitude. The fluidity of the micellar interior can also be determined by physical measurements by introducing either a fluorescent probe to follow the decay of fluorescence with time or a spin label to follow the linewidth of electron spin resonance.

Box 12.6 **Thermodynamics of micelle formation**

The treatment in this box follows Tanford (1973). We wish to calculate the chemical potential of an amphiphilic molecule in the micellar phase, $\mu^\circ_{\text{micelle}}$, with respect to the aqueous phase, μ°_{w}. If we write

$$\mu^\circ_{\text{micelle}} - \mu^\circ_{\text{w}} = RT \ln X_{\text{w}} + RT \ln f_{\text{w}}$$

we neglect the contribution of the micelle–solvent mixing entropy to the free energy. A micelle can be considered as a distinct thermodynamic entity because it is found experimentally to be formed and to be stable only with a minimum number of amphiphilic molecules. If the size of the micelle is ignored and if it contains m amphiphilic molecules on average, the cratic contribution per mole of the amphiphile is $1/m$ times the contribution per mole of micelle. Since the molar fraction X_{micelle} of amphiphiles incorporated in the micelle is m times the molar fraction of micelles, we have that

$$\mu_{\text{micelle}} = \mu^\circ_{\text{micelle}} + (RT/m) \ln(X_{\text{micelle}}/m)$$

Since, at equilibrium:

$$\mu_{\text{micelle}} = \mu_{\text{w}} = \mu^\circ_{\text{w}} + RT \ln X_{\text{w}} + RT \ln f_{\text{w}}$$

we obtain

$$\mu^\circ_{\text{micelle}} - \mu^\circ_{\text{w}} = RT \ln X_{\text{w}} + RT \ln f_{\text{w}} - (RT/m) \ln(X_{\text{micelle}}/m)$$

The concept of a CMC, i.e. a critical value of the concentration above which only micelles exist, is only valid in the limit when $m \to \infty$. In fact, for a large enough m, there is almost a phase transition between the molecular solution and the micelle since it occurs over such a narrow range of concentrations (see Fig. 12.6). Under these conditions, X_w could be replaced by [CMC] and the last term on the right could be neglected (m being large). This gives us

$$\mu^\circ_{\text{micelle}} - \mu^\circ_w = RT \ln[\text{CMC}] + RT \ln f_w$$

For solutions of surfactants, the CMC is a physical quantity as important as the melting point or boiling point is for a pure liquid. It can be determined experimentally in many ways: dye binding, electrochemical methods, optical methods using rotatory power, light scattering, refractometry or fluorescence, etc.

The CMC depends on the size and chemical nature of the non-polar chain in the amphiphilic molecule. For example, there is a linear relationship between ln[CMC] and the number n_c of carbon atoms in the aliphatic chain, so that

$$\mu^\circ_{\text{micelle}} - \mu^\circ_w = a - bn_c$$

The values of b cover only a narrow range, but those of a depend largely on the type of hydrophilic head.

13

Hydration of proteins

The configuration of a peptide chain in a given medium has long been known to depend on the thermodynamic equilibrium between several types of energy. These are related to the solvation mechanisms of different types of side chain (polar and non-polar, charged or neutral), which we have so far studied separately.

We first make an overall assessment of the interaction forces by considering four issues:

1. By minimizing contacts between water and aliphatic or aromatic side chains, the solvation processes involve hydrophobic interactions similar to those occurring in micelles. From a simple thermodynamic point of view, there are essentially two elementary amounts of work to be considered: one, $a_P \, dV_P$, due to the change in the total volume V_P of the protein, and the other, $\sigma \, dA$, due to the surface tension forces acting on the total accessible protein area A. From the general expression for dG, we have

$$a_P = \partial G/\partial V_P = (\partial G/\partial P)(\partial P/\partial V_P) = V_P \, \partial P/\partial V_P = \beta_P^{-1}$$

   where β_P is the compressibility of the protein. Hence:

$$\beta_P^{-1} \, dV_P - \sigma \, dA = 0$$

   Both $\beta_P{}^{-1}$ and σ behave as thermodynamic potentials and the Gibbs–Duhem relationship with constant T and P can be applied to give

$$V_P \, d\beta_P^{-1} - A \, d\sigma = 0$$

   and thus

$$V_P \delta\beta_P/\beta_P^2 + A\delta\sigma = 0$$

   This means that the protein is compressed in such a way as to minimize the interfacial energy σ (neglecting the change in volume of the water).

2. The interactions between charges are reduced by the high dielectric constant of water (see section 15.3.2).
3. Water may fill the cavities within the protein and stabilize structures by forming a network of hydrogen-bonded molecules.

4. Water will take part in the mechanisms of the interactions between proteins and other molecules. The release of bound water molecules is accompanied by an increase in their number of degrees of freedom and hence an increase in entropy, which in most cases forms the main part of the interaction free energy (*entropy-driven* reactions). The entropy term $-T\Delta S$ is predominant in the expression for the change in free energy.

These are general points, and we now examine a series of models put forward to explain and predict the effect of hydration on the conformation adopted by proteins and on their stability in solution. We then consider more thoroughly the structure of the network of water molecules bound to the protein.

13.1 Micellar model

A protein molecule containing amphiphilic amino acids can be represented by a micellar model. It can be viewed as an amphiphilic polymer folded like a micelle so as to bring hydrophilic amino acids (Thr, Ser, Asn, Gln) or the polar heads of amphiphiles (Glu^-, Asp^-, Lys^+, Arg^+) into contact with the solvent. By assigning each amino acid a van der Waals volume (as listed in Table 13.1), it is possible to calculate the ratio P of the volumes of polar to those of non-polar amino acids.

A spherical protein of known radius R can be pictured as being formed from two concentric layers: an outer one of thickness d ensuring solvation and an inner one of radius $R-d$ containing hydrophobic groups. The ratio of the volumes of the outer and inner layers for the sphere is

$$P_s = [R^3/(R-d)^3] - 1$$

Assuming a reasonable value for d of about 4 Å, a comparison of P and P_s shows that:

(1) if $P < P_s$, a monomeric protein is unstable and aggregation will tend to occur;

(2) if $P > P_s$, the protein can adopt an ellipsoidal shape.

Although this micellar model of a protein is crude, it does give some idea of the behaviour in solution. As Perutz (1978) picturesquely put it: 'Proteins are like wax inside and soap on the outside. When heated, wax melts and proteins unfold . . .'.

Table 13.1 Volumes of amino acids

Amino acid	Molar volume (ml g^{-1})	Amino acid	Molar volume (ml g^{-1})
Arg	125.7	Ala	32.2
Asp	58.4	Val	86.3
Glu	85.5	Leu	113.4
Lys	121	Ile	113.4
Ser	36	Phe	136.6
Thr	63.1	Try	175.5
Tyr	139	Pro	81
Gly	5.1	Cys	55

Table 13.2 Van der Waals radii (Å)

Tetrahedral C	2.0	Tetrahedral N	2.0
Trigonal C	1.7	Trigonal N	1.7
O (in C=O)	1.4	Divalent S	1.85
O (in OH)	1.6	S (in SH)	2.0
O (in COOH)	1.5		

X-ray diffraction studies of the tertiary structures of proteins known at present show that the distribution of amino acids does not rigidly follow the rule that the hydrophobic ones are inside and the hydrophilic ones outside.

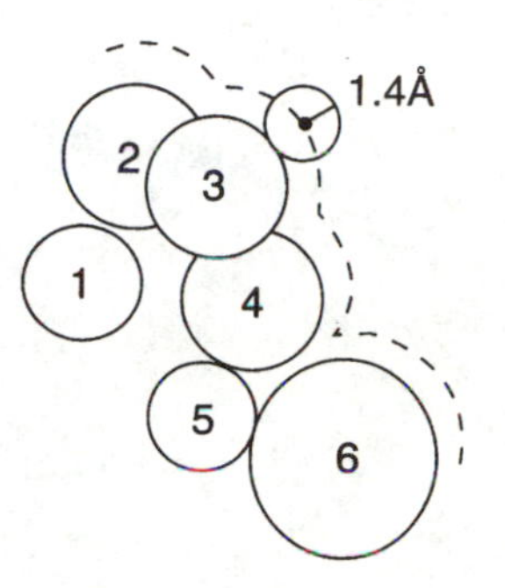

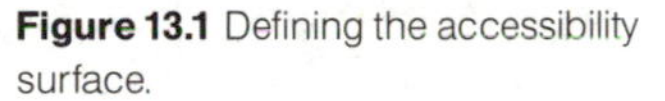
Figure 13.1 Defining the accessibility surface.

13.2 Solvent accessibility

Instead of defining a scale of hydrophobicity using criteria such as the partition coefficients between the aqueous and organic phases or their position inside or outside the protein, we can define a *solvent accessibility* from tertiary structures. Thousands of three-dimensional protein structures have now been determined from X-ray and NMR analyses, and reports on new structures are being published continuously.

A quantitative criterion for the accessibility of a protein to a water molecule can be established as follows (Lee and Richards, 1971). Each atom is represented by a sphere of radius R equal to its van der Waals radius (Table 13.2). The protein surface is defined by the envelope of all the atomic spheres, and another sphere of radius 1.4 Å representing the water molecule is rolled over the surface so defined. The centre of the rolling sphere will describe a surface that partially reflects the irregular shape of the van der Waals envelope (see Fig. 13.1, where atom 5 is not 'seen' by the water molecule). The area A of the surface so described is the *accessibility area*, and this can be calculated as shown in Box 13.1, together with the percentage accessibility of a given atom.

From an analysis of the solvent-accessible surfaces for 4110 residues in 23 proteins (Rose *et al.*, 1985), histograms can be drawn giving the percentages of each type of residue that are buried to varying degrees, from being fully accessible (0 per cent buried, at the origin), to being 100 per cent buried (Fig. 13.3, in which the expressions in Box 13.1 were used). From these distributions, we can calculate the mean accessible area $\langle A \rangle$ in a protein for each residue X. A reference value A^0 is taken to be the accessible area for the residue X in the tripeptide Gly-X-Gly.

The quantity $A^0 - \langle A \rangle$ is the mean buried area and $f = 1 - \langle A \rangle / A^0$ is the mean fraction of area lost when passing from an unfolded structure like the tripeptide to the folded protein. If A^0 is plotted against $A^0 - \langle A \rangle$, the amino acids arrange themselves along three straight lines characteristic of three families: hydrophobic (Ala, Cys, Val, Ile, Leu, Phe, Trp), moderately polar (Gly, Ser, Thr, His, Tyr) and highly polar (Pro, Asp, Asn, Glu, Gln, Arg). The glycine point is common to all three lines (Fig. 13.4): in other words, there is a linear relationship of the type $ax + b$ between A^0 and the buried area x, implying the existence of a threshold b beyond which $a = 1$ for hydrophobic residues, $a = 0.83$ for moderately polar residues and

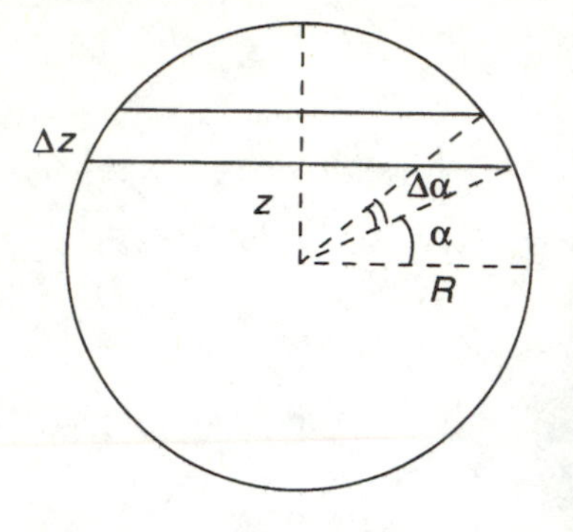

Figure 13.2 Calculation of the area of the accessibility surface.

Box 13.1 Calculation of the accessibility area

This box follows Lee and Richards (1971). The calculation of A makes use of computer-generated maps of parallel cross-sections separated by Δz (Fig. 13.2). The angle at the centre subtended by the arc cut off by Δz is $\Delta\alpha$.

The elementary surface area $\mathrm{d}S$ defined by $\mathrm{d}z$ and $\mathrm{d}\alpha$ is $2\pi R^2 \sin\alpha\,\mathrm{d}\alpha$ and since $z = R\cos\alpha$, $\mathrm{d}z = -R\sin\alpha\,\mathrm{d}\alpha$, and hence $\mathrm{d}S = 2\pi R\,\mathrm{d}z$. This applies to all the small but finite increments ΔS and Δz, so that

$$\Delta S = 2\pi R \Delta z$$

The circumference of the slice of thickness Δz is $2\pi(R^2 - z^2)^{1/2}$, and if only an arc of length L is exposed, this represents a fraction f given by

$$f = L/2\pi(R^2 - z^2)^{1/2}$$

The exposed surface of the slice ΔA is therefore given by:

$$\Delta A = f\Delta S = RL\Delta z/(R^2 - z^2)^{1/2}$$

The slicing process means that we have to define the thickness of the $(i+1)$th slice in terms of the ith by

$$\Delta z_{i+1} = \Delta z_i/2 + \Delta' z$$

where $\Delta' z$ is the smaller of the two quantities $\Delta z_i/2$ and $R - z_i$. The total accessible area at the centre of the water molecule is

$$A = \sum_i R(R^2 - z_i^2)^{-1/2} \Delta z_i L_i$$

The lengths L_i of the various circular arcs are determined for each of the i slices. After calculating A, the percentage accessibility for each atom of radius R is given by $100A/4\pi R^2$.

$a = 0.6$ for polar residues. Lysine appears to be the most highly solvent-accessible of all residues. There is also a linear relationship between A^0 (or $A^0 - \langle A\rangle$) and the free energy ΔG of the transfer from water to organic solvent.

This definition of accessibility not only enables us to take into account the exact protein structure, but it also allows us to compare the values determined for each amino acid with previous criteria for hydrophobicity. Conversely, the results can be used to refine the structure of a protein or simply as a guide in the search for the three-dimensional folding of a peptide chain.

Note on accessibility area

The accessibility area A is of course much greater than that of the ellipsoid enveloping the protein, but is much smaller than the surface area attributed to an extended chain. In the latter case, the total area $A_{\mathrm{tot}} = 1.45M$, whereas for small globular proteins $A = 11.1M^{2/3}$ (where A_{tot} and A are in Å^2 and M is the molecular weight in amu or daltons). Thus, for a protein with $M = 15\,000$, $A/A_{\mathrm{tot}} = 0.3$. If the same protein were represented instead by an equivalent sphere of surface

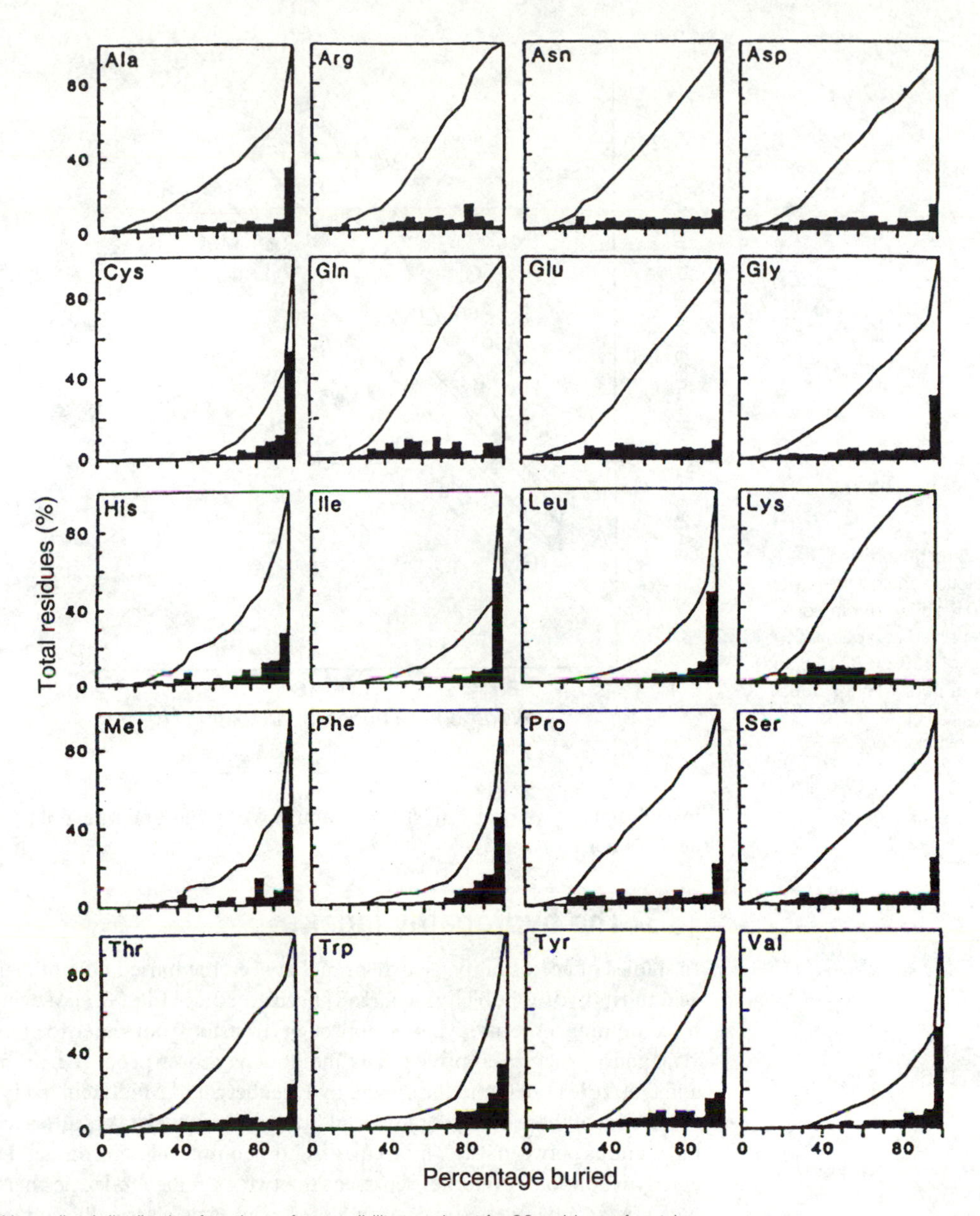

Figure 13.3 Normalized distribution functions of accessibility to solvent for 20 residues of proteins of known structure. Each function is in fact a histogram showing the percentage of residues of that type that are: fully accessible (origin, 0 per cent buried), 5 per cent buried, 10 per cent buried, etc. The full line is the integrated curve. (From Rose *et al.*, 1985. Reprinted with permission from *Science*, vol. 229, pp. 834–8 and with the authors' permission. © 1985 American Association for the Advancement of Science.)

area A_s, we should have $A_s = 4.84M^{2/3}$, where the factor 4.84 is $4\pi(3/4\pi)^{2/3}$, and hence $A/A_s = 2.3$. The concept of an accessibility area thus allows us to characterize protein folding in a quantitative way. Such a refinement of the description means that we must look again at the simplistic idea of 'all polar outside' and 'all non-polar

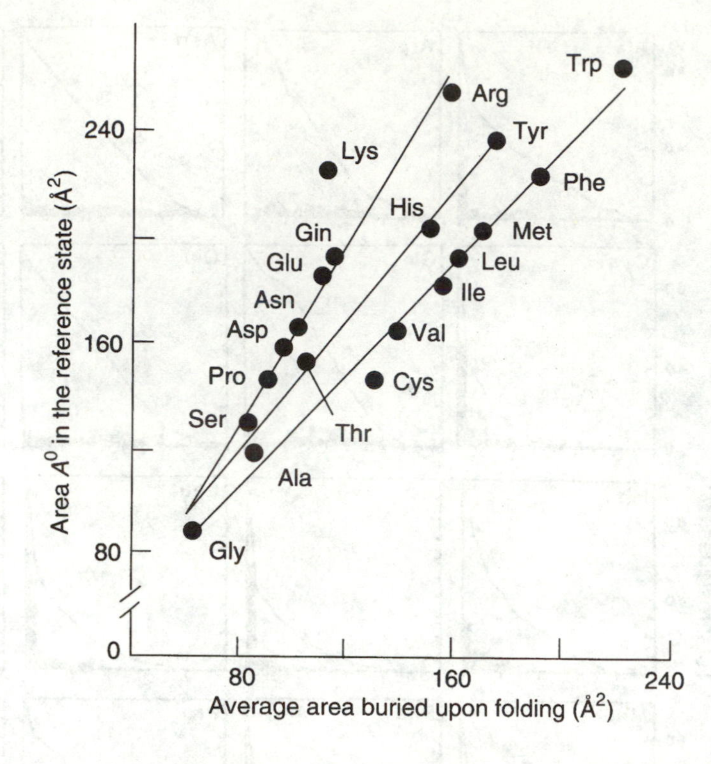

Figure 13.4 Mean buried area $A^0 - \langle A \rangle$ after folding plotted against the reference area A^0 for 20 residues. (From Rose *et al.*, 1985. Reprinted with permission from *Science*, vol. 229, pp. 834–8 and with the authors' permission. © 1985 American Association for the Advancement of Science.)

inside'. In fact, we find almost equal numbers of polar and non-polar atoms in the accessible area.

13.3 The hydropathy index

In studies of accessibility, a correlation between the buried area of a given amino acid and its hydrophobicity has already been noted (see Fig. 13.3). We can go further by combining a thermodynamic approach (transfer from water to an organic solvent) and a structural approach. Two methods have been proposed, the first by Kyte and Doolittle (1982) and the second by Eisenberg and MacLachlan (1986).

The first of these involves the idea of *hydropathy*, which takes into account all the interactions between water molecules and other molecules in contact. Hydrophilic and hydrophobic properties represent the two extreme tendencies here. A *hydropathy index*, characteristic of each amino acid, was classically defined from the energy of the transfer of the molecule from water to another solvent, such as ethanol or benzene. However, specific bindings with this solvent can be established that depend on the molecule in question. Kyte and Doolittle propose taking as their scale the energy transfer from liquid water to water vapour and apply this model to amino acids.

Because the calculation of the hydropathy index is largely empirical owing to the approximations involved, we give no details of the method used here but merely give a table of the indices estimated for each amino acid (Table 13.3).

Hydropathy maps can then be constructed by plotting a mean index against the amino acid number in the protein sequence, the mean index being obtained by averaging the indices of a given number of successive amino acids (usually 7–11) in

Table 13.3 Hydropathy indices

Ile	Val	Leu	Phe	Cys
4.5	4.2	3.8	2.8	2.5
Met	Ala	Gly	Thr	Ser
1.9	1.8	−0.4	−0.7	−0.8
Trp	Tyr	Pro	His	Gln
−0.9	−1.3	−1.6	−3.2	−3.5
Asn	Glu	Asp	Lys	Arg
−3.5	−3.5	3.5	−3.9	−4.5

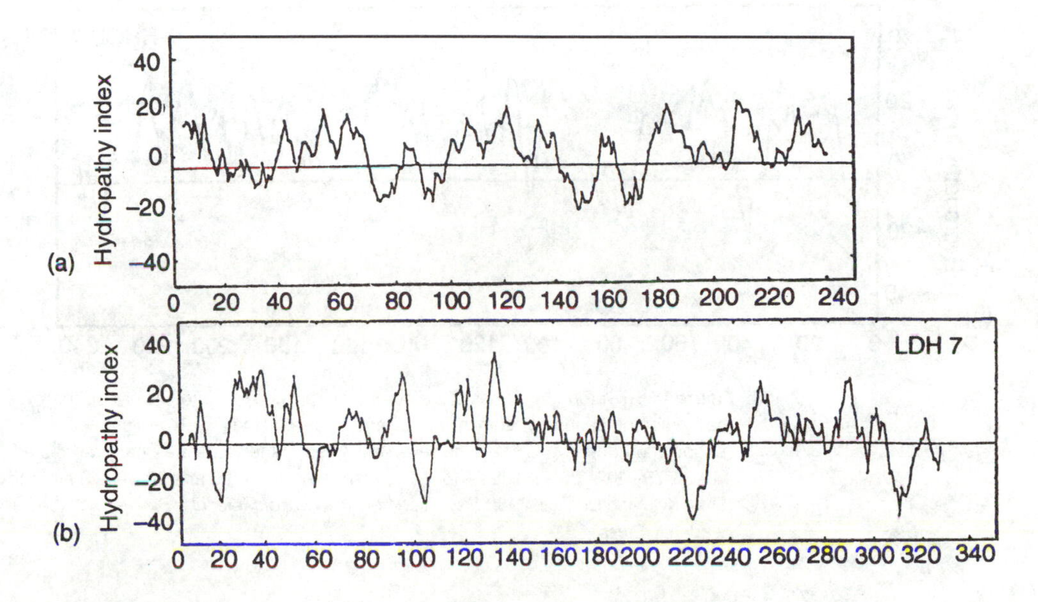

Figure 13.5 Hydropathy profiles of soluble proteins: (a) bovine chymotrypsinogen (mean over nine amino acids); (b) dogfish lactate dehydrogenase (mean over seven amino acids). (From Kyte and Doolittle, 1982. Reprinted from *Journal of Molecular Biology* with the permission of Academic Press.)

order to yield a curve smooth enough to be useful. The patterns obtained for proteins with a known three-dimensional structure are in agreement with information deduced from these structures (Fig. 13.5). In the case of membrane proteins (Fig. 13.6), there is a clear distinction between membrane-bound regions and membrane-crossing sequences (transmembrane helices). A final point is that an interesting result is obtained by comparing the mean hydropathy of 84 enzymes: membrane proteins cannot be fully distinguished from soluble proteins, and amongst soluble proteins the hydropathy values are almost independent of size and shape (Fig. 13.7). This means that the size, the shape and even the membrane characters of a protein cannot be deduced from its sequence. Only the hydropathy map is informative.

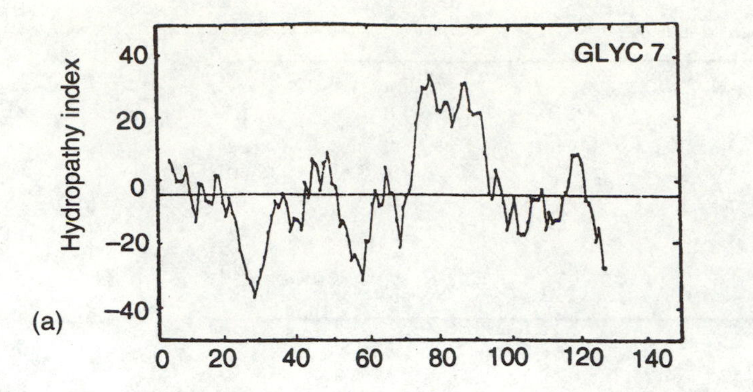

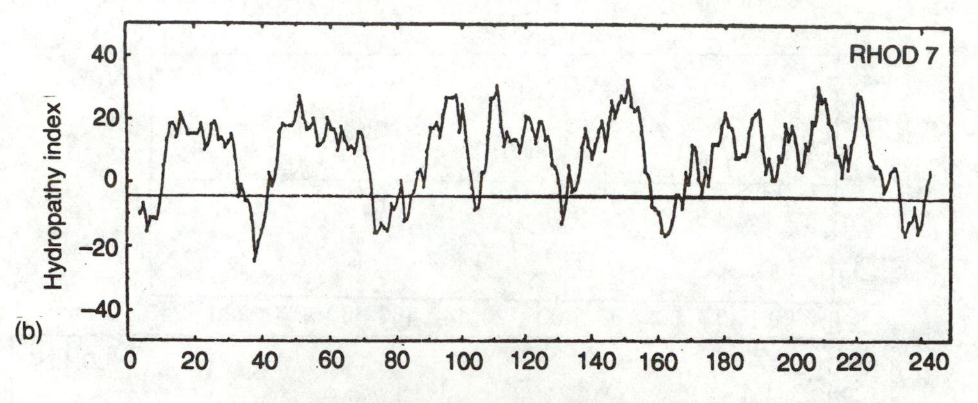

Figure 13.6 Hydropathy profiles of membrane proteins: (a) erythrocyte glycophorin – the membrane-spanning segment can clearly be seen in the 75–94 region; (b) bacteriorhodopsin – five of the seven transmembrane shafts are clearly delineated (10–34, 44–68, 78–102, 105–131, 134–158). The other two (177–199 and 200–224) are less distinct. (From Kyte and Doolittle, 1982. Reprinted from *Journal of Molecular Biology* with the permission of Academic Press.)

13.4 Protein hydration and stability

During the formation of a protein from a peptide chain or in the equilibrium between the protein and its denatured form, the change in free energy arises mainly from two sources:

(1) the energy associated with the transfer of hydrophobic side chains from water to the interior of the protein;

(2) the solvation energy of the hydrophilic groups lost when the groups are inside the protein.

The final shape of the protein in a given medium depends on the delicate equilibrium between these two contributions to the free energy.

Eisenberg and MacLachlan (1986) have developed a new method (see Box 13.2) of obtaining a quantitative estimate of the stability of proteins either modified or wrongly folded (after denaturation and renaturation). While the energy difference between native and denatured protein is about 100 kcal mol^{-1} for one with a hundred residues, differences of between 20 and 30 kcal mol^{-1} may occur between the correct structure and the less stable wrongly folded ones.

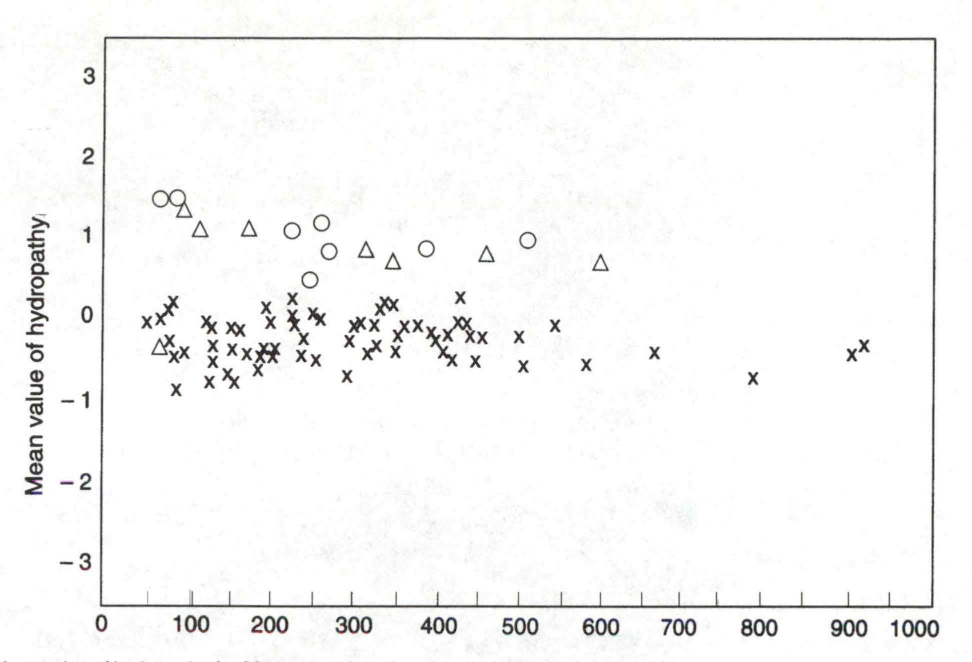

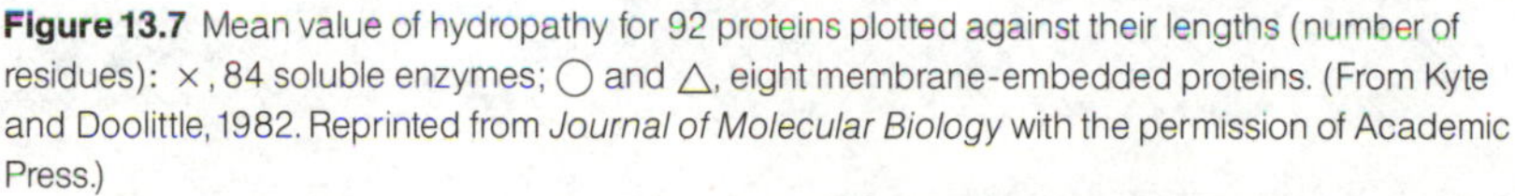
Figure 13.7 Mean value of hydropathy for 92 proteins plotted against their lengths (number of residues): ×, 84 soluble enzymes; ○ and △, eight membrane-embedded proteins. (From Kyte and Doolittle, 1982. Reprinted from *Journal of Molecular Biology* with the permission of Academic Press.)

Box 13.2 **Calculation of the protein stabilization energy**

The method depends on two main assumptions:

- The free energy associated with the transfer of an amino acid from the solvent to the interior of the protein is the sum of all the free energies related to the atoms contained in it.
- These free energies are linear functions of the accessible areas of the various atoms.

For each atom i, we define an atomic solvation parameter $\Delta\sigma_i$ such that the energy associated with the transfer of this atom from inside the protein to the solvent is

$$\Delta G_i = \Delta\sigma_i A_i$$

where A_i is the accessible area of the solvent.

For any amino acid X, $\Delta G_X = \sum_{\text{atoms } i \text{ of } X} A_i \Delta\sigma_i$. The atoms, apart from H, which is not taken into account, are divided into five classes: C, N or O, O^-, N^+, and S. Each of these provides a term; for C, for example, this is $\Delta G_C = \Delta\sigma_C \sum_{C \text{ atoms } i \text{ of } X} A(C_i, X)$ where $A(C_i, X)$ is the accessible area of the carbon atom in the standard conformation of amino acid X.

We then add the ΔG values for all the atoms in the amino acid (O^- in Asp and Glu; N^+ in Arg, Lys and His; S in Cys and Met; etc.), so that the 20 amino acids are replaced by five $\Delta\sigma$ parameters. These five values are obtained from the amino acid transfer energy, on the one hand, and from the values of A_X, on the other, i.e. from the accessible area of X in Gly-X-Gly, by using a least-squares method to obtain the best fit for

$$\Delta G_X = \Delta G_C + \Delta G_{N/O} + \Delta G_{O^-} + \Delta G_{N^+} + \Delta G_S$$

In this way, we obtain the following values (in cal mol^{-1} Å^{-2}):

$$\Delta\sigma_C = 16 \pm 2 \qquad \Delta\sigma_{N/O} = -6 \pm 4 \qquad \Delta\sigma_{O^-} = -24 \pm 10$$

$$\Delta\sigma_{N^+} = -50 \pm 9 \qquad \Delta\sigma_S = 21 \pm 10$$

From these, it is possible to estimate the energy contribution to protein folding due to solvation. If A_i is the accessible area of atom i and A_i' is the same quantity in an arbitrary reference state, the folding energy ΔG_r with respect to the reference state is

$$\Delta G_r = \Delta\sigma_C \sum_{\text{C atoms i}} (A_i - A_i') + \Delta\sigma_{N/O} \sum_{\text{N/O atoms i}} (A_i - A_i')$$

$$+ \Delta\sigma_{O^-} \sum_{\text{O}^- \text{ ions i}} (A_i - A_i') + \Delta\sigma_{N^+} \sum_{\text{N}^+ \text{ ions i}} (A_i - A_i') + \Delta\sigma_S \sum_{\text{S atoms i}} (A_i - A_i')$$

Table 13.4 Hydration of side chains

	Number	Number of bound H_2O	H_2O per side chain	H_2O per H bond site
Asp	119	225	1.9	0.47
Glu	92	191	2.1	0.52
Lys	135	107	0.8	0.26
Arg	72	88	1.2	0.24
Ser	176	117	0.66	0.22
Thr	140	114	0.8	0.27
Asn	133	119	0.9	0.22
Gln	94	109	1.2	0.29
Tyr	125	92	0.7	0.37
His	57	58	1	0.51
Trp	34	11	0.3	0.32

Source: From Baker and Hubbard (1984).

In the case of haemerythrin (113 residues), removal of the 606 C and three S atoms from the solvent contributes -159 kcal mol^{-1}. On the other hand, energy must be supplied to bury partially in the protein 297 N and O atoms (25 kcal mol^{-1}), 17 O^- ions (12 kcal mol^{-1}) and 21 N^+ ions (11 kcal mol^{-1}). The contribution from hydrophobic groups is therefore the major one in ensuring the stability of the tertiary structure. The $\Delta\sigma_C$ and $\Delta\sigma_S$ terms (see Box 13.2) almost exclusively represent the decrease in entropy of the water when the hydrophobic groups are solvated. These results strongly support what has been said above (section 5.6.3) about protein forming.

13.5 **Structural aspect**

In previous sections, the methods for calculating the hydration of proteins and its role in their structure and stability depended mainly on thermodynamic considerations (transfer energy) associated with accessibility parameters. The latter are determined from tertiary protein structures obtained from X-ray and neutron diffraction experiments on crystals, both of which are increasingly able to give accurate positions for the water molecules.

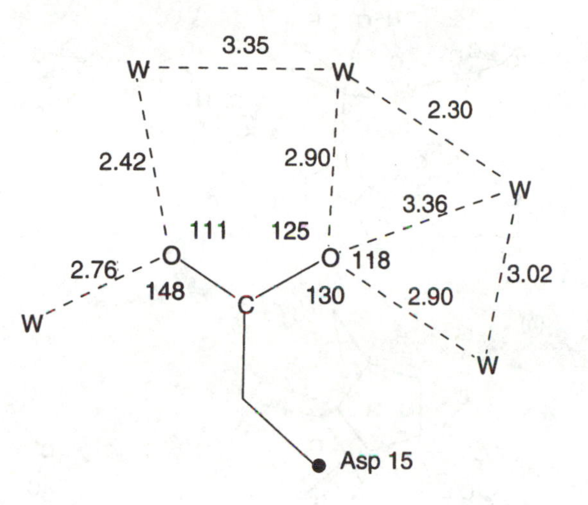

Figure 13.8 Example of a highly hydrated group: Asp 15 in the actinidin molecule, whose COO⁻ group is surrounded by five water molecules, four of which are at normal hydrogen bond distances. (From Baker and Hubbard, 1984. Reprinted from *Progress in Biophysics and Molecular Biology*, vol. 44, pp. 97 – 179, © 1984, with kind permission from Elsevier Science Ltd., The Boulevard, Langford Lane, Kidlington, Oxford OX5 1GB, UK.)

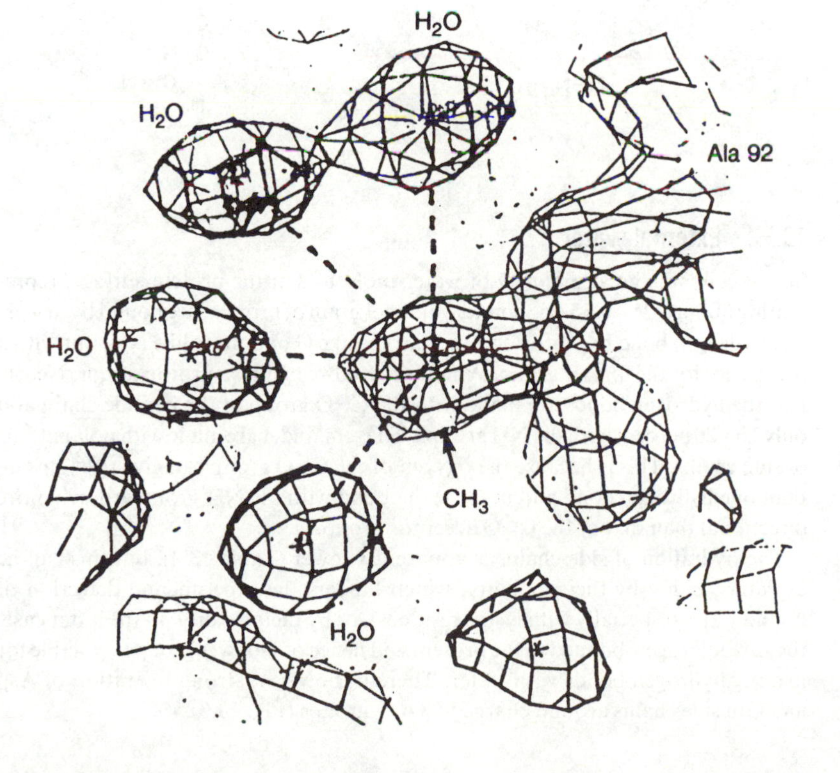

Figure 13.9 In this electron density map of part of the lysozyme molecule, four water molecules inside the protein can clearly be seen forming a semicircle around the Ala 92 side chain. Dotted lines connect water molecules to the methyl group. (From Blake *et al.*, 1983. Reprinted from *Journal of Molecular Biology* with the permission of Academic Press.)

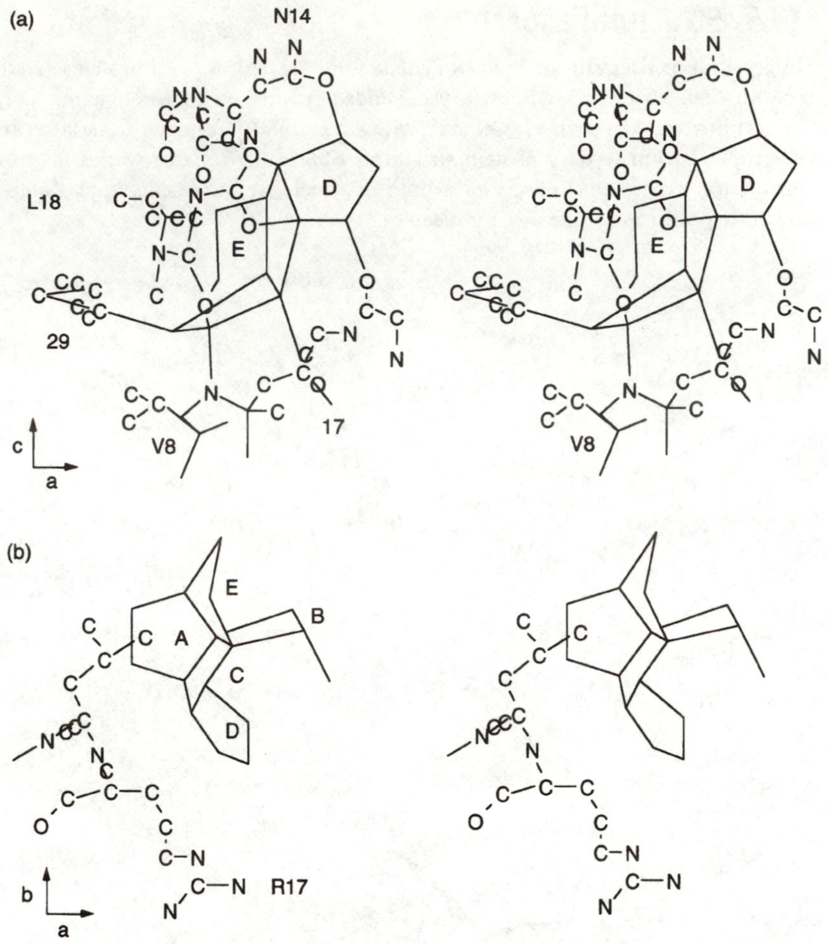

Figure 13.10 Stereo diagrams (to be observed directly by making the axes of the two eyes converge) of pentagons of water molecules bound to crambin in a region where there is a hydrophobic contact between two molecules. (a) The five hydrogen bonds with the α-helix at the top are all formed with the C=O groups of the C terminal part. Of the three hydrogen bonds made with the α-helix at the bottom, two are formed with N–H and one with C=O. (b) Pentagons A, C and E crown one of the carbon atoms of the L18 side chain, and pentagon D lies above the CH_2 group of R17. (From Teeter, 1984. Reprinted from *Proceedings of the National Academy of Sciences, USA* with the author's permission.)

13.5.1 External water

Statistically, the great majority of water molecules at the protein surface, representing about 0.25 – 0.35 g of bound water per gram of protein, are bound by at least one hydrogen bond to proton acceptor atoms like O or donors like NH, but with a preference for the former group. Water is therefore mainly a proton donor. Nearly half the hydrogen bonds are made with the C=O group of the peptide chain and only 10 – 20 per cent with the NH groups. The remainder are made with polar atoms of side chains. This is because the oxygen of the C=O group can give rise to more than one hydrogen bond and because the bond with the NH group is much more directional than that of the C=O acceptor group.

The hydration of side chains is apparently lower (Table 13.4), but this can be explained either by their mobility, which hinders the experimental detection of bound water (especially with lysine residues), or by their rigidity. In the latter case, they are hydrogen-bonded to the protein and hence offer fewer groups available for making hydrogen bonds with water. There is, however, strong hydration of Asp and Glu side chains around charged COO^- groups (Fig. 13.8).

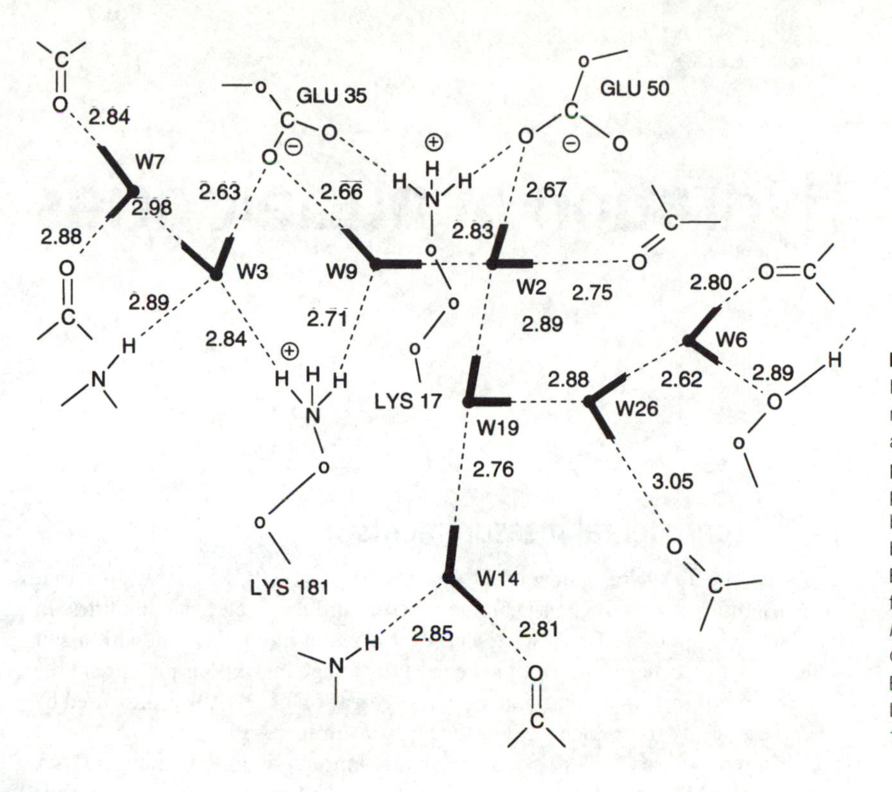

Figure 13.11 A network of eight hydrogen bonds is formed by eight water molecules between two domains of the actinidin molecule. The orientations of the protons are assumed to be the most probable ones. Note the bridge formed by two water molecules (W3 and W9) between Lys 181 and Glu 35. (From Baker and Hubbard, 1984. Reprinted from *Progress in Biophysics and Molecular Biology*, vol. 44, pp. 97 – 179, © 1984, with kind permission from Elsevier Science Ltd., The Boulevard, Langford Lane, Kidlington, Oxford OX5 1GB, UK.)

13.5.2 **Internal water**

The water molecules most strongly bound to a protein are those buried inside, occupying small internal cavities. These occur even in small proteins. For example, there are four water molecules for BPTI (58 residues), three for cytochrome c (103 residues) and four for lysozyme (129 residues) (Fig. 13.9; see also Fig. 13.10). From one to four hydrogen bonds may be involved in the interaction between a water molecule and the protein. Because of its size and because it can be both a donor and an acceptor, the water molecule can form a bridge between two groups, even in the case of COO^- and NH_3^+ (Fig. 13.11).

These internal water molecules make a considerable contribution to structural stability. In particular, when C=O and N−H groups cannot form hydrogen bonds with their respective partners, they may rely on water molecules to play this role. They can also create a locally polar medium around charged groups or form an H_3O^+ ion to compensate for a COO^- ion (see Part IV).

14

Hydration of nucleic acids

14.1 Experimental measurements

Ever since the first X-ray studies of DNA fibres in the 1960s, we have known that the A, B and C forms (Fig. 14.1) are obtained at different relative humidities. In these studies, stretched fibres were placed in a vessel in equilibrium with a salt solution, whose concentration could be used to change the vapour pressure at the temperature of the experiment. Later on, the hydration of DNA was measured by other methods, two of which yield particularly accurate results.

1. Infrared spectroscopy is used to measure the amount of bound water per DNA molecule as a function of the activity of the water, and to monitor its progressive dehydration and rehydration. Bound water is detected by its characteristic absorbance A at $3400\,\mathrm{cm}^{-1}$. From a curve of A plotted against relative humidity (Fig. 14.2), we can derive another curve showing the amount of bound water

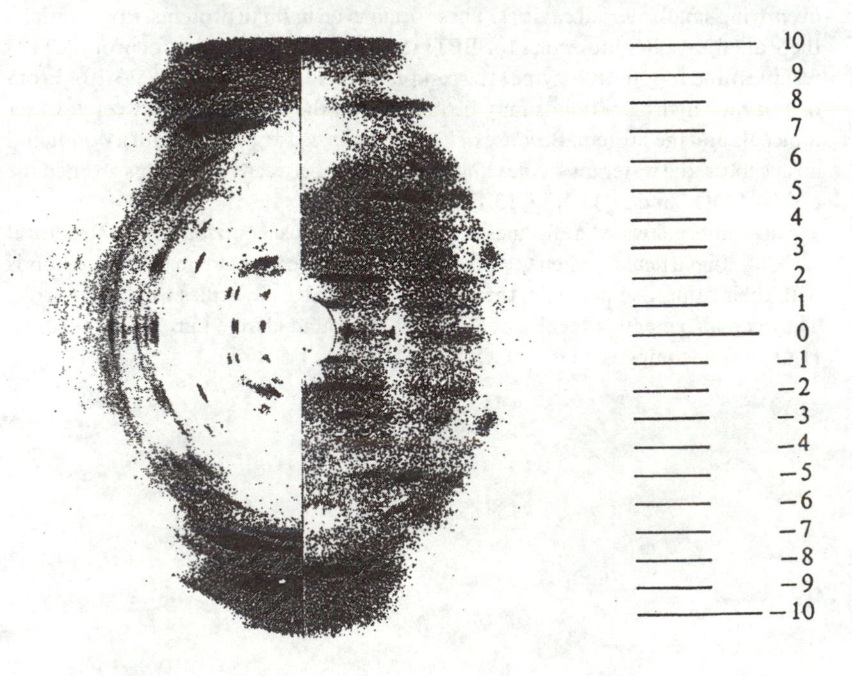

Figure 14.1 X-ray diffraction patterns of DNA fibres in the A form (left-hand diagram) and the B form (right-hand diagram). The layer lines are numbered from the equatorial line with index 0.

plotted against relative humidity (Fig. 14.3). It can be seen that the absorption and desorption curves are not exactly superimposed on each other, implying a kind of hysteresis in the hydration process. Another point to note is the last two or three water molecules, which are considered to be most strongly bound to phosphate groups. Falk *et al.* (1963) proposed some possible binding sites for the water

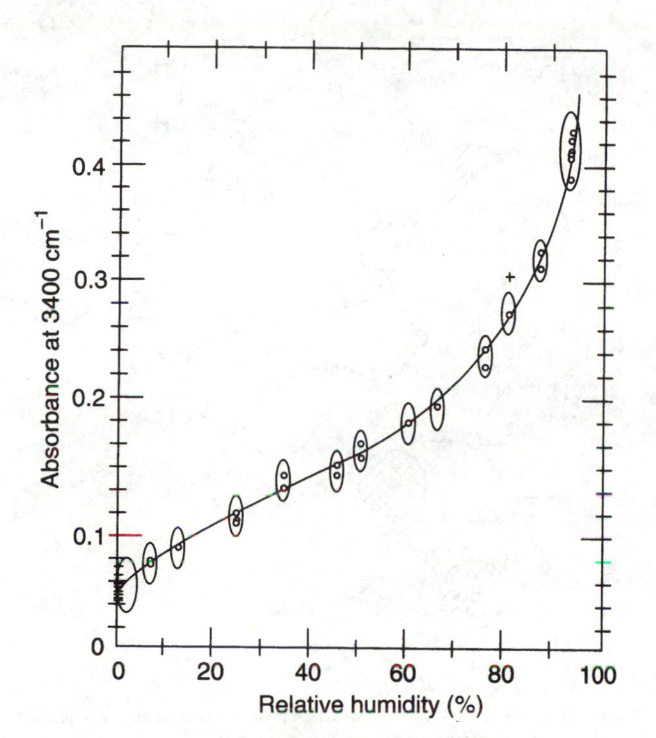

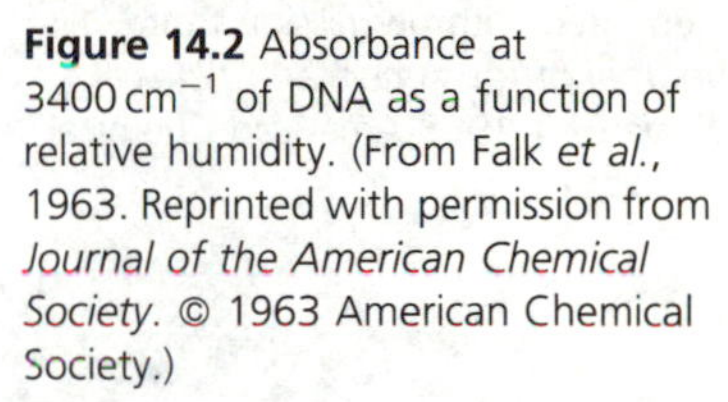

Figure 14.2 Absorbance at 3400 cm^{-1} of DNA as a function of relative humidity. (From Falk *et al.*, 1963. Reprinted with permission from *Journal of the American Chemical Society*. © 1963 American Chemical Society.)

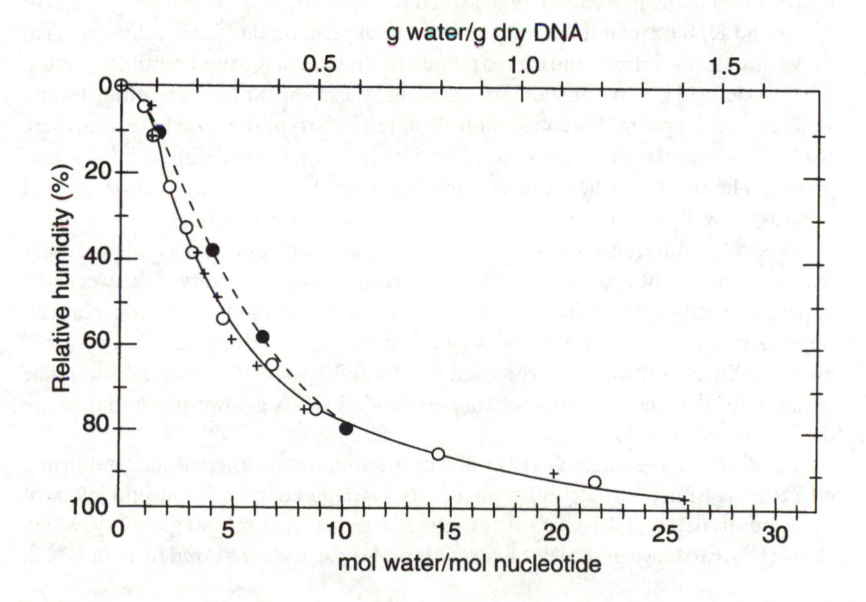

Figure 14.3 Adsorption (○) and desorption (●) of water for calf thymus DNA; and adsorption of water (+) for salmon sperm. (From Falk *et al.*, 1962. Reprinted with permission from *Journal of the American Chemical Society*. © 1962 American Chemical Society.)

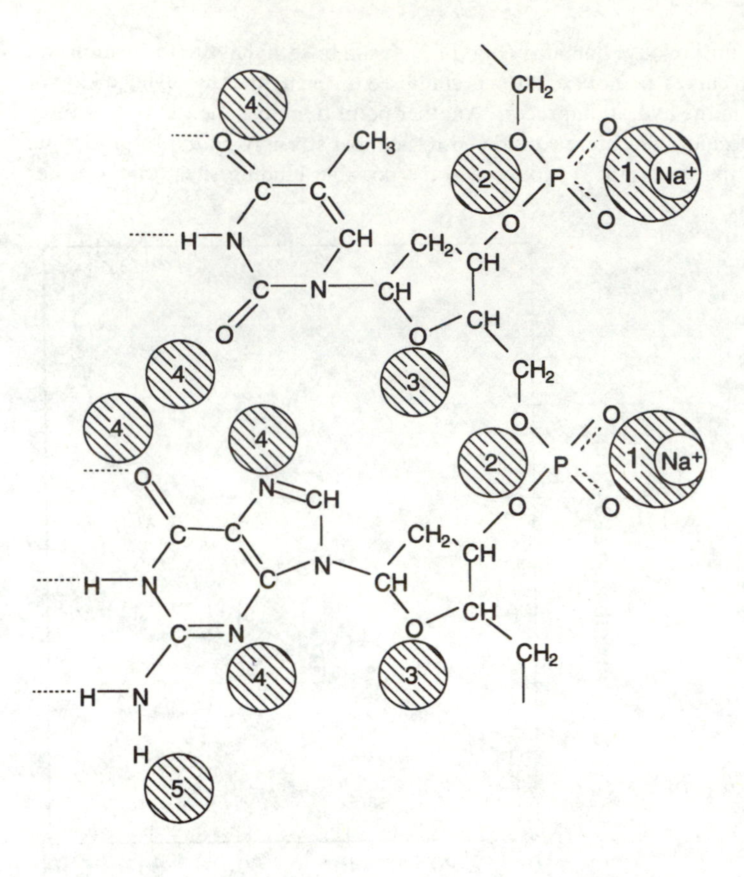

Figure 14.4 Possible sites for water molecules around pTpG: sites 1 and 2 are around phosphate groups, sites 3 around sugars and sites 4 around bases. (From Falk *et al.*, 1963. Reprinted with permission from *Journal of the American Chemical Society*. © 1963 American Chemical Society.)

molecules (Fig. 14.4). The number of bound water molecules per nucleotide was estimated to be about 20, of which 10 formed a first hydration shell.

2. Using density gradient ultracentrifugation, the apparent density of DNA can be measured in the presence of high salt concentrations. If we compare the salts of Li, Cs and K, the apparent density appears to depend on the water activity (a_w) in the various salt solutions and can be related to changes in the preferential solvation of DNA defined by a parameter Γ representing the number of preferentially bound water molecules per mole of nucleotide. Whatever the type of anion or cation, there is a one-to-one relationship between Γ and a_w (Fig. 14.5). By definition, $\Gamma \rightarrow \infty$ as $a_w \rightarrow 1$. The most significant values are therefore those obtained at low a_w, and these agree well with those obtained from infrared spectroscopy.

However, a difference between the AT and GC pairs appears as regards their degree of preferential solvation. A linear relationship is observed between the apparent density of DNA in a CsCl gradient and its GC content. The anion plays an important role since, for a given AT/GC ratio, the apparent density in a CsCl gradient differs considerably from that in a Cs_2SO_4 gradient. Finally, it should be pointed out that the hydration of single-stranded DNA is lower than that of the double-stranded molecule.

In addition to the selective effect of cations and anions, there is a destabilizing effect (depression of the melting point of DNA) triggered by high concentrations of some anions (ClO_4^-, CCl_3COO^-) due to the large decrease in water activity, which favours the existence of the least hydrated species, i.e. the denatured form of DNA.

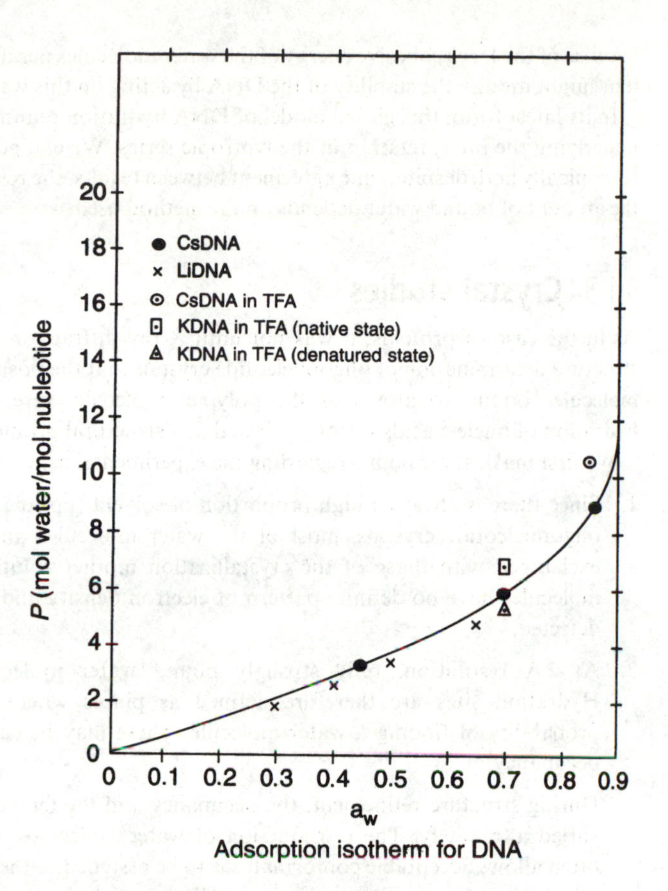

Figure 14.5 Preferential hydration of DNA and the adsorption isotherm, deduced from apparent densities using ultracentrifugation (TFA = trifluoroacetate). (From Tunis and Hearst, 1968. Reprinted from *Biopolymers* with the permission of John Wiley & Sons. © 1968 John Wliey & Sons.)

14.2 Thermodynamic model

In order to provide an overall explanation of all the experimental results on the hydration of DNA, Sinanoglu and Abdulnur (1964) put forward the theory that to dissolve DNA it is necessary to create a cavity in the solvent to accommodate the double helix.

The free energy of the stabilization in water of the double helix (H) with respect to that of the single strand (S) is the sum of three terms.

(1) the free-energy difference between the creation of a cylindrical cavity for H and of spherical cavities for the bases of the single strand – the difference varies considerably between solvents, being $-37\,\text{kcal mol}^{-1}$ per base pair for water and $-11\,\text{kcal mol}^{-1}$ per base pair for ethanol;

(2) the interaction of the base with the solvent in the two structures – in this case, there is little variation with the solvent, being $44.5\,\text{kcal mol}^{-1}$ per base pair for water, 50 for glycerol and 41 for ethanol;

(3) changes in the base–base and phosphate–phosphate interactions due to the presence of the solvent, which are almost independent of the solvent.

The first term is predominant: the stabilizing power of water arises from the large amount of energy required to create a cavity in it. The chain takes up a configuration such that the smallest surface area of the bases is exposed to the solvent because of the large cohesive energy of the water molecules per unit volume. Some ions might modify the stability of the DNA by acting on this water cohesion.

In its latest form, this global model of DNA hydration reminds us of the ideas underlying the interpretation of the lyotropic series. We can go no further macroscopically and, despite some agreement between results, the results of measuring the amount of bound water depends on the method used.

14.3 Crystal studies

As in the case of proteins, it was not until X-ray diffraction was used for the structure determination of oligonucleotide crystals that the positions of the water molecules bound to atoms of the polymer molecule were determined. The hydration of nucleic acids was then viewed as a structural problem.

We first make some points regarding the experimental aspect:

1. Since there is always a high proportion of solvent (up to 75 per cent) in the oligonucleotide crystals, most of the water molecules and ions are being exchanged with those of the crystallization mother solution. These water molecules have no definite pattern of electron density and hence cannot be detected.
2. At 2 Å resolution, only strongly bound water molecules are visible. Hydration sites are therefore defined as places where there is a high probability of finding a water molecule. These may be called sites of high occupancy.
3. During structure refinement, the occupancy and the temperature factor are varied alternately. The introduction of water molecules into the structure often allows acceptable conformations to be assigned to the sugar molecules. In other words, the first hydration shell plays a decisive role in maintaining the structure and stability of nucleic acids.
4. As for proteins, the water molecule acts mainly as a proton donor in the hydrogen bonds formed with the electronegative atoms in nucleic acids. These are principally the oxygens O^- bound to the phosphates, the O3′ and O5′ oxygens of the phosphodiester chain and the O4′ oxygen of the sugar. In addition, bridges formed by water molecules are established between these hydrated groups and the nitrogen or oxygen atoms of the bases.
5. Whereas the hydration of proteins is of central importance to the definition and stability of their tertiary structure, the interactions between water and nucleic acids are the physico-chemical basis of the polymorphism of DNA.

The first results were obtained for the self-complementary dodecamer (Drew and Dickerson, 1981):

CGCGAATTCGCG
GCGCTTAAGCGC

For the 24 nucleotides, the locations of about 80 water molecules were found, of which 60 are associated with the phosphodiester backbone. Helical geometry is favourable to the building of water 'bridges' between sites, which may be situated in the same residue, or in two adjacent residues of the same strand, or in two residues each in a different strand.

The hydration pattern is not the same in the two grooves (Fig. 14.6). Apart from the phosphate groups, which in all cases are the most highly hydrated, only one partially ordered layer of water molecules associated with N and O atoms (hydrogen bond acceptors) can be seen in the major groove, while there are two layers in the minor groove. The first forms bridges between O2 of T and N3 of A.

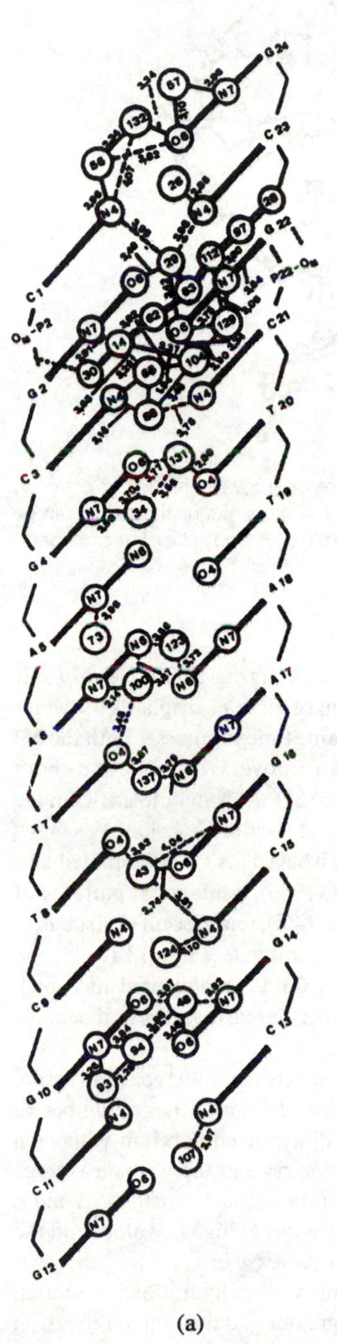

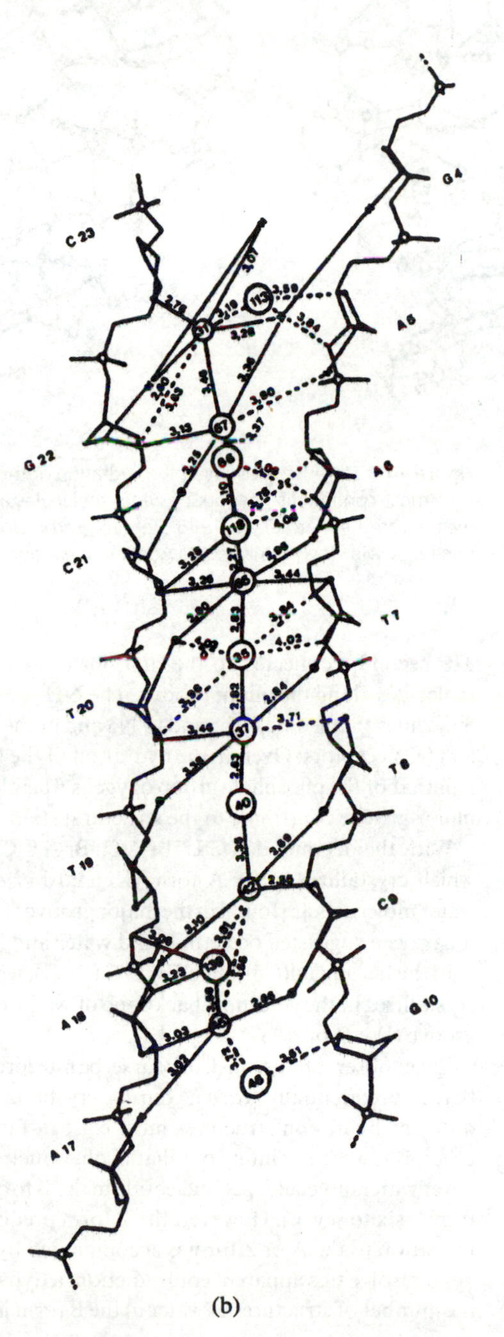

Figure 14.6 (a) Schematic view of the major groove of a B double-helix dodecamer, showing the first hydration shell. At the top, between the P2 and P22 phosphates, is the spermine molecule needed for crystallization. In all, there are 20 water molecules bound to N and O atoms, i.e. a little less than two per base pair. (b) Schematic view of the minor groove in the same oligomer. The hydration backbone and the hydrogen bonds between water molecules and neighbouring atoms are clearly visible. (From Kopka *et al.*, 1983. Reprinted from *Journal of Molecular Biology* with the permission of Academic Press.)

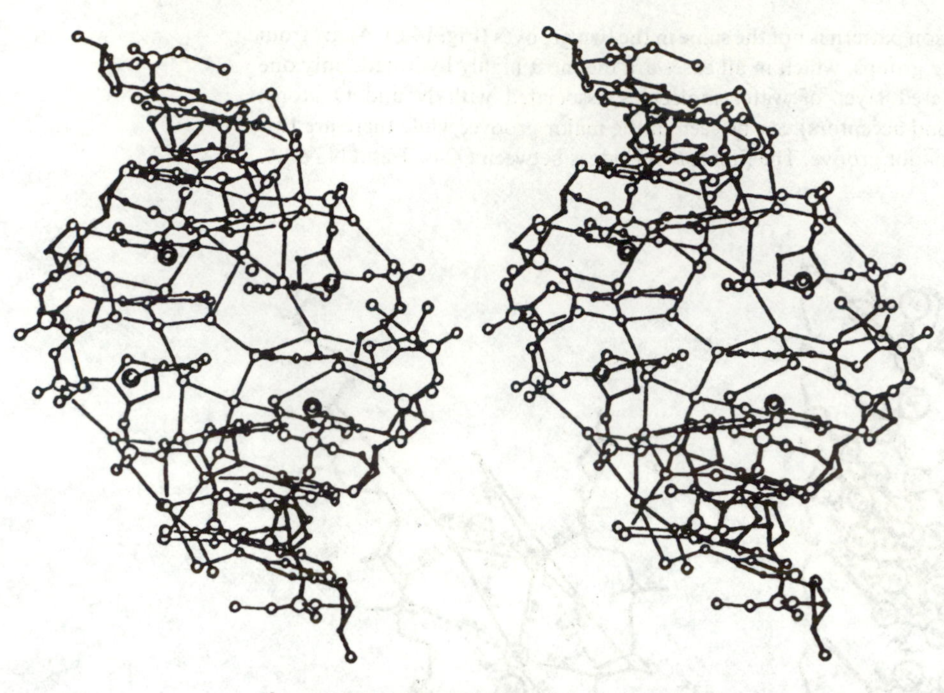

Figure 14.7 Stereoscopic view of the hydration of the major groove in the A-form octamer. An almost continuous network of water molecules and some pentagonal patterns can be seen. (From Kennard *et al.*, 1986. Reprinted from *Journal of Biomolecular Structure and Dynamics* with the permission of Adenine Press and the authors.)

The second, connected to the first, makes a kind of zigzag backbone of water molecules along the minor groove. The NH_2 group of guanine displaces the water molecules that would be bound to N3 and at the same time suppresses both the N3 and O2 acceptors. Overall, the hydration of the two grooves is the same and about half that of the phosphate group oxygens. It is clear that the water molecules in the minor groove contribute to the structural stability of the double helix.

With the octamer d(C G U5Br A U5Br A C C), where U is C5 brominated and which crystallizes in the A form (Kennard *et al.*, 1986), pentagonal patterns of water molecules are found in the major groove (Fig. 14.7), reminiscent of structures it has been suggested occur in liquid water and in crambin (see Fig. 13.10).

In the hexamer d(C5Br G C5Br G C5Br G), where C is C5 brominated and which crystallizes in the Z form, a backbone of water molecules also occurs in the minor groove (Westhof, 1987) (Fig. 14.8).

These observations must, of course, be interpreted as revealing a type of structure that is preferentially 'frozen' during crystallization. In fact, a large number of different hydration structures must exist in fast dynamic equilibrium with each other. It is worth pointing out that in oligonucleotide crystals there are on average fewer water molecules per nucleotide in the B form (about four) than in the A and Z forms (six to seven). However, the B form predominates at high a_w values and the transition to the A or Z form is accompanied by a decrease in a_w.

To resolve this apparent contradiction, a dynamic view of hydration is essential: the number of structures of water in the B form is greater and their interconversion

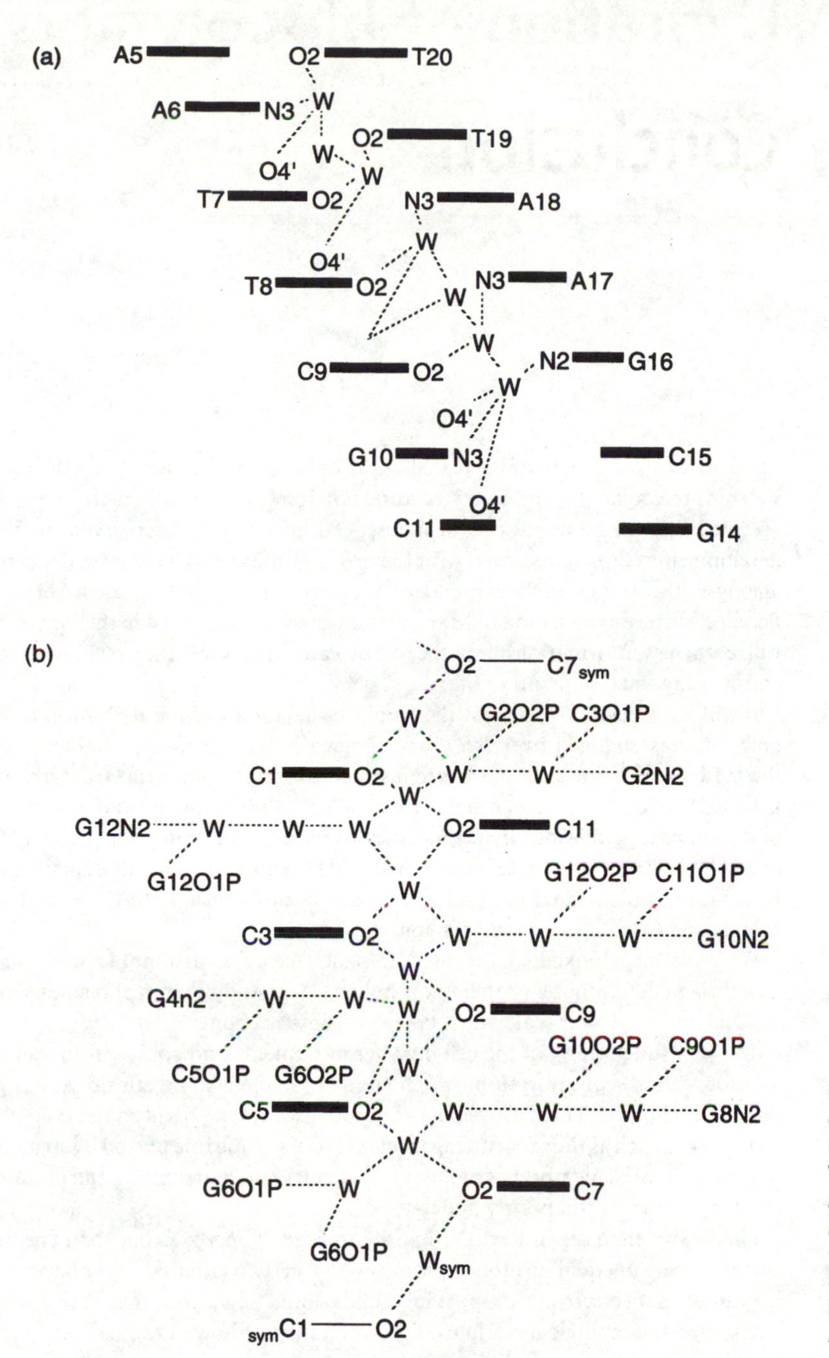

Figure 14.8 Hydration of the minor groove. Comparison between (a) the B form and (b) the Z form. (From Westhof, 1987. Reprinted from *International Journal of Biological Macromolecules*, vol. 9, pp. 186–92, © 1987 with kind permission from Elsevier Science NL, Sara Burgerhartstraat 25, 1055 KV Amsterdam, The Netherlands.)

is faster than in the A and Z forms. As a result, it is probably incorrect to interpret the B → A or B → Z transitions as arising solely from a change in the structure of the hydration shells. Work on the Brillouin scattering of phonons in the 4–80 GHz range (Tao *et al.*, 1987) has shown that measured relaxation times correspond to the dynamics of hydration shells.

Part III Hydration of biopolymers

Conclusion

The comparative brevity of Part III should not be taken as a signal that the role of water in molecular or membrane structures is in some way peripheral. It is generally accepted that all known forms of life depend heavily on water. Even prebiotic developments (those concerned with the origin of life) must have been dependent amongst other things on the existence of organic chemistry in the presence of water. Perhaps a future exploration of Mars, where water occurs at least in the form of ice, will reveal new information about the role of water in building the first biopolymers and creating the first forms of life.

It will have been noticed that the emphasis in studies of the hydration of biopolymers has changed over the years. Originally limited to a global and often thermodynamic approach, more recent studies are concerned with structures. Our knowledge of the location of water molecules in the three-dimensional structures of biopolymers is increasing all the time. In many cases, water can even be considered as an intrinsic part of a protein or nucleic acid structure. As a result, it plays a role both in molecular dynamics and in interaction mechanisms, but one that also complicates calculations using dynamic models.

We have also looked at the predominant role of hydrophobic–hydrophilic equilibria in determining protein morphology. Any study of morphogenesis, on a molecular or a cellular scale, must take water into account.

The very functioning of the cell involves movements and collisions in aqueous solutions whose concentration is high enough for them to be considered as gels rather than liquids. These movements cannot, however, be reduced to percolation mechanisms such as those occurring inside gels *in vitro*: the role played by structures like microtubules, membranes or the cytoskeleton in the marshalling and motion of macromolecules is still poorly understood.

Finally, we shall see in Part IV that the storage of energy other than chemical energy needs unequal proton concentrations in two adjacent compartments. Instead of using electron transport in a solid conducting medium as in the field of electronics, the biological world uses proton transport in water.

Part III Hydration of biopolymers

References and further reading

Water

Bernal, J.D. and Fowler, R.H. (1933). A theory of water and ionic solution, with particular reference to hydrogen and hydroxyl ions. *Journal of Chemical Physics*, **1**, 515–48.

Cantor, C.R. and Schimmel, P.R. (1980). *Biophysical Chemistry*, part III, pp. 1340–8. Freeman, San Francisco.

Edsall, J.T. and Wyman, J. (1958). *Biophysical Chemistry*, vol. I, p. 263. Academic Press, New York.

Eisenberg, D. and Kauzmann, W. (1969). *The structure and properties of water*. Oxford University Press, Oxford.

Frank, H.S. and Wen, W.Y. (1957). Structural aspects of ion–solvent interaction in aqueous solutions: a suggested picture of water structure. *Discussions of the Faraday Society*, **24**, 133–40.

Franks, F. (1972–9). *Water, a comprehensive treatise*, vols 1–6. Plenum Press, New York.

Grunwald, E. (1986). Model for the structure of the liquid water network. *Journal of the American Chemical Society*, **108**, 5719–26.

Gurney, R.W. (1953). *Ionic processes in solution*. McGraw-Hill, New York.

Hofmeister, F. (1888). Archiv für experimentelle Pathologie und Pharmakologie, **24**, 247.

Kay, R.L. and Evans, D.F. (1966). The effect of solvent structure on the mobility of symmetrical ions in aqueous solution. *Journal of Physical Chemistry*, **70**, 2325–35.

Lumry, R. and Rajender, S. (1970). Enthalpy–entropy compensation phenomena in water solutions of proteins and small molecules. A ubiquitous property of water. *Biopolymers*, **9**, 1125–227.

Narten, A.H. and Levy, H.A. (1969). Observed diffraction pattern and proposed models of liquid water. *Science*, **165**, 447–54.

Nemethy, G. and Scheraga, H.A. (1962). Structure of water and hydrophobic bonding in proteins. (1) A model for the thermodynamic properties of liquid water. *Journal of Chemical Physics*, **36**, 3382–400.

Rashin, A. and Honig, B. (1985). Re-evaluation of the Born model of ion hydration. *Journal of Physical Chemistry*, **89**, 5588–93.

Saenger, W. (1987). Structure and dynamics of water surrounding biomolecules. *Annual Review of Biophysics and Biophysical Chemistry*, **16**, 93–114.

Samoilov, O.Y. (1965). *Structure of aqueous electrolyte solutions and the hydration of ions*. Consultants Bureau, New York.

Tanford, C. (1973). *The hydrophobic effect*. Wiley, New York.

Walrafen, G.E. (1968). Raman spectral studies of HDO in H_2O. *Journal of Chemical Physics*, **48**, 244–51.

Yang, D.S.C., Sax, M., Chakrabartty, A. and Hew, C.L. (1988). Crystal structure of an antifreeze polypeptide and its mechanical implications. *Nature*, **333**, 232–7.

Proteins

Baker, E.N. and Hubbard, R.E. (1984). Hydrogen bonding in globular proteins. *Progress in Biophysics and Molecular Biology*, **44**, 97–119.

Blake, C.C.F., Pulford, W.C.A. and Artymiuk, P.J. (1983). X-ray studies of water in crystals of lysozyme. *Journal of Molecular Biology*, **167**, 693–723.

Bull, H.B. and Breese, K. (1968). Protein hydration. (1) Binding sites. (2) Specific heat of egg albumin. *Archives of Biochemistry and Biophysics*, **128**, 488–502.

Connolly, M.L. (1983). Solvent-accessible surfaces of proteins and nucleic acids. *Science*, **221**, 709–13.

Eisenberg, D. and McLachlan, A.D. (1986). Solvation energy in protein folding and binding. *Nature*, **319**, 199–203.

Fischer, H.F. (1964). A limiting law relating the size and shape of protein molecules to their composition. *Proceedings of the National Academy of Sciences, USA*, **51**, 1285–91.

Kauzmann, W. (1959). Some factors in the interpretation of protein denaturation. *Advances in Protein Chemistry*, **14**, 1–63.

Kyte, J. and Doolittle, R.F. (1982). A simple method for displaying the hydropathic character of a protein. *Journal of Molecular Biology*, **157**, 105–32.

Lee, B.K. and Richards, F.M. (1971). The interpretation of protein structures: estimation of static accessibility. *Journal of Molecular Biology*, **55**, 379–400.

Perutz, M.F. (1978). Electrostatic effects in proteins. *Science*, **201**, 1187–91.

Privalov, P.L. and Gill, S.J. (1988). Stability of protein structure and hydrophobic interaction. *Advances in Protein Chemistry*, **39**, 191–234.

Rose, G.D., Geselowitz, A.R., Lesser, G.L., Lee, R.H. and Zehfus, M.H. (1985). Hydrophobicity of amino acid residues in globular proteins. *Science*, **229**, 834–8.

Teeter, M.M. (1984). Water structure of a hydrophobic protein at atomic resolution. Pentagon rings of water molecules in crystals of crambin. *Proceedings of the National Academy of Sciences, USA*, **81**, 6014–8.

von Hippel, P., Peticolas, W., Schack, L. and Karlson, L. (1973). Model studies on the effect of neutral salts on the conformational stability of biological macromolecules. *Biochemistry*, **12**, 1256–63.

Nucleic acids

Drew, H.R. and Dickerson, R.E. (1981). Structure of B-DNA dodecamer. III: Geometry of hydration. *Journal of Molecular Biology*, **151**, 535–56.

Falk, M., Hartmann, K.A. and Lord, R.C. (1962). Hydration of DNA. I: A gravimetric study. *Journal of the American Chemical Society*, **84**, 3843–6.

Falk, M., Hartmann, K.A. and Lord, R.C. (1963). Hydration of DNA. II: An infra-red study. III: A spectroscopic study of the effect of hydration on the structure of DNA. *Journal of the American Chemical Society*, **85**, 387–94.

Kennard, O., Cruse, W.T.B., Nachman, J., Prange, T., Shakked, Z. and Rabinovich, D. (1986). Ordered water structure in an A-DNA octamer at 1.7 Å. *Journal of Biomolecular Structure and Dynamics*, **3**, 623–47.

Kopka, M.L., Fratini, A.V., Drew, H.R. and Dickerson, R.E. (1983). Ordered water structure around a B-DNA dodecamer. *Journal of Molecular Biology*, **163**, 129–46.

Sinanoglu, O. and Abdulnur, S. (1964). Hydrophobic stacking of bases and the solvent denaturation of DNA. *Photochemistry and Photobiology*, **3**, 333–42.

Subramanian, P.S. and Beveridge, D.L. (1989). A theoretical study of the hydration of canonical B d(GCCGAATTCGCG): Monte Carlo simulation and comparison with crystallographic ordered water sites. *Journal of Biomolecular Structure and Dynamics*, **6**, 1093–1122.

Tao, N.J., Lindsay, S.M. and Rupprecht, A. (1987). The dynamics of the DNA hydration shell at gigahertz frequencies. *Biopolymers*, **26**, 171–88.

Tunis, M.J.B. and Hearst, J.E. (1968). On the hydration of DNA. (1) Preferential hydration and stability of DNA in concentrated trifluoroacetate solution. (2) Base composition dependence of the net hydration of DNA. *Biopolymers*, **6**, 1325–53.

Wang, J.H. (1955). The hydration of DNA. *Journal of the American Chemical Society*, **77**, 258–60.

Westhof, E. (1987). Hydration of oligonucleotides in crystals. *International Journal of Biological Macromolecules*, **9**, 186–92.

Westhof, E. (1988). Water: an integral part of nucleic acid structure. *Annual Review of Biophysics and Biophysical Chemistry*, **17**, 125–44.

Part IV

Biopolymers as polyelectrolytes

15

Charge distributions: general laws

15.1 Introduction

Biological structures such as biopolymers, membranes or organelles normally contain a large number of charged groups. Examples are the COO^- groups in the side chains of aspartic and glutamic acids in proteins, in polar heads like $COO^-{-}(CH_2)_n{-}CH_3$ of fatty acids and in some sugars; the positively charged groups in the side chains of basic amino acids like $-NH_3^+$ in lysine, $NH_2{-}C{-}NH_2^+$ in arginine or $=NH^+$ in histidine; the PO_4^- groups in nucleic acids and phospholipids; and the PO_4^{2-} groups at the ends of chains or in certain compounds such as nucleoside monophosphates, diphosphates or triphosphates (Figs 15.1 and 15.2).

Biopolymers and cellular components therefore behave electrically as sets of point charges at fixed positions on molecules and migrating with them. In addition, there will of course be all the free ions present in the cell or the biological medium (such as Na^+, K^+, Mg^{2+}, Cl^-, CH_3COO^-, PO_4, PO_4^{2-}, etc.), which move along with their hydration shell. The charges on these ionized groups are all multiples of the elementary charge e ($= 1.6 \times 10^{-19}$ C) and will be denoted by $q = ze$, where z is a positive or negative integer for cations and anions respectively.

Acid–base equilibria apply to many of these groups. Their charge depends on the pH of the medium in which they occur and will normally be given by their pK_D (or $-\log K_D$) where K_D is the dissociation constant of the acid form defined by $K_D = [A][H^+]/[AH]$ in the equilibrium $AH \rightleftharpoons A^- + H^+$.

Alongside these ionized groups of charge ze, there are also groups that are partially charged by polarization: the distortion of the electron cloud around the molecule. The water molecule, for example, has a higher mean electron density around the O atom, giving it a slightly negative charge, while the two H atoms have a slightly positive charge. Because of its symmetry, the geometry of the water molecule means that it can be represented electrically by a model consisting of two charges $-\delta e$ and $+\delta e$ (with $\delta < 1$) separated by a distance x (see section 11.1). The C–N peptide bond is similar, with an excess negative charge on N and O and an excess positive charge on H and C, so that it can also be represented electrically by two charges q a distance x apart. In all the many examples of this type, the charge

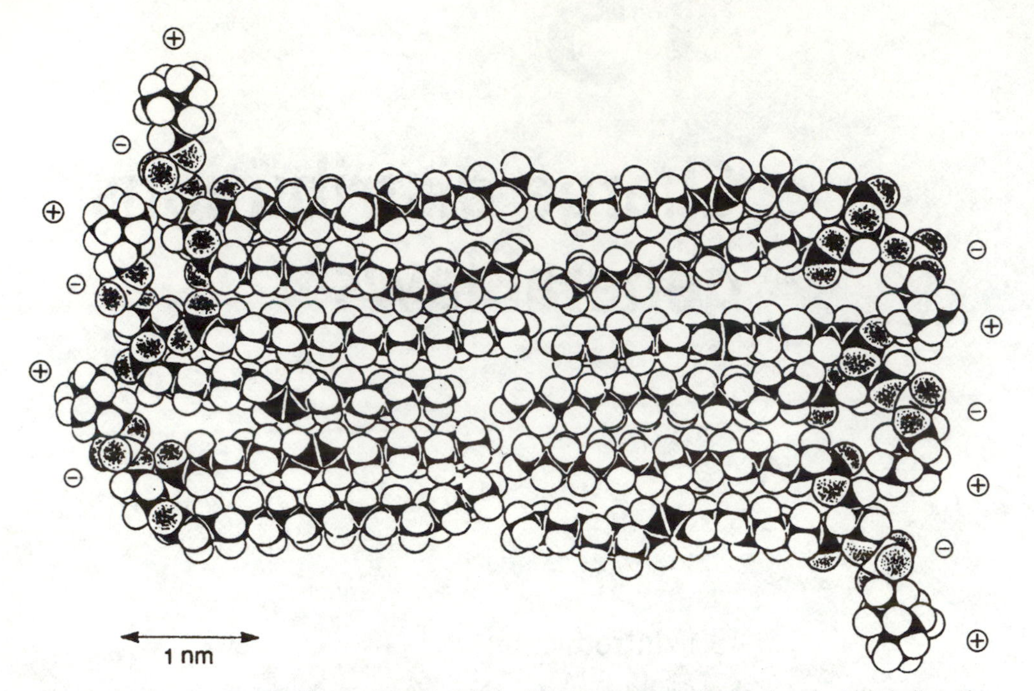

Figure 15.1 Model of a lipid bilayer consisting of phosphatidyl choline. The negative charges of the phosphate groups and the positive charges of choline quaternary ammonium are indicated. (From *Biochemistry*, 3rd edn, by L. Stryer. © 1988 by Lubert Stryer. Used with permission of W.H. Freeman and Company.)

distribution is defined by a quantity $p = qx$ known as the *dipole moment*, expressed in Cm. Thus $p = 6.11 \times 10^{-30}$ Cm for water and 11.55×10^{-30} Cm for the CO–NH group.

The dipole moment is in fact a vector quantity $\boldsymbol{p}$ with its origin at the negative charge, with a direction from the negative to the positive charge and with a magnitude equal to qx. In the water molecule it is directed along the symmetry axis, and in the CO–NH group it is parallel to the CO and NH bonds.

When peptide chains are organized into a regular structure such as α-helices or β-sheets, the vector addition of all the small constituent dipole moments can result in a large resultant moment. The α-helix, for example, behaves as a dipole consisting of two charges $\pm e/2$ located at the ends of the helix (Fig. 15.3).

Charges like these are present in most biopolymers and membranes, and more generally in subcellular structures. They create a complex electric potential and electric field distribution in the ionic medium of the cell. Any charge q situated at a point where the electric potential is Ψ has an energy $W = q\Psi$, and if the electric field at the point is $\boldsymbol{E}$ it experiences a force $\boldsymbol{F} = q\boldsymbol{E}$. The electric energy W is part of the energy of association between two molecules, while the electric force determines the speed and direction of their migration. Coulomb interactions thus play an important role both in the formation of structures and in transport processes within the cell or across membranes.

We describe the behaviour of the mobile charges, with their various magnitudes and physical dimensions, and of the dipoles by using the *electrostatic laws* governing stationary charge distributions together with the *laws of statistical mechanics* governing the motion of a set of particles undergoing Brownian motion under

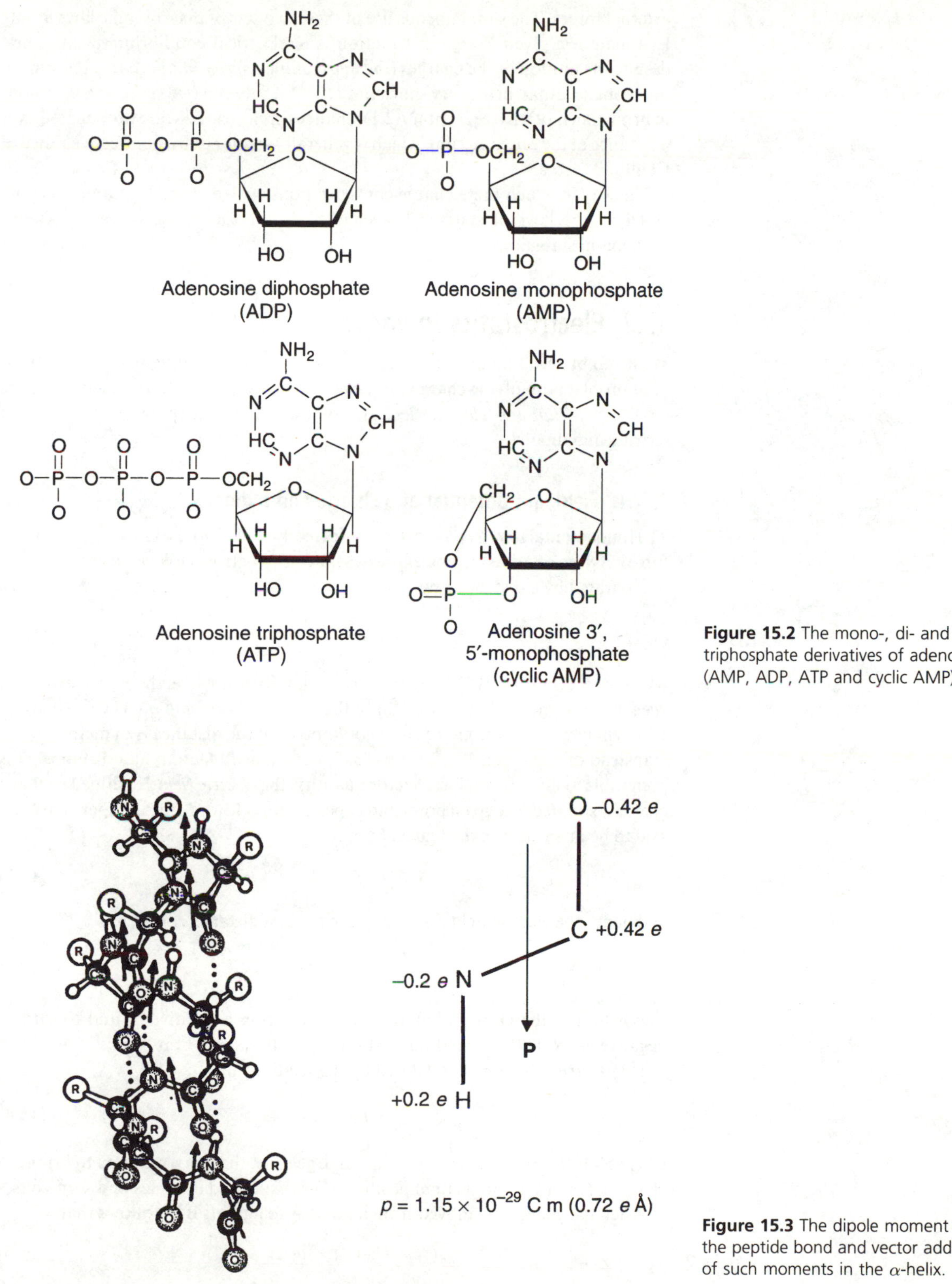

Figure 15.2 The mono-, di- and triphosphate derivatives of adenosine (AMP, ADP, ATP and cyclic AMP).

Figure 15.3 The dipole moment of the peptide bond and vector addition of such moments in the α-helix.

external forces. The simultaneous use of these two sets of macroscopic laws is only legitimate if a given charge distribution is in electrical equilibrium at all times, despite the motion of the charges (the *quasi-static approximation*). It is relevant to point out here that, in the very short time (10^{-14} s) between two successive collisions in Brownian motion (see section 7.2.1), changes in electric field are propagated over a distance of the order of 3 μm, which is generally greater than the effective range of Coulomb forces.

The way in which these charge configurations evolve in space and time is governed by the laws of statistical mechanics, in this case by classical Maxwell–Boltzmann statistics.

15.2 Electrostatics *in vacuo*

We begin by recalling the main results in electrostatics, dealing first with the simplest problems involving charges *in vacuo*. These results enable us to determine the *electric potential* and *electric field* at any point due to a spatial distribution of charges and dipoles.

15.2.1 Field and potential of a charge and a dipole

The fundamental law, *Coulomb's law*, is formally similar to the law of gravitational force between two masses, and expresses the force $\boldsymbol{f}$ between two *point charges* q and q' separated by a distance r by

$$\boldsymbol{f} = qq'\boldsymbol{u}/4\pi\varepsilon_0 r^2 \tag{15.1}$$

where $\varepsilon_0 = (1/36\pi) \times 10^{-9}$ in SI units is a quantity known as the *permittivity of a vacuum*, and where $\boldsymbol{u}$ is a unit vector in the direction from q to q'. The force in eqn 15.1 is in newtons if the charge is in coulombs and the distance is in metres.

Instead of using eqn 15.1 directly, as if dealing with 'action at a distance', it is preferable to introduce a local vector quantity, the *electric field* $\boldsymbol{E}$, defined so that a charge q situated at a given point and experiencing a force $\boldsymbol{f}$ due to other charges is said to be in an electric field given by

$$\boldsymbol{f} = q\boldsymbol{E} \tag{15.2}$$

and hence the electric field due to a charge q' at a distance r from it is

$$\boldsymbol{E} = \boldsymbol{u}q'/4\pi\varepsilon_0^2 \tag{15.3}$$

The field $\boldsymbol{E}$ is therefore radial, directed away from a positive q' and towards a negative q'. Note that the SI units of the quantity $\varepsilon_0 E$ are $\mathrm{C\,m^{-2}}$.

The *electric potential* Ψ is defined by the relationship

$$E_{\mathrm{P}} = q\Psi \tag{15.4}$$

where E_{P} is the potential energy of the charge q, i.e. the work required to bring the charge from infinity to its final position. E_{P} is expressed in joules and Ψ in volts.

When the charge q is moved from a point A to a point B, it follows that

$$E_{\mathrm{P}}(\mathrm{A}) - E_{\mathrm{P}}(\mathrm{B}) = q(\Psi_{\mathrm{A}} - \Psi_{\mathrm{B}}) \tag{15.5}$$

The left-hand side is the work W required to move the charge from A to B, and hence

$$W_{A\to B} = q(\Psi_A - \Psi_B) \tag{15.6}$$

In mechanics, the work done by a force $\boldsymbol{f}$ along a path can be written as $\int_A^B \boldsymbol{f}\cdot \mathrm{d}\boldsymbol{r}$, where $\mathrm{d}\boldsymbol{r}$ is the elementary vector displacement. In the present case this becomes $\int_A^B q\boldsymbol{E}\cdot \mathrm{d}\boldsymbol{r}$, and from eqn 15.6 we then have

$$\Psi_A - \Psi_B = \int_A^B \boldsymbol{E}\cdot \mathrm{d}\boldsymbol{r} = -(\Psi_B - \Psi_A) \tag{15.7}$$

If E_x, E_y and E_z are the usual three Cartesian components of $\boldsymbol{E}$, eqn 15.7 becomes

$$\Psi_B - \Psi_A = -\int_A^B (E_x\,\mathrm{d}x + E_y\,\mathrm{d}y + E_z\,\mathrm{d}z) \tag{15.8}$$

and the differential form of this is

$$E_x = -\partial\Psi/\partial x \qquad E_y = -\partial\Psi/\partial y \qquad E_z = -\partial\Psi/\partial z \tag{15.9}$$

The three partial derivatives are the components of the vector gradient of Ψ, so that eqn 15.9 in vector form is

$$\boldsymbol{E} = -\mathrm{grad}\,\Psi \tag{15.10}$$

The negative sign is conventional: the field is directed from a lower to a higher potential, whereas the gradient of the potential is in the opposite direction. Eqns 15.9 and 15.10 show that the electric field will be expressed in $\mathrm{Vm^{-1}}$.

For a point charge, the field and potential at a distance r depend only on r, so that

$$E_r = -\mathrm{d}\Psi/\mathrm{d}r$$

and from eqn 15.3 it follows that

$$\Psi = q/4\pi\varepsilon_0 r \tag{15.11}$$

assuming that $\Psi \to 0$ as $r \to \infty$.

For a *dipole* consisting of a pair of charges $\pm q$ separated by a distance x, the dipole moment in vector form is $\boldsymbol{p} = qx\boldsymbol{u}$ where $\boldsymbol{u}$ is a unit vector with a direction from $-q$ to $+q$. The electric potential at a point P (Fig. 15.4) due to the dipole is the sum of the potentials due to the two charges:

$$\Psi_P = (q/r_1 - q/r_2)/4\pi\varepsilon_0 = q(r_2 - r_1)/4\pi\varepsilon_0 r_1 r_2 \tag{15.12}$$

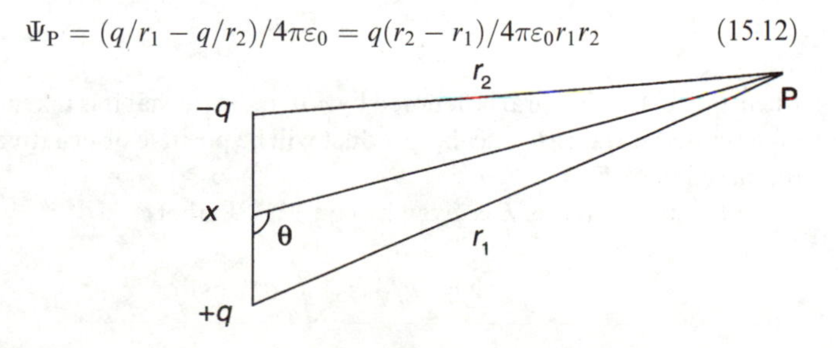

Figure 15.4 Geometry for calculating the electric field due to a dipole.

If $x \ll r$ (ideal dipole), $r_2 - r_1 = x\cos\theta$ and hence

$$\Psi_P = q(r_2 - r_1)/4\pi\varepsilon_0 r_1 r_2 \approx qx\cos\theta/4\pi\varepsilon_0 r^2$$

or

$$\Psi_P \approx p\cos\theta/4\pi\varepsilon_0 r^2 \tag{15.13}$$

The potential no longer varies as r^{-1} but as r^{-2} and depends on the angle θ. The electric field is no longer radial as it is for a point charge, but has both a radial and transverse component given by $E_r = -\partial\Psi/\partial r$ and $E_\theta = -(1/r)\,\partial\Psi/\partial\theta$. Equation 15.13 then gives

$$E_r = 2p\cos\theta/4\pi\varepsilon_0 r^3 \qquad \text{and} \qquad E_\theta = p\sin\theta/4\pi\varepsilon_0 r^3 \tag{15.14}$$

Now consider a dipole placed in an electric field, so that its two charges are at points with different potentials Ψ_+ and Ψ_-. Its potential energy from eqn 15.4 is given by $W = q(\Psi_+ - \Psi_-)$. The bracketed factor can be expressed in terms of the potential gradient along the direction of the dipole moment given by the unit vector $\boldsymbol{u}$, and is therefore $x\boldsymbol{u}\cdot \text{grad}\,\Psi$. The field $\boldsymbol{E}$ at the dipole is given by eqn 15.10, so that with $\boldsymbol{p} = qx\boldsymbol{u}$ we have

$$W = -\boldsymbol{p}\cdot\boldsymbol{E} = -pE\cos\theta \tag{15.15}$$

where θ is the angle between $\boldsymbol{p}$ and $\boldsymbol{E}$. The energy is therefore a minimum when the dipole moment is parallel to the field. For any other orientation, the dipole experiences a torque Γ tending to turn it into the minimum energy position. This torque is due to the forces $q\boldsymbol{E}$ and $-q\boldsymbol{E}$ acting on the charges of the dipole, which produce a couple of moment

$$\Gamma = \boldsymbol{p}\times\boldsymbol{E} \tag{15.16}$$

where Γ is in N m, $\boldsymbol{p}$ is in C m and $\boldsymbol{E}$ in V m^{-1}.

15.2.2 Gauss's theorem and continuous distributions of charge

Interactions between charged molecules can only be investigated if the electric potential around each of them is known. More generally, in a solution containing ions and charged polymers, it is possible to define a potential $\Psi(x, y, z)$ that must obey certain universal laws.

Consider a point charge q within an arbitrary closed surface S. Each elementary area of the surface dS is associated with a unit vector $\boldsymbol{u}_n$ normal to the element and directed from inside S to outside. A new physical quantity, the electric *flux* Φ, is defined by

$$\Phi = \oint \boldsymbol{E}\cdot\boldsymbol{u}_n\,\mathrm{d}S \tag{15.17}$$

where the surface integral is denoted by $\oint$ to indicate that it is taken over the whole of the closed surface. The scalar product will be positive or negative depending on the direction of $\boldsymbol{E}$.

For the point charge, $\boldsymbol{E}$ is given by eqn 15.3, so that

$$\Phi = (q/4\pi\varepsilon_0)\oint \mathrm{d}S\cos\theta/r^2 \tag{15.18}$$

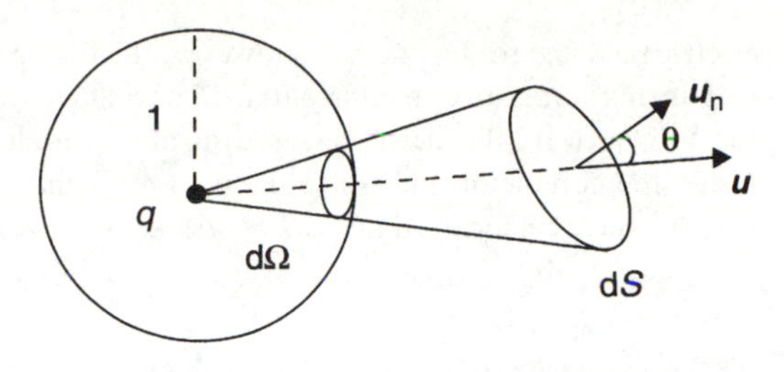

Figure 15.5 Definition of a solid angle.

where θ is the angle between $\boldsymbol{u}$ and $\boldsymbol{u}_n$. If a sphere of unit radius centred on q is drawn (Fig. 15.5), the cone with q at its apex and base dS has a *solid angle* $d\Omega$ (in steradians) at the centre, defined by

$$d\Omega = dS\cos\theta/r^2 \tag{15.19}$$

which is identical to the quantity after the surface integral sign in eqn 15.18. Integration over the whole closed surface yields the total solid angle subtended at the centre of the sphere, i.e. 4π. It then follows from eqn 15.18 that

$$\Phi = q/\varepsilon_0 \tag{15.20}$$

known as *Gauss's theorem*. This is a direct consequence of the $1/r^2$ variation of the electric field due to a point charge q at a distance r from it.

When there are several charges q_1, q_2, etc., within the same closed surface, the fluxes from each simply add and hence in general

$$\Phi = \sum_i q_i/\varepsilon_0 \tag{15.21}$$

Now consider a *spherical* charge distribution consisting of a charge q distributed uniformly over the surface of a sphere of radius R. The electric field due to this will be radial by symmetry. The flux of E over the surface of a sphere of radius $r > R$ is the product of E and the surface area $4\pi r^2$. Applying Gauss's theorem to the sphere of radius R gives us $4\pi r^2 E = q/\varepsilon_0$, so that

$$E_r = q/4\pi\varepsilon_0 r^2$$

as if the whole of the charge were concentrated at the centre of the sphere of radius R. This explains why the point charge model is used so frequently.

We now consider a *cylindrical* charge distribution with a charge λ per unit length distributed uniformly over the surface of a cylinder of infinite length whose radius is R. Once again, the electric field by symmetry is radial (directed away from the axis of the cylinder). If the electric field at a distance $r > R$ from the axis is E, Gauss's theorem applied to a cylindrical surface coaxial with the charge and of radius r and length L gives $2\pi rLE = \lambda L/\varepsilon_0$, since there is no flux over the ends. Hence

$$E = \lambda/2\pi\varepsilon_0 r \tag{15.22}$$

which is the same as the field created by a filament of charge along the axis of the cylinder. A cylindrical charge distribution produces the same field as a linear distribution along the axis, a model used for some linear polyelectrolytes.

Finally, consider a charged conductor of any shape with a surface charge density σ but containing no charge inside the surface. Any closed surface lying entirely within the conductor contains no charge and hence, by Gauss's theorem, the electric field is zero. Just outside, however, the field E will be normal to the surface and its flux across an elementary area $\mathrm{d}S$ parallel to the surface will be $E\,\mathrm{d}S$. This is the only outward flux across a closed surface bounded by $\mathrm{d}S$ just outside and just inside the conductor and, since the charge in the volume thus defined is $\sigma\,\mathrm{d}S$, Gauss's theorem means that $E\,\mathrm{d}S = \sigma\,\mathrm{d}S/\varepsilon_0$, so that

$$E = \sigma/\varepsilon_0 \qquad (15.23)$$

is the electric field immediately outside the surface of a charged conductor. There is a discontinuity in the field when crossing such a charged surface.

15.2.3 Divergence and Poisson's equation

Equations 15.20 and 15.21 apply to finite volumes of space and we need to find an equivalent expression allowing us to use the law locally at any point. For this purpose, a new quantity is defined: the *divergence* of the electric field, div $\boldsymbol{E}$. The treatment outlined in Box 15.1 shows that the differential expression of

Box 15.1 Definition of divergence

Consider an elementary volume in Cartesian coordinates (Fig. 15.6) with edges $\mathrm{d}x$, $\mathrm{d}y$, $\mathrm{d}z$ parallel to the coordinate axes. To apply Gauss's theorem to the volume, we have to evaluate the total outward flux of $\boldsymbol{E}$ across the bounding surface. First, take the two areas ABCD and A′B′C′D′. The outward flux across ABCD is simply $E_x\,\mathrm{d}y\,\mathrm{d}z$ since only the x component of $\boldsymbol{E}$ contributes to such a flux. Across A′B′C′D′, the x component of $\boldsymbol{E}$, E'_x, is inward and since we have $E'_x = E_x - (\partial E_x/\partial x)\,\mathrm{d}x$, the resultant outward flux from the two faces is $+(\partial E_x/\partial x)\,\mathrm{d}x\,\mathrm{d}y\,\mathrm{d}z$. Similar expressions are obtained for the other pairs of faces, so that the total outward flux of $\boldsymbol{E}$ is

$$\mathrm{d}\Phi = (\partial E_x/\partial x + \partial E_y/\partial y + \partial E_z/\partial z)\,\mathrm{d}x\,\mathrm{d}y\,\mathrm{d}z$$

The quantity in brackets defines the *divergence* of the vector $\boldsymbol{E}$, abbreviated to div $\boldsymbol{E}$. Since $\mathrm{d}x\,\mathrm{d}y\,\mathrm{d}z$ is the value of the elementary volume $\mathrm{d}V$:

$$\mathrm{d}\Phi = \mathrm{div}\,\boldsymbol{E}\,\mathrm{d}V$$

Equation 15.20, however, shows that $\mathrm{d}\Phi = \mathrm{d}q/\varepsilon_0$, where $\mathrm{d}q$ is the total charge in the volume $\mathrm{d}V$. It follows that div $\boldsymbol{E} = (\mathrm{d}q/\mathrm{d}V)/\varepsilon_0$. Because $\mathrm{d}q/\mathrm{d}V$ is the volume density of charge ρ, we have finally

$$\mathrm{div}\,\boldsymbol{E} = \rho/\varepsilon_0$$

as in eqn 15.24.

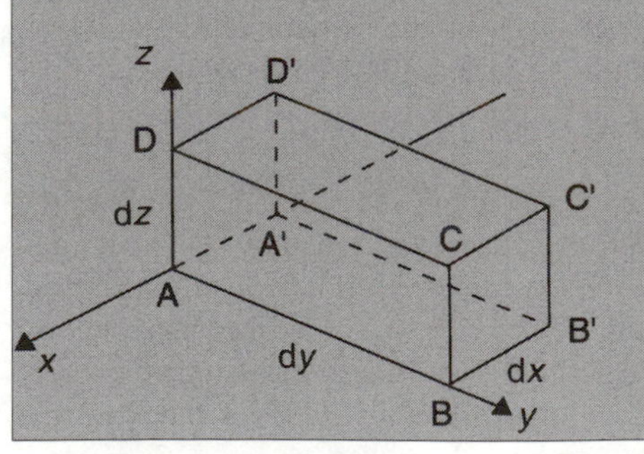

Figure 15.6 Elementary volume in the derivation of Poisson's equation.

Gauss's theorem is

$$\operatorname{div} \boldsymbol{E} = \rho/\varepsilon_0 \tag{15.24}$$

where ρ is the volume charge density at a point.

Equation 15.24 is the local differential form of Gauss's theorem, i.e. expressing the relationship at any point between the charge density and the scalar quantity $\operatorname{div} \boldsymbol{E}$. The validity of this is independent of any surface. Formally, electricity is being treated as a kind of 'fluid' filling space, with the charges being the 'sources' of the fluid: their value and spatial distribution thus determine the electric field at all points.

If eqn 15.24 is written in Cartesian coordinates by using the expression for $\operatorname{div} \boldsymbol{E}$ (see Box 15.1), and if we express the components of $\boldsymbol{E}$ as $E_x = -\partial\Psi/\partial x$, etc. (eqn 15.9), then

$$\partial^2\Psi/\partial x^2 + \partial^2\Psi/\partial y^2 + \partial^2\Psi/\partial z^2 = -\rho/\varepsilon_0 \tag{15.25}$$

where the left-hand side is a scalar quantity known as the *Laplacian* of Ψ and is written $\nabla^2\Psi$. Finally, therefore:

$$\nabla^2\Psi = -\rho/\varepsilon_0 \tag{15.26}$$

which is *Poisson's equation*. From a given charge distribution expressed as $\rho(x, y, z)$, the potential function $\Psi(x, y, z)$ can in principle be calculated provided the conditions to be satisfied at geometrical boundaries (the boundary conditions) are known. Solutions of Poisson's equation for various geometries have long been worked out, particularly for the case $\rho = 0$ (Laplace's equation). The equations can also be solved numerically by computer.

Using the expression $\boldsymbol{E} = -\text{grad}\,\Psi = -\nabla\Psi$, where ∇ is a vector operator with Cartesian components $\partial/\partial x$, $\partial/\partial y$, $\partial/\partial z$, Poisson's equation can also be written in the form

$$\nabla \cdot (\nabla\Psi) = -\rho/\varepsilon_0$$

15.3 Electrostatics in a medium

15.3.1 Polarization and dielectric constant

Throughout section 15.2, it has been assumed that both the charges and the points at which the field and potential were calculated were all in a vacuum. This is clearly not typical of the situation in biology, where all cellular activities occur either in an aqueous medium or inside membranes, i.e. in a medium similar to that of a liquid hydrocarbon. In such a medium, the behaviour of molecules subjected to an electric field depends on whether or not they have a permanent dipole moment, i.e. whether or not they are *polar*.

Polar molecules, like those of water, tend to become oriented in the direction of the field as discussed above when considering eqns 15.15 and 15.16. This is what happens near an ion, for example: the radial field of a cation or anion assumed to be spherical will align the dipoles of the water molecules and create a layer of partially immobilized water around the ion (a hydration shell). Similar behaviour is observed with ordered peptide chains or side chains of proteins. If, on the other hand, the molecules have no permanent moment, they become polarized and acquire an induced dipole moment.

Whether the dipole moments are permanent or induced, the overall effect is the same: they align themselves preferentially in the direction of the electric field and the medium becomes *polarized*. Such a medium is known as a *dielectric*, and is to be contrasted with a conductor in which charges are free to move through the atomic lattice. The electric moment per unit volume of the medium defines its *polarization* $\boldsymbol{P}$, so that, if there are N atoms or molecules per unit volume each with a dipole moment $\boldsymbol{p}$, then $\boldsymbol{P} = \mathrm{N}\boldsymbol{p}$.

Assuming that $\boldsymbol{P}$ is proportional to the electric field $\boldsymbol{E}$, then

$$\boldsymbol{P} = \varepsilon_0 \chi \boldsymbol{E} \tag{15.27}$$

and since both $\boldsymbol{P}$ and $\varepsilon_0\boldsymbol{E}$ have the same dimensions (SI units $\mathrm{C\,m^{-2}}$), the quantity χ is dimensionless. It describes the 'response' of the dielectric to the electric field and is called the *electric susceptibility*.

Now consider a small rectangular slab of dielectric with a thickness d and a surface area S perpendicular to an applied electric field $\boldsymbol{E}$ as in Fig. 15.7a. Because $\boldsymbol{P}$ is parallel to $\boldsymbol{E}$, the total electric moment of the slab is PSd in magnitude. The same electric moment would be produced if there were charges equal to PS appearing on the two areas S (Fig. 15.7a) since they would be a distance d apart. If the polarization were uniform, such charges would indeed appear and they would have a surface density σ_P numerically equal to P itself. It is assumed that this can be generalized to any volume of dielectric and that any elementary volume of such a dielectric would have a surface charge density equal to the component of $\boldsymbol{P}$ in the direction of the normal to the surface.

Suppose the field $\boldsymbol{E}$ is produced by a pair of parallel conducting plates carrying surface charge densities $\pm\sigma$ as in Fig. 15.7b. If the dielectric slab is placed between the plates as shown, the polarization charges appearing on its surfaces will partly

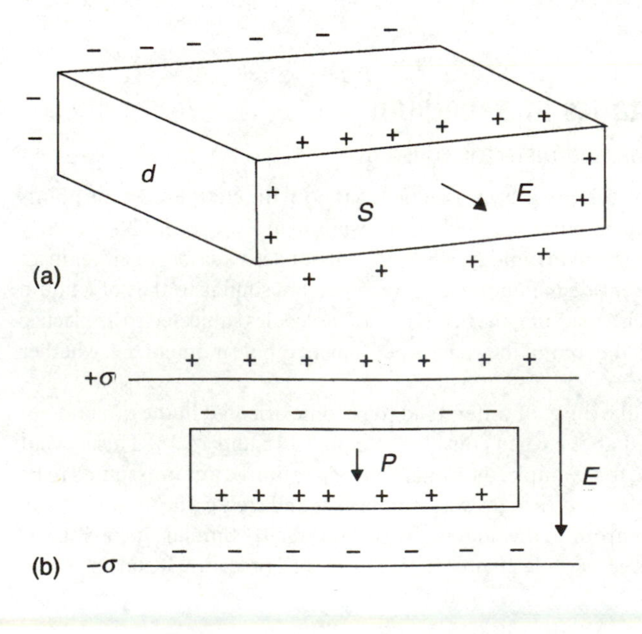

Figure 15.7 Polarization in a dielectric slab.

compensate for the electric field due to the conducting plates. There will be an effective surface charge of density σ_{eff} given by $\sigma - P$. According to eqn 15.23, therefore, the field in the dielectric will be $E = (\sigma - P)/\varepsilon_0$, so that

$$\sigma = \varepsilon_0 E + P \tag{15.28}$$

By analogy with $\sigma_P = P$, a new vector is defined by

$$D = \varepsilon_0 E + P \tag{15.29}$$

known as the *electric displacement*. The importance of this is that it is unaffected by the presence of the dielectric, since in the above case $D = \sigma$ numerically, and we can show more generally that

$$\sigma = \boldsymbol{D} \cdot \boldsymbol{u}_n \tag{15.30}$$

where $\boldsymbol{u}_n$ is a unit vector normal to the surface of a conductor immersed in a dielectric. The total free charge Q on the surface of a conductor will thus be

$$Q = \oint \sigma \, dS = \oint \boldsymbol{D} \cdot \boldsymbol{u}_n \, dS \tag{15.31}$$

the right-hand side being the flux of the vector $\boldsymbol{D}$ over the conducting surface. This equation is the same as Gauss's theorem in a vacuum, which can now be applied to the electric displacement when the conductor is in the presence of dielectrics. Note that the charge on the left-hand side of eqn 15.31 includes only the free charges on the conductor and not the polarization charges. When there are no free charges, $\oint \boldsymbol{D} \cdot \boldsymbol{u}_n \, dS$ is zero and the vector $\boldsymbol{D}$ is then said to have a conservative flux.

If eqn 15.17 is valid and is substituted into eqn 15.27, we obtain

$$\boldsymbol{D} = \varepsilon_0 (1 + \chi) \boldsymbol{E} = \varepsilon \varepsilon_0 \boldsymbol{E} \tag{15.32}$$

where the dimensionless quantity $\varepsilon = 1 + \chi$ is the *dielectric constant* and is generally greater than 1. The quantity $\varepsilon\varepsilon_0$ is called the *permittivity* of the medium and has the same units as ε_0.

Returning to the example of the dielectric slab in Fig. 15.7, the charge density σ_P on each face of the dielectric (equal to P) is given by $\sigma - \varepsilon_0 E$ from eqn 15.28. However, $\sigma = D = \varepsilon\varepsilon_0 E$ and hence

$$\sigma_P = \varepsilon_0 (\varepsilon - 1) E$$

Furthermore, if ε is constant, eqn 15.31 may be written as

$$Q = \varepsilon \varepsilon_0 \oint \boldsymbol{E} \cdot \boldsymbol{u}_n \, dS \tag{15.33}$$

showing that the flux of $\boldsymbol{E}$ through a closed surface is no longer Q/ε_0 as in eqn 15.20, but is now $Q/\varepsilon\varepsilon_0$.

The effect of the dielectric is therefore simply to divide the electric field (and the potential) by ε. In particular, the Coulomb interaction, as regards both the force and the energy, is divided by ε.

15.3.2 Water as a dielectric

The above effect explains why water ($\varepsilon \approx 80$ at 20°C) is such a good solvent of ionic crystals. Take the case of an Na^+F^- crystal in which the inter-ionic distance is 2.31 Å. The energy associated with the attractive force between Na^+ and F^- is given by $W = e^2/4\pi\varepsilon_0 d = 9.97 \times 10^{-19}$ J. Since $k_B T$ at 20°C is 4.95×10^{-21} J, the thermal energy is negligible compared with the inter-ionic energy.

When water is present, assuming the same inter-ionic distance and neglecting any change in dielectric constant near the ion (see section 16.2.2), the energy is reduced by a factor of 80 and becomes of the same order as $k_B T$. The crystal dissolves into a solution of free Na^+ and F^- ions, which, once solvated and surrounded by their own hydration shells, can no longer come close enough together to re-form an ion pair.

In fact, as we shall see later, the role played by water in electrostatic interactions is a little more subtle than that, for two reasons:

1. The dielectric constant ε is a property of a continuous medium, but its value depends on the molecular structure of the same medium. Local variations occur in the structure, so that on a molecular scale ε may not be constant or isotropic, and its very definition is problematic.
2. The high dielectric constant is the result of two competing processes: the orientation of the molecule in an electric field and the disorder due to thermal agitation, which produces a reduction of ε with increasing temperature. Both processes warrant a more thorough study.

Orientation of water molecules

Even in a strong external electric field, molecular dipoles are only partially aligned. The local organization of liquid water (see section 11.3) shows that there is short-range order around each molecule reminiscent of the tetrahedral order in ice I. Each water molecule can therefore be assumed (Kirkwood, 1939) to be subjected to the strong electric field E_i of neighbouring dipoles, which induces a dipole moment αE_i in the water molecule, α being its polarizability. The molecule will thus have a mean total moment $p' = p + \alpha E_i$, greater than the permanent moment $p = 6.11 \times 10^{-30}$ C m that is characteristic of the isolated molecule.

Moreover, the calculation of the dielectric constant (the 'response' of the water to the electric field) must take into account a *correlation parameter g*. This arises because each water molecule does not just become aligned in isolation but also influences a cluster of hydrogen-bonded neighbouring molecules. The overall effect of this is that

$$\varepsilon = p'^2 gn/2\varepsilon_0 k_B T$$

where n is the number of molecules per unit volume. Taking $g = 2.6$ at 20°C and $p' \approx 8.2 \times 10^{-30}$ C m, the calculated values of ε are found to be within a few per cent of the experimental values.

Thermal disorder

A rise in temperature not only increases the rotational Brownian motion of water molecules but also shortens the lifetime of the tetrahedral structure. This results in a decrease in correlation as measured by g. The lower values for p' and g lead to a more complex variation of ε with temperature than a simple $1/T$ law.

All this is only valid for a static or low-frequency electric field. At high frequencies, the water molecules cannot 'follow' the forced motion imposed by the field. If the field is represented by $E_0 \exp(i\omega t)$, the dispersion of the dielectric constant (i.e. its variation with frequency) can be shown to be

$$\varepsilon = \varepsilon_\infty + (\varepsilon_0 - \varepsilon_\infty)/[1 + (\omega\tau)^2]$$

where ε_0 and ε_∞ are the dielectric constants at zero and infinite frequencies, respectively, and τ is a *dielectric relaxation time*, found to be of the order of 10^{-11} s (10 ps) with ε_∞ of about 4 or 5. At a frequency of 10^{11} Hz, therefore, ε is nearly 6, and at a frequency of 10^{12} Hz the dielectric constant has almost reached its limiting value.

Box 15.2 considers the overall effect of thermal disorder on the orientation of dipoles in an external electric field, using a treatment due to Debye.

Box 15.2 **Orientation of dipoles in an external electric field**

An electric field $\boldsymbol{E}$ exerts a torque on a dipole moment $\boldsymbol{p}$ tending to align it in the direction of the field. The same applies to molecules like those of water which possess dipole moments. Molecular collisions, on the other hand, tend to oppose the tendency and produce disorder in the system.

To calculate the overall effect of the two processes, consider a dipole making an angle θ with the field and thus having an energy $-pE\cos\theta$. Since all values of θ are possible, we first calculate the number of dipoles with values of θ between θ and $\theta + \mathrm{d}\theta$ using the following geometry:

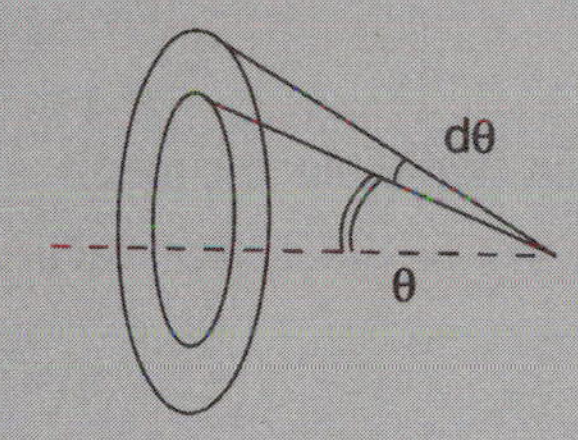

The dipoles are located at the centre of a sphere of unit radius, and those required lie between the cones with semi-vertical angles θ and $\theta + \mathrm{d}\theta$. The area between the two cones cut out of the spherical surface is $2\pi\sin\theta\,\mathrm{d}\theta$ and the required number is proportional to this. These have energy $-pE\cos\theta$, so that the partition function is

$$Z = \int_0^\pi [\exp(pE\cos\theta/k_\mathrm{B}T)]2\pi\sin\theta\,\mathrm{d}\theta$$

apart from a multiplying constant. For simplicity, we put the quantity $pE/k_\mathrm{B}T = x$, and to carry out the integration we change the variable by putting $pE\cos\theta/k_\mathrm{B}T = u$. Since $\sin\theta\,\mathrm{d}\theta = -\mathrm{d}u/x$,

$$Z = -(2\pi/x)\int_{+x}^{-x} \mathrm{e}^u\,\mathrm{d}u = (4\pi\sinh x)/x$$

The resultant electric moment of all the dipoles is directed along $\boldsymbol{E}$ by symmetry and is equal to the mean value $\langle p\rangle$ of the components $p\cos\theta$ of all the dipoles along $\boldsymbol{E}$. Hence

$$\langle p\rangle = (2\pi/Z)\int_0^\pi [\exp(pE\cos\theta/k_\mathrm{B}T)]\,p\cos\theta\sin\theta\,\mathrm{d}\theta$$

Using the same substitutions as above gives

$$\langle p \rangle = -(2\pi/Zx^2)p \int_{+x}^{-x} ue^u \, du$$

Integration by parts gives $\int ue^u \, du = ue^u - \int e^u \, du$ and after some manipulation this gives

$$\langle p \rangle = (4\pi p/Zx^2)(x \cosh x - \sinh x)$$

Since $Z = (4\pi \sinh x)/x$, the final expression for $\langle p \rangle$ is

$$\langle p \rangle = p\mathcal{L}(x)$$

where $\mathcal{L}(x)$ is the *Langevin function* $\coth x - 1/x$.

In general, the electric energy $pE \ll k_B T$, so that $x \ll 1$. For small x, $\coth x = 1/x + x/3$, so that $\mathcal{L}(x) \approx x/3 = pE/3k_B T$ and finally

$$\langle p \rangle = p^2 E/3k_B T$$

15.3.3 Boundary between two dielectric media

The local equation (eqn 15.24) is modified in a medium of dielectric constant ε to

$$\text{div } \boldsymbol{E} = \rho/\varepsilon\varepsilon_0 \tag{15.34}$$

Many media are heterogeneous with varying values of ε. We shall take such heterogeneity into account by assuming a set of volumes set alongside each other, each with its own uniform value of ε. The boundaries between media are important features in biological systems. For example, a protein in an aqueous medium consists mainly of aliphatic and aromatic side chains of low dielectric constant (between 2 and 4) whereas the external medium has a dielectric constant close to 80. Similarly, the interior of a planar lipid bilayer has a much lower dielectric constant than that of the aqueous medium in contact with it.

Equation 15.34 applies in each of the media concerned. In passing from one medium (ε_1) to an adjacent one (ε_2), we cross a boundary that in general may have a free surface charge density of σ. If the fields normal to the surface on the two sides of the boundary are $E_{1,n}$ and $E_{2,n}$, the application of Gauss's theorem to an elementary volume defined by the surface element of area dS gives

$$(\varepsilon_2 E_{2,n} - \varepsilon_1 E_{1,n}) \, dS = \sigma \, dS/\varepsilon_0 \tag{15.35}$$

Expressing eqns 15.34 and 15.35 in terms of potential Ψ allows us to describe the system by two equations

$$\nabla^2 \Psi = -\rho/\varepsilon\varepsilon_0 \tag{15.36a}$$

$$\varepsilon_2 (d\Psi/dn)_2 - \varepsilon_1 (d\Psi/dn)_1 + \rho/\varepsilon_0 = 0 \tag{15.36b}$$

the first of these being a generalization of eqn 15.26, and the second being another way of writing eqn 15.35, where $d\Psi/dn$ is the normal component of the electric field.

Functions $\Psi(x, y, z)$ that are solutions of eqns 15.36 have several important properties. One of more particular interest to us is the following: if the potential is

Box 15.3 **Electric images for a plane surface**

Consider a plane surface between two media of dielectric constants ε_1 and ε_2, and a point charge q placed at a distance x from the plane in medium 1:

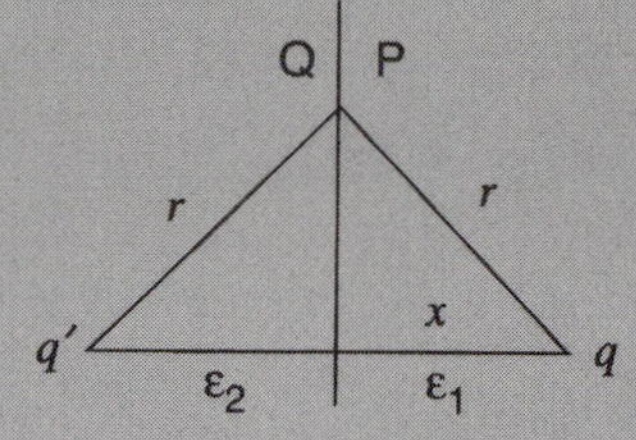

At any point P on the surface but just in medium 1, the potential may be considered as arising from q together with a fictitious charge q' at a point in medium 2 that would be the mirror image in the plane of the point occupied by q if the surface were a plane mirror. Then

$$\Psi_P = (q + q')/4\pi\varepsilon_1\varepsilon_0 r$$

At a point Q on the surface but just in medium 2, the potential is that of a fictitious charge q'' at the point occupied by q:

$$\Psi_Q = q''/4\pi\varepsilon_2\varepsilon_0 r$$

Since there are no free charges on the plane surface, there must be no discontinuity in Ψ, so that

$$(q + q')/\varepsilon_1 = q''/\varepsilon_2$$

The continuity of the electric displacement at the surface means that

$$\varepsilon_1(\mathrm{d}\Psi_P/\mathrm{d}x)_{x=0} = \varepsilon_2(\mathrm{d}\Psi_Q/\mathrm{d}x)_{x=0}$$

and hence

$$q - q' = q''$$

Combining the two relationships between the qs gives

$$q' = (\varepsilon_1 - \varepsilon_2)q/(\varepsilon_1 + \varepsilon_2) \qquad q'' = 2\varepsilon_2 q/(\varepsilon_1 + \varepsilon_2)$$

and if $\varepsilon_2 > \varepsilon_1$, then q' and q are of opposite signs.

specified over a closed surface enclosing all the charges, its value is uniquely defined at all external points. This is precisely the problem posed by the presence of charges inside a molecule like a protein. The specified conditions are known as *Dirichlet conditions* and the problem has led to many searches for mathematical functions (spherical harmonics, Bessel functions, Legendre functions, etc.) allowing boundary conditions to be written for special types of surface while at the same time satisfying Laplace's equation.

One particularly important result is that the potential Ψ can be written as the sum of the potential created by the charges within the surface and a *reaction potential* arising from the polarization of the external medium (Friedman, 1975). To a first approximation this reaction potential can be replaced by one created by fictitious point charges called *electric images*. Boxes 15.3 to 15.6 show how this concept can be used to solve some simple problems.

Box 15.4 **Dielectric sphere in a uniform field**

A sphere of radius a and dielectric constant ε_2 is immersed in a homogeneous medium of dielectric constant ε_1:

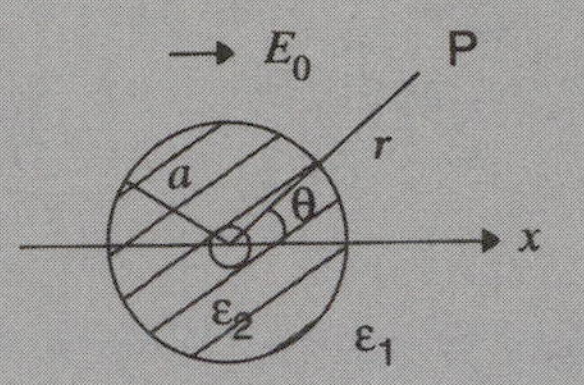

If E_0 is the uniform external field along the x direction, then $\Psi_0 = -E_0 x$ or $Er \cos\theta$.

Let Ψ_1 and Ψ_2 be the respective potentials outside and inside the sphere. Potential Ψ_1 can be considered as equal to the sum of Ψ_0 and the potential of a dipole corresponding to a distribution of positive and negative charges over the surface of the sphere. Since the dipole moment is proportional to the volume and its field varies as $1/r^3$, we can assume that Ψ_1 has the form

$$\Psi_1 = \Psi_0(1 - Aa^3/r^3)$$

which satisfies the condition that $\Psi_1 \to \Psi_0$ when $r \to \infty$.

The other potential has the form

$$\Psi_2 = \Psi_0(1 - A)$$

and the boundary condition $\Psi_1(a) = \Psi_2(a)$ is satisfied.

The continuity of the vector displacement $\boldsymbol{D}$ can be written

$$\varepsilon_1(\mathrm{d}\Psi_1/\mathrm{d}r)_{r=a} = \varepsilon_2(\mathrm{d}\Psi_2/\mathrm{d}r)_{r=a}$$

leading to

$$\varepsilon_1(1 + 2A) = \varepsilon_2(1 - A)$$

and hence

$$A = (\varepsilon_2 - \varepsilon_1)/(\varepsilon_2 + 2\varepsilon_1)$$

The potential Ψ_2 inside the sphere is therefore

$$\Psi_2 = \Psi_0 3\varepsilon_1(\varepsilon_2 + 2\varepsilon_1)$$

so that the uniform field inside the sphere is

$$E_2 = E_0 3\varepsilon_1(\varepsilon_2 + 2\varepsilon_1)$$

- If $\varepsilon_2 \gg \varepsilon_1$, then

$$E_2 \approx 3\varepsilon_1 E_0/\varepsilon_2$$

The internal field is much less than E_0 because of the depolarizing field produced by the charge distribution on the surface of the sphere. An extreme case is that when $\varepsilon_2 \to \infty$, for which $E_2 = 0$ and $\Psi_2 = 0$ (the sphere is an earthed conductor).

- If $\varepsilon_2 \ll \varepsilon_1$, then

$$E_2 \approx 3E_0/2$$

An example of this would be a protein in an electrolyte subjected to an electric field.

Box 15.5 **Dipole at the centre of a dielectric sphere**

We have seen that the α-helices of proteins behave as dipoles, and it is important to find out what role dipoles may play in interaction mechanisms. The following problem is relevant to this.

A homogeneous sphere of radius a and dielectric constant ε_2 is placed in a medium of dielectric constant ε_1. A dipole of moment $\boldsymbol{p}$ along the x axis is at the centre of the sphere. The potential Ψ_0 due to the dipole is given by eqn 15.13 with $\varepsilon_2\varepsilon_0$ instead of ε_0:

$$\Psi_0 = (p\cos\theta)/4\pi\varepsilon_2\varepsilon_0 r^2$$

The internal and external potentials can be shown to be

$$\Psi_1 = \Psi_0(1 - Aa^3/p)$$
$$\Psi_2 = \Psi_0(1 - Ar^3/p)$$

where A is a constant to be determined, and Ψ_1 is indeed equal to Ψ_2 when $r = a$.

The boundary condition that the normal component of $\boldsymbol{D}$ is continuous gives

$$\varepsilon_1(\mathrm{d}\Psi_1/dx)_{r=a} = \varepsilon_2(\mathrm{d}\Psi_2/\mathrm{d}x)_{r=a}$$

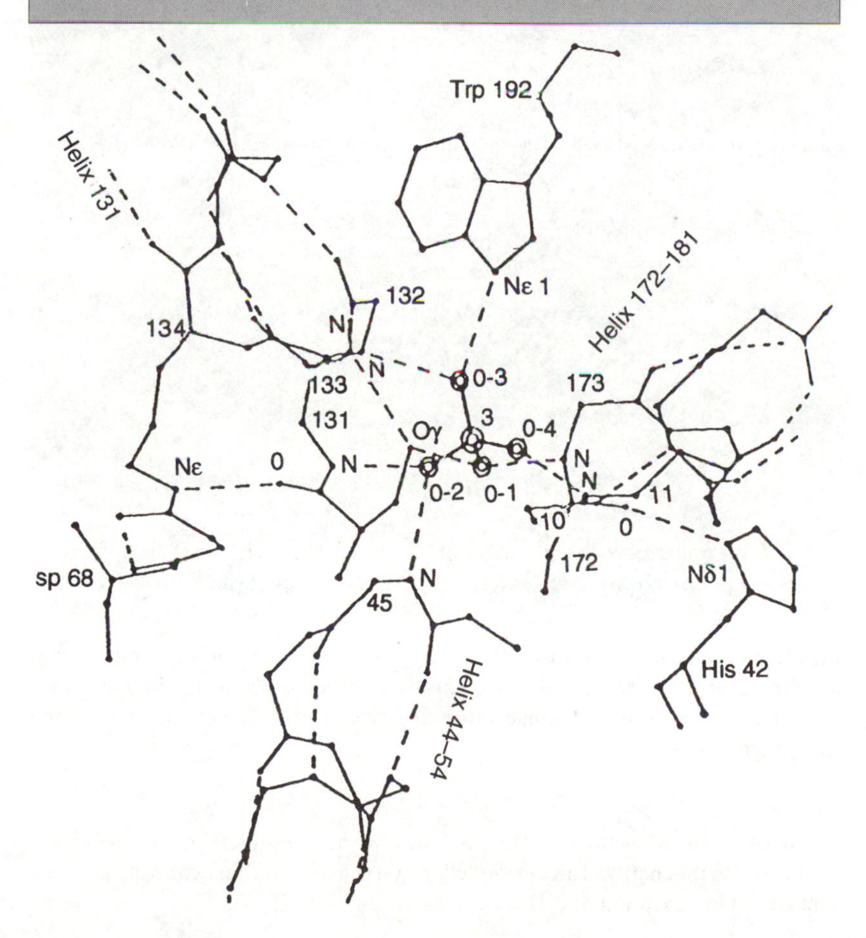

Figure 15.8 Three α-helices and seven hydrogen bonds keep the SO_4^{2-} ion in position in the protein to which it is bound. (From Quiocho *et al.*, 1987. Reprinted with permission from *Nature*, vol. 329, pp. 561–4 and with the authors' permission. © 1987 Macmillan Magazines Ltd.)

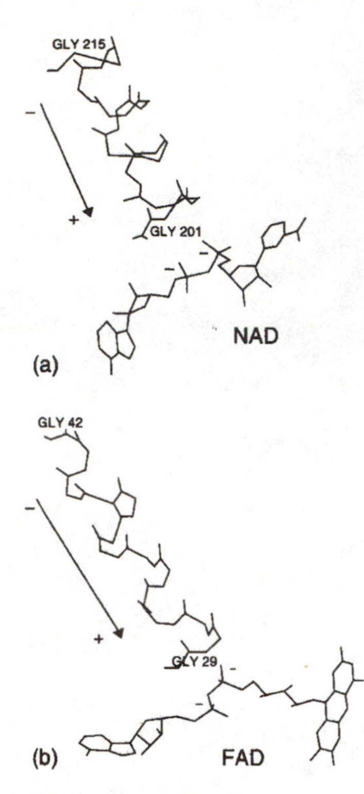

Figure 15.9 The role of α-helix dipoles in positioning negatively charged substrates: (a) nicotinamide adenine dinucleotide in alcohol dehydrogenase; (b) flavin adenine dinucleotide in glutathione reductase. (From Hol, 1985. Reprinted from *Progress in Biophysics and Molecular Biology*, vol. 45, pp. 149–95, © 1985, with kind permission from Elsevier Science Ltd., The Boulevard, Langford Lane, Kidlington, Oxford OX5 1GB, UK.)

Recalling that $d/dx(1/r^2) = -2x/r^4$ and $x = r\cos\theta$, we obtain

$$A = 2p(\varepsilon_1 - \varepsilon_2)/a^3(2\varepsilon_1 + \varepsilon_2)$$

whence

$$\Psi_1 = 3\Psi_0\varepsilon_2/(2\varepsilon_1 + \varepsilon_2)$$

Note that when $\varepsilon_2 \ll \varepsilon_1$ (e.g. protein in water), $A = p/a^3$ and $\Psi_1 = 0$. A dipole such as an α-helix inside the protein has a very limited influence outside, but may have a role in specific interactions within enzyme molecules (see Figs 15.8 and 15.9).

Box 15.6 **Cavity inside a dielectric**

This problem considers a simplified model of a protein in solution. Consider an arbitrary charge distribution of total charge Q inside a cavity of dielectric constant ε_1 itself located in an infinite medium of dielectric constant ε. Outside the cavity, Gauss's theorem applies to the flux of the electric field

$$\oint \varepsilon \boldsymbol{E} \cdot \boldsymbol{u}_n \, dS = Q/\varepsilon_0 \tag{1}$$

and in addition

$$\varepsilon_1 \boldsymbol{E}_1 \cdot \boldsymbol{u}_n = \varepsilon \boldsymbol{E} \cdot \boldsymbol{u}_n \tag{2}$$

The surface density of charge σ on the cavity surface satisfies the relationship

$$(\boldsymbol{E} - \boldsymbol{E}_1) \cdot \boldsymbol{u}_n = \sigma/\varepsilon_0 \tag{3}$$

Combining eqns 2 and 3:

$$\boldsymbol{E} \cdot \boldsymbol{u}_n(1 - \varepsilon/\varepsilon_1) = \sigma/\varepsilon_0 \tag{4}$$

and hence

$$\sigma = \varepsilon_0 \boldsymbol{E} \cdot \boldsymbol{u}_n(\varepsilon_1 - \varepsilon)/\varepsilon_1 \tag{5}$$

and substitution in eqn 1 gives

$$\oint \sigma \, dS = (\varepsilon_1 - \varepsilon)Q/\varepsilon\varepsilon_1 = (1/\varepsilon - 1/\varepsilon_1)Q \tag{6}$$

which determines the total polarization charge on the cavity surface. If $\varepsilon > \varepsilon_1$, the total polarization charge has the opposite sign to that of Q.

Finally, we consider the transfer of a charge carried by an ion of radius r from medium 1 (ε_1) to medium 2 (ε_2). This is a process occurring in ionization mechanisms and in ion exchange through a membrane. The expression for the transfer energy

$$\Delta W = q^2(1/\varepsilon_1 - 1/\varepsilon_2)/8\pi\varepsilon_0 r \tag{15.37}$$

was first established by Born. In Box 15.7, we give three completely equivalent ways of calculating the energy of an ion modelled by a sphere of radius r carrying a charge q immersed in a medium of dielectric constant ε.

Box 15.7 Energy of an ion in solution

1. We first calculate the work W required to bring the total charge to the surface of the ion from infinity (zero of potential) by infinitesimal elements dq. Since the potential $\Psi(r)$ at the surface of the ion is $q/4\pi\varepsilon\varepsilon_0 r$,

$$W = \int_0^q \Psi(r)\,\mathrm{d}q = q^2/8\pi\varepsilon\varepsilon_0 r \qquad (1)$$

and this is equal to the energy of the ion.

2. The spherical ion can be considered as a conductor of electric capacitance C and hence with a potential q/C. The work required to increase the charge by dq is $\mathrm{d}W = q\,\mathrm{d}q/C$ and the total work required to charge the sphere completely is

$$W = \int_0^q q\,\mathrm{d}q/C = q^2/2C$$

The capacitance of a conducting sphere of radius r in a medium of dielectric constant ε is $C = 4\pi\varepsilon\varepsilon_0 r$, which again gives eqn 1 for W.

3. The ion creates an electric field E at all points around it, which produces an energy density $w = \frac{1}{2}\varepsilon\varepsilon_0 E^2$ throughout the dielectric medium. The total energy stored in space can be calculated by considering a spherical shell centred on the sphere with radius R and thickness dR. The volume of the shell is $4\pi R^2\,\mathrm{d}R$, so that the stored energy it contains is the product of this and the energy density at $r = R$, using the electric field of $q/4\pi\varepsilon\varepsilon_0 R^2$. Integrating from r to ∞:

$$W = \int_r^\infty 4\pi R^2 w\,\mathrm{d}R = (q^2/8\pi\varepsilon\varepsilon_0)\int_r^\infty \mathrm{d}R/R^2 = q^2/8\pi\varepsilon\varepsilon_0 r$$

as before.

The work required to transfer an ion of charge q from a medium ε_1 to another ε_2 is thus

$$\Delta W = (q^2/8\pi\varepsilon_0 r)(1/\varepsilon_1 - 1/\varepsilon_2)$$

15.3.4 Poisson's equation and its consequences

Poisson's equation *in vacuo* (eqn 15.26) was obtained by combining the local flux theorem (eqn 15.23) with the relationship between field and potential (eqns 15.9 and 15.10). In a dielectric, Gauss's theorem applies to the vector displacement $\boldsymbol{D}$, and the local form of this is

$$\mathrm{div}\,\boldsymbol{D} = \rho$$

or

$$\mathrm{div}(\varepsilon\boldsymbol{E}) = \rho/\varepsilon_0$$

which generalizes eqn 15.24 to any material medium. If we put $\boldsymbol{E} = -\text{grad}\,\Psi$ or $\boldsymbol{E} = -\nabla\Psi$, we obtain

$$\nabla \cdot (\varepsilon \nabla \Psi) + \rho/\varepsilon_0 = 0 \tag{15.38}$$

a generalized form of Poisson's equation in any dielectric medium where ε may be a function of position in the medium.

As we mentioned at the beginning of this chapter, it is the combination of eqn 15.38 and the laws of statistical mechanics that enables us to deal with the problem of polyelectrolytes in solution. There are two other aspects to be considered alongside the general approach.

1. Measurable quantities (field and potential) can only be space and time averages. In particular, the *electrical neutrality* of the medium must be maintained: in other words, an elementary volume must contain on average an equal number of positive and negative charges.
2. There are fluctuations around these average quantities that become greater as the volume under consideration becomes smaller. At a given time, therefore, a small volume could have a non-zero net charge even though its time average will always be zero. The importance of these fluctuations, in field and potential as well as in charge, will be considered later. The fluctuations evolve rapidly with time and this dynamic aspect of the problem should not be forgotten.

If an attempt is made to describe phenomena and carry out calculations solely on a microscopic instead of a macroscopic scale, then the molecular mechanics demands that we consider an assembly of charges and dipoles moving in a given volume. We know that the laws of electrostatics are valid down to distances much smaller than interatomic distances, so that in this case we can use the same type of combination as before: between these electrostatic laws and, this time, the laws of *molecular dynamics*.

Before tackling the question of polyelectrolytes, we first consider the apparently simpler case of solutions of electrolytes in which almost all biological events take place. This will familiarize the reader with the arguments used later and will introduce some parameters that will be useful in studying polyelectrolytes.

16

Electrolytic solutions and the Debye–Hückel theory

An ion in solution can no longer be treated as an isolated charge. Since there must on average be electrical neutrality at any point in the solution, an ion will tend to be surrounded preferentially by ions of opposite sign. Underlying the general appearance of a uniform charge distribution and a mean zero charge in a large enough volume, there are appreciable local fluctuations of charge and potential.

To study this in more detail, we take an arbitrary ion in the solution and, using its centre as origin, we find an analytical expression for the electric potential Ψ around it produced by all the neighbouring ions. Other observable macroscopic quantities can be derived from Ψ.

16.1 Differential equation for the potential: assumptions and approximations

The laws of electrostatics are applied by using Poisson's equation for the relationship between the charge density ρ and the potential Ψ:

$$\nabla^2\Psi = -\rho/\varepsilon\varepsilon_0 \tag{16.1}$$

where $\nabla^2\Psi$ is the Laplacian of Ψ, ε_0 is the permittivity of a vacuum and ε is the dielectric constant of water, assumed to be a continuous homogeneous medium. The charge distribution is assumed to be fixed in the electrostatic sense over a short time interval.

The number of ions of type i per unit volume, n_i°, depends on the concentration of the electrolyte. If $z_i e$ is the ionic charge then electrical neutrality demands that

$$\sum z_i n_i^\circ = 0 \tag{16.2}$$

The Debye–Hückel theory is based on the following two assumptions and two approximations.

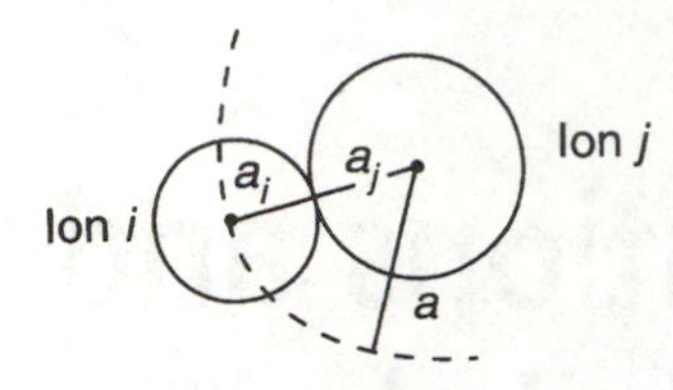

Figure 16.1 The distance of closest approach between ions i and j.

Assumption 1

Around any ion of type j, the charge distribution is assumed to be *spherically symmetrical*. In particular, $\rho_j(r)$ obeys

$$\int_a^{\infty} 4\pi r^2 \rho_j(r)\,\mathrm{d}r = -z_j e \tag{16.3}$$

which in fact expresses the electrical neutrality: the total charge of all the ions around the jth is equal and opposite to the charge on j. The quantity a is the distance of closest approach to the ion, equal to the sum of the radii a_i and a_j of the ith and jth ions (Fig. 16.1). The sphere of radius a around ion j is the volume from which the centres of gravity of other ions are excluded.

This approach corresponds to the assumption in the Debye–Hückel model that each spherical ion may be replaced by a point charge at its centre. This assumption, although valid from an electrostatic viewpoint, neglects ionic volumes and is only justified when the mean inter-ionic distance is large compared with their size (dilute solutions). The term *ionic radius* merits further examination (see Box 16.1).

Box 16.1 **Ionic radius**

The study of ions in solution raises the problem of what is meant by their size. It would be possible to adopt the radius r_c as given by crystallographic examination of crystals containing the ions. Table 16.1 gives values of r_c for the commonest ions.

It has been suggested that, to take hydration into account, the so-called Stokes radius r_S could be used, as defined by the coefficient of friction $6\pi\eta_0 r_S$ for a spherical particle moving in a fluid of viscosity η_0. Unfortunately, a correction factor has to be introduced because Stokes' law is no longer valid when the size of water molecules becomes comparable to that of the moving particles. Examples of the values of r_S are:

Ion	Li^+	Na^+	Mg^{2+}	Ca^{2+}	Zn^{2+}
r_S (Å)	3.7	3.3	4.4	4.2	4.4

The difference between r_c and r_S has often been interpreted as being due to the hydration shell around the solvated ion accompanying its motion. Rashin and Honig (1985) proposed replacing the ionic radius with that of a cavity defined as the sphere containing a negligible proportion of the solvent electron density (Fig. 16.2). This model and its consequences have already been analysed in detail in section 12.2.2 of Part III.

Table 16.1 Crystalline radii of ions

Ion	Li^+	Na^+	K^+	NH_4^+	Cs^+	
r_c (Å)	0.6	0.95	1.33	1.48	1.69	
Ion	Mg^{2+}	Ca^{2+}	Zn^{2+}	Mn^{2+}	Fe^{2+}	Co^{2+}
r_c (Å)	0.65	0.99	0.74	0.80	0.96	0.74
Ion	F^-	Cl^-	Br^-	I^-		
r_c (Å)	1.36	1.81	1.95	2.16		

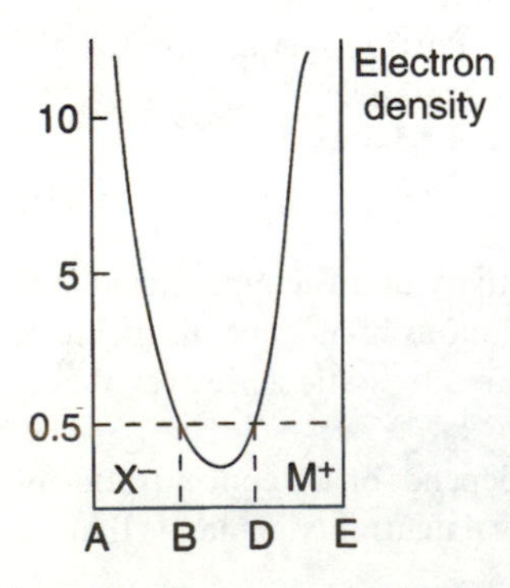

Figure 16.2 Electron density distribution in an alkali halide.

Assumption 2

The variation in charge density $\rho_j(r)$ is assumed to follow the Boltzmann distribution law. This states that the number n_i of ions around j is given by

$$n_i = n_i^\circ \exp[-w_{ij}(r)/k_B T] \tag{16.4}$$

and hence

$$\rho_j(r) = \sum_i e z_i n_i = \sum_i n_i^\circ z_i \exp[-w_{ij}(r)/k_B T] \tag{16.5}$$

where $w_{ij}(r)$ is the potential energy corresponding to the mean force exerted between the ions i and j.

The number n_i° is effectively the concentration of the salt solution, while the difference $n-n_i^\circ$ is the average local excess or deficiency of an ion (cation or anion) at a point where the potential energy is w_{ij}. The Boltzmann factor is the statistical weight multiplying the ionic concentration to account for mean local electric interactions between ions.

Approximation 1

The factor w_{ij} is replaced by the potential energy of an ion i located at a point where there is a potential Ψ_j due to ion j. This is a mean energy because it corresponds to the force acting between i and j after averaging over the $N-2$ other ions in all possible positions. Approximately, therefore,

$$w_{ij}(r) \approx z_i e \Psi_j(r) \tag{16.6}$$

Combining eqns 16.5 and 16.6, and using the expression for the Laplacian in spherical coordinates to satisfy the symmetry of Assumption 1, we obtain the so-called Poisson–Boltzmann equation:

$$r^{-2}\,\mathrm{d}/\mathrm{d}r(r^2\,\mathrm{d}\Psi/\mathrm{d}r) = (\bar{\varepsilon}\varepsilon_0)^{-1} \sum_i e z_i n_i^\circ \exp[-z_i e \Psi(r)/k_B T] \tag{16.7}$$

Approximation 2

Equation 16.7 does not have an analytical solution, so the last factor of the form $\exp(-y)$ is replaced by $1-y$, which is equivalent to assuming that $y \ll 1$ or that

$$z_i e \Psi / k_B T \ll 1 \tag{16.8}$$

This approximation is not made, as is sometimes said, to satisfy the superposition theorem of electrostatics, which depends on the linear form of the equations and not on a linear relationship between ρ and Ψ. There is, however, a contradiction when the non-linearized form of the Poisson–Boltzmann equation is used. Approximation 1 amounts to neglecting all interactions between groups of three, four, five, etc., ions and only considering those between pairs. This is only valid at very dilute concentrations when the number of clusters of more than two ions is negligible. The linearization of the Debye–Hückel theory, assuming that $z_i e \Psi \ll k_B T$, is equivalent to assuming that there is sufficient dilution to increase the mean distance between ions sufficiently to produce a low potential. Approximation 2 is then consistent with Approximation 1 and makes the Debye–Hückel theory a coherent one.

Equation 16.5 then becomes

$$\rho_j = \sum n_i^\circ e z_i - \sum n_i^\circ z_i e z_i \Psi e / k_B T \tag{16.9}$$

The first term on the right is zero (eqn 16.2), and substituting the rest in eqn 16.7 gives

$$r^{-2}\, \mathrm{d}/\mathrm{d}r(r^2\, \mathrm{d}\Psi/\mathrm{d}r) = \left(e^2 \sum n_i^\circ z_i^2 / \varepsilon\varepsilon_0 k_B T\right)\Psi = K^2\Psi \tag{16.10}$$

where

$$K^2 = e^2 \sum n_i^\circ z_i^2 / \varepsilon\varepsilon_0 k_B T \tag{16.11}$$

In this expression, the factor $e^2/\varepsilon\varepsilon_0 k_B T$ has the dimensions of a length, while n_i° is the reciprocal of a volume. Therefore $1/K$ has the dimensions of a length and is known as the *Debye length*. Its physical significance will be discussed later.

Moreover, apart from a numerical factor, $\sum n_i^\circ z_i^2$ is the ionic strength I of the medium, so that K varies as $I^{1/2}$. Note that $I = \frac{1}{2}\sum c_i z_i^2$ where c_i is the molar concentration (in moles of the free ionic species i), which, in some cases, has to be calculated by taking into account the incomplete dissociation of the salt in experimental conditions. In SI units, where the unit of volume is the m^3 or 10^3 litres, we have $n_i^\circ = 6.023 \times 10^{26} c_i$.

16.2 Solutions and consequences

The distance of closest approach $a = a_j + a_i$ introduces an 'internal' region $(a_j < r < a_j + a_i)$ where there is no charge and hence where

$$r^{-2}\, \mathrm{d}/\mathrm{d}r(r^2\, \mathrm{d}\Psi/\mathrm{d}r) = 0$$

which gives

$$\Psi_{\mathrm{int}} = -C/r + D \tag{16.12}$$

C and D being two constants of integration determined below. For $r > a$, the solution $\Psi(r)$ of eqn 16.10 is a modified Coulomb potential, which can be written as

$$\Psi(r) = u(r)/r \tag{16.13}$$

Substituting eqn 16.13 into eqn 16.10 gives

$$\mathrm{d}^2u/\mathrm{d}r^2 - K^2 u = 0$$

the standard solution of which is

$$u = A\exp(-Kr) + B\exp(Kr)$$

Since the potential must vanish as $r \to \infty$, only the first term on the right is retained, so that

$$\Psi(r) = Ar^{-1}\exp(-Kr) \tag{16.14}$$

The constant A is determined by using Gauss's theorem, which means that the flux of the electric field $E = -\mathrm{d}\Psi/\mathrm{d}r$ across the sphere of radius a is equal to the ionic charge divided by $\varepsilon\varepsilon_0$:

$$4\pi a^2 E(a) = z_j e/\varepsilon\varepsilon_0$$

Replacing $E(a)$ by its value obtained by differentiating eqn 16.14:

$$A = z_j e \exp(Ka)/4\pi\varepsilon\varepsilon_0(1 + Ka)$$

and

$$\Psi_{\text{ext}}(r) = z_j e \exp(Ka)\exp(-Kr)/4\pi\varepsilon\varepsilon_0(1 + Ka)r \qquad (16.15)$$

When $r = a$ there is no discontinuity in Ψ and $\mathrm{d}\Psi/\mathrm{d}r$ so that

$$\Psi_{\text{int}} = \Psi_{\text{ext}} \qquad \text{and} \qquad (\mathrm{d}\Psi/\mathrm{d}r)_{\text{int}} = (\mathrm{d}\Psi/\mathrm{d}r)_{\text{ext}}$$

These two conditions determine the values of C and D:

$$C = -z_j e/4\pi\varepsilon\varepsilon_0 \qquad \text{and} \qquad D = -Kz_j e/4\pi\varepsilon\varepsilon_0(1 + Ka)$$

so that finally

$$\Psi(r)_{\text{int}} = (z_j e/4\pi\varepsilon\varepsilon_0)[1/r - K/(1 + Ka)] \qquad (16.16)$$

16.2.1 Electrostatic shielding

We now examine the physical meaning of the two expressions for the potential. In eqn 16.16, the first term is the potential that the ion z_j alone would create (the Coulomb potential). The second term of opposite sign is the potential created by the ionic atmosphere around the ion z_j. In eqn 16.15, the exponential factor $\exp(-Kr)$ modifies the rate of change of Ψ with r compared to the $1/r$ Coulomb potential. The function $\Psi(r)$ decreases more rapidly with increasing r, so that its difference with $1/r$ will be greater when K, the ionic strength, is higher. Ions in solution form a kind of *electrostatic screen.*

Another way of looking at the expression for Ψ_{ext} in eqn 16.15 is to introduce an effective dielectric constant ε_{eff} depending on r and given by $\varepsilon_{\text{eff}} = \varepsilon\exp(Kr)$. This relationship or similar empirical expressions for $\varepsilon(r)$ have been used to calculate interactions between discontinuous charge distributions. However, this purely formal definition can often lead to inconsistencies when ε_{eff} is used in both microscopic and macroscopic descriptions of the solvent.

The problem emerges more clearly if we examine in more detail the calculation of the dielectric constant near an ion.

16.2.2 The dielectric constant of water near an ion

The calculation of the polarization of water molecules in the hydration shell and in the strong electric field that occurs near an ion involves the Langevin function (see Box 15.2). If we wish to define an apparent dielectric constant ε_{app} as a function of the distance r from the ion, we have to calculate the derivative $\mathrm{d}D/\mathrm{d}E$ at each point. Debye's calculation leads to a non-analytical expression for ε_{app} involving derivatives of the Langevin function, so that the dielectric constant can only be calculated numerically. Moreover, ε_{app} is no longer a scalar quantity since radial and transverse components exist. Numerical calculations of both components reveal a sigmoidal type of variation with r. As a result, empirical functions of this type have

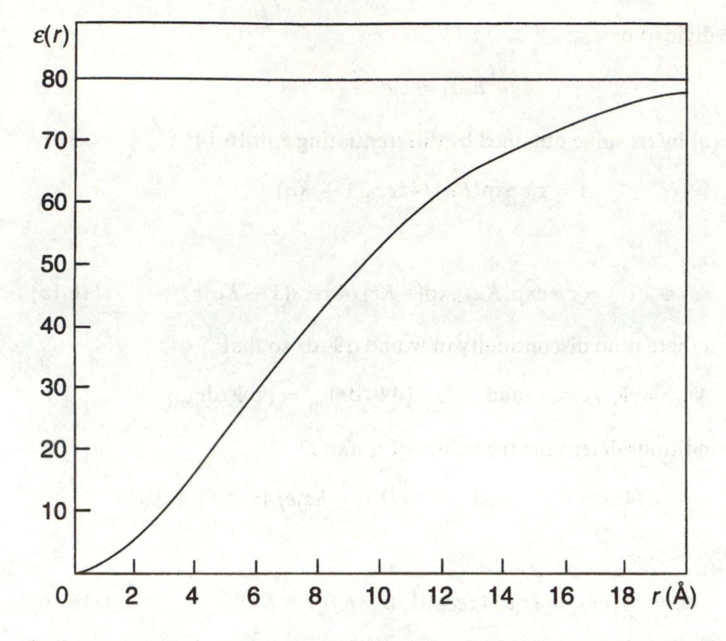

Figure 16.3 Variation of the dielectric constant of water with distance from a monovalent ion. A linear approximation can be used over a considerable region of the curve. (From Mehler and Eichele, 1984. Reprinted with permission from *Biochemistry*. © 1984 American Chemical Society.)

been proposed, an example of which is that of Mehler and Eichele (1984):

$$\varepsilon(r) = A + B/[1 + k\exp(-\lambda Br)]$$

When $r \to \infty$, $\varepsilon(r) = A + B = \varepsilon_{\text{water}}$. The function therefore depends on only the three parameters A, k and λ, with $B = 78.5 - A$ at 25°C. Taking $A = -8.55$, $k = 7.78$ and $\lambda = 3.63 \times 10^{-3}$ gives a sigmoidal curve to represent $\varepsilon(r)$ (see Fig. 16.3). This type of function enables observed pK changes in titratable groups of proteins to be predicted. Other simpler expressions have been proposed, in particular linear functions like $\varepsilon = ar$, which corresponds to the linear region of the curve in Fig. 16.3.

In all cases, these variations of ε with r near an ion raise the question of the validity of the electrostatic laws near a point charge in a highly polarizable medium. This also leads to doubts about the validity of the results obtained using the Debye–Hückel model. It should be pointed out that in a 0.1 M 1 : 1 salt there are 12.04×10^{25} ions per m^3, i.e. one ion on average in a volume of 0.83×10^4 Å^3. That is equivalent to an ion at the centre of a cube with an edge of 20.2 Å or of a sphere of radius 14 Å. In both cases, the use of the curve for $\varepsilon(r)$ shows that the limiting value ε_0 is almost attained.

The quantity $y = e\Psi/k_{\text{B}}T$ calculated at this distance and under these conditions has a value lying between 0.05 and 0.1, thus justifying *a posteriori* the approximation $\exp(-y) \approx 1 - y$. This once again shows the self-consistency of the Debye–Hückel theory valid for dilute electrolyte solutions.

It is worth pointing out that under physiological conditions (0.15 M), the approximation is still acceptable. However, this is no longer true for the so-called *halophilic* organisms living in extremely saline conditions (3–4 M KCl for example). Study of their function requires a new approach to the theory of electrolytes.

16.2.3 Ionic distribution

According to eqn 16.9, the charge density $\rho(r)$ is proportional to $\Psi(r)$, and thus to $r^{-1}\exp[-K(r-a)]$ (eqn 16.14). The charge dq in an elementary spherical shell of

Table 16.2 Values of $1/K$ for various ionic strengths

I (mol l^{-1})	10^{-4}	5×10^{-4}	10^{-3}	5×10^{-3}
$1/K$ (Å)	305	136	96.5	43
I (mol l^{-1})	10^{-2}	5×10^{-2}	10^{-1}	0.2
$1/K$ (Å)	30.5	13.6	9.6	6.8

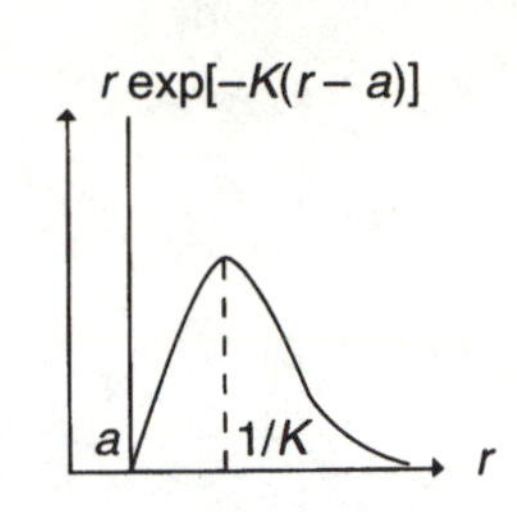

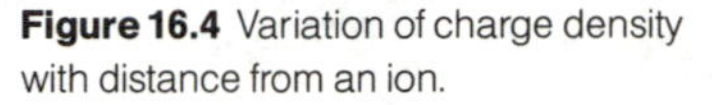
Figure 16.4 Variation of charge density with distance from an ion.

volume $4\pi r^2\,dr$ varies as $r\exp[-K(r-a)]$. This function, plotted against r in Fig. 16.4, passes through a maximum when $r = 1/K$. As a result, $1/K$ is often called the 'thickness' of the ion cloud around each ion. For a 1:1 electrolyte at 20°C, the values of $1/K$ (with $\varepsilon = 80$) for various ionic strengths are given in Table 16.2.

To build up the charge distribution formed by ion j in the presence of other ions, the charge $z_j e$ must be brought up from infinity. Suppose at a given time that the ion has the charge λz_j where $0 < \lambda < 1$, and then bring an element of charge $dq = z_j e\,d\lambda$ from infinity to $r = a_j$. The work done in the whole charging process is the sum of the elements of work $\Psi_{\text{int}}\,dq$, i.e.

$$W = z_j^2 e^2 \int_0^1 \lambda\,d\lambda/4\pi\varepsilon\varepsilon_0 a - z_j^2 e^2 \int_0^1 \lambda\,d\lambda/4\pi\varepsilon\varepsilon_0(1 + Ka)$$
$$= z_j^2 e^2/8\pi\varepsilon\varepsilon_0 a - z_j^2 e^2 K/8\pi\varepsilon\varepsilon_0(1 + Ka) = W_1 + W_2 \qquad (16.17)$$

The second term W_2 is the work required to 'create' an ion j in the presence of all the other ions in the system. In a way, it represents the ionic interaction term and can be related to the *activity coefficient* γ_j of ion j through

$$W_2 = k_B T \ln \gamma_j$$

and hence

$$\ln \gamma_j = -z_j^2 e^2 K/8\pi\varepsilon\varepsilon_0 k_B T(1 + Ka) \qquad (16.18)$$

16.3 Limits to the validity of the Debye–Hückel theory

All the calculations carried out above describe in principle the behaviour of very dilute solutions of electrolytes. Two questions then suggest themselves:

1. What is the upper limit of concentration for the validity of the Debye–Hückel theory?
2. Is there a theory applicable to high concentrations?

A great deal of work has been done in attempts to answer these questions, but it is the advent of computers and the use of the Monte Carlo method that have enabled precise tests of the validity of theories to be carried through. The Monte Carlo method studies the motion of an equal number of cations and anions in a limited cubic volume. The movement of an ion is determined by random selection of possibilities (hence the name of the method) and the modified distribution is examined after a large number of selections. In particular, it is possible to determine the radial distribution function $g_{ij}(r)$ and compare it with its theoretical value.

In the Debye–Hückel approximation ($z_i e \Psi_j / k_B T \ll 1$), we have

$$g_{ij}(r) \approx 1 - z_i e \Psi_j / k_B T \tag{16.19}$$

and replacing $\Psi_j(r)$ using eqn 16.15 gives us

$$g_{ij}(r) = 1 - X_{ij}$$

with

$$X_{ij} = z_i z_j e^2 \exp(Ka)\exp(-Kr)/4\pi\varepsilon\varepsilon_0(1 + Ka)r \tag{16.20}$$

a dimensionless quantity that remains unchanged if i and j are interchanged. The Debye–Hückel approximation in fact amounts to keeping only the first term in the expansion of $\exp(-z_i e \Psi_j / k_B T)$. A better approximation is obtained by retaining the second term as well, when the radial distribution function then becomes, from eqn 16.20:

$$g_{ij}(r) = 1 - X_{ij} + X_{ij}^2/2 \tag{16.21}$$

A priori, it seems contradictory to keep only one term in the expansion to solve the differential equation, then obtain $\Psi(r)$ and finally introduce this value of $\Psi(r)$ into a two-term expansion to get $g_{ij}(r)$. The contradiction disappears for 1 : 1 electrolytes since the second term in the expansion, by eqn 16.5, vanishes in the calculation of ρ_j.

Figure 16.5 shows the various approximations to g_{ij} given by eqns 16.19 and 16.21, as well as the complete form $g_{ij}\exp(-X_{ij})$ where X_{ij} is the parameter defined in eqn 16.20. It can be seen that there is excellent agreement between the complete form and points calculated using the Monte Carlo method. The discrepancy is greater using eqn 16.19 than it is with eqn 16.21, but beyond 6 Å even the Debye–Hückel approximation correctly describes the true situation when $a = 4$ Å. The activity coefficient γ given by eqn 16.18 is valid only for very low molar concentrations. The correction to be made so that it corresponds to the Monte Carlo calculations and so that it is valid between 0 and 1 M is more empirical than in the case of g_{ij}. It leads to a complicated formula involving the molar concentration and the ionic strength.

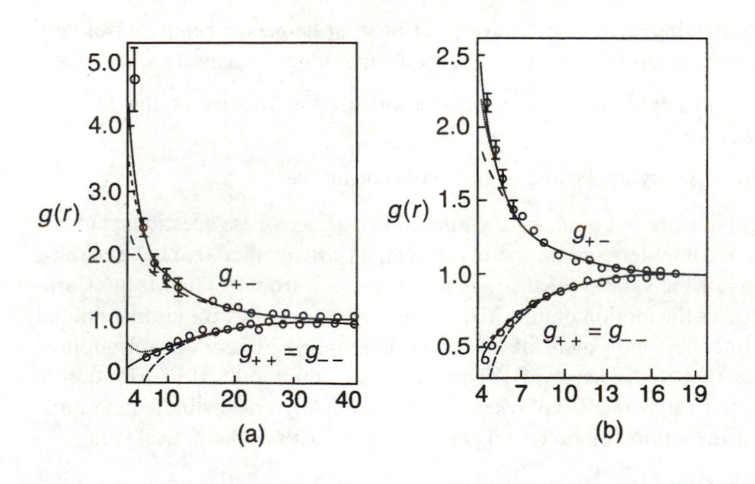

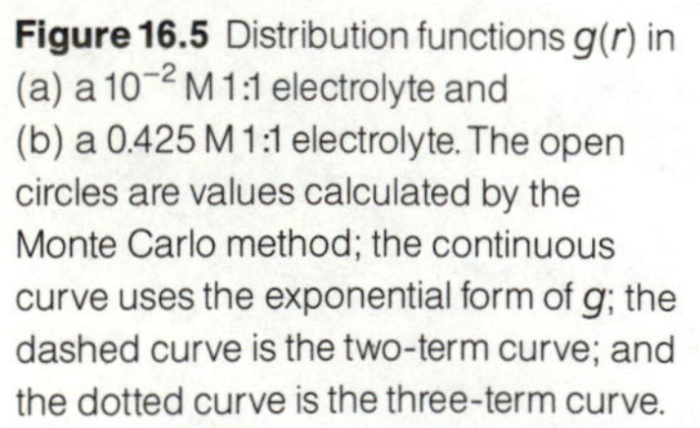
Figure 16.5 Distribution functions $g(r)$ in (a) a 10^{-2} M 1:1 electrolyte and (b) a 0.425 M 1:1 electrolyte. The open circles are values calculated by the Monte Carlo method; the continuous curve uses the exponential form of g; the dashed curve is the two-term curve; and the dotted curve is the three-term curve.

One of the best methods (Olivares and McQuarrie, 1975) takes as the radial distribution function

$$g_{ij}(r) = g_{ij}^{\mathrm{HS}}(r)\exp[-\zeta_{ij}(r)]$$

where $g_{ij}^{\mathrm{HS}}(r)$ is the radial distribution function in a *hard-sphere* model (see section 2.5.2) and ζ_{ij} is a complex potential, normalized to satisfy the condition of electrical neutrality. For low concentrations, it takes the form of a Debye–Hückel potential

$$\zeta_{ij}(r) = (z_i z_j e^2/4\pi\varepsilon\varepsilon_0 k_{\mathrm{B}} Tr)\exp(-Kr)$$

Under these conditions, there is excellent agreement between the values of the activity coefficient γ determined by the Monte Carlo method and that calculated from the above expressions.

16.4 **Ion pairs**

At the time of Arrhenius's studies on the dissociation of ions in solution, Bjerrum took a different view: that the distances between ions in a concentrated solution are small enough on average for associations to be created between ions of opposite charge. In principle, an *ion pair* of this kind can be produced when the distance between two elementary charges $+e$ and $-e$ is such that the electrostatic energy of attraction is equal to $k_{\mathrm{B}}T$. This gives rise to the definition of the *Bjerrum length* l_{B} by

$$l_{\mathrm{B}} = e^2/4\pi\varepsilon\varepsilon_0 k_{\mathrm{B}} T \tag{16.22}$$

so that at 20°C and with $\varepsilon = 80$, we find that $l_{\mathrm{B}} = 7.12$ Å.

Ion pairs in normal electrolytic solutions have a very short lifetime and very rarely occur. In polyelectrolytes, on the other hand, they are entropically much more likely to appear, and here we can observe a transient association between the charges carried by the macromolecule and the counter-ions. The Bjerrum length is another quantity, along with the Debye length $l_{\mathrm{D}} = 1/K$, that defines a scaling factor for polyelectrolytes.

17

Polyelectrolyte solutions

The term *polyelectrolyte* describes polymers carrying charges of the same sign. A good example is provided by nucleic acids, in which each phosphate group with a pK of about 1 is negatively charged in physiological conditions. Other examples are the acid polysaccharides carrying COO^- or SO_3^{2-} groups and basic oligomers like the polyamines (e.g. spermine and spermidine).

There are also biopolymers carrying both positive and negative charges, known as *polyampholytes*, proteins being the largest and most illustrative of these. In some cases there is an excess of positive charges, as in histones and protamines where the presence of a large number of lysines and arginines give a pH_i (isoelectric point) higher than 9. Proteins, on the other hand, have a net negative charge and may have a pH_i lower than 7 and even as low as 4.

17.1 General properties of polyelectrolytes

17.1.1 Charge accumulation

In polyelectrolytes, unlike simple electrolytes, charges are bound to a polymer chain, thus limiting their degrees of freedom and producing accumulations of charge at certain points in the solution. This leads us to expect strong fields and high potential in the vicinity of a polyelectrolyte.

In a polyampholyte, a local accumulation of positive charges may occur in one region of the biopolymer while one of negative charges may occur in another region. Such anisotropic charge distributions are often found in proteins and produce intramolecular domains with distinct properties.

17.1.2 Fluctuations

A given volume of a solution containing ions and polyelectrolytes must contain an equal number of positive and negative charges to satisfy electrical neutrality, as in ordinary electrolytes. Fluctuations still occur but in polyampholytes they also exist in the macromolecular domain itself. In other words, a collection of charges in a protein, say, will at a given time have a net charge that may differ by one or two units from that corresponding to equilibrium at the pH of the solution. These internal charge fluctuations could play a major role in association processes.

17.1.3 Counter-ions

In order to maintain electrical neutrality, the volume around the polyelectrolyte must contain ions of opposite sign to that of the polyelectrolyte. These compensating ions are known as *counter-ions* and will be cations such as Na^+, K^+ and Mg^{2+} in the case of nucleic acids or acid polysaccharides, and anions in the case of polyamines. Conversely, ions of the same sign or *co-ions* are excluded from the immediate neighbourhood. In other words, the ionic concentration near the polyelectrolyte may be very different from that in the bulk solution.

The attraction between counter-ions and the polymer is particularly significant when the former are H^+ ions. The charge distribution on a polyelectrolyte and more especially on a polyampholyte produces local pH values very different from the average pH of the solution. Reagent orientation and recognition specificity can be explained by such intramolecular micro-environments.

17.1.4 Polyelectrolyte conformation

The repulsive and attractive forces between charges will play a role in the conformation adopted by the polyelectrolyte. Two simplifying assumptions, the energy independence of bonds and the symmetry of the potential, are no longer valid. Coulomb $1/r^2$ forces are long-range directed forces and it is no longer possible to treat each bond as independent of the others, while there is no reason why they should have rotation symmetry about the valence bond. Charge repulsion will favour extended conformations in which the ratio r^2/na^2 is increased, all the more so for lower ionic strengths. The length L of the polyelectrolyte chain can therefore be altered either by changing the number of ions in solution or by changing the values of the charges and/or their distribution in the chain. Conversely, any mechanical strain (stretching, bending, twisting) produced in a charged chain could alter its charge density and a term representing the change in free electrostatic energy will have to be included in the mechanical equations for the deformation. For example, in transitions between different DNA structures (double to single strand, B to Z, etc.), there is a change in the charge distribution. The number of counter-ions and the electric potential are therefore modified, and this explains why transition parameters depend on ionic conditions.

Similarly, if there is repulsion between the phosphates, the polymer is very rigid and the basic proteins will help the folding and packing of the long thread of DNA by reducing its local rigidity. In tRNA, the charge density and high potential near the molecule create regions of strong interaction with divalent ions like Mg^{2+}. In proteins, the repulsion or attraction between charges on the surface and through the interface between two dielectric media plays a role in the folding of globular proteins.

17.1.5 Transport phenomena

During migration, polyelectrolytes and polyampholytes are always surrounded by their ionic atmosphere. Relaxation and dragging effects therefore occur because of the different mobilities of the various species. The energy loss due to the viscosity of the medium will be supplemented by a term due to the displacement of charges in the electric potential created by other charges.

We now move from such generalizations to a more quantitative treatment of polyelectrolytes in solution.

17.2 Activity of a polyelectrolyte solution

The effect of the strong interactions between polyelectrolytes and ions is to reduce the activities of the polyelectrolyte and the counter-ions and to increase that of the co-ions. As we shall see, even though we remain at an elementary level, several experimentally measurable quantities are related directly to these changes in activity.

17.2.1 Osmotic coefficient

Consider a salt solution with n_s cations and anions per unit volume, in which n_p molecules of a polyion with ν negative charges are dissolved (along with a corresponding number of counter-ions). Cations come from both the salt and the counter-ions, while anions come only from the salt. Hence:

$$n_+ = n_s + n_p\nu \qquad \text{and} \qquad n_- = n_s \tag{17.1}$$

The osmotic pressure Π of this solution, assumed dilute, is

$$\Pi = k_B T\phi(n_+ + n_-) = k_B T\phi(n_p\nu + 2n_s) \tag{17.2}$$

where ϕ by definition is the osmotic coefficient, i.e. the ratio of Π to a value Π_0 corresponding to n_+ and n_- free ions in solution:

$$\Pi_0 = (2n_s + n_p\nu)k_B T \tag{17.3}$$

Since a fraction of the $n_p\nu$ counter-ions remain bound to the polyelectrolyte, we shall have $\phi < 1$.

The osmotic pressure is the negative rate of change of the free energy with the volume V of the solution, and we can thus consider the difference $\Pi - \Pi_0$ as coming from the additional electric free energy G_e, so that

$$k_B T(\phi - 1)(2n_s + n_p\nu) = -(\partial G_e/\partial V)_{T,Np,Ns} \tag{17.4}$$

where $N_p = n_p V$ and $N_s = n_s V$.

As $n_s \to 0$, i.e. as the solution becomes salt-free, ϕ tends to a limiting value ϕ_0 known as the *free counter-ion fraction*.

17.2.2 Donnan equilibrium

Consider two compartments A and B separated by a membrane through which only small molecules like water and ions can pass. The same salt solution is placed in both compartments but a polyelectrolyte is added to B:

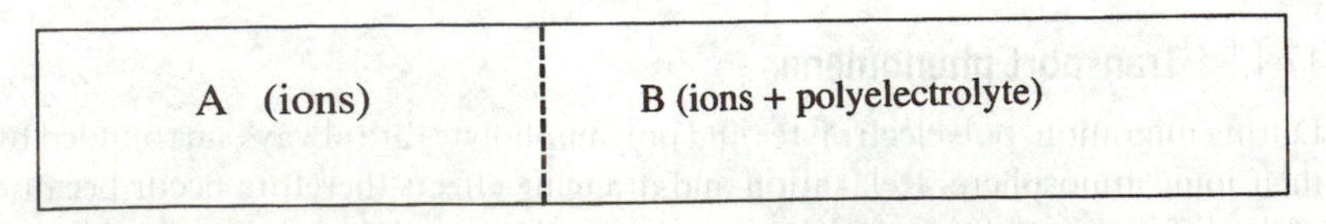

The equilibrium between cations and anions is disturbed by this and a new distribution of concentrations is required.

The chemical potentials of the salt(s) are the same on the two sides of the membrane since there is equilibrium across it:

$$\mu_s^A = \mu_s^B$$

Note that this equation is not valid for each ion taken separately since it does not constitute a phase (see Box 17.1). Neglecting the work of compression $V\,dp$, where V is the partial molar volume of the salt, we can write:

$$\mu = \mu_0 + RT \ln a$$

where a is the activity, and hence

$$a_s^A = a_s^B \tag{17.5}$$

However, electrical neutrality must be maintained in each compartment. Letting m be the molar concentration, and assuming a 1:1 salt for simplicity with the polyion carrying ν charges (negative say), we have:

- in A

$$m_+^A = m_-^A \tag{17.6a}$$

- in B

$$m_-^B + \phi \nu m_P = m_+^B \tag{17.6b}$$

where m_P is the molar concentration of the polyelectrolyte.

The activity of the salt can be expressed in terms of that of the ions by

$$a_s = (\gamma_+ m_+)(\gamma_- m_-)$$

where γ is the activity coefficient ($0 < \gamma < 1$) when m is in mol l^{-1}. A mean value γ is defined by

$$\gamma^2 = \gamma_+ \gamma_-$$

and hence in each compartment

$$a_s^A = \gamma_A^2 m_+^A m_-^A \qquad \text{and} \qquad a_s^B = \gamma_B^2 m_+^B m_-^B \tag{17.7}$$

and, substituting the neutrality equations (eqns 17.6) into eqns 17.7,

$$(m_+^A)^2 = (\gamma_A/\gamma_B)^2 (m_+^B)^2 (1 - \phi \nu m_P / m_+^B) \tag{17.8a}$$

$$(m_-^A)^2 = (\gamma_A/\gamma_B)^2 (m_-^B)^2 (1 + \phi \nu m_P / m_-^B) \tag{17.8b}$$

When $\phi \nu m_P \ll m_+^B$, and assuming that $\gamma_A \approx \gamma_B$, we obtain as a first approximation

$$m_+^A \approx m_+^B - \phi \nu m_P/2 + (\phi \nu m_P)^2 / 8 m_+^B \tag{17.9a}$$

$$m_-^A \approx m_-^B + \phi \nu m_P/2 + (\phi \nu m_P)^2 / 8 m_-^B \tag{17.9b}$$

In the new equilibrium, the cation concentration is clearly lower in A than in B, whereas the anion concentration is higher. This is *Donnan equilibrium*, generally characterized by a coefficient Γ_D defined by

$$\Gamma_D = \lim_{\nu m \to 0} (m_+^B - m_+^A)/\nu m_P = \phi/2 \tag{17.10}$$

Box 17.1 **Activity of an ion**

The activity of an ion in solution can be measured directly using selective electrodes. However, it is impossible to 'isolate' thermodynamically a solution formed from one cation or one anion. The chemical potential can only be defined for a salt, the classical expression for this being

$$\mu = \mu^\circ + RT \ln a$$

where a is the activity of the salt.

Dissociation, assumed to be complete, of the salt in water gives ν_1 moles of cations with valence z_1 and ν_2 moles of anion with valence z_2 per mole of the salt, with the condition for electrical neutrality:

$$\nu_1 z_1 + \nu_2 z_2 = 0$$

We can therefore write

$$\mu = \nu_1 \mu_1 + \nu_2 \mu_2 \qquad \text{and} \qquad \mu^\circ = \nu_1 \mu_1^\circ + \nu_2 \mu_2^\circ$$

where μ_1, μ_1° and μ_2, μ_2° are the potentials for cations and anions respectively. Putting

$$\mu_1 = \mu_1^\circ + \ln a_1 \qquad \text{and} \qquad \mu_2 = \mu_2^\circ + \ln a_2$$

we obtain

$$\mu = \nu_1(\mu_1^\circ + RT \ln a_1) + \nu_2(\mu_2^\circ + RT \ln a_2) = \mu^\circ + RT(\ln a_1^{\nu_1} + \ln a_2^{\nu_2})$$

Comparing this with the definition of μ gives

$$\ln a = \ln a_1^{\nu_1} + \ln a_2^{\nu_2}$$

i.e.

$$a = a_1^{\nu_1} a_2^{\nu_2}$$

For a 1:1 salt, we then have

$$a = a_1 a_2$$

As in the case of osmotic pressure, Γ can be determined experimentally from a Donnan equilibrium and its value compared with the theoretical value given by eqn 17.10. In cases where there are no bound counter-ions, $\phi = 1$ and $\Gamma = \frac{1}{2}$. For a DNA in a low-ionic-strength medium, it is found that $\Gamma = -0.1$, which implies that about 80 per cent of the counter-ions behave as if they were bound to the polyelectrolyte.

17.2.3 **Titration curves**

Consider first the classical case of a proton binding to a base A^- (in the generalized meaning of the term) to give an acid AH:

$$A^- + H^+ \rightleftharpoons AH \tag{17.11}$$

The association constant K_a is defined by

$$K_a = [AH]/[A^-][H^+] \quad (17.12)$$

If θ is the fraction of A^- with a bound proton, then $\theta = [AH]/([A^-]+[AH])$ and hence

$$\theta/(1-\theta) = K_a[H^+] \quad (17.13)$$

This can be generalized to a set of n identical functions (acidic or basic) located on the same macromolecule. In that case, θ will be the fraction of polyion sites with a bound proton. If ν is the mean number of bound protons, $\theta = \nu/n$, and eqn 17.13 becomes

$$\nu/(n-\nu) = K_a[H^+] \quad (17.14)$$

or the two equivalent expressions

$$\nu/[H^+] = K_a(n-\nu) \quad (17.15)$$

$$\nu = K_a n[H^+]/(1 + K_a[H^+]) \quad (17.16)$$

Depending on the type of graphical representation, any of the three expressions in eqns 17.14, 17.15 or 17.16 may be used. For example, the standard titration curve uses eqn 17.16 to plot ν against $\mathrm{pH} = -\log[H^+]$.

Very often, however, eqn 17.11 is considered as a reaction in which AH dissociates. A degree of dissociation is then defined by

$$\alpha = [A^-]/([A^-] + [AH]) = 1 - \theta \quad (17.17)$$

with a dissociation constant

$$K_d = K_a^{-1}$$

Equation 17.12 then becomes

$$(1-\alpha)/\alpha = K_a^{-1}[H^+] \quad (17.18)$$

With the usual notation $\mathrm{pH} = -\log[H^+]$ and $\mathrm{p}K_a = -\log K_a$, eqn 17.18 becomes

$$\mathrm{pH} = \mathrm{p}K_a + \log[\alpha/(1-\alpha)] \quad (17.19)$$

In fact, eqns 17.18 and 17.19 were obtained by assuming not only identical protonation sites but values of K_a (or K_d) that are independent of the charge on the polyion. The latter assumption is not generally valid.

If we put

$$\Delta G_0 = -RT\ln K_a = RT\ln K_d \quad (17.20)$$

then, with the definitions of $\mathrm{p}K_a$ or $\mathrm{p}K_d$, we have

$$\Delta G_0 = 2.3RT\mathrm{p}K_a = -2.3RT\mathrm{p}K_d \quad (17.21)$$

When the site is on a polyion, additional work W_{el} is required to bind the proton, work that is done against the polyion potential created by all the charged groups, so that $W_{el} = e\Psi(a)$, where $\Psi(a)$ is the potential at the surface of the polyion. For one mole:

$$\Delta G = \Delta G_0 + NW_{el}$$

Hence there is an apparent pK_a or pK_d given by

$$pK_a' = pK_a + 0.434e\Psi/k_B T \tag{17.22a}$$

or

$$pK_d' = pK_d - 0.434e\Psi/k_B T \tag{17.22b}$$

By combining eqns 17.19 and 17.22, it can be seen that the quantity pH − log[$\alpha/(1-\alpha)$] is only equal to the intrinsic pK_a characterizing the isolated group when $\Psi \rightarrow 0$. All the above treatment assumes that the charged group is accessible to the solvent.

17.3 Modelling

In fact, since the Debye–Hückel theory, the theories proposed for both electrolytes and polyelectrolytes have developed along two completely different paths.

In the *continuum model* (Chapter 18), the Poisson–Boltzmann equation is used with a continuous medium characterized by a dielectric constant and obeying the classsical laws of electrostatics. Depending on the system geometry and the approximations made, this type of approach leads to either analytical or numerical solutions. Analytical solutions are convenient in that many properties of the solutions can be determined from values of the potential (e.g. thermodynamic properties, the stability and conformational changes in biopolymers, ligand binding to macromolecules, or transport properties). However, the drawback of these is that they only apply to simple geometries and often require approximations that limit their range of application. Numerical solutions enable the Poisson–Boltzmann equation to be solved whatever the geometry of the system and also offer the possibility of checking the validity of the approximations made in the first approach involving analytical solutions. However, they are still dependent on the initial assumptions, which will be explained at the beginning of the next chapter.

In the discontinuous or *molecular model* (Chapter 20), the problem is treated as one of a distribution of ions, water molecules and charged groups on the macromolecule whose interactions are to be calculated. This approach is similar to that of molecular dynamics and it uses the same methods. It therefore starts with a three-dimensional structure as given by X-ray diffraction for example, and then introduces thermal motion and laws of interactions that are as precise and complete as possible. This approach still lacks a completely satisfactory model for water, particularly in the presence of ions: it is an n-body problem with a slow convergence of numerical solutions because of the long-range $1/r$ Coulomb potential. The computer time required is already considerable with the macromolecule alone, but with a 'physiological' Debye length of about 8 Å, the hydration of the macromolecule adds a large number of water molecules to the problem.

The diagrams in Fig. 17.1 (Nakamura, 1996) indicate the differences between the continuum and molecular models.

The two above approaches can be combined by introducing a *mixed model* (Chapter 19) in which the atomic description of the molecules is retained but in

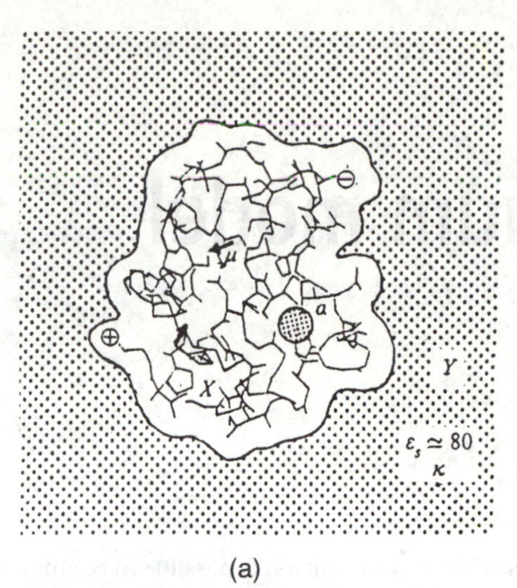

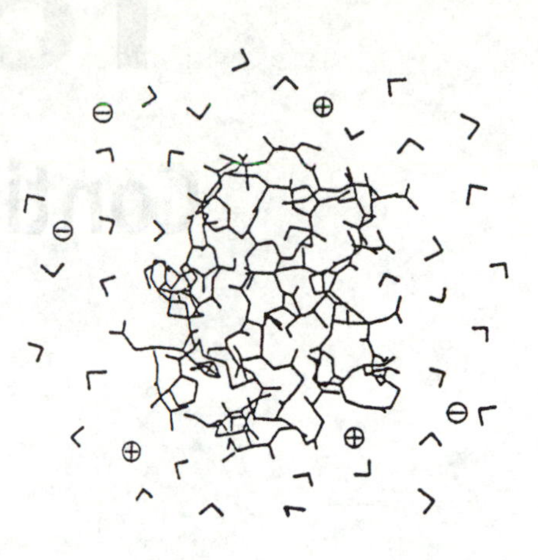

Figure 17.1 Two different models of a protein – solvent system. (a) The macroscopic or continuum model. A protein region (X) is surrounded by a continuum solvent region (Y), which has the characteristic high dielectric constant ε (≈80) and the Debye ionic shielding factor κ. In the region X, there are point charges (+ and −), permanent dipoles at polar groups and atoms having electronic polarizability α. (b) The microscopic or molecular model. A protein molecule is surrounded by small ions (+ and −) and many water molecules indicated by the symbol ∧. (From Nakamura, 1996. Reprinted from *Quarterly Review of Biophysics* with the permission of Cambridge University Press.)

which the solvent is treated as a continuous medium. For example, a protein is modelled as a medium with a low dielectric constant immersed in a medium of high dielectric constant, but the charge distribution in the protein is defined by the actual positions of the protein atoms. Because of recent developments in numerical methods, this approach is at present receiving considerable attention, and 'potential maps' around biopolymers are appearing in the literature in ever greater numbers. These are increasingly being used to derive their interactions either with small ligands (enzyme substrates) or with other charged biopolymers (e.g. specific associations between proteins and nucleic acids).

In the next three chapters, we give examples of each approach. In each case, complementary views of polyelectrolytes will gradually emerge and throw more and more light on the problem, eventually leading to a generally accepted picture.

18

Continuum model

We have already discussed the way in which it is possible to combine the laws of electrostatics, applicable in general to a fixed charge distribution, with a statistical description of the liquid medium in which the charged ions and molecules are moving. The success of the Debye–Hückel theory applied to electrolytes justifies this approach.

18.1 Establishing the Poisson–Boltzmann equation

In a continuum model, the polyelectrolyte is represented by a volume within which neither the dielectric constant nor the charge distribution will concern us. It is bounded by a charged closed surface in contact with an electrolyte whose dielectric constant is assumed to be time-invariant and uniform.

In this case, Poisson's equation takes the form:

$$\nabla^2\Psi = -\rho/\varepsilon\varepsilon_0 \tag{18.1}$$

where ρ is the volume charge density at any point in the electrolyte medium and ϵ is the dielectric constant of the medium.

The charge density ρ results only from the presence of counter-ions and co-ions. These are assumed to have a distribution obeying a Boltzmann type of variation in the potential $\Psi(r)$, where r is the distance in the polyelectrolyte from a plane, or from the centre for a sphere, or from the axis for a cylinder. This means that in a solution containing n_s 'molecules' of salt, the number n_1 of counter-ions with charge z_1 at a distance r is given by

$$n_1(r) = n_s \exp(z_1 e\Psi/k_B T) \tag{18.2a}$$

since z_1 and Ψ are of opposite signs, and the number n_2 of co-ions is

$$n_2(r) = n_s \exp(-z_2 e\Psi/k_B T) \tag{18.2b}$$

since z_2 and Ψ have the same sign. It follows that

$$\rho(r) = n_s e[-\exp(z_1 e\Psi/k_B T) + \exp(-z_2 e\Psi/k_B T)] \tag{18.3}$$

With a 1:1 salt, $z_1 = z_2 = 1$, and so

$$\rho(r) = -2n_s e \sinh(e\Psi/k_B T) \tag{18.4}$$

Using this in eqn 18.1 yields the *Poisson–Boltzmann equation*:

$$\nabla^2\Psi = 2n_s e \sinh(e\Psi/k_B T)/\varepsilon\varepsilon_0 \tag{18.5}$$

This can be put into a more compact form by introducing two dimensionless variables y and x:

$$y = e\Psi/k_B T \tag{18.6}$$

i.e. the ratio between the electrostatic and thermal energies, and

$$x = Kr \tag{18.7}$$

i.e. the ratio between the distance r and the Debye length $1/K$, where

$$K^2 = (2n_s e^2)/(\varepsilon\varepsilon_0 k_B T)$$

From eqns 18.6 and 18.7, we clearly have that $\nabla^2\Psi(r) = K^2\nabla^2 y(x)$, and substitution in eqn 18.5 gives the reduced form of the Poisson–Boltzmann equation as

$$\nabla^2 y = \sinh y \tag{18.8}$$

The solution of this equation depends on the boundary conditions given by the model adopted for the polyelectrolyte. For simple geometrical models (plane, sphere, cylinder), the Laplacian is written in the appropriate coordinates as follows.

- For *plane symmetry*, Ψ is a function only of the distance r from the plane and hence

  $$\nabla^2\Psi = \mathrm{d}^2\Psi/\mathrm{d}r^2$$

  and hence

  $$\nabla^2 y = \mathrm{d}^2 y/\mathrm{d}x^2 \tag{18.9}$$

- With *spherical symmetry*, Ψ is a function only of r (spherical equipotential surfaces), so that

  $$\nabla^2\Psi = r^{-2}\,\mathrm{d}(r^2\,\mathrm{d}\Psi/\mathrm{d}r)/\mathrm{d}r$$

  and hence

  $$\nabla^2 y = x^{-2}\,\mathrm{d}(x^2\,\mathrm{d}y/\mathrm{d}x)/\mathrm{d}x \tag{18.10}$$

- With *cylindrical symmetry*, Ψ depends only on the distance r from the axis of the cylinder and is constant along any line parallel to the axis, so that

  $$\nabla^2\Psi = r^{-1}\,\mathrm{d}(r\,\mathrm{d}\Psi/\mathrm{d}r)/\mathrm{d}r$$

  and hence

  $$\nabla^2 y = x^{-1}\,\mathrm{d}(x\,\mathrm{d}y/\mathrm{d}x)/\mathrm{d}x \tag{18.11}$$

The generalized Poisson–Boltzmann equation (eqn 18.8) has analytical solutions (when the potential is expressible as a function of parameters arising from the medium and the model) in only two cases: the plane and the cylinder in the presence of counter-ions only.

The simplified models of plane, sphere and cylinder are particularly useful. In the first place, they are good approximations to real systems: the charged plane for a membrane, the sphere for a protein or an approximately spherical assembly, and the cylinder for a linear polyelectrolyte. This simplified approach, which gives analytical expressions for Ψ, can be used for comparison with the results from numerical methods, and a certain number of general properties can also be obtained from these simplified models.

18.2 Examples of modelling

18.2.1 The non-linear Poisson–Boltzmann equation

For a charged cylinder surrounded only by counter-ions, analytical solutions have been obtained (Alfrey *et al.*, 1951; Fuoss *et al.*, 1951). The expressions yielded in this case are not easy to use, but they showed for the first time that there is a large concentration of counter-ions near the charged cylinder–a model for DNA, as we shall see later.

The charged plane, on the other hand, leads to a relatively simple expression for the potential (Zimm and Le Bret, 1983) of the form

$$y = 2\ln[(1+\gamma e^{-x})/(1-\gamma e^{-x})] \tag{18.12}$$

where $\gamma = \tanh(y_0/4)$, y_0 being the reduced value of the potential at the surface of the charged plane (see Box 18.1). The purely numerical value γ is related to the *surface charge density* on the plane, σ, by

$$\gamma = -KA + (1 + K^2A^2)^{1/2} \tag{18.13}$$

(see Box 18.2) where the quantity A, with the dimensions of a length, is given by

$$A = 2\varepsilon\varepsilon_0 k_B T/e\sigma \tag{18.14}$$

A set of curves $y(x)$ can be plotted for various values of the parameters σ and K, giving the variation of potential near a charged plane.

18.2.2 The linearized Poisson–Boltzmann equation

Spherical model

There is no general analytical solution with spherical symmetry, but if it is assumed that $y \ll 1$ as in the Debye–Hückel theory, then $\sinh y$ can be replaced by y (or e^y by $1+y$ and e^{-y} by $1-y$), and the Poisson–Boltzmann equation in the Debye–Hückel approximation becomes

$$d^2y/dx^2 + 2x^{-1}\,dy/dx - y = 0 \tag{18.15}$$

The approximation can be checked for spherical macro-ions by comparing the values of e^y and $1+y$. The difference is 1 per cent for $y = 0.15$ and reaches 10 per cent for $y = 0.5$. At $T = 293$ K, these y values correspond to potentials of 2.53 and 12.6 mV respectively. The Debye–Hückel approximation can thus still be used in this range of potentials at the surface of a polyelectrolyte or a polyampholyte.

Soon after the Debye–Hückel theory was published, the biochemist Linderstrøm-Lang (1924) had the idea of using the expression for the internal potential to

Box 18.1 The Poisson–Boltzmann equation for a charged plane

This box follows Zimm and Le Bret (1983). For a plane, the reduced Poisson–Boltzmann equation is

$$\nabla^2 y = \mathrm{d}^2 y/\mathrm{d}x^2 = \sinh y \tag{1}$$

Multiplying both sides of eqn 1 by $\mathrm{d}y/\mathrm{d}x$ gives

$$(\mathrm{d}y/\mathrm{d}x)(\mathrm{d}^2 y/\mathrm{d}x^2) = (\mathrm{d}y/\mathrm{d}x)\sinh y$$

or

$$\tfrac{1}{2}\mathrm{d}(\mathrm{d}y/\mathrm{d}x)^2/\mathrm{d}x = \mathrm{d}(\cosh y)/\mathrm{d}x \tag{2}$$

and a first integration of this gives

$$\tfrac{1}{2}(\mathrm{d}y/\mathrm{d}x)^2 = \cosh y + \text{constant} \tag{3}$$

When $y = 0$, $\mathrm{d}y/\mathrm{d}x = 0$, so that the constant is -1 and the right-hand side of eqn 3 can be written as $2\sinh^2(y/2)$. Hence

$$\mathrm{d}y/\mathrm{d}x = \pm 2\sinh(y/2)$$

Only the minus sign is an acceptable solution since y is a decreasing function of x, and so

$$\mathrm{d}y/\mathrm{d}x = -2\sinh(y/2) \tag{4}$$

which can be written in the form

$$\mathrm{d}(y/2)/\sinh(y/2) = -\mathrm{d}x$$

and this can be integrated by putting $u = \exp(y/2)$ so that $\mathrm{d}u/u = \mathrm{d}y/2$. The left-hand side then becomes $2\,\mathrm{d}u/(u^2-1)$ or $\mathrm{d}u[1/(u-1)-1/(u+1)]$, which is equal to $-\mathrm{d}x$. This can be integrated directly to give

$$\ln[(u-1)/(u+1)] = -x + \text{constant}$$

or

$$\ln[(\mathrm{e}^{y/2}-1)/(\mathrm{e}^{y/2}+1)] = -x + \text{constant} \tag{5}$$

When $x = 0$, $y = y_0$ at the surface and the constant is therefore equal to $\ln[(\mathrm{e}^{y_0/2}-1)/(\mathrm{e}^{y_0/2}+1)]$. Substituting this in eqn 5 and rearranging, we obtain

$$\mathrm{e}^{y/2} = (1+\gamma \mathrm{e}^{-x})/(1-\gamma \mathrm{e}^{-x}) \tag{6}$$

by putting

$$\gamma = (\mathrm{e}^{y_0/2}-1)/(\mathrm{e}^{y_0/2}+1) = \tanh(y_0/4) \tag{7}$$

This leads to the final expression for the reduced potential:

$$y = 2\ln[(1+\gamma \mathrm{e}^{-x})/(1-\gamma \mathrm{e}^{-x})] \tag{8}$$

If the potential at the surface is known, γ can be calculated from eqn 7 and $y(x)$ can then be obtained from eqn 8.

Box 18.2 **Distribution of ions near a charged plane**

The charge density (charge per unit area) $\sigma(r)$ at a distance r from the plane is given by

$$\sigma(r) = \int_0^r \rho(r)\,\mathrm{d}r = -\varepsilon\varepsilon_0 \int_0^r (\mathrm{d}^2\Psi/\mathrm{d}r^2)\,\mathrm{d}r \quad (1)$$

Since $\mathrm{d}^2\Psi/\mathrm{d}r^2 = (K^2 k_B T/e)\,\mathrm{d}^2y/\mathrm{d}x^2$, we have (from eqns 18.8 and 18.9: $\mathrm{d}^2y/\mathrm{d}x^2 = \sinh y$):

$$\sigma(r) = -(k_B T \varepsilon\varepsilon_0 K^2/e) \int_0^r \sinh y\,\mathrm{d}r \quad (2)$$

Since $\mathrm{d}y = -2\sinh(y/2)\,\mathrm{d}x$, we have $\mathrm{d}r = -\mathrm{d}y/2K\sinh(y/2)$ and with $\sinh y = 2\sinh(y/2)\cosh(y/2)$, eqn 2 becomes

$$\sigma(r) = (2k_B T\varepsilon\varepsilon_0 K/e) \int_0^r \cosh(y/2)\,\mathrm{d}(y/2)$$
$$= (-2k_B T\varepsilon\varepsilon_0 K/e)[\sinh y(r)/2 - \sinh y(0)/2] \quad (3)$$

When $r \to \infty$, $y(r) \to 0$ and $\sigma(\infty) \to$ the total charge σ per unit area carried by the plane, i.e.

$$\sigma = [2k_B T\varepsilon\varepsilon_0 K \sinh(y_0/2)/e \quad (4)$$

Using $\sinh(y_0/2) = 2\gamma/(1-\gamma^2)$, we obtain

$$\sigma = 4k_B T\varepsilon\varepsilon_0 K\gamma/e(1-\gamma^2) \quad (5)$$

which enables us to calculate γ as a function of σ by solving the quadratic equation in γ:

$$\gamma = -KA + (1 + K^2A^2)^{1/2} \quad (6)$$

where $A = 2\varepsilon\varepsilon_0 k_B T/e\sigma$.

We can use eqns 3 and 4 to find the ratio $\sigma(r)/\sigma$, with the help of eqn 6 that gives us $\sinh(y/2) = 2\gamma\mathrm{e}^{-x}/(1-\gamma^2\mathrm{e}^{-2x})$:

$$\sigma(r)/\sigma = 1 - (1-\gamma^2)\mathrm{e}^{-x}/(1-\gamma^2\mathrm{e}^{-2x}) = 1 - (1-\gamma^2)\mathrm{e}^{x}/(\mathrm{e}^{2x}-\gamma^2) \quad (7)$$

If γ from eqn 6 is substituted in this, we obtain finally

$$\sigma(r)/\sigma = \{1 - 2KA[(1+K^2A^2)^{1/2} - KA]\}\mathrm{e}^{Kr}/[\mathrm{e}^{2Kr} - 1 + 2KA(1+K^2A^2)^{1/2}] \quad (8)$$

When $K \to 0$ (infinite dilution), with A and r remaining constant, the expression in eqn 8 tends to a finite limit

$$\sigma(r)/\sigma \to 1 - 1/(1 + r/A) \quad (9)$$

In particular, when r becomes large, $\sigma(r)/\sigma \to 1$. Condensation of the counterions now occurs whatever the value of r since, in offering an unlimited volume of the solvent to the ions ($K \rightarrow 0$), the major part of the compensating charge remains near the charged plane.

The quantity r/A occurring in eqn 9 is quite simply the ratio between the distance r from the charged plane and A, which can be considered as the Debye length $1/K$ divided by $\sinh(y_0/2)$. The expression for A can also be related to the surface charge density on the plane as follows: if A is expressed in terms of the Bjerrum length $l_B = e^2/4\pi\varepsilon\varepsilon_0 k_B T$, we find

$$A = e/2\pi\sigma l_B \tag{10}$$

However, e/σ is the mean area S occupied by each elementary charge on the charged plane, so that

$$A = S/2\pi l_B \tag{11}$$

Since $l_B \approx 7.1$ Å at 20°C, the length A is related directly to the charge density on the plane. If, for example, $S \approx 90$ Å^2, $A \approx 2$ Å: at a distance $r = 8$ Å from the plane, the compensating charge already amounts to 80 per cent of the charge on the plane.

describe what happens at the surface of a protein assumed to be spherical. We now examine this simplified model (Fig. 18.1).

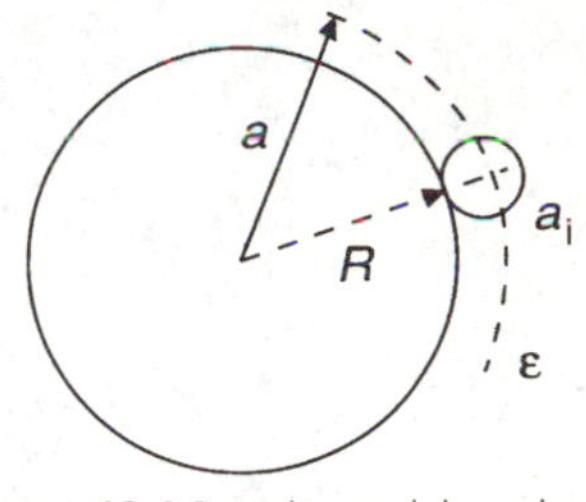

Figure 18.1 Protein model: exclusion sphere with adjacent ion.

The protein is represented by a sphere impervious to the solvent having a charge ze equal to the net charge of the protein and distributed uniformly over the surface. The Debye–Hückel theory valid for small ions can be applied to this macro-ion. Let a be the radius of the exclusion sphere and R the radius of the sphere equivalent to the protein, so that $a = R + a_i$, where a_i is the mean radius of the ions present. If the dielectric constant of water is ε, the potential near the protein ($R < r < a$) is given by an expression similar to eqn 16.16 calculated for Debye–Hückel theory:

$$\Psi = (ze/4\pi\varepsilon\varepsilon_0)[R^{-1} - K/(1 + Ka)] \tag{18.16}$$

The dissociation of protonated groups in the protein is governed by the classical equation relating the pH, the dissociation coefficient α and the *apparent* pK of the group in its environment:

$$\text{pH} = \text{p}K_{\text{app}} + \log[\alpha/(1 - \alpha)]$$

The difference between pK_{app} and the intrinsic pK, pK_{int}, has already been seen to depend on the work required to bring the proton to its site within the potential of the protein. It was found (eqn 17.22) that

$$\text{p}K_{\text{app}} = \text{p}K_{\text{int}} - 0.434e\Psi/k_B T$$

and hence for each ionizable group

$$\log[\alpha_i/(1 - \alpha_i)] = \text{pH} - (\text{p}K_{\text{int}})_i + 0.434e\Psi/k_B T \tag{18.17}$$

The titration curve is calculated by adopting the following values for pK_{int}:

Asp	Glu	αCOO$^-$	αNH$_3^+$	His	εLys	Arg	Tyr
4.0	4.4	3.8	7.5	6.3	10.4	12.5	10.2

For each pH value, the dissociation fraction α_i is calculated by assuming first that $z = 0$. From this α_i, a value for z is determined, which is then introduced to calculate

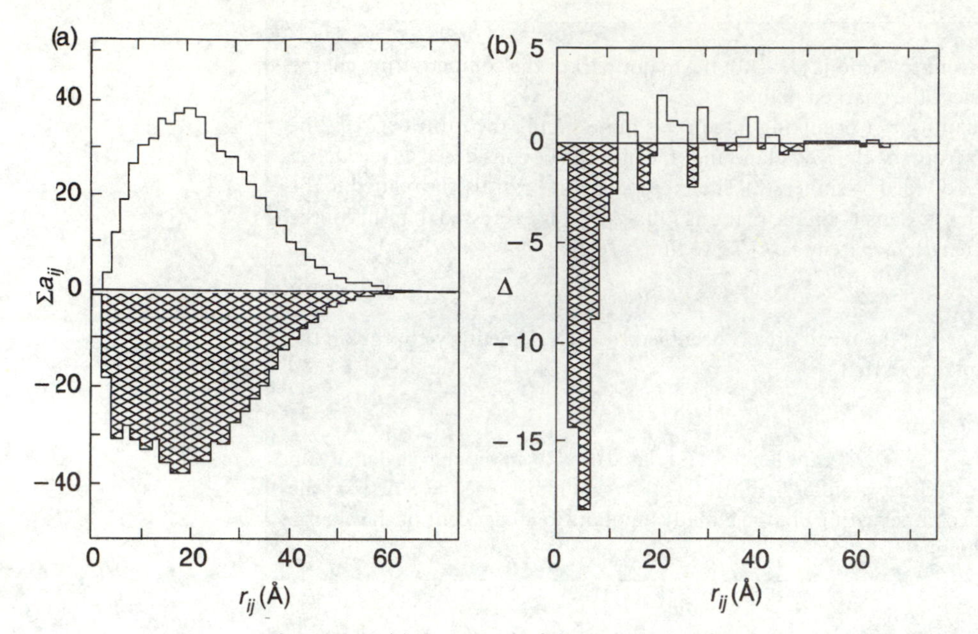

Figure 18.2 (a) Histograms of the distribution of distances r_{ij} between 44 592 pairs of charges in 14 proteins: top – charges of the same sign; bottom – charges of opposite sign. (b) Difference between the two histograms in (a). (From Wada and Nakamura, 1981. Reprinted with permission from *Nature*, vol. 293, pp. 757–8 and with the authors' permission. © 1981 Macmillan Magazines Ltd.)

a new value of α, and so on. This iteration process converges and gives z as a function of pH, i.e. a titration curve.

The relative success of this simplistic protein model depends on several factors.

1. It is found experimentally that acidic, basic and neutral regions in the protein titration curve correspond respectively to the titration ranges of acidic, basic and neutral amino acids. The variations in pK are never very great, which makes it appear that the assumption of their purely electrostatic origin is valid.

2. Modelling a protein by a charged spherical surface is equivalent to saying that all the protein charges are on or near the surface. A study of 36 globular proteins with known structures has shown that 95 per cent of the charges are indeed accessible to the solvent.

3. In a study of the distribution of charged groups in a protein, Wada and Nakamura (1981) drew up histograms for over 44 000 pairs of charges $e_i e_j$ separated by a distance r_{ij} selected from 14 different proteins (Fig. 18.2). Two almost symmetrical curves were found depending on whether $e_i e_j / r_{ij}$ is positive (repulsive) or negative (attractive). In the diagram showing the difference between the two histograms, a sharp minimum (attractive) occurs for $r_{ij} = 5 \pm 1.5$ Å, which means that on average any charge in the protein is surrounded by charges of opposite sign. This is a result reminiscent of the Debye–Hückel theory, with the difference that rearrangements of charge cannot occur in a protein as they can in a solution. This uniform distribution can thus be considered as partly reflecting the folding pattern of the peptide chain. Of course, this averaged model does not rule out the existence of highly anisotropic charge distributions in some proteins, as we shall see later.

Charged line model

Several polyelectrolytes (linear acidic polysaccharides, linear DNAs) behave in solution like rigid cylinders with a length that depends on the structure of the polymer and the parameters of the medium such as pH, temperature and ionic strength. To a first approximation, the equipotential surfaces around such a molecule will also be cylindrical and will correspond to those of an imaginary linear distribution of charge along the axis characterized by a mean distance b along the axis between two successive elementary charges e.

Two comments must be made at the outset about this. The first is that such an apparently crude model is justified insofar as any discrete charge distribution can be replaced, as in electrostatics, by a continuous distribution of charge density $\rho(r)$. At a point outside the volume of the polyelectrolyte, it is as if it were a charged conducting surface with a simple geometry (sphere, cylinder, ellipsoid, etc.).

The other comment concerns the implicit assumption of a linear distribution and hence one of infinite length. What physical meaning can be attributed to this mathematical fiction? One scaling factor in the frame of reference of the polyelectrolyte is the Debye length K^{-1}. Another, related to the rigidity of the chain, is the persistence length L_P. If $K^{-1} > L_P$ (at low ionic strengths), long-range interactions between charged groups can no longer be neglected. This amounts formally to the introduction of an 'end-effect' or to the replacement of the cylindrical model by an ellipsoidal one, which actually gives an equipotential surface like that due to a charged segment of finite length. Conversely, if K^{-1} is smaller than the local radius of curvature, the chain configuration no longer has a role to play.

The charged line model can therefore be considered as valid for polymers that are rigid in the given medium. For example, in 1 mM NaCl, DNA has an L_P of about 1000 Å, whereas K^{-1} is about 100 Å. It should also be pointed out that an increase in ionic strength not only reduces K^{-1} but also does the same to L_P because the flexibility is increased (L_P for DNA in 0.1 mM NaCl is no more than about 500 Å).

Before solving the Poisson–Boltzmann equation, we introduce a new parameter specific to linear distributions.

A charge Ne is distributed uniformly over the whole curved surface of a cylinder of radius a, length $l \gg a$ and surface potential Ψ_0. To simplify the problem, we assume that the cylinder is immersed in a 1:1 electrolyte with a dielectric constant ε. Let n be the number of ions of the electrolyte per unit volume far from the cylinder (Fig. 18.3). To obtain the boundary conditions at the surface of the cylinder, apply Gauss's theorem to the flux of the electric field E across the surface, i.e.

$$2\pi a l E = Ne/\varepsilon\varepsilon_0$$

Using eqn 18.7, $a = x_0/K$, so that

$$E = KeN/2\pi x_0 l\varepsilon\varepsilon_0 \tag{18.18}$$

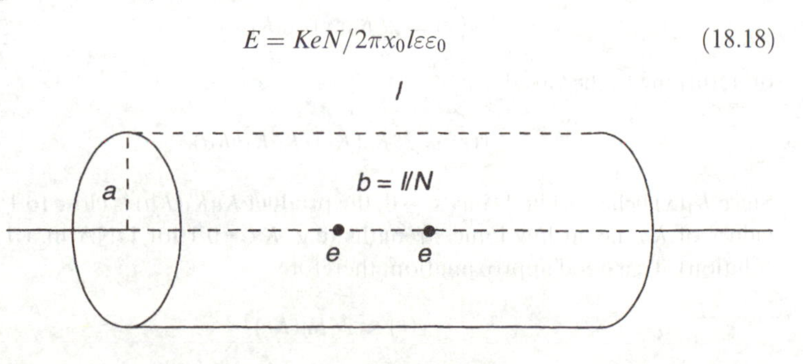

Figure 18.3 Cylindrical model for a linear polyelectrolyte.

Because the surface charge distribution is equivalent to a linear distribution along the axis of the cylinder (see section 15.2.2), we can define $b = 1/N$ as the mean distance along the axis between two elementary charges.

Using the elementary relationship between E and $\mathrm{d}y/\mathrm{d}x$, i.e. $E = -(Kk_B T\,\mathrm{d}y/\mathrm{d}x)/e$, and substituting this value for E in eqn 18.18, we obtain

$$-x_0(\mathrm{d}y/\mathrm{d}x)_{x=x_0}/2 = e^2/4\pi\varepsilon\varepsilon_0 b k_B T \tag{18.19}$$

The right-hand side of this equation is a dimensionless quantity called the *charge parameter* ξ, which characterizes the polyelectrolyte modelled as a charged cylinder. Hence

$$\xi = -(x_0/2)(\mathrm{d}y/\mathrm{d}x)_{x=x_0} \tag{18.20}$$

A comparison between eqns 18.19 and 16.22 shows the following simple relationship between ξ and l_B:

$$\xi = l_B/b \tag{18.21}$$

In the Debye–Hückel approximation it is assumed that $e\Psi \ll k_B T$ or $y \ll 1$, so that $\sinh y$ can be replaced by y. Poisson's equation in cylindrical coordinates is then

$$x^{-1}\,\mathrm{d}/\mathrm{d}x(x\,\mathrm{d}y/\mathrm{d}x) = y \tag{18.22}$$

(remembering that $x = Kr$, where r is the distance from the axis of the cylinder). On the surface of the cylinder $x_0 = Ka$ and $y_0 = e\Psi_0/k_B T$, and eqn 18.22 becomes

$$\mathrm{d}^2y/\mathrm{d}x^2 + x^{-1}\,\mathrm{d}y/\mathrm{d}x - y = 0 \tag{18.23}$$

which is a differential equation with solutions known as *modified Bessel functions of zero order* denoted by $K_0(x)$ (see Mathematical Appendix). The solution of eqn 18.23 is

$$y = CK_0(x) \tag{18.24}$$

Modified Bessel functions do not oscillate and behave like e^{-x} as $x \to \infty$. From eqn 18.24 and the properties of Bessel functions, we have

$$\mathrm{d}y/\mathrm{d}x = -CK_1(x) \tag{18.25}$$

where $K_1(x)$ is the modified Bessel function of the first order. Combining eqns 18.20, 18.24 and 18.25 gives

$$y(x) = 2\xi K_0(x)/x_0 K_1(x_0) \tag{18.26}$$

or, returning to the variable r,

$$y(r) = 2\xi K_0(Kr)/KaK_1(Ka) \tag{18.27}$$

Since $K_1(x)$ behaves like $1/x$ as $x \to 0$, the product $KaK_1(Ka)$ is close to 1 for small values of Ka, i.e. at low ionic strengths (e.g. $Ka = 0.1$ for DNA in a 1 mM salt solution). To a good approximation, therefore

$$y(r) \approx 2\xi K_0(Kr) \tag{18.28}$$

A table of $K_0(x)$ and $K_1(x)$ values is given in the Mathematical Appendix for $x = 0.02$ to $x = 4$. Note that the approximation $xK_1(x) = 1$ is within 5 per cent of its true value for $x = 0.2$.

When the charge density of the polyelectrolyte is high, the charge parameter ξ is also high, so that $y(r)$ is much greater than 1, which contradicts the initial assumption made in solving the Poisson–Boltzmann equation. Taking double-stranded DNA as an example: with $b = 1.7$ Å, we have $\xi = 4.2$ at 20°C. Values of x greater than 2 are required to obtain y values smaller than 1. For DNA molecules, the Debye–Hückel approximation is thus only valid at high ionic strengths and at large enough distances. In 0.1 M NaCl for example ($K^{-1} = 9.8$ Å), the distance must be much larger than twice the radius of the helix.

Returning to eqn 18.28, the Bessel function $K_0(Kr)$ at low ionic strengths ($Kr \to 0$) behaves like $-\ln(Kr)$ and hence

$$\Psi(r) = (-e/2\pi\varepsilon\varepsilon_0 b)(\ln K + \ln r) \tag{18.29}$$

The second term ($\ln r$) in the second brackets corresponds to the potential due to the line charge and the first term ($\ln K$) to the potential due to the counter-ions. This equation, in cylindrical coordinates, is similar to eqn 16.16 giving the internal potential $\Psi_{\text{int}}(r)$ in a system with spherical symmetry.

18.3 Numerical solutions

The Poisson–Boltzmann equation for a line charge without any approximations can only be solved by numerical methods: we can cite the older work by Kotin and Nagasawa (1962) and the more recent approach of Guéron and Weisbach (1980). In all cases, numerical values of y are obtained as a function of x, allowing potential curves to be plotted around the polyelectrolyte of reduced radius $x_0 = Ka$. The results obtained show that the xy plane can be divided into three regions (Fig. 18.4),

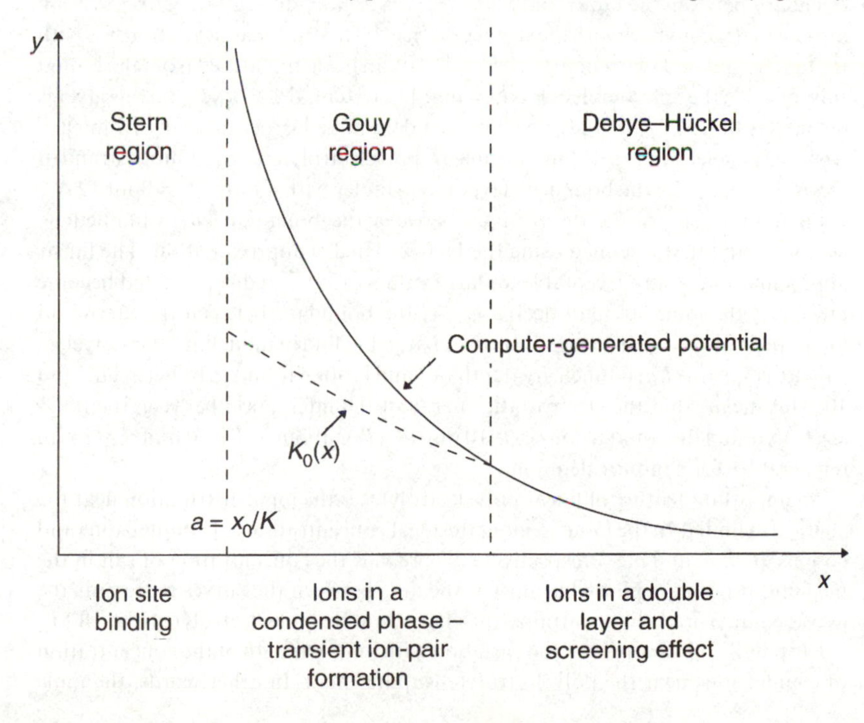

Figure 18.4 Potential variations near a linear polyelectrolyte. The full curve shows the numerically calculated potential. The dashed curve is the Bessel function $K_0(x)$. (From Schellman and Stigter, 1977. Reprinted from *Biopolymers* with the permission of John Wiley & Sons. © 1977 John Wiley & Sons.)

each corresponding to a domain over which the various approximations made in solving the equations are valid (Schellman and Stigter, 1977).

1. In the *Debye–Hückel region*, the solution of the differential equation is eqn 18.24:

$$y = CK_0(x)$$

$K_0(x)$ does not oscillate but decreases regularly with x and behaves similarly to the function $Ax^{-1}\exp(-x)$ obtained with spherical symmetry. The ions in this region are treated as point charges forming a typical double layer with a screening effect reflected in the behaviour of $K_0(x)$.

2. In the *Gouy region*, the cylindrical symmetry of the problem and the point charge model are still valid but y is no longer much less than 1, and the Poisson–Boltzmann equation must be solved without any approximation. A condensed phase of ions occurs near the polyelectrolyte, while transient ion pairs may be created between the polyelectrolyte charges and the counter-ions.

3. In the *Stern region*, the cylindrical symmetry disappears and the structural and geometrical factors have to be taken into consideration. In addition, the dimensions of the ions and the possible strong binding of some ions to specific sites with partial dehydration of the ion must also be included. In other words, a region close to the polyelectrolyte, such as the two grooves in a double-stranded DNA, can no longer be dealt with by a continuum model.

What values of x form the boundaries between these three regions? Between the Debye–Hückel and Gouy regions, the boundary may be defined as the value of x for which $K_0(x)$ approximates to the exact numerical solution for y/C. The boundary between the Gouy and Stern regions is more difficult to define. We can assume that it coincides with the surface defined by hydrodynamic friction, i.e. with the first boundary layer where the free solvent can be distinguished from the bound solvent when the macromolecule is moving. The potential $\Psi = k_B T/e$ on this layer is defined as the *zeta-potential*, which can be determined experimentally from electrophoretic measurements on the linear polyelectrolyte. For double-stranded DNA, for example, the boundary layer is a cylinder with a radius of about 12 Å.

Figure 18.4 shows the discrepancy between the potential $\Psi(x)$ obtained by computer and that obtained using the Debye–Hückel approximation. The linear approximation is only acceptable for large values of x, i.e. at distances that become greater as the ionic strength decreases. At the boundary between the Stern and Gouy regions, i.e. at the surface of the charged cylinder modelling the polyelectrolyte, $y(x_0)$ or $y(Ka)$ is much greater than 1 and is often found to be between 5 and 10. This means that the electrostatic energy of the interaction between the polyelectrolyte and the counter-ions is 5–10 times $k_B T$ or about 3–6 kcal mol^{-1} at room temperature for a monovalent ion.

An important feature of linear polyelectrolytes is the ionic distribution near the charged cylinder. In the Gouy region, the local concentrations of counter-ions and co-ions are $n_s e^y$ and $n_s e^{-y}$ respectively, where n_s is the concentration of salt in the medium. If both are plotted against x, the area between the curves represents the excess counter-ions that neutralize the charges on the polyelectrolyte (Fig. 18.5).

Keeping ξ and x_0 constant, the variation with ionic strength of the concentration of counter-ions near the polyelectrolyte is quite small. In other words, the ionic

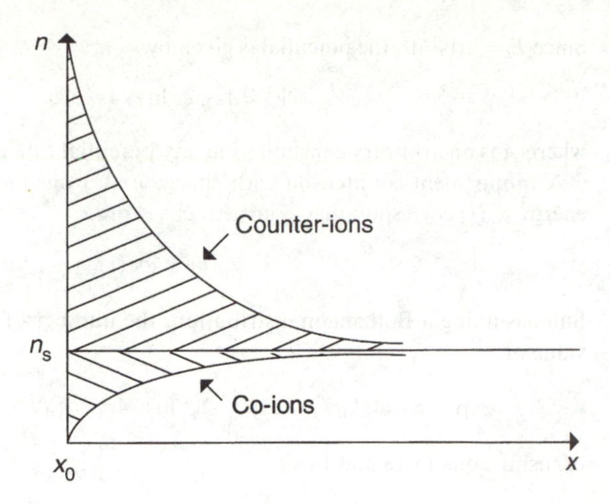

Figure 18.5 Distribution of ions near a charged cylinder.

atmosphere around the polyelectrolyte is not much affected by changes in ionic strength. For DNA, for example, a change from 10^{-3} M to 0.1 M of a 1:1 medium like NaCl causes the concentration of counter-ions (Na^+) around the phosphate groups to vary by only about 10 per cent. The counter-ion concentration stays high in any case (of the order of several mol l^{-1}) and cannot be compared with the bulk concentration far from the DNA molecule. This clearly reveals the ion-trapping behaviour of polyelectrolytes.

18.4 Condensation model for a charged line

Another way (Manning, 1969) of dealing with the problem of the charged line starts with the same assumptions as those made in using the Poisson–Boltzmann equation, namely:

(1) the solvent is a continuum with a dielectric constant ε;

(2) the ions are treated as point charges but as a whole they are considered to be represented by a charge density $\rho(r)$;

(3) the polyion is modelled by a line charge of infinite length characterized by a charge parameter ξ as defined above.

Instead of establishing the general equation, we simply consider how the third assumption affects the form of the potential.

18.4.1 Concept of condensation

We have seen previously (see section 18.2.2) how to calculate the field E using Gauss's theorem. Taking the case of an equipotential cylinder of radius r, cut a slice of length b containing only one elementary charge e, when the only flux to be considered is that of the radial field. Gauss's theorem then gives us

$$2\pi rbE = e/\varepsilon\varepsilon_0$$

so that

$$E = e/2\pi\varepsilon\varepsilon_0 b \tag{18.30}$$

Since $E = -\mathrm{d}\Psi/\mathrm{d}r$, the potential is given by

$$\Psi = A - 2e \ln r/4\pi\varepsilon\varepsilon_0 b \tag{18.31}$$

where A is an arbitrary constant as in any potential function.

A monovalent counter-ion with charge e at a distance r acquires a potential energy $E_p(r)$ corresponding to an attractive force:

$$E_p(r) = e\Psi(r)$$

Still assuming a Boltzmann distribution, the number of ions will depend on the value of

$$\exp\left[-E_p(r)/k_B T\right] = \exp\left(-2e^2 \ln r/4\pi\varepsilon\varepsilon_0 k_B T b\right) \exp\left(-eA/k_B T\right)$$

or, using eqns 18.18 and 18.19,

$$\exp\left[-E_p(r)/k_B T\right] = \exp\left(-2\xi \ln r\right) \exp\left(eA/k_B T\right) = W_0 r^{-2\xi} \tag{18.32}$$

where $W_0 = \exp(eA/k_B T)$. This gives the probability of finding an ion with charge e at a distance r from the line charge forming the model. The number of counter-ions inside a cylinder of radius r and of unit length is proportional to the integral

$$\int_0^{r_0} W_0 r^{-2\xi} 2\pi r \, \mathrm{d}r = 2\pi W_0 \int_0^{r_0} r^{1-2\xi} \, \mathrm{d}r \tag{18.33}$$

The integral diverges at $r = 0$ if $\xi > 1$, and this has a physical interpretation: the ionic distribution around the polyelectrolyte is unstable and the only way of solving the problem is to reduce the charge of the polyelectrolyte by *condensing* some of the counter-ions on to it so as to make $\xi = 1$. The fraction of charge neutralized is therefore

$$(\xi - 1)/\xi = 1 - 1/\xi \tag{18.34}$$

A more general expression for an ion of charge ze is

$$(\xi - 1)/\xi = 1 - 1/z\xi \tag{18.35}$$

The condensation process can be explained by the formation of *ion pairs* (Bjerrum) between the charges carried by the polyelectrolyte and the counter-ions. However, this is not a static phenomenon in which each cation finds a specific binding site: on the contrary, it is a dynamic process with each cation exploring all possible sites very quickly, so that at any time some of the polyelectrolyte charge is neutralized and brings ξ back to its critical value of 1.

In an analogy with the thermodynamic treatment of phase equilibria, the condensed counter-ions are closely comparable to a condensed phase (Oosawa, 1971) in which the cohesive forces come from charges bound to the polyelectrolyte. The free counter-ions then correspond to a kind of 'vapour' phase in equilibrium with the condensed phase.

Consider the example of double-stranded DNA at 20°C with $b = 1.7$ Å for the B structure and hence $\xi = 4.2$ and $1 - 1/\xi = 0.76$. This means that on average about three phosphate groups out of four are 'neutralized' by counter-ions, thus forming around the double helix a kind of condensed layer in which the local ionic concentration is much higher than that in the solution. From the outside, DNA behaves like a polyelectrolyte with an apparent charge density only about a quarter of that corresponding to the phosphate distribution on the double helix. In addition, because the counter-ions are rapidly exploring all the phosphate groups, we can say that the time-averaged net charge of each phosphate group is

$$q = e/z\xi \qquad (18.36)$$

18.4.2 Applications of the model

The condensation model, in providing simple analytical expressions for the potential, enables us to calculate many properties of polyelectrolytes.

Application 1

Assuming that the Debye–Hückel approximation can be applied to the partially neutralized line charge, we find a potential similar to that of eqn 18.29: the potential due to the charged line is added to that of the counter-ions. From this, we can calculate (see Box 18.3) the fraction ϕ of free ions in solution, i.e. the free counter-ions remaining after removing the 'condensed' ions and those bound in the classical Debye–Hückel double layer.

Table 18.1 gives some parameters calculated from Manning's theory for cases in which the ionic strength of the medium is practically negligible.

Application 2

A simple decrease in ionic strength cannot on its own return the condensed ions near the polyelectrolyte to the solution, but there are two important cases in which such a change may occur.

The first is when there is a structural change in the polyelectrolyte that alters its charge parameter ξ. An example of this is the transition between double-stranded and single-stranded DNA. The value of ξ for the double helix is much greater than that for the single strand, and it is the same for the condensation of counter-ions. Counter-ions will thus be released during the transition from double helix to single strand. The process is somewhat more subtle than this because the released ions have more degrees of freedom than bound ones and this gives rise to an increase in entropy during the transition. We shall see later that we must also take into account changes in the rotational degrees of freedom of the water molecule.

The second case arises because a charged ligand will compete with counter-ions when binding to the charged groups of the polyelectrolyte. The binding is non-specific and is governed mainly by the Coulomb attraction between two macromolecules of opposite sign. The polyelectrolyte is assumed to be modelled by a line charge. The ligands may have various sizes: for example, if the polyelectrolyte is a DNA, the same approach may be used either for a small molecule with one or two charges or for a protein. To begin with, any ligand that triggers a conformational

Table 18.1 Results from Manning's theory

Parameter	$\xi < 1$	$\xi > 1$
effective ξ	ξ	1
fraction condensed	0	$1-\xi^{-1}$
fraction of free ions (in infinitely dilute solution)	$1-\xi/2$	$1/2\xi$

Box 18.3 **Consequences of the condensation model**

In this box, we shall be concerned with the calculation of the osmotic coefficient, the fraction of free ions and the ionic concentration near the line charge.

Once condensation has occurred and ξ is reduced to 1, the Debye–Hückel approximation can be assumed to be valid because of the reduction in potential around the polyelectrolyte and can be used to study the distribution of non-condensed counter-ions.

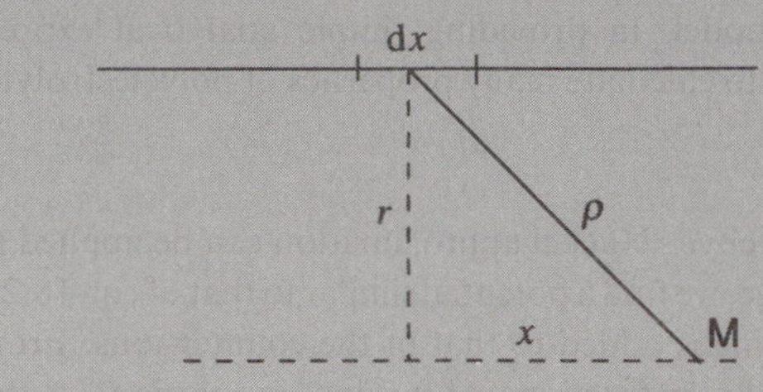

An elementary segment dx of the line charge creates a potential at a point M at a distance ρ that varies as $\rho^{-1}\exp(-K\rho)$. Since $\rho^2 = x^2 + r^2$, we have

$$\Psi(r) = (e/4\pi\varepsilon\varepsilon_0 b)\int_{-\infty}^{+\infty} \exp[-K(x^2+r^2)^{1/2}]/(x^2+r^2)^{1/2}\,\mathrm{d}x$$

where K is the usual Debye–Hückel parameter given by

$$K^2 = e^2\sum n_i^\circ z_i^2/\varepsilon\varepsilon_0 k_\mathrm{B}T$$

Integration leads to

$$\Psi(r) = eK_0(Kr)/2\pi\varepsilon\varepsilon_0 b \tag{1}$$

where K_0 is a modified Bessel function of zero order. As $Kr \to 0$, i.e. as the ionic strength becomes negligible, $K_0(Kr)$ behaves like $-\ln(Kr/2)$ and eqn 1 becomes the same as eqn 18.29.

The work W required to form the ionic atmosphere of the polyion can also be calculated from

$$W = \int \Psi' n_\mathrm{p}\,\mathrm{d}e \tag{2}$$

where Ψ' is the first term on the right-hand side of eqn 18.29 and n_p is the net number of charges carried by the polyion. Using the definition of ξ and integrating gives

$$W/k_\mathrm{B}T = (-\xi n_\mathrm{p}\ln K)/2$$

If there are N polyions in a volume V of solution, the total free electrostatic energy F_e is

$$F_\mathrm{e}/Vk_\mathrm{B}T = (-\xi N n_\mathrm{p}\ln K)/2V = (\xi c_\mathrm{e}\ln K)/2 \tag{3}$$

where c_e is the concentration of the polyion charged groups in the volume V.

This quantity can be used to predict several properties, in particular the fraction of ions that are free in the thermodynamic sense, i.e. those involved in creating osmotic pressure. Since the previous calculations assumed a salt-free solution ($K \to 0$), it will be assumed that the only ions present in the medium are counter-ions having a concentration $c_e = Nn_p/V$. The theoretical value of the osmotic pressure Π produced by these ions in solution (without a polyion) would be

$$\Pi_{th} = k_B T c_e \tag{4}$$

An *osmotic coefficient* ϕ can be defined from the experimental value of Π (with the polyion) by

$$\Pi_{exp} = k_B T \phi c_e \tag{5}$$

The difference $\Delta\Pi = \Pi_{exp} - \Pi_{th}$ depends on the free energy in a way given by the classical relationship

$$\Delta\Pi = -\partial F_e / \partial V \tag{6}$$

Using eqn 3 and remembering that $K^2 = e^2 c_e / \varepsilon\varepsilon_0 k_B T = e^2 N n_p / \varepsilon\varepsilon_0 k_B T V$ for a 1:1 electrolyte, the derivative in eqn 6 is easily obtained and gives

$$\Delta\Pi = -\xi k_B T c_e / 2 \tag{7}$$

Equations 4 and 5 also mean that $\Delta\Pi = (\phi - 1) k_B T c_e$, so that

$$\phi = 1 - \xi/2 \tag{8}$$

which is the fraction of thermodynamically free ions. This type of calculation is carried out with the Debye–Hückel approximation and is only significant if the approximation is valid. In the condensation model this means when $\xi \leq 1$.

What happens for a polyelectrolyte with $\xi > 1$? We saw that condensation means that a fraction $1 - 1/\xi$ of the polyelectrolyte charges are neutralized. In that case, the net number of charges on the polyion is no longer n_e but is n_e/ξ so that the effective value of ξ is brought back to 1. If ϕ' is the osmotic coefficient, then

$$k_B T \phi' c_e = k_B T \phi_{\xi=1} c_e / \xi \tag{9}$$

Equation 8 shows that $\phi_{\xi=1} = \frac{1}{2}$, so that

$$\phi' = 1/2\xi \tag{10}$$

change upon binding (bending, perturbations introduced by an intercalating agent in the case of DNA, etc.) will be excluded. The equilibrium between the free polyelectrolyte P, the free ligand L and the complex LP can be written

$$L + P \rightleftharpoons LP$$

The observed equilibrium constant K_{obs} is defined by

$$K_{obs} = [\mathrm{LP}]/[\mathrm{L}][\mathrm{P}]$$

In fact, the protein binding process involves a certain number m of ionic bonds between the charged groups of the polyelectrolyte and the charged groups of opposite sign carried by the ligand. At each bond, there is a release of counter-ions previously associated with the charged group (both those which are condensed and those present in the Debye atmosphere). At the same time, there is a release of counter-ions associated with the ligand. Consider, for example, a negatively charged polyelectrolyte. Let f be the fraction of bound counter-cations M^+ per charged group of the polyelectrolyte and let a be the number of anions X^- released from the positively charged groups of the ligand in the binding process. The equilibrium can be written

$$L + P \rightleftharpoons LP + mfM^+ + aX^- \tag{18.37}$$

The corresponding constant K_0 characterizes a thermodynamic equilibrium in which the concentrations of all the species present are taken into account. Assuming that the electrolyte in which the process occurs is M^+X^-, with $[M^+] = [X^-]$, we obtain

$$K_0 = [LP][M^+]^{mf}[X^-]^a/[L][P] = K_{obs}[M^+]^{mf+a}$$

and hence

$$\ln K_{obs} = \ln K_0 - (mf + a)\ln[M^+] \tag{18.38}$$

Since K_0 does not depend on $[M^+]$, taking the differential leads to

$$\mathrm{d}\ln K_{obs}/\mathrm{d}\ln[M^+] = -(mf + a) \tag{18.39}$$

As a result, when $\ln K_{obs}$ (or $\log K_{obs}$) is plotted against $\ln[M^+]$ (or $\log[M^+]$), a straight line is obtained whose slope is $-(mf + a)$ (Fig. 18.6). In many cases, $a \ll mf$

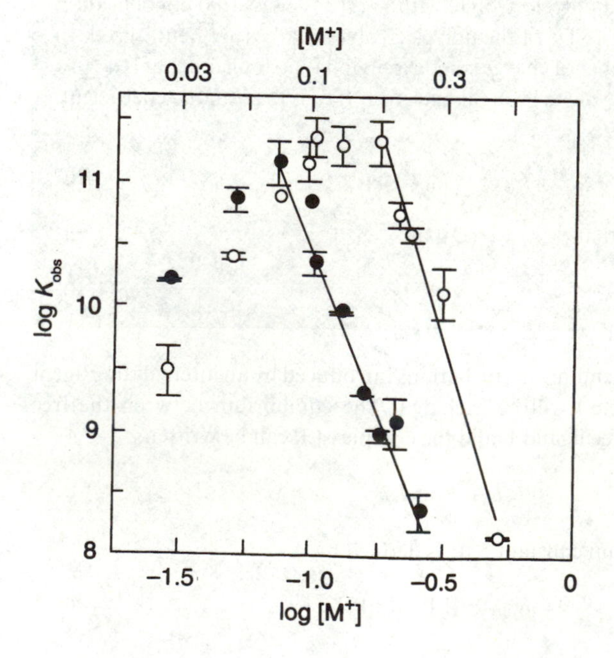

Figure 18.6 Variation with ionic strength of the association constant of the endonuclease *Eco*RI with the dodecamer CGCGAATTCGCG at pH 6 and 24°C: ○ = wild type, ● = mutant in which Arg187 is replaced by Ser. The slopes of the lines give eight and six ion pairs respectively. The difference corresponds to the disappearance of two arginines in the *Eco*RI dimer interacting with the oligonucleide. The decrease in K_{obs} below 0.05 M KCl is explained by protein aggregates that cannot bind the dodecamer. (From Jen-Jacobson *et al.*, 1983. Reprinted from *Journal of Biological Chemistry*, vol. 258, p. 14642, with the permission of the author.)

and the number m of interacting groups can be obtained provided f can be determined from the polyelectrolyte structure (for example, $f = 1-1/2\xi$ in Manning's model).

The relationship in eqn 18.39 has been widely used, but recent theoretical and experimental work (Sharp *et al.*, 1995) has shown that it is in practice only valid for small-sized charged ligands (dyes, antibiotics, etc.). We shall see later that a charged protein binding to a nucleic acid, for example, plays an important role in the variation of K_{obs} with ionic strength.

18.5 Comparison between counter-ion condensation (CC) and Poisson–Boltzmann (PB) models

The two models use the same assumptions:

1. A charged line of infinite length represents a polyelectrolyte with cylindrical symmetry and a charge parameter ξ.
2. The solvent is a continuous medium with a dielectric constant ε.
3. Ions are point charges whose dimensions and solvation can be neglected and whose distribution in space can be represented by a continuous charge density.
4. There are purely Coulomb interactions between ion and polyion.

The essential difference between the two models lies in the meaning given to the concept of bound ions.

The Poisson–Boltzmann approach

Here, bound ions are in equilibrium with free ions according to the classical law

$$(1-\alpha)/\alpha = Kc(a)$$

where α is the fraction of free ions, K the association constant and $c(a)$ the concentration of free ions on the polyion surface. We saw that

$$c(a) = c_0 \exp(y_a)$$

where c_0 is the counter-ion concentration far from the polyion and y_a is the reduced potential at the surface of the polyion. In this simple form, an important property is revealed: the true affinity as measured by K accounts for the high counter-ion concentration near the polyion. For example, when $y_a = 8$ ($e\Psi = 8k_BT$), we have $c(a) = 3000c_0$. Had the association constant been calculated with c_0 as the reference value, it would have been 3000 times as great.

The Manning model

Here, there is a dual limit on the validity: not only is the line of infinite length but the result is in principle only valid at a vanishing free ion concentration. The electric instability causes a condensation, which brings the value of ξ back to 1, implying that the fraction θ of condensed ions is equal to $1-\xi^{-1}$. Moreover, the volume occupied by condensed counter-ions is assumed to be a characteristic of the model. In that case, there cannot be a classical equilibrium between the concentrations of bound and free ions since the latter can tend to zero without altering the condensed ion concentration.

The binding energy between the condensed ions and the charges carried by the polyelectrolyte increases as the negative logarithm of the free ion concentration (see eqn 3 in Box 18.3).

Comparison

The two theoretical approaches can be compared in the first place by assuming the validity of the Poisson–Boltzmann equation and examining whether this approach can predict a phenomenon similar to condensation. The calculation has been carried out for a plane and a cylinder.

In the case of the plane, we have seen (Box 18.2) that an analytical solution exists for $\Psi(r)$ and that there is indeed a condensation mechanism, i.e. that a large fraction of counter-ions remain in the immediate vicinity of the charged plane when the total ion concentration tends to zero. For a cylinder, an analytical expression with $\xi > 1$ can only be found when the concentration of ions of the same sign around the polyion is negligible. We then find (Le Bret and Zimm, 1984) that the radius $R_{\theta(0)}$ of the cylinder enclosing the ion fraction $\theta(0) = 1 - \xi^{-1}$ increases indefinitely when $K \rightarrow 0$ (high dilution). It should, however, be noted that when $1/K \rightarrow \infty$, $R_{\theta(0)}$ increases like $(1/K)^{1/2}$. Yet if we seek the same limit of R_f for $\theta < \theta(0)$, we find a finite limit. For example, in the case of DNA ($\xi = 4.2$) and with $a = 1.25$ nm, we find at infinite dilution that 60 per cent of the charge is neutralized by counter-cations within a cylinder of radius $R \approx 3.5$ nm.

Moreover, the calculation shows that R_f is almost unchanged when the salt concentration of the medium increases by several orders of magnitude. In other words, the comparison between the Poisson–Boltzmann and Manning approaches leads to a better definition of the word 'condensation'.

In the immediate vicinity of the polyelectrolyte molecule, there is a true *condensed phase* of counter-ions, which, on average, remains strongly bound and which 'neutralizes' a large proportion of the charges carried by the polyelectrolyte. The concentration of this condensed phase is much higher than that of the solvent and changes little with the ionic strength. The cylindrical volume enclosing this condensed phase is, however, not constant and increases with the Debye length $1/K$ since R increases as $K^{1/2}$.

It is clear that the two approaches, the solution of the PB equation and the charged line condensation model, do not lead to exactly the same results when the ionic strength is no longer negligible, as is the case in physiological conditions for example, where it is often about 0.15 M. Much work has shown that the use of the Poisson–Boltzmann equation is the only valid approach in this case. However, we should realize that neither model takes account of a factor that cannot be neglected: the variation in the dielectric constant of water near a charge (see section 16.2.2).

The problem arose of how to interpret two calculations carried out with the charged line model of DNA (Fenley *et al.*, 1990). It was first seen that, in accordance with the assumptions made, the values found for $\theta = 1 - \xi^{-1}$ are no longer correct if the line is shorter than the Debye length corresponding to the ionic strength. Empirically, θ is found to vary as N^{-1}, where N is the number of phosphate groups. However, when an attempt is made to make the model more realistic by replacing the charged line by charges arranged on a double helix, the result is unexpected: the bound fraction $\theta(N)$ decreases as the salt concentration increases. When a function $\varepsilon(r)$ is introduced in place of the usual dielectric constant, the value of θ once again becomes independent of the salt concentration.

It is as if the naive charged line model used to represent a linear polyelectrolyte, without taking into account any internal order in the solvent, apparently offers a better interpretation of the facts than a more realistic model that neglects the organization of water molecules.

Note that, apart from this corrective effect of $\varepsilon(r)$, neither theoretical approach takes into account the size of ions or the repulsive potential between counter-ions.

To sum up, the study of polyelectrolytes using the continuum model yields several important results:

1. There is indeed an excess of counter-ions in the immediate vicinity of the polyion, but the quantitative features of the phenomenon are still not accounted for.
2. The charge parameter seems to be a significant variable in describing the behaviour of a linear polyelectrolyte.
3. The charged plane, which can be dealt with by a rigorous analytical treatment, is still a widely used model except when specific interactions depend on the geometry of the charge distributions over the plane.
4. The analytical solutions obtained are often used as the starting point for more sophisticated numerical models whose use we shall be examining a little later.

19

Mixed models

A continuum model cannot give a precise value for a local potential since the exact position of the charges is not taken into account, nor can it predict the path by which a charged molecule reaches its specific site. It is also difficult to predict local conformational changes caused by interactions between charges.

For all these reasons, new models have been developed that take into account the molecular structure and the detailed charge distribution. Two technological advances have contributed to the success of these new approaches:

(1) the increasing number of three-dimensional structures of proteins and nucleic acids now available using X-ray diffraction by crystals or NMR studies in solution;

(2) the astonishing increase in computer power, as regards both speed and memory, which has led to the development of new algorithms.

Because of the difficulty in dealing with purely molecular models, we begin in this chapter with an examination of 'mixed' methods, which combine a microscopic description of the charges carried by polyelectrolytes with a continuum model of the medium formed by water molecules and ions. For the time being, the latter will be described statistically. We start by recalling the form of the Poisson–Boltzmann equation.

19.1 The generalized Poisson–Boltzmann equation

Two approximations were made in establishing eqn 18.8:

1. The dielectric constant is uniform over the whole volume outside the charged surface representing the polyelectrolyte.
2. There are no charges other than mobile charges.

Neither of these can be retained if the exact three-dimensional structure of the polyelectrolyte is to be involved. The non-uniformity of ϵ requires the more general form of Poisson's equation to be used:

$$\nabla \cdot [\varepsilon(r)\nabla\Psi(r)] = -\rho(r)/\varepsilon_0 \tag{19.1}$$

while charges bound to the polyelectrolyte can be included by adding a charge distribution $\rho_1(r)$ independent of that of the mobile charges.

We may still use the reduced potential y defined by eqn 18.6, the term $2n_se/\epsilon\epsilon_0$ in eqn 18.5 becoming $2n_se^2/\epsilon_0k_BT$, i.e. $K^2\epsilon$. However, the dimensionless variable

$x = Kr$ cannot be used, so that we end up with

$$\nabla \cdot [\varepsilon(r)\nabla\Psi(r)] - \varepsilon K^2 \sinh y + \rho_1(r)/\varepsilon_0 = 0 \qquad (19.2)$$

which can be considered as the most general form of the Poisson–Boltzmann equation and which, of course, can in general only be solved numerically. Before we examine this, however, we consider a fairly old mixed model (Tanford and Kirkwood, 1957) devised for proteins, but one which could not take advantage of any structural knowledge since the first three-dimensional structure, that of myoglobin, did not appear until 1961. This is again a spherical model like that of Linderstrøm-Lang (1924) but one that takes into account the non-homogeneity of the dielectric medium and the presence of point charges inside the sphere.

19.2 Tanford and Kirkwood's model

This was the first model intended to predict the shape of a protein titration curve by describing the protein more realistically than in the Linderstrøm-Lang model. The charged groups have definite positions in the molecule (not necessarily on its surface). The protein molecule remains impervious to the solvent and retains its spherical shape with the same parameters R and a. Inside the molecule, the dielectric constant ε_{int} is assumed to be uniform and to lie between 2 and 10. The charges are assumed to be point charges, a simplification that does not alter the calculation of their (repulsive or attractive) interactions but precludes any calculation of self-energy terms.

The main parameters in this model are the distances between charges and the distance of each charge from the protein surface. Let z_ie and z_je be the charges at points i and j at distances r_i and r_j from the centre O of the sphere (Fig. 19.1). Assuming that the zero of energy is that of a completely uncharged protein, the work done in charging is

$$W = (e^2/8\pi\varepsilon_0)\sum_i\sum_j z_iz_j[(A_{ij} - B_{ij})/R - C_{ij}/a] \qquad (19.3)$$

with the following definitions of A_{ij}, B_{ij} and C_{ij}:

- $A_{ij} = R/\varepsilon_{\text{int}}r_{ij}$, the term in W giving the work done in introducing the charge into an infinite medium of dielectric constant ε_{int}.
- B_{ij} is given by

$$B_{ij} = \frac{1}{\varepsilon_{\text{int}}}\sum_{n=0}^{\infty}\frac{(n+1)(\varepsilon - \varepsilon_{\text{int}})}{(n+1)\varepsilon + n\varepsilon_{\text{int}}}\left(\frac{r_ir_j}{R^2}\right)^n P_n(\cos\theta_{ij}) \qquad (19.4)$$

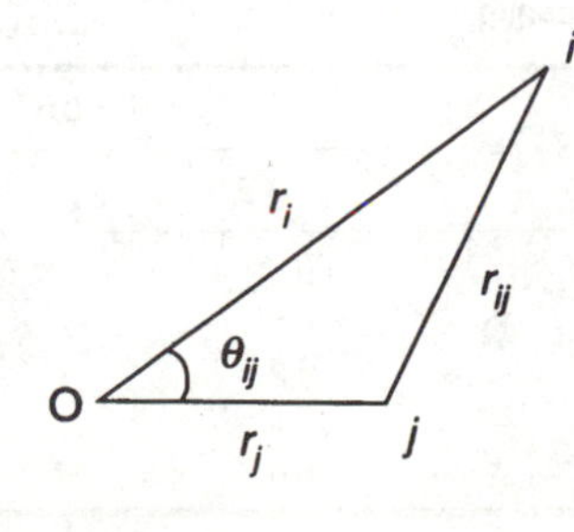

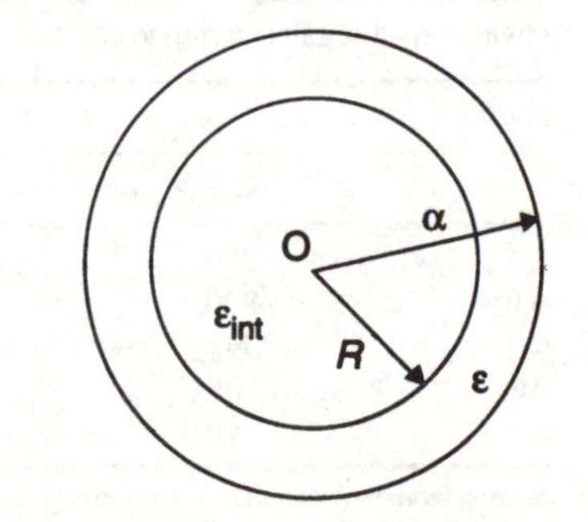

Figure 19.1 Geometry for Tanford and Kirkwood's model.

the term corresponding to the internal field resulting from the polarization of the surface separating two media with different dielectric constants, with spherical boundary conditions, P_n being Legendre polynomials (see Mathematical Appendix).

- C_{ij} represents the interaction with solvent ions in the Debye–Hückel approximation. Its first term is $Ka/\varepsilon(1+Ka)$, i.e. the term in the Debye–Hückel theory. When $Ka < 0.5$ (low ionic strength), C_{ij} can be limited to this first term. When $Ka \to 0$ (salt-free solution), then $C_{ij} \to 0$.

For point charges, the term $A_{ii} \to \infty$ and has no physical meaning. The same applies to B_{ii} when $r = R$ (charges on the protein surface).

To obtain the self-energy term, we can calculate B_{ii} when a single charged group is inside the protein of radius R provided it is not on the surface, i.e. $r < R$. In that case, B_{ii} can be replaced to a good approximation by the first two terms of eqn 19.4, i.e. $n = 0$ and $n = 1$ with $P_n = 1$ in both cases. We then find that

$$B_{ii} \approx (1/\varepsilon_{\text{int}})[(\varepsilon - \varepsilon_{\text{int}})/\varepsilon + 2(r^2/R^2)(\varepsilon - \varepsilon_{\text{int}})/(2\varepsilon + \varepsilon_{\text{int}})] \quad (19.5)$$

Substituting this in W and remembering that $\varepsilon_{\text{int}} \ll \varepsilon$, we obtain at zero ionic strength

$$W \approx -(e^2/8\pi\varepsilon_0)(1/\varepsilon_{\text{int}} - 1/\varepsilon)(1/R + r^2/R^3) \quad (19.6)$$

The term in $1/R$ is the energy required to surround a charged sphere of radius R and dielectric constant ε_{int} with a medium of dielectric constant ε. The term in r^2/R^3 is the energy of a dipole of length r inside a sphere of radius R. As a first approach, it is appropriate to separate the self-energy terms (discussed later) to leave the following for zero ionic strength:

$$W = \sum_i z_i^2 e^2 (A_{ii} - B_{ii})/8\pi\varepsilon_0 R + \sum\sum w_{ij} z_i z_j$$

with

$$w_{ij} = (e^2/8\pi\varepsilon_0 R)(A_{ij} - B_{ij}) \quad (19.7)$$

Calculation of $(A_{ij} - B_{ij})^{-1}$, which can be considered as an effective dielectric constant ε_{eff}, yields the results shown in Table 19.1. The values of ε_{eff} are almost a linear function of $\cos\theta_{ij}$. For points on the surface ($r/R = 1$), there is little variation with ε_{int}. As soon as the charges are inside the protein, both the distance d from the surface and the value of ε_{int} have a considerable effect.

Table 19.1 Values of $\varepsilon_{ij} = (A_{ij} - B_{ij})^{-1}$ for various values of the parameters and for pairs of charges at the same distance $r_i = r_j = r$ from the centre

$\cos\theta_{ij}$	$r/R = 1$		$r/R = 0.5$	
	$\varepsilon_{\text{int}} = 2$	$\varepsilon_{\text{int}} = 10$	$\varepsilon_{\text{int}} = 2$	$\varepsilon_{\text{int}} = 10$
−1	250	233	9	34
−0.5	200	185	6.5	25.5
0	145	139	4.2	17.4
0.5	86	85	2.1	9.5
0.9	28.5	29.2	0.6	2.9

Source: From Tanford and Kirkwood (1957).

Placing charges at 1 Å from the surface and dipoles at 1.5 Å, the Tanford–Kirkwood model gives quite a good representation of the protein titration curves by using the following relationship to calculate the new pK values of the ionizable groups:

$$\mathrm{p}K_i = \mathrm{p}K_0 - (2.3k_\mathrm{B}T)^{-1} \sum w_{ij}/z_i \tag{19.8}$$

where pK_0 is the value measured in a solution of the amino acid carrying the ionizable group in question.

When the three-dimensional structure of the protein is known, the positions of the charges with respect to the surface are also known. It is then possible to take into account their accessibility to the solvent by using the criterion of Lee and Richards (1971), in which a sphere modelling the water molecule is 'rolled' over the protein surface (see section 13.2).

If SA_i is the accessible fraction of site j ($0 < SA_j < 1$), w_{ij} in eqn 19.7 can be replaced by

$$w'_{ij} = w_{ij}(1 - SA_{ij}) \tag{19.9}$$

where SA_{ij} is an average taken over the two groups i and j. This is equivalent to defining a new effective dielectric constant

$$\varepsilon_\mathrm{eff} = [w_{ij}(1 - SA_{ij})r_{ij}]^{-1} \tag{19.10}$$

In principle, therefore, it is possible to determine a titration curve from the protein structure and compare it with an experimental curve. The agreement is remarkably good in spite of the arbitrary nature of the procedure and the inconsistency in envisaging a correction in the accessibility for the quantity A_{ij}. In Fig. 19.2, theoretical curves plotting $\varepsilon_\mathrm{eff}{}^{-1} = w_{ij}r_{ij}$ against r_{ij} are shown for various values of ionic strength. The dots represent the values of w'_{ij} calculated from eqn 19.9 for all pairs of sites i and j in ribonuclease (Matthew, 1985).

However, there is still a problem. In the Tanford–Kirkwood theory, where the charges are assumed to be point-like, transfer of charge from vacuum to the medium cannot be included. In fact, transfer occurs from the water to the inside of the protein. If the point charge is replaced by a charged sphere, the energy of transfer from a medium of dielectric constant ε_2 to one of dielectric constant $\varepsilon_1 \ll \varepsilon_2$ is given by the calculation already undertaken in Box 15.7:

$$W_\mathrm{transfer} = z^2e^2(1/\varepsilon_2 - 1/\varepsilon_1)/8\pi\varepsilon_0 r$$

where r is the radius of the charged sphere and ze is its charge. For $z = 1$, the following formula gives W in kcal mol^{-1} when r is in Å:

$$W_\mathrm{transfer} \approx 166(1/\varepsilon_2 - 1/\varepsilon_1)/r \tag{19.11}$$

Taking $\varepsilon_2 = 80$ and values of ε_1 equal in turn to 2, 4 and 10, we find that $W = 81/r$, $38.4/r$ and $14.5/r$ respectively. For an ion of radius $r = 2$ Å, this gives energies of 40.5, 19.7 and 7.2 kcal mol^{-1}. A change of one unit in pK (Δp$K = 1$) corresponds to $2.3RT$ or 1.38 kcal, so that we should expect dramatic changes in pK of the order of 29, 14 and 5 units respectively. More precisely, since the equilibrium is in favour of the neutral species in the medium with the lower dielectric constant, carboxylic acids would only be dissociated at very high pH while NH_3 groups would only be

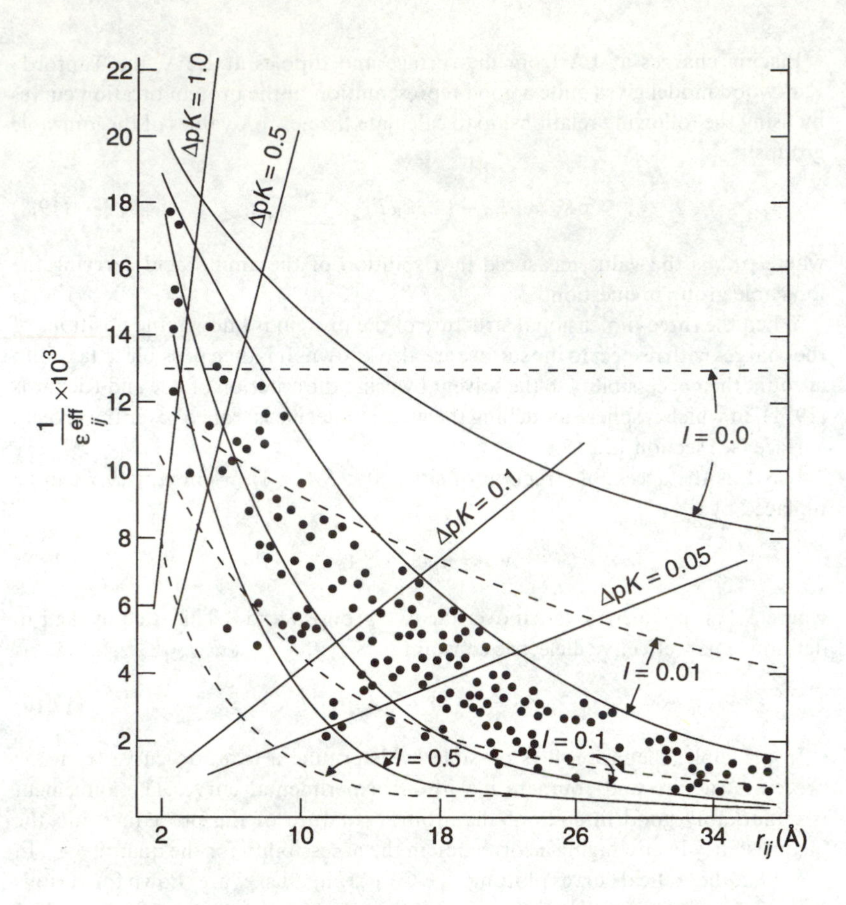

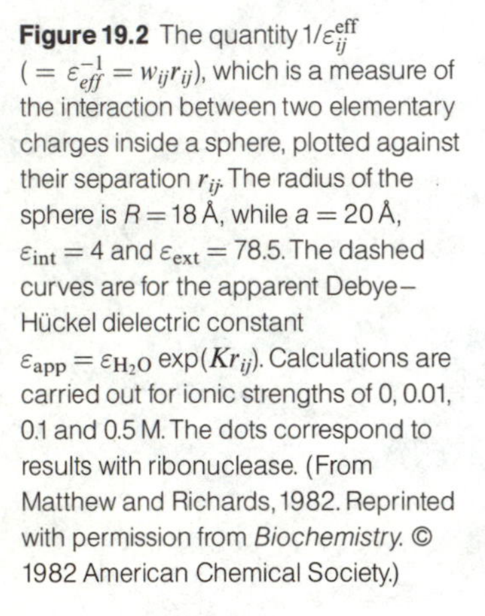

Figure 19.2 The quantity $1/\varepsilon_{ij}^{\text{eff}}$ ($= \varepsilon_{eff}^{-1} = w_{ij}r_{ij}$), which is a measure of the interaction between two elementary charges inside a sphere, plotted against their separation r_{ij}. The radius of the sphere is $R = 18$ Å, while $a = 20$ Å, $\varepsilon_{\text{int}} = 4$ and $\varepsilon_{\text{ext}} = 78.5$. The dashed curves are for the apparent Debye–Hückel dielectric constant $\varepsilon_{\text{app}} = \varepsilon_{\text{H}_2\text{O}} \exp(Kr_{ij})$. Calculations are carried out for ionic strengths of 0, 0.01, 0.1 and 0.5 M. The dots correspond to results with ribonuclease. (From Matthew and Richards, 1982. Reprinted with permission from *Biochemistry*. © 1982 American Chemical Society.)

formed at very low pH. Because this is never found, some explanation is needed and three reasons for it can be offered.

1. The interior of a protein is a medium of high dielectric constant. Such an explanation poses the problem of what exactly we mean by the dielectric constant of a protein. If only the electronic and atomic polarizabilities are used to account for ε, we should expect ε_{int} to be about 2. Yet because of the internal packing, a protein is very similar to a crystalline medium and ε_{int} might be close to 4 as it is for crystalline polyamides. In fact, ε_{int} must be considered as an anisotropic quantity that may vary from one point to another. Work by Antosiewicz *et al.* (1994), using quite a different approach that we shall look at later, shows that a value of 20 for the dielectric constant of a protein appears to give the best agreement between calculation and experiment.

2. Hydrogen bonds are formed locally and their energy may compensate for that needed to introduce the charge into a medium of low dielectric constant. When a proton is transferred from one medium to another (e.g. from COOH to O), *ab initio* calculations (Scheiner and Hillenbrand, 1985) show that the energy depends strongly on the respective geometries of the two groups and more precisely on the angle θ between C–O and O· · ·O, which is a measure of the distortion of the hydrogen bond. If $\theta = 0$, the C–O· · ·H–O form is preferred over C–OH· · ·O= accompanied by a change in the pK of COO^-. The difference of 7.5 kcal in favour of the first form is due both to the small distortion of the hydrogen bond and to the charge–dipole interaction between the C–O dipole and the charge carried by H–O.

Table 19.2 Frequency of buried ionized groups

Amino acid	Frequency	Amino acid	Frequency
Glu	15/368	His	15/368
Asp	34/411	N-terminal	2/38
Lys	5/525	C-terminal	0/38
Arg	4/242		

Source: From Rashin and Honig (1984).

This type of calculation remains valid inside a protein and shows that, under certain geometrical conditions, the respective pK values of the two groups can apparently be reversed.

3. The local dielectric environment directly responsible for the change in pK may be much smaller in volume than that of the dielectric separating the two interacting charges. In other words, the local presence of a few water molecules would be enough to maintain a pK close to pK_{int}. We should point out, however, that a polar environment does not necessarily imply a high dielectric constant: if polar groups are immobilized by structural constraints, their contribution to ε_{int} is small precisely because the high dielectric constant of liquid water arises partly from the concerted rotation of several water molecules. It is to avoid this problem that the Tanford–Kirkwood model assumes that the charges are close to the surface. From an analysis of 35 proteins with a known tertiary structure (Rashin and Honig, 1984), it is possible to draw up Table 19.2, giving the frequencies with which an ionized group (NH_3^+ or COO^-) is buried in the protein and not accessible to the solvent. More than two groups on average per protein are found to be buried, forming either salt bridges (6/36) or more probably hydrogen bonds.

19.3 **Other models**

The same kind of approach was adopted with DNA (Matthew and Richards, 1984), making the assumption, apparently verified by calculation, that the change from a spherical to a cylindrical geometry does not alter the distribution of counter-ions around the polyelectrolyte. The free electrostatic energy G_{el} is calculated from the positions r_i and the charges z_i (expressed as an algebraic number of elementary charges e) of the phosphate groups by summing all the interactions between them:

$$G_{\text{el}} = \sum_i \sum_j w_{ij}(1 - SA_{ij})z_i z_j \tag{19.12}$$

with $SA_{ij} \approx 0.75$ for phosphates and $z_i = z_j = -1$. We then seek the electric energy of a distribution of ions at points k near the phosphate groups, and similarly we can write:

$$G_q = \sum_i w_{iq}(1 - SA_{iq} z_i) \tag{19.13}$$

where w_{iq} is the coupling term between the phosphate groups and the ion q, and where SA_{iq} lies in the range 0.48 to 0.55.

To find the mean occupation z_k of each site k, an association constant for the ion at its site is defined by

$$K_k = \exp(G_q/k_{\text{B}}T)$$

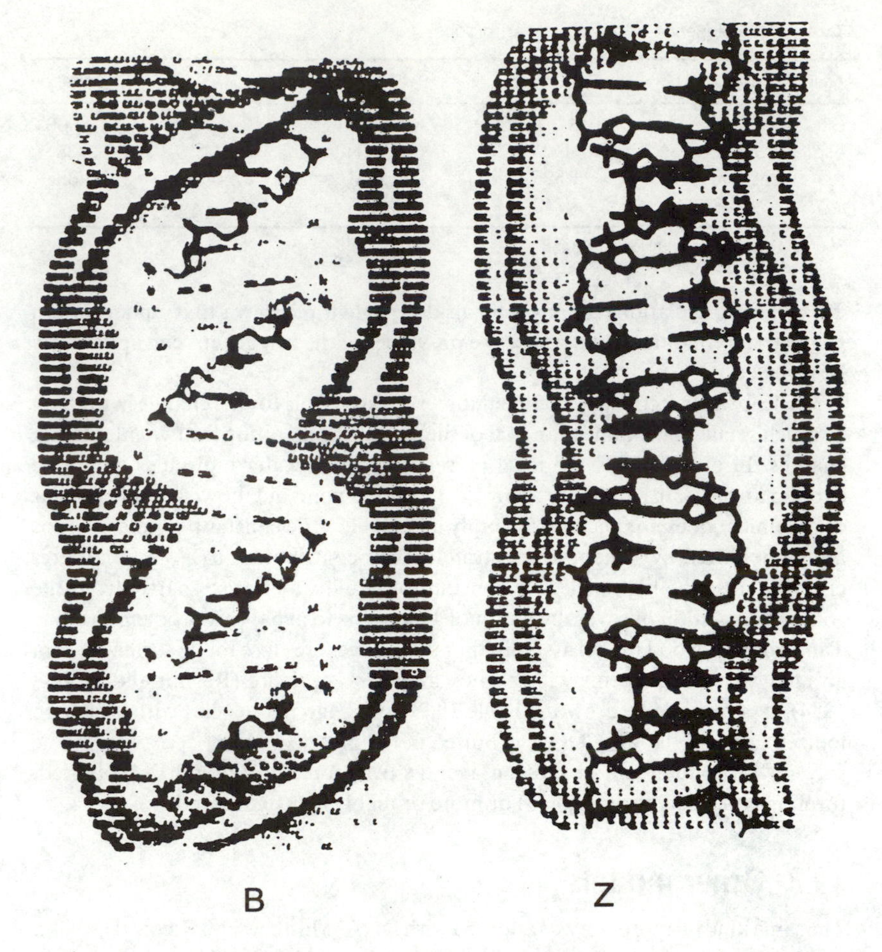

Figure 19.3 Equipotential surfaces for energies of $3k_{\mathrm{B}}T$ and $2k_{\mathrm{B}}T$ around B-DNA and Z-DNA at an ionic strength of 0.1 M without bound cations. (From Matthew and Richards, 1984. Reprinted from *Biopolymers* with the permission of John Wiley & Sons. © 1984 John Wiley & Sons.)

so that

$$z_k = K_k[q]/(1 + K_k[q]) \tag{19.14}$$

where $[q]$ is the free ion concentration. Since several binding sites are available (chosen to have an energy greater than $3k_{\mathrm{B}}T$), their number is determined by iteration. Equipotential lines can be obtained in this way with B-DNA or Z-DNA (Fig. 19.3).

Another method derived from that of Tanford and Kirkwood has been developed by Gilson *et al.* (1985). To avoid the criticism of the Tanford–Kirkwood model (neglecting self-energies when using point charges), the charge is now represented by a sphere of radius 2 Å. Inside the sphere, the dielectric constant is taken to be low ($\varepsilon_{\mathrm{int}} = 2$ to 4) when the charge is within the protein and high ($\varepsilon_{\mathrm{int}} = 80$) when it is in the solvent. Without entering into details of the calculations, we simply give the results shown in Figs 19.4 and 19.5.

Several points need to be made about this:

1. In the interaction energy between two charges, the polarization term is about the same order of magnitude as that due to the purely Coulomb interaction.
2. The self-energy term decreases rapidly when the charge gets near the protein surface (from 40 to 20 kcal mol^{-1} for a protein of radius 20 Å).

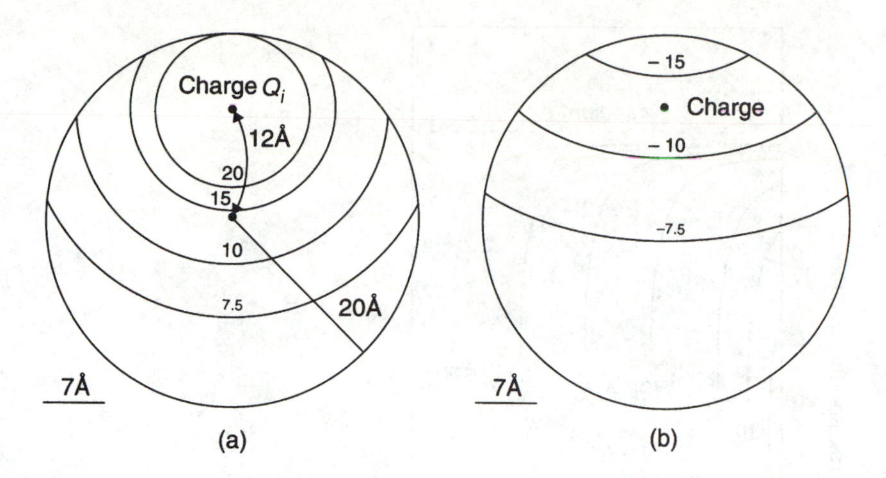

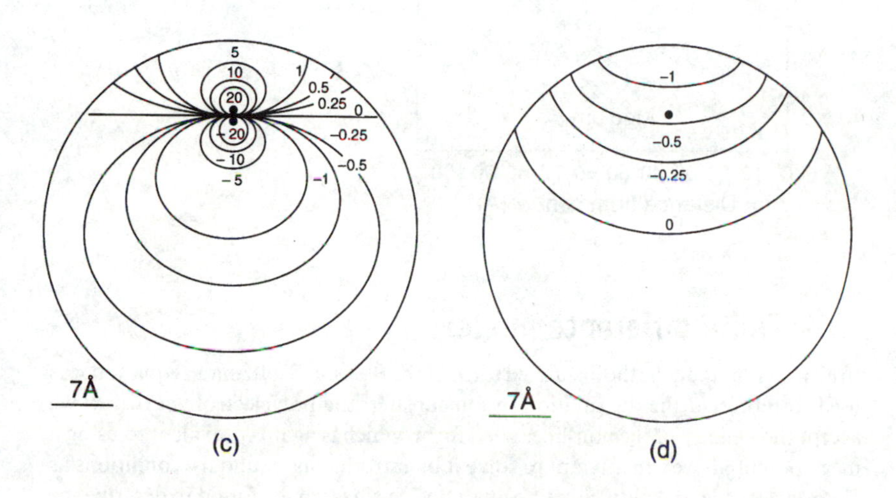

Figure 19.4 Charge–charge and charge–dipole interactions in a spherical protein immersed in water. (a) Coulomb interaction energy between two charges, one at 12 Å from the centre and the other at variable distances. (b) Polarization interaction energy between the same two charges. (c) Coulomb interaction energy between a charge and a dipole consisting of two elementary charges, $-e$ at 11.5 Å and $+e$ at 12.5 Å. (d) Polarization interaction energy between the same dipole and one charge. (From Gilson *et al.*, 1985. Reprinted from *Journal of Molecular Biology* with the permission of Academic Press.)

3. The effective dielectric constant defined above is not a simple function of geometrical parameters. It would be possible to adopt a constant value (but different from 2) at the centre of the protein and one that varies with distance near the surface.

4. The calculation of the forces exerted between the charges clearly shows the importance of the polarization forces in altering both the magnitude and direction of the resultant force. It is therefore incorrect to include only the purely Coulomb forces in calculations of the conformation and in those of molecular dynamics.

5. The solvent with electrolytes can be considered to be almost like a conductor ($\varepsilon \to \infty$). In other words, the electric field lines are concentrated in the solvent and the definition of ε_{eff} is almost devoid of interest.

6. There is no reason why the dielectric constant inside the protein should be high even if polar groups are present, since the rotational mobility of the dipoles (which explains the high value for water) is low inside the protein.

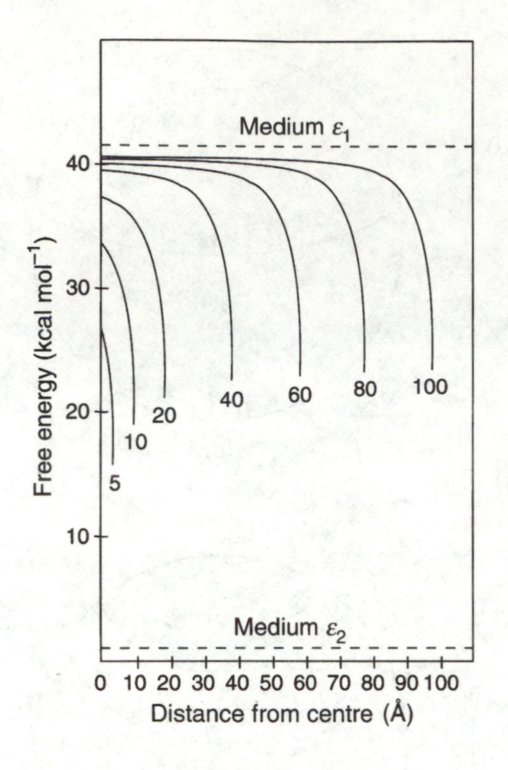

Figure 19.5 Free energy in kcal mol^{-1} plotted against distance from the centre in Å for a charge of radius 2 Å buried inside proteins with radii between 5 and 100 Å. The upper and lower dashed horizontal lines show the Born energies of the charge deep in the protein and water respectively. (From Gilson *et al.*, 1985. Reprinted from *Journal of Molecular Biology* with the permission of Academic Press.)

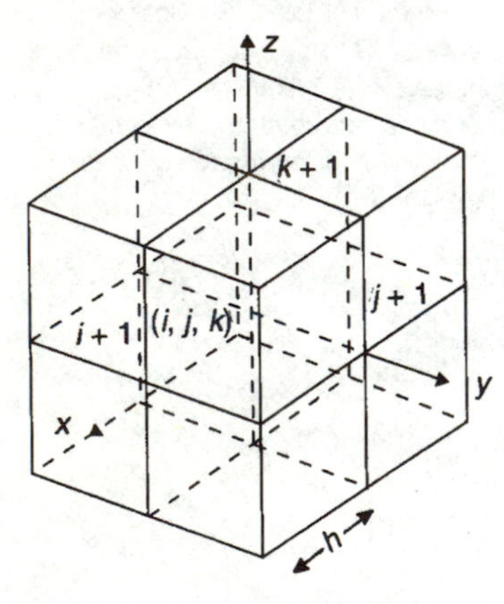

Figure 19.6 Three-dimensional grid for the finite-difference method.

19.4 Finite-difference model

In all previous models, the linearized form of the Poisson–Boltzmann equation was used to represent the distribution of ions around the polyelectrolyte. If now we accept the validity of the non-linearized form, which is again equivalent to using a mean potential, we can attempt to solve it by introducing boundary conditions as close as possible to reality. Since no analytical surface can be found to describe the exact shape of a biopolymer, the interface between the inside and outside of the polymer is represented by a grid of points that fits the shape of the polymer as exactly as possible (Warwicker and Watson, 1982). The interior and the solvent are also defined by a three-dimensional network of points whose spacings are defined in Cartesian coordinates.

The solution of Poisson's equation

$$\nabla \cdot (\varepsilon \nabla \Psi) + \rho/\varepsilon_0$$

involves calculating Ψ and ρ at each point of the network by replacing derivatives with finite differences. It should be pointed out that the function $\varepsilon(r)$ in the generalized equation in fact reflects the change at the boundary between the polymer and the solvent from a medium of low dielectric constant to one of high dielectric constant.

Each of the eight small parallelepipeds surrounding a point (i, j, k) is given a dielectric constant. The potential at (i, j, k) depends on the potential at the six nearest points and the eight values of ε chosen from among the *two* values ε_{int} and ε (Fig. 19.6).

We have the following relationships:

$$dx^2 = dy^2 = dz^2 = h^2$$

and

$$d\Psi = \Psi_{ijk} - \Psi_{i'j'k'}$$

with the coordinates of the six points

$$i' = i \pm 1 \qquad j' = j \qquad k' = k$$

$$i' = i \qquad j' = j \pm 1 \qquad k' = k$$

$$i' = i \qquad j' = j \qquad k' = k \pm 1$$

In addition, Gauss's theorem means that the integral of div($\varepsilon\nabla\Psi$) over a volume can be replaced by that of the gradient of $\varepsilon\nabla\Psi$ over a surface. Replacing grad Ψ by finite differences means that Poisson's equation now becomes

$$\sum_{i'j'k'}(\Psi_{ijk} - \Psi_{i'j'k'})\varepsilon_{i'j'k'} = -\rho_{ijk}h^2/\varepsilon_0 \tag{19.15}$$

where $\varepsilon_{i'j'k'}$ is the mean value of the four dielectric constants surrounding the line joining the points (i, j, k) and (i', j', k'). When one atom of the molecule with known crystallographic coordinates covers the centre of a parallelepiped with its van der Waals radius, we take ε_{int} as the value.

The solution starts with an arbitrary distribution of charges and potentials (given, for example, by a continuum model) and an iterative process enables the charges and potentials to be calculated at each point. The method converged very slowly and the computation times were excessive, but a considerable improvement has been made (Holst *et al.*, 1994) by carrying out the iteration over grids with decreasing spatial resolution by using an incomplete solution of Newton's algorithm. Since then, increasing numbers of maps of the potential $\Psi(r)$ around polyelectrolytes of known three-dimensional structures have been published.

Proteins

Distributions of potential and charge around proteins involve two types of function: the recognition of a nucleic acid or of another protein, and guidance by diffusion in an electric force field of a substrate towards the active site of an enzyme. Figs 19.7 and 19.8 illustrate these two aspects. The first gives the potential map for the β sub-unit of the *E. coli* DNA polymerase III (Kong *et al.*, 1992). In spite of the net negative charge ($-22e$) of the dimer, at the centre of the ring formed by it there is a positive potential favouring the interaction with DNA. The second figure shows that the approach of O_2^- towards the active site of superoxide dismutase is favoured by zones of positive potential whereas the net charge of the protein is negative (Getzoff *et al.*, 1983). Combining directed mutagenesis with the computation of potential maps gives access, for the first time, to the mechanisms of action of an increasing number of proteins.

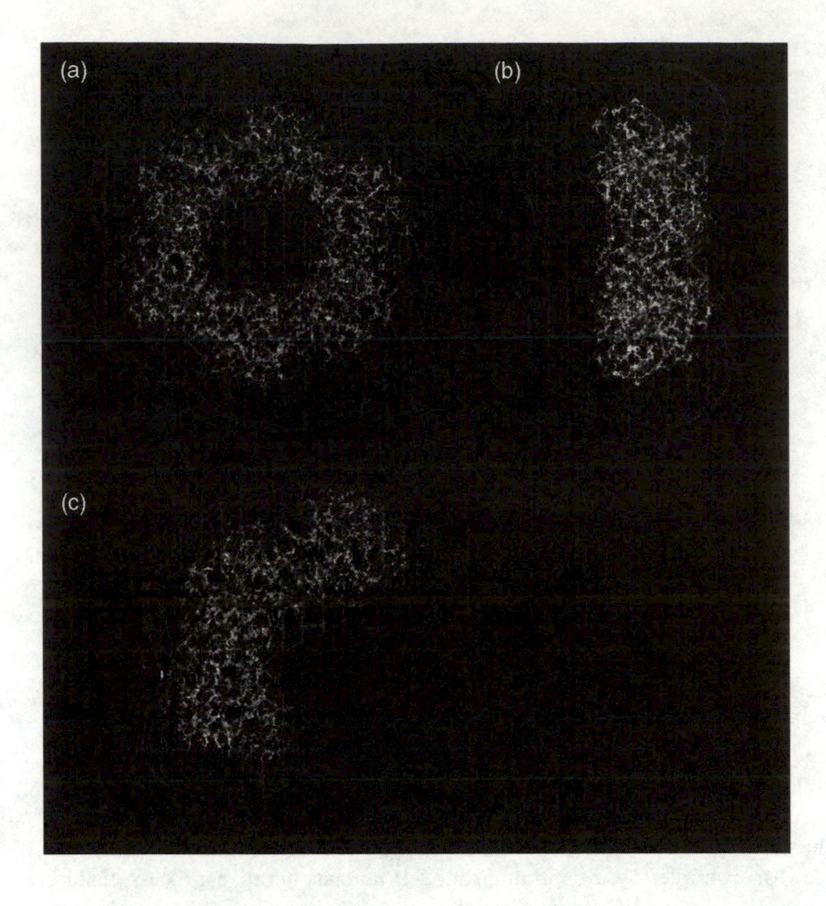

Figure 19.7 Electrostatic potential maps for the β sub-unit dimer of *E. coli* pol III and an isolated monomer. The calculation was carried out assuming a uniform dielectric constant of 80 for the solvent and 2 for the protein interior. The mesh contours represent electrostatic potential energies of $\pm 2.5 k_B T/e$. Two orthogonal views of the electrostatic potential of the dimer are shown in A and B, and C is the potential for an isolated monomer. (From Kong *et al.*, 1992. Reprinted from *Cell*, vol. 69, with the permission of Cell Press and with the authors' permission. © 1992 Cell Press.)

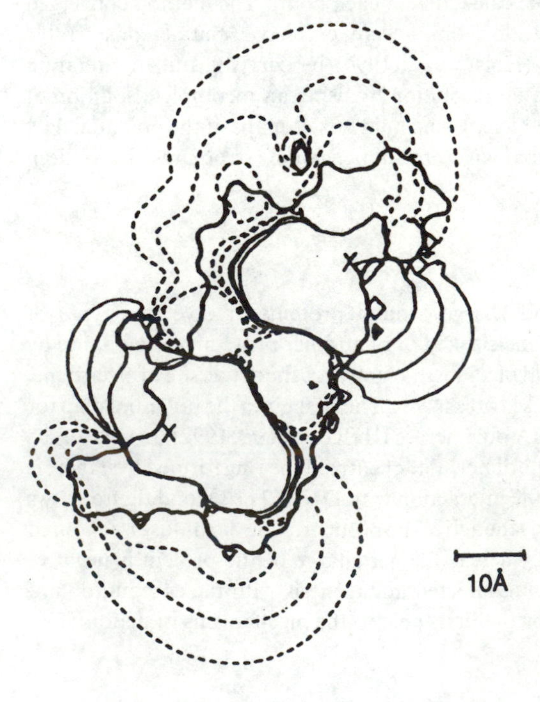

Figure 19.8 Cross-section through the dimeric molecule of superoxide dismutase in the region of the active site (the two copper atoms are indicated by crosses). The calculation was carried out with $\varepsilon_{int} = 2$ and $\varepsilon_{ext} = 80$ in a salt-free solution ($K^{-1} \rightarrow \infty$). Equi-energy contours are drawn for $k_B T/2$, $k_B T$ and $2k_B T$, with a dashed line for negative potentials and a full line for positive potentials. Note the attractive region around the active site to accommodate the negatively charged substrate (O^{2-}). (From Klapper *et al.*, 1986. Reprinted from *Proteins* with the permission of John Wiley & Sons. © 1973 John Wiley & Sons.)

Nucleic acids

An initial approach (Klein and Pack, 1983) applying the method to nucleic acids did not solve the Poisson–Boltzmann equation, but wrote the potential Ψ as the sum of Ψ^{C}, created by charges bound to the polyion depending only on its position in space, and Ψ^{S}, created by the distribution of counter-ions and co-ions. The latter are placed arbitrarily on a lattice, at each point i of which

$$\Psi_i^{C} = \sum_{k=1}^{K} (q_k/\varepsilon r_{ik})$$

(the polyion possessing K charges). The following relationship exists between ρ and Ψ^{S}:

$$\rho = 2Ne \sinh(e\Psi^{S}/k_{B}T)$$

Each of the j points in the external medium is associated with a volume V_j and a mean charge density ρ_j. The system of two equations

$$\Psi_i = \Psi_i^{C} + \sum (q_j/\varepsilon r_{ij}) \tag{19.16}$$

$$q_i = \rho_i V_i = V_i Ne[\exp(-e\Psi_i/k_{B}T) - \exp(e\Psi_i/k_{B}T)] \tag{19.17}$$

is solved by an iterative process, which converges only slowly because the values of q_i have to be slightly modified at the same time. Although the Poisson–Boltzmann equation does not appear explicitly in the calculation, the characteristic approximation is again obtained by replacing the potential of mean force by the mean Coulomb potential.

When the value of Ψ^{S} becomes close to that of the unperturbed potential Ψ^{C}, the perturbations of Ψ_i occur in exponential form in q_i, and these fluctuations may seriously upset the internal consistency of the process.

In the case of DNA, Klein and Pack (1983) made the calculation on a 30 bp portion of the double helix with Arnott coordinates (Arnott, 1970). A residual charge determined by molecular orbital calculations was placed on each atom. The lattice contained 4300 points up to 12 Å and 6100 points between 12 and 60 Å.

If we compare the two potentials created at the position of each DNA atom, we find that the negative potential due to the charge distribution on the phosphate groups is largely compensated by the positive potential due to the counter-ions. The differences between the gradient of these two potentials, on the other hand, are reflected by non-zero electrostatic forces acting on the various atoms (Figs 19.9 and 19.10).

If the volume occupied by the ions is taken into account by introducing a short-range repulsive potential, the volume of the condensed layer clearly increases. Peaks of positive charge density can be seen to appear in the minor groove of the B form.

The electrostatic stabilization of a nucleotide by the ionic environment is almost unchanged by the introduction of an ionic volume. It is the structure much more than the nature of the forces that modulates the ionic distribution around the DNA. However, the calculations do not yet take into account the place and role of water molecules, and for the time being use only regular 'canonical' forms to represent the double helix.

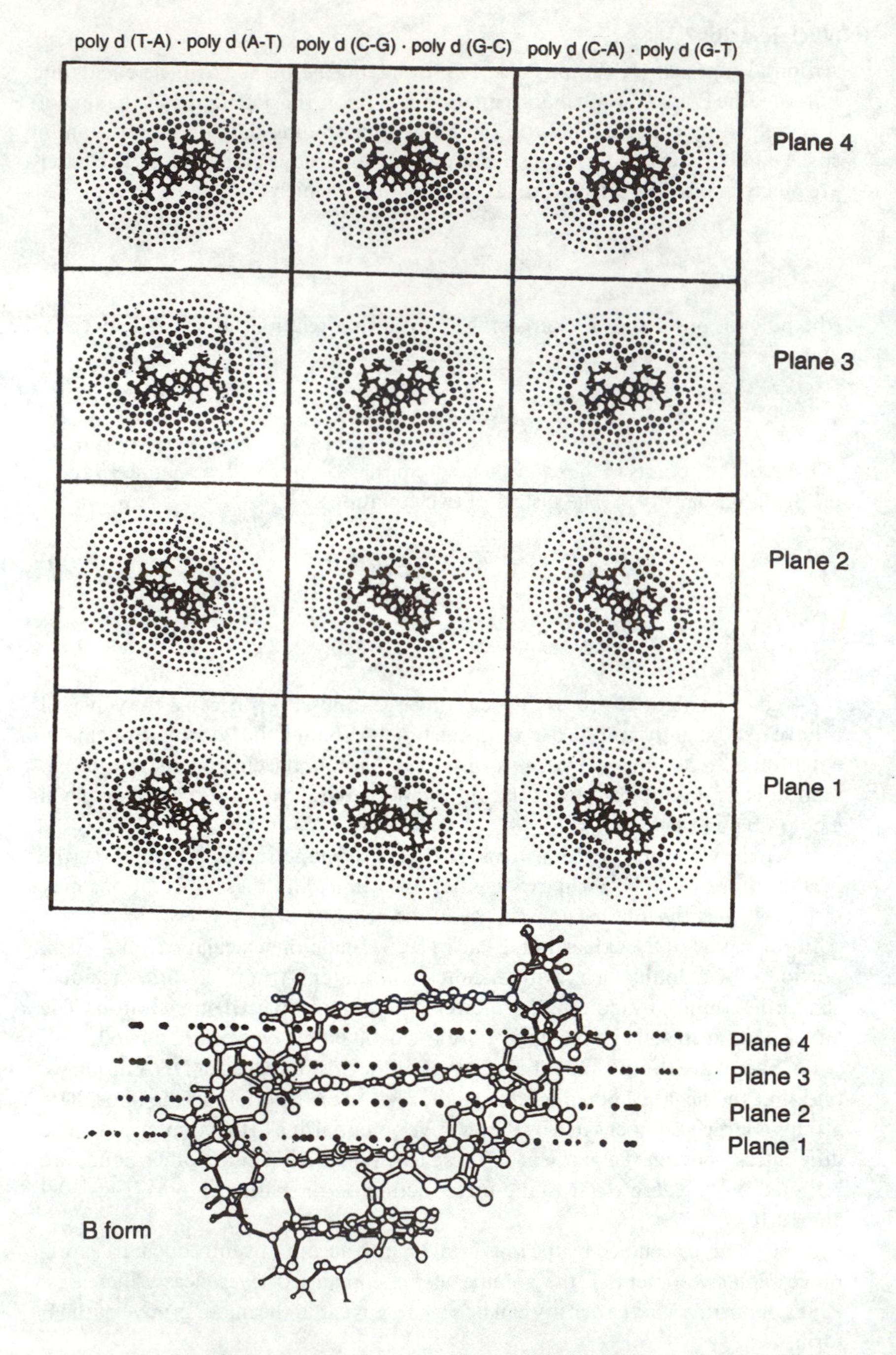

Figure 19.9 Equi-energy contours in four cross-sections of a B-type double helix for the polymers indicated at the top of the diagram. (From Pack *et al.*, 1986. Reprinted from *Nucleic Acids Research* by permission of Oxford University Press and with the authors' permission.)

The method has been applied (Lamm and Pack, 1990) to the calculation of the H ion concentration around poly(dG) · poly(dC). It is found that the pH distribution around the polynucleotide is far from being uniform, as can be seen from the curves in Fig. 19.11.

The problem has more recently been taken up again (Misra *et al.*, 1994a; Sharp, 1995; Sharp *et al.*, 1995; Sharp and Honig, 1995) by using methods of solving

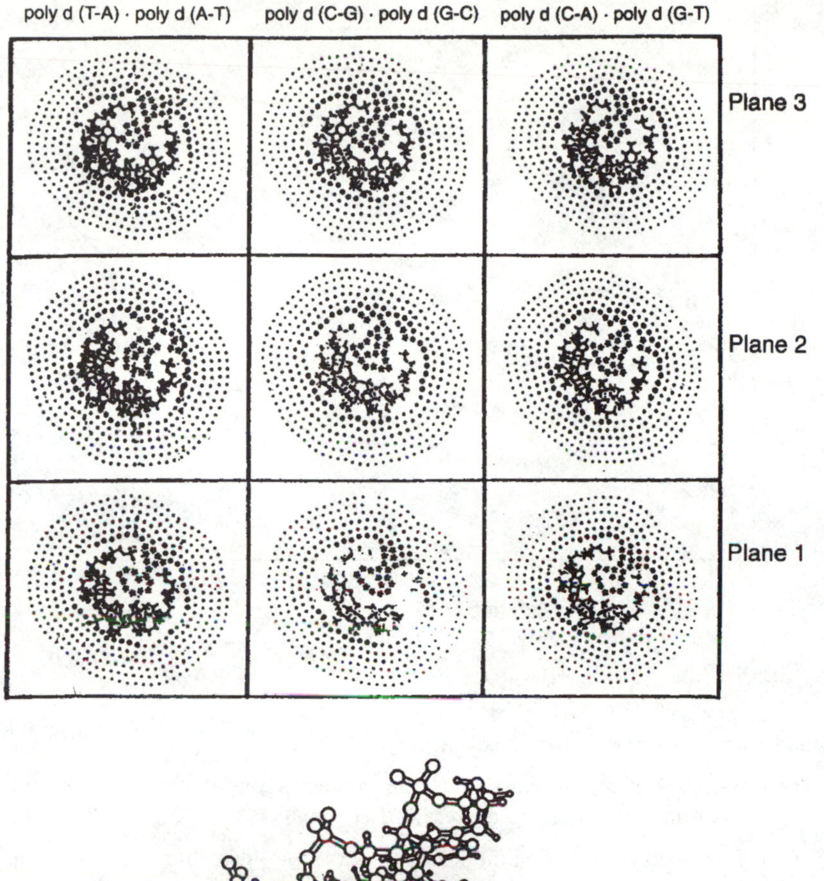

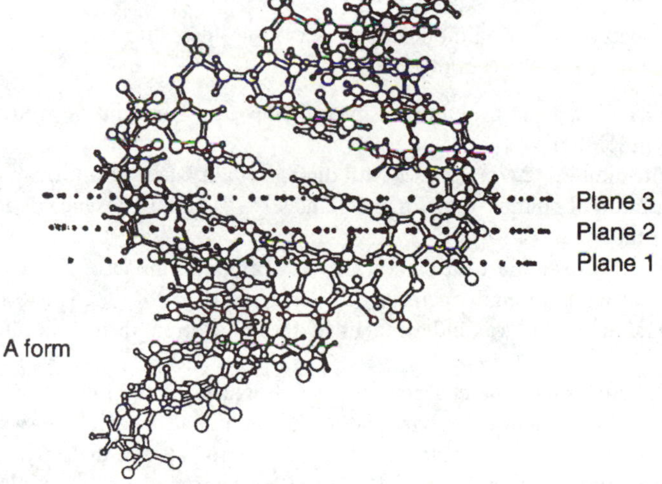

Figure 19.10 Equi-energy contours in three cross-sections of an A-type double helix for the same polymers as in Fig. 19.9. (From Pack *et al.*, 1986. Reprinted from *Nucleic Acids Research* by permission of Oxford University Press and with the authors' permission.)

the Poisson–Boltzmann equation on a three-dimensional grid. The procedure involves calculating the variation ΔG_e in the electrical free energy of the polyelectrolyte when it changes from its neutral to its charged state in the presence of external ions. The values of ΔS and ΔH can be calculated from ΔG_e. The entropy

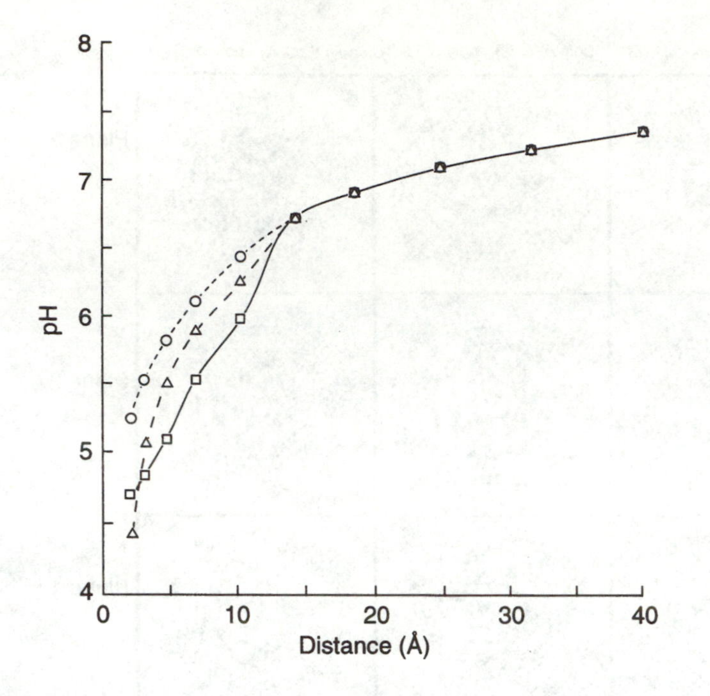

Figure 19.11 Mean value of pH plotted against the distance *r* from the surface of the cylinder representing the polyelectrolyte: □, major groove; △, minor groove; ○, elsewhere. Above 15 Å (■), because of the helical shape, the effect of the grooves is no longer detectable. (From Lamm and Pack, 1990. Reprinted from *Proceedings of the National Academy of Sciences, USA* with the authors' permission.)

change arises from two effects near the polyelectrolyte:

(1) a potential effect, which causes the condensation of counter-ions and reduces their translational degrees of freedom;

(2) a field effect, which immobilizes water molecules and reduces their rotational degrees of freedom.

These results allow us to evaluate the models previously used in the study of nucleic acids in solution.

1. At infinite dilution, ΔG_e increases with dimensionality: as a result, there is a total condensation of counter-ions for the plane ($\Gamma \rightarrow 0$) and an absence of condensation for the sphere ($\Gamma \rightarrow -0.5$).

2. The cylinder gives almost the same value as Manning's model ($\Gamma = -0.07$). As soon as the ionic strength increases, Γ decreases (e.g. to -0.25 in 0.1 M NaCl), whereas the value was independent of the ionic strength in Manning's theory.

3. The total entropy of the charged cylinder decreases as the ionic strength increases. On the other hand, the part due to the rotational behaviour of water increases from one-third of the total entropy in 10^{-3} M to half of it in 0.1 M.

4. When we compare (Sharp *et al.*, 1995) the binding of charged ligands to DNA, the result is different for the same total charge ($2e$) depending on whether it is a model sphere of radius 10 Å, an elongated molecule such as dapi (4′,6-diamidino-2-phenylindole) or a protein such as the λ-repressor, for which the net charge represents the balance between 22 negative charges and 24 positive charges. For the first two, the slope of the graph of $\log K$ against $\log[\mathrm{M}]$ (see eqn 18.39) is close to -1.9 and is little different from that deduced from Manning's theory (-1.76). For the protein, on the other hand, the slope of -4.7 is

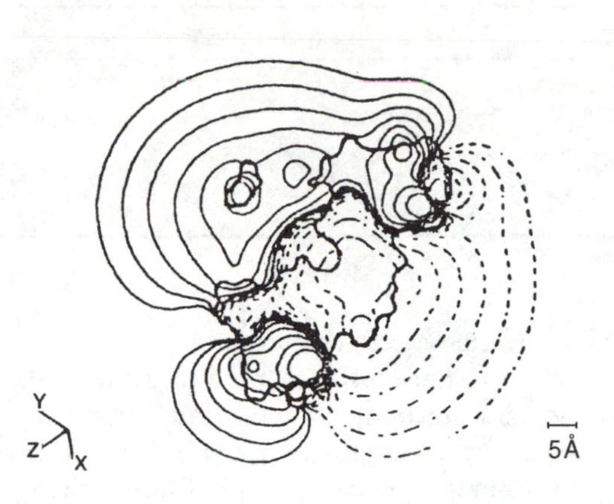

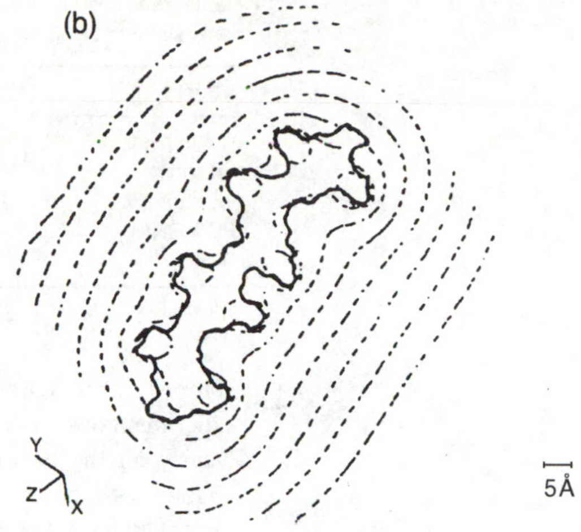

(c)

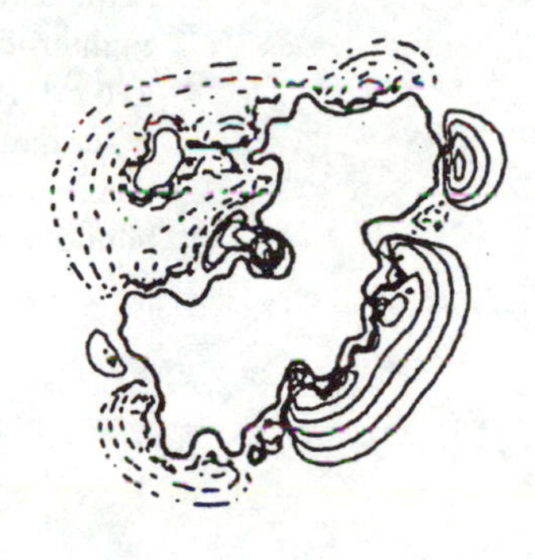

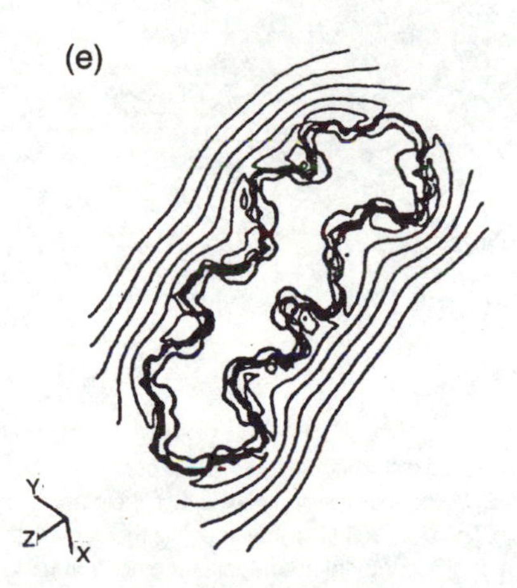

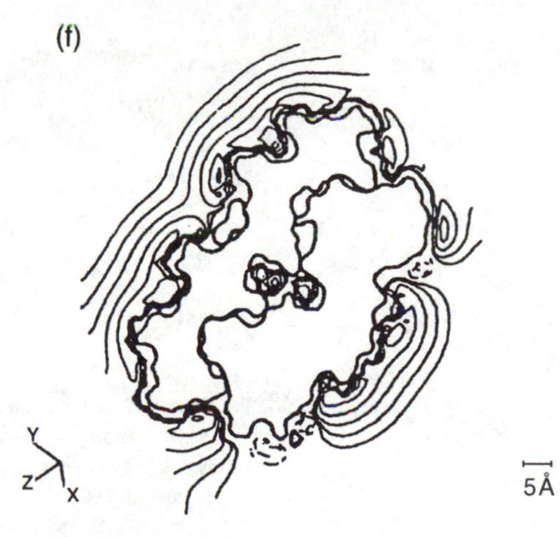

Figure 19.12 See p. 376 for caption.

Table 19.3 Entropic and enthalpic contributions to the variation in association constant with ionic strength at 25°C; the values in parentheses in the last column are experimental

Model	Enthalpy	Cratic $-T\Delta S$	Dielectric $-T\Delta S$	Total
Condensation	−0.084	1.51	0.332	1.76
PB sphere	−3.43	1.59	3.74	1.9
PB dapi/DNA	−3.34	1.47	3.82	1.95 (2.1)
PB λ-DNA	−1.17	0.41	5.5	4.7 (4.5–5.2)

Source: From Sharp *et al.* (1995).

completely different but is nevertheless closely comparable with that determined experimentally. In fact, there is an anisotropic charge distribution in the protein. Nearly all the positive charges are located on the 'face' of the repressor in contact with the DNA and almost neutralize the negative charges carried by the phosphate groups. The maps of the potential and ionic distribution in Fig. 19.12 throw light on this phenomenon.

Moreover, the entropic contribution due to the rotational degrees of freedom of the water molecules accounts for 90 per cent of the total change in entropy. Whereas in the condensation model the increase in entropy produced by the departure of counter-ions formed the main part of the interaction energy (Anderson and Record, 1982), it is in fact found that it is the change in rotational entropy of the water that is the main factor in the interaction energy (Table 19.3). Unexpectedly, this change in entropy arises from the variation of the electric energy of the non-bound protein with ionic strength.

Figure 19.12 Electrostatic potential and ion distributions around λ-repressor/DNA complex. A slice is taken through the repressor (a, d), DNA (b, e) or the complex (c, f) in the plane of the approximate two-fold symmetry axis of the protein. The bold solid line indicates the molecular surfaces of the protein and DNA. Diagrams (a) – (c) depict isopotential contours ranging from 1/8 to 8 times $k_B T$ in factors of 2. Diagrams (d) and (e) depict contours of constant net mobile ion charge density ranging from 1/8 to 8 times $k_B T$ in factors of 2. Positive contours are solid, and negative contours are dashed. (From Sharp *et al.*, 1995. Reprinted from *Biopolymers* with the permission of John Wiley & Sons. © 1995 John Wiley & Sons.)

20

Molecular models

20.1 Monte Carlo method

As the name of this method indicates, the displacements of the particles forming a canonical ensemble in a limited space are determined from a random series of numbers. This is used to simulate Brownian motion in the presence of the field of the interaction forces between the particles. From this apparently chaotic situation, order is gradually introduced into the initial 'solution'.

The modelling of ions around a polyelectrolyte generally involves representing the latter by a geometrical surface (sphere or cylinder) with well-defined fixed coordinates. In the box enclosing the polyion, a given number of ions (counter-ions and co-ions) with a mean diameter of about 6 Å are placed at random. One ion i with coordinates x_i^0, y_i^0, z_i^0 is chosen arbitrarily and given a new position x_i', y_i', z_i', where $x_i' = x_i^0 + \rho_x h$, h being the size of the increment and ρ_x being a randomly selected number between -1 and $+1$. The same operation is carried out on y and z and the energy E' of the new configuration is calculated from the general expression

$$E' = \sum_{i=1}^{N}\sum_{j=1}^{N} q_i q_j/\varepsilon r_{ij} + \sum_{i=1}^{N}\sum_{k=1}^{N} q_i q_k/\varepsilon r_{ik}$$

where the first term is due to the Coulomb interaction between ions and the second term is due to the interaction between the ions (i or j) and the charges carried by the stationary polyion. This expression can be supplemented by a repulsive energy to take into account the size of the ions or by a constraint preventing the distance between two ions from being smaller than the sum of their radii. The size of the polyion is similarly taken into account.

The new configuration is accepted or rejected depending on the sign of $\Delta E = E' - E^0$, where E^0 is the energy of the initial configuration. If $\Delta E < 0$, the motion and the configuration are accepted. If $\Delta E > 0$, the configuration is accepted if a randomly selected number between 0 and 1 is smaller than $\exp(-\Delta E/k_B T)$. It is found in general that 70 per cent of the movements are accepted.

The Monte Carlo process must satisfy two criteria:

1. *Convergence*. Every 5000 configurations, for example, a quantity such as the number of counter-ions per polyion charge is calculated and after a given time this quantity must converge to a steady value while remaining subject to fluctuations depending on the value of h.

2. *Validity*. The same number of selections is made but with different initial configurations in order to check that the result is independent of the initial charge configuration, i.e. of $\rho^0(r)$.

The Monte Carlo method has been applied to DNA (Le Bret and Zimm, 1984; Mills *et al.*, 1986; Paulsen *et al.*, 1988). It enables comparisons to be made with the application of the Poisson–Boltzmann equation, which is valid only when the size of the ions is negligible, as would be expected from the assumption of point charges. In fact, there is a kind of compensation between the size of the ions (excluded volume) and the electrostatic correlations, which causes the results from the Monte Carlo method and the Poisson–Boltzmann treatment to be in agreement. Because of the computing time required, the number of ions to be considered (i.e. the ionic strength of the medium) remains low: of the order of a few $\mathrm{mM\,l^{-1}}$.

On the other hand, the approximation of replacing the cylindrical charge distribution with a charged line appears to be entirely justified, since no modification in the ion distribution is found when changing from one model to the other.

20.2 Molecular dynamics

This method has already been described in section 10.2. In the polyelectrolyte field, it has mostly been applied to proteins in solution. In other words, the trajectories of the protein atoms are calculated by starting from a structure established by X-ray diffraction or NMR and solving Newton's equations. The expressions for the forces, however, must include those exerted between the charges carried by the protein and the ions present in the solvent medium as well as the dipoles of the water molecules. Although electrostatic forces are long-range, the computational demands make it essential to introduce a cut-off distance beyond which they are assumed to vanish. This approximation can lead to absurd results. One improvement is to preserve charge–charge interactions for the nearest atoms without cut-off. The more distant water molecules or ions are replaced by volume elements containing a charge or dipole moment whose volume increases with distance from the protein. The results obtained with a molecule of lysozyme (1301 atoms) immersed in 4797 water molecules are very satisfactory. One problem remains: the electronic polarizability is not taken into consideration in the force fields used in molecular dynamics.

There is an essential difference between the two methods. The Monte Carlo method assumes the existence of a medium with a given uniform dielectric constant but avoids the formalism and assumptions inherent in the Poisson–Boltzmann equation. Molecular dynamics makes no assumptions either about the mean potential or about the value of the dielectric constant. Although it is at present very limited, molecular dynamics will undoubtedly be the preferred method in future for studying the properties of polyelectrolytes as long as the power and speed of computers gives it the chance to fulfil its promise.

21

Cellular permeability and ion transport

21.1 Introduction

The discovery that the cell is the basic unit of any living system dates back to the seventeenth century. It may form the whole organism as in prokaryotes or unicellular eukaryotes. As we 'rise' through the evolutionary scale of the eukaryotes, the number of cells and their diversity related to their functional specificity both increase. However, there is a blueprint common to all these cases, particularly the existence of a membrane defining the internal volume of the cell as opposed to the external medium. Exchanges of matter, energy and information between the cell and the outside necessarily take place through the membrane, and as a result its structure plays a decisive role. In this chapter, we regard the membrane simply as a barrier that water and ions must cross, and as a result we shall attempt to discover the laws and general properties of ion transport without entering into details of any molecular mechanisms involved. Some idea of the importance of the problem can be gained by considering the values of ionic concentration on each side of a membrane given in Table 21.1.

However, a naive electrostatic approach immediately leads to a structural problem. The simplest membrane is known to consist of a lipid double layer and may be considered from an electric standpoint as a sheet with a dielectric constant ε_m of about 2 separating two aqueous media with a high dielectric constant ε_s (about 80). We have already seen (Box 15.7) that the energy W of transfer of an ion with charge ze (assumed to be a small charged sphere of radius r) from an aqueous

Table 21.1 Ionic concentration in mammalian cells (in mM at pH 7.4)

Ion	Intracellular	Extracellular
Na^+	5–15	145
K^+	140	5
Mg^{2+}	30	1–2
Ca^{2+}	1–2	2.5–5
H^+ (pH 7.4)	4×10^{-5}	$4 \times 10^{-}5$
Cl^-	4	110

medium to the interior of the membrane is given by

$$W = z^2e^2(1/\varepsilon_\mathrm{m} - 1/\varepsilon_\mathrm{s})/8\pi\varepsilon_0 r$$

We also saw that W was of the order of $100k_\mathrm{B}T$ for ions such as Na^+ or K^+. Such a potential barrier is inconsistent with experimental values of electric currents through membranes, so that specific transport mechanisms must exist. Before attempting to study these, we recall the general laws governing electric currents in a conductor, where the moving charges are no longer electrons but are ions.

21.2 **Ionic conductivity**

The classic statement of Ohm's law relating the current I through a conductor to the potential difference V between its ends is $V = RI$, where R is the resistance of the conductor. In ionic media, the conductance Λ is used in preference to R, where $\Lambda = 1/R$. When I is expressed in amps and V in volts, R is in ohms and Λ in siemens (symbol S).

Suppose that a current passing through a membrane is measured and found to be 10 pA (10^{-12} A) at a potential difference of 100 mV: the conductance would be 100 pS or 10^{-10} S. If the current were due to a flux of monovalent ions each with a charge $e = 1.6 \times 10^{-19}$ C, then it would involve the flow of about 70 million ions per second, sufficient to change the ionic concentration in a bacterium, say, by about 1 mM in 0.1 s.

In general, an electric current I can be considered as the flux through the cross-section S of the conductor of a vector $\boldsymbol{j}$ called the *current density*, so that

$$I = \int_S \boldsymbol{j} \cdot \boldsymbol{u}_\mathrm{n}\, \mathrm{d}S \tag{21.1}$$

where $\boldsymbol{u}_\mathrm{n}$ is a unit vector perpendicular to the elementary area $\mathrm{d}S$. Ohm's law can then be written in the form

$$\boldsymbol{j} = \sigma \boldsymbol{E} \tag{21.2}$$

where σ is the *conductivity* of the medium and $\mathbf{E}$ the electric field.

We now simplify matters by taking the case of a cylindrical conductor of cross-section S and length l with the same current density at every point. Equations 21.1 and 21.2 then show that

$$I = \sigma E S$$

and, since the potential gradient $E = V/l$,

$$I = \sigma S V/l \tag{21.3}$$

showing that the conductance and resistance are given by

$$\Lambda = \sigma S/l \qquad \text{and} \qquad R = \rho l/S \tag{21.4}$$

where $\rho = 1/\sigma$ is the *resistivity* of the medium. The SI unit of ρ is the Ω m and of σ is the S m^{-1}.

The conductivity can be related directly to the movement of charges in the medium. If n is the number of ions per unit volume, each with charge ze, and if $\boldsymbol{v}$ is their velocity, then

$$\boldsymbol{j} = nze\boldsymbol{v} \tag{21.5}$$

The velocity v has a limiting value (the terminal velocity) reached when the two forces acting on the ion, the electric force zeE and a frictional force kv, are equal and opposite, giving $v = zeE/k$. For motion through a fluid, $k = f\eta_0$, where f is a coefficient of friction depending on the geometry of the ion and η_0 is the viscosity of the medium. If the ion is a sphere of radius r, then $f = 6\pi r$ (Stokes' law).

Equating the two expressions for $\boldsymbol{j}$ in eqns 21.2 and 21.5, and using the expression for v, we obtain

$$\sigma = n(ze)^2/f\eta_0 \tag{21.6}$$

so that the experimental quantity σ may be related directly to properties connected with the microscopic description of ion transport.

Consider the Na^+ ion ($z = 1$) whose translational coefficient of diffusion $D_T = k_B T/f\eta_0$ is about $1.35 \times 10^{-9}\ m^2\ s^{-1}$, and assume a concentration of 0.1 M, i.e. about 6×10^{25} ion m^{-3}. This gives $\sigma = 0.52\ S\ m^{-1}$. To pass through a membrane of thickness $l = 5$ nm by means of an ion channel, the cross-section S of the channel would have to be of the order of $10^{-18}\ m^2$ or 100 Å^2 to explain the conductance of 100 pS as used in the previous example.

21.3 Transport mechanisms in ion channels

Hodgkin and Huxley (1952) showed that there is a linear relationship between the current I through a membrane and the transmembrane potential V. A total conductance can thus be defined as the sum of a whole series of independent elementary conductances. The current through the membrane can also be regarded as being composed of a set of independent ionic processes through what are called *ion channels*. Each of these is a kind of electrical short-circuit between the inside and outside of the membrane.

Three properties can be attributed to an ion channel:

1. It is a *gate* that controls the opening and closing of the channel and governs its accessibility to ions.
2. Each ion channel has its own conductance corresponding to ion transport.
3. Ions pass through the channel as a result of a driving force defined as the gradient of the electrochemical potential.

In nearly half a century since then, each of these has undergone considerable development.

1. A large variety of ion channels exist and fall into two main groups. In the first group, the opening is triggered by the potential difference between the inside and outside of the membrane. The ion channel is then described by the ion that passes selectively through it, so that we speak of Na^+, K^+, Ca^{2+}, Cl^-, etc., channels. In the second group, the opening is triggered by an effector acting on a receptor, and we speak of a channel with an acetylcholine receptor, with a serotonin receptor, etc. If an Na^+ ion passes through this type of channel, it does not trigger the opening, whereas acetylcholine, which can trigger the opening, does not pass into the channel.

2. With the so-called *patch-clamp* method (Fig. 21.1), it is now possible to measure the current passing through a single channel (Fig. 21.2). It is also possible to isolate the protein assembly forming the channel and insert it into a phospholipid vesicle while maintaining its function and its specificity (Fig. 21.3). The currents measured are found to be of the order of picoamperes, with conductances of several tens or hundreds of picosiemens. The accuracy of the measurements is limited by high-frequency electrical noise. Filtering this out is only effective if the channel opening time is not too short (e.g. 100 ms), otherwise the current through the channel must be greater than 0.05 pA to be detected.

3. Two theoretical models have been developed to explain the experimental results. The first combines a diffusion process with an electric force originating in the field due to the distribution of charges. This is the Nernst–Planck theory, which regards the neighbourhood of the membrane and the channel as a continuous medium. In the second theory, due to Eyring, the pathway followed by the ion is regarded as a crossing of several potential barriers whose position and height are determined by the structure of the proteins forming the channel. The selectivity could be due, for example, to electrostatic interactions between the ion crossing the membrane and the charges or dipoles located around the channel.

For the time being, we shall not examine the structure of an ion channel on the molecular scale, but will simply look for a general scheme capable of explaining and predicting how such a channel functions. Before considering such an approach, which is mainly a thermodynamic one, we review some of the concepts and results related to the electrochemical potential.

21.3.1 Electrochemical potential

In an open system with several components, as is normally encountered in chemistry and biology, the free-energy function G depends not only on pressure P and

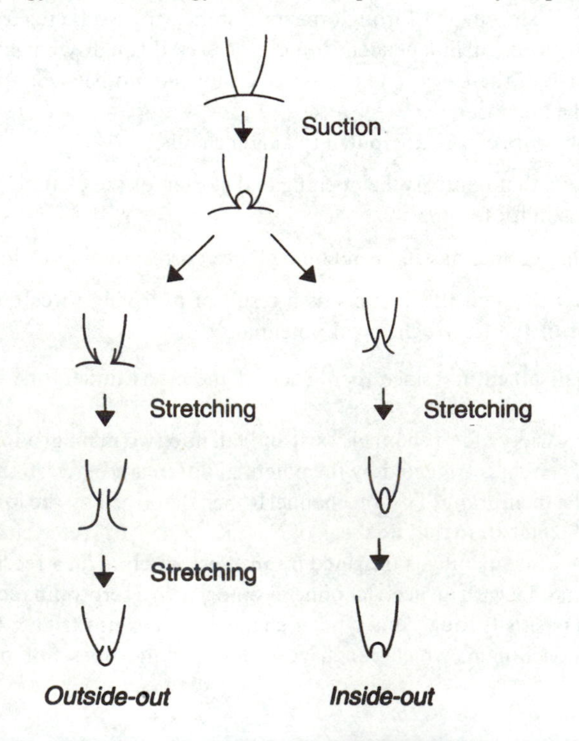

Figure 21.1 Schematic diagram of the patch-clamp technology.

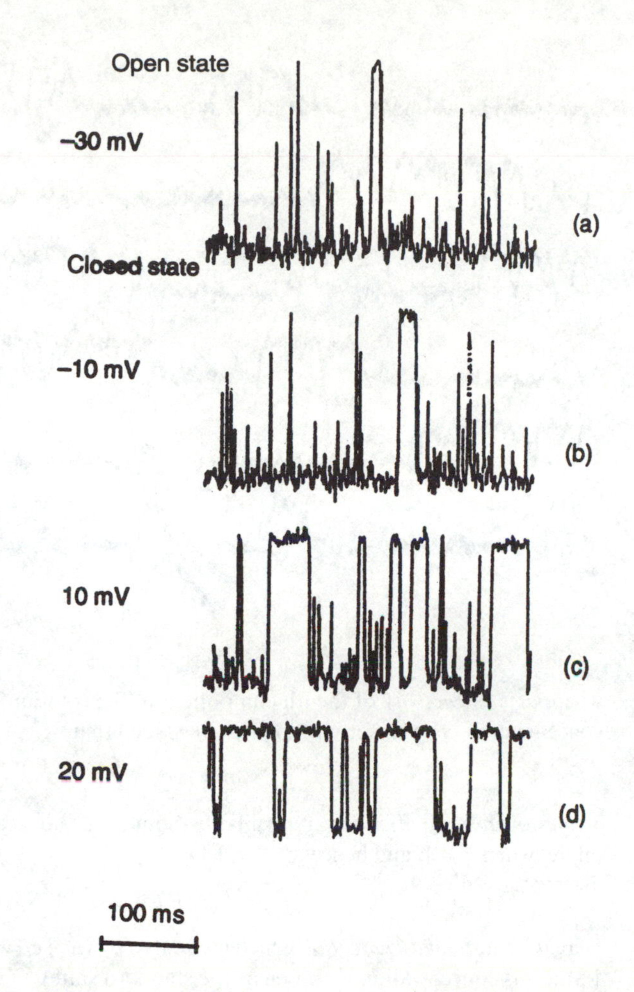

Figure 21.2 Recordings of a single channel in chicken embryo ventricle cells. (a) to (c) The system is inside-out with a resting potential of about −50 mV and a pipette potential of −20, −40 and −60 mV respectively. (d) An outside-out recording with a pipette potential of +20 mV. From this, it is possible to deduce (see Fig. 21.4) that the potential Ψ_m varies from −30 to +20 mV when going from (a) to (d). Channel opening induces an increase in the signal. The open states are interrupted by the flickering from a large number of short closures.

temperature T but on the number of moles of each component present in the system. For an elementary change in G, we can write

$$\mathrm{d}G = (\partial G/\partial P)_{T,n}\,\mathrm{d}P + (\partial G/\partial T)_{P,n}\,\mathrm{d}T + \sum(\partial G/\partial n_i)_{P,T,n_j \neq n_i}\,\mathrm{d}n_i \qquad (21.7)$$

where n is the sum of the number of moles $n_1, n_2, \ldots, n_i, n_j, \ldots$ of each species present. The first term on the right is equal to $V\,\mathrm{d}P$, and the second term is simply $-S\,\mathrm{d}T$. The third term is the change in free energy of the system when only the number of moles of component i varies. The partial derivative $(\partial G/\partial n_i)_{P,T,n_j \neq n_i}$ is called the partial molar free energy or simply the *chemical potential* of component i, usually denoted by μ_i. At constant temperature and pressure, therefore, the whole of eqn 21.7 may be replaced by

$$\mathrm{d}G = \sum \mu_i\,\mathrm{d}n_i \qquad (21.8)$$

By analogy with what is done in the case of a mixture of ideal gases, we write μ_i for any component i as

$$\mu_i = \mu_i^\circ + RT\ln a_i \qquad (21.9)$$

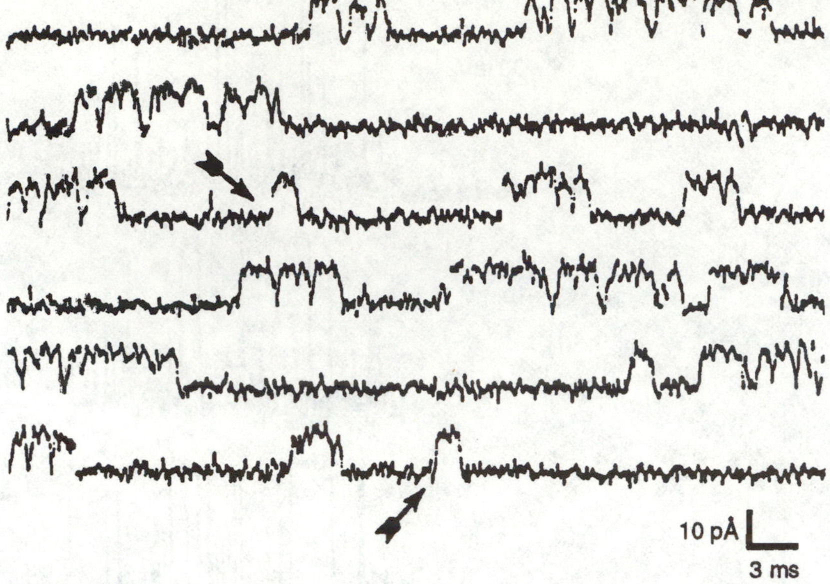

10 pÅ
3 ms

Figure 21.3 Recording of opening and closing in a single channel reconstituted in a lipid bilayer. The potential is held at zero and frequencies above about 10 kHz are filtered out.

where a_i is the *activity* of the ith component in the solution, related to the molar concentration c_i by an activity coefficient γ_i such that

$$a_i = \gamma_i c_i$$

Tables of these coefficients are available. Solutions may often be assumed to be dilute, when $\gamma = 1$, and hence

$$\mu_i = \mu_i^\circ + RT \ln c_i$$

Here $\mu_i{}^\circ$ is a *standard potential*, which is equal to μ_i when $c_i = 1$ (a concentration of 1 M if this concentration is chosen as the standard state).

However, in most open biological systems, the transport of matter is often accompanied by a transfer of charge (of protons, ions or charged molecules) and this must be taken into account in the energy balance. Consider for the moment a particular charged molecule or ion with a charge ze. When it is situated where the potential is Ψ, it has an energy $ze\Psi$. For one mole of such molecules or ions, this energy is given by $zNe\Psi$ or $zF\Psi$, where F is the faraday, a charge of about 96 500 C. The *electrochemical potential* is defined as the sum of the previously defined chemical potential and the molar electric energy:

$$\mu = \mu^\circ + RT \ln c + zF\Psi \qquad (21.10)$$

Now consider two compartments A and B separated by a membrane that allows charged species to pass through it. In each compartment we have

$$\mu_A = \mu_A^\circ + RT \ln c_A + zF\Psi_A$$

$$\mu_B = \mu_B^\circ + RT \ln c_B + zF\Psi_B$$

The standard state may be chosen so that $\mu_A^\circ = \mu_B^\circ$. Then the change in free energy ΔG per mole transported from A to B is numerically equal to the difference in electrochemical potentials:

$$\Delta G = \mu_B - \mu_A = RT \ln(c_B/c_A) + zF\Delta\Psi \tag{21.11}$$

where $\Delta\Psi = \Psi_B - \Psi_A$. Because the transport of charge induces a change of potential in each compartment, an equilibrium is established for which $\Delta G = 0$, and this leads to the two equivalent relationships

$$c_B = c_A \exp(-zF\Delta\Psi/RT) \tag{21.12}$$

$$\Delta\Psi = -(RT/zF)\ln(c_B/c_A) = \Psi_B - \Psi_A \tag{21.13}$$

For example, $\Delta\Psi < 0$ means that $c_B > c_A$ and, for $z = 1$, we have $\Delta\Psi = -120$ mV if $c_B \approx 100c_A$.

Equation 21.13 is the *Nernst equation*, which can also be obtained by applying Maxwell–Boltzmann statistics to a molecule of charge ze placed in a potential Ψ.

21.3.2 **Nernst–Planck model**

Ion flux is interpreted as being caused by the action of a generalized force equal to the negative gradient of the electrochemical potential. Assuming that the ions are moving along the x axis in a direction perpendicular to the membrane, the force X on a molar scale is given by

$$X = -\,\mathrm{d}\mu/\mathrm{d}x = -(RT\,\mathrm{d}c/c\,\mathrm{d}x + zF\,\mathrm{d}\Psi/\mathrm{d}x) \tag{21.14}$$

The ion is also subjected to a frictional force F_f, which depends on a geometrical factor f, the viscosity of the medium and the velocity v:

$$F_f = f\eta_0 v \tag{21.15}$$

The quantity $f\eta_0$ is usually expressed in terms of the translational coefficient of diffusion D_T through the relationship $D_T = k_B T/f\eta_0$, so that on a *molecular* scale:

$$f\eta_0 = k_B T/D_T \tag{21.16}$$

The limiting velocity v is reached when the frictional force $X_f = NF_f$ on the *molar* scale is equal and opposite to X, and hence

$$v = XD_T/RT \tag{21.17}$$

The ion flux J (the number of ions per m^2 per second) crossing the membrane is related to v by

$$J = cv \tag{21.18}$$

where c is the molar concentration of ions. From eqns 21.14, 21.17 and 21.18, we have

$$J = -D_T[\mathrm{d}c/\mathrm{d}x + (zFc/RT)\,\mathrm{d}\Psi/\mathrm{d}x] \tag{21.19}$$

Ion fluxes J_i exist for each component i. When the above equation is supplemented by the local form of Poisson's equation in a single variable x:

$$\mathrm{d}/\mathrm{d}x(\varepsilon\,\mathrm{d}\Psi/\mathrm{d}x) = -(e/\varepsilon_0)\sum_i c_i z_i \tag{21.20}$$

we obtain the two basic equations of the Nernst–Planck theory.

In both equations, Ψ is known as the *potential of mean force* and takes account of all the interactions (ion–ion and ion–channel) and of the external applied potential (the transmembrane potential). The potential at any instant fluctuates about this mean value. When moving in the channel, an ion remains in equilibrium as far as short-range forces are concerned. In other words, the range of the electric forces due to charges and dipoles in the channel is large compared with the mean free path of the ion.

Under steady-state conditions ($J = 0$), eqn 21.19 can be integrated to give the potential Ψ_m inside the membrane (the membrane potential) taking the potential in the external solution as zero:

$$\Psi_m = (k_B T/ze)\ln(c_O/c_I)$$

where c_O is the ion concentration outside and c_I its concentration inside the cell. This, of course, is equivalent to the classical Nernst equation (eqn 21.13), which shows that a selective membrane plays the role of an electric cell whose electromotive force depends on the ratio of concentrations on either side of the membrane. If, for example, we take data for the K^+ ion from Table 21.1, we find that, at 20°C, $\Psi_m = -84\,\text{mV}$, which represents the resting potential of a potassium ion selective membrane.

Away from the equilibrium state, a general solution to eqn 21.19 can be found by means of a procedure used by Kramers in which both sides are multiplied by $\exp(ze\Psi/k_B T) = \exp(y)$ with the notation used previously. This gives

$$(-J/D_T)e^y = e^y(\,dc/dx + c\,dy/dx) = d(c\,e^y)/dx \tag{21.21}$$

In this form, the equation can be integrated over the thickness of the membrane a:

$$\int_0^a (-J/D_T)e^y\,dx = \int_0^a d(c\,e^y) \tag{21.22}$$

Using the *steady-state* approximation, $\partial c/\partial t = 0$, which means that $\partial J/\partial x = 0$, i.e. J is constant and independent of x. If D_T is also independent of x, then

$$J = -D_T[c_I \exp(ze\Psi_m/k_B T) - c_O] \Big/ \int_0^a e^y\,dx \tag{21.23}$$

This equation for the flux can yield some simple results if additional conditions are introduced. It should be pointed out, however, that the equation is only valid for a single ion. If another ion were present in the channel, the potential created by this second ion would have to be taken into account. This amounts to what Hodgkin and Huxley (1952) call the *principle of independence*, which is valid only at low concentrations.

Equation 21.23 can be used to follow the flux of an isotope when its concentration is low enough not to modify c. If the isotope (denoted by *) is introduced into the external medium, the corresponding flux $J^*_{O\to I}$ is (with $c_I = 0$):

$$J^*_{O\to I} = D_T c_O \Big/ \int_0^a e^y\,dx$$

If the isotope is introduced into the internal medium, we then have (with $c_O = 0$):

$$J^*_{I\to O} = D_T c_I \exp(ze\Psi_m/k_B T) \Big/ \int_0^a e^y \, dx$$

The flux ratio

$$J^*_{I\to O}/J^*_{O\to I} = (c_I/c_O)\exp(ze\Psi_m/k_B T)$$

is characteristic of the Nernst equation. However, for K^+ channels it is found experimentally that $J^*_{I\to O}/J^*_{I\to O}$ varies as $[(c_I/c_O)\exp(ze\Psi_m/k_B T)]^{n'}$ where n' is a number between 2 and 2.5. This result can be interpreted as a correlated motion of ion clusters of mean size n'. It follows that there are at least two ions migrating together in the channel.

Equation 21.23 may also be used if assumptions are made about the distribution of the potential in the channel. We examine two models.

First model

This assumes that a constant potential exists within the membrane, which means according to eqn 21.20 that the number of charges within the membrane is negligible. Hence $d(\varepsilon \, d\Psi/dx) = 0$, and after integration this gives

$$\Psi(x) = x\Psi_m/a \tag{21.24}$$

Using this in eqn 21.23, we obtain

$$J = D_T[c_O - c_I \exp(ze\Psi_m/k_B T)] \Big/ \int_0^a \exp(ze\Psi_m/ak_B T)\, dx$$

and after the integration in the denominator:

$$J = (D_T ze\Psi_m/ak_B T)[c_O - c_I \exp(ze\Psi_m/k_B T)]/[\exp(ze\Psi_m/k_B T) - 1] \tag{21.25}$$

Since the current $I = zeJ$, this equation enables us to find forms of the relationship between the current and the membrane potential.

If a uniform gradient is assumed through the membrane and if we take the case of the usual three monovalent ions Na^+, K^+ and Cl^- ($z = \pm 1$), eqn 21.25 can be written for each ion j as

$$J_j = P_j y(c_{O_j} - c_{I_j} e^y)/(e^y - 1) \tag{21.26}$$

where $y = e\Psi_m/k_B T$ and $P_j = D_{T,j}/a$, which measures the *permeability* of the membrane to the ion j. P is expressed in $m\,s^{-1}$, i.e. it is equivalent to a velocity. It can be determined by measuring the accumulation of a radioactive solute over time.

Moreover, the equilibrium of the ion fluxes in the steady state means that $I_{Na} + I_K + I_{Cl} = 0$, i.e. given the relationship between I and J, that

$$J_{Na} + J_K + J_{Cl} = 0 \tag{21.27}$$

and eqns 21.26 and 21.27 can be combined to obtain the constant membrane potential Ψ_m resulting from the steady state. This yields the so-called *Goldman*

Box 21.1 Establishment of Goldman's equation

Starting from the flux given by eqn 21.26 for each ion, with $y = e\Psi_m/k_BT$ and $P = D_T/a$ and noting that $z = +1$ for Na and K and -1 for Cl, we obtain

$$J_{total} = y[P_{Na}(c_{O,Na} - c_{I,Na}e^y) + P_K(c_{O,K} - c_{I,K}e^y)]/(e^y - 1) - y[P_{Cl}(c_{O,Cl} - c_{I,Cl}e^{-y})]/(e^{-y} - 1) = 0$$

Multiplying the term relating to the Cl ion by e^y, we have

$$J_{total} = y[P_{Na}(c_{O,Na} - c_{I,Na}e^y) + P_K(c_{O,K} - c_{I,K}e^y) - e^y P_{Cl}(c_{O,Cl} - c_{I,Cl}e^{-y})]/(e^{-y} - 1) = 0$$

The expression within square brackets must be zero, so that

$$e^y = (P_{Na}c_{O,Na} + P_K c_{O,K} + P_{Cl}c_{I,Cl})/(P_{Na}c_{I,Na} + P_K c_{I,K} + P_{Cl}c_{O,Cl})$$

and this leads to eqn 21.28.

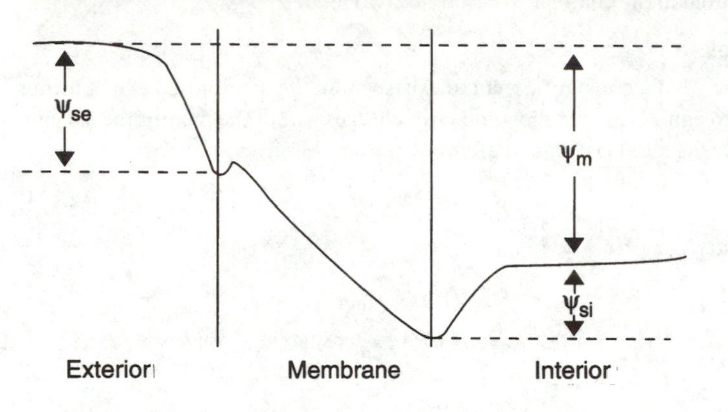

Figure 21.4 Potential variation across a membrane: Ψ_{se} is the potential difference between the external surface and the solvent; Ψ_{si} is the potential difference between the internal surface and the internal medium; Ψ_m is the membrane potential.

equation in which the subscript O indicates outside and I indicates inside:

$$\Psi_m = (RT/F)\ln\{[P_{Na}(\mathrm{Na})_O + P_K(\mathrm{K})_O + P_{Cl}(\mathrm{Cl})_I]/[P_{Na}(\mathrm{Na})_I + P_K(\mathrm{K})_I + P_{Cl}(\mathrm{Cl})_O]\} \quad (21.28)$$

(see Box 21.1). When the membrane is only permeable to a single ion, it is clear that this equation reduces to the Nernst equation. According to the concentrations given in Table 21.1, the equilibrium potential is positive for Na^+ but negative for K^+ and Cl^-. For Ψ_m to be negative, the permeability P_{Na} merely has to be small compared with the other two. For example, in the membrane of a squid axon, the resting potential is determined by the permeability to potassium ions, whereas in frog muscle the dominant term is the permeability to the chloride ion. This simple calculation also shows that the membrane potential can only be kept constant if active ionic transport processes are operating. This is the role played by ion pumps.

The above discussion can be summarized by representing the potential function as in the diagram of Fig. 21.4. It is important, however, to recall the assumptions forming the basis of this calculation:

1. The electric field is uniform, i.e. there is a linear variation of the potential through the membrane (Fig. 21.4). In fact, the interior of an ion channel is unlikely to have a homogeneous uniform electric field because charges and dipoles occur throughout the ion pathway.

2. The value of D_T is constant and hence so is the permeability inside the membrane. It is possible to introduce into P a partition coefficient β ($P = \beta D_T/a$), but it is not obvious that D_T will be the same all along the channel because of its geometry.

3. There is no coupling between fluxes of different ions or between the flux of an ion and that of the solvent. In other words, the theory does not lie within the realm of the thermodynamics of irreversible processes. A more careful approach using Onsager's laws must be attempted. Moreover, each sheet of the membrane is surrounded by an ionic atmosphere, which, to a first approximation, can be treated with the Debye–Hückel model.

Second model

Here, an attempt is made to involve the channel structure by assuming a potential distribution inside the membrane that is no longer a simple linear function of distance but takes into account the geometry and the charge distribution in the ion channel. The simplified model shown in Fig. 21.5 resembles a double-ended funnel and enables three parts to be distinguished. In the 'entrance vestibule' the ion coming from outside loses all or part of its hydration shell, which it recovers in the 'exit vestibule'. Between the entrance and exit there is a kind of 'queue' of ions waiting to pass through. The selectivity occurs through steric hindrance and a potential barrier. In $\Psi(x)$, therefore, the linear form of eqn 21.24 and a step function of height Ψ_0 and width $d = x_2 - x_1$ are superimposed to obtain the scheme shown in Fig. 21.6.

To integrate the denominator of eqn 21.25, there are three regions to be considered:

- region $0 < x < x_1$ for which

$$y(x) = x\Psi_m/a \tag{21.29a}$$

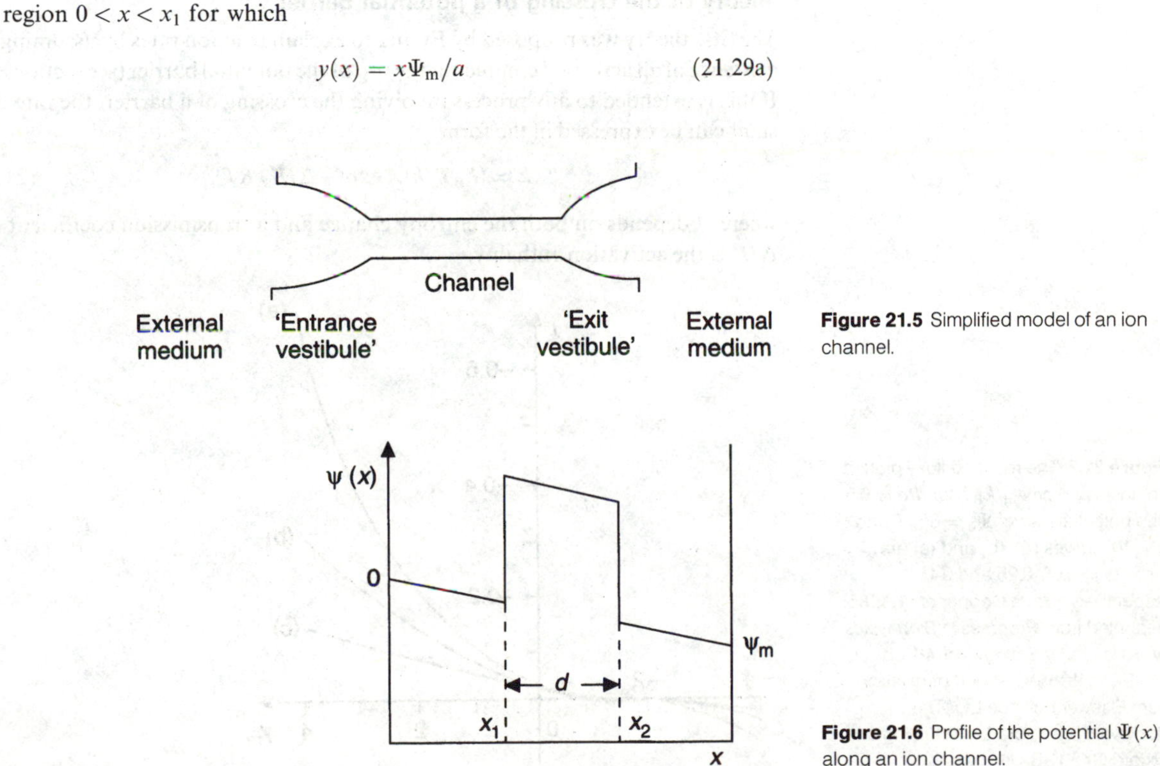

Figure 21.5 Simplified model of an ion channel.

Figure 21.6 Profile of the potential $\Psi(x)$ along an ion channel.

- region $x_1 < x < x_2$ for which
$$y(x) = x\Psi_m/a + \Psi_0 \tag{21.29b}$$
- region $x_2 < x < a$ for which
$$y(x) = x\Psi_m/a \tag{21.29c}$$

and by introducing the dimensionless variables

$$j = Ja/c_O D_T \qquad \eta = c_I/c_O \qquad y_m = ze\Psi_m/k_B T \tag{21.30}$$

we obtain after the integration:

$$j = y_m(1 - \eta \exp y_m)/\{\exp y_m - 1 + \exp(x_1 y_m/a)[1 - \exp(dy_m/a)](1 - \exp\Psi_0)\} \tag{21.31}$$

When $\Psi_0 \to 0$ or when $d \to 0$, eqn 21.31 of course gives eqn 21.25. If the potential barrier has the same width as the channel so that $d/a = 1$ and $x_1 = 0$, we obtain the simple expression

$$j = y_m(1 - \eta \exp y_m)/[\exp\Psi_0(\exp y_m - 1)] \tag{21.32}$$

Comparing eqns 21.32 and 21.25, we see that the presence of a constant potential Ψ_0 all along the channel is equivalent to reducing the coefficient D_T by a factor of $\exp\Psi_0$ or to introducing a partition coefficient between c_O and c_I. The calculated curves for the current I plotted against the membrane potential Ψ_m agree well with the experimental results (Fig. 21.7).

Theory of the crossing of a potential barrier

The first theory was proposed by Eyring to explain reaction rates by assuming the existence of an activated complex at the top of the potential barrier (see section 8.4). If this is extended to any process involving the crossing of a barrier, the rate constant can be expressed in the form

$$k = (k_B T/h)A\exp(-\Delta H^*/RT) \tag{21.33}$$

where A depends on both the entropy change and a transmission coefficient and ΔH^* is the activation enthalpy.

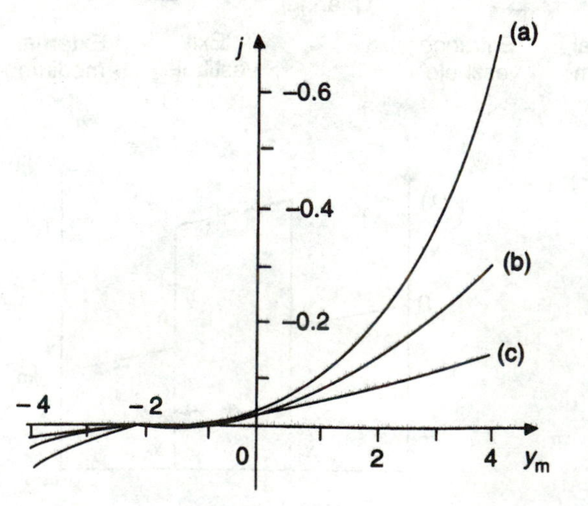

Figure 21.7 The reduced flux j plotted against $y_m = ze\Psi_m/k_B T$ for $d/a = 0.5$. The potential barrier $\Psi_0 = 6k_B T$ and $\eta = 10$. Curves (a), (b) and (c) are for $x_1/a = 0.05, 0.25$ and 0.45 respectively. (From Cooper *et al.*, 1985. Reprinted from *Progress in Biophysics and Molecular Biology*, vol. 46, pp. 51–96, © 1985, with kind permission from Elsevier Science Ltd., The Boulevard, Langford Lane, Kidlington, Oxford OX5 1GB, UK.)

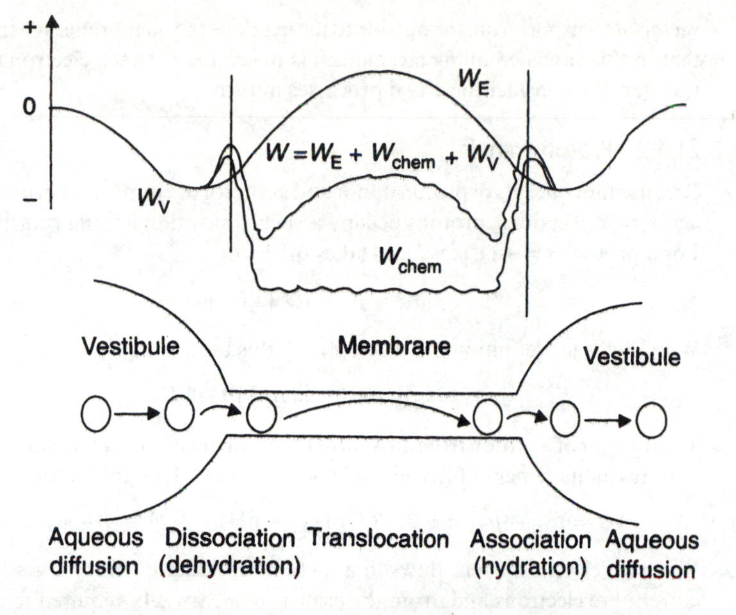

Figure 21.8 Schematic diagram of the potential energy profile along the ion pathway and the main stages in its transport along the channel. (From Eisenman and Dani, 1987. Reprinted with permission from *Annual Review of Biophysics and Biophysical Chemistry*, vol. 16 and with the authors' permission. © 1987 Annual Reviews Inc.)

In fact, in a viscous medium, the rate of potential barrier crossing has been shown by Kramers to be

$$k = (mD_T\omega_A\omega_C/2\pi k_B T)\exp(-E/k_B T) \qquad (21.34)$$

where ω_A and ω_C are the natural vibrational frequencies of the system in state A (the start) and state C (top of the barrier), and E is the potential energy ($= mV_C$) of the particle at the top of the barrier. These relationships were established in Part II (see section 8.4).

The flux J can be calculated from eqns 21.33 and 21.34 provided the parameters of the potential barrier and the concentrations at both ends of the channel are known. A possible energy diagram is shown in Fig. 21.8.

The end result is usually a set of differential equations that cannot always be solved analytically. If, for example, three sites are available in the ion channel and if all can be occupied, the various transitions between states (four states and 24 rate constants) can be represented as in Fig. 21.9.

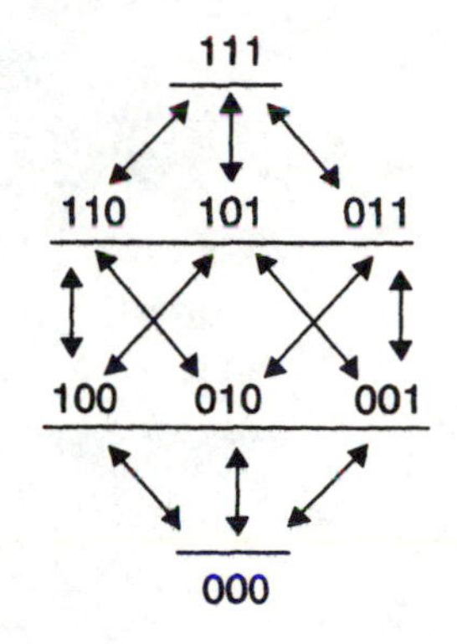

Figure 21.9 Possible transitions between four states.

The kinetic approach leads to equations similar to the Michaelis–Menten equation. With only one type of ion, saturation occurs; with two different ions, there is competition and behaviour similar to that of a competitive inhibitor.

21.4 Functioning of an ion pump

21.4.1 Introduction

As well as ion channels, there are *ion pumps* formed from proteins capable of transporting cations. With Na^+ and K^+ pumps, energy is required to transfer Na^+ from inside to outside and to transfer K^+ in the opposite direction, since the transfer is made against the concentration gradient. An ATP molecule is needed to phosphorylate the membrane enzyme and to trigger a conformational change between two structures each offering a greater affinity for one of the ions than the other.

The flux in ion pumps is about 300 ions per second even though ion channels can transport between 10^6 and 10^7 ions in the same time. In *proton pumps* there is *vectorial* transport from the outside to the inside of the membrane. We shall see later that in this case a coupling mechanism is operating between electron and proton transfer. We consider these two processes in turn.

21.4.2 Proton transfer

Because the concepts of proton donor and acceptor are important in understanding acid – base reactions, protons occupy a special position among transported ions. For a proton ($z = +1$), eqn 21.10 takes the form

$$\mu_{[\mathrm{H}^+]} = \mu^\circ + RT\ln[\mathrm{H}^+] + F\Psi \qquad (21.35)$$

With the usual definition $\mathrm{pH} = -\log[\mathrm{H}^+]$, this becomes

$$\mu_{[\mathrm{H}^+]} = \mu^\circ - 2.3RT\,\mathrm{pH} + F\Psi \qquad (21.36)$$

The transfer of a proton from an internal compartment to an external compartment requires an increment of free energy given by eqns 21.11 and 21.36:

$$\Delta G = \mu_{\mathrm{ext}} - \mu_{\mathrm{int}} = -2.3RT(\mathrm{pH}_{\mathrm{ext}} - \mathrm{pH}_{\mathrm{int}}) + F(\Psi_{\mathrm{ext}} - \Psi_{\mathrm{int}}) \qquad (21.37)$$

When an electric current flows in a metallic conductor, the charges forming the current are electrons and, from the electrical energy ΔW required to move 1 mol of electrons round the circuit (i.e. a charge $Ne = F$), an *electromotive force* is defined by

$$e_{\mathrm{M}} = \Delta W/F$$

By analogy with this, when the charges forming the current are protons, we can define what is called a *proton motive force* e_{P} by using eqn 21.37:

$$e_{\mathrm{P}} = -2.3RT(\mathrm{pH}_{\mathrm{ext}} - \mathrm{pH}_{\mathrm{int}})/F + (\Psi_{\mathrm{ext}} - \Psi_{\mathrm{int}}) \qquad (21.38)$$

where the term $\Psi_{\mathrm{ext}} - \Psi_{\mathrm{int}}$ is simply the potential difference across the membrane. Another way of putting it is that, while $\Delta W/F$ is an electron-induced potential difference, $\Delta G/F$ is a proton-induced potential difference.

The comparison between the proton motive force e_{P} and the electromotive force e_{M} can be carried further. In classical electricity, an electric generator (dry cell, accumulator, dynamo, etc.) creates a potential gradient causing a current of electrons to flow in a closed metallic circuit or a current of ions in an electrolytic tank. In a living cell, the generator has to maintain a potential gradient originating from protons forming a kind of energy reservoir to supply all the cellular processes that consume energy. This is the role of the proton pump. To obtain some idea of orders of magnitude, a ratio of concentrations across the membrane of 1 : 10 for a monovalent ion like H^+ requires a potential difference to be maintained given by $2.3RT/F$. This is equal to $2.3k_{\mathrm{B}}T/e$ or 59 mV at 25°C. Transporting 1 mol of H^+ across this potential difference requires an energy of $2.3RT$ or 5.7 kJ. What is the source of this energy? A remarkable idea put forward by Mitchell (1961) was to consider the source as arising from the successive transfer of electrons into a metabolic redox system. We study this coupling between electron and proton transfer in two stages.

21.4.3 Proton transfer and electron transfer

A comparison is first made between redox systems involving electron transfer and acid–base transformations involving proton transfer.

In redox systems, the donor is a reducing agent and the acceptor is an oxidizing agent. The redox potential Ψ is given by the Nernst equation in a form similar to that of eqn 21.13:

$$\Psi = \Psi_0 + 2.3(RT/nF)\log([\text{acceptor}]/[\text{donor}]) \quad (21.39)$$

where n is the number of transferred electrons.

In acid–base systems, the donor is an acid and the acceptor is a base, and the *Henderson–Hasselbach* relationship applies:

$$\text{pH} = \text{p}K_\text{D} + \log([\text{acceptor}]/[\text{donor}]) \quad (21.40)$$

The coupling between the proton and electron transfer can be shown to be equivalent to the formation of one water molecule:

$$H_2 \rightarrow 2H^+ + 2e^-$$

$$\tfrac{1}{2}O_2 + 2e^- \rightarrow O^{2-}$$

$$O^{2-} + 2H^+ \rightarrow H_2O$$

If the system is so constructed as to allow the free circulation of protons in an aqueous solution around two metal electrodes, then a fuel cell is created (Fig. 21.10). The flux of hydrogen around one electrode creates an excess of electrons and the electrode potential becomes negative. The oxygen flux around the other electrode creates a deficiency of electrons and the electrode potential becomes positive. The two electrodes together form a classical source of e.m.f. and, when connected by a wire, they cause an electron current to flow in the wire, supplied by the fluxes of the two gases and the redox reaction.

If, on the other hand, the system is built so as to allow a free circulation of electrons, then unequal proton concentrations will develop in the two compartments of the aqueous solution. A *unidirectional* flow of protons is established if the two compartments are connected through a membrane for example (Fig. 21.11).

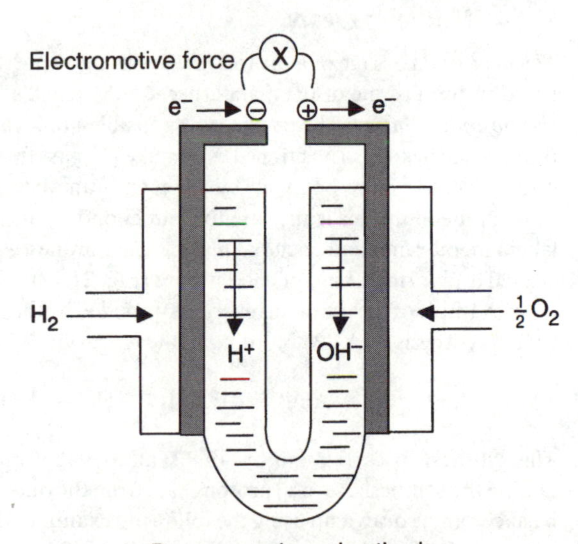

Figure 21.10 Principle of the fuel cell for generating electricity. The electromotive force depends on the redox potential.

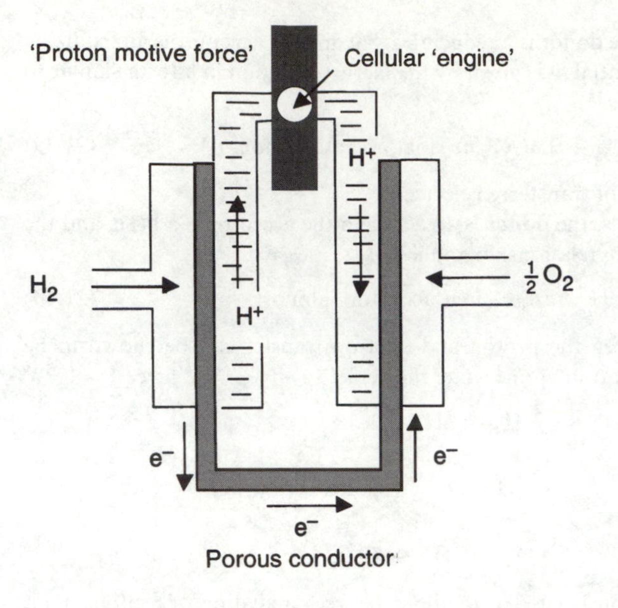

Figure 21.11 Principle of the proton current generator in a living cell. The 'proton motive force' depends on the redox chain along which electrons are transferred.

Clearly, the flow of electrons cannot be considered as equivalent to that occurring in a metallic conductor. In the enzymic complex responsible for the redox process, the electron is transferred from one enzymic site to the next in space, as was shown in the first crystal structure of a photosynthetic receptor. The transfer times that could be measured are of the order of 10^{-9} to 10^{-11} s.

Because the phospholipid membrane offers a 'resistance' to H^+ transfer, it can support a proton gradient and a potential difference of several hundred millivolts. This gradient is maintained by proton pumps, which, for example, eject a proton to the outside and capture an internal proton. Although this appears to be a proton current, it is in fact a *translocation* of protons rather than a true transport.

21.4.4 Mitchell's theory

This *vectorial* passage of a proton (always in the same direction for a given membrane system) is one of the main aspects of Mitchell's (1961) theory or so-called *chemiosmotic theory*. The enzymes responsible are located in the membrane and their active sites are so positioned as to make H^+ pass from the inside to the outside. The internal volume bounded by the membrane becomes more basic than the external medium: this is the case for mitochondrial membranes, chloroplast thylakoid membranes and bacterial plasmic membranes. A proton motive force e_P is created and forms a kind of energy store (Fig. 21.12).

An example of the use of this energy is provided by the oxidative phosphorylation for the synthesis of ATP according to the reaction

$$ADP^{3-} + PO_4H_2^{2-} + H^+ \rightarrow ATP^{4-}$$

This synthesis is carried out by ATP synthases that may be reversible ATPases. During the synthesis, several protons pass from the outside to the inside. An energy balance can be drawn up using the following example. If in eqn 21.38 we assume a potential gradient of one unit and a potential difference across the membrane of

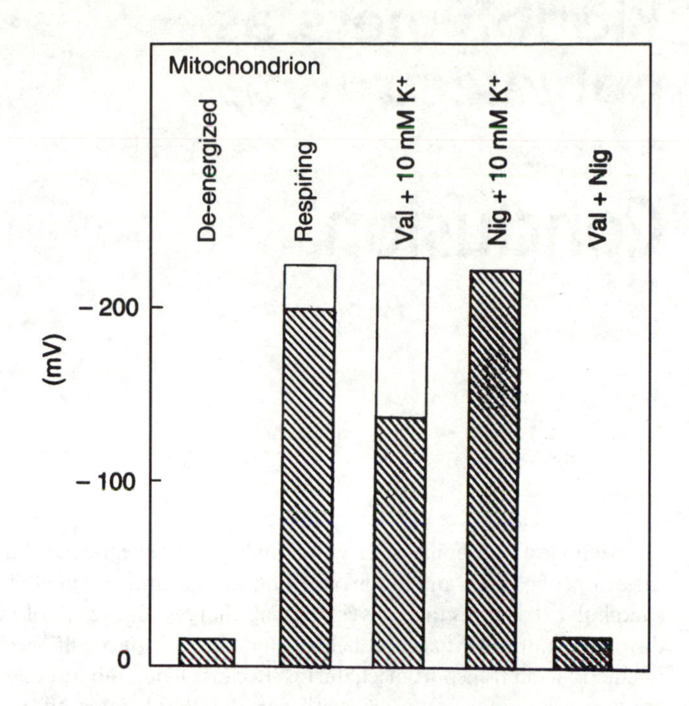

Figure 21.12 Ionophores modulate the composition of the proton potential. For a respiring mitochondrion, the proton motive force $e_p = -220$ mV is almost equal to $\Delta\Psi$. Valinomycin, in the presence of 10 mM K^+, enhances ΔpH but reduces $\Delta\Psi$ without changing e_p. Nigericin reduces ΔpH to zero but enhances $\Delta\Psi$. Both ionophores together enable K^+ ions to be transported through the membrane. This is equivalent to proton conduction, i.e. to the disappearance of both ΔpH and $\Delta\Psi$. Cross-hatching indicates $\Delta\Psi$; the open parts of the bars indicate ΔpH. (From *The Vital Force: A Study of Bioenergetics* by F.M. Harold. © 1986 by W.H. Freeman and Company. Used with permission.)

about 100 mV, we obtain $e_P \approx 160$ mV. One mole of protons crossing the membrane corresponds to an energy of about 15.4 kJ. Since 40 kJ are needed to synthesize one mole of ATP, about three protons must enter the membrane for each ATP molecule created.

It should be pointed out that the coupling is not local: the whole of the internal compartment defined by the membrane supplies the necessary energy. In other words, in the case of phosphorylation for example, the synthesis of ATP should be independent of the way in which the proton motive force is generated. It might in fact be produced in artificial vesicles by means of an electrochemical gradient between the outside and inside: ATP synthases embedded in the vesicles can then synthesize ATP molecules.

Part IV Biopolymers as polyelectrolytes

Conclusion

From an electrical point of view, the living cell appears as a huge assembly of molecules of all sizes and of ions carrying charges and moving through or around subcellular structures themselves carrying charges. The result of this is a complex distribution of potentials and fields, which in their turn will become involved in interaction and transport mechanisms. Because Coulomb forces vary as $1/r^2$ their effect is long-range compared with van der Waals interactions and hydrogen bonds. A problem with this kind of distribution cannot be tackled directly, and that is why this Part of the book began with a study of the fields and potentials in simple models: ionic solutions, the charged plane, sphere and cylinder. Laws had to be discovered and parameters defined that could in turn be used in a more detailed approach to the problems. Physical quantities like the Debye length, the screening factor and the charge parameter are examples of well-defined and measurable quantities. It then proved possible, by using the classical physics of continuous media, to characterize polyelectrolytes, to predict their behaviour and begin to examine their interactions. The price to be paid for this, of course, is an over-simplification of models, whether looking for solutions of the Poisson–Boltzmann equation or using Manning's theory. Despite their approximations and limitations, the models provide a good overall picture of polyelectrolytes.

This initial approach, however, while essential in clarifying ideas and concepts, is not enough to provide us with a detailed description of the electrical environment of a biopolymer. Present methods of representation favour a description of the charge distribution on an atomic scale provided the tertiary structure of the macromolecule is known with high resolution. 'Mixed' methods already make it possible to draw maps of the potential and field around an increasing number of proteins. This should lead to a better understanding of the process of enzyme–substrate interaction and, undoubtedly in the near future, of the passage of an ion through an ion channel or the functioning of an ion pump. To go any further, two questions must be tackled:

1. How can we take into account at any instant the internal dynamics of the biopolymer by calculating the interactions between charges carried by the macromolecule and all the ions moving in the solvent?
2. How can we represent on an atomic scale all the polarization forces reflected on a macroscopic scale by the existence of a dielectric constant?

The solution of these two problems will then allow molecular dynamics to be the preferred method for all polyelectrolyte studies. This is also the way in which we can deal with the conformational changes of a protein caused by the phosphorylation of serine, threonine or tyrosine: these seem to play a decisive role in controlling cellular division and perhaps in the storage of information in the brain. In the variety of problems discussed so far, an approach using a combination of thermodynamics and electrostatics has already given way to a structural and dynamic approach. This has been developed above for the case of hydration. Does this mean that all the models and methods of calculation based on the Poisson–Boltzmann equation will become obsolete? Quite apart from their pedagogical value, it is very probable that they may be the only accessible methods for calculating potentials and fields in large complexes such as the ribosome or the nucleosome, and thus even more so for the cell as a whole.

Part IV Biopolymers as polyelectrolytes

References and further reading

General

Stryer, L. (1988). *Biochemistry*, 3rd edn. Freeman, San Francisco.

DNA

Alfrey, T., Berg, P.W. and Morawetz, H. (1951). The counterion distribution in solutions of rod-shaped polyelectrolytes. *Journal of Polymer Science*, **7**, 543–7.

Anderson, C.F. and Record, M.T. (1982). Polyelectrolyte theories and their applications to DNA. *Annual Review of Physical Chemistry*, **33**, 191–222.

Arnott, S. (1970). The geometry of nucleic acids. *Progress in Biophysics and Molecular Biology*, **21**, 265–319.

Fenley, M.O., Manning, G.S. and Olson, W.K. (1990). Approach to the limit of counterion condensation. *Biopolymers*, **30**, 1191–203.

Fixman, M. (1979). The Poisson–Boltzmann equation and its application to polyelectrolytes. *Journal of Chemical Physics*, **70**, 4995–5005.

Friedman, H.L. (1975). Image approximation in the reaction field. *Molecular Physics*, **29**, 1533–43.

Friedman, R.A.G. and Manning, G. (1984). Polyelectrolyte effects on site-binding equilibria with application to the intercalation of drugs into DNA. *Biopolymers*, **23**, 2671–714.

Fuoss, R.M., Katchalsky, A. and Lifson, S. (1951). The potential of an infinite rod-like molecule and the distribution of the counter ions. *Proceedings of the National Academy of Sciences*, **37**, 579–89.

Gilson, M.K. (1995). Theory of electrostatic interactions in macromolecules. *Current Opinion in Structural Biology*, **5**, 216–23.

Guéron, M. and Weisbach, G. (1979). Polyelectrolyte theory. The polarizability of the counterion sheath. *Journal of Physical Chemistry*, **83**, 1991–8.

Guéron, M. and Weisbach, G, (1980). Polyelectrolyte theory. I: Counterion accumulation, site-binding and their insensitivity to polyelectrolyte shape in solutions containing finite salt concentrations. *Biopolymers*, **19**, 353–82.

Honig, B. and Nicholls, A. (1995). Classical electrostatics in biology and chemistry. *Science*, **268**, 1144–9.

James, A.E. and Williams, D.A. (1985). Numerical solutions of the Poisson–Boltzmann equation. *Journal of Colloid and Interface Science*, **107**, 44–59.

Jen-Jacobson, L., Kurpiewski, M., Lesser, D., Grable, J., Boyer, H.W., Rosenberg, J.M. and Greene, P. (1983). Coordinate ion pair formation between *Eco*RI endonuclease and DNA. *Journal of Biological Chemistry*, **258**, 14638–46.

Kirkwood, J. C. (1939). The dielectric polarization of polar liquids. *Journal of Chemical Physics*, **7**, 911–9.

Klein, B.J. and Pack, G.R. (1983). Calculations of the spatial distribution of charge density in the environment of DNA. *Biopolymers*, **22**, 2331–52.

Klein, B.J., Anderson, C.F. and Record, M.T. (1981). Comparison of Poisson–Boltzmann and condensation model expressions for the colligative properties of cylindrical polyions. *Biopolymers*, **20**, 2263–80.

Kotin, L. and Nagasawa, M. (1962). Chain model for polyelectrolyte. VII: Potentiometric titration and ion binding in solutions of linear polyelectrolytes. *Journal of Chemical Physics*, **36**, 873–9.

Lamm, G. and Pack, G.R. (1990). Acidic domains around nucleic acids. *Proceedings of the National Academy of Sciences, USA*, **87**, 9033–6.

Lampert, M.A. (1985). The non-linear Poisson–Boltzmann equation. *Nature*, **315**, 159.

Lampert, M.A. and Crandall, R.S. (1980). Non-linear Poisson–Boltzmann theory for polyelectrolyte solutions: the counterion condensate around a line charge as a δ-function. *Chemical Physics Letters*, **72**, 481–6.

Le Bret, M. and Zimm, B. (1984). Distributions of counterions around a cylindrical polyelectrolyte and Manning's condensation theory. *Biopolymers*, **23**, 287–312.

Lifson, S. and Katchalsky, A. (1954). The electrostatic free energy of polyelectrolyte solutions. II: Fully stretched macromolecules. *Journal of Polymer Science*, **13**, 43–55.

Lohman, T.M., de Haseth, P.L. and Record, M.T. (1980). Pentalysins–DNA interactions: a model for the general effects of ion concentrations on the interaction of proteins with nucleic acids. *Biochemistry*, **19**, 3522–30.

Manning, G.S. (1969). Limiting laws and counterion condensation in polyelectrolyte solutions. I: Colligative properties. II: Self-diffusion of the small ions. *Journal of Chemical Physics*, **51**, 924–38.

Manning, G.S. (1977). Limiting laws and counterion condensation in polyelectrolyte solutions. IV: The approach to the limit and the extraordinary stability of the charge fraction. *Biophysical Chemistry*, **7**, 95–102.

Manning, G.S. (1978). The molecular theory of polyelectrolyte solutions with application to the electrostatic properties of polynucleotides. *Quarterly Review of Biophysics*, **11**, 179–246.

Manning, G.S. (1979). Counterion binding in polyelectrolyte theory. *Accounts of Chemical Research*, **12**, 443–9.

Matthew, J.B. and Richards, F.M. (1982). Anion binding and pH-dependent electrostatic effects in ribonuclease. *Biochemistry*, **21**, 4989–99.

Matthew, J.B. and Richards, F.M. (1984). Differential electrostatic stabilization of A, B and Z forms of DNA. *Biopolymers*, **23**, 2743–59.

Mills, P., Anderson, C.F. and Record, M.T. (1986). Grand canonical Monte Carlo calculation of thermodynamic coefficients for a primitive model of DNA salt solution. *Journal of Physical Chemistry*, **90**, 6541–8.

Misra, V.K., Sharp, K.A., Friedman, R.A. and Honig, B. (1994a). Salt effects on ligand–DNA binding. *Journal of Molecular Biology*, **238**, 245–63.

Misra, V.K., Hecht, J.L., Sharp, K.A., Friedman, R.A. and Honig, B. (1994b). The λcI repressor and *Eco*RI endonuclease. *Journal of Molecular Biology*, **238**, 264–80.

Nakamura, H. (1996). Roles of electrostatic interaction in proteins. *Quarterly Review of Biophysics*, **29**, 1–90.

Olivares, W. and McQuarrie, D.A. (1975). On the theory of ionic solutions. *Biophysical Journal*, **15**, 143–62.

Oosawa, F. (1971). *Polyelectrolytes*. Dekker, New York.

Orttung, W.H. (1978). Direct solution of the Poisson equation for biomolecules of arbitrary shape. Polarizability density and charge distribution. *Annals of the New York Academy of Sciences*, **303**, 22–37.

Pack, G.R. and Klein, B.J. (1984). Generalized Poisson–Boltzmann calculation of the distribution of electrolyte ions around the B and Z conformers of DNA. *Biopolymers*, **23**, 2801–23.

Pack, G.R., Wong, L. and Prasad, C.V. (1986). Counterion distribution around DNA: variation with conformation and sequence. *Nucleic Acids Research*, **14**, 1479–93.

Paulsen, M.D., Anderson, C.F. and Record, M.T. (1988). Counterion exchange reactions on DNA: Monte Carlo and Poisson–Boltzmann analysis. *Biopolymers*, **27**, 1249–65.

Rashin, A. and Honig, B. (1985). Re-evaluation of the Born model of ion hydration. *Journal of Chemical Physics*, **89**, 5588–93.

Saito, M. (1994). Molecular dynamics simulations of proteins in solution; artifacts caused by the cut-off appoximation. *Journal of Chemical Physics*, **101**, 4055–64.

Schellman, J.A. and Stigter, D. (1977). Electrical double layer, zeta potential and electrophoretic charge of double-stranded DNA. *Biopolymers*, **16**, 1415–34.

Sharp, K.A. (1995). Polyelectrolyte electrostatics: salt dependence, entropic and enthalpic contributions to the free energy in the non-linear Poisson–Boltzmann model. *Biopolymers*, **36**, 227–43.

Sharp, K.A. and Honig, B. (1995). Salt effects on nucleic acids. *Current Opinion in Structural Biology*, **5**, 323–8.

Sharp, K.A., Friedman, R.A., Misra, V., Hecht, J. and Honig, B. (1995). Salt effects on polyelectrolyte–ligand binding: comparison of Poisson–Boltzmann and limiting law/counterions binding models. *Biopolymers*, **36**, 245–62.

Soumpasis, D.M. (1984). Statistical mechanics of the B–Z transition of DNA: contribution of diffuse ionic interactions. *Proceedings of the National Academy of Sciences, USA*, **81**, 5116–20.

Stigter, D. (1975). The charged colloidal cylinder with a Gouy double layer. *Journal of Colloid and Interface Science*, **53**, 296–306.

Stigter, D. (1978). A comparison of Manning's polyelectrolyte theory with the cylindrical Gouy model. *Journal of Physical Chemistry*, **82**, 1603–6.

Wensel, T.G., Meares, C.F., Vlachy, V. and Matthew, J.B. (1986). Distribution of ions around DNA, probed by energy transfer. *Proceedings of the National Academy of Sciences, USA*, **83**, 3667–71.

Wilson, R.W., Ray, D.C. and Bloomfield, V.A. (1980). Comparison of polyelectrolyte theories of the binding of cations to DNA. *Biophysical Journal*, **30**, 317–26.

Zimm, B.H. and Le Bret, M. (1983). Counterion condensation and system dimensionality. *Journal of Biomolecular Structure and Dynamics*, **1**, 461–71.

Proteins

Antosiewicz, J., McCammon, J.A. and Gilson, M.K. (1994). Prediction of pH dependent properties of proteins. *Journal of Molecular Biology*, **238**, 415–36.

Barlow, D.J. and Thornton, J.M. (1986). The distribution of charged groups in proteins. *Biopolymers*, **25**, 1717–33.

Getzoff, E.D., Tainer, J.A., Weiner, P.K., Kollman, P.A., Richardson, J.S. and Richardson, D.C. (1983). Electrostatic recognition between superoxide and copper, zinc superoxide dismutase. *Nature*, **306**, 287–90.

Gilson, M.K. and Honig, B. (1988). Calculation of the total electrostatic energy of a macromolecular system: solvation energies, binding energies and conformational analysis. *Proteins*, **4**, 7–18.

Gilson, M.K., Rashin, A., Fine, R. and Honig, B. (1985). On the calculation of electrostatic interactions in proteins. *Journal of Molecular Biology*, **183**, 503–16.

Hol, W.G.J. (1985). The role of the α-helix dipole in protein structure and function. *Progress in Biophysics and Molecular Biology*, **45**, 149–95.

Hol, W.G.J., van Duijnen, P.T. and Berendsen, H.J.C. (1978). The α-helix dipole and the properties of proteins. *Nature*, **273**, 443.

Holst, M., Kozack, R.E., Saied, F. and Subramaniam, S. (1994). Protein electrostatics: rapid multigrid-based Newton algorithm for solution of the full non-linear Poisson–Boltzmann equation. *Journal of Biological Structure and Dynamics*, **11**, 1437–45.

Honig, B., Hubbell, W.L. and Flewelling, R.F. (1986). Electrostatic interactions in membranes and proteins. *Annual Review of Biophysics and Biophysical Chemistry*, **15**, 163–93.

Imoto, T. (1983). Electrostatic free energy of lysozyme. *Biophysical Journal*, **44**, 293–8.

Klapper, I., Haegstrom, R., Fine, R., Sharp, K.A. and Honig, B. (1986). Focusing of electric fields in the active site of Cu–Zn superoxide dismutase: effects of ionic strength and amino acid modification. *Proteins*, **1**, 47–59.

Kong, X.-P., Onrust, R., O'Donnell, M. and Kuriyan, J. (1992). Three-dimensional structure of the β subunit of *E. coli* DNA polymerase III holoenzyme: a sliding DNA clamp. *Cell*, **69**, 425–37.

Lee, B.K. and Richards, F.M. (1971). The interpretation of protein structures: estimation of static accessibility. *Journal of Molecular Biology*, **55**, 379–400.

Linderstrøm-Lang, K. (1924). On the ionization of proteins. *Compte Rendu des Travaux du Laboratoire de Carlsberg*, **15**, 1–29.

Matthew, J.B. (1985). Electrostatic effects in proteins. *Annual Review of Biophysics and Biophysical Chemistry*, **14**, 387–417.

Mehler, E.L. and Eichele, G. (1984). Electrostatic effects in water-accessible regions of proteins. *Biochemistry*, **23**, 3887–91.

Paul, C.H. (1982). Building models of globular protein molecules from their amino acid sequences. I :Theory. *Journal of Molecular Biology*, **155**, 53–62.

Perutz, M.F. (1978). Electrostatic effects in proteins. *Science*, **201**, 1187–91.

Quiocho, F.A., Sack, J.S. and Vyas, N.K. (1987). Stabilization of charges on isolated ionic groups sequestered in proteins by polarized peptide units. *Nature*, **329**, 561–4.

Rashin, A. and Honig, B. (1984). On the environment of ionizable groups in globular proteins. *Journal of Molecular Biology*, **173**, 515–21.

Rogers, N.K. (1986). The modelling of electrostatic interactions in the function of globular proteins. *Progress in Biophysics and Molecular Biology*, **48**, 37–66.

Rogers, N.K. and Sternberg, M.J.E. (1984). Electrostatic interactions in globular proteins: different dielectric models applied to the packing of α-helices. *Journal of Molecular Biology*, **174**, 527–42.

Scheiner, S. and Hillenbrand, E.A. (1985). Modification of pK values caused by change in H-bond geometry. *Proceedings of the National Academy of Sciences, USA*, **82**, 2741–5.

Sheridan, R.P., Levy, R.M. and Salemme, F.R. (1982). α-helix dipole model and electrostatic stabilization of 4-α-helical proteins. *Proceedings of the National Academy of Sciences, USA*, **79**, 4545–9.

Tanford, C. (1957). Theory of protein titration curves. II: Calculations for simple models at low ionic strength. III: The location of electrostatic charges in Kirkwood's model of organic ions. *Journal of the American Chemical Society*, **79**, 5340–52.

Tanford, C. and Kirkwood, J.G. (1957). Theory of protein titration curves. I: General equations for impenetrable spheres. *Journal of the American Chemical Society*, **79**, 5333–9.

Vanbelle, D., Couplet, I., Prevost, M. and Wodak, S. (1987). Calculations of electrostatic properties in proteins. Analysis of contributions from induced protein dipoles. *Journal of Molecular Biology*, **198**, 721–35.

Wada, A. and Nakamura, H. (1981). Nature of the charge distribution in proteins. *Nature*, **293**, 757–8.

Warshel, A. and Levitt, M. (1976). Theoretical studies of enzymic reactions: dielectric, electrostatic and steric stabilization of the carbonium ion in the reaction of lysozyme. *Journal of Molecular Biology*, **103**, 227–49.

Warshel, A., Russell, S.T. and Churg, A.K. (1984). Macroscopic models for studies of electrostatic interaction in proteins limitation and applicability. *Proceedings of the National Academy of Sciences, USA*, **81**, 4785–9.

Warwicker, J. and Watson, H.C. (1982). Calculation of the electric potential in the active site cleft due to α-helix dipoles. *Journal of Molecular Biology*, **157**, 671–9.

Yang, A.S. and Honig, B. (1993). On the pH dependence of protein stability. *Journal of Molecular Biology*, **231**, 459–74.

Ion channels

Catterall, W.A. (1988). Structure and function of voltage sensitive ion channels. *Science*, **242**, 50–61.

Chan, D.Y.C. and Halle, B. (1984). The Smolukowski–Poisson–Boltzmann description of ion diffusion at charged interfaces. *Biophysical Journal*, **46**, 387–407.

Cooper, K., Jakobsson, E. and Wolynes, P. (1985). The theory of ion transport through membrane channels. *Progress in Biophysics and Molecular Biology*, **46**, 51–96.

Eisenman, G. and Dani, J.A. (1987). An introduction to molecular architecture and permeability of ion channels. *Annual Review of Biophysics and Biophysical Chemistry*, **16**, 205–26.

Fishman, H.M. (1985). Relaxations, fluctuations and ion transfer across membranes. *Progress in Biophysics and Molecular Biology*, **46**, 127–62.

Harold, F.M. (1986). *The vital force: a study of bioenergetics.* Freeman, New York.

Hodgkin, A.L. and Huxley. A.F. (1952). A quantitative description of membrane current and its application to conduction and excitation in nerves. *Journal of Physiology*, **117**, 500–44.

Jan, L.Y. and Jan, Y.N. (1989). Voltage sensitive ion channels. *Cell*, **56**, 13–25.

Levitt, D.G. (1985). Strong electrolyte continuum theory solution for equilibrium profiles, diffusion, limitation and conductance in charged ion channels. *Biophysical Journal*, **48**, 19–31.

Levitt, D.G. (1986). Interpretation of biological ion channel flux data. Reaction rate versus continuum theory. *Annual Review of Biophysical Chemistry*, **15**, 29–57.

Levitt, D.G. (1991). General continuum theory for multi-ion channel. (1) Theory. (2) Application to acetylcholine channel. *Biophysical Journal*, **59**, 271–88.

Mitchell, P. (1961). Coupling of phosphorylation to electrons and hydrogen transfer by a chemiosmotic type of mechanism. *Nature*, **191**, 144–8.

Polymeropoulos, E.E. and Brickmann, J. (1985). Molecular dynamics of ion transport through transmembrane model channels. *Annual Review of Biophysical Chemistry*, **14**, 315–30.

Stevens, C.F. (1984). Biophysical studies of ion channels. *Science*, **225**, 1346–50.

Part V

Association between molecules

Introduction

When studied on a molecular scale, most biological phenomena are found to involve the sometimes transient collision, interaction and association between two molecules. Such encounters occur in all enzyme – substrate reactions and in the mode of action of many molecules that are mutagenic, carcinogenic or antibiotic, or more generally act as selective effectors on specific targets (steroid hormones, oligopeptides, neurotransmitters). They also occur in the binding of divalent metallic cations (Mg^{2+}, Zn^{2+}, Mn^{2+}, Fe^{2+}, etc.), triggering structural changes in macromolecules or membranes, and in the specific recognition between two macromolecules (antigen – antibody, repressor – DNA, etc.) or between several macromolecules as in the replication, repair or transcription processes with DNA. Collisions between macromolecules and their association are, moreover, essential features in the building of subcellular structures either stable (chromatin, ribosomes, mitochondria, cytoskeleton, nucleoskeleton, etc.) or transient (microtubules, endocytotic vesicles, etc.), all of which are essential to cell life.

Considerable advances have occurred since the 1980s in the experimental structure determination of some macromolecular complexes, such as the antigen – antibody complex, several viruses, the repressor – operator specific complex, the polymerase – DNA association, and so on. We shall see later how these structural studies yield precise data essential to an understanding of the process of specific recognition, but do not make it possible to formulate any general laws. There appears to be no recognition 'code' operating when biopolymers interact: they simply seem to make use of the various modes of interaction listed in Part I of this book. Moreover, there are 'assembly processes' that occur in the formation of some transient complexes without the slightest relationship existing between the process and the function. If a clever 'trick' has been discovered, it is used wherever possible until a cleverer one is found. The aim of current work is thus simply to catalogue these interactions and to reveal the morphogenesis, similar in some respects to a kind of systematics at the molecular level. It does not bring into play any new forces or energies, but merely reveals a particularly effective form of 'tinkering' in Nature.

Accordingly, only a limited time will be spent on describing such interacting systems; instead, we shall concentrate on the search for models capable of representing different types of interaction. For equilibrium processes, these will be based on thermodynamics and statistics; for reaction rates, on kinetics and dynamics. Specificity will always appear as a competition between several reaction processes and not as an all-or-none mechanism.

In all these phenomena, the number and variety of sites, the association constants and the changes in thermodynamic variables characteristic of the association are quantities essential to a description and understanding of the equilibrium process. They must, of course, be supplemented by studies of reaction kinetics and conformational dynamics. Our initial study is restricted to equilibrium, and here it is essential to use a theoretical approach as simple and general as possible, allowing a definition of parameters common to all the processes, even if it means using a more elaborate and complex description in some special cases.

22

Equilibrium studies

22.1 Two simple molecules

Consider the bimolecular association between a molecule A defined as a *ligand* and another molecule M capable of binding only one A molecule. The reaction $M + A \rightleftharpoons MA$ is characterized by an *association constant* K:

$$K = [MA]/[M][A] \tag{22.1}$$

or a *dissociation constant* $K_d = K^{-1}$. The latter is often used in enzymology, but we prefer to use the former because the process involved is one of association.

For a given concentration of M and A, the solution will contain A, M and MA molecules. In this population, a mean fraction ν of bound A molecules can be defined as the ratio of the concentration of the complex MA to the total concentration of M, i.e.

$$\nu = [MA]/([M] + [MA]) \tag{22.2}$$

so that $0 < \nu < 1$. Replacing [MA] using eqn 22.1 gives

$$\nu = K[A]/(1 + K[A]) \tag{22.3}$$

The quantity ν can have two other meanings:

1. Out of all the possible sites provided by all the M molecules, ν is the fraction of occupied sites.
2. If all the M molecules were examined one after another, some of them would be free and others would be found as the complex MA. The quantity ν is then the probability of finding A bound to a randomly selected molecule M.

Equation 22.3 can be rewritten to express K in terms of ν:

$$K[A] = \nu/(1 - \nu) \tag{22.4}$$

or

$$\log K + \log[A] = \log[\nu/(1 - \nu)] \tag{22.5}$$

The quantity $-\log K = \log K_d$ is simply the pK and, by analogy, we can put $-\log[A] = \text{pA}$. Equation 22.5 then becomes

$$\text{pA} = \text{p}K - \log[\nu/(1 - \nu)] \tag{22.6}$$

In the case of electrolytes, A is simply the proton H^+ and we generally use the dissociation constant K_d of the acid MH together with a dissociated fraction α. We then have $\alpha = 1 - \nu$ and eqn 22.6 becomes the classical relationship

$$\mathrm{pH} = \mathrm{p}K_d + \log[\alpha/(1-\alpha)] \tag{22.7}$$

If $\alpha = \nu = \frac{1}{2}$, then eqns 22.6 and 22.7 show that pA = pK or pH = pK_d.

22.2 A ligand and a macromolecule

A macromolecule P may offer a number of binding sites, say n, to a ligand A. We then have to deal with a *multiple equilibrium* since the reaction in eqn 22.1 may occur at each site.

22.2.1 Equivalent sites

Assume first that the n sites on P are equivalent, i.e. having the same reactivity for the ligand. The ligand occupation rate of P is defined as the number of A molecules bound per molecule of P. As in the case of one site, eqns 22.2 and 22.3 still apply for each of the n sites. Because there are n sites per macromolecule, the mean fraction of occupied sites, ν^*, is n times the expression in eqn 22.3, i.e.

$$\nu^* = nK[\mathrm{A}]/(1 + K[\mathrm{A}]) \tag{22.8}$$

this time with $0 < \nu^* < n$.

This basic relationship for multiple equilibrium with n equivalent sites per macromolecule can be represented graphically in three different ways, called *binding isotherms* since the experiments are generally carried out at constant temperature. Each method enables us to determine the two parameters n and K characterizing the process.

Bjerrum plot

Here, ν^* is plotted against log[A], producing a graph (Fig. 22.1) similar to a titration curve in which the number of bound protons is plotted against $\mathrm{pH} = -\log[\mathrm{H}^+]$. The curve behaves asymptotically as $\nu^* \to n$ and there is a point of inflection when $\log[\mathrm{A}] = -\log K$.

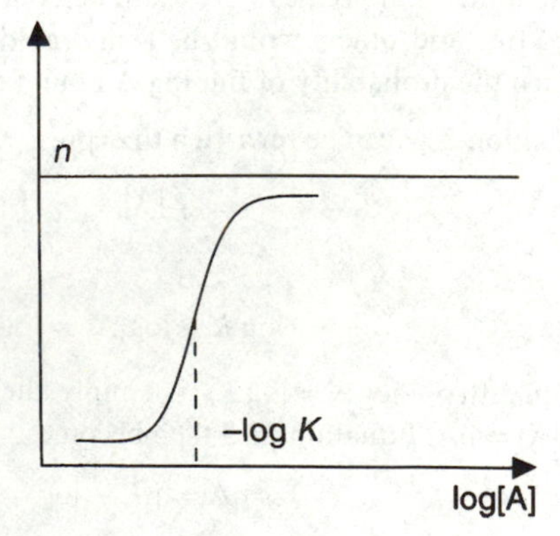

Figure 22.1 The Bjerrum plot.

Reciprocal plot

Taking the reciprocal of both sides of eqn 22.8 gives

$$1/\nu^* = 1/n + 1/nK[\mathrm{A}] \qquad (22.9)$$

Plotting $1/\nu^*$ against $1/[\mathrm{A}]$ gives a straight line of slope $1/nK$ and intercept on the vertical axis of $1/n$ (Fig. 22.2).

Scatchard plot

Equation 22.8 can also be written in the form

$$(n/\nu^*) - 1 = 1/K[\mathrm{A}]$$

or

$$\nu^*/[\mathrm{A}] = K(n - \nu^*) \qquad (22.10)$$

This is the form generally used, and here $\nu^*/[\mathrm{A}]$ is plotted against ν^* to give a straight line of slope $-K$ and intercept nK (Fig. 22.3).

Note that with the reciprocal plot, high values of [A] are needed to obtain an accurate value for the intercept, while the Scatchard plot requires small values of ν^* and thus of [A].

Before resuming the analysis of multiple equilibrium by looking at more complex examples, it is worth demonstrating the formal analogy between the treatment of a

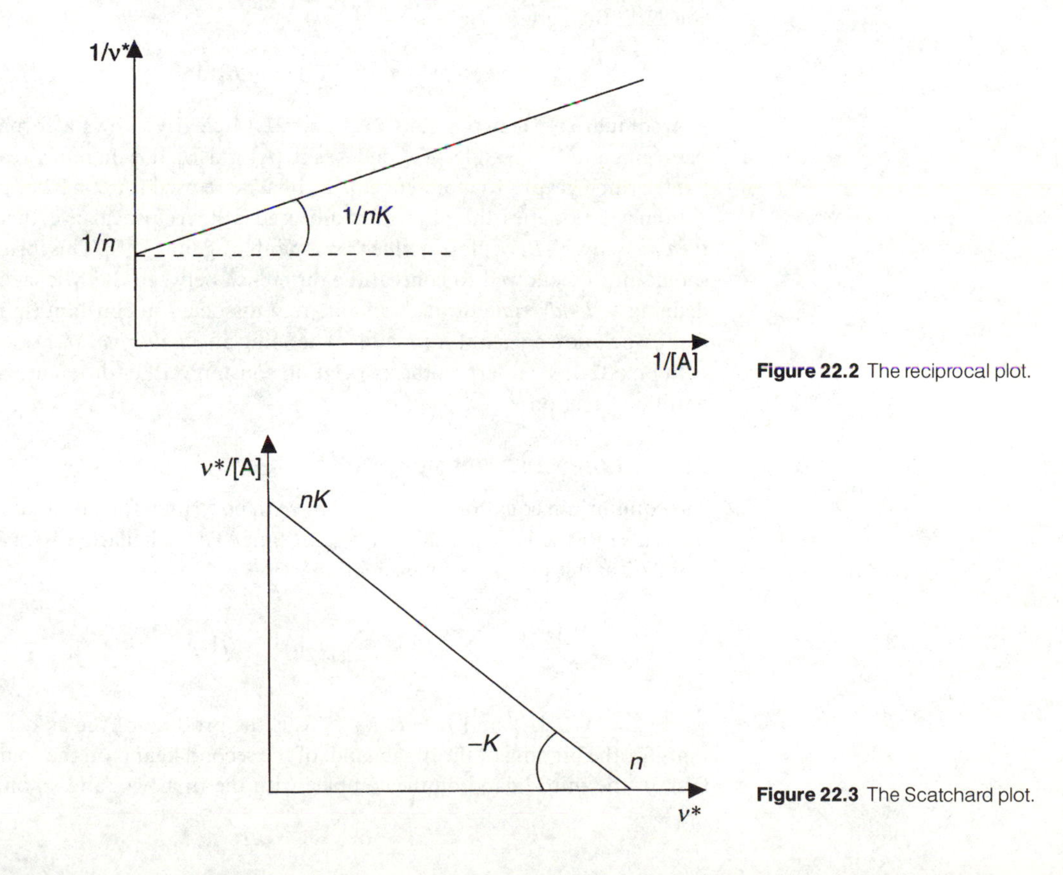

Figure 22.2 The reciprocal plot.

Figure 22.3 The Scatchard plot.

system with n sites and enzyme kinetics in a steady state. In the latter case, we have

$$\mathrm{E} + \mathrm{S} \underset{k_2}{\overset{k_1}{\rightleftharpoons}} \mathrm{ES} \underset{k_3}{\rightarrow} \mathrm{P} + \mathrm{E}$$

in which k_1, k_2 and k_3 are rate constants for the two successive reactions. The rate of formation of the enzyme–substrate complex is given by

$$\mathrm{d[ES]/d}t = k_1[\mathrm{E}][\mathrm{S}] - (k_2 + k_3)[\mathrm{ES}]$$

In the steady state, i.e. with an excess of substrate, $\mathrm{d[ES]/d}t = 0$ and hence

$$[\mathrm{E}][\mathrm{S}]/[\mathrm{ES}] = (k_2 + k_3)/k_1 = K_\mathrm{M} \tag{22.11}$$

where K_M is the Michaelis constant. If $[\mathrm{E_T}]$ is the total enzyme concentration, then

$$[\mathrm{E_T}] = [\mathrm{E}] + [\mathrm{ES}] \tag{22.12}$$

and using this to replace [E] in eqn 22.11:

$$[\mathrm{ES}] = [\mathrm{S}][\mathrm{E_T}]/([\mathrm{S}] + K_\mathrm{M})$$

The rate at which the product P is created, v, is

$$v = k_3[\mathrm{ES}] = k_3[\mathrm{E_T}][\mathrm{S}]/([\mathrm{S}] + K_\mathrm{M}) \tag{22.13}$$

As [S] tends to infinity, v tends to the limiting rate $v_\mathrm{M} = k_3[\mathrm{E_T}]$, which corresponds to the substrate binding to each active site of the enzyme. Using this value of v_M in eqn 22.13 then gives

$$v = v_\mathrm{M}[\mathrm{S}]K_\mathrm{M}^{-1}/(1 + K_\mathrm{M}^{-1}[\mathrm{S}]) \tag{22.14}$$

A comparison between eqns 22.8 and 22.14 clearly shows a formal analogy between v and v^*, v_M and n, K_M^{-1} and K, and [A] and [S]. It is therefore reasonable to use the same graphs to represent eqn 22.14 as were used for eqn 22.8.

In enzyme kinetics, the graph normally used is the reciprocal plot, then known as the *Lineweaver–Burk* plot, in which $1/v$ is plotted against $1/[\mathrm{S}]$. This formal analogy should not be allowed to conceal the difference between the Michaelis constant defining a *steady state* through an apparent dissociation constant (in mol) and a true association constant K (in mol^{-1}) defining an *equilibrium state*.

In Box 22.1, we interpret the association constant rather differently in terms of partition functions.

22.2.2 Non-equivalent sites

The equilibrium equations in this case have to be written for the binding of ligand molecules to the polymer taken one at a time. The calculation is carried out in Box 22.2 and leads to the final equation

$$\nu^* = \sum_{i=0}^{n} iL_i[\mathrm{A}]^i \Big/ \sum_{i=0}^{n} L_i[\mathrm{A}]^i = \partial \log \Sigma / \partial \log [\mathrm{A}] \tag{22.15}$$

where $\Sigma = \sum_i L_i[\mathrm{A}]^i$ and $L_i = K_1 K_2 \ldots K_i$ is the product of the association constants for the binding of the first ligand, of the second ligand on the complex to the first, of the third ligand on the complex with the first two, and so on. This very

Box 22.1 **Association constants and partition function**

Statistical thermodynamics can be used to interpret the association constant somewhat differently. Consider a linear sequence of n sites, of which ν are occupied. All the sites are assumed to be independent and equivalent. If q_1 is the partition function of a ligand bound to a site, the partition function Z of the ν occupied sites will be

$$Z = q_1^{\nu} n!/\nu!(n-\nu)!$$

where the binomial coefficient is the number of ways of placing ν objects into n boxes. The general relationship

$$\mu_1^{\mathrm{b}} = -RT\,\partial(\ln Z)/\partial\nu$$

exists between the chemical potential μ_1^{b} of the bound ligand and the partition function Z. Using Stirling's approximation that $n! \approx n\ln n - n$, we obtain

$$\mu_1^{\mathrm{b}} = RT\ln[\nu/q_1(n-\nu)]$$

The bound and free ligands can be regarded as forming two distinct phases in equilibrium. If a is the activity of the free ligand (often the same as its concentration), we have in the solvent phase

$$\mu_1^{\mathrm{f}} = \mu_0 + RT\ln a$$

where μ_1^{f} is the chemical potential of the free ligand. Since $\mu_1^{\mathrm{b}} = \mu_1^{\mathrm{f}}$, we have

$$\nu/a(n-\nu) = q_1\exp(\mu_0/RT)$$

which is also valid for the mean value ν^*. Comparison of this with eqn 22.10 gives the following relationship between K, μ_0 and q_1:

$$K = q_1\exp(\mu_0/RT)$$

However, there is also the classical relationship between the partition function q_0 of the free ligand and μ_0:

$$\mu_0 = -RT\ln q_0$$

and hence

$$K = q_1/q_0$$

i.e. the association constant is the ratio of the partition functions for the free and bound ligands respectively.

general expression is of no practical use and is only of value in studying the interaction free energy. Several of its properties are, however, interesting from a practical point of view.

1. Experimentally, the curve of $\log[\nu^*/(n-\nu^*)]$ plotted against $\log[\mathrm{A}]$ (the *Hill plot*, Fig. 22.4), is almost linear over a considerable range of $\log[\mathrm{A}]$ values. For n equivalent sites, eqn 22.10 shows that the Hill plot is a straight line of slope 1 given by

$$\log[\nu^*/(n-\nu^*)] = \log K + \log[\mathrm{A}]$$

Box 22.2 **Calculation of a multiple equilibrium**

For the first molecule, we have

$$[PA]/[P][A] = K_1 \quad (1)$$

where [PA] is the concentration of the complex consisting of one macromolecule and one ligand bound to any one of the n sites. In the next step, we have

$$[PA_2]/[PA][A] = K_2 \quad (2)$$

where $[PA_2]$ is the concentration of the complex consisting of a macromolecule with two bound ligands, etc.

In total, the fraction ν^* of bound ligand is

$$\nu^* = \{[PA] + 2[PA_2] + \ldots + (n-1)[PA_{n-1}] + n[PA_n]\}/ \\ ([P] + [PA] + [PA_2] + \ldots + [PA_n]) \quad (3)$$

Introducing eqns 1, 2 and so on into this equation, and dividing numerator and denominator by P, we obtain

$$\nu^* = (K_1[A] + 2K_1K_2[A]^2 + \ldots + nK_1K_2\ldots K_n[A]^n)/ \\ (1 + K_1[A] + K_1K_2[A]^2 + \ldots + K_1K_2\ldots K_n[A]^n) \quad (4)$$

To understand the significance of the constants K_1, K_2, etc., we look again at the case of equivalent sites with an intrinsic association constant K. The constants $K_1, K_2, \ldots$ are then apparent constants in which a probability factor is included. For K_1 is the number of ways in which the ligand A is bound to one of the n sites, i.e. $K_1 = nK$. Similarly, K_1K_2 is the number of ways in which two ligands can be randomly bound to two of the n sites:

$$K_1K_2 = n(n-1)K^2/2$$

and since $K_1 = nK$, $K_2 = (n-1)K/2$. Extending this argument, the general expression for $K_1K_2\ldots K_i$ is

$$n!K^i/i!(n-i)!$$

Substituting these values for K_1, etc., in eqn 4, and remembering that the factors $n!/i!(n-i)!$ are simply the binomial coefficients, we arrive at

$$\nu^* = nK[A](1 + K[A])^{n-1}/(1 + K[A])^n = nK[A]/(1 + K[A])$$

which is eqn 22.8.

Equation 4 can be expressed in a more compact form by defining quantities $L_i = K_1K_2\ldots K_i$ with the convention that $L_0 = 1$. We then have

$$\nu^* = \sum_{i=0}^{n} iL_i[A]^i / \sum_{i=0}^{n} L_i[A]^i \quad (5)$$

or

$$\nu^* = \partial \log \Sigma / \partial \log [A]$$

where $\sum = \sum_i L_i[A]^i$.

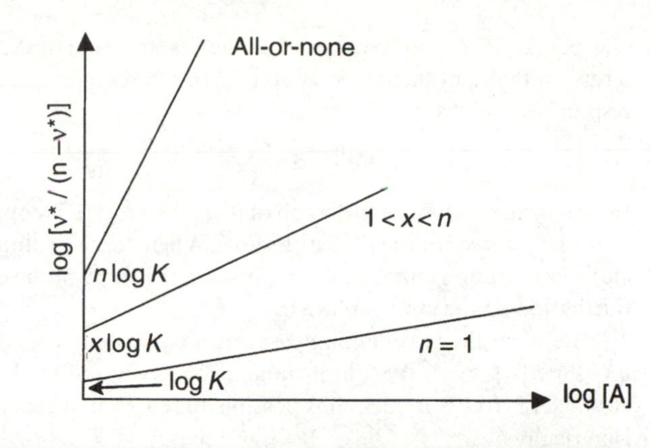

Figure 22.4 The Hill plot.

Box 22.3 All-or-none process

In this case only two states are possible: either no site is occupied or all sites are occupied. Equation 4 in Box 22.2 then takes the simple form:

$$\nu^* = nK_1K_2\ldots K_n[\mathrm{A}]^n/(1 + K_1K_2\ldots K_n[\mathrm{A}]^n) = n(K[\mathrm{A}])^n/(1 + K[\mathrm{A}]^n)$$

and hence

$$\nu^*/(n - \nu^*) = (K[\mathrm{A}])^n$$

is the new equation of the isotherm.

In the Scatchard plot, we have

$$\nu^*/[\mathrm{A}] = K\nu^{*(n-1)/n}(n - \nu^*)^{1/n}$$

This isotherm passes through a maximum when $\nu^* = n-1$ (as can be shown by equating to zero the derivative with respect to ν^*).

In the Hill plot, we have

$$\log[\nu^*/(n - \nu^*)] = n\log K + n\log[\mathrm{A}]$$

i.e. a straight line of slope n.

If there is cooperative binding, the association constant for the second ligand molecule is greater than that for the first, that for the third is greater than that for the second, and so on. The slope of the Hill plot is then always greater than 1. Let us take as an example an all-or-none process: as soon as one ligand is bound to a site of the macromolecule P, all the sites on P will be saturated before another macromolecule could bind a ligand. In such a case, the fraction ν represents a heterogeneous population of macromolecules without A and of macromolecules saturated with A.

This means that, in the expression for ν^*, only the last term remains in the numerator and denominator (see Box 22.3) and hence

$$\nu^* = n(K[\mathrm{A}])^n/\{1 + (K[\mathrm{A}])^n\}$$

so that $\nu^*/(n-\nu^*) = (K[\mathrm{A}])^n$ giving a Hill plot with a slope of n.

2. If in eqn 22.15 we differentiate ν^* with respect to log[A], we obtain

$$\mathrm{d}\nu^*/\mathrm{d}\log[\mathrm{A}] = A\left[\sum i^2L_iA^{i-1}/\sum L_iA^i - \sum L_iA^{i-1}\sum iL_iA^i/\left(\sum L_iA^i\right)^2\right]$$

$$= \sum i^2L_iA^i/\sum L_iA^i - \left(\sum iL_iA^i/\sum L_iA^i\right)^2 \qquad (22.16)$$

where [A] has been replaced by A on the right.

In eqn 22.15, ν^* can be regarded as the first moment of the function $\sum L_i A^i$. The terms on the right-hand side of eqn 22.16 are therefore equal to $\langle \nu^{*2} \rangle$ and $\langle \nu^* \rangle^2$ respectively, so that

$$d\nu^* / d \log [A] = \langle \nu^{*2} \rangle - \langle \nu^* \rangle^2 \quad (22.17)$$

In this form, the slope of the graph of ν^* against log[A] is equal to the *fluctuation* of ν^* or the *variance* of the ν^* distribution. When A is a proton, the ν^* versus log[A] curve is a titration curve, and for a protein the slope of the curve gives the charge fluctuation on the polyampholyte.

3. Between the two limiting cases of n equivalent sites described by $\nu^*/[A] = nK/(1 + K[A])$ and the all-or-none process described by $\nu^*/[A] = nK[A]^{n-1}/(1 + K^n[A]^n)$ (see Box 22.3), it is possible to represent a general type of association empirically by

$$\nu^* / [A] = nK[A]^{x-1} / (1 + K^x [A]^x)$$

with $1 < x < n$. The parameter x is called the *Hill coefficient* and a graph of ν^* against log[A] is a sigmoid curve.

4. If the n sites are distributed over several families of identical sites, say n_1 with constant K_1, n_2 with constant K_2, etc., and with $n = \sum n_j$, then ν^* can be written as the sum

$$\nu^* = \sum_j n_j K_j [A] / (1 + K_j [A]) \quad (22.18)$$

in which j takes on as many values as there are independent families of identical sites. When there are only two or three families and when their respective association constants differ considerably, it might be possible to distinguish the various parts of the binding isotherm experimentally and to derive the values for the n_j and K_j, albeit at the expense of a considerable amount of computer time.

22.3 Binding energy

The association constant K and the change in standard free energy ΔG are related by the classical equation

$$\Delta G = -RT \ln K \quad (22.19)$$

At 20°C, each 10-fold increase in K corresponds to an increase in ΔG of about 1.35 kcal mol^{-1}.

In studying any interaction, we have to examine the variation of K with temperature, i.e. to establish a set of isotherms at temperatures compatible with the existence of the molecules involved. Replacing ΔG in eqn 22.19 by $\Delta H - T\Delta S$,

$$\ln K = -\Delta H / RT + \Delta S / R$$

and hence, assuming ΔH and ΔS to be independent of temperature:

$$d \ln K / dT = \Delta H / RT^2$$

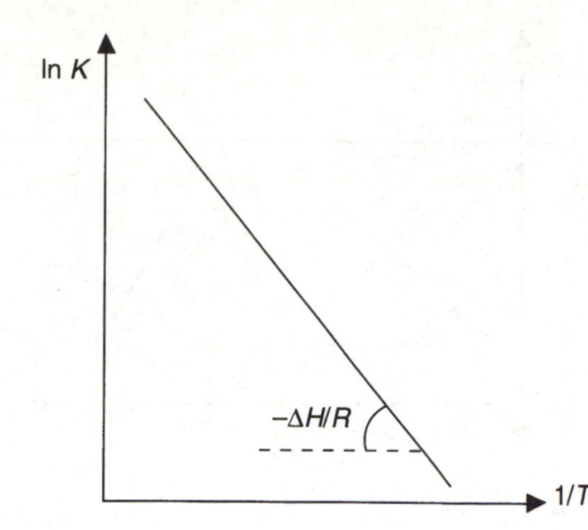

Figure 22.5 Graph of ln K against $1/T$ giving ΔH and ΔS.

or, more usually,

$$d \ln K/d(1/T) = -\Delta H/R \tag{22.20}$$

This leads to the standard graph of ln K plotted against $1/T$ (Fig. 22.5), giving a straight line of slope $-\Delta H/R$ from which ΔS can be calculated.

These thermodynamic quantities are characteristic of the association process. For example, a ΔH close to zero and a positive ΔS accompany a purely electrostatic binding. Such an increase in entropy, which allows the reaction to take place, generally comes from the release of ions and water molecules (see Part IV). In passing from the bound to the free state, they gain more degrees of freedom and this increases the entropy of the system.

In some cases, the free energy of binding is a function of ν^*: in other words, the sites are not energetically equivalent. The simplest way of changing ΔG is to assume a linear variation and write

$$\Delta G = \Delta G_0 + w\nu^* \tag{22.21}$$

Because ΔG is negative, the absolute value of the energy will increase if $w < 0$ and decrease if $w > 0$. In the first case, K will increase with ν^* and the process is said to be *cooperative*: the presence of an already bound ligand favours the binding of the next one. On the other hand, if $w > 0$, the process is called *anticooperative*. These effects may originate either from an interaction between bound ligands or from a structural change in the polymer induced by the ligand binding, which modifies one of the parameters (n, K) of the association.

Using eqn 22.19, eqn 22.21 can be written

$$K = K_0 \exp(-w\nu^*/RT) \tag{22.22}$$

and, substituting this in eqn 22.10 of the Scatchard plot, we obtain:

$$\nu^*/[A] = K_0(n - \nu^*) \exp(-w\nu^*/RT) \tag{22.23}$$

The derivative $d(\nu^*/[A])/d\nu^*$ is zero when

$$\nu^* = n + RT/w \tag{22.24}$$

and for a cooperative process ($w < 0$), the isotherm passes through a maximum at this value of ν^*, which is less than n (Fig. 22.6).

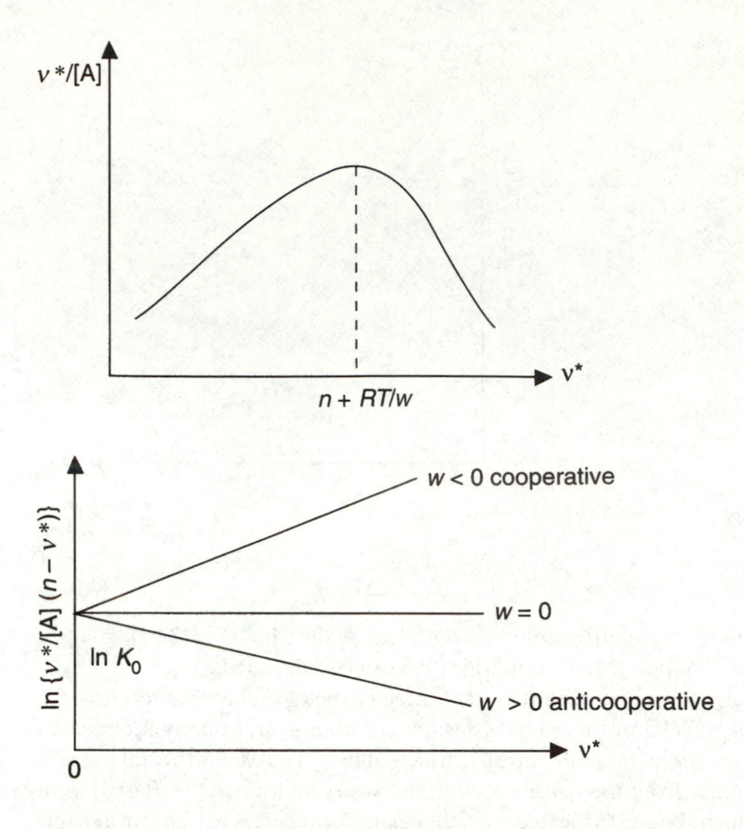

Figure 22.6 Scatchard isotherm for a cooperative process.

Figure 22.7 Cooperative and anticooperative processes.

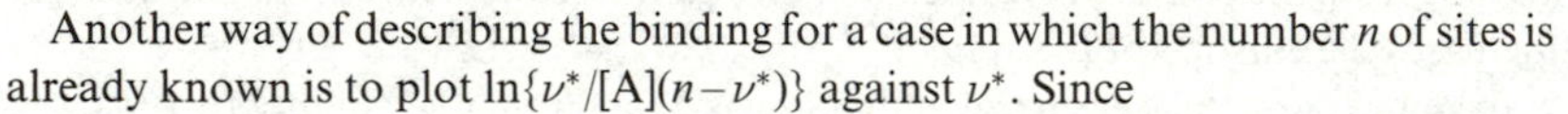

Another way of describing the binding for a case in which the number n of sites is already known is to plot $\ln\{\nu^*/[\mathrm{A}](n-\nu^*)\}$ against ν^*. Since

$$\ln[\nu^*/[\mathrm{A}](n-\nu^*)] = \ln K_0 - w\nu^*/k_{\mathrm{B}}T \qquad (22.25)$$

the linear isotherm given by this has a positive slope when the process is cooperative and a negative slope when it is anticooperative (Fig. 22.7).

The general case of the free energy of binding for any system of sites is discussed in Box 22.4.

22.4 Large ligands

For a discussion of this topic see McGhee and von Hippel (1974). When a ligand is large enough to cover several binding sites or when it modifies the local structure of the polymer in such a way as to prevent some neighbouring sites being occupied, the above arguments and calculations must be looked at again.

To begin with, we simplify the problem by making three assumptions:

1. The polymer is assumed to have a linear lattice of N repeated units (residues), where $N \to \infty$. This type of model is very appropriate for the binding of large ligands to DNA, particularly for that of many proteins that play a part in expressing and regulating the genetic message.
2. There is polar binding of the ligands to the lattice sites, i.e. binding in the same direction.
3. The ligand covers n successive residues, e.g. a protein covering several phosphate groups in the DNA molecule.

Box 22.4 **Binding free energy: the general case**

A generalization to any system of sites can be carried out using eqn 22.18 (Edsall and Wyman, 1958). The chemical potential of species PA_i is

$$\mu_i = \mu_i^\circ + RT\ln[PA_i]$$

and that of the free ligand is

$$\mu_A = \mu_A^\circ + RT\ln[A]$$

According to eqn 22.15, the fraction f_i of molecules in the state PA_i is L_iA^i/Σ with $f_0 = 1/\Sigma$ by convention.

To simplify the matter, consider one mole of P. The free energy before the interaction is

$$G_1 = \mu_0^\circ + RT\ln[P_T] + \nu(\mu_A^\circ + RT\ln[A]) + G_0 \quad (1)$$

where G_0 is the excess free energy of the ligand and of the other species in solution (ions, for example), where

$$P_T = \sum_{i=0}^{n}[PA_i] = [P]\Sigma$$

and where $\nu = \sum_1^n if_i$ is the number of moles of bound ligand at equilibrium.

After the interaction, the energy is

$$G_2 = f_0(\mu_0^\circ + RT\ln[P]) + \sum_1^n f_i(\mu_i^\circ + RT\ln[PA_i]) + G_0$$

However, $f_0 = 1 - \sum_i^n f_i$ and so

$$G_2 = \mu_0^\circ + RT\ln[P] + \sum_1^n f_i(\mu_i^\circ + RT\ln[PA_i] - \mu_0^\circ + RT\ln[PA_i]) + G_0 \quad (2)$$

The difference ΔG between eqns 1 and 2 is therefore

$$\Delta G = RT\ln([P]/[P_T]) + \sum f_i(\mu_i^\circ - \mu_0^\circ - i\mu_A^\circ) + RT\ln([PA_i]/[P][A_i]) \quad (3)$$

The quantity $\mu_i^\circ - \mu_0^\circ - i\mu_A^\circ$ is the free energy of formation of PA_i from P and the free ligand, and is therefore $-RT\ln L_i$. Since the last term on the right-hand side of eqn 3 is $+RT\ln L_i$, the two terms cancel and we end up with the important relationship

$$\Delta G = -RT\ln([P]/[P_T]) = -RT\ln\Sigma \quad (4)$$

and, comparing this with equations in statistical thermodynamics, this shows that Σ behaves like a partition function. To bring out the significance of eqn 4, consider once again the simple case of n independent sites for which $\Sigma = (1 + K[A])^n$. This gives us

$$\Delta G = -nRT\ln(1 + K[A]) \quad (5)$$

which, of course, differs from eqn 22.19 giving the *standard free energy* independent of the free ligand concentration. The quantity ΔG given by eqn 4 can be determined experimentally. Combining eqns 4 and 22.15, we have

$$\partial\Delta G/\partial\ln[A] = -RT\partial\ln\Sigma/\partial\ln[A] = -\nu^* RT$$

and hence

$$\Delta G = -RT \int_0^A \nu^* \, d(\ln[A]) \qquad (6)$$

In the Bjerrum plot of ν^* against ln[A], ΔG in eqn 6 is given by the area between the curve and the horizontal axis.

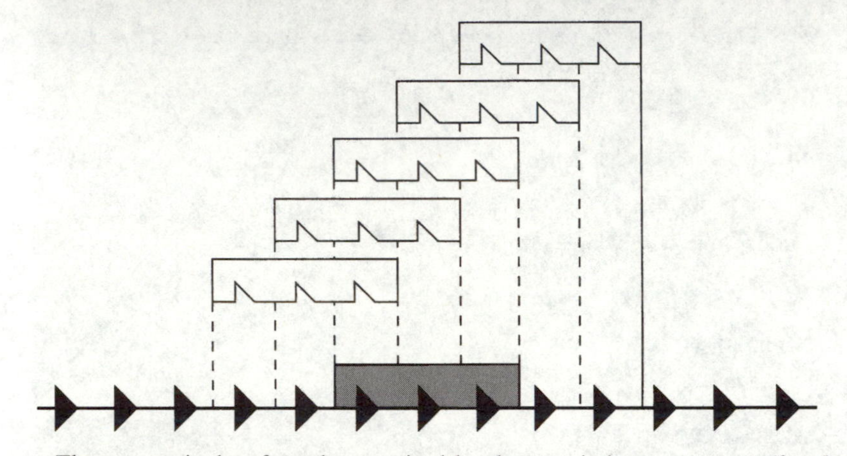

Figure 22.8 The binding of a ligand covering n residues excludes $2n-1$ potential sites. In the example shown with $n=3$, five possibilities are excluded. (From McGhee and von Hippel, 1974. Reprinted from *Journal of Molecular Biology* with the permission of Academic Press.)

The process is therefore characterized by the association constant K (for the ligand to its site) and the value of n (which is not related in any way to the number of sites defined in section 22.2). The concentration of the free ligand will still be denoted by [A] but ν^* will now be the ratio of the concentration of bound ligand to the total concentration of residues.

It would be possible to take the naive view that the number of available sites is N/n and that therefore the number of sites available to the ligand at all times is N/n minus the number of bound ligands. This argument is incorrect for two reasons. Firstly, in a completely free lattice of N residues, the number of binding sites is not N/n: these sites are not 'booked in advance' but any residue can form the beginning of a site. There are thus $N-n+1$ *potential* sites ($N-n$ residues at the beginning of a site plus the last n residues for the $(N-n+1)$th site). Thus, at the start of the binding process, the value of K would be greatly underestimated. Secondly, when a ligand is bound, a minimum of n residues are excluded when there is a chance of finding exactly one free space of n residues between two occupied sites. On the other hand, the first bound ligand molecule excludes $2n-1$ residues (Fig. 22.8). For the following ones, if a 'hole' of g residues is free, there are two possibilities: either $g > n$ and there are $g-n+1$ sites, or $g < n$ and there are no sites. The lattice is therefore very difficult to saturate. The process is reminiscent of parking a car at the kerbside: to reverse into a parking space requires a free space at least equal to the length of the car (Fig. 22.9).

Scatchard's equation (eqn 22.10), with N as the number of sites, can be written in the form:

$$(\nu^*/N)/[A] = K(1 - \nu^*/N) \qquad (22.26)$$

In the case now under discussion, it is ν^*/N that will be ν, i.e. the *binding density* (moles of bound ligands/moles of all the residues). The binding isotherm thus has

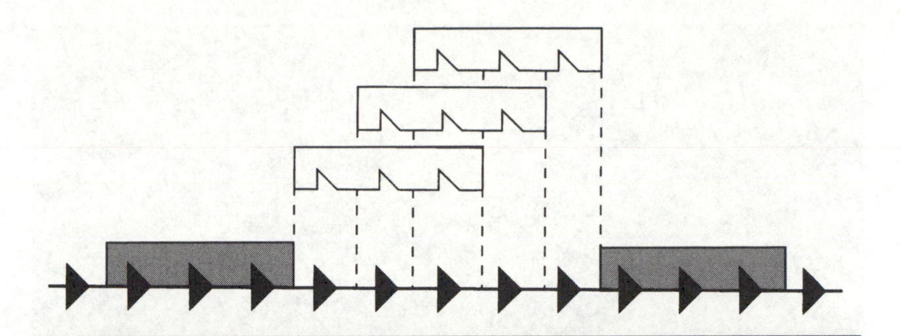

Figure 22.9 A 'hole' of g residues provides $g-n+1$ potential sites. In the example shown with $g=5$ and $n=3$, there are three possible sites. (From McGhee and von Hippel, 1974. Reprinted from *Journal of Molecular Biology* with the permission of Academic Press.)

Box 22.5 **Calculation of the binding isotherm for a large ligand**

We start by defining the probability P_g of finding g residues in the free space between two ligands. The mean number s of free sites per empty space is then

$$s = \sum_{g=n}^{N} (g-n+1)P_g \tag{1}$$

If B ligands are bound, there are $B+1$ free spaces with $g \geq 0$. The available density of binding sites is therefore

$$(B+1)\sum_{g=n}^{N} (g-n+1)P_g \tag{2}$$

If V denotes an 'empty' residue (not covered) and O an 'occupied' residue (covered), we must have four types of probability as we move in a given direction along the lattice of sites:

- (VV) = probability of two contiguous empty residues
- (VO_1) = probability of an occupied residue to the right of an empty one
- $(O_1O_2)\ldots(O_{n-1}O_n)$ = probability that two contiguous residues are occupied (this can only occur inside the n residues covered by the ligand)
- (O_nV) = probability of an empty residue to the right of a covered one.

We have the obvious relationships:

- $(VV)+(VO_1)=1$ (to the right of an empty residue)
- $(O_nV)+(O_nO_1)=1$ (to the right of a bound ligand)
- $(O_1O_2)=(O_2O_3)= \ldots =(O_{n-1}O_n)=1$ (by definition)
- $(VO_2)=(VO_3)= \ldots =0$ (impossibility)

The probability P_g can be calculated as

$$P_g = (O_nV)\,(VV)^{g-1}\,(VO_1) \tag{3}$$

where (VO_1): beginning of the right ligand; $(VV)^{g-1}$: $g-1$ contiguous empty sites; (O_nV): end of the left ligand.

We now select a point on the isotherm where B ligands are bound ($\nu = B/N$). There are Bn (i.e. $nN\nu$) covered residues and hence a fraction $n\nu N/N = n\nu$ of covered sites. The free fraction $1-n\nu$ is the probability that a randomly selected site is free. The probability that the next residue to the right of the first is free is also $1-n\nu$. But this second residue can be selected in two ways:

- the first residue is free – probability (VV)
- the first residue is the last covered by the ligand – probability (O_nV)

so that

$$(1-n\nu)(VV)+\nu(O_nV)=1-n\nu$$

With no cooperativity, we must also have $(O_nV)=(VV)$ and $(O_nO_1)=(VO_1)$, and hence

$$(VV)=(O_nV)=(1-n\nu)/[1-(n-1)\nu]$$

and because $(O_nO_1)=1-(O_nV)$, we have

$$(O_nO_1)=(VO_1)=\nu/[1-(n-1)\nu] \tag{4}$$

Substituting this in eqn 3 gives us

$$P_g=\nu(1-n\nu)^g/[1-(n-1)\nu]^{g+1}$$

Using this in eqn 1 and making N tend to infinity, we obtain a series of geometric terms and their derivatives, giving

$$s=(1-n\nu)n/\nu[1-(n-1)\nu]^{n-1}$$

Since $B\approx B+1$, we obtain (available density $=Bs/N=\nu s$):

$$\nu/[A]=K(1-n\nu)^n/[1-(n-1)\nu]^{n-1} \tag{5}$$

as in eqn 22.27.

the new form:

$$\begin{aligned}&\text{(binding density)/(free ligand concentration)}\\&=K\text{(available density of binding sites)}\end{aligned}$$

and the problem is to calculate the available density as shown in Box 22.5. The final result is

$$\nu/[A]=K(1-n\nu)\{(1-n\nu)/[1-(n-1)\nu]\}^{n-1} \tag{22.27}$$

When this more elaborate form of the isotherm is used for $n=1$, the usual form of Scatchard's equation $\nu/[A]=K(1-\nu)$ is of course obtained: this is similar to eqn 22.10 because $\nu=\nu^*/N$ (or ν^*/n in the notation of section 22.2).

An example of a case where $n=2$ is the intercalation of a planar aromatic dye between two base pairs. One base pair is then regarded as a residue and the intercalation excludes the neighbouring base pair from dye binding. Putting $n=2$ in eqn 22.27 yields the expression in eqn 4.34 derived directly in Part I.

In the general case, the intercept on the vertical axis of the Scatchard isotherm is K. The isotherm crosses the ν axis at $1/n$ (saturation) (Fig. 22.10). It is in fact curved and changes only slightly with ν in this region, so that it is very difficult to determine n. Physically, the end of binding can be regarded as an accumulation of 'empty' spaces of length less than n. To fill them, the already bound ligands must be moved

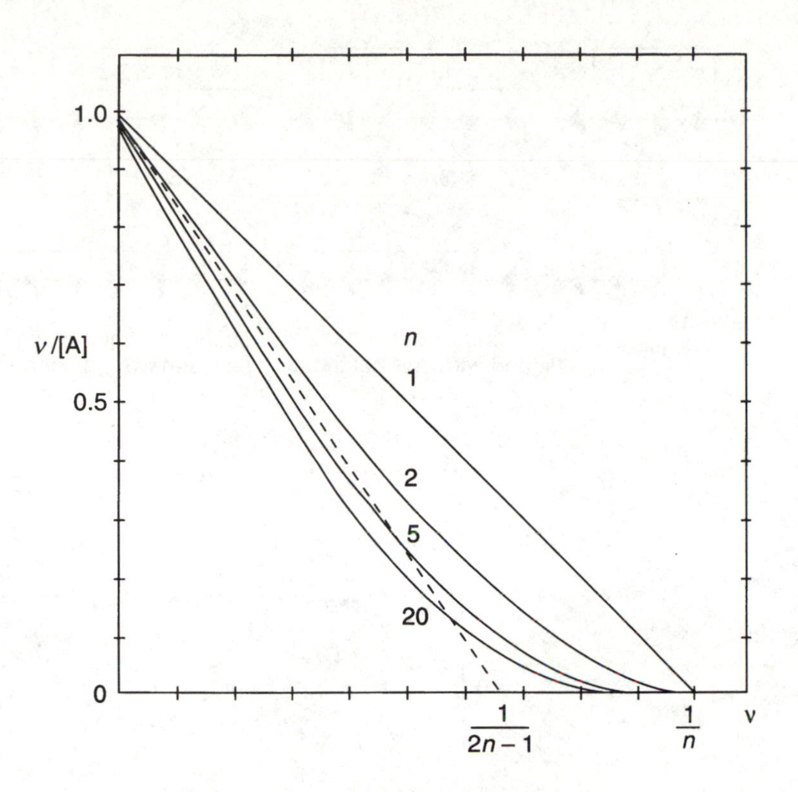

Figure 22.10 Scatchard isotherms calculated from eqn 22.27, with K taken arbitrarily as equal to 1 $\mathbf{M}^{-1}$ and with the values of n shown on the curves. The intercalation process (non-cooperative) is generally well represented by the curve for $n = 2$, and the dashed line is the tangent to this curve at $\nu = 0$. (From McGhee and von Hippel, 1974. Reprinted from *Journal of Molecular Biology* with the permission of Academic Press.)

and redistributed over the lattice of sites. This operation 'costs' entropy and saturation can only be achieved by greatly increasing the free ligand concentration with all the parasitic effects that might then occur (aggregation, precipitation, non-specific binding).

As $\nu \to 0$, the limiting value of the derivative $d(\nu/[A])/d\nu$ is

$$-K(2n-1) \tag{22.28}$$

and no longer $-K$ as in the case of the Scatchard isotherm. If, as a result, the presence of a large ligand is neglected, the value of K will be overestimated. It may be that, for experimental reasons, sufficiently high values of ν are not available, preventing any observation of the curvature of the isotherm. There will be a tendency to extrapolate the straight line passing through the first few points. According to eqn 22.28, the line will cross the ν axis at a value of $1/(2n-1)$ (see Fig. 22.10). For $n = 4$, say, with the isotherm regarded as a straight line, the value of K obtained would be seven times as great as it should be, and an apparent density of 1/7 would seem to indicate that one residue out of seven is capable of binding a ligand.

The calculation of isotherms just developed can be extended to cooperative binding. In this case, the cooperativity comes only from interaction between a bound ligand molecule and the nearest bound molecules. Three possibilities exist, each corresponding to a different value of the association constant (Fig. 22.11):

	Situation	Constant	Mean no. of bound ligands
(0)	Ligand alone on isolated site	K	$B_0 = K[A]$
(1)	Ligand with one neighbour	$K\omega$	$B_1 = K\omega[A]$
(2)	Ligand between two neighbours	$K\omega^2$	$B_2 = K\omega^2[A]$

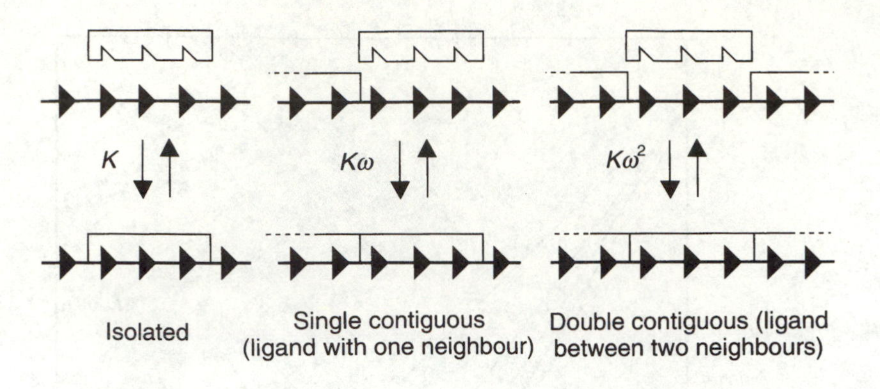

Figure 22.11 Definition of the three possible types of ligand binding sites when only nearest-neighbour interactions are considered. (From McGhee and von Hippel, 1974. Reprinted from *Journal of Molecular Biology* with the permission of Academic Press.)

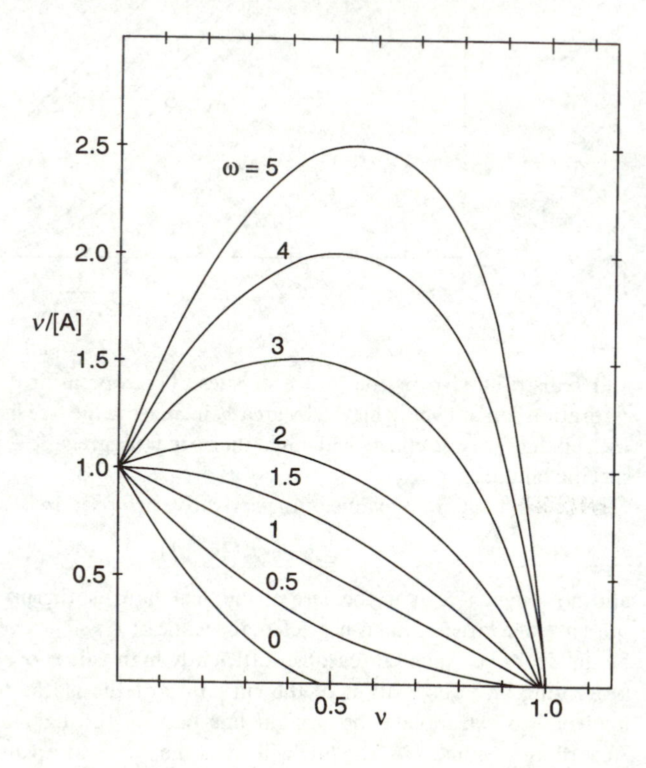

Figure 22.12 Scatchard plot for cooperative binding of a ligand with $n = 1$ and $K = 1\ \mathrm{M}^{-1}$ for ω ranging from 0 to 5. (From McGhee and von Hippel, 1974. Reprinted from *Journal of Molecular Biology* with the permission of Academic Press.)

The parameter ω is the equilibrium constant of the process in which a bound ligand is carried from situation (0) to situation (1) or from (1) to (2). The binding is cooperative if $\omega > 1$ and anticooperative if $\omega < 1$. The binding isotherm depends on the three parameters K, n and ω.

The calculation, which will not be given in detail here, yields the following complex equation for the isotherm

$$\nu/[\mathrm{A}] = \frac{K(1 - n\nu)[(2\omega - 1)(1 - n\nu) + \nu - R]^{n-1}[1 - (n+1)\nu + R]^2}{[2(\omega - 1)(1 - n\nu)]^{n-1}[2(1 - n\nu)]^2} \qquad (22.29)$$

where $R = \{[1 - (n+1)\nu]^2 + 4\omega\nu(1 - n\nu)\}^{1/2}$.

In spite of its formidable appearance, eqn 22.29 can easily be used by drawing a set of isotherms for different values of ω and comparing them with experimental curves (Fig. 22.12). There are several points to be made about this:

1. The intercept on the vertical axis is always equal to K and the intersection with the horizontal axis always occurs at $1/n$.
2. The slope at the origin is $K(2\omega - 2n - 1)$. It may be positive, negative or even zero when $\omega = n + \frac{1}{2}$.
3. In some cases, compensation may occur between the effect of the size and the cooperativity, giving the isotherm a quasi-linear appearance.
4. The importance of ω can be illustrated as follows. Assume a ratio of only 2 between the two association constants K corresponding to the binding of the same ligand to a lattice of sites on two different biopolymers. If in fact ω is large, the binding process has a tendency to form 'clusters' of molecules on the receptor. If, for example, 12 ligand molecules are bound side by side in this way, the ratio of the two binding constants is no longer 2 but $2^{12} = 4096$. Cooperativity can thus lead to specificity. A good illustration of this is the strongly cooperative binding of the protein RecA to a single strand of DNA during a recombination or repair process.

23

Kinetic study and facilitated diffusion

23.1 Introduction

The kinetics of the association between proteins and nucleic acids may involve a choice between two extreme models:

1. The rate of association is diffusion-controlled, in which case association takes place every time a collision occurs between the two reacting molecules.
2. The rate of association is limited by the reaction itself, either because of the chemical reactivity or because conformational changes are needed before the reaction can start.

We shall return later to the question of distinguishing between these two models. For the moment, we consider the first model and show that we are dealing with *facilitated diffusion*, because of a change of dimensionality during the diffusion process.

The first experimental data on the binding of the *lac* repressor to the *lac* operator gave a high value for the association constant, and particularly for the rate constant (of the order of $2 \times 10^{10}\,\mathrm{M}^{-1}\,\mathrm{s}^{-1}$). To interpret this value, we can use the expression obtained in the theory of a purely diffusion-controlled reaction (see eqn 8.17) for the rate constant k_a:

$$k_a = 4\pi Kfb(D_R + D_O)10^3 N$$

where K is a steric factor, f is an electrostatic factor, b is the interaction distance, D_R and D_O are the translational coefficients of diffusion for the repressor and operator respectively, and N is Avogadro's number. Giving K and f their maximum values of 1, assuming that $b = 50$ Å (the sum of the radii of the double helix and the repressor) and that $D_R = 5 \times 10^{-11}\,\mathrm{m}^2\,\mathrm{s}^{-1}$ and $D_O = 10^{-13}\,\mathrm{m}^2\,\mathrm{s}^{-1}$, we find that $k_a = 2 \times 10^9\,\mathrm{m}^{-1}\,\mathrm{s}^{-1}$, only a tenth of the experimental value. To account for this, a collision process different from classical three-dimensional diffusion is needed, and there are two important clues that point the way:

1. It has been shown that the efficiency of a collision as a result of diffusion is increased by reducing the number of dimensions, i.e. by compelling the molecule to undertake a two-dimensional (surface) or one-dimensional (line) random walk.

2. With the DNA concentration generally used (10^{-12} M), it can be assumed that on average one DNA molecule is present in a cube of side 10 μm. Since the radius of gyration of the DNA used is about 0.6 μm, there is a local concentration of segments in the 'internal' space of the DNA much higher than that in the intermolecular space. In other words, the probability of a collision between protein and DNA is much higher when the protein is 'confined' inside the coil.

These two clues have led to the development of a new model for the *in vitro* association of protein with DNA (Berg *et al.*, 1981; Winter *et al.*, 1981; von Hippel and Berg, 1986).

23.2 Models of DNA–repressor recognition

We must first distinguish clearly between several different mechanisms (Fig. 23.1).

Microscopic dissociation and reassociation

After dissociating from its complex with DNA, a protein molecule remains close to its previous binding site and can very quickly be bound again. It is, however, far enough away to allow dissociated counter-ions to recondense on to phosphate groups (see Part III). These local changes take place in a time much shorter than the residence time of the protein in the immediate neighbourhood of the DNA chain. The protein can thus 'sample' several nearby sites by means of this microscopic dissociation process.

If the protein is confined between two DNA segments that are close to each other, it can also jump from one segment to the other in a *translocation* process.

Intersegment transfer

The movement of the protein from one segment to another close to the first no longer involves Brownian motion as it did previously, but occurs as a result of the

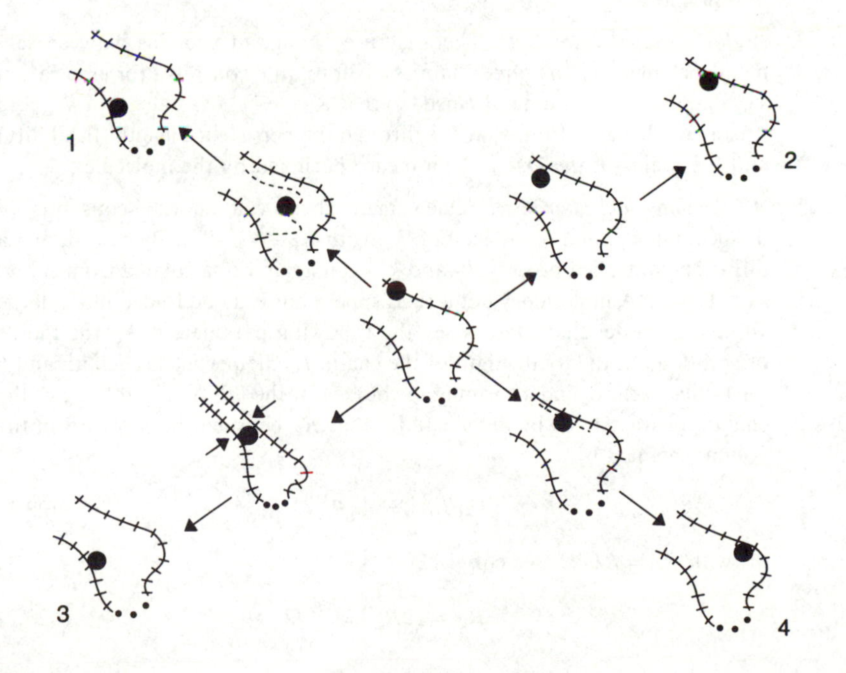

Figure 23.1 The various processes by which a protein is transferred from one DNA binding site (at the centre) to another: (1) three-dimensional diffusion; (2) local jump (microscopic association–dissociation); (3) intersegment transfer (fluctuation of the DNA); (4) sliding along the DNA. (From Berg *et al.*, 1981. Reprinted with permission from *Biochemistry*. © 1981 American Chemical Society.)

internal dynamics of the DNA chain, which, at a given moment, may be doubly bound to the protein.

Sliding

The protein remains non-specifically bound to DNA (purely Coulomb binding) and slides along an equipotential surface of the polyelectrolyte. This is a one-dimensional Brownian motion accompanied from the beginning by the simultaneous release of several counter-ions 'ahead' and the condensation of the same number of counter-ions 'astern'.

The recognition of a specific site (O, for operator) would be achieved through a non-specific intermediate binding state (P) according to the following reaction scheme:

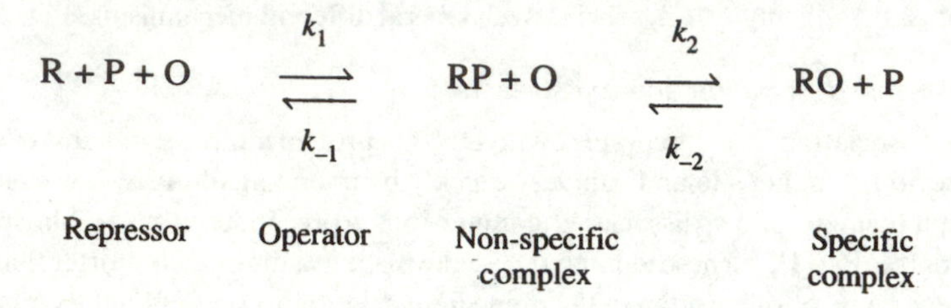

where k_1 and k_{-1} are true rate constants corresponding to a classical collision between two molecules, while k_2 and k_{-2} are rate constants for the transfer from one site to the other.

It is then possible to describe three processes by specific dissociation and association constants.

1. *At the microscopic level*: Let k_i be the rate constant of association and λ that of dissociation, so that the equilibrium constant for the non-specific process is $K_{RP} = k_i/\lambda$.
2. *At the intramolecular level*: There is an exchange of proteins between segments belonging to the same chain. As before, rate constants for association k_{ass} and dissociation Λ are defined, so that $K_{RP} = k_{ass}/\Lambda$. Values for k_{ass} and Λ can be calculated from k_i and λ through the persistence length (flexibility) and the radius of the DNA chain (space occupied by the molecule).
3. *At the intermolecular level*: Once again there will be rate constants of dissociation k_{-1} and of association k_1, with $K_{RP} = k_1/k_{-1}$. By considering a diffusion-controlled process, k_1 and k_{-1} can again be calculated from Λ and k_{ass}. The DNA molecule is not a rigid sphere but is a coil folded into a large volume, a model characterized by the following parameters: R_G the radius of gyration, $2L$ the total length of the chain, L_P the persistence length and b the radius of the cylinder forming the model for the DNA. The folding of the chain is characterized by the mean distance R_C between the segments in the volume, defined by

$$2\pi L R_C^2 = 4\pi R_G^3/3 \tag{23.1}$$

or, with $R_G^2 = LL_P/3$ (see eqn 4.12),

$$R_C = (2R_G L_P)^{1/2}/3 \tag{23.2}$$

The probability that the protein moves across the volume without any collisions with the DNA is far from negligible. In fact, it is found that

$$k_1 = 4\pi M^{-1} D R_G [1 - (\tanh u)/u]$$

where M is the number of non-specific sites per chain, and

$$u^2 = 3L/R_G[\ln(R_C/b) + 2\pi Dl/k_i] \qquad (23.3)$$

l being the length of a base pair. In the expression for k_1, the bracketed term is the probability that the protein binds to the DNA chain instead of freely diffusing across the volume. For a double-stranded DNA this term increases from 0.44 to 0.7 when the number of base pairs increases from 2000 to 50 000.

Returning to the initial reaction scheme, we have at equilibrium

$$[RP]/[R] = k_1[P_T]/k_{-1} = [P_T]K_{RP}$$

where $[P_T]$ is the total DNA concentration (because [P], the concentration of non-specific sites, is almost equal to $[P_T]$). Similarly:

$$[RO]/[R][O] = k_2 K_{RP}/k_{-2} = K_{RO}$$

Such a scheme is no longer valid if $k_{-1} \ll k_2[O]$. The protein then remains attached to the non-specific sites and there are no more exchanges with the specific site. However, the measured quantity is the ratio of the concentration of bound repressor molecules to the total concentration of free repressor molecules:

$$K_{RO}^{obs} = [RO]/([RP] + [R])[O] = k_2 K_{RP}/k_{-2}(1 + [P_T]K_{RP}) \qquad (23.4)$$

The *effective association constant* k_a can then be calculated, assuming that $k_{-2} = 0$ in the reaction scheme. We then have

$$k_a = k_2[P_T]K_{RP}/(1 + [P_T]K_{RP} + k_2[O_T]/k_{-1}) \qquad (23.5)$$

where $[O_T]$ is the concentration of operator sites expressed in base pairs.

Similarly, an *effective dissociation constant* k_d can be calculated assuming that $k_1 = 0$, i.e. under conditions such that the protein cannot reassociate after it has dissociated from DNA. This is equivalent experimentally to adding enough 'cold' DNA containing the operator. It is found that

$$k_d = k_2[P_T]/(1 + k_2[O_T]/k_{-1} + k_{-2}[P_T]/k_{-1}) \qquad (23.6)$$

It is thus generally possible to obtain theoretical expressions for k_a or k_d as functions of K_{RP} depending on the model for the elementary process: jump, transfer or sliding. Figure 23.2 shows the theoretical curves for the variation of k_a with K_{RP}.

From these calculations, it ought to be possible to observe large changes in the kinetics of association and dissociation by varying either the non-specific affinity (e.g. by changing the ionic strength) or the length of the DNA containing the specific site, in this case the operator.

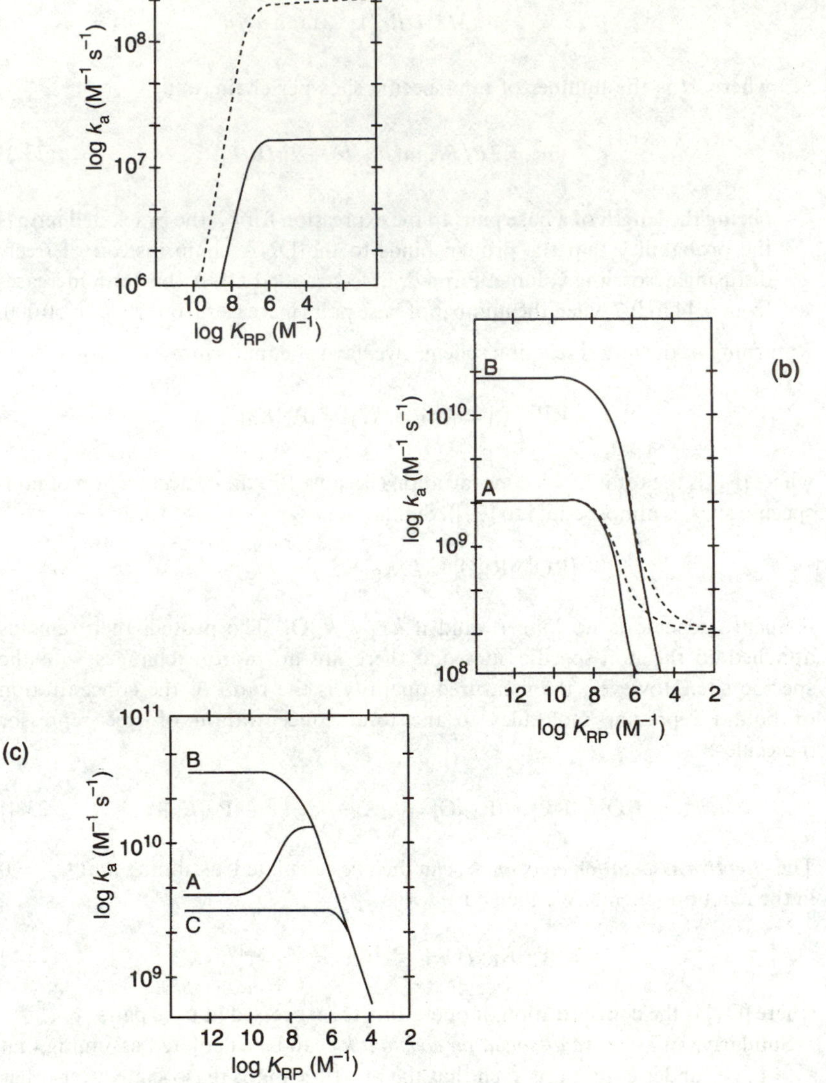

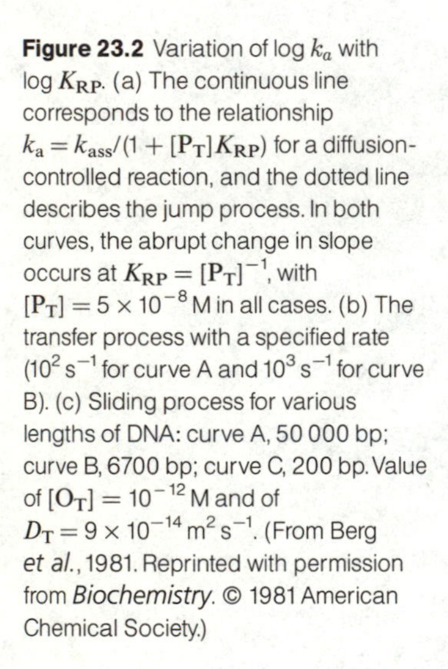

Figure 23.2 Variation of log k_a with log K_{RP}. (a) The continuous line corresponds to the relationship $k_a = k_{ass}/(1 + [P_T]K_{RP})$ for a diffusion-controlled reaction, and the dotted line describes the jump process. In both curves, the abrupt change in slope occurs at $K_{RP} = [P_T]^{-1}$, with $[P_T] = 5 \times 10^{-8}$ M in all cases. (b) The transfer process with a specified rate (10^2 s^{-1} for curve A and 10^3 s^{-1} for curve B). (c) Sliding process for various lengths of DNA: curve A, 50 000 bp; curve B, 6700 bp; curve C, 200 bp. Value of $[O_T] = 10^{-12}$ M and of $D_T = 9 \times 10^{-14}$ m^2 s^{-1}. (From Berg *et al.*, 1981. Reprinted with permission from *Biochemistry*. © 1981 American Chemical Society.)

23.3 Results

Experimental curves showing the variation of log k_a with log[KCl] are plotted in Fig. 23.3. It is immediately clear that

(1) the variation of log k_a with log[KCl] is not monotonic;

(2) k_a decreases as the size of the DNA molecules decreases;

(3) the increase in k_a between 0.2 and 0.1 M follows a corresponding increase in K_{RP}, i.e. in the lifetime of the non-specific complex.

A comparison between theoretical and experimental curves together with an interpretation of the data clearly shows that sliding is the dominant mechanism.

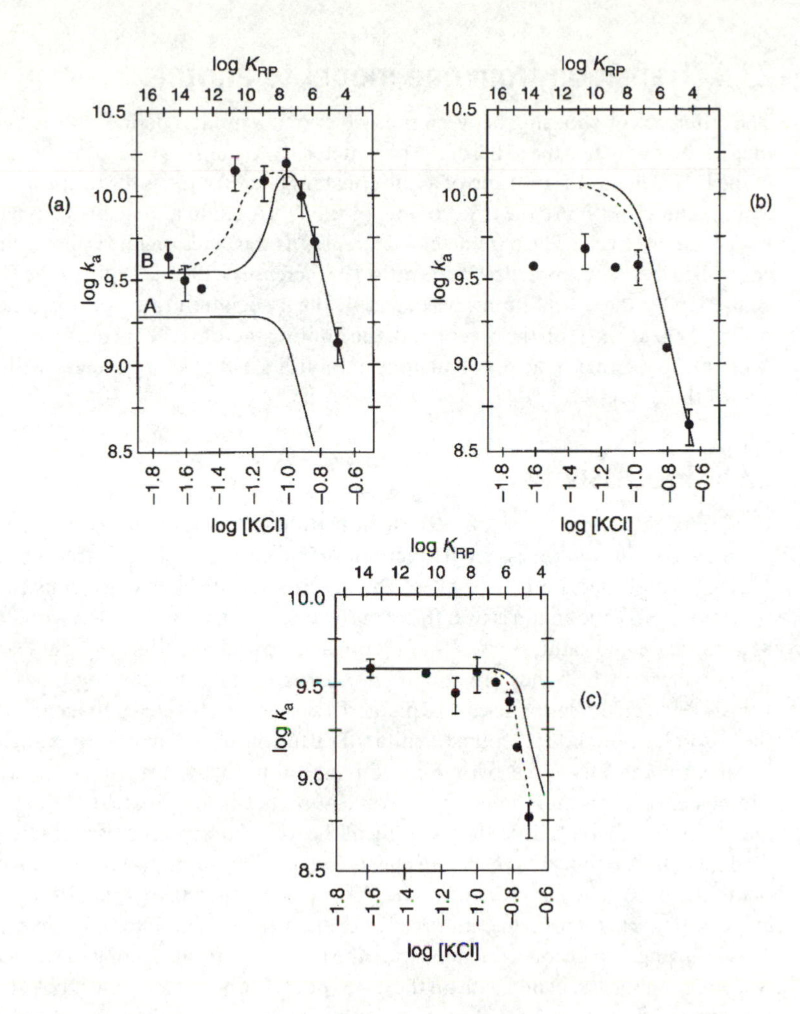

Figure 23.3 Variation of log k_a with log[KCl] when an RO complex is formed in the case of the *lac* repressor. Experimental points are indicated with error bars. The curves are theoretical curves calculated using the values of K_{RP} along the top. (a) $D_T = 5 \times 10^{-11}\,m^2\,s^{-1}$, DNA with 50 000 bp, $R_G = 5500$ Å, $[O_T] = 10^{-12}$ M; curve A is for intersegment transfer and curve B is for the sliding mechanism. Over the dotted portion, K_{RP} is supposed to depend less on the ionic strength. (b) $D_T = 5 \times 10^{-11}\,m^2\,s^{-1}$, DNA with 6700 bp, $R_G = 2100$ Å, $[O_T] = 0.5 \times 10^{-12}$ M; the curve is for the sliding mechanism. (c) $D_T = D_R + D_O = 6.5 \times 10^{-11}\,m^2\,s^{-1}$, DNA with 175 bp. The curves are for the sliding mechanism with $D_T = 9 \times 10^{-14}\,m^2\,s^{-1}$ (continuous curve) and $D_T = 3 \times 10^{-14}\,m^2\,s^{-1}$ (dotted curve). (From Winter *et al.*, 1981. Reprinted with permission from *Biochemistry*. © 1981 American Chemical Society.)

The influence of the ionic strength is mainly exerted through the rapid increase of the dissociation rate constant k_{-1} with ionic strength. At 0.075 M, the probability of a non-specifically bound repressor molecule finding the operator before dissociating from DNA is about 75 per cent. At 0.2 M this probability falls to 1 per cent. It should, however, be pointed out that with too small an ionic strength the dissociation rate becomes so low that the repressor remains on a non-specific site for too long. The optimum value occurs when k_{-1} is of the same order of magnitude as the DNA concentration.

For small pieces of DNA, the transfer rate constant is much greater than k_{-1}. The limiting step then becomes the classical diffusion process.

The one-dimensional diffusion constant D_1 can be measured and is found to be $9 \times 10^{-14}\,m^2\,s^{-1}$. The use of $x^2 = 2Dt$ shows that an average of 1000 base pairs are explored per second.

23.4 Transition from one model to another

The criterion for choosing between the two extreme models defined at the beginning of this chapter is the value of the parameter $x = bk \ln V/v$, where b is the radius of the DNA chain, k is the ratio of a rate constant to a flux and is the reciprocal of a length, and v and V are the true volume of the DNA chain and its hydrodynamic volume respectively. The parameter k decreases as the reaction rate is increasingly controlled by the reaction itself. The ratio V/v increases with the length of the DNA chain. Under these conditions, when $x \gg 1$, the association rate is diffusion-controlled. When $x \ll 1$, on the other hand, the limiting factor is the chemical reaction. Most protein–nucleic acid interactions exhibit a kinetics in agreement with the first of these models ($x \gg 1$).

23.5 Processivity

Facilitated diffusion, which has been demonstrated *in vitro*, exists *in vivo* for many enzymes carrying out the essential functions of replication, repair and transcription of DNA. Unlike most other enzymes, which are dissociated from their substrate as soon as the enzymic action is over (*distributive* enzymes), these function with DNA as a substrate and catalyse a series of identical chemical reactions without leaving the polymer to which they are bound (*processive* enzymes). Work by Dowd and Lloyd (1990) on the endonuclease of phage T4 showed by directed mutagenesis that there is indeed, in the bacterium, facilitated diffusion of the enzyme enabling it to localize on the DNA the pyrimidine dimers induced by ultraviolet radiation. Moreover, it seems that there is good correlation between the extent of the repair and the processivity. A little later, a study of the replication mechanism of DNA by pol III, a DNA polymerase of *E. coli* consisting of 16 sub-units, revealed the role of one of them (β) in the processivity of the polymerase (Stukenberg *et al.*, 1991; Kong *et al.*, 1992). This β protein dimerizes to form a ring with an internal diameter of 35 Å allowing it to become threaded over the DNA and to slide along the molecule while remaining associated with all the other pol III sub-units. A new DNA strand is synthesized at a rate of 750 nucleotides per second, the polymerase remaining bound to the DNA over a distance of more than 10^5 nucleotides! Directed mutagenesis makes it possible to render a distributive enzyme like ribonuclease processive (del Cardayré and Raines, 1994). It is apparently sufficient to fulfil two conditions: firstly, that the substrate must provide a repetitive structural motif, in this case a poly(rA), and secondly that there must be more than one protein site that is firmly bound to the repetitive motif.

In eukaryotes, the DNA is embedded in a compact chromatin structure that makes the existence of a sliding mechanism difficult if not impossible. Nevertheless, the three-dimensional structure of a complex of the human polymerase β with DNA indicates that it is entirely compatible with a sliding mechanism (Pelletier *et al.*, 1996).

23.6 Guided diffusion of a substrate (channelling)

When two enzymic processes are carried out by the same protein, it may happen that the product X from the first reaction becomes the substrate for the second. For maximum efficiency of the overall reaction, X will be transported between the two reaction sites by guided diffusion. One example of this is that of bacterial tryptophan synthase, which catalyses the synthesis of tryptophan from indole-3-glycerol phosphate (IGP) and serine. The enzyme is an $\alpha-\beta-\beta-\alpha$ tetramer of two sub-units α and β. The transformation of IGP into indole is catalysed by α and the condensation of indole with serine is catalysed by β. Studies using fast-reaction kinetics (Anderson *et al.*, 1991) can be interpreted by assuming that the indole diffuses quickly inside the protein. A determination of the three-dimensional structure (Hyde *et al.*, 1988) confirms this hypothesis by showing that there is a hydrophobic tunnel about 25 Å in diameter joining the two sub-units.

Another example (Knighton *et al.*, 1994) is provided by the three-dimensional structure of a dual-function enzyme, dihydrofolate reductase–thymidilate synthase (DHFR-TS) found in some eukaryotes. The TS part of the enzyme catalyses the change of dUMP to dTMP with release of dihydrofolate (FH_2). The DHFR part reduces the FH_2 to tetrahydrofolate. To move from one reaction site to the other over a distance of 70 Å, there is guided diffusion of the FH_2 (with a charge $-2e$) along a positively charged surface consisting of lysines and arginines.

A final example concerns the movement of a substrate towards the active site of the enzyme, involving the O_2^- radical and superoxide dismutase (SOD), whose potential map can be found in Fig. 19.8. SOD has a very high rate constant ($10^9\ \mathrm{M}^{-1}\ \mathrm{s}^{-1}$) while the substrate has to react with a copper ion occupying no more than 0.1 per cent of the surface area of the protein. Molecular dynamics simulations (Sharp *et al.*, 1987; Sines *et al.*, 1990) can be computed from the potential map. The superoxide radical-ion is represented by a sphere of radius 1.8 Å and its trajectory is calculated using the following relationship applied to the three spatial directions:

$$x_i(t+\Delta t) = x_i(t) - (\partial\Psi/\partial x_i)D\Delta t/k_B T + n_i(2D\Delta t)^{1/2} \qquad (i = 1, 2, 3)$$

which describes the Brownian motion of the ion in the gradient of the electric potential Ψ, n_i being a small random number and the time interval Δt being of the order of 10^{-13} s. Several thousand trajectories were calculated, from which the probability of the substrate encountering the active site could be determined. The association rate constant k_1 is derived from this and is found to be almost equal to that obtained with the Cu^{2+} ion alone: the small size of the target was in some way compensated by the focusing of the electric field lines in the channel leading to the active site.

24

Mechanisms and specificity of recognition

In all the many examples studied so far as illustrations of the different types of interaction mentioned in the Introduction to this Part, the forces involved are exactly the same as those examined in Chapter 2, i.e. van der Waals forces, hydrogen bonds and Coulomb interaction forces. Specific recognition and association arise, not from new types of force, but from the spatial distribution of bonds, which ensures that the two surfaces coming into contact are complementary. The specificity also arises from the kinetics of the formation and breakdown of the associations that occur. The energies of the bonds in such associations are all very similar and are therefore not significant here: the total free-energy change, often large, comes from the addition of all the contributions from bond energies.

During the various evolutionary stages, innumerable 'processes' must have been tried out. When some of them proved to be particularly efficient, they were retained and used in a large number of systems. This chapter thus analyses a few of these processes and is, in a way, a catalogue of models similar to one that an architect might offer!

24.1 Calcium transport and the EF hand

Because of the considerable involvement of the Ca^{2+} ion in a large number of biological processes, particularly in muscular contraction, it is essential that this ion should be movable as part of a complex that is both stable and easily available. Dozens of proteins are capable of forming such a complex, with the great variation of binding sites perhaps reflecting the wide diversity of functions. Amongst all the proteins, there is one class, exemplified particularly by troponin C and calmodulin, that uses a special structure to bind with the Ca^{2+} ion: two perpendicular α-helices E and F are connected through a large loop, which 'winds up' around the calcium ion. This so-called *helix–loop–helix* (*HLH*) *structure* is often pictured as a hand (the *EF hand*) whose thumb and forefinger point along the two helices and in which the Ca atom coincides with the clenched middle finger (Fig. 24.1). Around the Ca ion there are generally seven ligands arranged in a double pentagon-based pyramid, one apex of which is occupied by a water molecule, the others by oxygen atoms in the side chains of Asp, Glu, Asn or Tyr.

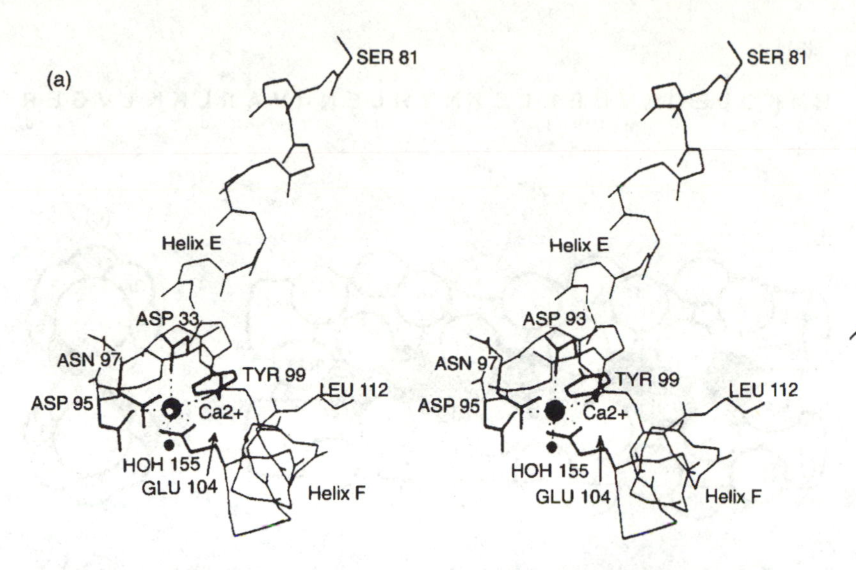

Figure 24.1 (a) Stereo diagrams of site 3 in calmodulin with the two helices E and F and the hepta-coordinated Ca^{2+} ion. (From Strynadka and James, 1989. Reproduced with permission from *Annual Review of Biochemistry*, vol. 58 and with the authors' permission. © 1989 Annual Reviews Inc.) (b) The EF hand.

24.2 Protein dimerization and leucine zippers

The idea of creating a dimer between two proteins by intertwining two α-helices was proposed by Crick as early as 1953. In his model, the hydrophobic 'knobs' of one α-helix were housed in the 'crevices' of the other. After a similar mechanism was discovered in several transcription factors, a *zipper model* was proposed that made it possible to create a protein dimer, i.e. a structure with two-fold symmetry capable of recognizing a double-stranded DNA with the same symmetry. The yeast GCN4 factor has received the most attention. In its sequence of 33 amino acids, a leucine occurs at regular intervals of seven residues (Fig. 24.2). X-ray studies showed that the two α-helices taking part in the zipper were parallel (the same $N \rightarrow C$ orientation). The structure was later refined to 1.8 Å resolution (Fig. 24.3).

The two α-helices form a quarter-turn of a left-handed superhelix with a pitch of 181 Å. The axes of the two helices are 9.3 Å apart and make an angle of 18° with each other. The whole structure can still be represented by a ladder with twisted uprights and rungs made of the side chains of the interface. The leucine side chains do not interpenetrate each other but are side by side as in a 'handshake'. The role of the leucine may be explained by the presence of a branched aliphatic chain with a tighter packing than a linear chain. Coulomb interactions between K15 and E20′, between E22 and K27, and between K27 and E22′ supplement the hydrophobic interactions characteristic of all zipper interfaces.

It is the combination of these two types of interaction (hydrophobic and Coulomb) that is typical of the leucine zipper and may well explain the specific formation of a heterodimer between the two oncogenes *jun* and *fos* rather than a homodimer.

(a)

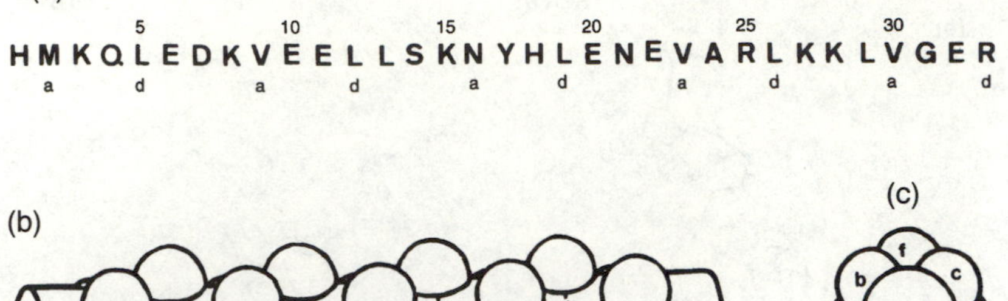

(b)

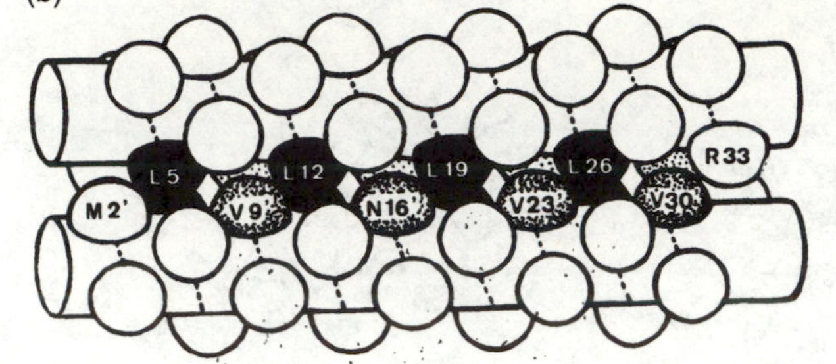

(c)

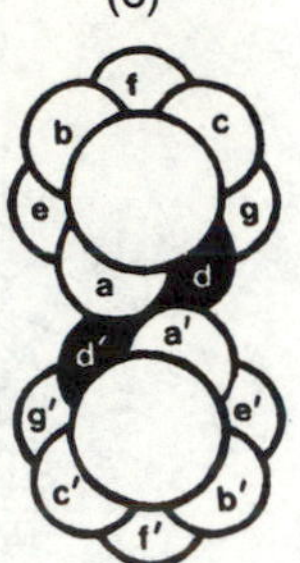

(d)

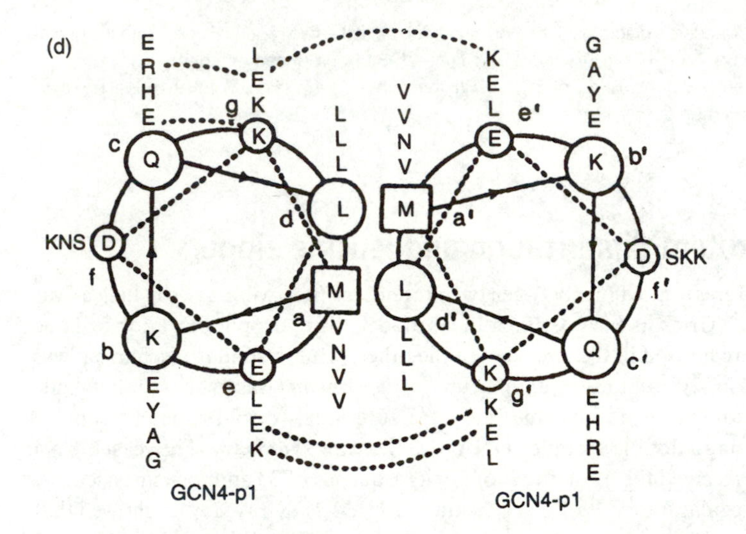

Figure 24.2 The leucine zipper of GCN4 modelled by two intertwined helices. (a) Sequence: leucines are located at 'd' and hydrophobic or neutral groups at 'a'. (b) Side view of the junction showing interactions between the leucines of one α-helix and the hydrophobic groups of the other. (c) End view. (d) Distribution over a helical wheel of α-helix residues numbered from 'a' to 'g' as in (c). Dotted lines indicate ionic bonds, which contribute to the stability of the dimer. ((a – c) From Hu *et al.*, 1990. Reprinted with permission from *Science*, vol. 250, pp. 1400 – 3 and with the authors' permission. © 1990 American Association for the Advancement of Science. (d) From O'Shea *et al.*, 1991. Reprinted with permission from *Science*, vol. 254, pp. 539 – 44 and with the authors' permission. © 1991 American Association for the Advancement of Science.)

24.3 Recognition of nucleic acids by proteins

In the first two examples (trapping of a Ca^{2+} ion and protein dimerization), van der Waals forces, Coulomb forces and hydrogen bonds were all involved. It is the same with protein – nucleic acid interactions.

24.3.1 Interacting groups in proteins

Among the amino acids whose side chains can play a role in selective recognition, we can distinguish three groups.

1. The basic groups (arginine, lysine and histidine), for which a large electrostatic contribution can be expected, are involved. Calculations of interaction energies (Kumar and Govil, 1984) take into account attractive forces (Coulomb, polarization, dispersion) and short-range repulsion. Figure 24.4 illustrates how ion pairs

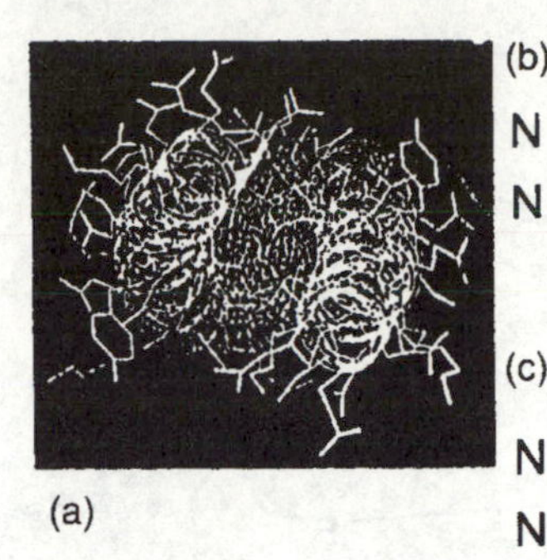

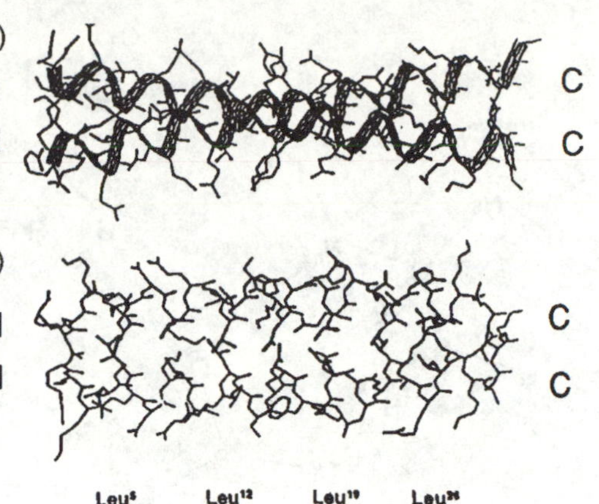

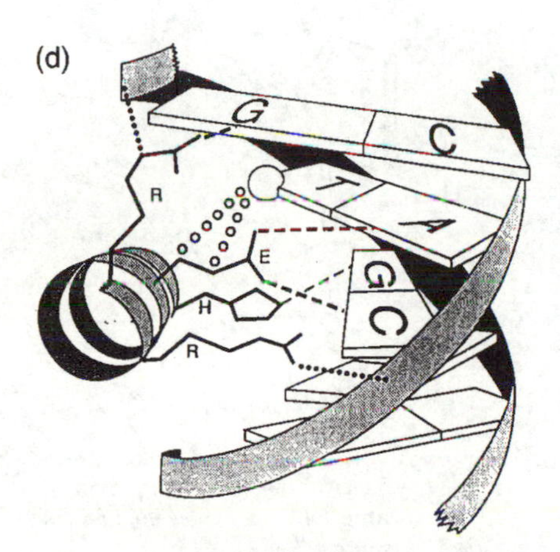

Figure 24.3 Different aspects of the superhelical structure. (a) Viewed from the N-terminal end. Note the bending of the two α-helices with a quarter-turn of the superhelix in total. (b) Side view showing the intersection of the two α-helices at 18°. (c) Side view perpendicular to (b) showing the interaction of side chains. (d) Detailed interactions with DNA of the basic region of GCN4. The α-helix is shown in the major groove. All types of bond occur: salt bridge, hydrogen bonds, van der Waals interactions. In the last-named, a major role is played by contacts between side chains of Ala and methyl groups of T. ((a–c) From O'Shea *et al.*, 1991. Reprinted with permission from *Science*, vol. 254, pp. 539–44 and with the authors' permission. © 1991 American Association for the Advancement of Science. (d) From Ellenberger, 1994. Reprinted from *Current Opinion in Structural Biology* with the permission of Current Biology Ltd.)

may be formed. The interaction energies listed in Table 24.1 (calculations unfortunately carried out without water) are for some of the bonds formed preferentially.

2. Independently of the peptide chain, several side chains can form hydrogen bonds with bases or base pairs and a few examples are shown in Fig. 24.5. Calculations of association energies show that they are higher for single bases than for base pairs. Glu and Asp have a greater affinity for a GC pair than for an AT (AU) pair. In all cases, however, the energy is three to four times as great with Arg^+ as with the other amino acids.

With arginine, another interaction model has been proposed involving hydrogen bonds between Arg and the phosphate groups. There are five donor groups on the arginine: the four hydrogens of the two NH_2 groups (N_η) and the hydrogen of the NH group (N_ε) (Fig. 24.6). This type of interaction could occur between the protein and RNA, the arginine forming a kind of bridge between two phosphate groups.

3. Lastly, another type of protein–nucleic acid interaction involves the aromatic rings of the three amino acids Tyr, Trp and Phe. The stacking energy between these aromatic rings and the base, for a well-defined geometry, is comparable to that of several hydrogen bonds. The major contribution comes from the dispersive term. Some examples are shown in Fig. 24.7.

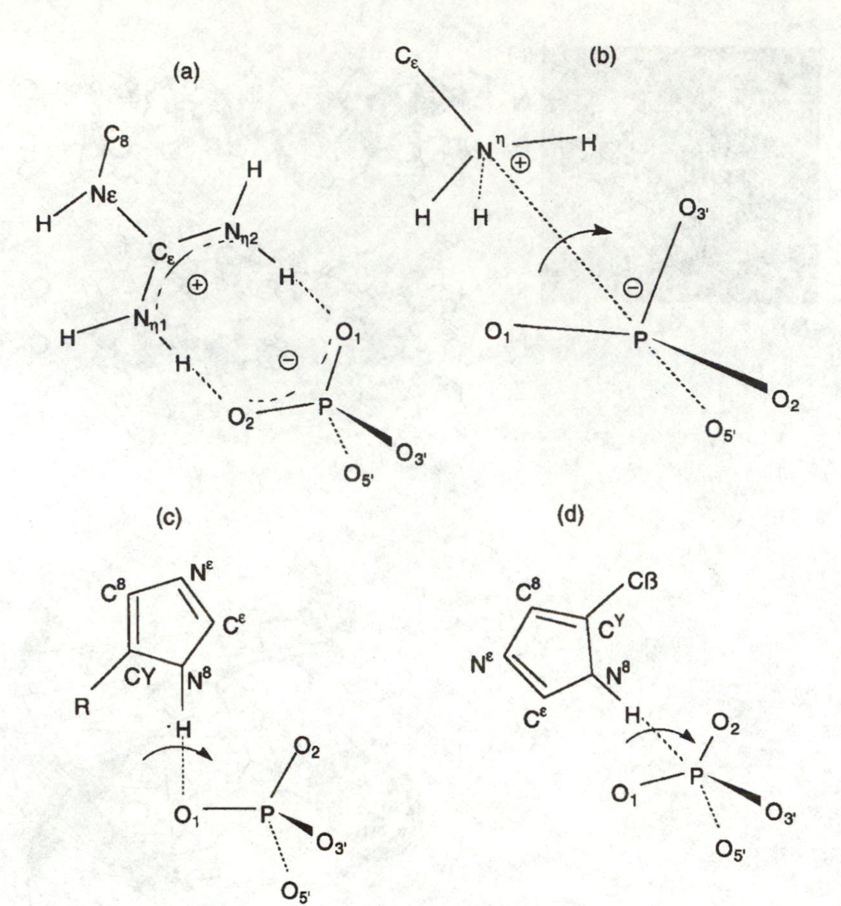

Figure 24.4 Coulomb interactions between the phosphate group and basic amino acids: (a) arginine, (b) lysine and (c, d) histidine. (From Kumar and Govil, 1984. Reprinted from *Biopolymers* with the permission of John Wiley & Sons. © 1984 John Wiley & Sons.)

Table 24.1 Interaction energies of basic groups with nucleic acids (kcal mol^{-1})

	A-RNA	A-DNA	B-DNA
Arg	96.4	98.6	93.2
Lys	130.4	128.9	122.6
His	114.8	112.2	107.0

24.3.2 Interacting groups in nucleic acids

DNA is accessible from the two grooves characteristic of the B form, and this allows a 'direct' recognition of base pairs and hence of the sequence. Because of the atoms available (N, O and H), only proton donor–acceptor (D–A) pairs can be formed. Each base pair can therefore be described by a succession of 'A's and 'D's going from one edge of the groove to the other. Four distinct positions are then found in the major groove and three in the minor groove (Table 24.2).

It is clear from this table that the minor groove cannot be a region where discrimination is possible between the four types of pairing within a sequence. As for the major groove, a single hydrogen bond is not enough to distinguish between, for example, TA and GC or between AT and CG. If, on the other hand, a second hydrogen bond is formed, the protein can specifically recognize each one of the four arrangements of base pairs. The two hydrogen bonds can be formed with a single amino acid, e.g. Arg with its two NH_2 donor groups, or Asn or Gln with their donor – acceptor pair O=C–NH–. Such recognition patterns are in fact found in the repressor – operator complex in which Arg is bound to G and Gln is bound to A, as could be predicted from Table 24.2.

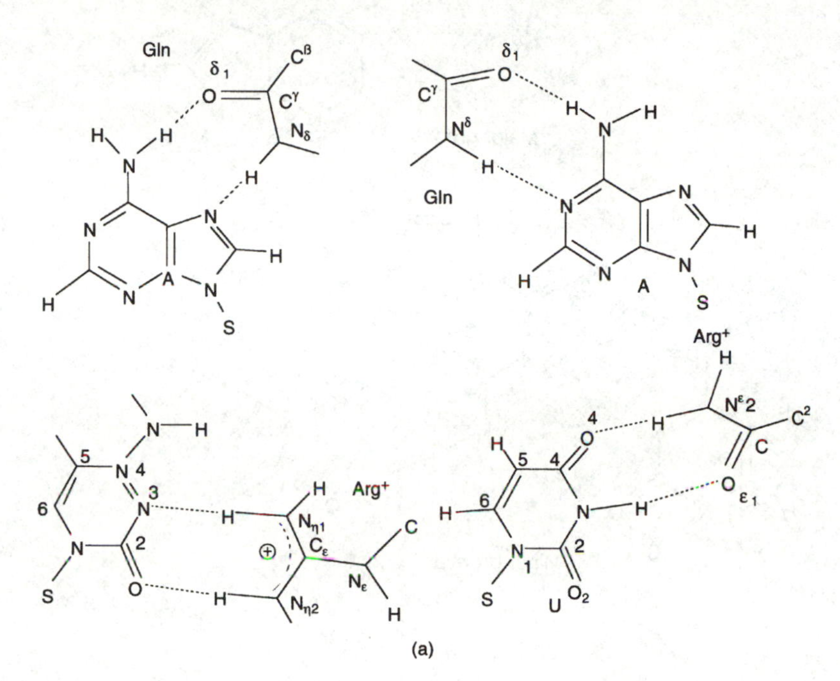

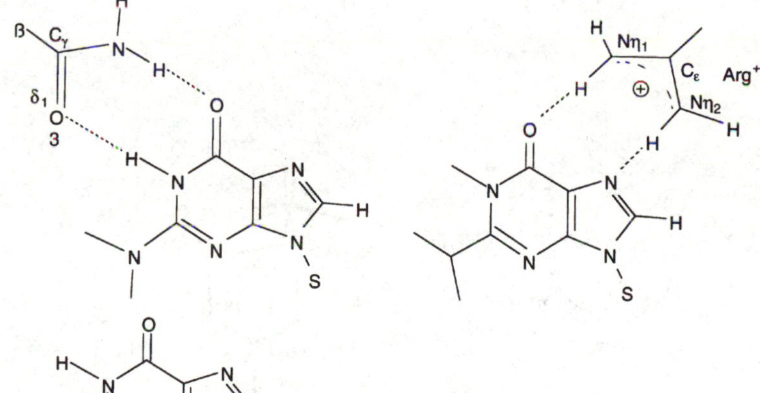

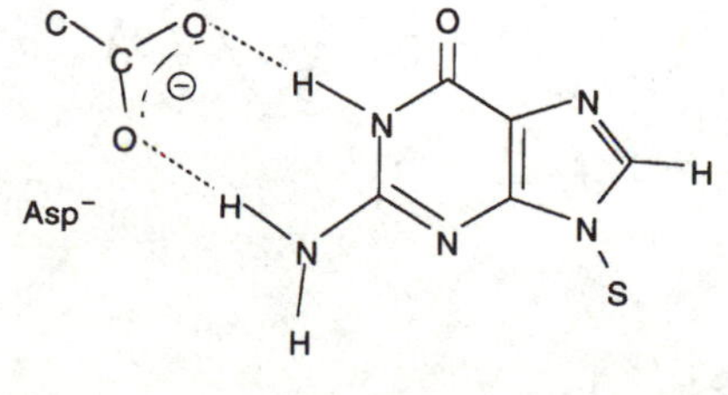

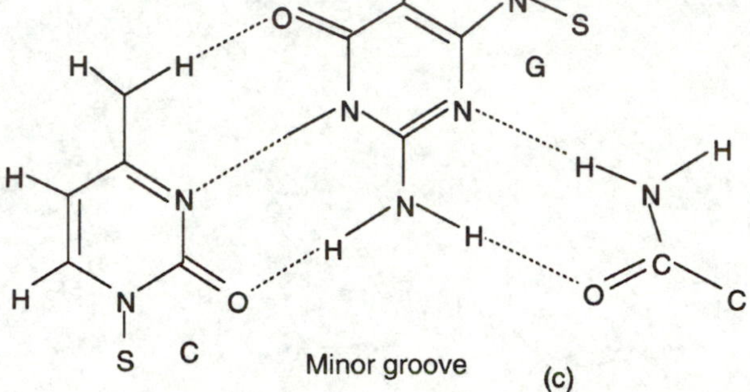

Figure 24.5 Formation of hydrogen bonds between bases and side chains of amino acids. (a) Gln with A and U, Arg with C. (b) G with the CONH group of Asn and Gln, and with Arg and Asp. (c) Base pairs AU and CG with the amide group of Gln or Asn. (From Kumar and Govil, 1984. Reprinted from *Biopolymers* with the permission of John Wiley & Sons. © 1984 John Wiley & Sons.)

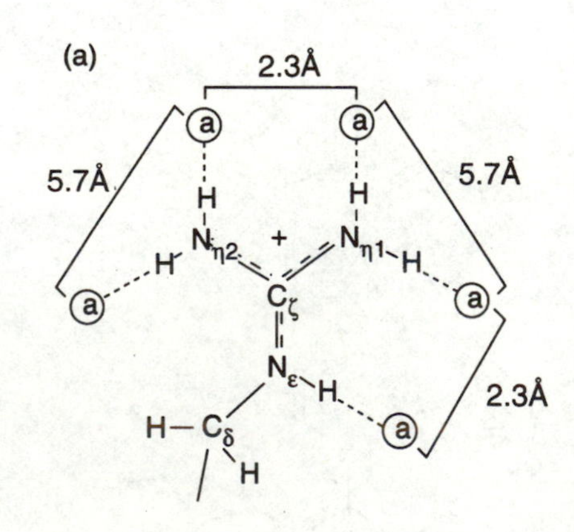

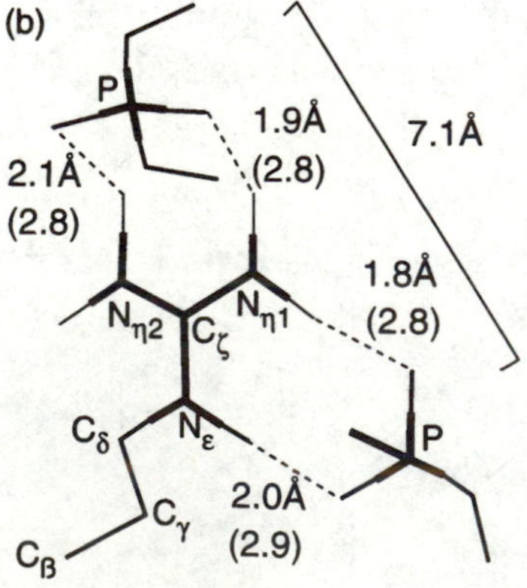

Figure 24.6 Models of hydrogen bonds between phosphate groups and arginine. (a) Geometry of possible hydrogen bonds with five donor groups. (b) Possible bridging role of arginine. (From Calnan *et al.*, 1991. Reprinted with permission from *Science*, vol. 252, pp. 1167–71 and with the authors' permission. © 1991 American Association for the Advancement of Science.)

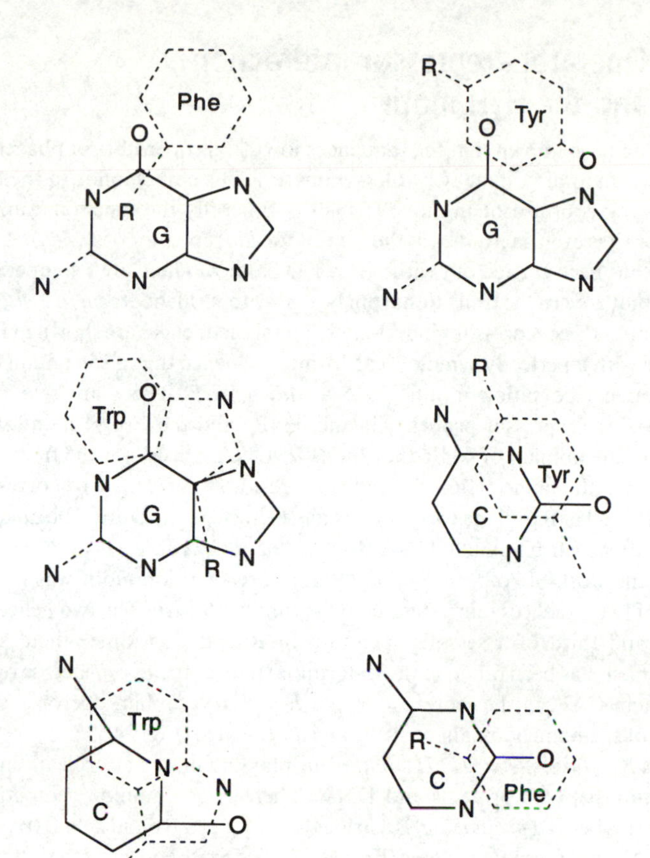

Figure 24.7 Stacking of aromatic side chains with G and C bases. (From Kumar and Govil, 1984. Reprinted from *Biopolymers* with the permission of John Wiley & Sons. © 1984 John Wiley & Sons.)

Table 24.2 Distribution of donor and acceptor groups in B-DNA

Pair	Major groove				Minor groove		
	(1)	(2)	(3)	(4)	(1)	(2)	(3)
AT	A (N7)	D (N6)	A (O4)	(Me)* (C5)	A (N3)	D (C2)	A (O2)
TA	(Me) (C5)	A (O4)	D (N6)	A (N7)	A (O2)	D (C2)	A (N3)
GC	A (N7)	A (O6)	D (N4)		A (N3)	D (N2)	A (O2)
CG		D (N4)	A (O6)	A (N7)	A (O2)	D (N2)	A (N3)

*(Me) denotes the methyl group on the C3 of thymine.

24.4 Operator–repressor interaction and the HTH motif

It has long been known that the sequences in various operators of phages or bacteria have an approximate two-fold symmetry. This corresponds to the two-fold symmetry axis of a protein dimer, which is generally the functional unit of the repressor (it is even a tetramer in the case of the *lac* repressor of *E. coli*). A detailed analysis has been carried out with phages 434 and λ. There are two operators O_R and O_L that govern the transition from lysogeny to virulence (*genetic switch*). Each operator has three repressor binding sites. The six sequences are shown in Fig. 24.8, together with a perfectly symmetrical 14-mer sequence that also contains the two palindromic recognition motifs ACAA and TTGT. There are two repressor molecules: the repressor proper (λR and 434R) and the Cro protein (λCro and 434Cro). The domain of 434R that binds to DNA consists of the first 69 amino acids, hence the name RI-69. In addition, residues 3–71 of Cro correspond to residues 1–69 of R and the two proteins have a similar structure. The dimer is only formed after their binding to DNA (Figs 24.9 and 24.10).

With the work of Anderson *et al.* (1981), a recognition motif was found consisting of two α-helices joined by a chain forming a β-turn. The two helices are the second and third of a set of five found in R and Cro. Since then, the same arrangement has been found in the N-terminal part of the *E. coli* lactose operon, in the protein CAP, in the repressor of the *E. coli* tryptophan operon, and in the protein of the homeodomain of *Antennapedia* (Antp) (Fig. 24.11).

The *helix–turn–helix* (*HTH*) *motif* thus plays an important role in this type of recognition between proteins and DNA. There is no common geometry but in broad terms helix 2 (*positioning*) is astride the major groove and helix 3 (*recognition*) fits roughly into the major groove (Fig. 24.12). A precise code of recognition cannot be defined. However, when the complexes RI-69/OR3 and RI-69/OR1 are compared, differences in the structure of one of the two halves of the operator appear which are correlated with the change from TTGT to CTGT. Differences of contact could be reflected in differences of constraint imposed by the protein on the DNA.

Site	Approximate 2-fold symmetry	$K_{dissociation}$ R1-69	$K_{dissociation}$ Cro
OR1	TACAAGAAAGTTTGTT ATGTTCTTTCAAACAA	1	2.5
OR2	AACAAGATACATTGTA	4	6
OR3	CACAAGAAAAACTGTA	4	1
OL1	TACAAGGAAGATTGTA	2.5	2
OL2	AACAATAAATATTGTA	2.5	3
OL3	AACAATGGAGTTTGTT	11	9
14-mer	ACAATATATATTGT	9	10

1 corresponds to 2×10^{-8} M

Figure 24.8 Sequences of phage 324 operators. (From Harrison and Aggarwal, 1990. Reprinted with permission from *Annual Review of Biochemistry*, vol. 59 and with the authors' permission. © 1990 Annual Reviews Inc.)

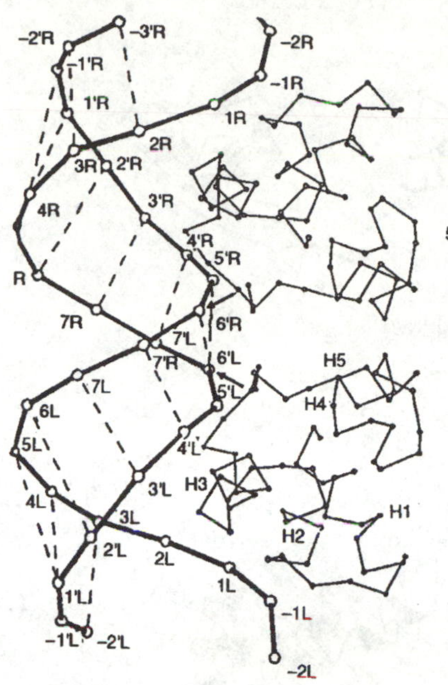

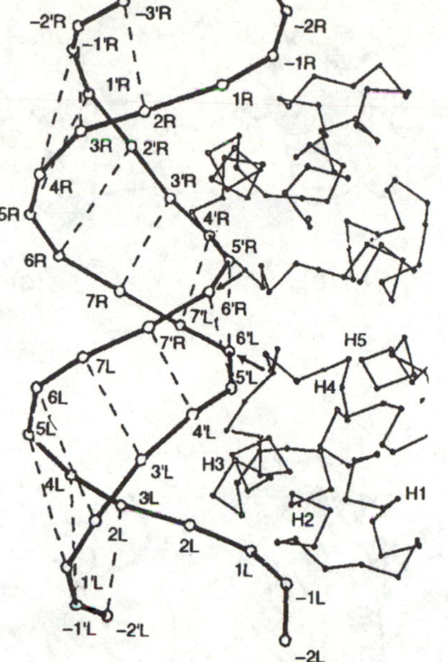

Figure 24.9 Stereo diagram of the 20 bp oligonucleotide interacting with the RI-69 dimer. In the middle of the operator the width of the minor groove is reduced to about 9 Å but recovers its normal value of 14 Å at both ends. The operator site OR1 is 14 bp long and is divided into two half-sites R and L, each 7 bp long. The phosphates are therefore numbered from 1 to 7 in one strand and from 1′ to 7′ in the other, with index R or L according to which half-site they belong. (From Aggarwal *et al.*, 1988. Reprinted with permission from *Science*, vol. 242, pp. 899–907 and with the authors' permission. © 1988 American Association for the Advancement of Science.)

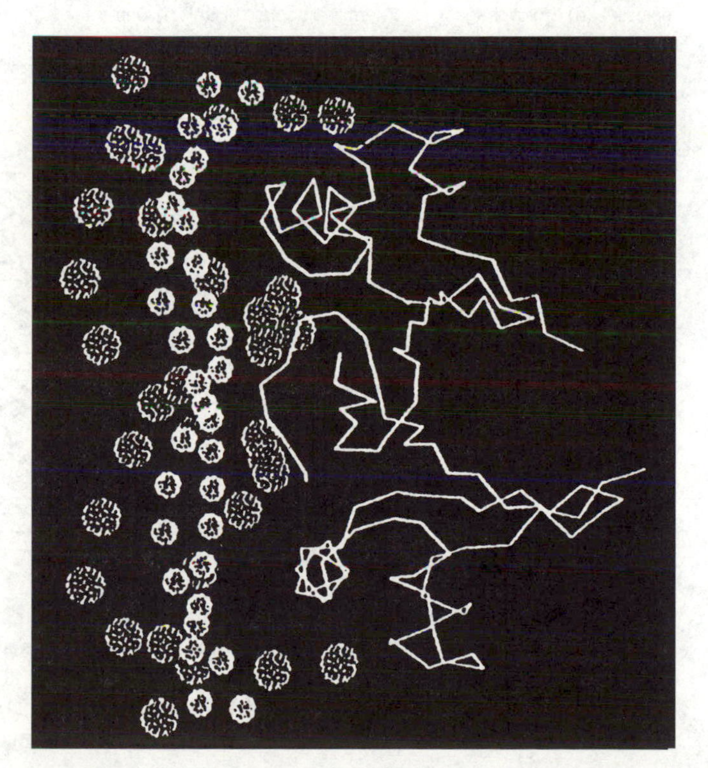

Figure 24.10 The complex of the Cro repressor with DNA. The protein is represented by its α-carbon backbone, the DNA by phosphates (grey spheres) and the bottom of the two grooves (small white spheres). One of the recognition helices (at the bottom of the figure) has its axis perpendicular to the plane of the figure. The second helix is rotated by about 35–50° with respect to the first. (From Brennan *et al.*, 1990. Reprinted from *Proceedings of the National Academy of Sciences, USA* with the authors' permission.)

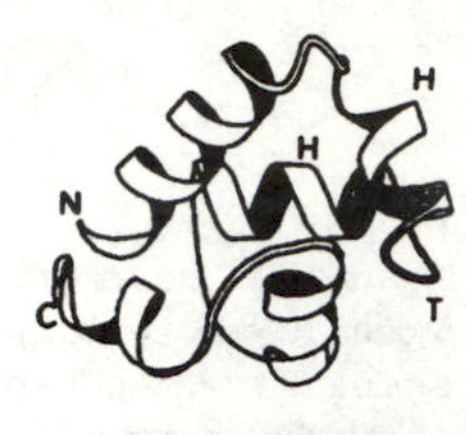

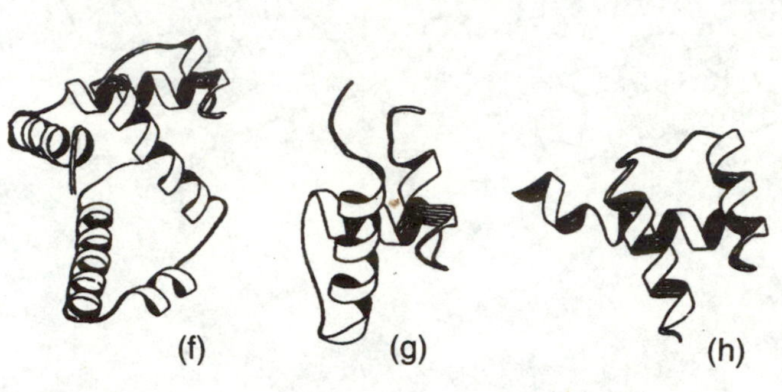

Figure 24.11 Eight examples of helix – turn – helix motifs oriented in the same way, indicated in (d). (a) Domain RI-69 of phage 434 repressor. (b) Cro of phage 434. (c) N-terminal domain of phage λ repressor. (d) Cro of phage λ. (e) C-terminal domain of CAP. (f) One of the sub-units of Trp repressor of *E. coli.* (g) Head-piece of *lac* repressor. (h) Homeodomain of the gene *Antennapedia* (Antp) of *Drosophila.* (From Harrison and Aggarwal, 1990. Reprinted with permission from *Annual Review of Biochemistry*, vol. 59 and with the authors' permission. © 1990 Annual Reviews Inc.)

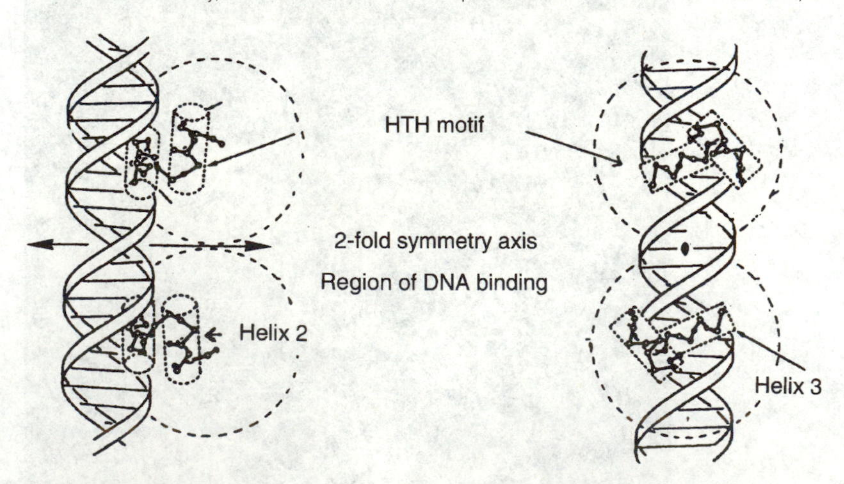

Figure 24.12 One of the binding modes of the HTH motif to DNA. The left-hand diagram is a view perpendicular to the two-fold symmetry axis. The right-hand diagram is a view along the symmetry axis. The dotted spheres indicate the volumes occupied by the repressor dimers. (From Matthews, 1988. Reprinted with permission from *Nature*, vol. 335, pp. 294 – 5 and with the author's permission. © 1988 Macmillan Magazines Ltd.)

In the case of the Trp repressor, there is a rigid central core containing A, B, C and F helices and flexible 'reading heads' formed from D and E helices arranged in an HTH motif (Fig. 24.13).

In CAP, E helices (two) and F helices (three) form the HTH motif. In this case, it is the C-terminal domain that interacts with DNA and the N-terminal domain that

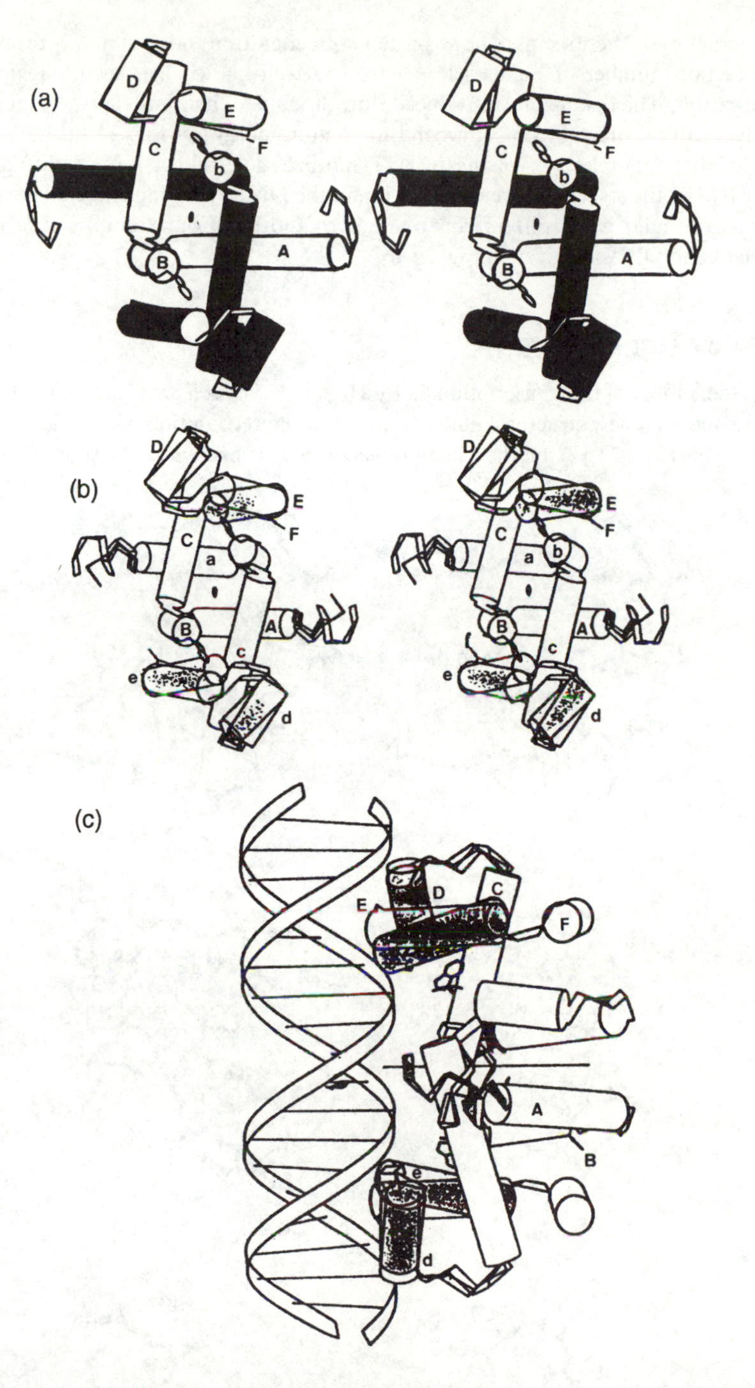

Figure 24.13 The repressor of the *E. coli* Trp operon. (a) Stereo diagram of the protein with the two intertwined subunits, one in white, the other in black. (b) Change in structure induced by tryptophan binding. The positions of the D and E helices are indicated in grey in the aporepressor (without Trp). (c) DNA binding. Reading heads of the D and E helices only fit into the major groove when tryptophan is bound to the repressor. (From Luisi and Sigler, 1990. Reprinted from *Biochimica et Biophysica Acta*, vol. 1048, pp. 113–26, © 1987 with kind permission from Elsevier Science NL, Sara Burgerhartstraat 25, 1055 KV Amsterdam, The Netherlands.)

contains the binding site of cyclic AMP.Homeodomains are small proteins, encoded by early genes, which regulate the embryogenesis of *Drosophila*. Detailed descriptions of the specific interactions that are established between groups of atoms can be found in the published literature but no general rule can at present be given. However, an analysis of the interactions is particularly significant: it reveals all the possibilities previously described in the study of protein–nucleic acid interactions. There seems to be no particular recognition code but simply the use of a certain number of bonds selected from a well-known list, according to the situation. The distortions introduced into the DNA chain only have to create a favourable geometry to maximize the interaction energy (see Figs 24.14 and 24.15).

In all the complexes involving the HTH motif, as determined by X-ray diffraction or NMR, the structure reveals distortion in the DNA (bending and change in the torsion angle), which differs from one case to another. Four examples of proteins that bend DNA are given in Fig. 24.16.

24.5 Zinc fingers

In their study of the transcription factor IIIA of *Xenopus laevis*, Miller *et al.* (1985) discovered a new structural motif involved in the recognition of nucleic acids by proteins (Fig. 24.17). Of the 344 amino acids in this protein, 264 (from 13 to 276)

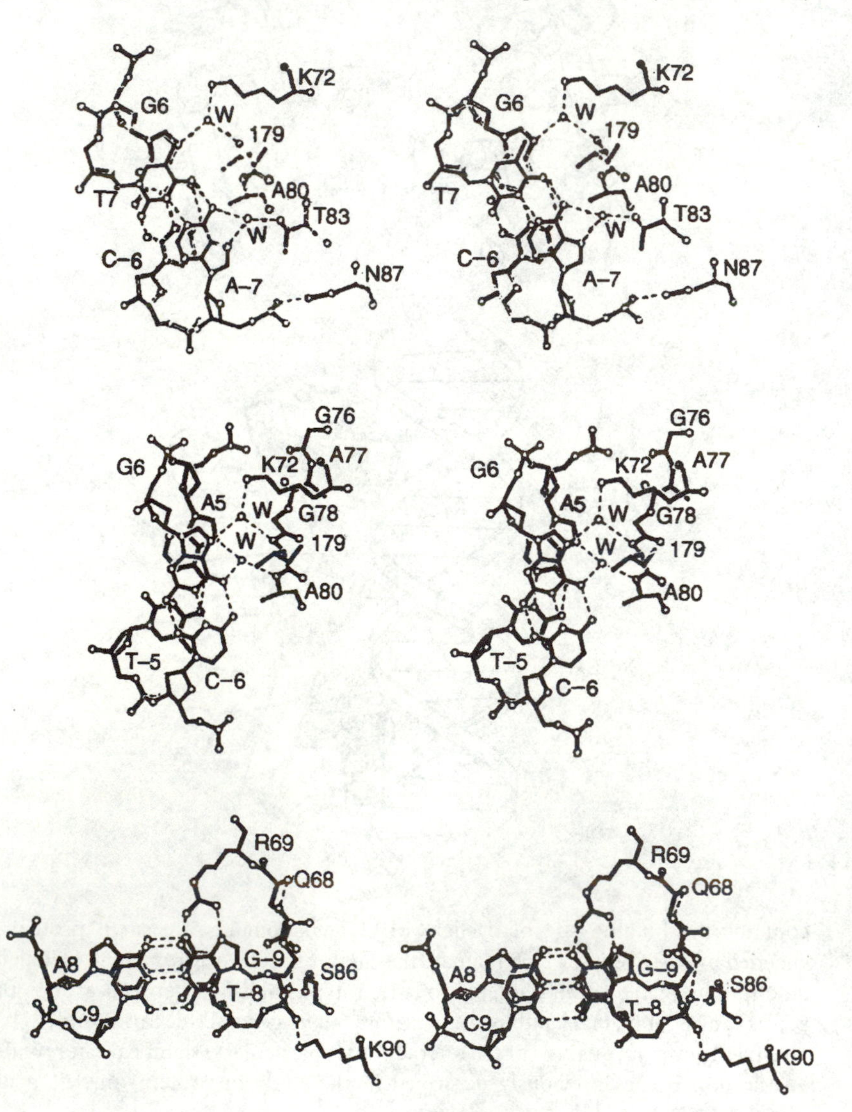

Figure 24.14 Stereo diagrams of interactions between base pairs and amino acids in the complex of the Trp repressor with its operator. Note the structural role of several water molecules (W). (From Luisi and Sigler, 1990. Reprinted from *Biochimica et Biophysica Acta*, vol. 1048, pp. 113–26, © 1987 with kind permission from Elsevier Science NL, Sara Burgerhartstraat 25, 1055 KV Amsterdam, The Netherlands.)

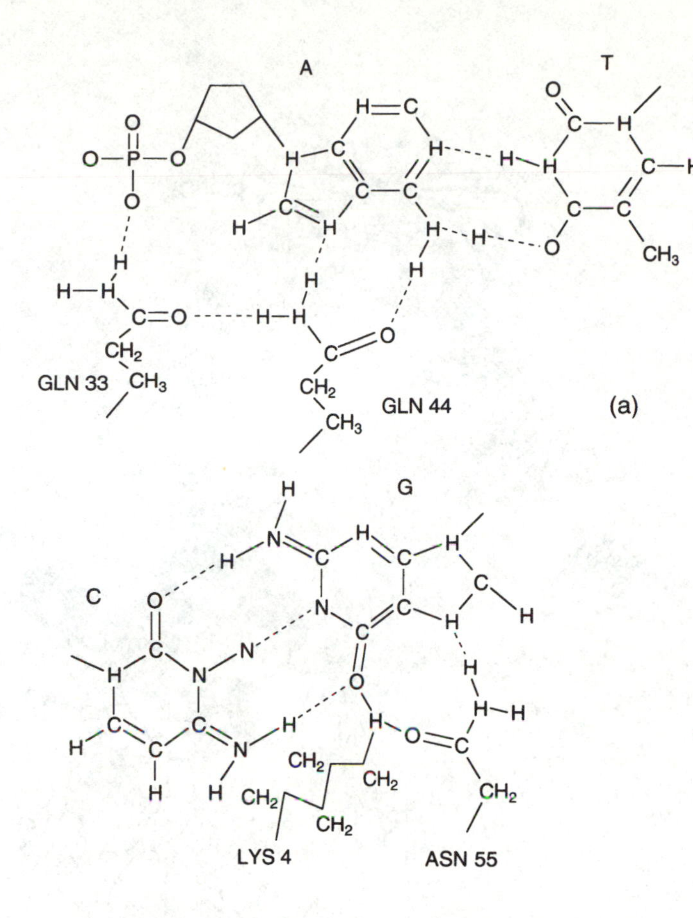

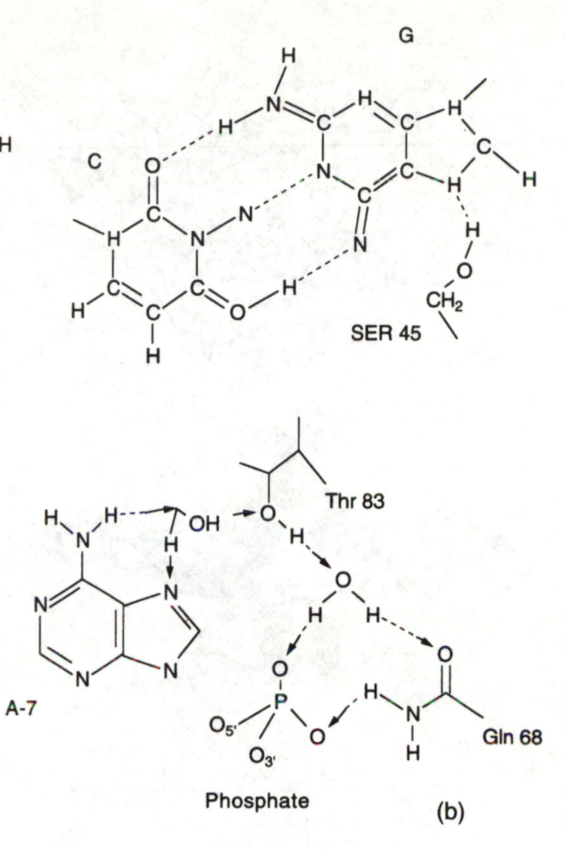

Figure 24.15 Examples of protein – DNA interactions. (a) Complex of the λ repressor. (From Jordan and Pabo, 1988. Reprinted with permission from *Science*, vol. 242, pp. 893–9 and with the authors' permission. © 1988 American Association for the Advancement of Science.) (b) Complex of the Trp repressor. Water molecules are involved as donors or acceptors of hydrogen bonds. The arrows indicate the acceptor groups. (From Luisi and Sigler, 1990. Reprinted from *Biochimica et Biophysica Acta*, vol. 1048, pp. 113–26, © 1987 with kind permission from Elsevier Science NL, Sara Burgerhartstraat 25, 1055 KV Amsterdam, The Netherlands.)

form a series of nine loops each of about 30 amino acids containing two cysteines (2C) and two histidines (2H) with a zinc atom forming a tetrahedral complex with these four amino acids. Since then, the same structural motif (*zinc finger*) has been found in many proteins that bind DNA, in some cases with a zinc complex having four cysteines (instead of 2C and 2H). This leads to a distinction between two families of proteins with zinc fingers, one with the 2C2H complex, the other with an xC complex, x being not less than 4.

The structure of the zinc finger in the first family was determined using NMR. It is hairpin-shaped, with the two cysteines in the loop being linked to an α-helix containing the two histidines (Fig. 24.18). Hydrophilic (S, T) and charged (K, R) side chains are on the same side of the α-helix and must be involved in the interaction of the zinc finger protein with DNA. In a more refined model, the helix is

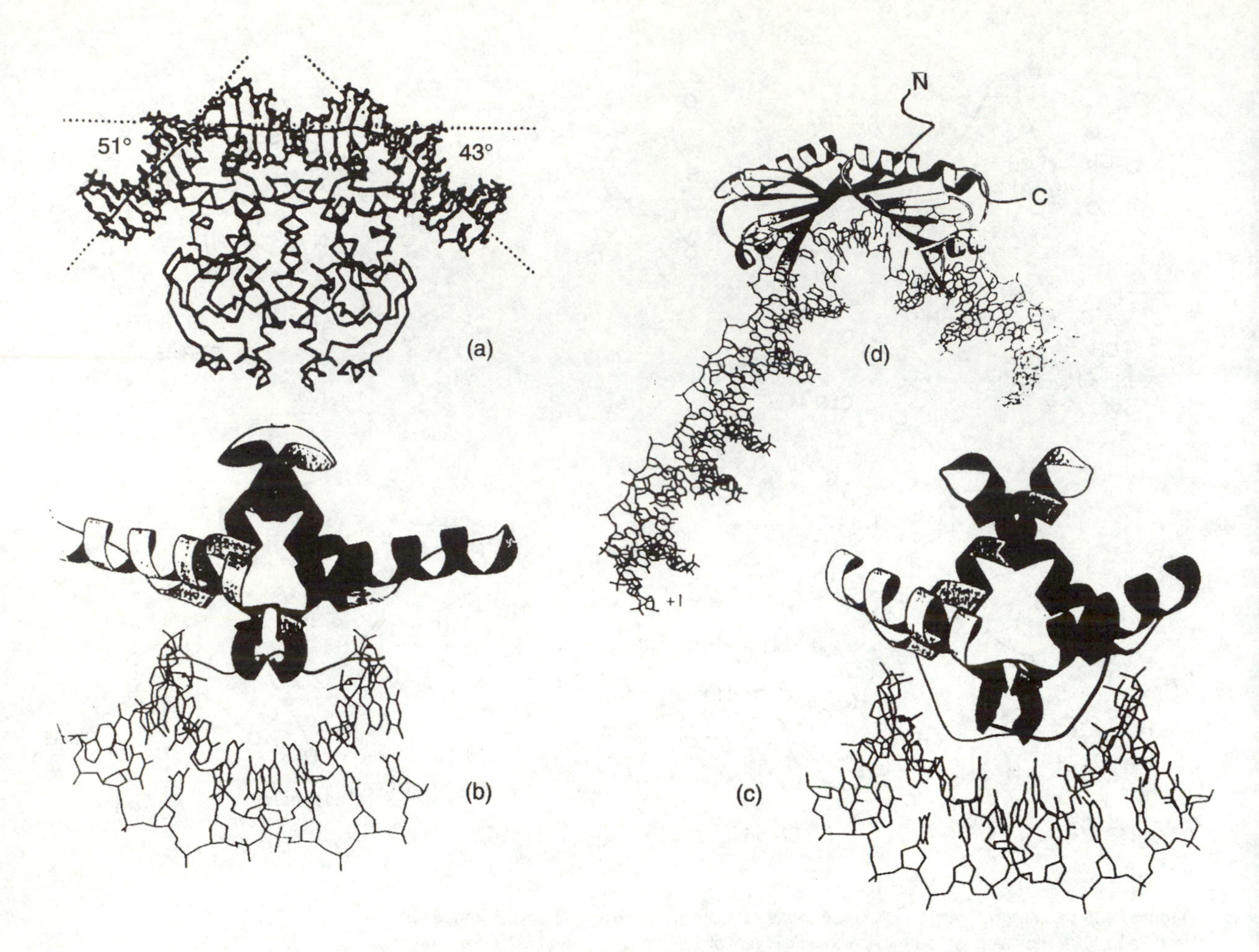

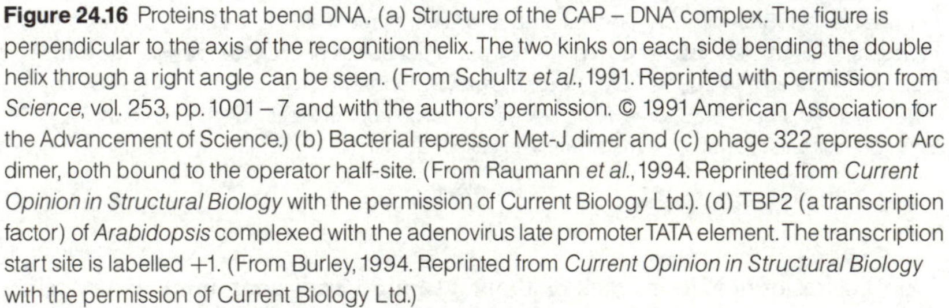

Figure 24.16 Proteins that bend DNA. (a) Structure of the CAP – DNA complex. The figure is perpendicular to the axis of the recognition helix. The two kinks on each side bending the double helix through a right angle can be seen. (From Schultz *et al.*, 1991. Reprinted with permission from *Science*, vol. 253, pp. 1001 – 7 and with the authors' permission. © 1991 American Association for the Advancement of Science.) (b) Bacterial repressor Met-J dimer and (c) phage 322 repressor Arc dimer, both bound to the operator half-site. (From Raumann *et al.*, 1994. Reprinted from *Current Opinion in Structural Biology* with the permission of Current Biology Ltd.). (d) TBP2 (a transcription factor) of *Arabidopsis* complexed with the adenovirus late promoter TATA element. The transcription start site is labelled +1. (From Burley, 1994. Reprinted from *Current Opinion in Structural Biology* with the permission of Current Biology Ltd.)

located in the major groove, the recognition sequence consisting of GGG or GCG triplets (Fig. 24.19).

In a recent review (Choo and Klug, 1997), a degenerate recognition code is proposed within the zinc finger protein family (Fig. 24.20), each base being capable of recognition by a limited number of amino acids. Its existence undoubtedly arises from the position of the zinc fingers lying almost perpendicular to the plane of the bases. The amino acids responsible for the recognition are about 3.4 Å apart, which is in agreement with the axial distance between base pairs and a 1:1 stoichiometry. Two points should be made about this important class of proteins

(a)

```
    1      8      13   17     23   26    30
   TGEK*PYVC..DGCDKRFTKK..LK*RH..*.H

1 (MGEKALPVVYKR)                          12
1          YICSFADCGAAYNKNWKLQ*AHLC*KH    37
2  TGEK*PFPCKEEGCEKGFTSLHHLT*RHSL*TH     67
3  TGEK*NFTCDSDGCDLRFTTKANMK*KHFNRFH     98
4  NIKICVYVCHFENCGKAFKKHNQLK*VHQF*SH    129
5  TQQL*PYECPHEGCDKRFSLPSRLK*RHEK*VH    159
6  AG--*-YPCKKDDCCSFVGKTWTLYLKHVAECH    188
7  QD--*LAVC--DVCNRKFRHKDYLR*DHQK*TH    214
8  EKERTVYLCPRDGCDRSYTTAFNLR*SHIQSFH    246
9  EEQR*PFVCEHAGCGKCFAMKKSLE*RHSV*VH    276
                                        277
(DPEKRKL*KEKCPRPKRSLASRLTGYIPPKSKEKNA  311
SVSGTEKTDSLVKNKPSGTETNGSLVLDKLTIQ)     344
```

(b)

Figure 24.17 (a) Sequence of the transcription factor IIIA of *Xenopus*. The nine repeating units have been aligned to show the homology and the location of the two cysteines (2C) and two histidines (2H). The box at the top corresponds to a consensus sequence of a 30-residue zinc finger. (b) Folding model of two successive zinc fingers. The conserved amino acids are circled. (From Klug and Rhodes, 1987. Reprinted from *Trends in Biochemical Science*, vol. 12, pp. 464–69, with permission from Elsevier Trends Journals. © 1987 Elsevier Science Ltd.)

(Choo and Klug, 1997):

1. The stereochemical complementarity of the amino acids and bases (Fig. 24.21) is somewhat similar to that encountered in the formation of the triple helix (compare Fig. 24.21 with Fig. 4.35).
2. Instead of the dimeric form adopted by most proteins in specific interaction with a palindromic domain of the DNA double helix, these monomeric zinc finger proteins have an α-helix that can bend into segments of three or four base pairs and then fit the geometry of the major groove with ease. In this case, as in many others, the DNA double helix is also distorted to facilitate the interaction.

24.6 Polymerase–nucleic acid interactions

The hand model proposed above (section 24.1) to represent the interaction between a protein and a Ca^{2+} ion used only three fingers (thumb, forefinger and middle finger) and provided a picture of the internal structure of a protein. Another hand model can be used to represent the binding of a DNA or RNA polymerase to DNA. This time the hand is a half-closed right hand containing the DNA in its palm. Preserving the analogy, the three polymerase domains are called fingers (F), palm (P) and thumb (T).

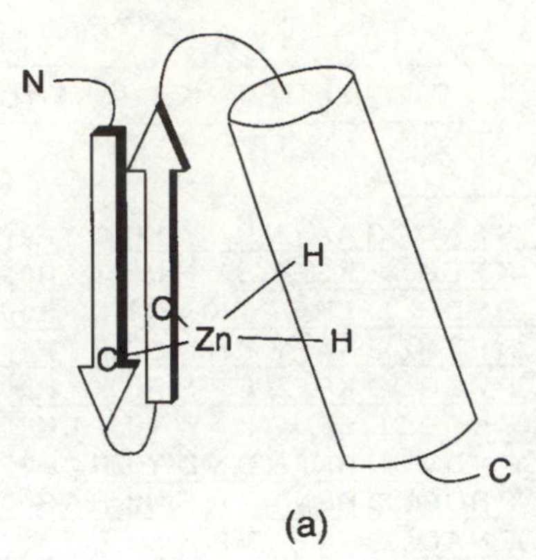

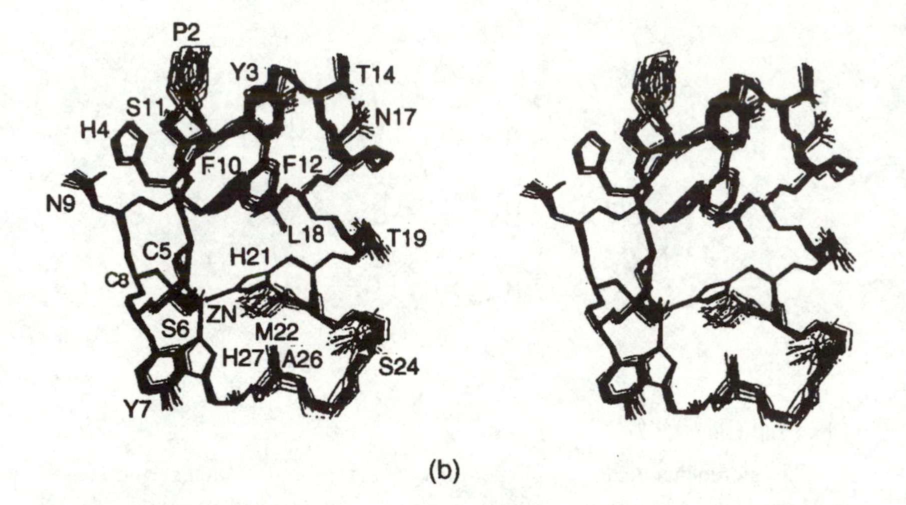

Figure 24.18 (a) Diagrammatic structure of a zinc finger with the α-helix represented by a cylinder. (b) Stereo diagram of the structure of a zinc finger with the sequence:

R P Y H C S Y C N F
1 5 10
S F K T K G N L T K
15 20
H M K S K A H S K K
25 30

A tetrahedral complex is formed between the zinc atom and C5, C8, H21 and H27 residues. The α-helix extends from T14 to S24. The thick line represents the superposition of 40 different structures each satisfying NMR constraints. (From Omichinski *et al.*, 1990. Reprinted with permission from *Biochemistry*. © 1990 American Chemical Society.)

Four polymerases conforming to this model are shown in Fig. 24.22 (Arnold *et al.*, 1995): the Klenow fragment of *E. coli* polymerase, the phage T7 RNA polymerase, HIV-1 reverse transcriptase and the DNA polymerase polβ. It is significant that similar overall conformations are found for the four proteins, all of which must interact with double-stranded DNA. Before this 'hand' model can be generalized to many other polymerases, however, more structural information is needed.

The three types of recognition process between proteins and nucleic acids just described in sections 24.4 to 24.6 will undoubtedly be extended by the discovery of new modes of recognition. Although there is only a limited list of short-range interactions between atomic groups for very different overall conformations, it is enough to allow a large number of combinations.

24.7 Specificity of recognition

So far, we have been dealing merely with a classification of interactions related to a restricted number of atomic combinations. There are, however, some general rules governing the specificity of recognition.

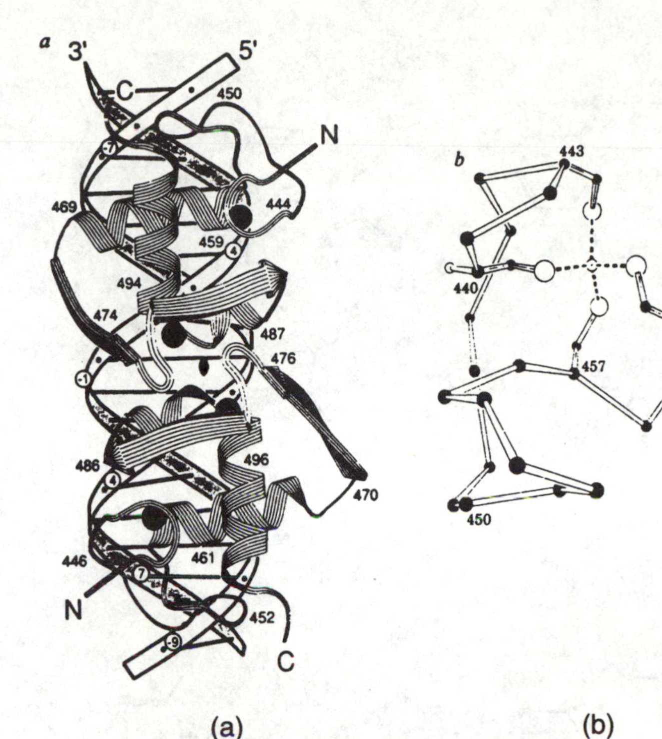

Figure 24.19 Interaction with DNA of a glycocorticoid receptor. The protein domain (440 – 525) that binds DNA contains two zinc fingers and is dimerized. (a) View of the complex along the two-fold axis of the dimer. Zinc atoms are represented by discs. (b) Detail of an *S*-configuration of one of the zinc fingers, with the four cysteines 440, 443, 457 and 460. (From Luisi *et al.*, 1991. Reprinted with permission from *Nature*, vol. 352, pp. 497 – 505 and with the authors' permission. © 1991 Macmillan Magazines Ltd.)

POSITION IN TRIPLET

NUCLEOTIDE	5′	Mid.	3′
G	Arg 6 Ser 6 / Asp 2* Thr 6 / Asp 2*	His 3	Arg –1 / Asp 2
A		Asn 3	Gln –1 / Ala 2
T	Ser 6 / Asp 2* Thr 6 / Asp 2*	Ala 3 Ser 3 Val 3	Asn –1 Gln –1 / Ser 2
C		Asp 3 Leu 3 Thr 3 Val 3	Asp –1

Figure 24.20 Elements of a consensus zinc finger DNA recognition code. Cognate amino acids and their positions in the zinc finger α-helix are listed in a matrix relating the four bases to each position of a DNA triplet. Cooperating or interdependent cognate amino acids are separated by a solidus (/). The aspartate residues that help specify 5′G or 5′T are from an adjacent zinc finger and are marked with an asterisk. (From Choo and Klug, 1997. Reprinted from *Current Opinion in Structural Biology* with the permission of Current Biology Ltd.)

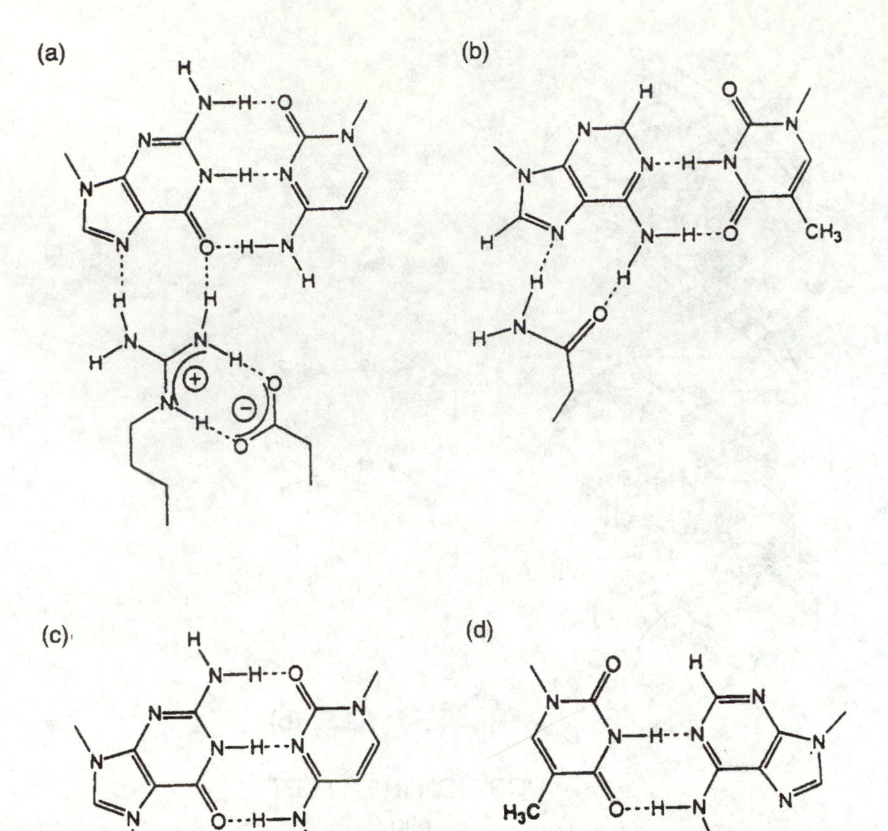

Figure 24.21 Illustration of some intermolecular contacts responsible for zinc finger specificity. (a) Interaction between arginine and guanine showing buttressing of Arg(−1) by Asp(2). (b) Interaction between asparagine and adenine. (c) Possible interaction between histidine and guanine. (d) Possible interaction between alanine and thymine. Note that all the above interactions occur in the plane of the paper. (From Choo and Klug, 1997. Reprinted from *Current Opinion in Structural Biology* with the permission of Current Biology Ltd.)

At an elementary level, the specificity of interaction between protein and nucleic acid can be quantified by finding the minimum number n of nucleotides necessary to define a unique site in the genome. Taking the *E. coli* genome with 4×10^6 base pairs as an example, and assuming an equal probability (1/4) of finding any of the four bases in a nucleotide, then a unique site on this genome entails an $n \approx 11$. It is thus unnecessary to know a large number of nucleotides at the beginning of an unknown gene in order to retrieve it unambiguously in the now established *E. coli* genome. This model can be used to its limit by determining the difference in free energy between binding to a specific and a non-specific site (see Boxes 24.1 and 24.2). The ratio of association constants found in this way is in good agreement with experiment.

To analyse this problem more thoroughly, Berg and von Hippel (1987) developed an interesting theoretical approach. Although we shall not go into details of the calculations, it is important to explain the underlying ideas and assumptions.

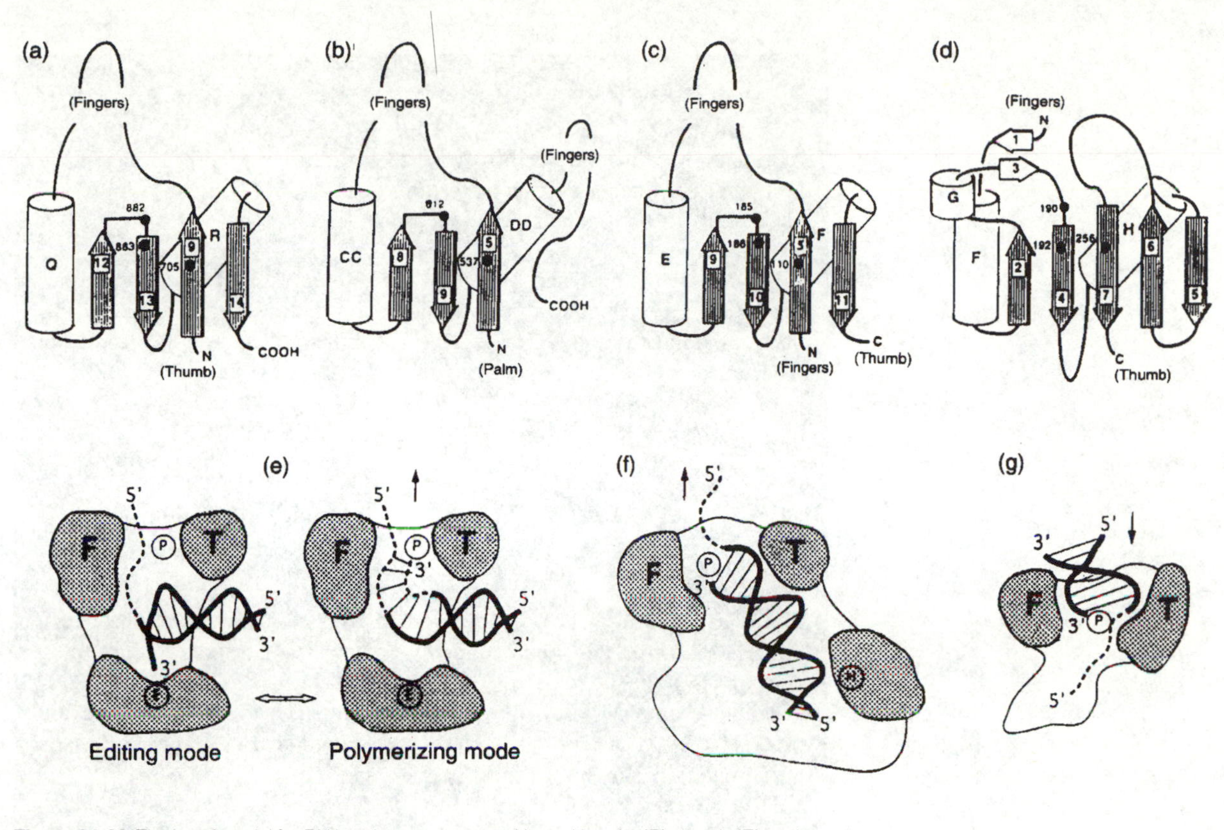

Figure 24.22 The hand model for DNA-polymerase recognition, with palm (P), thumb (T) and fingers (F): (a) Klenow fragment; (b) T7 DNA polymerase; (c) HIV-1 reverse transcriptase; and (d) polβ. Representation of the structurally conserved palm subdomains: in (a), (b) and (c), the interrupted loop at the top of the diagrams represents departures of the main chain from the palm into the finger subdomain; in (d), the palm subdomain is topologically different. (e – g) Assumed binding modes in the protein – DNA complex. DNA strands observed in the crystal structure are drawn as solid black lines. (From Arnold *et al.*, 1995. Reprinted from *Current Opinion in Structural Biology* with the permission of Current Biology Ltd.)

Following experimental studies of the DNA sequences recognized either by repressors (*operator* sites) or by RNA polymerase (*promoter* sites), biochemists introduced the idea of a *consensus sequence*. This defines a sequence of base pairs that is always or almost always present at the same place in the recognition site without the need to consider the local DNA structure associated with this sequence. Assuming that the interaction energy of the protein with each sequence is known, this is equivalent to considering a system with a set of energy states at constant temperature and attempting to determine the occupancy of each state (microcanonical ensemble). Three assumptions are made:

1. A range of interaction energy can be defined within which all the sequences that can interact with the protein are taken into account.
2. The number of such sequences is large and all possible sequences have the same probability.

Box 24.1 Protein–DNA recognition

A protein has to recognize a group of n nucleotides among the N ($N \gg n$). Assuming an equal probability for the occurrence of the four bases, the probability of finding that such a group exists is $(1/4)^n$. For this group of n nucleotides to be unique among all the N, we must have $(1/4)^n = 1/N$ or $n = \ln N/\ln 4$.

Let ΔG_1 be the free-energy change (almost entirely electrostatic in origin) accompanying the binding of the protein to any part of the DNA and ΔG_2 be that for binding to the specific site. The probability p_1 that the non-specific site is occupied is

$$p_1 = [(N-n)/N]\exp(-\Delta G_1/RT)/[(N-n)/N]\exp(-\Delta G_1/RT) + (n/N)\exp(-\Delta G_2/RT)$$

Let $\Delta G_2 - \Delta G_1 = \Delta G$ ($\Delta G < 0$), and then, neglecting n in comparison with N:

$$p_1^{-1} \approx (n/N)\exp(-\Delta G/RT)$$

If, for example, $p_1 = 10^{-2}$ in order to favour the specific binding, then $\Delta G = -RT\ln(100N)\ln 4/\ln N$, and this free energy is shared among the n nucleotides of the site.

If, as an example, we now take an operator with $n = 10$ and $N = 10^6$, we find that $\Delta G \approx -9.7\,\text{kcal mol}^{-1}$ or $-16RT$. Less than 1 kcal mol^{-1} per nucleotide is thus enough to ensure the specificity.

The extra energy required to make the binding specific is provided by several bonds of low energy formed after the repressor has been docked at its site by means of some non-specific Coulomb interactions. This approximate calculation ignores two important points:

- The different bases do not have equal probabilities, particularly in bacterial genomes where the GC/AT equilibrium is not observed.
- More subtly, the specificity must be measured for a given recognition protein by the competition between the unique site and 'pseudo-sites' differing by only a few bases (see Box 24.2).

3. Each base pair b ($b = 0, 1, 2, 3$) located in a position i of the site ($i = 1, 2, 3, \ldots, s$, where s is the size of the site) contributes $\varepsilon_{ib}k_BT$ to the total interaction energy between the protein and the site.

To a first approximation, these individual contributions are assumed to be independent and therefore additive. If the base pair is 'good', i.e. if the binding energy is the largest, then we can take $\varepsilon = 0$ and $b = 0$ ($\varepsilon_{i0} = 0$). With this convention, the ε_{ib} are dimensionless numbers. The quantity $\varepsilon_{ib}k_BT$ will measure the decrease in binding free energy when the 'good' base is replaced by another one, so that the ε_{ib} appear as *local energies of discrimination*. This is equivalent to allocating to each base pair b a probability factor $\exp(-\lambda\varepsilon_{ib})$, where λ is a constant to be determined but playing a role similar to $1/k_BT$ in classical Boltzmann statistics. For each sequence, a *discrimination energy* can thus be calculated by adding all the local energies. If, to a first approximation, it is assumed that all the base pairs are present in the genome

Box 24.2 Competition between specific and non-specific sites

Let the concentration of specific sites D_s in a DNA molecule be $[D_s]$ with an association constant K_s, and the concentrations of non-specific sites be $[D_1]$, $[D_2]$, ... with constants K_1, K_2, ... The theory of multiple equilibria shows that, if $[A_0]$ is the concentration of the free ligand (e.g. a protein), the concentration of the bound ligand $[A_L]$ is obtained by adding all the possible contributions:

$$[A_L] = \sum_i K_i[D_i][A_0]/(1 + K_i[A_0]) + K_s[D_s][A_0]/(1 + K_s[A_0])$$

If $\theta_s = [AD_s]/([D_s] + [AD_s])$ is the occupied fraction of D_s:

$$\theta_s = K_s[A_0]/(1 + K_s[A_0])$$

and hence $[A_0] = K_s^{-1}\theta_s/(1-\theta_s) = x/K_s$ where $x = \theta_s/(1-\theta_s)$ is a selectivity factor for the specific site. For example, if $\theta_s = 0.999$, then $x \approx 1000$.

With these new parameters, $K_s = x/[A_0]$ and $\theta_s = x/(1+x)$, the total concentration $[A_T] = [A_0] + [A_L]$ can be written as

$$[A_T] = x/K_s + x[D_s]/(1 + x) + \sum x[D_i]/(x + K_s K_i)$$

and a selectivity factor x can therefore be calculated from K_s, K_i and the concentrations $[D_s]$, $[D_i]$ and $[A_T]$.

If 'strong' pseudo-specific sites exist, then $K_s/K_i < x$. On the other hand, for all the 'weak' non-specific sites, $K_s/K_i > x$. If $x \gg 1$, an apparent association constant for binding to the specific site can be defined by

$$K_s^{app} = K_s/(1 + \sum K_i[D_i])$$

with the same probability, the probability f_{ib} that b is in position i is given by:

$$f_{ib}(E) = \exp(-\lambda\varepsilon_{ib}/[1 + \exp(-\lambda\varepsilon_{i1}) + \exp(-\lambda\varepsilon_{i2}) + \exp(-\lambda\varepsilon_{i3})] \qquad (24.1)$$

where λ is a constant determined by the choice of E, i.e. by the value arbitrarily given to the total discrimination energy. For example, if all the ε_i have the same value ε^*, then

$$f_{ib} = \exp(-\lambda\varepsilon^*)/[1 + 3\exp(-\lambda\varepsilon^*)]$$

and

$$E = 3\varepsilon^* s \exp(-\lambda\varepsilon^*)/[1 + 3\exp(-\lambda\varepsilon^*)]$$

which gives

$$\lambda = (\varepsilon^*)^{-1}\ln[3(\varepsilon^* s - E)/E]$$

and values of λ are generally found to lie between 0.5 and 1.5.

The probability f_{ib} can also be written in the form

$$f_{ib} = \exp(-\lambda\varepsilon_{ib})/4q_1 \tag{24.2}$$

where q_1 is the partition function such that

$$f_{i0}^{-1} = (4q_1) = 1 + \exp(-\lambda\varepsilon_1) + \exp(-\lambda\varepsilon_2) + \exp(-\lambda\varepsilon_3) \tag{24.3}$$

Hence

$$\lambda\varepsilon_{ib} = \ln(f_{i0}/f_{ib}) \tag{24.4}$$

Each sequence can thus be characterized by *information* I, which can be calculated from the observed frequencies with which base pairs are used. If E is the free energy of the sequence, it can be shown that $dI/dE = -\lambda$, an expression similar to the classical thermodynamic relationship $dS/dE = 1/T$. The quantity $-I$ is in fact proportional to the entropy of the sequence ($-I = k_B S$) and can be defined as a *selection entropy*.

Mutations in the sequence not changing the binding affinity lead to equiprobable sequences: these are similar to thermal transitions in a thermodynamic system. We can also calculate the specific part of the mean binding constant K_R for a randomly selected site. For example, if $\lambda = 1$, we have

$$K_R = K_0 \exp[-(E + I)]$$

where K_0 is the association constant for the reference specific site.

Some interesting results are obtained when this theory of specific recognition is applied to promoters. We first recall that there are at least two steps in the binding of RNA polymerase to a promoter: a 'closed' complex with a binding constant K_B and an 'open' complex obtained from the closed one with a transition rate constant k_2. The 'strength' of a promoter is usually defined as the product k_2K_B. A linear relationship has been found empirically between $\log(k_2K_B)$ and a 'homology index' defined from the difference between each promoter and the consensus sequence. Berg and von Hippel's (1987) theory enables the strength of the promoter to be predicted from physical quantities using eqn 24.4.

If n_{ib} is the number of times the base pair b is observed in position i and N is the total number of sites used in the sample, it can be shown that

$$f_{ib} = (n_{ib} + 1)/(N + 4) \tag{24.5}$$

In particular, in the absence of any information, $n_{ib} = 0$ and $N = 0$, so that $f_{ib} = 1/4$ and hence

$$\lambda\varepsilon_{i0} = \ln[(n_{i0} + 1)/(n_{ib} + 1)] \tag{24.6}$$

with a standard deviation that can be estimated.

Such a relationship could be used with promoters by introducing two parameters characterizing the recognition sequence of RNA polymerase: the base composition of the regions -10 and -35, and the length (expressed in base pairs) between these two regions. Assuming that these are independent, a summation over all the base

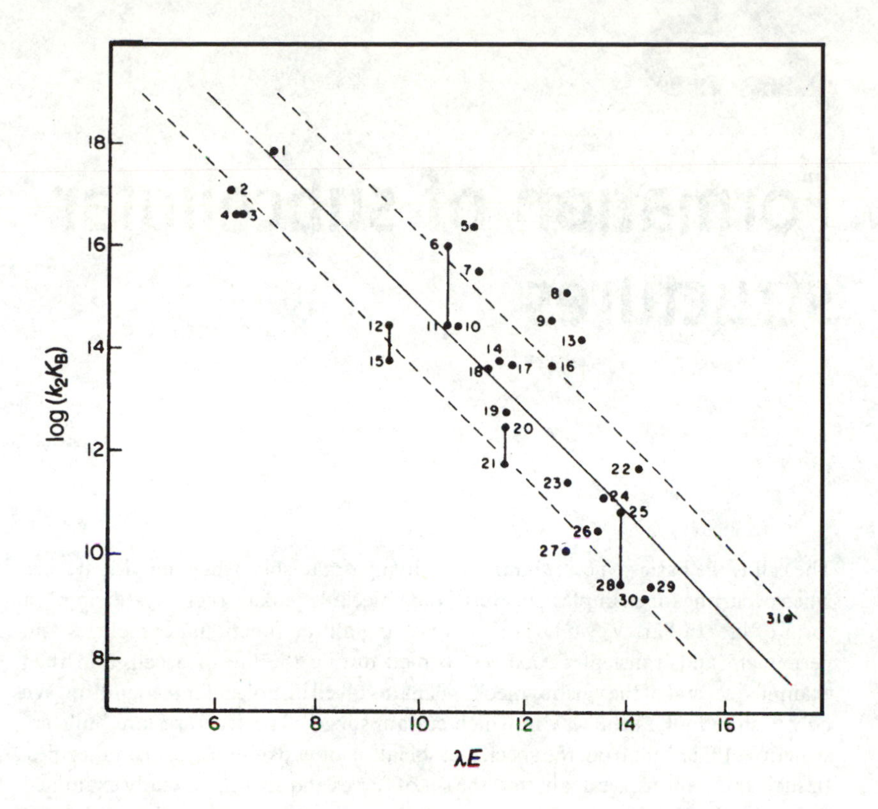

Figure 24.23 Linear relationship between $\ln(k_2K_B)$ and λE for a series of 31 promoters. The two dashed lines represent the ranges of uncertainty in both coordinates. (From Berg and von Hippel, 1987. Reprinted from *Journal of Molecular Biology* with the permission of Academic Press.)

pairs gives us

$$\lambda E = \sum_{i=1}^{s} \ln[(n_{i0}+1)/(n_{ib}+1)] + \ln[n(L_{opt})+1]/[n(L)+1] \qquad (24.7)$$

where $L_{opt} = 17$ and $n(L)$ is the number of times the length L is found in the sample of promoters being studied. This energy of discrimination must be a characteristic of the strength of promoters k_2K_B. When $\ln(k_2K_B)$ is plotted against λE for a series of 31 promoters, a linear relationship is found (Fig. 24.23).

It is meaningless to extrapolate to $E = 0$, for if k_2K_B is large enough the diffusion limit is replaced by a chemical reaction limit and the correlation line becomes a constant for small values of E. The good agreement between the classification of promoters obtained in this way and that proposed on the basis of simple rules of homology arises because the two quantities $\sum\ln[(n_{i0}+1)/(n_{ib}+1)]$ and $\sum(n_{i0}-n_{ib})$ are generally proportional.

25

Formation of subcellular structures

The cell is the basic structural unit of all living organisms, while individual cells, when occurring for example in bacteria and unicellular eukaryotes, are the simplest form of life. In Part V, we began by surveying all the functional complexes and permanent and transient structures formed during the life of a cell, and then examined several of the various mechanisms involved in molecular recognition. We now wish to look at the way in which certain subcellular structures are built up, structures that result from the specific association of two or more macromolecules. In such associations, we again find the set of forces and energies already examined in detail in Chapter 2: no new ones are involved. The complexity and specificity of subcellular forms are explained by the structure and complementary nature of their component macromolecules.

The shape of the cell, its plasticity (i.e. the extent to which its internal organization can change) and its movements mean that there must be a *cytoskeleton*–a complex network of proteins assembled together in the cytoplasm–modelling the bones and muscles of a vertebrate on the cellular scale. The great majority of these protein structures are either *actin filaments* formed by the polymerization of *G-actin* (a globular protein with a molecular mass of 41 800 Da) or *microtubules* arising from the polymerization of *tubulin* α and β forming a dimer with a mass of 100 000 Da. Table 25.1 compares the properties of these two modes of assembly.

Microtubules are found in stable structures like cilia and flagella, whose movements arise from interactions between the microtubule and *dynein*. Actin filaments occur in the microvilli of brush borders or in the stereoscopic cilia of the cells in the inner ear, where their movement is produced by interactions between actin and *myosin* as in muscular contraction. Beyond these organelles, actin and tubulin are part of some *labile structures*, like the mitotic spindle and the contractile ring that appear during cell division. Polymerization and depolymerization must take place rapidly: the dynamic nature of the assembly and disappearance of these transient structures is essential for cell growth.

25.1 Formation of polymeric structures

We now examine a few simple mechanisms in order to throw more light on the meaning of some of the parameters involved. The treatment closely follows that of Oosawa and Asakura (1975).

Table 25.1 Comparison of actin and tubulin

	Actin	Tubulin
Molecular weight (Da)	41 800	50 000 for α 50 000 for β
Type of protein	globular monomer	globular dimer
Related nucleotide (in the non-polymerized state)	ATP (one per monomer)	GTP (two per dimer)
Factors required for polymerization	Ca^{2+} or Mg^{2+}, NaCl	Mg^{2+}, chelating agent to remove Ca^{2+}, NaCl
Polymer	two-stranded helix, diameter 7 nm	hollow tube, diameter 25 nm, containing 13 protofilaments

25.1.1 Cycle of conformational change

In the conformational change between two forms A and B (section 8.1), a relaxation time τ was defined by

$$\tau^{-1} = k_1 + k_{-1}$$

(eqn 8.6), where k_1 and k_{-1} are the two rate constants for the A $\rightleftharpoons$ B transition. We now consider what happens in a steady state for which

$$\mathrm{d}[\mathrm{A}]/\,\mathrm{d}t = -k_1[\mathrm{A}] + k_{-1}[\mathrm{B}] = 0 \tag{25.1}$$

If $[\mathrm{A_T}]$ is the total concentration of component A, then

$$[\mathrm{B}] = [\mathrm{A_T}] - [\mathrm{A}]$$

and hence

$$-\,k_1[\mathrm{A}] + k_{-1}([\mathrm{A_T}] - [\mathrm{A}]) = 0$$

and

$$[\mathrm{A}] = k_{-1}[\mathrm{A_T}]/(k_1 + k_{-1}) \tag{25.2}$$

The equilibrium is dynamic. In other words, each molecule may oscillate between the two forms A and B, with the mean number of molecules in each state remaining constant. The rate at which A disappears is

$$k_1[\mathrm{A}] = k_1k_{-1}[\mathrm{A_T}]/(k_1 + k_{-1}) \tag{25.3}$$

The factor $k_1k_{-1}/(k_1 + k_{-1})$ expressed in s^{-1} is the number of cycles per second experienced by each molecule. The higher its value, the greater is the probability of a reaction with one of the two forms. A cycle of conformational change such as this may explain the mechanism of proton exchange and may also be involved in polymerization if it is only possible in one of the two forms of protein.

25.1.2 Polymerization at the end of a polymer

When a protein polymerizes to create a filamentary or tubular aggregate, the two main parameters, apart from temperature, pH and ionic strength, are the concentration [m] of free ends and the concentration c_1 of the monomer. The time variation of c_1 takes the form

$$\mathrm{d}c_1/\mathrm{d}t = -k_1 c_1[\mathrm{m}] + k_{-1}[\mathrm{m}] \tag{25.4}$$

where the first term on the right describes the polymerization (collision of the monomer with one end) and the second term the depolymerization. In the steady state, the length of the fibre remains constant because of a dynamic equilibrium between the growth and disappearance of the ends. This means that $k_1 c_1[\mathrm{m}] = k_{-1}[\mathrm{m}]$, each of the quantities being equal to the number of cycles per second characterizing each process. This is what happens, for example, in the polymerization of G-actin to form F-actin.

The polymerization process can be described as a multiple equilibrium between the monomer M and the *i*-mer *i*M:

$$[2\mathrm{M}] = K[\mathrm{M}][\mathrm{M}],$$
$$[3\mathrm{M}] = K[2\mathrm{M}][\mathrm{M}], \qquad \ldots, \qquad [\mathrm{iM}] = K[(i-1)\mathrm{M}][\mathrm{M}]$$

where the equilibrium constant K is assumed to be independent of the degree of polymerization. The concentration c_i of the *i*-mer is given by

$$c_i = K^{i-1}[\mathrm{M}]^i = (Kc_1)^i/K \tag{25.5}$$

If c_0 is the total concentration of protein molecules, then

$$c_0 = \sum_i ic_i = c_1/(1 - Kc_1)^2 \tag{25.6}$$

and the mean degree of polymerization $\langle i \rangle$ is

$$\langle i \rangle = \sum ic_i / \sum c_i = (1 - Kc_1)^{-1} \tag{25.7}$$

This simplified model does not generally represent the experimental behaviour. Most often, a helical polymer is formed only after a certain minimum size is achieved. In a tetramer, for example, if a bond is established between the first and fourth monomer with a conformational change, a helix can grow from an initial core consisting of the trimer. The growth of such a helical polymer can be calculated (see Box 25.1). Instead of eqn 25.6, we find

$$c_0 \approx c_1 + \sigma c_1/(1 - K_\mathrm{h} c_1)^2 \tag{25.8}$$

where K_h is the association constant for a monomer binding to the end of the helix and the quantity $\sigma = \gamma(K/K_\mathrm{h})^2$ is a nucleation parameter similar to that already encountered in the helix–coil transition (section 8.3). The factor $\gamma = \exp(-\Delta G^*/RT)$ is very small because ΔG^* (> 0) is the free energy required to distort the linear trimer, while $K \ll K_\mathrm{h}$, so that σ is only of the order of 10^{-8}.

If c_0 is increased while keeping σ and K_h constant (i.e. not altering the experimental conditions), then at the beginning $c_1 \approx c_0$. When c_1 becomes of the same order of magnitude as ${K_\mathrm{h}}^{-1}$, the second term on the right of eqn 25.8 becomes significant and the helix begins to grow. ${K_\mathrm{h}}^{-1}$ can thus be considered as a *critical*

Box 25.1 **Growth of a helical polymer**

Consider a linear tetramer that is distorted by the formation of a bond between the first and fourth monomer, thus triggering the growth of a helix consisting of an assembly of trimers. The concentration c_{3h} of the trimer in helical form is given by

$$c_{3h} = \gamma c_3 \quad (1)$$

where γ is a kind of nucleation factor ($\ll 1$) of the form $\exp(-\Delta G^*/RT)$, ΔG^* being the free-energy change (> 0) required to distort the trimer. From eqn 25.5:

$$c_3 = K^{-1}(Kc_1)^3$$

and hence

$$c_{3h} = \gamma K^{-1}(Kc_1)^3 \quad (2)$$

For the transition from the trimer to the tetramer, a monomer must be added with an association constant $K_h \gg K$, so that

$$c_{4h} = K_h c_{3h} c_1 = \gamma K^{-1} K_h c_1^4 \quad (3)$$

and more generally

$$c_{ih} = \gamma (K/K_h)^2 K_h^{-1} (K_h c_1)^i \quad (4)$$

This equation is similar to eqn 25.5 provided K is replaced by K_h and a multiplying factor $\sigma = \gamma(K/K_h)^2$ is introduced. The total concentration of helical polymers is

$$c_h = \sum_i i c_{ih} = \sum \sigma K_h^{-1} i (K_h c_1)^i \quad (5)$$

Here there is a sum of the form $\sum ix^i$ with $x = K_h c_1$ but from $i = 3$. Since $\sum ix^i = x(1-x)^{-2}$, we have

$$\sum_{i \geq 3}^{\infty} ix^i = x(1-x)^{-2} - (x + 2x^2)$$

and hence

$$c_h = \sigma c_1/(1 - K_h c_1)^2 - (\sigma c_1 + 2\sigma K_h c_1^2) \quad (6)$$

Since the total concentration in linear polymers and monomers is given by eqn 25.6, we finally obtain

$$c_0 = c_1/(1 - Kc_1)^2 + \sigma c_1/(1 - K_h c_1)^2 - (\sigma c_1 + 2\sigma K_h c_1^2) \quad (7)$$

Because $\sigma \ll 1$ and $K \ll K_h$, the third term on the right is negligible and the first is almost equal to c_1. Hence

$$c_0 \approx c_1 + \sigma c_1/(1 - K_h c_1)^2$$

concentration c_c similar to the critical micelle concentration defined in section 12.3.3. At concentrations higher than c_c, the excess monomers are converted into polymers and the graph representing the growth of helices is similar to that showing the formation of micelles.

This type of calculation can be carried out for geometrical shapes other than the helix: for example, for the spherical aggregates formed in some multimeric proteins. It can also be applied to two-dimensional layers formed artificially on membranes to create two-dimensional crystals for subsequent examination by electron microscopy.

25.1.3 **Copolymerization**

In many cases, proteins polymerize from two or more different monomeric molecules. A large number of polymeric proteins are known in which associations of two sub-units α and β are of the $\alpha_m\beta_n$ type. In flagella, for instance, the polymerization of *flagellin* involves different monomeric units, some of which give the flagellum a spiral shape while others keep it linear.

A simple treatment of copolymerization will be adopted here by a restriction to two monomers α and β whose respective association constants K_α and K_β are independent of polymer composition. This is equivalent to assuming that the α and β monomers retain their conformation when incorporated in the polymer.

Using a notation similar to that for a single monomer, let $c_{0\alpha}$ and $c_{0\beta}$ be the initial concentrations of α and β, and $c_{1\alpha}$ and $c_{1\beta}$ be the free monomer concentrations at any later time. It is then found (see Box 25.2) that

$$c_{0\alpha} = c_{1\alpha} + Ac_{1\alpha}K_\alpha/[1 - (c_{1\alpha}K_\alpha + c_{1\beta}K_\beta)]^2 \tag{25.9a}$$

$$c_{0\beta} = c_{1\beta} + Ac_{1\beta}K_\beta/[1 - (c_{1\alpha}K_\alpha + c_{1\beta}K_\beta)]^2 \tag{25.9b}$$

The critical condition for copolymerization is now

$$c_{1\alpha}K_\alpha + c_{1\beta}K_\beta = 1 \tag{25.10}$$

This is a generalization of the equation established above for helical polymerization ($c_1K_h = 1$). The examples chosen are simple but bring out two important characteristics of protein polymerization:

(1) the *cooperative aspect*, with nucleation centres and a critical concentration;

(2) the *dynamic aspect*, with a polymer continuously growing and vanishing and with fluctuations occurring in the monomer and polymer concentrations.

In spite of the great variety of binding energies between monomers, one essential feature emerges from all these processes: the role of *entropy*. If the free-energy change favours the formation of a polymer, it is almost entirely due to the corresponding increase in entropy ($\Delta S > 0$). The association of two monomers should be viewed as replacing an area of contact with the solvent by an area of contact between two proteins. Ions and water molecules are released during the binding process, and the increase in their degrees of freedom prevails over the ordering in the polymer to give a positive ΔS.

To go any further in studying the mechanism, a knowledge of the detailed structure of the monomer and its assembly is required. The only structural studies

Box 25.2 **Growth of a copolymer**

The concentration c_{lm} of a polymer consisting of l units of α and m units of β is

$$c_{lm} = Ac_m^i(c_{1\alpha}K_\alpha)^l(c_{1\beta}K_\beta)^m \quad (1)$$

where $l + m = i$, $c_m{}^i$ is the number of possible arrangements of the monomers (1 among i) and A is a constant that takes into account the nucleation mechanism.

The total concentration c_i of polymers is

$$c_i = \sum_{l+m=i} c_{lm}$$

i.e. from the binomial expansion:

$$c_i = A(c_{1\alpha}K_\alpha + c_{1\beta}K_\beta)^i \quad (2)$$

Since AK_α and AK_β are negligibly small compared with 1, the mean fractions $\langle l\rangle/i$ and $\langle m\rangle/i$ of α and β respectively in the i-mers are given by

$$\langle l\rangle/i = (c_{1\alpha}K_\alpha)/(c_{1\alpha}K_\alpha + c_{1\beta}K_\beta) \quad (3a)$$

$$\langle m\rangle/i = (c_{1\beta}K_\beta)/(c_{1\alpha}K_\alpha + c_{1\beta}K_\beta) \quad (3b)$$

The total amount of monomer in the polymer is

$$c_h = \sum ic_i = A\sum i(c_{1\alpha}K_\alpha + c_{1\beta}K_\beta)^i$$

or

$$c_h = A(c_{1\alpha}K_\alpha + c_{1\beta}K_\beta)[1 - (c_{1\alpha}K_\alpha + c_{1\beta}K_\beta)]^{-2}$$

In copolymers, there are $c_h\langle l\rangle/i$ α-monomers and $c_h\langle m\rangle/i$ β-monomers. The conservation laws are thus

$$c_{0\alpha} = c_{1\alpha} + c_h\langle l\rangle/i \qquad \text{and} \qquad c_{0\beta} = c_{1\beta} + c_h\langle m\rangle/i$$

or

$$c_{0\alpha} = c_{1\alpha} + Ac_{1\alpha}K_\alpha/[1 - (c_{1\alpha}K_\alpha + c_{1\beta}K_\beta)]^2 \quad (4a)$$

$$c_{0\beta} = c_{1\beta} + Ac_{1\beta}K_\beta/[1 - (c_{1\alpha}K_\alpha + c_{1\beta}K_\beta)]^2 \quad (4b)$$

and the critical condition for copolymerization becomes

$$c_{1\alpha}K_\alpha + c_{1\beta}K_\beta = 1 \quad (5)$$

carried out so far have dealt with:

(1) the assembly of protein molecules in tobacco mosaic virus (TMV) either alone or in the presence of viral RNA around which they form a helical jacket (Namba and Stubbs, 1986);

(2) the assembly of molecules of RecA, a protein involved in the recombination process but also in releasing the repression of the SOS system in bacteria.

In this latter case, it is the helical assembly of monomeric proteins around a single-stranded DNA fragment that is the essential process. It is accompanied by a change in the conformation of RecA, which becomes a protease capable of cleaving the repressor LexA common to all the operators of genes in the SOS system (Story *et al.*, 1992).

25.2 The nucleus and chromatin

The problem of packing DNA in the nucleus of a cell can be stated very simply: a thread with a length of 1.5 m and a diameter of 2 nm has to be accommodated in a 'box' with a volume of a few μm^3. Such compaction is achieved in several steps that we now attempt to analyse. In the first step, the DNA is wound around a protein core to form a *nucleosome* (Oudet *et al.*, 1975). This is a sub-unit of chromatin, which can be defined from an operational point of view as the result of a partial digestion of nuclear DNA by an endonuclease such as micrococcal nuclease. The size of the DNA fragment per sub-unit depends on the type of eukaryote, on the tissue and on the organ in question. It is generally in the range 160–220 bp. Another sub-unit can be defined either by a partial digestion or by a precise three-dimensional structure: this is the *core particle* (CP), considered as the elementary structural unit of chromatin. As the first neutron scattering experiments showed (Pardon *et al.*, 1975), the DNA is outside the CP and consists of a 146 bp fragment of the B form wrapped around a core of four histone pairs. The properties of the core particle are listed in Table 25.2.

The composition and structure of the H3 and H4 histones have hardly changed during the evolutionary process, suggesting that they play a vital structural role. A whole series of studies has indeed shown that the $(H3–H4)_2$ tetramer determines the position of the CP on the DNA, its stability and almost its geometry (Kornberg, 1977).

The first studies of CP crystals by X-ray diffraction (Richmond *et al.* 1984; at 7 Å resolution) showed that the histone octamer can be viewed as a kind of spool around which about 1.7 supercoils of double-stranded DNA are wrapped in a left-handed superhelix. The histones are relatively well structured inside the octamer, with about 50 per cent α-helices and few or no β-sheets. The DNA remains in the B form. The histone core is a distorted wedge-shaped disc with a diameter of 70 Å and a thickness of 55 Å.

The nucleosome as a whole has a symmetry axis formed by the axis of the DNA superhelix and a perpendicular two-fold axis. The three-dimensional structure of

Table 25.2 Properties of the core particle DNA fragment – length = 146 bp or 496 Å, mass = 96 360 Da Protein core – octamer of histones, two each of the following four:

Histone	Mass (Da)	Net positive charge*
H4	11 200	19 (17)
H3	15 300	20 (18)
H2A	14 000	21 (17)
H2B	13 800	22 (19)
Mass of the octamer	108 600 Da	
Total mass of core particle	204 960 Da	
Total positive charge		164 (142)†

*The bracketed number is for uncharged histidines. †Roughly half the negative charge of DNA.

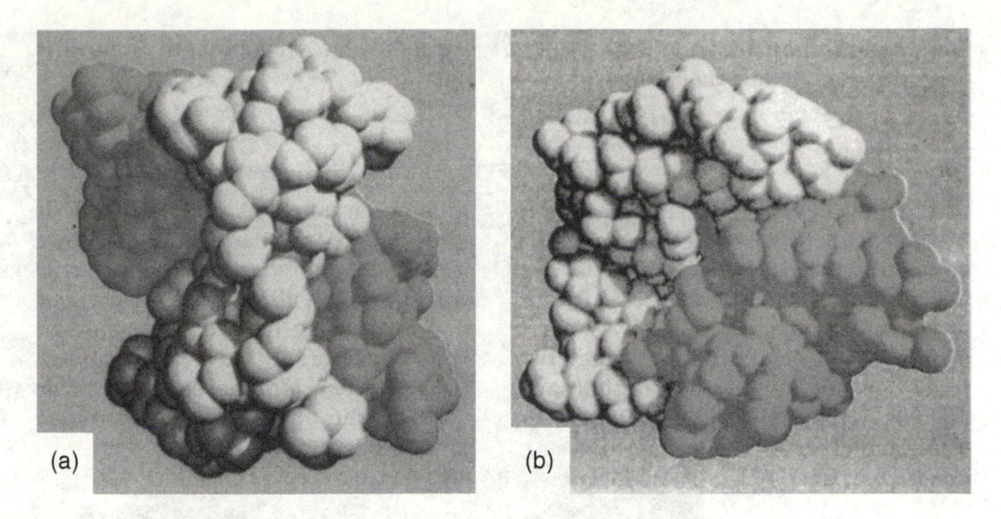

Figure 25.1 Structure of the histone octamer. (a) View along the two-fold axis: the axis of the superhelix is horizontal. (b) View along the axis of the superhelix with the two-fold axis horizontal. The octamer appears as a disc. View (b) is rotated through 90° with respect to (a), the rotation axis lying in the plane of the sheet. (From Arents *et al.*, 1991. Reprinted from *Proceedings of the National Academy of Sciences, USA* with the authors' permission.)

the octamer alone was later determined by X-ray diffraction at a resolution of 3.1 Å (Arents *et al.*, 1991; Arents and Moudrianakis, 1993). It is a tripartite assembly of the tetramer H3–H4 and the two dimers H2A–H2B forming a kind of trapezoidal wedge. The area of contact between the tetramer and the two dimers is particularly large because of the way they are assembled, a little like a handshake (Fig. 25.1).

The complete structure of the CP has now been obtained at 2.8 Å resolution by X-ray diffraction (Luger *et al.*, 1997). In previous structural studies, the DNA sequence in the CPs varied from one sample to another, and the histones also showed varying degrees of acetylation and phosphorylation, which meant that the material being used was not homogeneous and the resolution was therefore low. The much higher resolution of the more recent work is explained by the use of CPs artificially reconstructed (and hence absolutely identical) from a DNA fragment with a palindromic sequence (to comply with the two-fold pseudo-symmetry of the CP) and from recombinant histones expressed in bacteria and lacking post-translational modifications. The overall features of the whole assembly are confirmed: symmetry around the superhelix axis and a perpendicular two-fold axis, a superhelix with a diameter of about 42 Å and an irregular curvature, and histones highly ordered into α-helices. Without entering into the details of the structure, it is worth mentioning some of the results (see Figs 25.2 and 25.3).

1. A structural motif (a histone fold), formed from three α-helices joined by two loops as in α_1–L–α_2–L–α_3, is common to the four histones. The motif can be seen in the figure, α_2 being the longest helix in every case and the two short helices α_1 and α_3 being joined to α_2 by curved threads representing the loops L. It can be considered (Arents and Moudrianakis, 1993) as arising from a 'tandem' replication of the gene coding for the HTH motif, which we saw previously as one of the elements in protein–DNA recognition. The

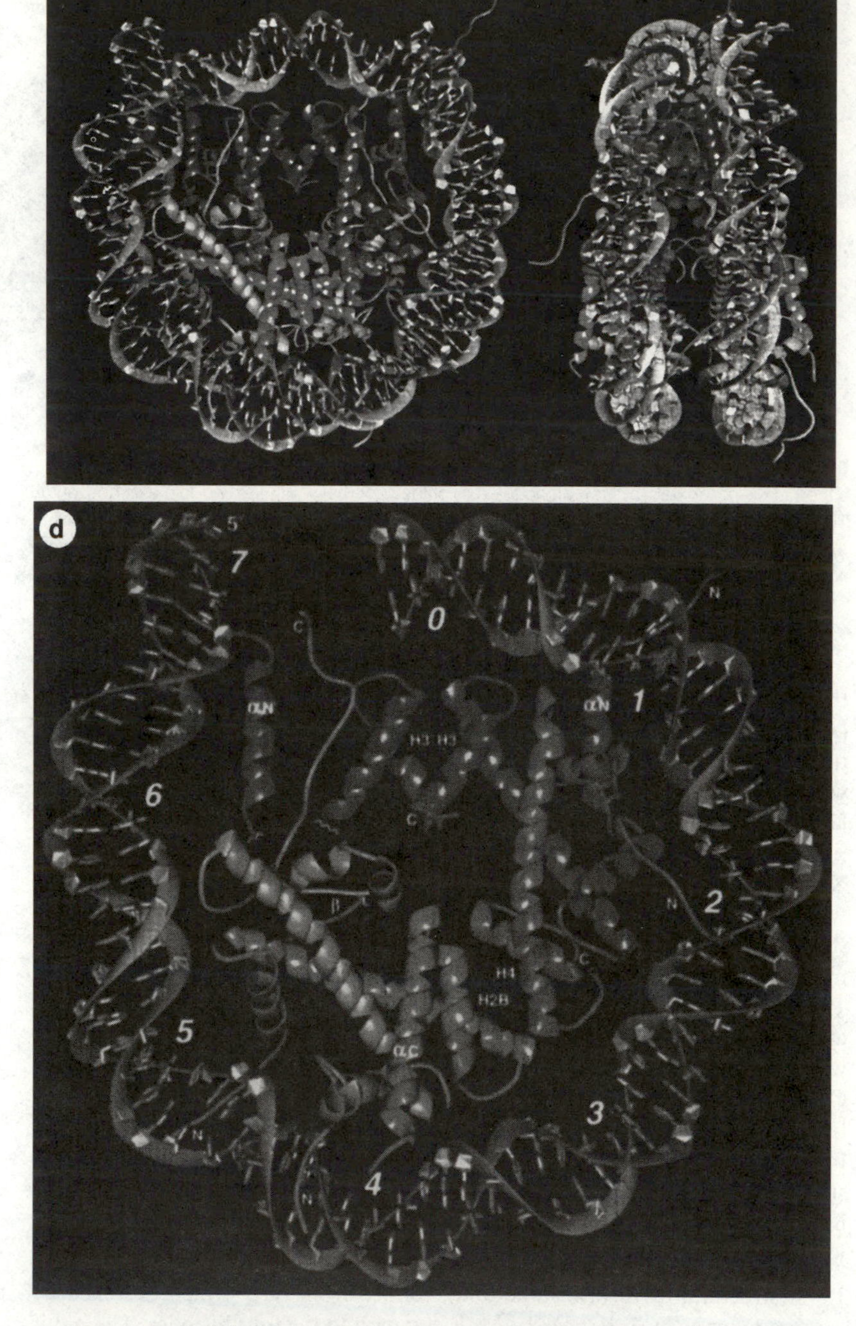

Figure 25.2 Two views of the core particle: on the left, along the superhelix axis; on the right, along a perpendicular axis. In both cases, the two-fold pseudo-symmetry axis is vertical. (From Luger *et al.*, 1997. Reprinted with permission from *Nature*, vol. 389, pp. 251–60 and with the authors' permission. © 1997 Macmillan Magazines Ltd.)

Figure 25.3 View of half the core particle (73 bp) with the same orientation as the left-hand diagram in Fig. 25.2. The number 0 indicates the base pair through which the pseudo-symmetry axis passes. The other numbers correspond to complete turns of the double helix. The motif $\alpha_1 - L_1 - \alpha_2 - L_2 - \alpha_3$ is indicated for each histone, α_2 being the longest helix in every case. (From Luger *et al.*, 1997. Reprinted with permission from *Nature*, vol. 389, pp. 251–60 and with the authors' permission. © 1997 Macmillan Magazines Ltd.)

N-terminal of the histone helices (i.e. the positive charge of the helix dipole) interacts with the negative charge of a phosphate group.

2. Coulomb interactions also arise through the insertion of the side chains of 14 arginines into the minor groove of the DNA. Hydrogen bonds and van der Waals contacts complete the interaction forces.

3. The N-terminal 'tails' of the H3 and H2B histones have no special structure and pass through the superhelix. This position outside the structured core favours the acetylation of lysines.
4. Despite the even number of nucleotides, the symmetry axis passes through a base pair, leaving 72 and 73 base pairs respectively on the two sides.
5. There are only 1.65 supercoils because 10 nucleotides at the two ends do not form part of the superhelix.
6. The mean twist angle of the DNA is 35.3°, or 10.2 bp per turn. This is greater than that of DNA alone in solution (34.3° or 10.5 bp per turn) and reflects the constraint exercised by the histone core.

Although this high-resolution structure for the CP of the nucleosome was determined on crystals of a reconstituted particle, it is very probable that a 'native' CP would have a very similar, if not identical, structure. However, the acetylation and phosphorylation of the histones may introduce large conformational changes in the N-terminal regions of H3 and H4 and may interfere with the interactions between nucleosomes, i.e. in the successive levels of order in chromatin.

The chain of nucleosomes is in fact ordered into a helix to give a fibre of diameter 300 Å, followed by successive levels of folding, ending up with a degree of compaction for the DNA of the order of 10^5 to 10^6 (although this structure has essentially only the status of a model at present).

To condense chromatin into chromosomes during cell division requires actin-like proteins to form a kind of 'scaffold', which helps to define the shape of the chromosome. We do not yet know the mechanisms underlying all these processes of organization, condensation and decondensation.

The situation is much simpler in bacterial chromosomes. Proteins with compositions different from histones are involved in the folding of DNA, its supercoiling and its compaction in the bacteria. In the case of the HU protein, for example, interactions between protein molecules *already bound to DNA* have been shown to impose the overall conformation of the bacterial chromosome.

25.3 **Viruses**

Since the famous experiment in which the tobacco mosaic virus (TMV) was reconstituted in a test tube, great advances have occurred in our knowledge of virus structure. In these nucleoprotein assemblies and unlike the nucleosome, the nucleic acid (RNA or DNA) is enclosed in a protein shell called a *capsid*. This is formed by assembling small protein sub-units. As in the case of nucleosome, therefore, it is the assembly of proteins that determines the morphology of the virus. In TMV, a cylindrical shape is obtained by the assembly of slightly distorted discs with a complicated RNA threading process (Lebeurier *et al*., 1977).

Spherical viruses have a protein shell formed by the assembly of a large number of identical sub-units completely enveloping the folded nucleic acid inside. The first electron microscope images obtained suggested an icosahedral shape, and this formed the basis of a theory for morphogenesis (Caspar and Klug, 1962). A regular icosahedron is a solid with cubic symmetry whose 20 faces are equilateral triangles the bases of which form convex regular pentagons. A maximum of three protein sub-units can be accommodated in each face if the equivalence of the faces is to be maintained, i.e. if there is the same type of contact between all the sub-units. Since

Table 25.3 Parameters of the pseudo-equivalent system and corresponding viruses

h	k	P	f	T	Virus*
0	1	1	1	1	ΦX174
			2	4	Nωv
			3	9	
			4	16	
			5	25	Adenovirus
1	1	3	1	3	CCMV and TYMV
			2	12	
1	2	7	1	7	HK97, SV40, papovavirus
1	3	13	1	13	Rotavirus
2	3	19	1	19	
1	4	21	1	21	

*Nωv, *Nudaurelia capensis* omega virus; CCMV, cowpea chlorotic mottle virus; HK97, Hong Kong 97 bacteriophage; ΦX174, bacteriophage; TYMV, turnip yellow mosaic virus; SV40, simian virus.

many viruses are formed by the assembly of a much larger number of protein subunits, complete equivalence cannot be sustained and is replaced by a *quasi-equivalence*, i.e. the virus is constructed from a flat lattice of triangles forming an *icosadeltahedron* having $20T$ faces, T being a *triangulation number* defined by

$$T = (h^2 + hk + k^2)f^2 = Pf^2$$

where f is any integer, and h and k are a pair of integers with no common factor. Table 25.3 lists the results obtained for various values of f, h and k.

The flat lattice can only be spatially folded by introducing a combination of $10(T-1)$ hexamers and 12 pentamers on the surface of the icosahedron. As an example, consider the particularly well-investigated case of CCMV (cowpea chlorotic mottle virus) (Johnson and Speir, 1997). The whole system involves 84 dimers and one dodecamer, or 180 asymmetrical units (nine units per face, i.e. $T = 3$) assembled into 12 pentamers and 20 hexamers. The contacts between the sub-units in a pentamer and in a hexamer are no longer exactly the same: there is only quasi-equivalence. This viral structure is reminiscent of the geodesic domes built by Buckminster Fuller. The idea put forward by Caspar and Klug (1962) is now, over 30 years later, generally accepted: specific interactions between atomic groups have been found in several cases of protein sub-units whose three-dimensional structures have been determined at high resolution (Johnson and Speir, 1997). Moreover, quasi-equivalence is explained on the molecular scale by the structural reorganization of a small part of the protein.

The DNA or RNA filament can be packed in phage heads. Except in TMV, where the charges in the cylindrical capsid force the RNA molecule to adopt a single-stranded helical structure, polyamines are always necessary for the tight packing of a highly negatively charged filament. The loss of entropy due to the ordering of the nucleic acid is largely compensated by the gain in entropy from the release of ions and water molecules during the assembly process.

In these varied examples (nucleosome and viruses), there is no unique scheme for assembling proteins but just two general principles:

1. It is the protein–protein interactions that determine the geometry of the system, to which the configuration of the nucleic acid must conform.

2. In all cases, the negative change in free energy in favour of the assembling process comes almost entirely from the positive entropy change.

25.4 **Ribosome**

A better understanding of the structure and morphogenesis of the ribosome must await its complete structure determination or at least that of one sub-unit. Nevertheless, some of the topological experiments using a physical approach (neutron scattering after selective deuteration of ribosomal proteins) or a chemical approach (use of bridging reagents) have clearly shown that in this case the tertiary structure of ribosomal RNA is important. It provides specific binding sites to ribosomal proteins, which then, *once bound to the RNA*, are able to interact and complete the morphogenesis of the ribosome (see section 4.7).

The principles stated in section 25.3 remain valid if the tertiary structure of ribosomal RNA is assumed to be as well determined as that of a protein, at least in regions outside the loops.

Part V Association between molecules

Conclusion

The final Part of this book has provided an outline of what may be called the second level in our description of the living world, that of the assembly of macromolecules to form functional structures present in the cell. The forces acting are the same as those involved in building polymers but the specificity of interactions, as found for example between proteins and nucleic acids, is not yet fully understood. A specific association may well be *observed* but it cannot be *predicted* from our knowledge of the structure and dynamics of the reactants. Recognition processes may be described but no rules of assembly can yet be stated. Some concepts such as the complementarity between two surfaces would be worth a more thorough study through modelling.

The biological order evident to any observer of the living world is not the result of assembling identical units as in a crystal, neither is it comparable to the assembly of protons and neutrons in an atomic nucleus or to that of the nuclei and electrons in an atom. More relevant is the process by which different types of atom are assembled in a molecule, where the decisive role and specificity of each atom can be recognized. In this case the assembling forces are simply the valence forces, which obey the laws of quantum mechanics.

The assembly of macromolecules is a much more subtle process, which at the same time is much richer in possibilities because both the building blocks and the forces involved come in such a large variety. Since the list of these forces is still the same as that used in the building of biopolymers, the great diversity of structures results mainly from the high specificity and the large range of macromolecular structure and dynamics. For example, although it is possible to define a certain charge configuration on the surface of two macromolecules, the first event in their mutual recognition will be the spatial matching of their charge distributions due to the long-range nature of Coulomb forces.

If it is not yet possible to state any rules for building a ribosome, a virus or a chromosome, how much more complex a problem it is when it is a question of understanding polyenzymatic systems such as those found in the replication, transcription and repair of DNA. Here, it is no longer a matter of the 'simple' building of a functional structure, but rather the regulation in space and time by means of reversible associations of a relatively rapid process (on the scale of milliseconds) that only tolerates a very small number of errors.

A similar problem exists in understanding the operation of the protein synthesis 'factory' that produces ribosomes. Even at this relatively simple level of cell functioning, a *systems* approach proves to be essential. In tackling this fascinating new area of biophysics, new models and new concepts will have to be developed alongside new experimental approaches.

Part V Association between molecules

References and further reading

Aggarwal, A.K., Rodgers, D.W., Drottar, M., Ptashne, M. and Harrison, S.C. (1988). Recognition of a DNA operator by the repressor of phage 434: a view at high resolution. *Science*, **242**, 899–907.

Alberts, B., Bray, D., Lewis, J., Raff., M., Roberts, K. and Watson, J.D. (1988). *Biologie Moléculaire de la Cellule*. Flammarion, Paris.

Anderson, K.S., Miles, E.W. and Johnson, K.A. (1991). Serine modulates substrate channeling in tryptophan synthase. *Journal of Biological Chemistry*, **266**, 8020–33.

Anderson, W.F., Ohlendorf, D.H., Takeda, Y. and Matthews, B.W. (1981). Structure of the cro repressor from bacteriophage λ and its interaction with DNA. *Nature*, **290**, 754–8.

Arents, G. and Moudrianakis, E.N. (1993). Topography of the histone octamer surface: repeating structural motifs utilized in the docking of nucleosomal DNA. *Proceedings of the National Academy of Sciences, USA*, **90**, 10489–93.

Arents, G., Burlinghame, R.W., Wang, B.C., Love, W.E. and Moudrianakis, E.N. (1991). The nucleosomal core histone octamer at 3.1 Å resolution: a tripartite protein assembly and a left-handed superhelix. *Proceedings of the National Academy of Sciences, USA*, **88**, 10148–52.

Arnold, E., Ding, J., Hughes, H.S. and Hostomsky, Z. (1995). Structures of DNA and RNA polymerases and their interactions with nucleic acid substrates. *Current Opinion in Structural Biology*, **5**, 27–38.

Berg, O.G. and Blomberg, C. (1977). Association kinetics with coupled diffusion. An extension to coiled-chain macromolecules applied to the *lac* repressor–operator system. *Biophysical Chemistry*, **7**, 33–9.

Berg, O.G. and von Hippel, P.H. (1985). Diffusion-controlled macromolecular interactions. *Annual Review of Biophysics and Biophysical Chemistry*, **14**, 131–60.

Berg, O.G. and von Hippel, P.H. (1987). Selection of DNA binding sites by regulatory proteins. Statistical mechanical theory and application to operators and promoters. *Journal of Molecular Biology*, **193**, 723–50.

Berg, O.G., Winter, R.B. and von Hippel, P.H. (1981). Diffusion-driven mechanisms of protein translocation on nucleic acids. (1) Models and theory. *Biochemistry*, **20**, 6929–48.

Brennan, R.G., Roderick, S.L., Takeda, Y. and Matthews, B.W. (1990). Protein–DNA conformational changes in the crystal structure of a λ cro-operator complex. *Proceedings of the National Academy of Sciences, USA*, **87**, 8165–9.

Burley, S.K. (1994). DNA binding motifs from eukaryote transcription factors. *Current Opinion in Structural Biology*, **4**, 3–11.

Calnan, B.J., Tidor, B., Biancalana, S., Hudson, D. and Frankel, A.D. (1991). Arginine-mediated RNA recognition: the arginine fork. *Science*, **252**, 1167–71.

Caspar, D.L.C. and Klug, A. (1962). Physical principles in the construction of regular viruses. *Cold Spring Harbor Symposia on Quantitative Biology*, **27**, 1–24.

Choo, Y and Klug, A. (1997). Physical basis of a protein–DNA recognition code. *Current Opinion in Structural Biology*, **7**, 117–25.

del Cardayré, S.B. and Raines, R.T. (1994). Structural determinants of enzymatic processivity. *Biochemistry*, **33**, 6031–7.

Dowd, D.R. and Lloyd, R.S. (1990). Biological significance of facilitated diffusion in protein–DNA interactions. *Journal of Biological Chemistry*, **265**, 3424–31.

Edsall, J.T. and Wyman, J. (1958). *Biophysical Chemistry*, vol. I, p. 263. Academic Press, New York.

Ellenberger. T. (1994). Getting a grip on DNA recognition: structure of the basic region leucine zipper and the basic region helix–loop–helix DNA binding domains. *Current Opinion in Structural Biology*, **4**, 12–21.

Freemont, P.S., Lane, A.N. and Sanderson, M.R. (1991). Structural aspects of protein–DNA recognition. *Biochemical Journal*, **278**, 1–23.

Harrison, S.C. and Aggarwal, A.K. (1990). DNA recognition by proteins with the helix–turn–helix motif. *Annual Review of Biochemistry*, **59**, 933–69.

Hu, J.C., O'Shea, E.K., Kim, P.S. and Sauer, R.T. (1990). Sequence requirements for coiled-coils: analysis with λ repressor–GCN4 leucine zipper fusions. *Science*, **250**, 1400–3.

Hyde, C.C., Ahmed, S.A., Padlan, E.A., Miles, E.W. and Davies, D.R. (1988). Three-dimensional structure of the tryptophan synthase α,β,multienzyme complex from *Salmonella typhimurium*. *Journal of Biological Chemistry*, **263**, 17857–71.

Johnson, J.E. and Speir, J.A. (1997). Quasi-equivalent viruses: a paradigm for protein assemblies. *Journal of Molecular Biology*, **269**, 665–75.

Jordan, S.R. and Pabo, C.O. (1988). Structure of the lambda complex at 2.5 Å resolution. Details of the repressor–operator interactions. *Science*, **242**, 893–9.

Klug, A. and Rhodes, D. (1987). Zinc fingers: a novel protein motif for nucleic acid recognition. *Trends in Biochemical Science*, **12**, 464–9.

Knighton, D.R., Kan, C.C., Howland, E., Janson, C.A., Hostomsky, Z., Welsch, K.M. and Matthews, D.A. (1994). Structure of and kinetic channelling in bifunctional dihydrofolate reductase–thymidylate synthase. *Nature Structural Biology*, **1**, 186–94.

Kong, X.-P., Onrust R., O'Donnell, M. and Kuriyan, J. (1992). Three-dimensional structure of *E. coli* DNA polymerase III holoenzyme: a sliding DNA clamp. *Cell*, **69**, 425–37.

Kornberg, R. (1977). Structure of chromatin. *Annual Review of Biochemistry*, **46**, 931–54.

Kumar, N.V. and Govil, G. (1984). Theoretical studies on protein–nucleic acid interactions. *Biopolymers*, **23**, 1979–2024.

Lebeurier, G., Nicolaieff, A. and Richards, K.E. (1977). Inside-out model for self-assembly of tobacco mosaic virus. *Proceedings of the National Academy of Sciences, USA*, **74**, 149–153.

Lohman, T.M. (1985). Kinetics of protein–nucleic acids interactions: use of salt effects to probe mechanism of interaction. *CRC Critical Review of Biochemistry*, **19**, 191–245.

Luger, K., Mäder, A.W., Richmond, R.K., Sargent, D.F. and Richmond, T.J. (1997). Crystal structure of the nucleosome core particle at 2.8 Å resolution. *Nature*, **389**, 251–60.

Luisi, B.F. and Sigler, P.B. (1990). The stereochemistry and biochemistry of the trp repressor–operator complex. *Biochimica et Biophysica Acta*, **1048**, 113–26.

Luisi, B.F., Xu, W.X., Otwinowski, Z., Freedman, L.P., Yamamoto, K.R. and Sigler, P.B. (1991). Crystallographic analysis of the interaction of the glucocorticoid receptor with DNA. *Nature*, **352**, 497–505.

McGhee, J.D. and von Hippel, P.H. (1974). Theoretical aspects of DNA–protein interactions: cooperative and non-cooperative binding of large ligands to a one-dimensional homogeneous lattice. *Journal of Molecular Biology*, **86**, 469–89.

Matthews, B.W. (1988). No code for recognition. *Nature*, **335**, 294–5.

Miller, J., McLachlan, A.D. and Klug, A. (1985). Repetitive zinc-binding domains in the protein transcription factor IIIA from *Xenopus* oocytes. *The EMBO Journal*, **4**, 1609–14.

Namba, K. and Stubbs, G. (1986). Structure of TMV at 3.6 Å resolution: implications for assembly. *Science*, **231**, 1401–6.

Omichinski, J.G., Clore, G.M., Appella, E., Sakaguchi, K. and Gronenborn, A.M. (1990). High-resolution three-dimensional structure of a single zinc finger from a human enhancer binding protein in solution. *Biochemistry*, **29**, 9324–34.

Oosawa, F. and Asakura, S. (1975). *Thermodynamics of the polymerization of proteins*. Academic Press, New York.

O'Shea, E.K., Klemm, J.D., Kim, P.S. and Alber, T. (1991). X-ray structure of the GCN4 leucine zipper, a two-stranded parallel coiled-coil. *Science*, **254**, 539–44.

Oudet, P., Gross-Bellard, M. and Chambon, P. (1975). Electron microscopy and biochemical evidence that chromatin structure is a repeating unit. *Cell*, **4**, 281–300.

Pardon, J.P., Worcester, D.L., Wooley, J.C., Tatchett, K., Van Holde, K.B. and Richards, B.M. (1975). Low-angle neutron scattering from chromatin subunit particle. *Nucleic Acids Research*, **2**, 2163–76.

Pelletier, H., Sawaya, M.R., Wolfle, W., Wilson, S.H. and Kraut, J., (1996). Crystal structure of human DNA polymerase β complexed with DNA: implications for catalytic mechanism, processivity and fidelity. *Biochemistry*, **35**, 12742–61.

Raumann, R.E., Brown, B.M. and Sauer, R. (1994). Major groove DNA recognition by β-sheets: the ribbon–helix–helix family of gene regulatory proteins. *Current Opinion in Structural Biology*, **4**, 36–43.

Richmond, T.J., Finch, J.T., Rushton, B., Rhodes, D. and Klug, A. (1984). Structure of the nucleosome core particle at 7 Å resolution. *Nature*, **311**, 532–7.

Schultz, S.C., Shields, G.C. and Steitz, T.A. (1991). Crystal structure of a CAP–DNA complex: the DNA is bent by 90°. *Science*, **253**, 1001–7.

Seeman, N.C., Rosenberg, J.M. and Rich, A. (1976). Sequence-specific recognition of double helical nucleic acids by proteins. *Proceedings of the National Academy of Sciences, USA*, **73**, 804–8.

Sharp, K., Fine, R. and Honig, B. (1987). Computer simulation of the diffusion of a substrate to an active site of an enzyme. *Science*, **236**, 1460–3.

Sines, J., Allison, S.A. and McCammon, J.A. (1990). Point charge distributions and electrostatic steering in enzyme/substrate encounter: Brownian dynamics of modified copper/zinc superoxide dismutases. *Biochemistry*, **29**, 9403–12.

Story, R.M., Weber, I.T. and Steitz, T.A. (1992). The structure of the *E. coli* RecA protein monomer and polymer. *Nature*, **355**, 318–25.

Strynadka, N.J.C. and James, M.N.G. (1989). Crystal structure of the helix–loop–helix calcium-binding proteins. *Annual Review of Biochemistry*, **58**, 951–81.

Stukenberg, P.T., Studwell-Vaughan, P.S. and O'Donnell, M. (1991). Mechanism of the sliding β-clamp of DNA polymerase III holoenzyme. *Journal of Biological Chemistry*, **266**, 328–34.

van Holde, K.E. (1988). *Chromatin*. Springer-Verlag, New York.

von Hippel, P.H. and Berg, O.G. (1986). On the specificity of DNA–protein interactions. *Proceedings of the National Academy of Sciences, USA*, **83**, 1608–12.

Winter, R.B. and von Hippel, P.H. (1981). Diffusion-driven mechanisms of protein translocation on nucleic acids. (2) The *E. coli* repressor–operator interactions: equilibrium measurements. *Biochemistry*, **20**, 6948–60.

Winter, R.B., Berg, O.G. and von Hippel, P.H. (1981). Diffusion-driven mechanisms of protein translocation on nucleic acids. (3) The *E. coli* repressor–operator interactions: kinetic measurements and conclusions. *Biochemistry*, **20**, 6961–77.

Mathematical appendix

A.1 Evaluation of some integrals

Integrals of the type $I(n) = \int_0^\infty e^{-x} x^n \, dx$ (n **a positive integer**)

Integration by parts is used:

$$\int_0^\infty e^{-x} x^n \, dx = -[x^n e^{-x}]_0^\infty + n \int_0^\infty e^{-x} x^{n-1} \, dx$$

The first term on the right is zero, so that the following recursion formula is obtained:

$$I(n) = nI(n-1)$$
$$I(n-1) = (n-1)I(n-2)$$
$$\dots$$
$$I(1) = I(0)$$

But $I(0) = \int_0^\infty e^{-x} \, dx = [-e^{-x}]_0^\infty = 1$, and hence

$$I(n) = n!$$

Integrals of the type $J(n) = \int_0^\infty \exp(-ax^2) x^n \, dx$ (n **a positive integer**)

A recursion method can also be used here by noting that $J(n)$ is the negative derivative with respect to a of $J(n-2)$:

$$\partial J(n-2)/\partial a = \int_0^\infty -x^2 \exp(-ax^2) x^{n-2} \, dx = -J(n)$$

so that

$$J(n) = -\partial J(n-2)/\partial a$$

and hence

$$J(n-2) = -\partial J(n-4)/\partial a$$

$$\dots$$

$$J(2) = -\partial J(0)/\partial a$$

To calculate $J(0) = \int_0^\infty \exp(-ax^2)\,dx$, put $u = xa^{1/2}$ so that $dx = a^{-1/2}\,du$ and $J(0) = a^{-1/2}\int_0^\infty \exp(-u^2)\,du$. The value of the latter integral is $\pi^{1/2}$, and hence

$$J(0) = (\pi/a)^{1/2}/2$$

and

$$J(2) = (\pi/a^3)^{1/2}/4$$

In addition,

$$J(1) = \int_0^\infty \exp(-ax^2)x\,dx = -(1/2a)[\exp(-ax^2)]_0^\infty = 1/2a$$

From this we obtain $J(3) = 1/2a^2$ and hence all the other values of $J(n)$ for odd n.

A.2 Legendre polynomials

A.2.1 Definition and notation

Consider the simplest charge distribution with axial symmetry consisting of a charge q located at a point M on the z axis at a distance ρ from the origin. Except for a numerical coefficient $1/4\pi\varepsilon\varepsilon_0$, the potential at a point P is $\Psi_P = q/r'$ where $r' = \text{MP}$, and this can be expressed in terms of ρ, r', $r = \text{OP}$ and θ, the angle between OM and OP:

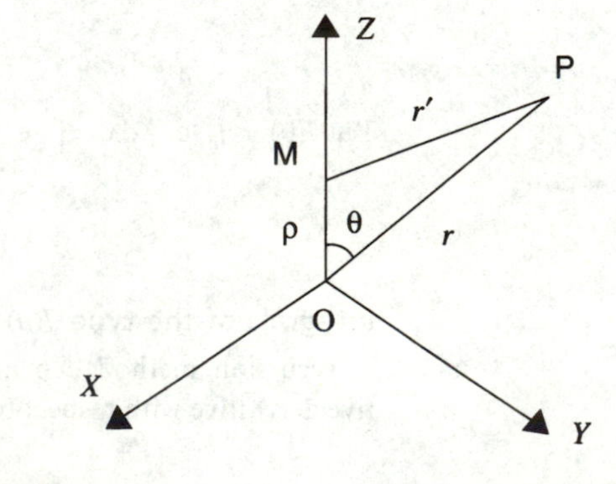

In the triangle OMP

$$r'^2 = r^2 + \rho^2 - 2\rho r\cos\theta$$

and hence

$$r' = r[1 - 2(\rho/r)\cos\theta + \rho^2/r^2]^{1/2}$$

so that

$$\Psi_P = (q/r)[1 - 2(\rho/r)\cos\theta + \rho^2/r^2]^{-1/2}$$

Assuming that $-2(\rho/r)\cos\theta + \rho^2/r^2 < 1$, Ψ_P can be expanded using the binomial theorem. The series converges, and rearranging terms in ascending order of powers of ρ/r gives

$$\Psi_P = (q/r)\sum_{n=0}^{\infty} P_n(\cos\theta)(\rho/r)^n$$

where $P_n(\cos\theta)$ are called Legendre polynomials. The first few terms are easy to calculate:

$$P_0 = 1, \qquad P_1 = \cos\theta, \qquad P_2 = (3\cos^2\theta - 1)/2$$

The rest of the terms are obtained from the recursion relation:

$$(n+1)P_{n+1} - [(2n+1)\cos\theta]P_n + nP_{n-1} = 0$$

and this gives

$$P_3 = (5\cos^3\theta - 3\cos\theta)/2$$
$$P_4 = (35\cos^4\theta - 30\cos^2\theta + 3)/8$$

It can also be shown that for two such polynomials of degrees l and m:

$$\int_{-1}^{+1} P_l(u)P_m(u)\,\mathrm{d}u = 0$$

where $u = \cos\theta$ (this can easily be verified for P_1 to P_4). In other words, Legendre polynomials define a family of orthogonal functions.

A.2.2 Potential due to a dipole

Consider a dipole consisting of two charges $\pm q$ at points M_1 and M_2 which are symmetrically positioned with respect to the origin (i.e. at $\pm\rho$ along the z axis). The potential at a point P is the sum of the potentials Ψ_1 and Ψ_2 at P due to the two charges:

$$4\pi\varepsilon_0\Psi_1 = (q/r)\sum_{n=0}^{\infty} P_n(\cos\theta)(\rho/r)^n$$

$$4\pi\varepsilon_0\Psi_2 = -(q/r)\sum_{n=0}^{\infty} P_n(\cos(\pi-\theta))(\rho/r)^n$$

$$= -(q/r)\sum_{n=0}^{\infty} P_n(-\cos\theta)(\rho/r)^n$$

Since the polynomials of even degree are also even functions, they cancel each other in the sum and what remains is

$$4\pi\varepsilon_0\Psi_P = 2(q/r)[P_1(\cos\theta)\rho/r + P_3(\cos\theta)\rho^3/r^3 + \cdots]$$

If $2q\rho = p$, the dipole moment, and if $\rho \ll r$, only the first term remains in Ψ_P so that

$$4\pi\varepsilon_0\Psi_P = (p\cos\theta)/r^2$$

since $P_1 = \cos\theta$.

A.2.3 Potential due to a quadrupole

In a linear quadrupole there are three charges: $-q$ at both M_1 and M_2 and $+2q$ at the midpoint. If M_1, M_2 and P are the same points as in section A.2.2, the potential at P due to the charge at the origin, Ψ_0, is given by $4\pi\varepsilon_0\Psi_0 = 2q/r$, while the potentials due to the other two charges are the same as Ψ_1 and Ψ_2 in section A.2.2 but are of the same sign. In the sum, only the even terms remain this time, while Ψ_0 cancels the two terms in P_0. Hence

$$4\pi\varepsilon_0\Psi_P = 2(q/r)[P_2(\cos\theta)\rho^2/r^2 + P_4(\cos\theta)\rho^4/r^4 + \cdots]$$

If the quadrupole moment is defined by

$$Q = 2q\rho^2$$

then for $\rho \ll r$,

$$4\pi\varepsilon_0\Psi_P = (Q/r^3)(3\cos^2\theta - 1)$$

Note that for a linear quadrupole of this type, both the total charge and the total dipole moment are zero.

A.2.4 Generalization and spherical functions

We now consider the general case of a charge placed at a point M defined by its spherical coordinates (ρ, θ', ϕ') and the potential Ψ_P it produces at a point P (r, θ, ϕ) expressed in terms of the coordinates of M and P. This is the problem that arises when we require the potential created by a charge distribution over a macromolecule:

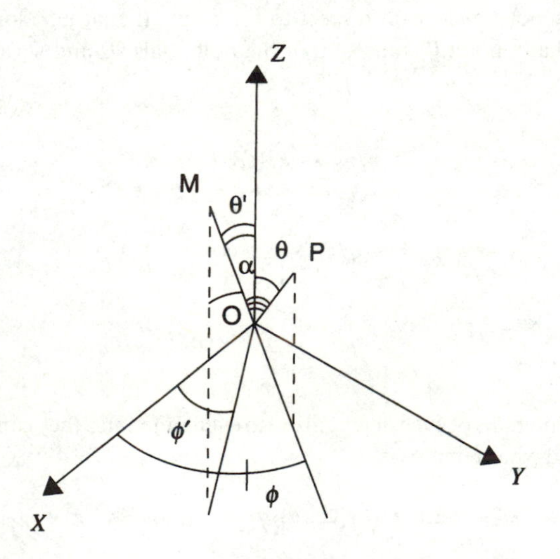

The Cartesian components of the vector displacement OM are $\rho \sin\theta' \cos\phi'$, $\rho \sin\theta' \sin\phi'$ and $\rho\cos\theta'$, while those of OP are $r\sin\theta\cos\phi$, $r\sin\theta\sin\phi$ and $r\cos\theta$. If the angle between OM and OP is α, the scalar product OM · OP is $\rho r\cos\alpha$. Using the identity for a scalar product $\boldsymbol{A}\cdot\boldsymbol{B} = A_xB_x + A_yB_y + A_zB_z$, it follows that

$$\begin{aligned}\cos\alpha &= \sin\theta\sin\theta'(\cos\phi\cos\phi' + \sin\phi\sin\phi') + \cos\theta\cos\theta' \\ &= \sin\theta\sin\theta'\cos(\phi-\phi') + \cos\theta\cos\theta'\end{aligned}$$

As in section A.2.2, the potential Ψ_P can then be expressed in terms of Legendre polynomials in which $\cos\alpha$ is the variable:

$$\Psi_P = (q/r)\sum_{n=0}^{\infty} P_n(\cos\theta\cos\theta' + \sin\theta\sin\theta'\cos(\phi-\phi'))(\rho/r)^n$$

It can be shown that Ψ_P can be put into the following form:

$$\Psi_P = (q/r)\sum_{n=0}^{\infty}\Bigg\{P_n(\cos\theta)P_n(\cos\theta') + \sum_{m=1}^{n}[(n-|m|)!/(n+|m|)!]y_n^m(\theta,\phi)y_n^m(\theta',\phi')\Bigg\}(P/r)^n$$

where the functions y are known as *associated spherical functions* with the following forms (putting $\cos\theta = u$):

$$y_n^m = (1-u^2)^{m/2}\,\mathrm{d}^m/\mathrm{d}u^m[P_n(u)]\mathrm{e}^{\mathrm{i}\phi}$$

so that

$$y_0^0 = 1, \qquad y_1^0 = \cos\theta, \qquad y_1^1 = \sin\theta\,\mathrm{e}^{\mathrm{i}\phi}$$

$$y_2^0 = (3\cos^2\theta - 1)/2, \qquad y_2^1 = 3\sin\theta\cos\theta\,\mathrm{e}^{\mathrm{i}\phi}$$

$$y_2^2 = 3\sin^2\theta\,\mathrm{e}^{2\mathrm{i}\phi}$$

A.3 Bessel functions

Bessel functions are solutions of Bessel's equation, originally introduced in the study of planetary motion:

$$\mathrm{d}^2y/\mathrm{d}x^2 + x^{-1}\,\mathrm{d}y/\mathrm{d}x + (1 - n^2/x^2)y = 0 \qquad \text{(MA.1)}$$

with $n \geq 0$.

Bessel functions occur in many physical problems that have cylindrical symmetry. They are classified according to the value of n into functions of zero order, first order, etc.

A.3.1 Bessel functions (of the first kind) of zero order

These are solutions of

$$\mathrm{d}^2y/\mathrm{d}x^2 + x^{-1}\,\mathrm{d}y/\mathrm{d}x + y = 0 \qquad \text{(MA.2)}$$

with the initial conditions

$$y(0) = 1 \quad \text{and} \quad y'(0) = 0 \tag{MA.3}$$

To find the solution of eqn MA.2, y is replaced by a power series in x:

$$y = a_0 + a_1x + a_2x^2 + \cdots + a_nx^n + \cdots$$

so that

$$\mathrm{d}y/\mathrm{d}x = a_1 + 2a_2x + 3a_3x^2 + \cdots + na_nx^{n-1} + \cdots$$
$$\mathrm{d}^2y/\mathrm{d}x^2 = 2a_2 + 6a_3x + \cdots + n(n-1)a_nx^{n-2} + \cdots$$

Substituting these in eqn MA.2 with the conditions of eqn MA.3 gives $a_0 = 1$, $a_2 = -1/4$ and shows that the odd terms are zero. Therefore y is an alternating series

$$y(x) = 1 - x^2/2^2 + x^4/2^4(2!)^2 - x^6/2^6(3!)^2 + \cdots + (-1)^n x^{2n}/2^{2n}(n!)^2 + \cdots \tag{MA.4}$$

This series defines the Bessel function of zero order $J_0(x)$. The ratio of the $(n+1)$th term to the nth term is $-x^2/(2n)^2$, which tends to zero as n tends to infinity. The series therefore converges and $J_0(x)$ and its derivatives are continuous functions.

If the two functions

$$J_0(x) = \sum_{k=0}^{\infty}(-1)^k x^{2k}/2k(k!)^2$$

$$\cos x = \sum_{k=0}^{\infty}(-1)^k x^{2k}/(2k!)$$

are compared, it can be seen that $J_0(x)$ is an oscillating but non-periodic function of x with a decreasing amplitude. As $x \to \infty$, $J_0(x)$ tends asymptotically to

$$(2/\pi x)^{1/2}\cos[x - (\pi/4)] \tag{MA.5}$$

A.3.2 Bessel functions (of the first kind) of first order

These are solutions of

$$\mathrm{d}^2y/\mathrm{d}x^2 + x^{-1}\,\mathrm{d}y/\mathrm{d}x + (1 - x^{-2})y = 0 \tag{MA.6}$$

with the initial conditions $y(0) = 0$ and $y'(0) = \frac{1}{2}$. Using the same method as in solving eqn MA.2, we find

$$J_1(x) = x/2 - x^3/2^2 4 + x^5/2^2 4^2 6 + \cdots = \sum_{k=0}^{\infty}(-1)^k x^{2k+1}/2^{2k+1}k!(k+1)! \tag{MA.7}$$

We now compare this with

$$\sin x = \sum_{k=0}^{\infty}(-1)^k x^{2k+1}/(2k+1)!$$

and see that $J_1(x)$ behaves like a sine function with a decreasing amplitude. Just as $\sin x = -\mathrm{d}(\cos x)/\mathrm{d}x$, we have

$$J_1(x) = -\,\mathrm{d}J_0(x)/\mathrm{d}x \qquad \text{(MA.8)}$$

A.3.3 Bessel functions of the second kind or Neumann functions

Function $J_0(x)$ is only one possible solution to eqn MA.2. In fact, as with any second-order differential equation, there must be another solution in the form of a function $u(x)$. Let $J_0(x) = v$ and substitute this in eqn MA.2, giving the following equations for u and v:

$$xu'' + u' + xu = 0 \qquad \text{and} \qquad xv'' + v' + xv = 0 \qquad \text{(MA.9)}$$

Multiplying the first equation by v and the second by u and subtracting, we obtain:

$$x(u''v - v''u) + u'v - v'u = 0$$

or

$$\mathrm{d}[x(u'v - v'u)]/\mathrm{d}x = 0$$

and hence

$$x(u'v - v'u) = B$$

where B is a constant of integration. Dividing throughout by v^2x gives

$$(u'v - v'u)/v^2 = \mathrm{d}/\mathrm{d}x(u/v) = B/v^2x$$

and hence

$$u/v = A + B\int \mathrm{d}x/v^2x$$

If v is now replaced by $J_0(x)$, then

$$u(x) = AJ_0(x) + BJ_0(x)\int \mathrm{d}x/J_0^2(x)x \qquad \text{(MA.10)}$$

Equation MA.4 can be used to integrate this and, after rearrangement, we have finally

$$u(x) = AJ_0(x) + B[J_0(x)\ln x + x^2/4 - 3x^4/128 + \cdots]$$

The *Neumann function of zero order* $N_0(x)$ is defined by putting $A = -2/\pi(\log 2 - \gamma)$ and $B = 2/\pi$ in the above equation, where γ is Euler's constant defined by

$$\gamma = \lim_{n\to\infty}\sum_{k=1}^{n}(1/k - \ln n) = 0.557\,7215\ldots \qquad \text{(MA.11)}$$

So finally:

$$N_0(x) = (2/\pi)[J_0(x)(\ln x/2 + \gamma)] + (2/\pi)(x^2/4 - 3x^4/128 + \cdots) \qquad \text{(MA.12)}$$

$N_0(x)$ is also an oscillating aperiodic function, defined only for $x > 0$. When $x \to 0$, $N_0(x) \to -\infty$ like $\log x$. When x becomes large, there is a simple asymptotic form reminiscent of eqn MA.5:

$$N_0(x) = (2/\pi x)^{1/2} \sin[x - (\pi/4)] \quad \text{(MA.13)}$$

A comparison of eqns MA.5 and MA.13 suggests combining the two asymptotic forms, giving

$$\text{asymptotic form of} \quad J_0(x) + \mathrm{i}N_0(x) = (2/\pi x)^{1/2} \exp[\mathrm{i}(x - \pi/4)] \quad \text{(MA.14a)}$$

$$\text{asymptotic form of} \quad J_0(x) - \mathrm{i}N_0(x) = (2/\pi x)^{1/2} \exp[-\mathrm{i}(x - \pi/4)] \quad \text{(MA.14b)}$$

From these two relationships, we can define new functions known as *Hankel functions*:

- Hankel function of the first kind of zero order

$$H_0^1(x) = J_0(x) + \mathrm{i}N_0(x) \quad \text{(MA.15a)}$$

- Hankel function of the second kind of zero order

$$H_0^2(x) = J_0(x) - \mathrm{i}N_0(x) \quad \text{(MA.15b)}$$

These definitions can, of course, be extended to Bessel and Neumann functions of the first, second, etc., orders to give

$$H_n^1(x) = J_n(x) + \mathrm{i}N_n(x) \qquad \text{and} \qquad H_n^2(x) = J_n(x) - \mathrm{i}N_n(x)$$

A.3.4 Modified Bessel functions

The functions $J_1(x)$ and $J_0(x)$ were respectively comparable to $\sin x$ and $\cos x$ with oscillating amplitudes. It is important to find Bessel functions that do not oscillate and that resemble $\sinh x$ and $\cosh x$ rather than $\sin x$ and $\cos x$. These are precisely the functions involved in the cylindrically symmetrical solutions of the Poisson–Boltzmann equation.

The functions $\sin x$ and $\cos x$ are solutions of the differential equation $\mathrm{d}^2y/\mathrm{d}x^2 + y = 0$, and $\sinh x$ and $\cosh x$ are solutions of $\mathrm{d}^2y/\mathrm{d}x^2 - y = 0$ (where x is replaced by $\mathrm{i}x$).

Bessel's equation (eqn MA.1) can be changed in the same way to give

$$\mathrm{d}^2y/\mathrm{d}x^2 + x^{-1}\,\mathrm{d}y/\mathrm{d}x - (1 - n^2/x^2)y = 0 \quad \text{(MA.16)}$$

whose solutions are linearly independent pairs of functions known as *modified Bessel functions*, denoted by $I_n(x)$ and $K_n(x)$. These are related to $J_n(x)$ and $H_n(x)$ through the easily derived relationships

$$I_n(x) = \mathrm{e}^{-\mathrm{i}n\pi/2} J_n(\mathrm{i}x) \quad \text{(MA.17a)}$$

$$K_n(x) = (\mathrm{i}\pi/2)\mathrm{e}^{\mathrm{i}n\pi/2} H_n(\mathrm{i}x) \quad \text{(MA.17b)}$$

In particular, the functions of zero order are $I_0(x)$ and $K_0(x)$, so that the function of zero order that is a general solution of eqn MA.16 is

$$y = AI_0(x) + BK_0(x)$$

where A and B are constants.

Table MA.1 Values of $K_0(x)$ and $K_1(x)$

x	$K_0(x)$	$K_1(x)$
0.02	4.0284	49.955
0.04	3.3365	24.923
0.06	2.9329	16.564
0.08	2.6475	12.374
0.10	2.4371	9.854
0.16	1.9674	6.053
0.20	1.7527	4.776
0.24	1.5798	3.919
0.30	1.3725	3.056
0.40	1.1145	2.184
0.50	0.9244	1.656
0.70	0.6605	1.050
1.00	0.421	0.602
1.50	0.213	0.277
2.00	0.114	0.140
3.00	0.035	0.040
4.00	0.011	0.0125

Neither $I_0(x)$ nor $K_0(x)$ oscillates, but as x increases $I_0(x)$ increases and $K_0(x)$ decreases. Since the solution of the Poisson–Boltzmann equation must lead to a potential decreasing with increasing x, the coefficient of $I_0(x)$ must be zero, so that

$$y = BK_0(x)$$

Two important properties of $K_0(x)$ are often used:

- As in the case of eqn MA.8, we have

$$K_1(x) = -\mathrm{d}/\mathrm{d}x[K_0(x)]$$

  If, therefore, $K_0(x)$ describes the potential, $K_1(x)$ describes the field. Table MA.1 lists the values of $K_0(x)$ and $K_1(x)$ in the range $0 < x < 4$.
- When $x \to 0$, the function $K_0(x)$ behaves like $-\ln x$.

A.4 Fourier transforms

Fourier transforms are widely used in contemporary science: in analyses of X-ray diffraction patterns, in pulsed NMR spectra, in random processes (particularly Brownian motion) and in many other areas of physics. The transform can be thought of as providing a mathematical correspondence between two physical spaces: one we wish to know about (molecular structure and dynamics) but not directly accessible to us, and another in which measurements are made. Biophysicists need to have a good knowledge of the concept and we thus approach it in stages.

A.4.1 Fourier series

The sound wave from a complex source, such as a musical instrument or human voice, is a periodic function of time. Fourier's theorem states that such a function is equivalent to the sum of harmonic waves whose frequencies are multiples of a fundamental frequency and whose amplitudes are appropriately chosen. The theoretically infinite number of terms in the sum forms a *Fourier series*. The same applies to any periodic function $A(t)$ with a period T, which can thus be written as

$$A(t) = a_0 + \sum_{n=1}^{\infty} [a_n \cos(2\pi nt/T) + b_n \sin(2\pi nt/T)] \qquad \text{(MA.18a)}$$

where

$$a_0 = (1/T) \int_0^T A(t)\,\mathrm{d}t \qquad \text{(MA.18b)}$$

$$a_n = (2/T) \int_0^T A(t) \cos(2\pi nt/T)\,\mathrm{d}t \qquad \text{(MA.18c)}$$

$$b_n = (2/T) \int_0^T A(t) \sin(2\pi nt/T)\,\mathrm{d}t \qquad \text{(MA.18d)}$$

Equations MA.18 are only valid under the following conditions (Dirichlet conditions):

(1) $A(t)$ is a continuous function almost everywhere in the interval $[0, T]$, i.e. it has only a finite number of discontinuities;

(2) $A(t)$ has a finite number of maxima and minima over the same interval.

The expressions in MA.18 can be compared with those giving the n rectangular components of an n-dimensional vector by realizing that the functions $\sin(2\pi nt/T)$ and $\cos(2\pi nt/T)$ are orthogonal and define basis functions similar to the unit vectors in an orthonormal coordinate system.

Replacing a function of time $A(t)$ by a spectrum of frequencies brings out a correspondence between the time scale in which $A(t)$ exists and a frequency scale, i.e. one that has the dimensions of a reciprocal of time ($2\pi nt$ being an angle in radians). We shall return to this correspondence later. Note, however, that if the variation in $A(t)$ is only given over a finite range of time (t_1, t_2) (i.e. $A(t) = 0$ for any $t < t_1$ or $t > t_2$), then the corresponding Fourier series has an infinite number of terms. So representing it by a finite number will never be exactly correct, but in all practical applications a sufficient number of terms is used to match the experimental accuracy with which $A(t)$ is known.

An example of this is the re-creation of a sound using a synchronizer which superimposes an appropriate number of harmonic oscillations with suitably chosen frequencies and amplitudes. Because the series is finite, the synthesis is never quite true to the original. However, the decomposition of the sound into a Fourier series is also the way the inner ear analyses it, and it is our own physiological limit that finally decides how true to the original is the synthesis. More generally, the agreement between a reconstituted signal and the original 'true' signal is said to be adequate if the differences are within experimental error.

Can this type of analysis be extended to non-periodic functions?

A.4.2 Definition of the Fourier transform

We first change eqns MA.18 in two ways:

1. The function of time $A(t)$ is replaced by a function $A(x)$ of any variable x defined over the interval $[-a, +a]$, with $T = 2a$. Equations MA.18 then become

$$A(x) = a_0 + \sum_{n=1}^{\infty}[a_n \cos(n\pi x/a) + b_n \sin(n\pi x/a)] \tag{MA.19a}$$

with

$$a_0 = (1/2a)\int_{-a}^{+a} A(x)\,\mathrm{d}x \tag{MA.19b}$$

$$a_n = (1/a)\int_{-a}^{+a} A(x)\cos(n\pi x/a)\,\mathrm{d}x \tag{MA.19c}$$

$$b_n = (1/a)\int_{-a}^{+a} A(x)\sin(n\pi x/a)\,\mathrm{d}x \tag{MA.19d}$$

2. The relationship $\mathrm{e}^{\mathrm{i}x} = \cos x + \mathrm{i}\sin x$ is used, and eqns MA.19 are written in complex notation as

$$A(x) = \sum_{-a}^{+a} C_n \exp(\mathrm{i}n\pi x/a) \tag{MA.20a}$$

$$C_n = (1/2a)\int_{-\infty}^{+\infty} A(x)\exp(-\mathrm{i}n\pi x/a)\,\mathrm{d}x \tag{MA.20b}$$

with

$$C_0 = a_0 \qquad \text{and} \qquad C_n = \tfrac{1}{2}(a_n - \mathrm{i}b_n)$$

Like all expressions involving functions of a complex variable, the physical significance can be appreciated as follows. Let $\omega_n = n\pi/a = 2n\pi/T$, and once again we have a sum of harmonic functions with an angular frequency ω_n. Since C_n is a complex quantity, we can write

$$C_n = \rho_n \exp(-\mathrm{i}\phi_n)$$

where ρ_n is the amplitude and ϕ_n the phase angle. Hence

$$A(x) = \sum_{-\infty}^{+\infty} \rho_n \exp(\omega_n x - \phi_n) \tag{MA.21}$$

A graph of ρ_n against ω_n (or $\nu_n = 2\pi/\omega_n$) gives an *amplitude spectrum* of $A(x)$ as an infinite series of discrete values of the frequency with an amplitude ρ_n. A graph of ϕ_n against ω_n or ν_n gives the *phase spectrum* of $A(x)$. These two spectra completely define the function $A(x)$ and represent its *Fourier analysis*. Reconstructing the function from the spectra is *Fourier synthesis*.

To extend the process to any function defined over all values of x from $-\infty$ to $+\infty$, a must be made to tend to infinity. Let $\delta s = 1/2a$ and $S = n\delta s = n/2a$. Substituting these in eqn MA.20b, the expression for C_n becomes

$$C_n = \delta s \int_{-a}^{+a} A(x) \exp(-2\pi i s x)\, dx = G(s)\delta s \qquad \text{(MA.22)}$$

where $G(s)$ is the value of the integral, which is a function of the new variable s. Substituting eqn MA.22 into MA.20a gives

$$A(x) = \sum_{-\infty}^{+\infty} G(s)\delta s \exp(2\pi i s x) \qquad \text{(MA.23)}$$

It can be shown that in the limit when $a \to \infty$, $A(x)$ becomes the integral

$$A(x) = \int_{-\infty}^{+\infty} G(s) \exp(2\pi i s x)\, ds \qquad \text{(MA.24)}$$

and eqn MA.22 similarly becomes

$$G(s) = \int_{-\infty}^{+\infty} A(x) \exp(-2\pi i s x)\, dx \qquad \text{(MA.25)}$$

The two functions $A(x)$ and $G(s)$ defined by eqns MA.24 and MA.25 are known as *Fourier transforms* (FT), written as

$$G(s) = \mathrm{FT}\{A(x)\} \qquad \text{(MA.26)}$$

A.4.3 **Properties of Fourier transforms**

Existence

The conditions for the existence of Fourier transforms are always satisfied when $A(x)$ and $G(s)$ describe physical quantities, but this is not necessarily so when using simplified mathematical models. For example, the $\sin x$ function has no Fourier transform since the integral does not converge for all values of s. In fact, of course, the sinusoidal model represented by $\sin x$ does not describe a real physical situation since it would correspond to a wave existing for an infinite time. However, it is possible to extend the concept of a function by passing to a limit, and this allows us to define the FT of $\sin x$ as we shall see later.

Written form

If the variable $u = 2\pi s$ is to be used instead, eqns MA.24 and MA.25 become

$$A(x) = (1/2\pi) \int_{-\infty}^{+\infty} G(u) \exp(iux)\, du \qquad \text{(MA.27a)}$$

$$G(s) = \int_{-\infty}^{+\infty} A(x) \exp(-iux)\, dx \qquad \text{(MA.27b)}$$

Contents

The value of each of these functions at the origin is equal to the contents (i.e. the total area under the curve) of the other. From eqns MA.24 and MA.25, we have

$$\text{for } s = 0: \qquad G(0) = \int_{-\infty}^{+\infty} A(x)\,\mathrm{d}x \tag{MA.28a}$$

$$\text{for } x = 0: \qquad A(0) = \int_{-\infty}^{+\infty} G(s)\,\mathrm{d}s \tag{MA.28b}$$

Parity and symmetry

If $A(x)$ is an even function of x, then $A(-x) = A(x)$. The exponential in eqn MA.25 is replaced by trigonometrical functions using $\exp(-2\pi \mathrm{i}sx) = \cos(2\pi sx) - \mathrm{i}\sin(2\pi sx)$. Since $\cos(2\pi sx) = \cos[2\pi s(-x)]$ and $\sin(2\pi sx) = -\sin[2\pi s(-x)]$, the sine terms cancel out and for any even function we are left with

$$G(s) = 2\int_{0}^{\infty} A(x)\cos(2\pi sx)\,\mathrm{d}x \tag{MA.29}$$

Similarly, for any odd function:

$$G(s) = 2\int_{0}^{\infty} A(x)\sin(2\pi sx)\,\mathrm{d}x \tag{MA.30}$$

Translation

To determine the FT of $A(x - x_0)$, denote the required function by $G'(s)$. Then

$$G'(s) = \int_{0}^{+\infty} A(x - x_0)\exp(2\pi \mathrm{i}sx)\,\mathrm{d}x$$

Writing x in the last two factors as $x = x - x_0 + x_0$, this becomes

$$G'(s) = \exp(2\pi \mathrm{i}sx_0)\int_{0}^{+\infty} A(x - x_0)\exp(2\pi \mathrm{i}s)\,\mathrm{d}(x - x_0)$$

i.e.

$$G'(s) = G(s)\exp(2\pi \mathrm{i}sx_0) \tag{MA.31}$$

A translation of x_0 in x-space is transformed into a rotation (i.e. a change of phase angle) in s-space. For a given x_0, the phase change increases as s increases.

A.4.4 The Dirac delta function

Returning to eqns MA.24 and MA.25, we introduce another variable into one of the integrals: in eqn MA.25, for example, we change x into y:

$$G(s) = \int_{-\infty}^{+\infty} A(y)\exp(-2\pi \mathrm{i}sy)\,\mathrm{d}y \tag{MA.32}$$

and substituting this in eqn MA.24 gives

$$A(x) = \int_{-\infty}^{+\infty} \exp(2\pi i s x)\, ds \int_{-\infty}^{+\infty} A(y) \exp(-2\pi i s y)\, dy \qquad \text{(MA.33)}$$

Reversing the order of integration, we have

$$A(x) = \int_{-\infty}^{+\infty} A(y)\, dy \int_{-\infty}^{+\infty} \exp[2\pi i s(x-y)]\, ds$$

The function $f(x-y)$ defined by the second integral has some strange properties. The integral can be written as the limit, when $a \to \infty$, of

$$\int_{-a}^{+a} \exp[2\pi i s(x-y)]\, ds \qquad \text{(MA.34)}$$

and this has the value

$$\{\exp[2\pi i a(x-y)] - \exp[-2\pi i a(x-y)]\}/2\pi i(x-y) = \sin[a(x-y)]/\pi(x-y) \qquad \text{(MA.35)}$$

Putting $u = x - y$, this is the function $\sin(au)/\pi u$, which is plotted against u in the graph below. It passes through a maximum whose value is a/π at $u = 0$ (i.e. $x = y$). The first two zeros of the function on both sides of $u = 0$ correspond to $x = \pm\pi/a$ and the width of the central peak can then be defined as $2\pi/a$.

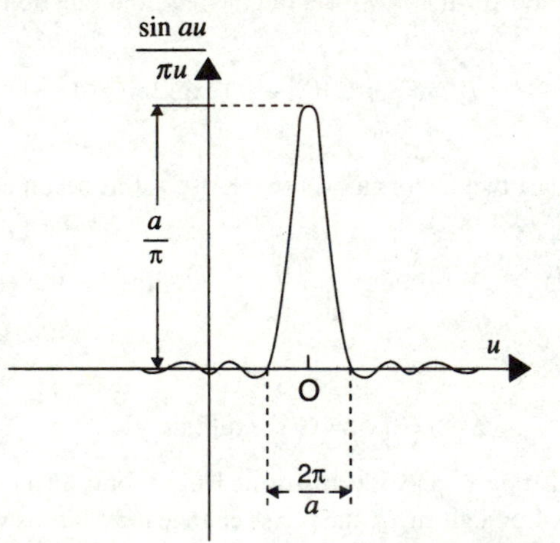

As a tends to infinity, the function also tends to infinity for $x = 0$ and the width tends to zero. The 'function' is thus zero everywhere except when $u = 0$ where it becomes infinite. It can also be shown that the area under the central peak given by

$$\int_{-a}^{+a} [\sin(ax)/\pi x]\, dx$$

tends to 1 when $a \to \infty$. Such a 'function' is known as the *Dirac delta function* or simply the *delta function* and is written as $\delta(x-y)$. It satisfies the following relationships:

$$\begin{aligned}
&\delta(x-y) = 0 \qquad &\text{if } x \neq y \\
&\delta(x-y) \to \infty \qquad &\text{if } x = y \\
&\int_{-\infty}^{+\infty} \delta(x-y)\,\mathrm{d}y = 1 \\
&\delta(x-y) = \delta(y-x)
\end{aligned} \qquad \text{(MA.36)}$$

For any function $f(x)$, the integral

$$\int_{-\infty}^{+\infty} f(x)\delta(x-y)\,\mathrm{d}y$$

is equal to $f(y)$, i.e. the value of the function for $x = y$. In particular

$$\int_{-\infty}^{+\infty} f(x)\delta(x)\,\mathrm{d}x = f(0)$$

What is the importance of the delta function? It is a way of representing mathematically an effect that is infinitesimally small in space or time. A point mass or point charge, an infinitesimally short pulse of light or energy are familiar concepts but do not correspond to anything real. Their importance arises from the fact that the response to such an infinitesimally short phenomenon cannot be distinguished from the response to a realizable physical pulse, given the resolving power of instruments.

According to the Dirichlet conditions, the delta function $\delta(x)$ is not in principle a Fourier transform. However, as in the case of the sine function, a careful passage to the limit enables Fourier transformation to be extended to this special type of function.

If, for example, $\delta(s)$ or $\delta(x)$ is introduced into eqns MA.24 or MA.25, we find once again that δ can be associated with an FT that is equal to 1 throughout the space in question.

If we consider the two functions $\delta(x+\frac{1}{2})$ and $\delta(x-\frac{1}{2})$, it can be seen immediately from eqn MA.25 that the functions $\exp(\mathrm{i}\pi s)$ and $\exp(-\mathrm{i}\pi s)$ respectively correspond to them. The additive property of Fourier transforms thus leads to the relationships

$$\mathrm{FT}\left\{\tfrac{1}{2}[\delta(x+\tfrac{1}{2}) + \delta(x-\tfrac{1}{2})]\right\} = \cos(\pi s) \qquad \text{(MA.37a)}$$

$$\mathrm{FT}\left\{\tfrac{1}{2}[\delta(x-\tfrac{1}{2}) - \delta(x+\tfrac{1}{2})]\right\} = \mathrm{i}\sin(\pi s) \qquad \text{(MA.37b)}$$

which specify the correspondence between the sine and cosine functions mentioned above and combinations of delta functions.

A.4.5 Examples of Fourier transforms

Table MA.2 gives some examples of functions that are Fourier transforms of each other. They include those that are often encountered in physical situations: (a) the

Table MA.2 Examples of Fourier transforms

$A(x)$	$G(s)$
Gaussian $\exp(-\pi x^2)$	Gaussian $\exp(-\pi s^2)$
Rectangular pulse $\Pi(x)$ $= 1$ for $\lvert x\rvert < \frac{1}{2}$ $= 0$ for $\lvert x\rvert > \frac{1}{2}$	$(\sin\pi s)/\pi s$
Exponential function $\exp - \lvert x\rvert$	Lorentzian $2/[1 + (2\pi s)^2]$
Step function $H(x)$ $= 1$ for $x \geq 0$ $= 0$ for $x < 0$	$\delta(s)/2 - \mathrm{i}/2\pi s$
Sampling function $\sum_{n=-\infty}^{+\infty} \delta(s-n)$	$\sum_{n=-\infty}^{+\infty} \delta(x-n)$

rectangular pulse $\Pi(x)$ has an FT $G(s)$ whose variation maps that in the optical diffraction pattern from a rectangular slit; (b) a decaying amplitude in space (like a damped harmonic oscillator) has an FT in the shape of a resonance curve in s-space (a Lorentzian curve); (c) a step function has an FT representing the response of a system to a sudden perturbation; (d) the sampling function in x-space has an FT that is also a sampling function but in s-space (e.g. the diffraction pattern of a diffraction grating).

A.4.6 Convolution integrals

We have seen that several physical techniques involve examining the response of a particular system to a short pulse of a physical quantity (electric field, magnetic field, light intensity, sound intensity, etc.). Take as an example the measurement of pulsed fluorescence: a solution of fluorescent molecules is subjected to a short flash of light and the time variation in the emitted fluorescence is analysed. However rapid it is, the flash cannot be described by $\delta(t)$. The light is produced over a very short but finite time interval and its intensity varies over this interval according to a certain function $L(t')$ characteristic of the method of producing the flash. For each value of this intensity, the fluorescence intensity examined during time t is a function $i_{\mathrm{F}}(t-t')$ proportional to $L(t')$. There is no fluorescence before the flash is triggered, but between its onset and extinction the exciting intensity will have varied like $L(t')$ and overall the measured fluorescence intensity is represented by the function

$$I_{\mathrm{F}}(t) = \int_0^t L(t')i_{\mathrm{F}}(t-t')\,\mathrm{d}t' \qquad \text{(MA.38)}$$

An integral of this sort is called a *convolution integral* of the functions L and i_{F}. In practice, the function $i_{\mathrm{F}}(t)$ must be extracted from the experimental quantity $I_{\mathrm{F}}(t)$

by using the function $L(t')$. The Fourier transform provides the method for this. Let

$$J(\nu) = \mathrm{FT}\{I_{\mathrm{F}}(t)\}, \qquad A(\nu) = \mathrm{FT}\{L(t)\}, \qquad B(\nu) = \mathrm{FT}\{I_{\mathrm{F}}(t)\}$$

where the frequency ν has the dimensions of the reciprocal of a time t. This means that

$$J(\nu) = \int_{-\infty}^{+\infty} \exp(2\pi i\nu t) I_{\mathrm{F}}(t)\,\mathrm{d}t = \int_{-\infty}^{+\infty} \exp(-2\pi i\nu t)\,\mathrm{d}t \int_{-\infty}^{+\infty} L(t') i_{\mathrm{F}}(t-t')\,\mathrm{d}t'$$

where the limits of integration can be extended to $-\infty$ and $+\infty$ since we know that the intensities are zero between $-\infty$ and 0 and between t and $+\infty$. Putting $z = t - t'$:

$$J(\nu) = \int_{-\infty}^{+\infty} \exp[2\pi i\nu(t'+z)]\,\mathrm{d}z \int_{-\infty}^{+\infty} L(t') i_{\mathrm{F}}(z)\,\mathrm{d}t'$$

and reversing the order of integration:

$$J(\nu) = \int_{-\infty}^{+\infty} \exp(2\pi i\nu t') L(t')\,\mathrm{d}t' \int_{-\infty}^{+\infty} \exp(-2\pi i\nu z) i_{\mathrm{F}}(z)\,\mathrm{d}z$$

i.e.

$$J(\nu) = A(\nu)B(\nu) \qquad \text{(MA.39)}$$

The convolution integral in eqn MA.38 has been converted into a simple product of two functions. The integration in time space has been transformed into a multiplication in frequency space. Equation MA.39 gives $B(\nu) = J(\nu)/A(\nu)$ and the inverse Fourier transform of $B(\nu)$ leads to the required function $i_{\mathrm{F}}(t)$.

More generally, a convolution integral appears when a physical phenomenon is detected by an apparatus with a characteristic 'scanning' function. An example is the slit function in spectroscopy. Such an integral also appears when a completely or partially random process described by a function $f(x)$ is sampled. For two values x and $x + x_0$ of the variable, the *autocorrelation function* C of the process is defined by

$$C = \int_{-\infty}^{+\infty} f(x_0) f(x_0 + x)\,\mathrm{d}x_0 \qquad \text{(MA.40)}$$

which appears as the convolution integral of f with itself. This is in fact a measure of a correlation between successive values of the random process defined by f when x changes with time. If $x = 0$, C becomes

$$C_{x=0} = \int_{-\infty}^{+\infty} |f(x_0)|^2\,\mathrm{d}x_0 \qquad \text{(MA.41)}$$

It can be shown that the value given by eqn MA.41 is the maximum value of C, which then decreases as x moves further away from x_0. In a completely random process, the correlation disappears and $C \to 0$ as $x \to \infty$.

Equation MA.40 can be considered as a convolution integral. If $F(s)$ is the Fourier transform of $f(x)$, then

$$C = \mathrm{FT}\{|F(s)|^2\} = \int_{-\infty}^{+\infty} |F(s)|^2 \exp(2\pi i x s)\,\mathrm{d}s \qquad \text{(MA.42)}$$

Note that in this equation any information about the phase in $F(s)$ is completely lost.

If $x = 0$, because the variables x and x_0 are interchangeable, we have that

$$\int_{-\infty}^{+\infty} |f(x)|^2\,\mathrm{d}x = \int_{-\infty}^{+\infty} |F(s)|^2\,\mathrm{d}s \qquad \text{(MA.43)}$$

a result known as *Parseval's theorem*.

For a detailed study of Fourier transforms, see R. N. Bracewell, *The Fourier transform and its applications*, 3rd edition, McGraw-Hill, New York (1986).

A.5 Wiener–Khinchin relations

Consider a *steady* random process described by the function $y(t)$. The correlation function for such a process, $G(\tau) = \langle y(t)y(t+\tau)\rangle$, is independent of t. Like any random function, $G(\tau)$ can be expressed as a Fourier integral:

$$G(\tau) = \int_{-\infty}^{+\infty} J(\nu)\exp(2\pi i\nu\tau)\,\mathrm{d}\nu \qquad \text{(MA.44)}$$

where $J(\nu)$ is called the *spectral density* of $y(t)$. The inverse Fourier transform of this is

$$J(\nu) = \int_{-\infty}^{+\infty} G(\tau)\exp(-2\pi i\nu\tau)\,\mathrm{d}\tau \qquad \text{(MA.45)}$$

Equations MA.44 and MA.45 are the *Wiener–Khinchin relations.*

$G(\tau)$ is a real function and also satisfies $G(\tau) = G(-\tau)$. To prove the latter, put $t_1 = t-\tau$, when the time-independent function $G(\tau)$ becomes

$$G(\tau) = \langle y(t_1)y(t_1+\tau)\rangle = \langle y(t-\tau)y(t)\rangle = G(-\tau)$$

The fact that $G(\tau)$ is real means that $G^*(\tau) = G(\tau)$ where * indicates the complex conjugate. It follows from eqn MA.45 that $J(\nu)$ has the same properties, i.e. $J^*(\nu) = J(\nu)$ and $J(-\nu) = J(\nu)$. These are proved as follows:

$$J^*(\nu) = \int_{-\infty}^{+\infty} G(\tau)\exp(2\pi i\nu\tau)\,\mathrm{d}\tau = \int_{-\infty}^{+\infty} G(\tau)\exp(-2\pi i\nu\tau)\,\mathrm{d}\tau = J(\nu)$$

and since

$$J(-\nu) = \int_{-\infty}^{+\infty} G(\tau)\exp(2\pi i\nu\tau)\,\mathrm{d}\tau$$

changing the sign of the variable ν, we obtain

$$J(-\nu) = \int_{-\infty}^{+\infty} G(\tau)\exp(-2\pi i\nu\tau)\,d\tau = J(\nu)$$

This result can be used by realizing that $G(0) = \langle y^2 \rangle$, so that eqn MA.44 gives

$$\langle y^2 \rangle = \int_{-\infty}^{+\infty} J(\nu)\,d\nu = 2\int_{0}^{+\infty} J_{+}(\nu)\,d\nu \tag{MA.46}$$

where $J_{+}(\nu)$ is the part of the function corresponding to positive values of ν. When ν is a frequency in the classical meaning of the term, $J_{+}(\nu)$ is the experimentally measured spectral density.

Index

Italic page numbers refer to illustrations.
Bold page numbers refer to tables.